Nutritional Epidemiology

FATTY ACIDS

NUTRITION ◊ AND ◊ HEALTH
Adrianne Bendich, Series Editor

Fatty Acids: *Physiological and Behavioral Functions*, edited by *David I. Mostofsky, Shlomo Yehuda, and Norman Salem Jr.*, 2001

Nutrition and Health in Developing Countries, edited by *Richard D. Semba and Martin W. Bloem*, 2001

Preventive Nutrition: *The Comprehensive Guide for Health Professionals*, Second Edition, edited by *Adrianne Bendich and Richard J. Deckelbaum*, 2001

Nutritional Health: *Strategies for Disease Prevention*, edited by *Ted Wilson and Norman J. Temple*, 2001

Clinical Nutrition of the Essential Trace Elements and Minerals: *The Guide for Health Professionals*, edited by *John D. Bogden and Leslie M. Klevey*, 2000

Primary and Secondary Preventive Nutrition, edited by *Adrianne Bendich and Richard J. Deckelbaum*, 2000

The Management of Eating Disorders and Obesity, edited by *David J. Goldstein*, 1999

Vitamin D: *Physiology, Molecular Biology, and Clinical Applications*, edited by *Michael F. Holick*, 1999

Preventive Nutrition: *The Comprehensive Guide for Health Professionals*, edited by *Adrianne Bendich and Richard J. Deckelbaum*, 1997

FATTY ACIDS
PHYSIOLOGICAL AND BEHAVIORAL FUNCTIONS

Edited by

DAVID I. MOSTOFSKY, PhD
Department of Psychology, Boston University, Boston, MA

SHLOMO YEHUDA, PhD
Psychopharmacology Laboratory, Bar Ilan University, Ramat Gan, Israel

NORMAN SALEM JR., PhD
National Institute on Alcohol Abuse and Alcoholism, Rockville, MD

Forewords by

RALPH HOLMAN, PhD
AND WILLIAM LANDS, PhD

HUMANA PRESS
TOTOWA, NEW JERSEY

© 2001 Humana Press Inc.
999 Riverview Drive, Suite 208
Totowa, New Jersey 07512

www.humanapress.com

All rights reserved. No part of this book may be reproduced, stored in a retrieval system, or transmitted in any form or by any means, electronic, mechanical, photocopying, microfilming, recording, or otherwise without written permission from the Publisher.

All papers, comments, opinions, conclusions, or recommendations are those of the author(s), and do not necessarily reflect the views of the publisher.

Cover design by Patricia F. Cleary.

Production Editor: Mark J. Breaugh.

For additional copies, pricing for bulk purchases, and/or information about other Humana titles, contact Humana at the above address or at any of the following numbers: Tel.: 973-256-1699; Fax: 973-256-8341; E-mail: humana@humanapr.com or visit our website at http://humanapress.com

This publication is printed on acid-free paper. ∞
ANSI Z39.48-1984 (American National Standards Institute) Permanence of Paper for Printed Library Materials.

Photocopy Authorization Policy:
Authorization to photocopy items for internal or personal use, or the internal or personal use of specific clients, is granted by Humana Press Inc., provided that the base fee of US $10.00 per copy, plus US $00.25 per page, is paid directly to the Copyright Clearance Center at 222 Rosewood Drive, Danvers, MA 01923. For those organizations that have been granted a photocopy license from the CCC, a separate system of payment has been arranged and is acceptable to Humana Press Inc. The fee code for users of the Transactional Reporting Service is: [0-89603-942-0/01 $10.00 + $00.25].

Printed in the United States of America. 10 9 8 7 6 5 4 3 2 1

Library of Congress Cataloging-in-Publication Data

Fatty acids : physiological and behavioral functions / edited by David I. Mostofsky,
Shlomo Yehuda, and Norman Salem.
      p. cm. -- (Nutrition and health)
   Includes bibliographical references and index.
   ISBN 0-89603-942-0 (alk. paper)
    1. Essential fatty acids in human nutrition 2. Fatty acids--Metabolism. 3. Essential fatty acids--Psychological aspects. I. Mostofsky, David I. II. Yehuda, Shlomo. III. Salem, Norman. IV. Nutrition and health (Totowa, N.J.)

QP752.E84 F38 2001
612.3'97--dc21

00-067293

Series Editor Page

The *Nutrition and Health* series of books have, as an overriding mission, to provide health professionals with texts that are considered essential because each includes: (1) a synthesis of the state of the science, (2) timely, in-depth reviews by the leading researchers in their respective fields, (3) extensive, up-to-date fully annotated reference lists, (4) a detailed index, (5) relevant tables and figures, (6) identification of paradigm shifts and the consequences, (7) virtually no overlap of information between chapters, but targeted, inter-chapter referrals, (8) suggestions of areas for future research, and (9) balanced, data-driven answers to patient/health professionals questions that are based upon the totality of evidence rather than the findings of any single study.

The series volumes are not the outcome of a symposium. Rather, each editor has the potential to examine a chosen area with a broad perspective, both in subject matter as well as in the choice of chapter authors. The international perspective, especially with regard to public health initiatives, is emphasized where appropriate. The editors, whose trainings are both research- and practice-oriented, have the opportunity to develop a primary objective for their book, define the scope and focus, and then invite the leading authorities from around the world to be part of their initiative. The authors are encouraged to provide an overview of the field, discuss their own research, and relate the research findings to potential human health consequences. Because each book is developed *de novo*, the chapters are coordinated such that the resulting volume imparts greater knowledge than the sum of the information contained in the individual chapters.

Fatty Acids: Physiological and Behavioral Functions edited by David I. Mostofsky, Shlomo Yehuda, and Norman Salem, clearly exemplifies the goals of the *Nutrition and Health* series. In fact, this volume is surely ahead of the curve with regard to awareness of the importance of fatty acids in virtually every aspect of human health and disease prevention. Two fatty acids are considered essential nutrients for humans: linoleic and linolenic acids. Thus, there is no question about the importance of dietary intake of adequate levels of these essential nutrients. However, the story of the changes in our food supply and the consequences of consumption by us, as well as our farm animals, of foods that no longer have the same balance of these essential fatty acids had yet to be captured in one authoritative, up-to-date volume until now.

Fatty Acids: Physiological and Behavioral Functions, edited by David I. Mostofsky, Shlomo Yehuda, and Norman Salem, has set the benchmark for providing the most critical data on fatty acids in the most accessible volume published to date. Understanding the metabolism of fatty acids and their roles in human health is certainly not simple and the terms used can often seem daunting; however, the editors and authors have focused on assisting those who are unfamiliar with this field in understanding the critical issues and important new research findings that can impact their fields of interest. Moreover, the two Forewords by the well-acknowledged leaders in the field, Drs. Ralph Holman and William Lands provide the historic perspective as well as a clear overview of the critical importance of fatty acid balance to human health.

Emphasis is placed on the physiological role of the two essential fatty acids, linoleic acid (n-6) and linolenic acid (n-3) and their metabolites—arachidonic acid (n-6), docosahexaenoic acid (n-3), and eicosapentaenoic acid (n-3). The uninitiated reader is clearly guided through the, at first, complex terminology that often makes fatty acid biology seem foreboding. By making the chapters accessible to all readers, the editors have worked to broaden the base of professionals that can know the importance of fatty acids first-hand. All cells contain fatty acids in their membranes, from the outer cell membrane to the inner membranes including mitochondria, endoplasmic reticulum and nuclear membranes. Thus it is easy to understand that the physiological roles of these ubiquitous molecules are critical to the understanding of human health by many researchers in different disciplines such as cardiovascular function, brain and retinal function, immune responses, nephrology, respiratory function, etc. Moreover, by also including the novel findings of the critical value of fatty acids to brain structure and the functionings of the mind, the reader is made aware of this exciting new area of research. Once the reader learns that docosahexaenoic acid makes up about 50% of the fatty acids in the developing brain and retina, it becomes obvious why this major reference volume devotes one-quarter of the book to the behavioral consequences of fatty acid status.

Fatty acids are also a source of energy for the body because these are sources of fat calories and the reader is guided through the dilemma that faces the neonate who has insufficient energy sources and must shift the balance of essential fatty acids needed for brain development to the more immediate need of energy for survival. The challenge is particularly relevant when a premature infant is not provided with sufficient fatty acid resources to continue the optimal development of its brain, retinas, and other vital organs, a process that requires high levels of long-chain fatty acids at that critical point in development. The provision of nutritional sources for premature infants has moved from the research bench to the political arena and has been a regulatory question for several years. These issues are touched upon in several of the chapters in this important volume.

It is not generally recognized that certain fatty acids in cell membranes affect the electrical conductance through the cells. The consequences of lower-than-recommended levels of n-3 fatty acids in the food supply may be one important factor in the development of arrythmias in individuals with this type of cardiac tissue dysregulation. Again, the many roles of fatty acids in human physiology are critically reviewed in this volume and the newest research is highlighted with a focus on communicating the totality of the evidence and the current level of progress in these new areas of therapeutic roles for fatty acids.

Drs. Mostofsky, Yehuda, and Salem have carefully chosen the very best researchers who can communicate the relevance of fatty acid biology to professionals who are not experts in this field. The authors have worked hard to make their information accessible to health professionals interested in public health, child health, nursing, pharmacy, psychology, as well as nutrition-related health professions.

In conclusion, *Fatty Acids: Physiological and Behavioral Functions* provides health professionals in many areas of research and practice the most up-to-date, well-referenced, and easy-to-understand volume on the importance of fatty acids for optimal human health. This volume will serve the reader as the authoritative resource in this field for many years to come.

Adrianne Bendich, PhD, FACN

FOREWORD

The relationship of physiological and *behavioral* functions to dietary levels of ω3 and ω6 essential nutrients are now being investigated intensely, whereas talk of such relationships was almost heresy a few years ago. The climate was similar 70 years ago, for the community of nutritional scientists did not immediately accept the concept of essential fatty acids when George and Mildred Burr proposed it in 1930. They had found that elimination of fat in the diet induced a dermatitis in rats, and that dietary linoleic and linolenic acids could prevent or correct the dermatitis. Arild Hansen, who had done his thesis with George Burr at the University of Minnesota, pursued the possibility of the essentiality in human infants, and put his findings into the medical literature in the 1930s through the 1950s. Both linoleic acid and linolenic acid were effective suppressants of dermatitis, and thus came to be considered equivalent. At Texas A & M in the 1950s, my student Carl Widmer and I were the first to show that linoleic acid is the precursor of arachidonic acid, and that linolenic acid is the precursor of eicosapentaenoic and docosahexaenoic acids in the rat. Now, 70 years later, we are treating neurological diseases with ω3 essential fatty acids. We have come a long way.

Because linolenic acid is much more subject to autoxidation than is linoleic acid, nutritionists and industrial laboratories attempted to eliminate linolenic acid from food formulations to minimize rancidity during storage. Autoxidation was the great enemy for designers of stable foods, and a lifetime of effort was required to breed linolenic acid out of soybeans sufficiently to make soybean oil a more stable and convenient component of industrial products intended for human consumption.

In the 1950s and 1960s, worldwide studies concluded that linoleate-rich dietary oils, such as corn oil and cottonseed oil, lowered the cholesterol level of human plasma, and therefore were advocated as preferred sources of polyunsaturated fatty acids. Hence, in the food industry, polyunsaturated began to mean two double bonds per molecule. More than two double bonds per molecule was associated with rapid unwanted rancidity. In the effort to minimize plasma cholesterol, food oils for humans became richer in linoleic acid (18:2ω6) and lower in linolenic acid (18:3ω3), but were also becoming deficient in essential ω3 polyunsaturates normally found in high levels in brain and nerve lipids.

Our studies in the 1960s revealed that with a constant dietary level of 18:3ω3, increasing dietary 18:2ω6 suppressed the metabolism of the 18:3ω3 to its more highly unsaturated products. Our later studies of EFA profiles in human health and in disease, revealed that in immune-deficiency diseases, and in diseases with neurological manifestations, low levels of ω3 polyunsaturated acids were found in plasma phospholipids. Our studies of several human populations revealed that Americans had the *lowest* levels of ω3 polyunsaturated acids in their plasmas. We have come to believe that low ω3 status is a feature of many diseases, and that the American public is chronically deficient in ω3, in comparison with other national populations.

We now realize that deficiencies of ω3 essential fatty acids are pandemic, especially in modern industrialized societies, and that this is an underlying cause of many burgeoning neurological diseases. The United States of America probably is the current leader in ω3 deficiencies. How can we reverse the trend? Our entire agricultural industry is currently dedicated to the production of crops and products low in ω3 and high ω6 essential fatty acids. Most of our food animals are fed in feed lots, largely on corn, which contains little ω3 fatty acids, but is rich in linoleic acid, a competitor for enzyme sites. Therefore, our major national sources of animal protein are now relatively ω3-deficient. We cannot cure our ω3 deficiency by eating ω3-deficient meat. However, we can replace mammalian meat by fish and enhance our intake of ω3 fatty acids. Present supplies of fish may not be sufficient to meet future demand. Current fish-farming practices must be modified to enhance the ω3 content of the fish, for we cannot cure our ω3 deficiencies with present-day corn-fed ω3-deficient fish.

Perhaps part of the solution to this national and worldwide problem, could be the insertion of genes for ω3 synthesis into corn, rather than trying to shift to new crop species for farmers and their animals. This one effort to enhance the ratio of the ω3/ω6 in corn could solve the problems related to many of our farm animals. Another solution could be to revive the soybean strains that we had 50 years ago. At this stage of the game, one cannot predict what the solution will be, but a solution must be found to eliminate our current pandemic ω3 deficiency.

Ralph T. Holman, PhD

Foreword

By the time that you read this foreword, essential fatty acids in your tissues will have already had profound effects on your body's development and health. Knowing that all that's past is prologue, the authors of this collection of reviews assembled knowledge from their past discoveries to set the stage for readers to anticipate another wave of discovery about essential fatty acids (EFA). For 70 years, a growing body of information illumined ways in which n-3 and n-6 fatty acids maintain health and also act in disease. The essential actions of these fatty acids in physiological and behavioral functions occur through three different modes of acyl chain interaction: specific lipid–protein actions in membrane function; specific lipid–protein actions inducing gene expression; and specific receptor-mediated eicosanoid signaling (*see* Figure 1). When any of these interactions is influenced differently by the n-3 or n-6 arrangement of double bonds in the essential acid, it produces an important consequence of daily food choice. Voluntary food choice is important in health maintenance because the relative abundance of n-3 and n-6 acids in each person's tissues depend on the daily supply. To help people decide whether they wish to make different food choices in the future, the authors address some very complex processes that underlie simple terms like "seizure threshold," "immune function," "retinal function," or "learning behavior." Even now, there are uncertainties about the degree to which the three modes of EFA action in Figure 1 mediate these phenomena. Throughout the 23 chapters and 2078 citations in this book, the authors give information on physiological consequences of dietary EFA supply to help readers evaluate the impact of n-3 and n-6 acid supplies on health maintenance and disease prevention.

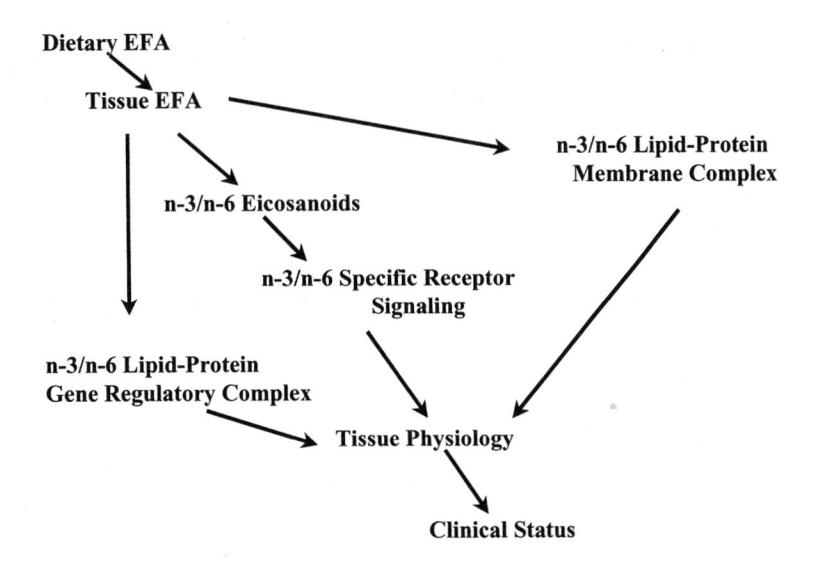

Fig. 1. Essential fatty acids in diets and disease.

THE "DRIFT" IN SUPPLIES OF EFA

For all essential nutrients, a dietary supply is the sine qua non of their action. Chapter 1 points to the general change in intake of EFA over the past centuries that has produced adverse effects on tissue abundances with resultant undesired physiologic and public health outcomes. Negative health consequences of this apparently accidental drift in relative intakes are now seen for different populations. Appendix I of Chapter 1 gives readers clear recommendations for corrective levels of dietary intakes of n-3 and n-6 acids. Chronic inattention to the simple principle that dietary supplies affect tissue abundances of EFA has also led to nearly all experimental animal models in drug development to have excessive n-6 eicosanoid signals with little moderation by n-3 eicosanoid signals. Such polarized experimental models are useful in developing patented pharmaceuticals for treating the consequences of excessive n-6 abundances in tissues. However, measurements in animals fed such imbalanced diets may give little insight into an effective nutritional strategy for preventing the onset of the pathology in the first place. Readers may want to consider the relative abundance of the essential fatty acids in the various foods they routinely eat. To help identify palatable foods that can maintain relative tissue levels of n-3 and n-6 highly unsaturated fatty acids (HUFA) at whatever level desired, readers can use a convenient interactive software program, KIM (Keep It Managed), that is accessible through the website http://ods.od.nih.gov/eicosanoids. Knowledge about the different supplies of EFA in common food servings (over 9000 are listed) seems certain to affect future voluntary dietary choices of well-informed people.

Competition in Maintaining Tissue EFA

The abundant dietary 18-carbon EFA compete vigorously for the limited space conserved for long-chain HUFA in tissue lipids. This competitive metabolism creates reciprocal changes between these acids in tissues. Also, when essential n-3 and n-6 supplies are limited, n-7 and n-9 HUFA accumulate in their place. Chapters 6 through 11 explore these competitions with a focus on healthy perinatal development. These chapters extend beyond the 40-yr-old quantitative evidence of hyperbolic competitive interactions between the n-3 and n-6 acids for elongation, desaturation, and incorporation into tissue lipids (Mohrhauer and Holman, 1963a,b). After reading these chapters, readers might enjoy re-examining that seldom-discussed evidence of competitive hyperbolic metabolic processes for both 18:2n-6 and 18:3n-3, which have midpoints near 0.1 percent of ingested calories (Mohrhauer and Holman, 1963a, b as confirmed by Lands, 1991). Clearly, the effective midpoint for maintaining tissue HUFA is far below the dietary supply now common in the United States.

Bioequivalence is an important concept addressed in several chapters because the highly conserved tissue HUFA are formed from 18-carbon homologs more abundant in the diet. The quantitative competitions among n-3 and n-6 HUFA in the liver, plasma, and visceral tissues parallel, but quantitatively differ from, those in the brain and retina where DHA dominates. Throughout this book, authors lead readers to recognize a special ability of brain and nervous tissue to sequester DHA, illustrating important brain/body differences in EFA dynamics. Tissue abundances of EFA maintained in response to dietary supplies are readily measured by gas chromatography, and extensive efforts were made to predict plasma proportions produced from dietary intakes (e.g., Lands et al., 1992). As a result, the proportion of plasma phospholipid total HUFA that is n-6 HUFA has a predictable relationship to the various dietary EFA supplies. This biomarker of intake (% n-6 HUFA in total HUFA) also relates to the probable intensity of n-6 eicosanoid signaling when the tissue is stimulated. However, despite some progress in estimating tissue EFA, estimating their probable actions along all three modes in Figure 1 remains a major challenge for authors and readers alike.

Preformed docosahexaenoate (DHA) has a 4- to 20-fold greater relative efficacy (or bioequivalence) over linolenic acid (LNA) as substrate for accumulating as brain DHA during perinatal development. Chapter 6 reviews the kinetic data from primates to strengthen the recommendation of including at least a modest supply of DHA in infant formulas to aid brain development. Discussion of the low levels of LNA in brain tissue may reflect the predominance of phosphoglycerides in brain lipids known to differ from triacylglycerols by accumulating linolenate but little LNA. We still have no explanation for how tissue triacylglycerols accumulate LNA while the metabolically related phosphoglycerides do not! Recent advances in evaluating the supply of DHA to the nervous system are noted in Chapter 7. The authors conclude that much ingested LNA is not available for synthesis of DHA, and that elongation and desaturation events in the liver must be accompanied by biosynthetic activity in brain and nervous tissue to maintain adequate DHA levels in those tissues.

Radiolabeled long-chain fatty acids help quantify EFA incorporation rates, turnover, and half-lives and imaging brain phospholipid metabolism in vivo. Chapter 8 notes that the rapid entry of plasma non-esterified fatty acids into brain acyl-CoA pools may be adequate to meet any increased neuronal demands as long as plasma DHA levels are sufficient. Half-lives for turnover of some fatty acids in phosphatidyl inositol and phosphatidyl choline of brain were around 3 h (although that for DHA in phosphatidyl choline was 22 h). The results indicate that incorporation of arachidonate into brain lipids is stimulated by the muscarinic agent, arecoline, and diminished with chronic lithium treatment. Chapter 9 focuses attention on how oxidation and the reuse of acetyl-CoA units (carbon recycling) diverts LNA from its elongation and desaturation to long-chain n-3 HUFA. Carbon recycling is described as a process that decreases the bioavailability of the n-3 fatty acids needed for neural development. Most dietary LNA is completely oxidized for energy even in rapidly growing young animals, and less than 10% seems available for DHA synthesis and esterification in the suckling rat. A three- to four-fold higher oxidation of LNA in the early postnatal period makes carbon recycling more active in neonates than adults.

Readers can find an interesting aspect of brain development in Chapter 10. Astrocytes may provide to neurons 22:6n-3 produced from the 20:5n-3 that had been made by cerebroendothelial cells from 18:3n-3 acquired from plasma. If multicellular transfers are needed to provide neuronal DHA, much more needs to be known about control of this intercellular transport. The authors illustrate ways in which interpreting the nonlinear responses of brain DHA to dietary abundances continues to challenge researchers studying DHA supply and conservation. Until those dynamics are better understood, the functional and behavioral consequences of these nonlinear responses seem certain to remain equally nonlinear and puzzling. Readers will find valuable quantitative insight in Chapter 11, which re-examines bioequivalence during the rapid regain of DHA in primate brain phospholipids during recovery from a dietary n-3 fatty acid deficiency. The extensive results with rhesus monkeys indicate the existence of mechanisms in the brain to conserve HUFA and to manage rapid, reciprocal competitive changes in n-3 and n-6 HUFA. The authors describe analyses of brain and retina EFA that parallel (but quantitatively differ from) analyses of the easily obtained plasma and red cells, which have some quantitatively predictable relations to dietary supplies (Lands et al., 1992). The overall results give clear support that the concepts in the preceding chapters about the reversible, reciprocal changes as n-3 and n-6 HUFA compete for limited space are likely transferrable to human conditions.

Tissue Physiology as Proof of EFA Importance

Vitamins and hormones were identified in the early 20th century by their impact on growth and physiology. Burr and Burr (1929, 1930) identified the n-3 and n-6 EFA when they restored normal physiology to young animals on fat-free diets. Poor overall growth, irregular ovulation, scaly skin, tail necrosis, renal degeneration, and water loss could then be better interpreted. In addition, subnormal testicular development was restored by either dietary EFA or by injected gonadotropin (Greenberg and Ershoff, 1951). Now we can ask which of the three modes in Figure 1 underlie EFA support of the needed pituitary hormone production. Similar questions can address the inadequate dermal integrity that led to greater water loss during EFA deficiency. When researchers used a water rationing protocol to study growth with EFA, a clear difference between n-3 and n-6 nutrients was seen (Thomassen, 1962) that was not apparent when the water supply and humidity were adequate (Burr et al., 1940). Overall, three general physiologic processes seem supported more effectively by n-6 than n-3 EFA: dermal integrity and water balance; renal function; parturition. Readers may see in Chapters 2, 22, and 23 tantalizing clues to mechanisms for those different physiologic outcomes. Eczema and watery stools were clear biomarkers of insufficient EFA for human infants, and they were 50% prevented by about 0.07 % calories as linoleate (Hansen et al., 1963). A later meta-analysis (Cuthbertson, 1976) noted that EFA symptoms in human infants are completely prevented by less than 0.5% of calories as linoleate. Such low thresholds are similar to the 0.3% calories of EFA that proved adequate for growing rats (Mohrhauer and Holman, 1963a,b), supporting the concept that EFA metabolism and physiology are quite similar in rats and humans (Lands et al., 1992).

To help readers interpret body fluid homeostasis, Chapter 22 describes possible roles for n-3 EFA in ways that extend beyond past results with water balance. In addition, new information on mediators of energy homeostasis extends beyond past results on modulating growth hormones and cytokines. Whenever physiologic processes are differentially influenced by n-3 and n-6 EFA, then possible strategies for preventive nutrition can be developed. Chapter 23 illustrates ways in which EFA fit into our expanding awareness about the "information traffic" that integrates the body's nervous system, immune system, and endocrine system to maintain health. The authors note that mechanisms for differential modulation of stress hormones or cytokines by n-3 and n-6 EFA may involve either membrane fluidity, oxidized eicosanoid signaling, or regulation of gene expression. Any successful approach to possibly preventing a disorder will need monitoring with biomarkers that reflect health imbalances prior to the full expression of the clinical disorder that requires treatment. As current choices of traditional foods are now being extended with new "functional foods" to help maintain a balance of n-3 and n-6 EFA, we can expect more nutritional efforts to prevent or diminish the severity of some diseases. Clinicians familiar with the symptoms and methods of treating disorders must share with nutritionists and dietitians an interest in finding useful biomarkers that can properly assess the success of prevention efforts.

Biomarkers of Complex EFA-mediated Events

Many chapters address specific processes that help readers evaluate biomarkers useful in health maintenance and disease prevention. For example, Chapter 3 explores ways that DHA suppresses the expression of VCAM-1, E-selectin, and ICAM-1 on the cell surface, and normal and neoplastic leukocytes exhibit diminished adhesion through the interactions of integrins and selectins. DHA also suppresses expression of major histocompatibility II molecules. In contrast, arachidonate increased adhesion of human blood leukocytes to endothelial cells. As in other chapters of this book, readers are challenged to discern which of the three modes of interaction in Figure 1 regulate the cellular events. In Chapter 4, the effects of essential fatty acids on shifting the voltage dependence of voltage-regulated ion channels are noted as affecting seizure thresholds in animals. The discussion of ion channels extends the important finding (Kang and Leaf, 1994) that many polyunsaturated fatty acid soaps can diminish arrhythmia, but arachidonate can also exacerbate it in a manner dependent on n-6 eicosanoid formation. Release of a mixture of HUFA from tissue phospholipids can give direct anti-arrythmogenic and indirect arrythmogenic actions (Li et al., 1997). The relative proportions of n-3 and n-6 HUFA that had accumulated in the tissue prior to a physiologic challenge are clearly important to tissue responsiveness.

Aging is accompanied by altered cytokine expression and release during a progressive shift in relative abundance of Th1, Th2, and CD^{5+} cells and decline in immune function, which Chapter 5 describes in detail. This chapter provides readers an opportunity to explore how the three different modes of EFA interactions in Figure 1 might participate in altering cytokine-mediated physiology in aging. This background may then be extended to the continually appearing reports on how EFA and their oxidized products regulate expression of genes for cytokines (e.g., Wallace et al., 2001) and metabolic mediators (e.g., Clarke, 2001). Chapter 14 examines disturbances of EFA metabolism during neural complications of diabetes to help readers interpret the reported decreases in arachidonoyl species of phospholipids. Eicosanoid imbalances were indicated for this condition when a cyclooxygenase inhibitor blocked the EFA-supported improvement in nerve conduction velocity and blood flow. Chapter 15 describes another disorder that affects EFA metabolism. It is linked to defects in protein import into peroxisomes by peroxins. The important role of peroxisome enzymes in providing DHA from its docosapentaenoate precursor gives a rationale for DHA therapy in these conditions, which is described in some detail. Chapter 16 extends beyond information in Chapter 4 as it describes diverse effects of fatty acids and ketones on neuronal excitability, exploring their implications for epilepsy and its treatment. The appearance of potentially toxic fatty acid ethyl esters as non-oxidative metabolites of ethanol is reviewed in Chapter 17.

Although the formation is not selective for EFA, these products represent an intriguing biomarker of alcohol exposure.

EFA in Vision, Learning, and Behavior

An outstanding integrated view of DHA actions in vision and brightness discrimination, ranging from acyl chain packing to learning events, is in the combined information of Chapters 2, 12, and 13. Chapter 2 describes how the active form of the visual G-protein-coupled receptor, metarhodopsin II, is modulated by membrane phospholipid acyl chain packing (sometimes referred to as "fluidity"). An increased bilayer area per headgroup of the polyunsaturated phospholipid is associated with more favorable kinetic coupling of the signaling components and is also linked with an increased permeability to water. The authors noted that di-18:3-PC is five times as permeable to water as 18:0,18:1-PC and 18:0,22:6-PC is four times as permeable as 18:0,18:1-PC. Readers may imagine hundreds of other G-protein-mediated systems in which the principles developed for DHA actions with rhodopsin may be extended to signaling systems throughout the body. This concept is advanced further in Chapter 12, which extends beyond phospholipid interactions with rhodopsin to describe detailed kinetics of electroretinogram waveforms and how they are used to explore possible roles of omega-3 polyunsaturated fatty acids in photo pigment activation kinetics. Careful interpretation of results on electrophysiologic signals from photoreceptors led the authors to suggest that DHA is not essential for neural function, but is needed to avoid subtle neural anomalies and produce optimal function. The important action of retinal pigmented epithelium in recycling photoreceptor components with interreceptor retinoid binding protein also depends on DHA levels, providing an indirect means by which DHA abundance can affect vision. Thus, DHA deprivation provides a puzzling mixture of neural impairments perhaps due to altered receptoral mechanisms.

Extending beyond previous information, Chapter 13 explores how performance in the brightness discrimination learning test is impaired by a relative n-3 EFA deficiency. Experimenters observed diet-dependent differences in behavior as a complex outcome of retinal function and conditioned appetitive behavior. In recovering full learning behavior, the experimenters showed an expected competition by high dietary linoleate, which decreased the efficacy of n-3 EFA supplements. Also, the observed decreased turnover of DHA-rich brain ethanolamine phospholipids during n-3-deficient conditions may functionally relate to decreased neurotrophin and synaptic vesicle densities in rat hippocampus, a brain area important to learning and memory. Thus, these three chapters on vision and learning introduce key concepts that may help interpret the behavioral and cognitive phenomena described in Chapters 18 through 21. Dramatic cross-national data in Chapter 18 associate seafood and n-3 EFA supply with psychiatric disorders including major depression, bipolar affective disorder, postpartum depression, hostility and homicide, and suicide. A more positive clinical efficacy of EFA over DHA raises the possibility that these disorders may have an imbalance in specific oxygenated eicosanoids that mediate receptor signaling or gene expression rather than having inadequate membrane DHA levels. Chapter 19 reviews possible disorders of phospholipid metabolism in schizophrenia, affective disorders, and neurodegenerative disorders, and Chapter 20 describes use of eicosapentaenoic acid as a potential new treatment for schizophrenia. Chapter 21 continues exploring the importance of DHA in optimal cognitive function by describing several rodent models that measure learning and motivation. The overall results support continued interest in providing some DHA to infants to ensure adequate neurologic development. Readers of this book will find many reasons to re-examine future food choices for themselves and their families.

William E. M. Lands, PhD

References

Burr GO, Burr MM. A new deficiency disease produced by rigid exclusion of fat from the diet. J Biol Chem 1929; 82:345–367.

Burr GO, Burr MM. On the nature and role of the fatty acids essential in nutrition. J Biol Chem 1930; 86:587–620.

Burr GO, Brown JB, Kass JP, Lundberg WO. Comparative curative values of unsaturated fatty acids in fat deficiency. Proc Soc Exp Biol Med 1940; 44:242–244.

Clarke SD. Polyunsaturated fatty acid regulation of gene transcription: A molecular mechanism to improve the metabolic syndrome. J Nutr 2001; 131:1129–1132.

Cuthbertson WFJ. Essential fatty acid requirements in infancy. Am J Clin Nutr 1976; 29:559–568.

Greenberg SM, Ershoff BH. Effects of chorionic gonadotropin on sex organs of male rats deficient in essential fatty acids. Proc Soc Exp Biol Med. 1951; 78:552–554.

Hansen AE, Wiese HF, Boelsche AN, Haggard ME, Adam DJD, Davis H. Role of linoleic acid in infant nutrition. Clinical and chemical study of 428 infants fed on milk mixtures varying in kind and amount of fat. Pediatrics 1963; 31:171–192.

Kang JX, Leaf A. Proc Nat Acad Sci 1994; 91(21):9886–9890

Lands WEM. Dose-response relationships for 3/6 effects. In Simoupoulos AP et al., eds. Health Effects of 3 Polyunsaturated Fatty Acids in Seafoods. World Review of Nutrition and Dietetics, Vol. 66, pp.177–194. Karger, Basel, 1991.

Lands WEM, Libelt B, Morris A, Kramer NC, Prewitt TE, Bowen P, Schmeisser D, Davidson MH, and Burns JH. Maintenance of lower proportions of n-6 eicosanoid precursors in phospholipids of human plasma in response to added dietary n-3 fatty acids. Biochem Biophys Acta 1992; 1180:147–162.

Li Y, Kang JX, Leaf A. Differential effects of various eicosanoids on the production or prevention of arrhythmias in cultured neonatal rat cardiac myocytes. Prostaglandins 1997; 54(2):511–530

Mohrhauer H, Holman, RT. Effect of linolenic acid upon the metabolism of linoleic acid. J Nutr 1963a; 81:67–74.

Mohrhauer H, Holman RT. The effect of dose level of essential fatty acids upon fatty acid composition of the rat liver. J Lipid Res 1963b; 4:151–159.

Thomassen HJ. Essential fatty acids. Nature 1962; 194:973

Wallace FA, Miles EA, Evans C, Stock TE, Yaqoob P, Calder PC. Dietary fatty acids influence the production of Th1- but not Th2-type cytokines. J Leukocyte Biol 2001; 69:449–457.

Preface

Among the major scientific research efforts of the recent period has been the recognition of the importance of the "essential fatty acids" (EFA). The profound effects of these special chemical entities, and equally profound effects of their deficit, are appreciated by a variety of disciplines, including (but not necessarily limited to) lipid biochemistry, physiology, nutrition, psychology, psychiatry, and, perhaps most intensely, by the neurosciences at large. Functions of the central nervous system, in particular, may be seriously compromised by deficits in the levels of these FA or the ratio (or balance) among major constituents. The role of the polyunsaturated fatty acids (PUFA) α-linolenic acid (LNA; omega-3; n-3; 18:3n-3) and linoleic acid (LA; omega-6; n-6; 18:2n-6) and their metabolites has generated the most exciting findings. Health and medical implications related to these FA extend to visual development in infants, cognitive and emotional development, immunological responses, and cardiovascular health. Several foci of interest are worth noting at this point; foci that are represented in the chapters that follow and that mirror the directions in the field of FA research.

The experimental study of FA deficit has been characterized by investigations that utilize food deprivation or restrictions on nutritional intake, and by designs that have provided for dietary supplementation of the FA and/or their metabolites (especially DHA and its precursors EPA and LNA). Metabolic studies continue to address many of the unexplained complexities associated with the behavior performance observations in the laboratory. Among the questions of interest are: How do the EFAs get into the brain and other organs? What is the basis for the apparent selectivity of various organs, cells, and subcellular organelles for particular lipids and FA? Why is DHA (docosahexaenoic acid; 22:6n-3) concentrated in the brain? How can the adult brain maintain its DHA even when there is little support in the diet? How much can the metabolism of the precursors of DHA (e.g., LNA, EPA, etc.) support DHA composition in the brain in comparison to the incorporation of preformed DHA taken in the diet? In addition to their basic science value, these issues have practical implications for public health policy, such as the design of infant formulas.

The studies of supplementation have drawn attention to peripheral effects, such as the beneficial consequences of DHA in reducing cardiovascular mortality, reduction of immune and inflammatory responses, and influences in the management of diabetes. Supplementation effects also continue to be studied in order to better delineate complex behavioral patterns, with some critical insight on aggression, as but one example, in human studies.

Deprivation of n-3 in animal research has often been concentrated on the F2 offspring where demonstrable impairments in visual function and nonvisual cognitive behaviors have repeatedly been observed. Similar outcomes in human infants have been reported, with a pronounced increase in the frequency of randomized control trials being reported in the literature. Infant behavior appears to suffer quite seriously at the hands of nutritional deprivation, with some long-term followup studies suggesting that the early deficits appear to be maintained with functional loss in later years. The reader will soon discover that differences among outcome studies may be attributable, in part or in total, to variations in the test designs used to assess physiological or behavioral function. Often the attempts to describe complex cognitive and emotional behaviors by use of learning and performance paradigms require a liberal interpretation of the results to support such assessments, which may be open to question or dispute.

Despite a number of known weaknesses, unexplained phenomena, and sorely needed pieces of information yet to be discovered, the present overview of activities in these areas allows one to justifiably conclude major advances in the chemistry and biochemistry of fatty acids have contributed to a considerable understanding about the metabolism and function of fatty acids and their impact on the physiology and behavior of whole organisms. The diversity of actions of fatty acids in many biological systems such as physiological, neurological, endocrinological, and immune begs for elucidation. The management of many chronic health issues will surely benefit from such knowledge in the near term. The purpose of *Fatty Acids: Physiological and Behavioral Functions* is to examine such a representative segment of the scientific aspects of this area, with topics ranging from molecular analyses to functional performance of physiological and cognitive behaviors. To assist the relative newcomer to the vocabulary of the field, we have provided a glossary at the end of the volume. Considerable additional helpful information is easily obtainable from many sources on the web, as even a brief search will indicate.

We hope that *Fatty Acids: Physiological and Behavioral Functions* will facilitate a consolidation of understanding among the separate disciplinary specialists, and will excite other investigators to enter this arena, so that even more dramatic advances and developments in chemistry, behavior, and health management will be forthcoming.

David I. Mostofsky, PhD
Shlomo Yehuda, PhD
Norman Salem Jr., PhD

CONTENTS

Series Editor Page v
Foreword by *Ralph Holman*, PhD vii
Foreword by *William Lands*, PhD ix
Preface xvii
Contributors xxiii

Part I Basic Mechanisms

1   Evolutionary Aspects of Diet: *Essential Fatty Acids* 3
    Artemis P. Simopoulos

2   Modulation of Receptor Signaling by Phospholipid Acyl Chain
    Composition 23
    Drake C. Mitchell and Burton J. Litman

3   Role of Docosahexaenoic Acid in Determining Membrane
    Structure and Function: *Lessons Learned from Normal
    and Neoplastic Leukocytes* 41
    Laura J. Jenski and William Stillwell

4   Effects of Essential Fatty Acids on Voltage-Regulated
    Ionic Channels and Seizure Thresholds in Animals 63
    Robert A. Voskuyl and Martin Vreugdenhil

5   Role of Dietary Fats and Exercise in Immune Functions
    and Aging 79
    Jaya T. Venkatraman and David Pendergast

Part II Phospholipid and Fatty Acid Composition and Metabolism

6   On the Relative Efficacy of α-Linolenic Acid and Preformed
    Docosahexaenoic Acid as Substrates for Tissue Docosahexaenoate
    During Perinatal Development 99
    Meng-Chuan Huang and J. Thomas Brenna

7   Recent Advances in the Supply of Docosahexaenoic Acid
    to the Nervous System 115
    Robert J. Pawlosky and Norman Salem Jr.

8   Quantifying and Imaging Brain Phospholipid Metabolism
    In Vivo Using Radiolabeled Long Chain Fatty Acids 125
    Stanley I. Rapoport

9   Carbon Recycling: *An Important Pathway in α-Linolenate
    Metabolism in Fetuses and Neonates* 145
    Stephen C. Cunnane

Part III DHA and CNS Development

10 Impact of Dietary Essential Fatty Acids on Neuronal Structure and Function *159*
M. Thomas Clandinin, R.A.R. Bowen, and Miyoung Suh

11 Dietary N-3 Fatty Acid Deficiency and its Reversibility: *Effects upon Brain Phospholipids and the Turnover of Docosahexaenoic Acid in the Brain and Blood* *177*
William E. Connor, Gregory J. Anderson, and Don S. Lin

12 The Role of Omega-3 Polyunsaturated Fatty Acids in Retinal Function *193*
Algis J. Vingrys, James A. Armitage, Harrison S. Weisinger, B.V. Bui, Andrew J. Sinclair, and Richard S. Weisinger

13 Brightness–Discrimination Learning Behavior and Retinal Function Affected by Long-Term α-Linolenic Acid Deficiency in Rat *219*
Harumi Okuyama, Yoichi Fujii, and Atsushi Ikemoto

Part IV Pathology

14 Disturbances of Essential Fatty Acid Metabolism in Neural Complications of Diabetes *239*
Joseph Eichberg and Cristinel Mîinea

15 Docosahexaenoic Acid Therapy for Disorders of Peroxisome Biogenesis *257*
Hugo W. Moser and Gerald V. Raymond

16 Effects of Fatty Acids and Ketones on Neuronal Excitability: *Implications for Epilepsy and its Treatment* *273*
Carl E. Stafstrom

17 Fatty Acid Ethyl Esters: *Toxic Nonoxidative Metabolites of Ethanol* *291*
Zbigniew M. Szczepiorkowski and Michael Laposata

Part V Psychiatry and Behavior

18 Omega-3 Fatty Acids and Psychiatric Disorders: *Current Status of the Field* *311*
Joseph R. Hibbeln and Norman Salem Jr.

19 Disorders of Phospholipid Metabolism in Schizophrenia, Affective Disorders, and Neurodegenerative Disorders *331*
David F. Horrobin

20 Eicosapentaenoic Acid: *A Potential New Treatment for Schizophrenia?* *345*
Malcolm Peet and Shaun Ryles

21 The Importance of DHA in Optimal Cognitive Function in Rodents *357*
Claus C. Becker and David J. Kyle

	22	The Role of Omega-3 Polyunsaturated Fatty Acids in Body Fluid and Energy Homeostasis 377
		Richard S. Weisinger, James A. Armitage, Peta Burns, Andrew J. Sinclair, Algis J. Vingrys, and Harrison S. Weisinger
	23	PUFA: *Mediators for the Nervous, Endocrine, and Immune Systems* 403
		Shlomo Yehuda, Sharon Rabinovitz, and David I. Mostofsky

Glossary 421
Index 425

CONTRIBUTORS

GREGORY J. ANDERSON • *Oregon Health Sciences University, Portland, Oregon*
JAMES A. ARMITAGE • *Department of Optometry and Vision Sciences, University of Melbourne, Melbourne, Victoria, Australia*
CLAUS C. BECKER • *Martek Biosciences Corporation, Columbia, Maryland*
R.A.R. BOWEN • *Nutrition and Metabolism Research Group, Department of Agricultural, Food and Nutritional Science and Department of Medicine, University of Alberta, Edmonton, Alberta, Canada*
J. THOMAS BRENNA • *Division of Nutritional Sciences, Cornell University, Ithaca, New York*
B.V. BUI • *Department of Optometry and Vision Sciences, University of Melbourne, Melbourne, Victoria, Australia*
PETA BURNS • *Howard Florey Institute, University of Melbourne, Melbourne, Victoria, Australia*
M. THOMAS CLANDININ • *Nutrition and Metabolism Research Group, Department of Agricultural, Food and Nutritional Science and Department of Medicine, University of Alberta, Edmonton, Alberta*
WILLIAM E. CONNOR • *Oregon Health Sciences University, Portland, Oregon*
STEPHEN C. CUNNANE • *Department of Nutritional Sciences, University of Toronto, Toronto, Ontario*
JOSEPH EICHBERG • *Department of Biology and Biochemistry, University of Houston, Houston, Texas*
YOICHI FUJII • *Faculty of Pharmaceutical Sciences, Nagoya City University, Nagoya, Japan*
JOSEPH R. HIBBELN • *National Institute on Alcohol Abuse and Alcoholism, National Institute of Health, Rockville, Maryland*
DAVID F. HORROBIN • *Laxdale LTD, Stirling, United Kingdom*
MENG-CHUAN HUANG • *Division of Nutritional Sciences, Cornell University, Ithaca, New York*
ATSUSHI IKEMOTO • *Faculty of Pharmaceutical Sciences, Nagoya City University, Nagoya, Japan*
LAURA J. JENSKI • *Department of Biological Sciences, Marshall University, Huntington, West Virginia*
DAVID J. KYLE • *Martek Biosciences Corporation, Columbia, Maryland*
MICHAEL LAPOSATA • *Division of Laboratory Medicine, Massachusetts General Hospital, Boston, Massachusetts*
DON S. LIN • *Oregon Health Sciences University, Portland, Oregon*
BURTON J. LITMAN • *Section of Fluorescence Studies, Laboratory of Membrane Biochemistry and Biophysics, National Institute on Alcohol Abuse and Alcoholism, National Institute of Health, Rockville, Maryland*
CRISTINEL MÎINEA • *Department of Biology and Biochemistry, University of Houston, Houston, Texas*
DRAKE C. MITCHELL • *Section of Fluorescence Studies, Laboratory of Membrane Biochemistry and Biophysics, National Institute on Alcohol Abuse and Alcoholism, National Institute of Health, Rockville, Maryland*

HUGO W. MOSER • *Kennedy Krieger Institute, Baltimore, Maryland*
DAVID I. MOSTOFSKY • *Department of Psychology, Boston University, Boston, Massachusetts*
HARUMI OKUYAMA • *Faculty of Pharmaceutical Sciences, Nagoya City University, Nagoya, Japan*
ROBERT J. PAWLOSKY • *FCL/Beltsville Human Nutrition Research Center, Beltsville, Maryland*
MALCOLM PEET • *Department of Psychiatry, The Longley Centre, University of Sheffield, Sheffield, United Kingdom*
DAVID R. PENDERGAST • *Department of Physiology and Biophysics, State University of New York at Buffalo, Buffalo, New York*
SHARON RABINOVITZ • *Psychopharmacology Laboratory, Bar Ilan University, Ramat Gan, Israel*
STANLEY I. RAPOPORT • *Brain Physiology and Metabolism Section, National Institute on Aging, National Institutes of Health, Bethesda, Maryland*
GERALD V. RAYMOND • *Kennedy Krieger Institute, Baltimore, Maryland*
SHAUN RYLES • *Department of Psychiatry, The Longley Centre, University of Sheffield, Sheffield, United Kingdom*
NORMAN SALEM JR. • *National Institute on Alcohol Abuse and Alcoholism, National Institute of Health, Rockville, Maryland*
ARTEMIS P. SIMOPOULOS • *The Center for Genetics, Nutrition and Health, Washington, DC*
ANDREW J. SINCLAIR • *Department of Food Science, RMIT University, Melbourne, Victoria, Australia*
CARL E. STAFSTROM • *Departments of Neurology and Pediatrics, University of Wisconsin, Madison, Wisconsin*
WILLIAM STILLWELL • *Department of Biology, Indiana University, Purdue University Indianapolis, Indianapolis, Indiana*
MIYOUNG SUH • *Nutrition and Metabolism Research Group, Department of Agricultural, Food and Nutritional Science and Department of Medicine, University of Alberta, Edmonton, Alberta, Canada*
ZBIGNIEW M. SZCZEPIORKOWSKI • *Blood Transfusion Service, Massachusetts General Hospital, Boston, Massachusetts*
JAYA T. VENKATRAMAN • *Department of Physical Therapy, Exercise and Nutrition Sciences, State University of New York at Buffalo, Buffalo, New York*
ALGIS J. VINGRYS • *Department of Optometry and Vision Science, University of Melbourne, Melbourne, Victoria, Australia*
ROBERT A. VOSKUYL • *Leiden/Amsterdam Center for Drug Research; Epilepsy Clinics Foundation of the Netherlands, Heemstede, The Netherlands*
MARTIN VREUGDENHIL • *Division of Neuroscience, Department of Neurophysiology, Medical School, University of Birmingham, Birmingham, United Kingdom*
HARRISON S. WEISINGER • *Department of Food Science, RMIT University, Melbourne, Victoria, Australia*
RICHARD S. WEISINGER • *Howard Florey Institute, University of Melbourne, Melbourne, Victoria, Australia*
SHLOMO YEHUDA • *Psychopharmacology Laboratory, Bar Ilan University, Ramat Gan, Israel*

I Basic Mechanisms

1 Evolutionary Aspects of Diet
Essential Fatty Acids

Artemis P. Simopoulos

1. INTRODUCTION

The interaction of genetics and environment, nature, and nurture is the foundation for all health and disease. This concept, based on molecular biology and genetics, was originally defined by Hippocrates. In the 5th century BC, Hippocrates stated the concept of positive health as follows:

> Positive health requires a knowledge of man's primary constitution [which today we call genetics] and of the powers of various foods, both those natural to them and those resulting from human skill [today's processed food]. But eating alone is not enough for health. There must also be exercise, of which the effects must likewise be known. The combination of these two things makes regimen, when proper attention is given to the season of the year, the changes of the winds, the age of the individual and the situation of his home. If there is any deficiency in food or exercise the body will fall sick.

In the last two decades, using the techniques of molecular biology, it has been shown that genetic factors determine susceptibility to disease and environmental factors determine which genetically susceptible individuals will be affected (Simopoulos and Childs, 1990; Simopoulos and Nestel, 1997; Simopoulos, 1999d). Nutrition is an environmental factor of major importance. Whereas major changes have taken place in our diet over the past 10,000 yr since the beginning of the Agricultural Revolution, our genes have not changed. The spontaneous mutation rate for nuclear DNA is estimated at 0.5% per million years. Therefore, over the past 10,000 yr there has been time for very little change in our genes, perhaps 0.005%. In fact, our genes today are very similar to the genes of our ancestors during the Paleolithic period 40,000 yr ago, at which time our genetic profile was established (Eaton and Konner, 1985). Genetically speaking, humans today live in a nutritional environment that differs from that for which our genetic constitution was selected. Studies on the evolutionary aspects of diet indicate that major changes have taken place in our diet, particularly in the type and amount of essential fatty acids (EFA) and in the antioxidant content of foods (Eaton and Konner, 1985; Simopoulos, 1991; Simopoulos, 1999a; Simopoulos, 1999c) (Table 1, Fig. 1). Using the tools of molecular biology and genetics, research is defining the mechanisms by which genes influence nutrient absorption, metabolism and excretion, taste perception, and degree of satiation, and the mechanisms by which nutrients influence gene expression.

From: *Fatty Acids: Physiological and Behavioral Functions*
Edited by: D. Mostofsky, S. Yehuda, and N. Salem Jr. © Humana Press Inc., Totowa, NJ

Table 1
Characteristics of Hunter–Gatherer and Western Diet and Lifestyles

Characteristic	Hunter–Gatherer diet and lifestyle	Western diet and lifestyle
Physical activity level	High	Low
Diet		
Energy density	Low	High
Energy intake	Moderate	High
Protein	High	Low–moderate
Animal	High	Low–moderate
Vegetable	Very low	Low–moderate
Carbohydrate	Low–moderate	Moderate
(rapidly absorbed)	Slowly absorbed	Rapidly absorbed
Fiber	High	Low
Fat	Low	High
Animal	Low	High
Vegetable	very low	Moderate to high
Total long chain n-6 + n-3	High (2.3 g/d)	Low (0.2 g/d)
Ratio n-6:n-3	Low (2.4)	High (12.0)
Vitamins (mg/d)	*Paleolithic period*	*Current US intake*
Riboflavin	6.49	1.34–2.08
Folate	0.357	0.149–0.205
Thiamin	3.91	1.08–1.75
Ascorbate	604	77–109
Carotene	5.56	2.05–2.57
(retinol equivalent)	(927)	—
Vitamin A	17.2	7.02–8.48
(retinol equivalent)	(2870)	(1170–429)
Vitamin E	32.8	7–10

Source: Modified from Simopoulos, 1999a.

Whereas evolutionary maladaptation leads to reproductive restriction (or differential fertility), the rapid changes in our diet, particularly the last 100 yr, are potent promoters of chronic diseases such as atherosclerosis, essential hypertension, obesity, diabetes, and many cancers. In addition to diet, sedentary lifestyles and exposure to noxious substances interact with genetically controlled biochemical processes, leading to chronic diseases. This chapter discusses evolutionary aspects of diet and the changes that have occurred in Western diets, due to the increase in omega-6 and decrease in omega-3 fatty acid intake from the large-scale production of vegetables oils, agribusiness, and modern agriculture, with emphasis on the balance of omega-6 : omega-3 fatty acids. The Appendix is a portion of the summary of The Workshop on the Essentiality of and Recommended Dietary Intakes (RDIs) for Omega-6 and Omega-3 Fatty Acids, held at the National Institutes of Health (NIH) in Bethesda, MD, USA, April 7–9, 1999 (Simopoulos, et al., 1999).

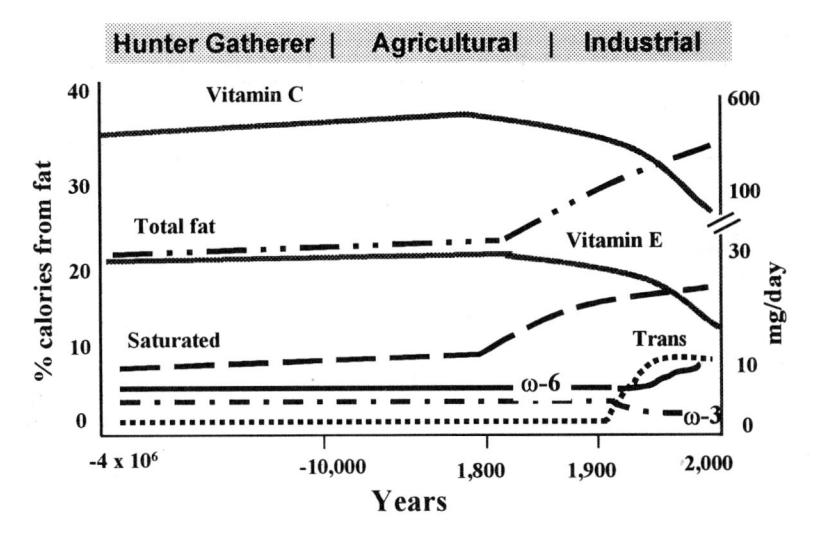

Fig. 1. Hypothetical scheme of fat, fatty acid (ω-6, ω-3, *trans* and total) intake (as percent of calories from fat) and intake of vitamins E and C (mg/d). Data were extrapolated from cross-sectional analyses of contemporary hunter–gatherer populations and from longitudinal observations and their putative changes during the preceding 100 yr. (Simopoulos, 1999a).

2. EVOLUTIONARY ASPECTS OF DIET

The foods that were commonly available to preagricultural humans (lean meat, fish, green leafy vegetables, fruits, nuts, berries, and honey) were the foods that shaped modern humans' genetic nutritional requirements. Cereal grains as a staple food are a relatively recent addition to the human diet and represent a dramatic departure from those foods to which we are genetically programmed and adapted (Cordain, 1999; Simopoulos, 1995a; Simopoulos, 1999d). Cereals did not become a part of our food supply until very recently—10,000 yr ago—with the advent of the Agricultural Revolution. Prior to the Agricultural Revolution, humans ate an enormous variety of wild plants, whereas, today, about 17% of plant species provide 90% of the world's food supply, with the greatest percentage contributed by cereal grains (Cordain, 1999; Simopoulos, 1995a; Simopoulos, 1999d). Three cereals, wheat, maize, and rice, together account for 75% of the world's grain production. Human beings have become entirely dependent on cereal grains for the greater portion of their food supply. The nutritional implications of such a high grain consumption upon human health are enormous. And yet, for the 99.9% of mankind's presence on this planet, humans never or rarely consumed cereal grains. It is only since the last 10,000 yr that humans consume cereals. Up to that time, humans were non-cereal-eating hunter–gatherers since the emergence of Homo erectus 1.7 million years ago. There is no evolutionary precedent in our species for grass seed consumption (Eaton and Konner, 1985). Therefore, there has been little time (<500 generations) since the beginning of the Agricultural Revolution 10,000 years ago to adapt to a food type which now represents humanity's major source of both calories and protein. Cereal grains are high in carbohydrates and omega-6 fatty acids, but low in omega-3 fatty acids and in antioxidants, particularly in comparison to green leafy vegetables. Recent studies show that low-fat/high-carbohydrate diets increase insulin resistance and hyperinsulinemia, conditions

Table 2
Late Paleolithic and Currently Recommended Nutrient Composition for Americans

	Late Paleolithic	Current recommendations
Total dietary energy (%)		
Protein	33	12
Carbohydrate	46	58
Fat	21	30
Alcohol	~0	—
P : S ratio[a]	1.41	1.00
Cholesterol (mg)	520	300
Fiber (g)	100–150	30–60
Sodium (mg)	690	1100–3300
Calcium (mg)	1500–2000	800–1600
Ascorbic acid (mg)	440	60

[a]P : S = polyunsaturated to saturated fat.
Source: Modified from Eaton et al., 1998.

that increase the risk for coronary heart disease, hypertension, diabetes, and obesity (Fanaian, et al., 1996; Simopoulos, 1994a; Simopoulos, 1994b).

A number of anthropological, nutritional, and genetic studies indicate that human's overall diet, including energy intake and energy expenditure, has changed over the past 10,000 yr, with major changes occurring during the past 150 yr in the type and amount of fat and in vitamins C and E intake (Eaton and Konner, 1985; Leaf and Weber, 1987; Simopoulos, 1995a; Simopoulos, 1995b; Simopoulos, 1998b; Simopoulos, 1999a; Simopoulos and Visioli, 2000) (Table 1, Fig. 1). Eaton and Konner (1985) have estimated higher intakes for protein, calcium, potassium, and ascorbic acid, and lower sodium intakes for the diet of the late Paleolithic period than the current US and Western diets. Most of our food is calorically concentrated in comparison with wild game and the uncultivated fruits and vegetables of the Paleolithic diet. Paleolithic man consumed fewer calories and drank water, whereas today most drinks to quench thirst contain calories. Today, industrialized societies are characterized by (1) an increase in energy intake and decrease in energy expenditure, (2) an increase in saturated fat, omega-6 fatty acids, and *trans* fatty acids, and a decrease in omega-3 fatty acid intake, (3) a decrease in complex carbohydrates and fiber, (4) an increase in cereal grains, but a decrease in fruits and vegetables, and (5) a decrease in protein, antioxidants, and calcium intake (Eaton and Konner, 1985; Eaton et al., 1988; Eaton et al., 1998; Simopoulos 1995c; Simopoulos, 1998b; Simopoulos 1999a) (Tables 1 and 2).

3. ESSENTIAL FATTY ACIDS AND THE OMEGA-6 : OMEGA-3 BALANCE

3.1. Large-Scale Production of Vegetable Oils

The increased consumption of omega-6 fatty acids in the last 100 yr is the result of the development of technology at the turn of the century that marked the beginning of the modern vegetable oil industry and the result of modern agriculture with the emphasis on grain feeds for domestic livestock (grains are rich in omega-6 fatty acids) (Kirshenbauer, 1960). The invention of the continuous screw press, named Expeller® by V.D. Anderson,

Table 3
Comparison of Dietary Fats

Dietary fat	Saturated fat (%)	Polyunsaturated fat LA (%)	LNA (%)	LA : LNA (%)	Monounsaturated fat (%)	Cholesterol (%)
		Fatty acid content normalized to 100%				
Flaxseed oil	10	16	53	(0.3)	20	0
Canola oil	6	22	10	(2.2)	62	0
Walnut oil	12	58	12	(4.8)	18	0
Safflower oil	10	77	Trace	(77)	13	0
Sunflower oil	11	69	—	(69)	20	0
Corn oil	13	61	1	(61)	25	0
Olive oil	14	8	1	(8.0)	77	0
Soybean oil	15	54	7	(7.7)	24	0
Margarine	17	32	2	(16)	49	0
Peanut oil	18	33	—	(33)	49	0
Palm oil[a]	51	9	0.3	(30)	39	0
Coconut oil[a]	92	2	0	(2.0)	7	0
Chicken fat	31	21	1	(21)	47	11
Lard	41	11	1	(11)	47	12
Beef fat	52	3	1	(3.0)	44	14
Butterfat	66	2	2	(1.0)	30	33

[a]Palm oil has arachidic of 0.2 and coconut oil has arachidic of 0.1.
Sources: Canola oil: data on file, Procter & Gamble. All others: Reeves JB, Weihrauch JL. Composition of Foods, Agriculture Handbook No. 8-4. US Department of Agriculture, Washington, DC, 1979.

and the steam-vacuum deodorization process by D. Wesson made possible the industrial production of cottonseed oil and other vegetable oils for cooking (Kirshenbauer, 1960). Solvent extraction of oilseeds came into increased use after World War I and the large-scale production of vegetable oils became more efficient and more economic. Subsequently, hydrogenation was applied to oils to solidify them. The partial selective hydrogenation of soybean oil reduced the α-linolenic (LNA) content of the oil while leaving a high concentration of linoleic acid (LA). The LNA content was reduced because LNA in soybean oil caused many organoleptic problems. It is now well known that the hydrogenation process and particularly the formation of *trans* fatty acids has led to increases in serum cholesterol concentrations, whereas LA in its regular state in oil is associated with a reduced serum cholesterol concentration (Emken, 1984; Troisi et al., 1992). The effects of *trans* fatty acids on health have been reviewed extensively elsewhere (Simopoulos, 1995c).

Since the 1950s, research on the effects of omega-6 polyunsaturated fatty acids (PUFA) in lowering serum cholesterol concentrations has dominated the research support on the role of PUFA in lipid metabolism. Although a number of investigators contributed extensively, the article by Ahrens et al. in 1954 and subsequent work by Keys et al. in 1957 firmly established the omega-6 fatty acids as the important fatty acid in the field of cardiovascular disease. The availability of methods for the production of vegetable oils and their use in lowering serum cholesterol concentration led to an increase in both the fat content of the diet and the greater increase in vegetable oils rich in omega-6 fatty acids (Table 3).

Table 4
Fatty Acid Content of Plants (mg/g Wet Weight)

Fatty acid	Purslane	Spinach	Buttercrunch lettuce	Red Leaf lettuce	Mustard
14:0	0.16	0.03	0.01	0.03	0.02
16:0	0.81	0.16	0.07	0.10	0.13
18:0	0.20	0.01	0.02	0.01	0.02
18:1n-9	0.43	0.04	0.03	0.01	0.01
18:2n-6	0.89	0.14	0.10	0.12	0.12
18:3n-3	4.05	0.89	0.26	0.31	0.48
20:5n-3	0.01	0.00	0.00	0.00	0.00
22:6n-3	0.00	0.00	0.001	0.002	0.001
Other	1.95	0.43	0.11	0.12	0.32
Total fatty acid content	8.50	1.70	0.60	0.702	1.101

Source: Modified from Simopoulos and Salem, 1986.

3.2. Agribusiness and Modern Agriculture

Agribusiness contributed further to the decrease in omega-3 fatty acids in animal carcasses. Wild animals and birds who feed on wild plants are very lean, with a carcass fat content of only 3.9% (Ledger, 1968) and contain about five times more PUFA per gram than is found in domestic livestock (Crawford, 1968; Eaton et al. 1998). Most importantly, 4% of the fat of wild animals contains eicosapentaenoic acid (EPA). Domestic beef contains very small or undetectable amounts of LNA because cattle are fed grains rich in omega-6 fatty acids and poor in omega-3 fatty acids (Crawford et al., 1969) whereas deer that forage on ferns and mosses contain more omega-3 fatty acids (LNA) in their meat. Modern agriculture with its emphasis on production has decreased the omega-3 fatty acid content in many foods. In addition to animal meats mentioned earlier (Crawford, 1968; Crawford et al., 1969; Eaton et al., 1998; Ledger, 1968), green leafy vegetables (Simopoulos and Salem, 1986; Simopoulos et al., 1992; Simopoulos et al., 1995), eggs (Simopoulos and Salem, 1989; Simopoulos and Salem, 1992), and even fish (van Vliet and Katan, 1990) contain less omega-3 fatty acids than those in the wild. Foods from edible wild plants contain a good balance of omega-6 and omega-3 fatty acids. Table 4 shows purslane, a wild plant, and compares it to spinach, red leaf lettuce, buttercrunch lettuce, and mustard greens. Purslane has eight times more LNA than the cultivated plants. Modern aquaculture produces fish that contain less omega-3 fatty acids than do fish grown naturally in the ocean, rivers, and lakes (Table 5). As can be seen from Table 6 comparing the fatty acid composition of egg yolk from free-ranging chickens in the Ampelistra farm in Greece and the standard US Department of Agriculture (USDA) egg, the former has an omega-6 : omega-3 ratio of 1.3, whereas the USDA egg has a ratio of 19.9 (Simopoulos and Salem, 1989; Simopoulos and Salem, 1992). By enriching the chicken feed with fishmeal or flax, the ratio of omega-6 : omega-3 decreased to 6.6 and 1.6, respectively (Simopoulos and Salem, 1992). Similarly, milk and cheese from animals that graze contain arachidonic acid (AA), EPA, and docosahexaenoic acid (DHA), whereas milk and cheese from grain-fed animals do not (Table 7) (Simopoulos, 1998b).

Table 5
Fat Content and Fatty Acid Composition of Wild and Cultured Salmon (Salmo Salar)

	Wild (n = 2)	Cultured (n = 2)
Fat (g/100 g)	10 ± 0.1	16 ± 0.6[a]
Fatty acids (g/100 g fatty acid)		
18:3n-3	1 ± 0.1	1 ± 0.1
20:5n-3	5 ± 0.2	5 ± 0.1
22:6n-3	10 ± 2	7 ± 0.1[b]
Other n-3 (18:4n-3 + 20:3n-3 + 22:5n-3)	3 ± 0.5	4 ± 0.1
18:2n-6	1 ± 0.1	3 ± 0.1
Other n-6 (20:4n-6 + 22:4n-6)	0.2 ± 0.1	0.5 ± 0.1
Total n-3	20 ± 2	17 ± 0.2
Total n-6	2 ± 0.1	3 ± 0.1[a]
Ratio of n-3 : n-6	11 ± 2	6 ± 0.1[b]

[a]Significantly different from wild, $p < 0.01$.
[b]Significantly different from wild, $p < 0.05$.
Source: Modified from van Vliet and Katan, 1990.

3.3. Imbalance of Omega-6 : Omega-3 and Its Biological Significance

It is evident that food technology and agribusiness provided the economic stimulus that dominated the changes in the food supply (Dupont et al., 1991; Hunter, 1989; Litin and Sacks, 1993; Raper et al., 1992). From per capita quantities of foods available for consumption in the US national food supply in 1985, the amount of EPA is reported to be about 50 mg/capita/d and the amount of DHA was 80 mg/capita/d. The two main sources of EPA and DHA are fish and poultry (Raper et al., 1992). It has been estimated that the present Western diet is "deficient" in omega-3 fatty acids with a ratio of omega-6 to omega-3 of 15–20 : 1, instead of 1 : ,1 as is the case with wild animals and presumably human beings (Crawford, 1968; Crawford et al., 1969; Eaton and Konner, 1985; Eaton et al., 1998; Ledger, 1968; Simopoulos, 1991; Simopoulos, 1999a; Simopoulos, 1999b; Simopoulos, 1999c). Before the 1940s, cod-liver oil was ingested mainly by children as a source of vitamins A and D, with the usual dose being a teaspoon. Once these vitamins were synthesized, the consumption of cod-liver oil was drastically decreased.

Thus, an absolute and relative decrease of the omega-6 : omega-3 fatty acid ratio has occurred in the food supply of Western countries over the last 100 yr (Tables 8–10). Eaton et al. (1998) have estimated the intake of EFA from animal and vegetable sources, assuming 35% of energy came from animals and 65% from plants. In the Late Paleolithic Period, Table 8 shows that the ratio of LA : LNA was 0.70, whereas the ratio of longer chain omega-6 : omega-3 was 1.79, with a total omega-6 : omega-3 ratio of 0.79. In the United States, again considering the same subsistence ratio of animal sources : plant sources of 35 : 65, the current diet would provide a ratio of 16.74 (Table 9), which is close to the estimates of 15–20:1 of other investigators (Simopoulos, 1991; Eaton et al, 1998). Considering other populations (Table 11), in Japan, the omega-6 : omega-3 ratio is 4 : 1 (Sugano and Hirahara, 2000), and in the United Kingdom, this ratio is 15 : 1, whereas 20 yr ago it was 10 : 1 (Sanders, 2000). Similar ratios have been suggested for northern Europe and Holland, with lower ratios in southern Europe because of a higher consumption of

Table 6
Fatty Acid Levels (mg/g Yolk) in Chicken Egg Yolks

Fatty acid	Greek egg	Supermarket egg	Fishmeal egg	Flax egg
Saturates				
14:0	1.1	0.7	1.0	0.6
15:0	—	0.1	0.3	0.2
16:0	77.6	56.7	67.8	58.9
17:0	0.7	0.3	0.8	0.5
18:0	21.3	22.9	23.0	26.7
Total	100.7	80.7	92.9	86.9
Monounsaturates				
16:1n-7	21.7	4.7	5.1	4.4
18:1	120.5	110.0	102.8	94.2
20:1n-9	0.6	0.7	0.9	0.5
24:1n-9	—	—	0.1	—
Total	142.8	115.4	108.9	99.1
n-6 Polyunsaturates				
18:2n-6	16.0	26.1	67.8	42.4
18:3n-6	—	0.3	0.3	0.2
20:2n-6	0.2	0.4	0.6	0.4
20:3n-6	0.5	0.5	0.5	0.4
20:4n-6	5.4	5.0	4.4	2.6
22:4n-6	0.7	0.4	0.3	—
22:5n-6	0.3	1.2	0.2	—
Total	23.1	33.9	74.1	46.0
n-3 Polyunsaturates				
18:3n-3	6.9	0.5	4.1	21.3
20:3n-3	0.2	—	0.1	0.4
20:5n-3	1.2	—	0.2	0.5
22:5n-3	2.8	0.1	0.4	0.7
22:6n-3	6.6	1.1	6.5	5.1
Total	17.7	1.7	11.3	28.0
P : S ratio	0.4	0.4	0.9	0.9
M : S ratio	1.4	1.4	1.2	1.1
n-6 : n-3 Ratio	1.3	19.9	6.6	1.6

P : S = polyunsaturates : saturates; M : S = monounsaturates : saturates.
Source: Modified from Simopoulos and Salem, 1992.
Note: The eggs were hard-boiled, and their fatty acid composition and lipid content were assessed as described elsewhere (Simopoulos and Salem, 1989). Greek eggs, free-ranging chickens; supermarket eggs, standard US Department of Agriculture eggs found in US supermarkets; fish meal eggs, main source of fatty acids provided by fish meal and whole soybeans; flax eggs, main source of fatty acids provided by flax flour.

olive oil instead of corn and safflower oils. In the past 20 yr, the ratio changed from 10 : 1 to 15 : 1 in England and northern Europe. The shift in the decrease in omega-3 fatty acid intake is reflected in the declining concentrations of DHA and rising concentrations of LA in human milk (Sanders, 2000). The traditional diet of Greece prior to 1960 had a ratio of total omega-6 : omega-3 fatty acids of 1–2 : 1 (Simopoulos 1999b; Simopoulos and Sidossis, 2000). In general, there are a few reliable estimates of the intake of longer chain omega-3 PUFA.

Table 7
Fatty Acid Content of Various Cheeses (per 100 g Edible Portion)

	2% Milk	Cheddar	American	Swiss	Greek myzithra	Greek feta
Total saturated fat (g)	1.2	21.00	19.69	16.04	9.30	7.20
12:0 (g)	<1	0.54	0.48	0.57	—	—
14:0 (g)	<1	3.33	3.21	2.70	1.90	1.60
16:0 (g)	<1	9.80	9.10	7.19	5.40	3.90
18:0 (g)	<1	4.70	3.00	2.60	2.00	1.70
Total monounsaturated fat (g)	1	9.99	8.95	7.05	3.90	3.00
Total polyunsaturated fat (g)	0.07	0.94	0.99	0.62	0.80	0.58
18:2 (g)	0.04	0.58	0.61	0.34	0.38	0.29
18:3 (g)	0.03	0.36	0.38	0.28	0.30	0.20
Arachidonic acid (mg)	—	—	—	—	14	10
Eicosapentaenoic acid (mg)	—	—	—	—	18	14
Docosapentaenoic acid (mg)	—	—	—	—	31	23
Docosahexaenoic acid (mg)	—	—	—	—	5.5	5.1
Total fat (g)	2.27	31.93	29.63	23.71	14.00	10.78

Note: Milk, cheddar, American, and Swiss from US Department of Agriculture Handbook No. 8; Greek myzithra and Greek feta from National Institute on Alcohol Abuse and Alcoholism analyses.
Source: From Simopoulos, 1998b.

A balance existed between omega-6 and omega-3 for millions of years during the long evolutionary history of the genus *Homo*, and genetic changes occurred partly in response to these dietary influences. During evolution, omega-3 fatty acids were found in all foods consumed: meat, wild plants, eggs, fish, nuts, and berries. Recent studies by Cordain et al. (1998) on the omega-3 fatty acid content of wild animals confirm the original observations of Crawford and Sinclair et al. (Crawford, 1968; Sinclair et al., 1982). Furthermore, rapid dietary changes over short periods of time as have occurred over the past 100–150 yr is a totally new phenomenon in human evolution.

Linoleic acid and LNA and their long-chain derivatives are important components of animal and plant cell membranes. When humans ingest fish or fish oil, the EPA and DHA from the diet partially replace the omega-6 fatty acids, especially AA, in the membranes of probably all cells, but especially in the membranes of platelets, erythrocytes, neutrophils, monocytes, and liver cells [reviewed in Simopoulos, 1991]. A diet that has a high ratio of omega-6 : omega-3 fatty acids has detrimental effects on eicosanoid metabolism and gene expression.

Because of the increased amounts of omega-6 fatty acids in the Western diet, the eicosanoid metabolic products from AA, specifically prostaglandins, thromboxanes, leukotrienes, hydroxy fatty acids, and lipoxins, are formed in larger quantities than those formed from omega-3 fatty acids, specifically EPA. The eicosanoids from AA are biologically active in very small quantities and, if they are formed in large amounts, they contribute to the formation of thrombus and atheromas, to allergic and inflammatory disorders, particularly in susceptible people, and to the proliferation of cells. Thus, a diet rich in omega-6 fatty acids shifts the physiological state to one that is prothrombotic and

Table 8
Estimated n-3 and n-6 Fatty Acid Intake in the Late Paleolithic Period (g/d)

Plants	
LA	4.28
LNA	11.40
Animals	
LA	4.56
LNA	1.21
Total	
LA	8.84
LNA	12.60
Plants and animals	
AA (n-6)	1.81
EPA (n-3)	0.39
DTA (n-6)	0.12
DPA (n-3)	0.42
DHA (n-3)	0.27
Ratios of n-6:n-3	
LA : LNA	0.70
AA+DTA : EPA+DPA+DHA	1.79
Total n-6 : n-3	0.79

LA, linoleic acid; LNA, linolenic acid; AA, arachidonic acid; EPA, eicosapentaenoic acid; DTA, docosatetranoic acid; DPA, docosapentaenoic acid; DHA, docosahexaenoic acid.
Note: Assuming an energy intake of 35 : 65 of animal : plant sources.
Source: Data from Eaton et al., 1998.

Table 9
Estimated n-3 and n-6 Fatty Acid Intake in Current Dietary Western Patterns (g/d)

LA	22.5
LNA	1.2
AA	0.6
EPA	0.05
DTA	—
DPA	0.05
DHA	0.08
Ratios of n-6 : n-3	
LA : LNA	18.75
AA+DTA : EPA+DPA+DHA	3.33
Total n-6 : n-3	16.74

LA, linoleic acid; LNA, linolenic acid; AA, arachidonic acid; EPA, eicosapentaenoic acid; DTA, docosatetranoic acid; DPA, docosapentaenoic acid; DHA, docosahexaenoic acid.
Note: Assuming an energy intake of 35 : 65 of animal : plant sources.
Source: Data from Eaton et al, 1998.

proaggregatory, with increases in blood viscosity, vasospasm, and vasocontriction and decreases in bleeding time. Bleeding time is decreased in groups of patients with hypercholesterolemia (Brox et al., 1983), hyperlipoproteinemia (Joist et al., 1979), myocardial

Table 10
Ratios of Dietary n-6 : n-3 Fatty Acids in the
Late Paleolithic Period and in Current Western Patterns (g/d)

	Paleolithic	Western
LA : LNA	0.70	18.75
AA+DTA : EPA+DPA+DHA	1.79	3.33
Total	0.79	16.74

Note: Assuming an energy intake of 35 : 65 of animal : plant sources.
Source: Data from Eaton et al., 1998.

Table 11
n-6 : n-3 Ratios in Various Populations

Population	n-6 : n-3	Ref.
Paleolithic	0.79	(Eaton et al., 1998)
Greece prior to 1960	1.00–2.00	(Simopoulos, 1999)
Current United States	16.74	(Eaton et al., 1998)
United Kingdom and northern Europe	15.00	(Sanders, 2000)
Japan	4.00	(Sugano and Hirahana, 2000)

Table 12
Ethnic Differences in Fatty Acid Concentrations in
Thrombocyte Phospholipids and Percentage of All Deaths From Cardiovascular Disease

	Europe and United States	Japan	Greenland Eskimos
Arachidonic acid (20:4n-6)	26	21	8.3
Eicosapentaenoic acid (20:5n-3)	0.5	1.6	8.0
Ratio of n-6 : n-3	50	12	1
Mortality from cardiovascular disease	45	12	7

Source: Data modified from Weber, 1989.

infarction, other forms of atherosclerotic disease, and diabetes (obesity and hypertriglyceridemia). Atherosclerosis is a major complication in non-insulin-dependent diabetes mellitus (NIDDM) patients. Bleeding time is longer in women than in men and longer in young than in old people. There are ethnic differences in bleeding time that appear to be related to diet. Table 12 shows that the higher the ratio of omega-6 : omega-3 fatty acids in platelet phospholipids, the higher the death rate from cardiovascular disease (Weber, 1989). Figure 2 shows that in India, the higher omega-6 : omega-3 fatty acid ratio in the diet led to a higher prevalence of type II diabetes in the population as a result of changes in the vegetable oil consumption of the population (Raheja, et al., 1993).

In the early stages of evolution, cellular growth and evolutionary success required that the developing organism respond to a myriad of environmental factors. In particular, the organism needed to be able to fulfill its nutrient needs and to develop a sense for nutrient deficiency and excess in order to turn on pathways of synthesis or storage. Thus, nutrient control of gene expression probably evolved as one of the earliest environmental sensor

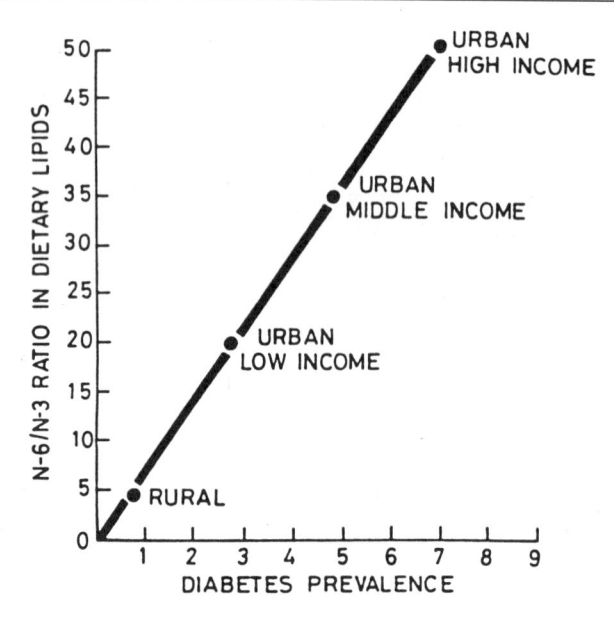

Fig. 2. Relation between *n*-6 : *n*-3 ratio in dietary lipids in Indian diet and prevalence of NIDDM. (Raheja et al., 1993).

mechanisms. Nutrient regulation of gene expression remains, today, a fundamental player in growth and development and is a key player in the development of nutritionally related chronic diseases such as cardiovascular disease, hypertension, diabetes, arthritis, and cancer. For this reason the role of EFA in gene expression is a major field of study. The phospholipid class and fatty acid composition and cholesterol content of biomembranes are critical determinants of physical properties of membranes and have been shown to influence a wide variety of membrane-dependent functions, such as integral enzyme activity, membrane transport, and receptor function. The ability to alter membrane lipid composition and function in vivo by diet, even when EFAs are adequately supplied, demonstrates the importance of diet in growth and metabolism (Galli, et al., 1971).

Complex interactions and displacements of the omega-3 and omega-6 fatty acids take place in plasma and cellular lipids after dietary manipulations. Early steps of cell activation, such as generation of inositol phosphates, are induced by dietary fatty acids (Galli et al., 1989). The effects of dietary fatty acids on the inositol phosphate pathway indicate that diet-induced modifications of PUFA at the cellular level affect the activity of the enzymes responsible for the generation of lipid mediators in addition to the formation of products (eicosanoids) directly derived from their fatty acid precursors. This shows that dietary fats affect key processes in cell function.

The role of omega-3 fatty acids in the control of gene expression is an area that is expected to expand over the next 5 yr as we begin to understand the role of nutrients in gene expression. It is known that nutrients, like hormones, influence and control gene expression and research is now providing more examples (Rucker and Tinker, 1986). Tables 13 and 14 summarize the effects of various PUFA on gene expression. Because of their coordinate or opposing effects, both classes of PUFA are needed in the proper amounts for normal growth and development. Although, so far, the studies in infants have concentrated on the effects of PUFA on retinal and brain phospholipid composition and

Table 13
Effects of Polyunsaturated Fatty Acids on Several Genes
Encoding Enzyme Proteins Involved in Lipogenesis, Glycolysis, and Glucose Transport

Function and gene	Ref.	Lino-lenic acid	a-Lino-lenic acid	Arachi-donic acid	Eicosa-penta-enoic acid	Docosa-hexaenoic acid
Hepatic cells						
Lipogenesis						
FAS	Clarke and Jump (1996)	↓	↓	↓	↓	↓
	Clarke and Jump (1993)					
	Clarke et al. (1990)					
	Clarke et al. (1977)					
S14	Clarke and Jump (1996)	↓	↓	↓	↓	↓
	Clarke and Jump (1993)					
	Clarke et al. (1990)					
	Clarke et al. (1977)					
SCD1	Ntambi (1991)	↓	↓	↓	↓	↓
SCD2	DeWillie and Farmer (1993)	↓	↓	↓	↓	↓
ACC	Clarke and Jump (1996)	↓	↓	↓	↓	↓
ME	Clarke and Jump (1996)	↓	↓	↓	↓	↓
Glycolysis						
G6PD	Jump and Clarke (1994)	↓				
GK	Jump and Clarke (1994)	↓	↓	↓	↓	↓
PK	Liimaatta et al. (1994)	—	↓	↓	↓	↓
Mature adiposites						
Glucose transport						
GLUT4	Tebbey et al. (1994)	—	—	↓	↓	—
GLUT1	Tebbey et al. (1994)	—	—	↑	↑	—

↓ = suppresses or decreases; ↑ = induces or increases.

IQ (Lucas et al., 1992; Eurocat Working Group, 1991), motor development is very much dependent on intermediary metabolism and on overall normal metabolism, both of which are influenced by fatty acid biosynthesis and carbohydrate metabolism.

The amounts of PUFA found in breast milk in mothers fed diets consistent with our evolution should serve as a guide to determine omega-6 and omega-3 fatty acid requirements during pregnancy, lactation, and infant feeding. Of interest is the fact that saturated, monounsaturated, and *trans* fatty acids do not exert any suppressive action on lipogenic or glycolytic gene expression, which is consistent with their high content in human milk serving primarily as sources of energy. Because nutrients influence gene expression and many chronic diseases begin *in utero* or in infancy, proper dietary intake of PUFA, even prior to pregnancy may be essential, as shown for folate deficiency in the development of neural tube defects (Eurocat Working Group, 1991).

A balance between the omega-6 and omega-3 fatty acids is a more physiologic state in terms of gene expression (Simopoulos, 1996), prostaglandin and leukotriene metabolism, and interleukin-1 (IL-1) production (Simopoulos, 1991). The current recommendation to substitute vegetable oils (omega-6) for saturated fats leads to increases in IL-1,

Table 14
Effects of Polyunsaturated Fatty Acids on Several Genes
Encoding Enzyme Proteins Involved in Cell Growth, Early Gene Expression,
Adhesion Molecules, Inflammation, β-Oxidation, and Growth Factors

Function and gene	Ref.	Linoleic acid	α-Linolenic acid	Arachidonic acid	Eicosapentaenoic acid	Docosahexaenoic acid
Cell growth and early gene expression						
c-fos	Sellmayer et al., 1996	—	—	↑	↓	↓
Egr-1	Sellmayer et al., 1996	—	—	↑	↓	↓
Adhesion molecules						
VCAM-1 mRNA[a]	De Caterina et al., 1996	—	—	↓	[b]	↓
Inflammation						
IL-1β	Robinson et al., 1996	—	—	↑	↓	↓
β-oxidation						
Acyl–CoA oxidase[c]	Clarke and Jump, 1996	↑	↑	↑	↑↑	↑
Growth factors						
PDGF	Kaminski, 1993	—	—	↑	↓	↓

Note: VCAM, vascular cell adhesion molecule; IL, interleukin; PDGF, platelet-derived growth factor. ↓ suppresses or decreases, ↑ induces or increases.

[a]Monounsaturated fatty acids (MONOs) also suppress VCAM1 mRNA, but to a lesser degree than does DHA. AA also suppresses to a lesser extent than DHA.

[b]Eicosapentaenoic acid has no effect by itself but enhances the effect of docosahexaenoic acid (DHA).

[c]MONOs also induce acyl-CoA oxidase mRNA.

prostaglandins, and leukotrienes, is not consistent with human evolution, and may lead to maladaptation in those genetically predisposed.

The time has come to return the omega-3 fatty acids into the food supply. Progress in this regard is being made (Simopoulos, 1998a; Simopoulos, 1999b; Simopoulos and Robinson, 1999). In the past, industry focused on improvements in food production and processing, whereas now and in the future, the focus will be on the role of nutrition in product development and its effect on health and disease (Simopoulos, 1998c).

4. CONCLUSION

Several sources of information suggest that humans evolved on a diet with a ratio of omega-6 to omega-3 fatty acids of approx 1, whereas today this ratio is approx 10–20 : 1, with an average of 16.74, indicating that Western diets are deficient in omega-3 fatty acids compared with the diet on which humans evolved and their genetic patterns were established. Because omega-3 fatty acids are essential in growth and development throughout the life cycle, they should be included in the diets of all humans. Omega-3 and omega-6 fatty acids are not intercovertible in the human body and are important components of practically all cell membranes. Whereas cellular proteins are genetically determined, the PUFA composition of cell membranes is to a great extent dependent on the dietary intake. Therefore, appropriate amounts of dietary omega-6 and omega-3 fatty acids need to be considered in making dietary recommendations, and these two classes of PUFA should be distinguished because they are metabolically and functionally distinct and have opposing physiological functions. Their balance is important for homeostasis and normal development.

REFERENCES

Ahrens, E.H., Blankenhorn, D.H., Tsaltas, T.T. Effects on human serum lipids of substituting plant for animal fat in the diet. Proc Soc Exp Biol Med 1954; 86:872–878.

Brox, J.H., Killie, J.E., Osterud B, Holme S, Nordoy, A.. Effects of cod liver oil on platelets and coagulation in familial hypercholesterolemia (type IIa). Acta Med Scand 1983; 213:137–144.

Clarke SD, Romsos DR, Leveille GA. Differential effects of dietary methyl esters of long chain saturated and polyunsaturated fatty acids on rat liver and adipose tissue lipogenesis. J Nutr 1977; 107:1170–1180.

Clarke SD, Armstrong MK, Jump DB. Nutritional control of rat liver fatty acid synthase and S14 mRNA abundance. J Nutr 1990; 120:218–224.

Clarke SD, Jump DB. Fatty acid regulation of gene expression: a unique role for polyunsaturated fats. In: Berdanier C, & Hargrove JL, eds. Nutrition and Gene Expression CRC, Boca Raton, FL, 1993, pp. 227–246.

Clarke SD, Jump DB. Polyunsaturated fatty acid regulation of hepatic gene transcription. Lipids 1996; 31:S7–S11.

Crawford MA. Fatty acid ratios in free-living and domestic animals. Lancet 1968; I:1329–1333.

Crawford MA, Gale MM, Woodford MH. Linoleic acid and linolenic acid elongase products in muscle tissue of Syncerus caffer and other ruminant species. Biochem J 1969; 115:25–27.

Cordain L, Martin C, Florant G, Watkins BA. The fatty acid composition of muscle, brain, marrow and adipose tissue in elk: evolutionary implications for human dietary requirements. In: Simopoulos AP, ed. The Return of ω3 Fatty Acids into the Food Supply. I. Land-Based Animal Food Products and Their Health Effects. World Review of Nutrition and Dietetics, Vol. 83, Karger, Basel, 1998, p. 225.

Cordain L. Cereal grains: humanity's double-edged sword. In: Simopoulos AP, ed. Evolutionary Aspects of Nutrition and Health. Diet, Exercise, Genetics and Chronic Disease. World Review of Nutrition and Dietetics, Vol. 84. Karger, Basel, 1999, pp. 19–73.

De Caterina R, Libby P. Control of endothelial leukocyte adhesion molecules by fatty acids. Lipids 1996; 31(Suppl):S57–S63.

DeWillie JW, Farmer SJ. Linoleic acid controls neonatal tissue-specific stearoyl- CoA desaturase mRNA levels. Biochim Biophy Acta 1993; 1170:291–295.

Dupont J, White PJ, Feldman EB. Saturated and hydrogenated fats in food in relation to health. J Am Coll Nutr 1991; 10:577–592.

Eaton SB, Konner M. Paleolithic nutrition. A consideration of its nature and current implications. N Eng J Med 1985; 312:283–289.

Eaton SB, Konner M, Shostak M. Stone agers in the fast lane: chronic degenerative diseases in evolutionary perspective. Am J Med 1988; 84:739–749.

Eaton SB, Eaton SB III, Sinclair AJ, Cordain L, Mann NJ. Dietary intake of long-chain polyunsaturated fatty acids during the Paleolithic. In: Simopoulos AP, ed. The Return of ω3 Fatty Acids into the Food Supply. I. Land-Based Animal Food Products and Their Health Effects. World Review of Nutrition and Dietetics vol. 83. Karger, Basel, 1998, pp. 12–23.

Emken EA. Nutrition and biochemistry of trans and positional fatty acid isomers in hydrogenated oils. Ann Rev Nutr 1984; 4:339–376.

Eurocat Working Group. Prevalence of neural tube defects in 20 regions of Europe and the impact of prenatal diagnosis 1980–86. J Epidemiol Community Health 1991; 45:52–58.

Fanaian M, Szilasi J, Storlien L, Calvert GD. The effect of modified fat diet on insulin resistance and metabolic parameters in type II diabetes. Diabetologia, 1996; 39(Suppl):A7.

Galli C, Mosconi C, Medini L, Colli S, Tremoli E. N-6 and N-3 fatty acids in plasma and platelet lipids, and generation of inositol phosphates by stimulated platelets after dietary manipulations in the rabbit. In: Galli C, Simopoulos AP, eds. Dietary ω3 and ω6 Fatty Acids: Biological Effects and Nutritional Essentiality. Life Sciences Series Vol. 171. Plenu, New York, 1989, pp. 213–218.

Galli C, Trzeciak HI, Paoletti R. Effects of dietary fatty acids on the fatty acid composition of brain ethanolamine phosphoglyceride. Reciprocal replacement of n-6 and n-3 polyunsaturated fatty acids. Biochim Biophys Acta 1971; 248:449–454.

Hunter JE. Omega-3 fatty acids from vegetable oils. In: Galli C, Simopoulos AP, eds. Dietary ω3 and ω6 Fatty Acids: Biological Effects and Nutritional Essentiality. Series A: Life Sciences. Plenum Press, New York, 1989, pp. 43–55.

Joist JH, Baker RK, Schonfeld, G. Increased in vivo and in vitro platelet function in type IV-hyperlipoproteinemia. Thrombos Res 1979; 15:95–108.

Jump DB, Clarke SD. Coordinate regulation of glycolytic and lipogenic gene expression by polyunsaturated fatty acids. J Lipid Res 1994; 35:1076–1084.

Kaminski WE, Jendraschak E, Kiefl R, von Schacky, C. Dietary omega-3 fatty acids lower levels of platelet-derived growth factor mRNA in human mononuclear cells. Blood 1993; 81:1871–1879.

Keys A, Anderson JT, Grande F. Serum cholesterol response to dietary fat. Lancet 1957; I:787.

Kirshenbauer HG. Fats and Oils. 2nd ed. Reinhold, New York, 1960.

Leaf A, Weber PC. A new era for science in nutrition. Am J Clin Nutr 1987; 45:1048–1053.

Ledger HP. Body composition as a basis for a comparative study of some East African animals. Symp Zool Soc London 1968; 21:289–310.

Liimatta M, Towle HC, Clarke SD, Jump DB. Dietary PUFA interfere with the insulin glucose activation of L-Type pyruvate kinase. Mol Endocrinol 1994; 8:1147–1153.

Litin L, Sacks F. Trans-fatty-acid content of common foods. N Eng J Med 1993; 329(26):1969–1970.

Lucas A, Morley R, Cole TJ, Lister G, Leeson-Payne C. Breast milk and subsequent intelligence quotient in children born premature. Lancet 1992; 339:261–264.

Ntambi JM. Dietary regulation of stearoyl–CoA desaturase I gene expression in mouse liver. J Biol Chem 1991; 267:10,925–10,930.

Raheja BS, Sadikot SM, Phatak RB, Rao MB. Significance of the n-6/n-3 ratio for insulin action in diabetes. In: Klimes I, Howard, BV, Storlien LH, Sevokova E, eds. Dietary Lipids and Insulin Action. New York Academy of Sciences, New York, 1993, pp. 258–271.

Raper NR, Cronin FJ, Exler J. Omega-3 fatty acid content of the US food supply. J Am Coll Nutr 1992; 11(3):304.

Robinson DR, Urakaze M, Huang R, Taki H, Sugiyama E, Knoell CT, et al. Dietary marine lipids suppress the continuous expression of interleukin 1B gene transcription. Lipids 1996; 31(Suppl):S23–S31.

Rucker R, Tinker D. The role of nutrition in gene expression: a fertile field for the application of molecular biology. J Nutr 1986; 116:177–189.

Sanders TAB. Polyunsaturated fatty acids in the food chain in Europe. Am J Clin Nutr 2000; 71(Suppl) 176S–178S.

Sellmayer A, Danesch U, Weber PC. Effects of different polyunsaturated fatty acids on growth-related early gene expression and cell growth. Lipids 1996; 31(Suppl):S37–S40.

Simopoulos AP, Salem N Jr. Purslane: a terrestrial source of omega-3 fatty acids. N Eng J Med 1986; 315:833.

Simopoulos AP, Salem N Jr. N-3 fatty acids in eggs from range-fed Greek chickens. N Eng J Med 1989; 321:1412.

Simopoulos AP, Childs B, eds, Genetic Variation and Nutrition. World Review of Nutrition and Dietetics. Karger, Basel, 1990, vol. 63.

Simopoulos A.P. Omega-3 fatty acids in health and disease and in growth and development. Am J Clin Nutr 1991; 54:438–463.

Simopoulos AP, Norman HA, Gillaspy JE, Duke JA. Common purslane: a source of omega-3 fatty acids and antioxidants. J Am Coll Nutr 1992; 11(4):374–382.

Simopoulos AP, Salem N Jr. Egg yolk as a source of long-chain polyunsaturated fatty acids in infant feeding. Am J Clin Nutr 1992; 55:411–414.

Simopoulos AP, Is insulin resistance influenced by dietary linoleic acid and trans fatty acids? Free Radical Biol Med 1994; 17(4):367–372.

Simopoulos AP, Fatty acid composition of skeletal muscle membrane phospholipids, insulin resistance and obesity. Nutr Today 1994; 2:12–16.

Simopoulos AP, Norman, H.A., Gillaspy, J.E. Purslane in human nutrition and its potential for world agriculture. In: Simopoulos AP, ed. Plants in Human Nutrition. World Review of Nutrition and Dietetics Vol. 77. Karger, Basel, 1995, pp. 47–74.

Simopoulos AP, ed, Plants in Human Nutrition. World Review of Nutrition and Dietetics. Karger, Basel, Vol. 77, 1995a.

Simopoulos AP. The Mediterranean diet: Greek column rather than an Egyptian pyramid. Nutr Today 1995b; 30(2):54–61.

Simopoulos AP. Trans fatty acids. In: Spiller GA, ed. Handbook of Lipids in Human Nutrition. CRC, Boca Raton, 1995c, pp. 91–99.

Simopoulos AP. The role of fatty acids in gene expression: health implications. Ann Nutr Metab 1996; 40:303–311.

Simopoulos AP, Nestel PJ, eds, Genetic Variation and Dietary Response. World Review of Nutrition and Dietetics Vol. 80, Karger, Basel, 1997.

Simopoulos AP, ed, The Return of ω3 Fatty Acids into the Food Supply. I. Land-Based Animal Food Products and Their Health Effects. World Review of Nutrition and Dietetics Vol. 83, Karger, Basel, 1998a.

Simopoulos AP. Overview of evolutionary aspects of ω3 fatty acids in the diet. In: Simopoulos AP, ed. The Return of ω3 Fatty Acids into the Food Supply. I. Land-Based Animal Food Products and Their Health Effects. World Review of Nutrition and Dietetics Vol. 83, Karger, Basel, 1998b, pp. 1–11.

Simopoulos AP. Redefining dietary recommendations and food safety. In: Simopoulos AP, ed. The Return of ω3 Fatty Acids into the Food Supply. I. Land-Based Animal Food Products and Their Health Effects. World Review of Nutrition and Dietetics Vo. 83, Karger, Basel, 1998c, pp. 219–222.

Simopoulos AP. Genetic variation and evolutionary aspects of diet. In: Papas AM, ed. Antioxidant Status, Diet, Nutrition, and Health. CRC, Boca Raton, 1999a, pp. 65–88.

Simopoulos AP. New products from the agri-food industry. The return of n-3 fatty acids into the food supply. Lipids 1999b; 34(Suppl):S297–S301.

Simopoulos AP. Evolutionary aspects of omega-3 fatty acids in the food supply. Prostaglandins, Leukotrienes Essential Fatty Acids 1999c; 60(5&6):421–429.

Simopoulos AP, ed., Evolutionary Aspects of Nutrition and Health. Diet, Exercise Genetics and Chronic Disease. World Review of Nutrition and Dietetics Vol. 84, Karger, Basel, 1999d.

Simopoulos AP, Robinson J. The Omega Diet. The Lifesaving Nutritional Program Based on the Diet of the Island of Crete. HarperCollins, New York, 1999.

Simopoulos AP, Leaf A, Salem N Jr. Essentiality of and recommended dietary intakes for omega-6 and omega-3 fatty acids. Ann Nutr Metab 1999; 43:127–130.

Simopoulos AP, Sidossis LS. What is so special about the traditional diet of Greece: the scientific evidence. In: Simopoulos AP, Visioli F, eds. Mediterranean Diets. World Review of Nutrition and Dietetics. Karger, Basel, 2000, pp. 24–42.

Simopoulos AP, Visioli F, eds. Mediterranean Diets. World Review of Nutrition and Dietetics Vol. 87, Karger, Basel, 2000.

Sinclair AJ, Slattery WJ, O'Dea K. The analysis of polyunsaturated fatty acids in meat by capillary gas-liquid chromotography. J Food Sci Agric 1982; 33:771–776.

Sugano M, Hirahara F. Polyunsaturated fatty acids in the food chain in Japan. American J Clin Nutr 2000; 71(Suppl):189S–196S.

Tebbey PW, McGowan KM, Stephens JM, Buttke TM, Pekata PH. Arachidonic acid down regulates the insulin dependent glucose transporter gene (Glut 4) in 3T3-L1 adipocytes by inhibiting transcription and enhancing mRNA turnover. J Biol Chem 1994; 269:639–644.

Troisi R, Willett WC, Weiss ST. Trans-fatty acid intake in relation to serum lipid concentrations in adult men. Am J Clin Nutr 1992; 56:1019–1024.

Van Vliet T, Katan MB. Lower ratio of n-3 to n-6 fatty acids in cultured than in wild fish. Am J Clin Nutr 1990; 51:1–2.

Weber PC. Are we what we eat? Fatty acids in nutrition and in cell membranes: cell functions and disorders induced by dietary conditions. Report No. 4, Svanoy Foundation, Svanoybukt, Norway, 1989, pp. 9–18.

APPENDIX I: RECOMMENDED DIETARY INTAKES FOR OMEGA-6 AND OMEGA-3 FATTY ACIDS

On April 7–9, 1999, an international working group of scientists met at the National Institutes of Health in Bethesda, Maryland (USA) to discuss the scientific evidence relative to dietary recommendations of omega-6 and omega-3 fatty acids (Simopoulos, et al., 1999). The latest scientific evidence based on controlled intervention trials in infant nutrition, cardiovascular disease, and mental health was extensively discussed. Tables A1 and A2 include the Adequate Intakes (AI) for omega-6 and omega-3 essential fatty acids for adult and infant formula/diet, respectively.

Adults. The working group recognized that there are not enough data to determine Dietary Reference Intakes (DRI), but there are good data to make recommendations for Adequate Intakes (AI) for Adults as shown in Table A1.

Pregnancy and Lactation. For pregnancy and lactation, the recommendations are the same as those for adults with the additional recommendation seen in footnote b of Table 2, that during pregnancy and lactation women must ensure a DHA intake of 300 mg/d.

Composition of Infant Formula/Diet. It was thought of utmost importance to focus on the composition of the infant formula considering the large number of premature infants around the world, the low number of women who breast-feed, and the need for proper nutrition of the sick infant. The composition of the infant formula/diet was based on studies that demonstrated support for both the growth and neural development of infants in a manner similar to that of the breast-fed infant (Table A2).

One recommendation deserves explanation here. After much discussion, consensus was reached on the importance of reducing the omega-6 polyunsaturated fatty acids (PUFAs) even as the omega-3 PUFAs are increased in the diet of adults and newborns for optimal brain and cardiovascular health and function. This is necessary to reduce adverse effects of excesses of arachidonic acid and its eicosanoid products. Such excesses can occur when too much LA and AA are present in the diet and an adequate supply of dietary omega-3 fatty acids is not available. The adverse effects of too much arachidonic acid and its eicosanoids can be avoided by two interdependent dietary changes. First, the amount of plant oils rich in LA, the parent compound of the omega-6 class, which is converted to AA, needs to be reduced. Second, simultaneously the omega-3 PUFAs need to be increased in the diet. LA can be converted to arachidonic acid and the enzyme, Δ-6 desaturase, necessary to desaturate it, is the same one necessary to desaturate ALA, the parent compound of the omega-3 class; each competes with the other for this desaturase. The presence of LNA in the diet can inhibit the conversion of the large amounts of LA in the diets of Western industrialized countries, which contain too much dietary plant oils rich in omega-6 PUFAs (e.g., corn, safflower, and soybean oils). The increase of LNA, together with EPA and DHA, and reduction of vegetable oils with high LA content, are necessary to achieve a healthier diet in these countries.

Table A1
Adequate Intakes[a] for Adults

Fatty acid	Grams/d (2000 kcal diet)	% Energy
LA	4.44	2.0
(upper limit)[b]	6.67	3.0
LNA	2.22	1.0
DHA + EPA	0.65	0.3
DHA to be at least[c]	0.22	0.1
EPA to be at least	0.22	0.1
TRANS-FA		
(upper limit)[d]	2.00	1.0
SAT		
(upper limit)[e]	—	< 8.0
MONOs[f]	—	—

[a]If sufficient scientific evidence is not available to calculate an Estimated Average Requirement, a reference intake called an Adequate Intake (AI) is used instead of a Recommended Dietary Allowance. The AI is a value based on experimentally derived intake levels or approximations of observed mean nutrient intakes by a group (or groups) of healthy people. The AI for children and adults is expected to meet or exceed the amount needed to maintain a defined nutritional state or criterion of adequacy in essentially all members of a specific healthy population; LA = linoleic acid; LNA = α-linolenic acid; DHA = docosahexaenoic acid; EPA = eicosapentaenoic acid; TRANS-FA = *trans* fatty acids; SAT = saturated fatty acids; MONOs = monounsaturated fatty acids.

[b]Although the recommendation is for AI, the Working Group felt that there is enough scientific evidence to also state an upper limit (UL) for LA of 6.67 g/d based on a 2000-kcal diet or of 3.0% of energy.

[c]For pregnant and lactating women, ensure 300 mg/d of DHA.

[d]Except for dairy products, other foods under natural conditions do not contain *trans* FA. Therefore, the Working Group does not recommend *trans* FA to be in the food supply as a result of hydrogenation of unsaturated fatty acids or high-temperature cooking (reused frying oils).

[e]Saturated fats should not comprise more than 8% of energy.

[f]The Working Group recommended that the majority of fatty acids are obtained from monounsaturates. The total amount of fat in the diet is determined by the culture and dietary habits of people around the world (total fat ranges from 15% to 40% of energy) but with special attention to the importance of weight control and reduction of obesity.

Table A2
Adequate Intake[a] for Infant Formula/Diet

Fatty Acid	Percent of fatty acids
LA[b]	10.00
LNA	1.50
AA[c]	0.50
DHA	0.35
EPA[d] (upper limit)	< 0.10

[a]If sufficient scientific evidence is not available to calculate an Estimated Average Requirement, a reference intake called an Adequate Intake (AI) is used instead of a Recommended Dietary Allowance. The AI is a value based on experimentally derived intake levels or approximations of observed mean nutrient intakes by a group (or groups) of healthy people. The AI for children and adults is expected to meet or exceed the amount needed to maintain a defined nutritional state or criterion of adequacy in essentially all members of a specific healthy population; LA = linoleic acid; LNA = α-linolenic acid; AA = arachidonic acid; DHA = docosahexaenoic acid; EPA = eicosapentaenoic acid; TRANS-FA = *trans* fatty acids; SAT = saturated fatty acids; MONOs = monounsaturated fatty acids.

[b]The Working Group recognizes that in countries like Japan, the breast milk content of LA is 6–10% of fatty acids and the DHA is higher, about 0.6%. The formula/diet composition described here is patterned on infant formula studies in Western countries.

[c]The Working Group endorsed the addition of the principal long-chain polyunsaturates, AA and DHA, to all infant formulas.

[d]EPA is a natural constituent of breast milk, but in amounts more than 0.1% in infant formula may antagonize AA and interfere with infant growth.

2
Modulation of Receptor Signaling by Phospholipid Acyl Chain Composition

Drake C. Mitchell and Burton J. Litman

1. INTRODUCTION

A guiding principle in the diverse investigations of biological molecules is that the functional and structural properties of macromolecular assemblies are determined by chemical and structural properties of the constituent molecules and the manner in which those molecules interact. In biological membranes, this requires an understanding of how membrane lipids, primarily phospholipids and cholesterol, and proteins interact with each other and among themselves, so as to carry out a wide range of biological functions. Most of the functions associated with biological membranes (e.g., signal transduction, ion movement, energy conversion, etc.) are carried out by membrane proteins. The phospholipids of neuronal and retinal cells are rich in highly unsaturated acyl chains, especially those of docosahexaenoic acid, 22:6n-3. A primary point of interest when considering receptor signaling is the role of highly unsaturated phospholipids and their effect on membrane protein function. Thus, the focus of this chapter will be on highly unsaturated acyl chains as components of phospholipids and their role in modulating membrane-associated signaling pathways. Therefore, the effects of polyunsaturated free fatty acids on the function of membrane proteins, such as L-type calcium channels (Kang & Leaf, 2000), γ-aminobutyric acid receptor (Nabekura et al., 1998), and voltage-gated potassium channels (Poling et al., 1996) will not be discussed.

Among the polyunsaturated fatty acid constituents of phospholipids, 22:6n-3 is among the most important. This fatty acid represents about 50% of the phospholipid acyl chain composition of the retinal rod outer segment (ROS) disk membrane (Stinson et al, 1991a) and about 40% of the acyl chains in synaptosomal membranes. It is also abundant in the membranes of neuronal and sperm cells (Salem, 1989). The importance of 22:6n-3 in retinal function is indicated by the difficulty in depleting this fatty acid in the retina by dietary manipulation (Stinson et al, 1991b; Bazan et al, 1993). The general importance of DHA in the nervous system is demonstrated by the observed visual and cognitive deficits in animals maintained on an n-3-deficient diet (for reviews, *see* Hamosh & Salem, 1998; Gibson & Makrides, 1998). Neuringer et al. (1984) have shown in nonhuman primates, that infants born to mothers raised on n-3-deficient diets have impaired visual response. Birch et al. (2000) have studied a group of preterm infants, which were fed formula, formula supplemented with 22:6n-3, or breast-fed. In cognitive and visual

From: *Fatty Acids: Physiological and Behavioral Functions*
Edited by: D. Mostofsky, S. Yehuda, and N. Salem Jr. © Humana Press Inc., Totowa, NJ

testing, the 22:6n-3 supplemented and breast-fed groups showed significantly better performance than the formula-fed group.

Several psychological disorders are currently being discussed in terms of the effect of the physical state of the membrane lipids on neurotransmitter receptor function. Hibbeln and Salem (1995) suggest that serotonin levels and membrane 22:6n-3 content are directly linked, whereby low 22:6n-3 yields low serotonin. This results in an individual being susceptible to depression or other affective diseases. These authors suggest that the depletion of 22:6n-3 induces a change in membrane physical properties, which, in turn, influences the function of either serotonergic receptors or serotonin reuptake systems.

In other literature, a conflict in the role of cholesterol in depression and suicide is evident. The rate of suicide and depression has been linked to total serum cholesterol levels. Suicide and violent behavior have been correlated with low serum cholesterol (Muldoon et al, 1990; Engelberg, 1992); however, a recent report indicates that the ratio of violent to nonviolent suicide rates correlates directly with total serum cholesterol (Tanskanen et al, 2000). The latter study suggests that the ratio of violent to nonviolent suicide rates, rather than the total suicide rate, might be a better correlative parameter with cholesterol levels.

The importance of investigating how cholesterol content and acyl chain composition alter the physical properties of membranes is highlighted by the functional deficits associated with 22:6n-3-deficient diets and the antisocial behavior associated with varied cholesterol levels. In the context of psychological disease and neurotransmitter receptor function, it is important to investigate how compositionally induced changes in membrane physical properties influence membrane-associated signaling processes.

The visual transduction pathway is the best characterized G-protein-coupled signal transduction system. Study of the visual receptor, rhodopsin, over the past several decades has made it the archetype of the growing superfamily of heptahelical G-protein-coupled receptors (reviewed in Litman & Mitchell, 1996a). The preeminent position of rhodopsin in this important superfamily will likely increase with the recent publication of the three-dimensional structure of rhodopsin (Palczewski et al., 2000). Many neurotransmitter receptors, as well as the olfactory and taste receptors, are members of this superfamily. Therefore, the effect of lipid membrane composition on various steps in visual signaling will be reviewed in some detail in this chapter. Given the similarity in mode of signaling, the observations made for the vision system should be of general applicability to other members of this receptor superfamily.

2. PROPERTIES OF POLYUNSATURATED PHOSPHOLIPID BILAYERS

In order to understand the biophysical mechanisms whereby polyunsaturated phospholipid acyl chains may alter membrane protein function, a number of investigators have examined model bilayer systems consisting of defined phospholipid composition. Measurements with a wide variety of techniques demonstrate that the introduction of a single *cis* double bond into a saturated acyl chain results in a large decrease in both acyl chain intramolecular order and the intermolecular acyl chain packing order in the liquid-crystalline or fluid phase. However, the effects of higher levels of unsaturation are not as widely agreed upon and appear to vary depending on the specific location of the double bonds and on whether one or both phospholipid acyl chains are unsaturated. Knowledge of the effects of acyl chain unsaturation on bilayer properties is especially incomplete for high levels of polyunsaturation (i.e., four or more double bonds). It does not appear that the effects of a single double bond on acyl chain packing lead to a general description that can be used to explain the effects of high levels of polyunsaturation on molecular order and bilayer properties. However, direct investigation of the properties of highly polyunsaturated bilayers with a variety of techniques has significantly advanced our understanding of these systems in recent years.

Among the wide range of bilayer properties that are determined by the degree of acyl chain unsaturation, special emphasis in this chapter will be given to bilayer properties that have been proven or postulated to modulate the activity of one or more membrane proteins. Widespread attention has been accorded three physical mechanisms wherein a compositionally derived membrane property is correlated with membrane protein function: these are curvature strain (Epand, 1998; Gruner 1985), membrane thickness (Killian, 1998), and acyl chain packing (Hazel, 1995; Litman & Mitchell, 1996b). The effect on bilayer properties of the interaction of cholesterol with polyunsaturated acyl chains will also be reviewed.

2.1. General Physical Properties

Phospholipids containing unsaturated acyl chains melt or undergo the gel to liquid-crystalline phase transition, T_m, at much lower temperatures than phospholipids with saturated acyl chains containing an equal number of carbon atoms. For *sn-1* saturated, *sn-2* unsaturated phospholipids, the first double bond in the *sn-2* chain produces the greatest depression in the T_m. The T_m for 16:0,18:0 phosphatidylcholine (PC) is 321 K, whereas the T_m for 16:0,18:1n9 is 271 K, and the T_m for 16:0,18:2n6 PC is 255 K (Hernandez-Borrell et al., 1993). This trend is also observed for PCs with an 18:0 chain in the *sn-1* position (Niebylski & Salem, 1994). A simple, elegant molecular model has been presented recently that explains the T_m values of *sn-1* saturated, *sn-2* unsaturated phospholipids in terms of the number and position of *cis* carbon–carbon double bonds (Huang and Li, 1999). For diunsaturated PCs, the incremental reduction in T_m with increased number of double bonds is also quite small; the T_m for di18:2n6 PC is 216 K, whereas for di22:6n3 PC, the T_m is 205 K (Kariel et al., 1991; Keough et al., 1987). The relatively similar values of the T_m for dienes and higher polyenes show that the energetic difference between the gel and liquid crystalline states for these lipids are comparable. However, as detailed below, levels of phospholipid acyl chain unsaturation beyond two double bonds leads to alteration of other important bilayer properties.

High levels of phospholipid acyl chain unsaturation produce lipid bilayers with unique mechanical or material properties. Micropipet pressurization measurements demonstrate that bilayers composed of symmetrically substituted polyunsaturated acyl chain phospholipids are more flexible than bilayers composed of *sn-1*-saturated, *sn-2*-monounsaturated phospholipids of the same thickness (Rawicz et al., 2000). Similar measurements demonstrate that the area expansion modulus of di20:4 PC is much lower than that of bilayers composed of saturated or monounsaturated PCs, indicating that high levels of unsaturation yield less cohesive acyl chain packing, which results in a reduction in the energy required for membrane expansion (Needham & Nunn, 1990). The elastic area compressibility modulus is an indicator of the relative amount of energy required to compress a bilayer. The elastic area compressibility modulus for 18:0,22:6 PC is half that of 18:0,18:1 PC, and in 18:0,22:6 PC, the 18:0 chain is less compressible than the 22:6 chain (Koenig et al., 1997). These results demonstrate that the presence of polyunsaturated acyl chains reduces the amount of energy required to elastically deform a phospholipid bilayer and imparts a higher degree of compressibility.

It was recently shown that for bilayers composed of phospholipids with 18-carbon chains, increasing acyl chain unsaturation increases water permeability to such an extent that di18:3 PC is five times as permeable to water as 18:0,18:1 PC (Olbrich et al., 2000). Increased water permeation with increased acyl chain unsaturation does not appear to require both chains to be unsaturated, as 18:0,22:6 PC is about four times as permeable as 18:0,18:1 PC, but only about 30% less permeable than di22:6 PC (Huster et al., 1997).

2.2. Bilayer Thickness

Several investigators have reported variation in bilayer thickness with changing acyl chain unsaturation. X-ray diffraction measurements of peak-to-peak headgroup spacing show that a di18:2n6 PC bilayer is 2 Å thinner than a di18:1n9 PC bilayer. However, in a di20:4 PC bilayer, the bilayer thickness is reduced by only 0.5 Å, to 34.4 Å, relative to a di18:2n6 PC bilayer (Rawicz et al., 2000). This is consistent with deuterium nuclear magnetic resonance (NMR) measurements on *sn-1* saturated, *sn-2* unsaturated PCs, which indicate that 18:0,22:6 PC is only 1 Å thinner than 18:0,18:1 PC and that bilayer thickness is not further reduced after more than three double bonds are introduced to the *sn-2* acyl chain (Holte et al., 1995). Both x-ray diffraction and NMR measurements demonstrate that variation in acyl chain unsaturation produces relatively small changes in bilayer thickness.

2.3. Curvature Strain

Biological membranes are lamellar bilayers and lipid mixtures, extracted from biological membranes, form liquid-crystalline-phase bilayers under physiological conditions (McElhaney, 1984). However, some of the phospholipid components of biological membranes consist of nonbilayer phases in a purified form (Cullis & De Kruijff, 1979). Phospholipids with a headgroup cross-sectional area that is smaller than the cross-sectional area of the volume occupied by the acyl chains do not pack well into a planar bilayer; thus, they tend to undergo a lamellar to nonbilayer or inverted hexagonal (HII) phase transition. The stress that nonbilayer preferring lipids create in planar biological membranes has been termed curvature stress or curvature strain (Epand, 1998). A phosphatidylethanolamine (PE) headgroup is much smaller than a PC head group; thus, the presence of PE in a membrane contributes to curvature strain. A good measure of the relative curvature strain introduced by phospholipids is the temperature of their lamellar to hexagonal phase transition, T_H, with a lower T_H indicating greater curvature strain. For phospholipids with a PE headgroup, T_H is reduced as the unsaturation of the acyl chains increases (Dekker et al., 1983). This is consistent with the results of NMR measurements, which show that the cross-sectional area of the volume occupied by the phospholipid acyl chains increases with unsaturation (Holte et al., 1995; 1996). Curvature strain is also sensitive to the position of the unsaturation, as T_H for di18:1n9 PE is 20°C lower than for either di18:1n6 PE or di18:1n11 PE (Epand et al., 1996). Thus, membrane curvature strain may be altered by changes in either phospholipid headgroup composition or acyl chain unsaturation.

2.4. Acyl Chain Packing

Most biophysical studies of the effects of highly polyunsaturated phospholipid acyl chains on membranes have focused on changes in acyl chain packing, sometimes referred to as "fluidity," a term which has no validity at the dimensions of the bimolecular leaflets forming a membrane. Changes in acyl chain packing are commonly characterized via changes in the steady-state anisotropy of a fluorescent probe like 1,3,5-diphenylhexatriene (DPH). A number of studies have demonstrated that a great deal more information about phospholipid acyl chain packing can be obtained from an analysis of the time-resolved decay of DPH fluorescence anisotropy in terms of an orientational distribution model (Mitchell & Litman, 1998a; Straume & Litman, 1987; Straume & Litman, 1988; van Ginkel et al., 1989). These studies show in detail that increased acyl chain polyunsaturation decreases the cohesion or tightness of acyl chain packing.

In *sn-1* saturated, *sn-2* unsaturated species, the orientation order decreases and the probe motional dynamics increase as the unsaturation of the *sn-2* chain is increased. The degree of orientation order drops dramatically in dipolyunsaturated species compared with PCs that contain a saturated *sn-1* chain and a polyunsaturated *sn-2* chain (Mitchell & Litman, 1998a). The reduction in acyl chain packing order upon going from a disaturated PC to a monounsaturated PC is well known, and symmetric, monounsaturated PCs such as di18:1n9 PC are often used in studies employing model membranes as representative of unsaturated phospholipid bilayers. However, the difference between a symmetrically substituted, highly polyunsaturated bilayer and a symmetrically substituted, monounsaturated bilayer is much greater than the difference between a monounsaturated bilayer and a bilayer where all of the acyl chains are saturated (Mitchell & Litman, 1998a).

The relative differences in acyl chain packing among highly unsaturated, monounsaturated, and saturated bilayers can be illustrated by comparing the angular orientations available to DPH (i.e., the DPH orientation distribution), for each bilayer. The relative probability of all DPH orientations, ranging from parallel to the bilayer normal (0°) to parallel to the plane of the membrane (90°) is derived from time-resolved measurements of DPH fluorescence. The 0° population is approximately parallel to the acyl chains, whereas the 90° population of DPH molecules is in the bilayer midplane (Mitchell & Litman, 1998a,b; Straume & Litman, 1987, 1988; van Ginkel et al., 1989). Comparisons of the DPH orientation probability for di14:0 PC with both di18:1 PC and di22:6 PC are shown in Fig. 1. The curves are the result of subtracting the orientation distribution for DPH in the di14:0 PC bilayer from the DPH orientation distribution in the unsaturated bilayer. Areas above the zero line indicate the range of angular orientations of DPH that are permitted to a greater extent in the unsaturated PCs than in di14:0 PC. Areas below the zero line indicate the range of angular orientations where DPH is has a higher probability of being located in di14:0 PC than it has being located in the unsaturated PCs. The curves in Fig. 1 show that the DPH population in the bilayer midplane, centered about 90°, is greater in both unsaturated bilayers than in di14:0 PC. However, the "difference curve" for di14:0 PC in Fig. 1 is the zero line, and the difference curve for di18:1 PC deviates much less from the zero line than the difference curve for di22:6 PC. This comparison shows that acyl chain packing in di18:1 PC is more similar to that found in di14:0 PC than it is to that found in di22:6 PC. In terms of important membrane properties, such as acyl chain packing, bilayer area per molecule, and water permeability, a bilayer consisting exclusively of monounsaturated acyl chains bears more resemblance to a saturated bilayer than it does to a bilayer containing highly polyunsaturated acyl chains, which is more representative of neuronal membranes.

In an early molecular modeling study, it was determined that the highest probability geometries for 22:6n-3 acyl chains were essentially linear, rigid configurations that would

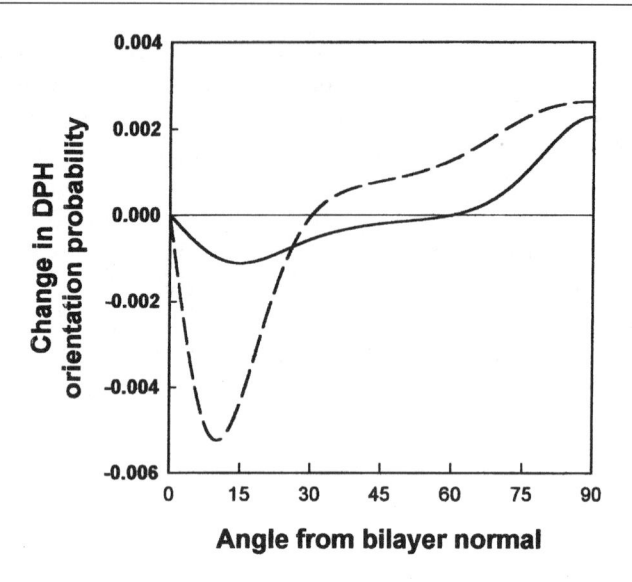

Fig. 1. Difference in orientation probability for the fluorescent membrane probe DPH in di14:0 PC compared to di18:1 PC (—) and di14:0 PC compared to di22:6 PC (---). Each curve is the result of subtracting the probability distribution for DPH orientation in the di14:0 PC from the probability distribution for DPH orientation in the unsaturated bilayer. Regions above the zero line correspond to a range of angular orientations that have a higher probability in the unsaturated bilayer than in di14:0 PC.

facilitate the formation of tightly packed two-dimensional arrays (Applegate & Glomset, 1986). However, the experimental evidence published to date, regarding either 22:6n-3 chain packing or conformation indicates that 22:6n-3 acyl chains have a high degree of flexibility and do not have a high probability of being in an extended linear conformation. A series of NMR measurements have shown that 22:6n-3 acyl chains are loosely packed in fluid-phase bilayers and individual 22:6n-3 acyl chains have very low orientational order (Holte et al., 1995; Holte et al., 1996; Huster et al., 1998). The area of the bilayer occupied by a phospholipid molecule is a strong indicator of the orderliness of acyl chain packing. The area per molecule for 18:0,22:6 PC is approx 12% higher than in 18:0,18:1 PC (Koenig et al., 1997), indicating a higher degree of disorder in the 18:0,22:6 PC. Monolayer studies show that, at the lateral surface pressure corresponding to biological membranes, the area per molecule for di22:6 PC is 50% greater than for 18:0,22:6 PC (Zerouga et al., 1995), which is consistent with greater bilayer free volume in a di22:6 PC bilayer than in a 16:0,22:6 PC bilayer, particularly in the bilayer midplane (Mitchell & Litman, 1998a).

Both increased temperature and increased acyl chain unsaturation introduce disorder in phospholipid acyl chain packing; thus, their effects on bilayer properties are often compared and discussed as being equivalent perturbations on the bilayer structure. However, detailed studies of the effect of these two factors on both intrachain order and acyl chain packing reveal significant differences. The difference between temperature-induced disorder and *sn-2* unsaturation-induced disorder of the saturated *sn-1* acyl chain is illustrated in Fig. 2A. The difference in NMR order-parameter data is plotted for each methyl group of the saturated 18:0 *sn-1* acyl chain. The two difference curves show that the disorder caused by exchanging 18:1n-9 at the *sn-2* position for 22:6n-3, occurs mainly

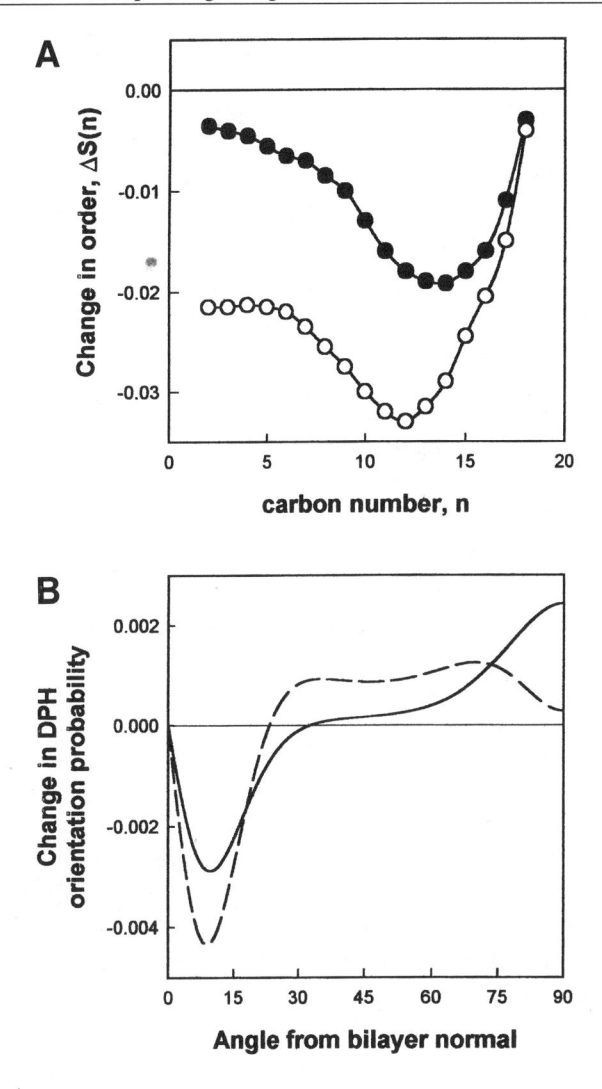

Fig. 2. Increased acyl chain unsaturation and increased temperature produce dissimilar increases in acyl chain disorder. (**A**) Difference order-parameter profiles, $\Delta S(n)$, of deuterium NMR measurements on perdeuterated 18:0 in the *sn-1* position as a function of changes in temperature and unsaturation at the *sn-2* position. ●: $\Delta S(n)$ between $18:0_{d35}$, 22:6 PC and $18:0_{d35}$, 18:1 PC; ○:$\Delta S(n)$ for $18:0_{d35}$,18:1 PC between 27°C and 47°C. Carbon atoms are numbered beginning at the glycerol backbone. (Data from Gawrisch and Holte, 1996; used by permission of K. Gawrisch). (**B**) Difference in orientation probability for the fluorescent membrane probe DPH. Orientation distribution of DPH in 16:0,18:1 PC at 40°C minus that in 20°C (—), and the distribution of di22:6 PC minus that of 16:0,18:1 PC at 20°C (---).

in the terminal half of the acyl chain, whereas the disorder induced by raising the temperature occurs over the entire chain.

Elevated temperature and increased unsaturation also have distinct effects on the average acyl chain packing. DPH orientation difference curves that compare the DPH orientation probability in 16:0,18:1 PC between 20°C and 40°C, and between 16:0,18:1 PC and di22:6 PC are shown in Fig. 2B. The solid curve shows that raising the temperature reduces the population oriented parallel to the acyl chains and increases the population in the bilayer midplane, oriented parallel to the bilayer surface. This shift indicates that an increase in temperature alters acyl chain packing in such a way that the bilayer midplane can accommodate a larger fraction of the DPH molecules. However, the distributions remain fairly narrow. The region below the zero line in the dashed curve shows that the presence of 22:6n-3 acyl chains also reduces the DPH population oriented parallel to the acyl chains. However, the dashed curve above the zero line shows that the presence of 22:6n-3 acyl chains produces a pronounced broadening of both orientational modes, rather than just a shift of DPH from the parallel mode to the perpendicular mode. The two comparisons in Fig. 2 demonstrate that elevated temperature and increased acyl chain unsaturation have distinct effects on both acyl chain disorder and average acyl chain packing and that the properties of highly unsaturated membranes are not equivalent to more saturated membranes at a higher temperature.

2.5. Interaction With Cholesterol

It is well established that cholesterol has a strong ordering effect on saturated phospholipid acyl chains in the fluid phase. This results in condensation of phospholipid monolayers and a reduction in enthalpy of the main gel to liquid-crystalline phase transition. These effects are reduced as *sn-2* acyl chain unsaturation is increased for phospholipids with a saturated *sn-1* chain (Hernandez-Borrell et al., 1993; Smaby et al., 1997), although the proximity of the double bonds to the headgroup is as important as the number of double bonds (Stillwell et al., 1994).

For all symmetrically substituted, unsaturated PCs, the chain-ordering effect of cholesterol is greatly reduced when compared with the corresponding *sn-1* saturated, *sn-2* unsaturated PC, and the effect of cholesterol decreases as the level of unsaturation increases (Mitchell & Litman, 1998b). A few studies have examined the effects of cholesterol in bilayers consisting of dipolyunsaturated PCs. All of these studies demonstrate that even at concentrations above 30 mol%, cholesterol has very little effect on the acyl-chain-packing properties of dipolyunsaturated bilayers. In both di20:4 PC and di22:6 PC, cholesterol has almost no effect on the gel–liquid-crystalline phase transition (Kariel et al., 1991), causes minimal change in acyl chain packing (Mitchell & Litman, 1998b), and causes only a small increase in the monolayer elastic area compressibility modulus (Smaby et al., 1997). In 18:0,18:1 PC bilayers, 50 mol% cholesterol increases the elastic area expansion modulus by 600 dyn/cm, whereas in di20:4 PC, 50 mol% cholesterol increases this parameter by only 50 dyn/cm (Needham & Nunn, 1990). The best explanation of these observation comes from a deuterium NMR study, which showed that cholesterol is soluble in di20:4n6 PC only to 15 mol% and that the molecular organization of cholesterol in this bilayer is profoundly different from that observed in *sn-1* saturated, *sn-2* polyunsaturated bilayers (Brzustowicz et al., 1999).

In recent years, much evidence has accumulated for lateral membrane domains that differ in their relative cholesterol content (Schroeder et al., 1995). In addition, it has been proposed that high levels of *sn-2* unsaturation may promote formation of microdomains, in which the saturated *sn-1* chains preferentially interact with each other (Litman et al., 1991). Several studies of cholesterol in bilayers containing high levels of polyunsaturation have reported evidence of lateral domains, which are driven by the preference of cholesterol for saturated acyl chains over polyunsaturated acyl chains (Huster et al., 1998; Mitchell & Litman, 1998b; Polozova & Litman, 2000; Zerouga et al., 1995). The recent

Chapter 2 / Modulation of Receptor Signaling

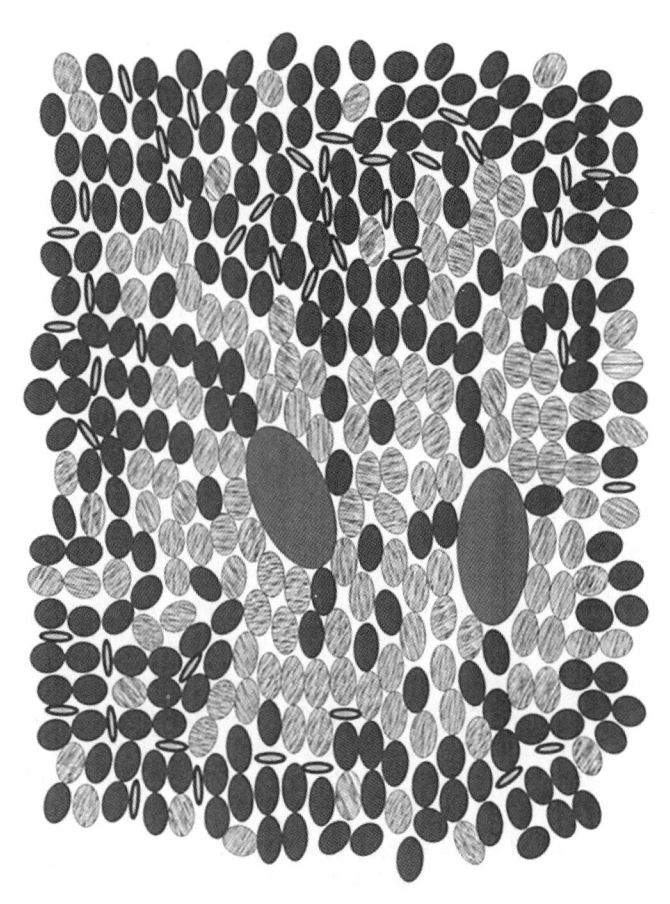

Fig. 3. Schematic representation of lateral domains in a bilayer consisting of di16:0 PC (dark ovals), di22:6 PC(striated ovals), cholesterol (small, light ovals) in a 7 : 3 : 3 ratio and rhodopsin (large gray ovals) at a 100 : 1 ratio of PC : rhodopsin. Rhodopsin is in a cluster, highly enriched in 22:6n-3 acyl chains, whereas cholesterol is mainly associated with the saturated 16:0 acyl chains. The enrichment of di22:6 PC in the cluster around rhodopsin is enhanced about six times relative to the bulk concentration. The cluster extends about three layers around rhodopsin.

work of Polozova and Litman (2000) is especially significant in terms of biological mechanisms, because it was found that in a mixed PC system composed of di16:0 PC and di22:6 PC, lateral domain behavior was observed only when both cholesterol and the integral membrane protein rhodopsin were included in the bilayer. A conceptual diagram of the proposed protein-containing microdomains is shown in Fig. 3. The observation that rhodopsin was essential for the formation of domains and showed a distinct preference for di22:6 PC indicates a mechanism whereby changes in either phospholipid acyl chain unsaturation or membrane cholesterol could control membrane domain formation and, thereby, integral membrane protein function.

3. EFFECTS OF ACYL CHAIN COMPOSITION ON MEMBRANE PROTEIN FUNCTION

Numerous studies have been published describing modulation of membrane protein function by changes in the degree of unsaturation of the phospholipid acyl chains. It is convenient to divide these studies into two types, those that investigated membrane protein function in natural membranes and those that examined the function of purified membrane proteins reconstituted with defined phospholipids. The literature abounds with examinations of the effects of diets or drugs on a wide range of physiological and behavioral outcomes. In this chapter, we are concerned with the relatively few studies that have examined the isolated function of one or more specific receptors and performed an analysis of the composition of the receptor's host membrane.

Several different receptor systems in the heart have been examined in this kind of detail. The amount of 22:6n-3-containing phospholipid in the sarcolemmal membranes of rats was elevated by injections of hydrocortisone, and this was accompanied by a downregulation of β-adrenergic receptors (Skuladottir et al., 1993). This finding is supported by the results of a detailed examination of 22:6n3-supplemented cardiomyoctes (Grynberg et al., 1995). Fatty acid enrichment produced cells in which 20% of the total fatty acids were 22:6n3. The α-adrenergic system was unaffected, but the β-adrenergic receptors had a decreased affinity for ligand. In a study of streptozotocin-induced diabetes mellitus, the sarcoplasic reticulum membranes of treated rats showed a loss of arachidonic acid, 20:4n6, acyl chains and an increase in 22:6n-3 (Kuwahara et al., 1997). This was accompanied by decreased sodium/potassium-ATPase activity and a reduction in calcium uptake.

A number of different types of membrane protein have been examined in reconstituted lipid vesicles containing high levels of acyl chain unsaturation, 20:4n6 or 22:6n3. These include sarcoplasmic reticulum calcium ATPase (Matthews et al., 1993), protein kinase C (PKC) (Giorgione et al., 1995; Slater et al., 1994;), gramicidin (Cox et al., 1992), and rhodopsin (Brown, 1994; Gibson & Brown, 1993; Litman & Mitchell, 1996b; Mitchell et al., 1990; Mitchell et al., 1992; O'Brien et al., 1977; Wiedmann et al., 1988). Higher levels of acyl chain unsaturation promote membrane protein function in all of these membrane proteins except calcium ATPase. Calcium ATPase was reconstituted into PCs with a 16:0 acyl chain at the *sn-1* position and 18:0, 18:1n9, 18:2n6, 20:4n6, or 22:6n3 acyl chain at the *sn-2* position. Enzyme function obtained with 18:1n9 or 18:2n6 at the *sn-2* position was more than 10 times higher than that obtained with 20:4n6 or 22:6n3 at the *sn-2* position (Matthews et al., 1993). In the other studies cited above, the highest level of protein function was obtained in the most highly unsaturated bilayer examined. However, there is no agreement among these studies regarding the physical property of the bilayer and the associated forces, which promote membrane protein function in highly unsaturated bilayers. The following subsections describe studies, wherein the authors have attributed their observations relative to membrane protein function to bilayer thickness, curvature stress, or acyl chain packing properties (e.g., acyl chain packing free volume).

3.1. Bilayer Thickness

Relatively large changes in membrane thickness have been demonstrated to alter the function of integral membrane proteins. An example of the magnitude of the change in membrane thickness needed to alter protein function is provided by studies of the sarcoplasmic reticulum calcium ATPase. Activity of this integral membrane protein in bilayers with symmetrically substituted, monounsaturated acyl chains with 16, 18, or 20 carbons is nearly constant. However, when the acyl chains are shortened to 14 carbons or lengthened to 22 carbons, activity is reduced by more than a factor of 3 (Lee, 1998).

This indicates that extreme changes in membrane thickness can alter protein function. However, the small changes in membrane thickness caused by physiologically relevant changes in acyl chain unsaturation are well within the range of constant calcium ATPase activity and seem unlikely to alter the function of other integral membrane proteins.

3.2. Curvature Strain

The activity of several different membrane proteins has been correlated with the propensity of the lipids composing the membrane to form nonlamellar structures or the curvature strain of the host bilayer (Epand, 1998; Li et al., 1995). Examples of alteration of membrane protein function via variable acyl chain composition, which were interpreted as changes in curvature stress, are provided by mitochondrial ubiquinol–cytochrome-*c* reductase and H^+ ATPase. When these protein are reconstituted into liposomes of di18:1n9-*cis* PC and di18:1n9-*trans* PE, their activity is increased as the percentage of di18:1n9-*cis* PE is raised, whereas the addition of di18:1n9-*trans* PE had no effect (Li et al., 1995). This demonstrated that the effect was not the result of the addition PE headgroups to the bilayer and suggested that it was related to the greater propensity of di18:1n9-*cis* to form a nonbilayer phase.

Epand and co-workers found that 22:6n-3 acyl chains produced the highest level of PKC function when incorporated into PE, but not PC, and the activity was correlated with increased partitioning of PKC to the membrane (Giorgione et al., 1995). Stubbs and co-workers found that PKC activity was optimal when the membrane had a combination of headgroup spacing and bilayer curvature, which could be obtained with a mixture of PEs and PCs containing a 22:6n3 acyl chain (Slater et al., 1994). The rate at which gramicidin converted from a nonchannel to a channel-forming conformation was highest in PC membranes, which contained 22:6n3 acyl chains or PE phospholipids, and it was proposed that this is the result of increased curvature stress (Cox et al., 1992).

A recent detailed study of calcium-ATPase function and phospholipid motion concluded that PE headgroups promote the activity of calcium ATPase by specific noncovalent interactions, rather than by bilayer curvature stress, as had been previously proposed (Hunter et al., 1999). In biological membranes, which contain both PE and PC and a range of acyl chain compositions, the effects of polyunsaturated acyl chains and PE headgroups on curvature stress can be difficult to assess, because of the complex influence the acyl chains of one phospholipid species exert on those of other species (Holte et al., 1996; Separovic & Gawrisch, 1996). Although quantitative correlation of protein function with curvature stress derived from acyl chain unsaturation remains difficult, there is ample evidence to suggest that the curvature stress induced by high polyunsaturated acyl chains could be functionally significant. The challenge is to measure both alterations in function and curvature stress in the same system, so as to allow a direct correlation between these membrane properties.

Rhodopsin, the light receptor in the G-protein-coupled visual transduction system, has been studied extensively in reconstituted systems. Light converts rhodopsin's antagonist, 11-*cis* retinal, to the agonist, all-*trans* retinal, resulting in the formation of activated receptor. The conformation of activated receptor, which binds the visual G protein, metarhodopsin II (MII), exists in equilibrium with an inactive conformation, metarhodopsin I (MI). Thus, the extent of functional activation is given by the equilibrium between MI and MII. The general observation is that the formation of MII is highest in bilayers composed of phospholipids with 22:6n3 acyl chains. In a series of experiments,

Brown and co-workers demonstrated that components that produce curvature strain could replace 22:6n3 acyl chains without compromising the extent of formation of MII. The ability of PE headgroups to support optimal rhodopsin function was analyzed in mixtures of di18:1 PC and di18:1 PE. MII formation in bilayers composed of 75% di18:1 PE and 25% di18:1 PC was found to be equivalent to that observed in bilayers where 50% of the acyl chains are 22:6n-3 (Brown, 1994). In a second set of measurements, it was found that diphytanoyl PC was as effective as di22:6 PC in promoting MII formation in PC/PE/PS mixtures, which mimicked the headgroup composition of the native rod outer segment disk membrane (Brown, 1994). Based on these findings, Brown has proposed that MII formation is facilitated by a lipid bilayer, which has curvature strain because the formation of MII results in a release of the curvature strain in the bilayer adjacent to rhodopsin (Brown, 1994). Although these studies provide a strong inference relative to the role of curvature strain in modulating MII formation, this interpretation suffers from a lack of direct measurements of curvature strain on the reconstituted membrane systems used in this study.

3.3. Acyl Chain Packing

An alternative explanation of the promotion of MII formation comes from direct measurements of acyl chain packing in rhodopsin-containing bilayers composed of a series PCs with varying levels of acyl chain unsaturation and cholesterol (Litman & Mitchell, 1996b; Mitchell et al., 1990; Mitchell et al., 1992). Acyl chain packing was characterized by analyzing the decay of fluorescence anisotropy of the membrane probe DPH in terms of the orientation distribution of DPH in the bilayer. The orientation distribution of DPH was summarized by a parameter, F_v, which is a measure of the difference between the DPH orientation distribution in the sterically restricted space of the bilayer and that which would be observed for an unrestricted free-tumbling DPH molecule (Mitchell & Litman, 1998a; Straume & Litman, 1987). F_v is positively correlated with the acyl chain packing free volume. Both MII formation and, F_v were measured in bilayers composed of a variety of PCs with and without cholesterol.

Examples of the effects of acyl chain composition and cholesterol on the MI–MII equilibrium constant, K_{eq}, are summarized in Fig. 4. The equilibrium constant for the formation of MII from its inactive precursor MI, K_{eq}, was measured in these rhodopsin-containing vesicles. For each acyl chain composition, K_{eq} was determined to be linearly correlated with F_v in a manner that was independent of cholesterol content (Mitchell et al., 1990; Mitchell et al., 1992). These correlations demonstrate that each acyl chain composition produces a unique correlation between MII formation and acyl chain packing that is not altered by cholesterol. The bars in Fig. 4 demonstrate that high levels of acyl chain unsaturation enhance MII formation. The linear correlations between K_{eq} and F_v demonstrate that this enhancement is related to the higher degree of disorder in acyl chain packing, resulting in increased acyl chain packing free volume. An unexplored question is whether the enhancement of MII formation by PE phospholipid groups is also correlated with acyl chain packing disorder.

In visual signal transduction, activation proceeds from the receptor, rhodopsin, to the effector, phosphodiesterase (PDE), via the visual G protein, G_t. Each MII sequentially binds and activates up to 100 G_t, thus MII–G_t binding initiates the first stage of signal amplification in the visual pathway. Litman et al. (2001) have studied the phospholipid acyl chain dependence of the kinetics of formation of both the MII conformation and the

Chapter 2 / Modulation of Receptor Signaling

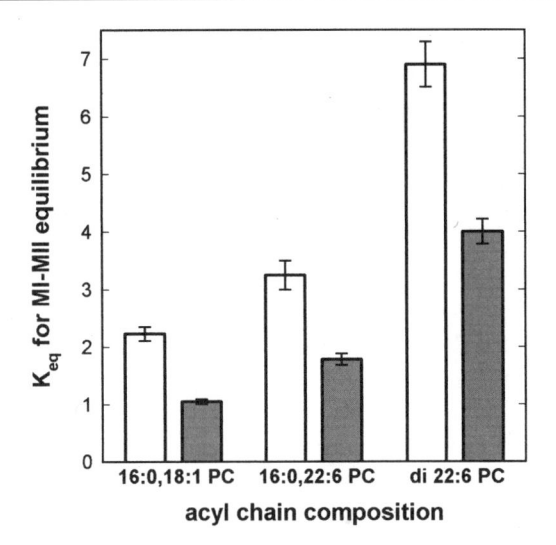

Fig. 4. Examples of the effects of acyl chain composition (white bars) and cholesterol (gray bars) on K_{eq} for the MI–MII equilibrium of photolyzed rhodopsin at 37°C. Higher values of K_{eq} correspond to higher equilibrium concentrations of MII, the state of photolyzed rhodopsin that participates in visual signal transduction by binding the visual G protein.

MII–G_t complex. The temporal nature of the interaction of MII and G_t is characterized by the ratio of the rate of formation of the MII–G_t complex divided by the formation rate of MII. This ratio varied from 1.39 to 4.95 in native disk membranes and 18:0,18:1 PC bilayers, respectively. In 18:0,22:6 PC bilayers, the ratio had an intermediate value of 3.46.

An important feature of 22:6n-3-containing bilayers is the ability to buffer the inhibitory effects of cholesterol. The inclusion of 30 mol% cholesterol in an 18:0,22:6 PC bilayer had relatively little effect on MII coupling to G_t, whereas this level of cholesterol in 18:0,18:1 PC bilayers resulted in a ratio of 9.1 and an increase in lag time for complex formation of 6.5-fold, relative to native disk membranes. Complex formation involves MII and G_t diffusion in the surface of the membrane. The increased lag time suggests that the diffusion process is dramatically slowed by the presence of cholesterol in 18:0,18:1 PC, whereas this process is relatively unaffected by cholesterol in 18:0,22:6 PC bilayers. A delay in the coupling of MII with G_t decreases the response time of the pathway. In addition to the kinetics of MII–G_t complex formation, a reduced binding affinity of MII to G_t was observed in 18:0,18:1 PC relative to 18:0,22:6 PC bilayers (Niu, Mitchell, and Litman, unpublished results). The addition of cholesterol reduced the binding affinity of MII for G_t to a greater degree in 18:0,18:1 PC bilayer than in 18:0,22:6 PC bilayers. Thus, signal amplification along the pathway will be reduced in less unsaturated bilayers. These data demonstrate explicitly that 22:6n-3-containing phospholipids can buffer the inhibitory effects of cholesterol in a signaling pathway and highlight the potential importance of 22:6n-3 acyl chains in optimizing both the response time and magnitude of response in signaling pathways.

In the visual pathway, the activity of the PDE is a measure of the integrated pathway activity. Litman et al. (2001) studied the phospholipid acyl chain dependence of the light-stimulated PDE activity. This study was carried out in reconstituted systems, which included G_t, PDE, and rhodopsin in unilamellar vesicles, whose phospholipid composition was either 16:0,18:1 PC or 16:0,22:6 PC. Each rhodopsin absorbing a photon is analogous to an agonist-bound receptor. A level of 1 in 1000 rhodopsin molecules activated by light produced 59% of the activity obtained in native disk membranes for rhodopsin in 16:0,22:6 PC bilayers and only 26% of disk activity for rhodopsin in 16:0,18:1 PC bilayers. Under conditions of saturating stimulation, 97% of the activity of native disk-membrane response was observed in the 16:0,22:6 PC vesicles system, whereas only 50% of the disk activity was seen in 16:0,18:1 PC vesicles. Here again, the system properties are optimized in 22:6n-3-containing bilayers.

The presence of lateral domains in di22:6 PC bilayers demonstrates an additional mechanism whereby 22:6n-3-containing bilayers can enhance signaling processes. If, in addition to rhodopsin, G_t and PDE also show preferential partitioning into regions rich in 22:6n-3, then lateral domain formation will increase the efficiency of association of these proteins by reducing the diffusion pathway for their interaction and increasing their effective concentration in the region of the microdomains.

4. SUMMARY

Highly polyunsaturated phospholipids produce membranes with several unique characteristics, and these characteristics are beneficial to many biological membrane functions. The high levels of 22:6n-3 acyl chains in the membrane phospholipids of the nervous system and retina suggest that these phospholipids play an important structural role. There is general agreement that the presence of polyunsaturated acyl chains in the phospholipids of membranes imparts a variety of unique features to these membranes. The specialized physical properties of highly polyunsaturated bilayers include increased bilayer area per headgroup, increased water permeability, higher degree of acyl chain flexibility, dynamics and disorder, sharply reduced interaction with cholesterol, and a tendency to enhance the formation of lateral domains. Although the actual forces that modulate protein function are still under investigation, it is clear that the unique properties imparted to biological membranes by the presence of polyunsaturated acyl chains and, in particular, 22:6n-3 chains play a fundamental role in determining membrane protein function.

The visual pathway, which is a prototypical G-protein-coupled receptor system, is one of the best characterized of this family of receptor systems. Various steps in this pathway are optimized in 22:6n-3-containing bilayers. The highest levels of MII formation in reconstituted systems are observed in 22:6n-3-containing bilayers (Litman & Mitchell, 1996b). The 22:6n-3-containing bilayers exhibited increased levels of MII–G_t complex formation and more favorable kinetic coupling of MII and G_t, relative to less unsaturated membrane systems (Litman et al., 2001). MII and G_t must interact rapidly to form a complex upon formation of MII in order to make signaling along the pathway efficient. As the bilayer acyl chains become less unsaturated, this process becomes delayed, introducing a lag time in the signaling process. Reduced affinity of the receptor, MII, for G_t will result in less amplification in the pathway, resulting from there being fewer G_ts activated. Observations made in reconstituted systems are in good agreement with electroretinogram (ERG) measurements made on n-3-deficient animals, where reduced signal amplitude and a lag time in signal development are seen. In view of the similarities between the visual system and other G-protein-coupled receptor pathways, the findings in the vision pathway ought to be applicable to neurotransmitter receptors in this superfamily. This extrapolation is supported by studies evaluating olfactory discrimination of rats raised on either an n-3-deficient or n-3-adequate diet. The n-3-adequate group made fewer errors in odor discrimination tests than the n-3-deficient group (Greiner et al, 1999). The olfactory bulb of rats raised on an n-3-deficient diet showed an 82% loss of 22:6n-3 relative to rats raised on an n-3-adequate diet. Olfactory and visual signaling are both G-protein-coupled receptor pathways and both are less sensitive in 22:6n-3-deficient animals.

The finding that 22:6n-3-containing bilayers buffer the inhibitory effects of cholesterol has strong implications for psychological disorders that are associated with variable levels of cholesterol. Under conditions of reduced levels of 22:6n-3 in n-3-deficiency and an increased cholesterol level, a reduced sensitivity in G-protein-coupled receptor signaling would be anticipated. Thus, the effectiveness of serotonin or other associated neurotransmitters would be decreased, potentially inducing some of the observed psychological dysfunction associated with cholesterol.

The studies described herein provide insight into the mechanism whereby 22:6n-3-containing phospholipids optimize membrane-associated signaling processes. Studies of both odor discrimination in n-3-deficient rats (Greiner et al, 1999) and visual deficits in n-3-deficient rhesus monkeys (Neuringer et al., 1984) and formula-fed infants (Birch et al., 2000) demonstrate a marked desensitization of two distinct G-protein-coupled signaling pathways to 22:6n-3 deficiency. It is anticipated that the sensitivity of G-protein-coupled receptor systems to levels of 22:6n-3 in membrane phospholipids will be of a general nature and this phenomenon may provide an explanation of the deficiencies in cognitive processes observed in n-3 deficiency.

REFERENCES

Applegate KR, Glomset JA. Computer-based modeling of the conformation and packing properties of docosahexaenoic acid. J Lipid Res 1986; 27:658–680.

Bazan NG, de Turco R, Gordon WC. Pathways for the uptake and conservation of docosahexaenoic acid in photoreceptors and synapses: biochemical and autoradiographic studies. Can J Physiol Pharmacol 1993; 71:690–698.

Birch EE, Garfield S, Hoffman DR, Uauy R, Birch DG. A randomized controlled trial of early dietary supply of long-chain polyunsaturated fatty acids and mental Development in infants. Dev Med Child Neurol 2000; 42:174–181.

Brown MF. Modulation of rhodopsin function by properties of the membrane bilayer. Chem Phys Lipids 1994; 73:159–180.

Brzustowicz MR, Stillwell W, Wassall SR. Molecular organization of cholesterol in polyunsaturated phospholipid membranes: a solid state 2H NMR investigation. FEBS Lett 1999; 451:197–202.

Cox KJ, Ho C, Lombardi V, Stubbs CD. Gramicidin conformational studies with mixed-chain unsaturated phospholipid bilayer systems. Biochemistry 1992; 31:1112–1117.

Cullis PR, de Kruijff B. Lipid polymorphism and the functional roles of lipids in biological membranes. Biochim Biophys Acta 1979; 559:399–420.

Engelberg H. Low serum cholesterol and suicide. Lancet 1992; 339:727–729.

Dekker CJ, Geurts van Kessel WS, Klomp JP, Pieters J, De Kruijff B. Synthesis and polymorphic phase behaviour of polyunsaturated phosphatidylcholines and phosphatidylethanolamines. Chem Phys Lipids 1983; 33:93–106.

Epand RM, Fuller N, Rand RP. Role of the position of unsaturation on the phase behavior and intrinsic curvature of phosphatidylethanolamines. Biophys J 1996; 71:1806–1810.

Epand RM. Lipid polymorphism and protein-lipid interactions. Biochim Biophys Acta 1998; 1376:353–368.

Gawrisch K, Holte LL. NMR investigations of non-lamellar phase promoters in the lamellar phase state. Chem Phys Lipids 1996; 81:105–116.

Gibson R, Makrides M. The role of long chain polyunsaturated fatty acids (LCPUFA) in neonatal nutrition. Acta Paediatr 1998; 87:1017–1022.

Gibson NJ, Brown MF. Lipid headgroup and acyl chain composition modulate the MI-MII equilibrium of rhodopsin in recombinant membranes. Biochemistry 1993; 32:2438–2454.

Giorgione J, Epand RM, Buda C, Farkas Y. Role of phospholipids containing docosahexaenoyl chains in modulating the activity of protein kinase C. Proc Natl Acad Sci USA 1995; 92:9767–9770.

Gruner SM. Intrinsic curvature hypothesis for biomembrane lipid composition: a role for nonbilayer lipids. Proc Natl Acad Sci USA 1985; 82:3665–3669.

Greiner RS, Moriguchi T, Hutton A, Slotnick BM, Salem N Jr. Rats with low levels of brain docosahexaenoic acid show impaired performance in olfactory-based and spatial learning tasks. Lipids 1999; 34:S239–243.

Grynberg A, Fournier A, Sergiel JP, Athias P. Effect of docosahexaenoic acid and eicosapentaenoic acid in the phospholipids of rat heart muscle cells on adrenoceptor responsiveness and mechanism. J Mol Cell Cardiol 1995; 27:2507–2520.

Hamosh M, Salem N Jr. Long-chain polyunsaturated fatty acids. Biol Neonate 1998; 74:106–120.

Hazel JR. Thermal adaptation in biological membranes: is homeoviscous adaptation the explanation? Ann Rev Phys 1995; 57:19–42.

Hernandez-Borrell J, Keough KM. Heteroacid phosphatidylcholines with different amounts of unsaturation respond differently to cholesterol. Biochim Biophys Acta 1993; 1153:277–282.

Hibbeln JR, Salem N Jr. Dietary polyunsaturated fatty acids and depression: When cholesterol does not satisfy. Am J Clin Nutr 1995; 62:1–9.

Holte LL, Peter SA, Sinnwell TM, Gawrisch K. ^{2}H nuclear magnetic resonance order parameter profiles suggest a change of molecular shape for phosphatidylcholines containing a polyunsaturated acyl chain. Biophys J 1995; 68:2396–2403.

Holte LL, Separovic F, Gawrisch K. Nuclear magnetic resonance investigation of hydrocarbon chain packing in bilayers of polyunsaturated phospholipids. Lipids 1996; 31:S199–S203.

Huang C, Li S. Calorimetric and molecular mechanics studies of the thermotropic phase behavior of membrane phospholipids. Biochim Biophys Acta 1999; 1422:273–307.

Hunter GW, Negash S, Squier TC. Phosphatidylethanolamine modulates Ca-ATPase function and dynamics. Biochemistry 1999; 38:1356–1364.

Huster D, Jin AJ, Arnold K, Gawrisch K. Water permeability of polyunsaturated lipid membranes measured by ^{17}O NMR. Biophys J 1997; 73:855–864.

Huster D, Arnold K, Gawrisch K. Influence of docosahexaenoic acid and cholesterol on lateral lipid organization in phospholipid mixtures. Biochemistry 1998; 37:17,299–17,308.

Kang JX, Leaf A. Prevention of fatal cardiac arrhythmias by polyunsaturated fatty acids. Am J Clin Nutr 2000; 71:202s–207s.

Kariel N, Davidson E, Keough KM. Cholesterol does not remove the gel–liquid crystalline phase transition of phosphatidylcholines containing two polyenoic acyl chains. Biochim Biophys Acta 1991; 1062:70–76.

Keough KM, Kariel N. Differential scanning calorimetric studies of aqueous dispersions of phosphatidylcholines containing two polyenoic chains. Biochim Biophys Acta 1987; 902:11–18.

Killian JA. Hydrophobic mismatch between proteins and lipids in membranes. Biochim Biophys Acta 1998; 1376:401–415.

Koenig BW, Strey HH, Gawrisch K. Membrane lateral compressibility determined by NMR and x-ray diffraction: effect of acyl chain polyunsaturation. Biophys J 1997; 73:1954–1966.

Kuwahara Y, Yanagishita T, Konno N, Katagiri T. Changes in microsomal membrane phospholipids and fatty acids and in activities of membrane-bound enzyme in diabetic rat heart. Basic Res Cardiol 1997; 92:214–222.

Lee AG. How lipids interact with an intrinsic membrane protein: the case of the calcium pump. Biochim Biophys Acta 1998; 1376:381–390.

Li L, Zheng LX, Yang FY. Effect of propensity of hexagonal II phase formation on the activity of mitochondrial ubiquinol-cytochrome c reductase and H(+)-ATPase. Chem Phys Lipids 1995; 76:135–144.

Litman BJ, Lewis EN, Levin IW. Packing characteristics of highly unsaturated bilayer lipids: Raman spectroscopic studies of multilamellar phosphatidylcholine dispersions. Biochemistry 1991; 30:313–319.

Litman BJ, Mitchell DC. Rhodopsin structure and function. In: Lee AG, ed, Biomembranes, Volume 2A Jai, Greenwich, CT, 1996, pp. 1–32.

Litman BJ, Mitchell DC. A role for phospholipid polyunsaturation in modulating membrane protein function. Lipids 1996; 31:S193–S197.

Litman BJ, Niu SL, Polozova A, Mitchell DC. The role of docosahexaenoic acid containing phospholipids in modulating G protein-coupled signaling pathways: Visual transduction. Mol Cell Neuorsci, In Press, 2001.

Matthews PL, Bartlett E, Ananthanarayanan VS, Keough KM. Reconstitution of rabbit sarcoplasmic reticulum calcium ATPase in a series of phosphatidylcholines containing a saturated and an unsaturated chain: suggestion of an optimal lipid environment. Biochem Cell Biol 1993; 71:381–389.

McElhaney RN. The structure and function of the Acholeplasma laidlawii plasma membrane. Biochim Biophys Acta 1984; 779:1–42.

Mitchell DC, Straume M, Miller JL, Litman BJ. Modulation of metarhodopsin formation by cholesterol-induced ordering of bilayer lipids. Biochemistry 1990; 29:9143–9149.

Mitchell DC, Straume M, Litman BJ. Role of sn-1-saturated, sn-2-polyunsaturated phospholipids in control of membrane receptor conformational equilibrium: effects of cholesterol and acyl chain unsaturation on the metarhodopsin I in equilibrium with metarhodopsin II equilibrium. Biochemistry 1992; 31:662–670.

Mitchell DC, Litman BJ. Molecular order and dynamics in bilayers consisting of highly polyunsaturated phospholipids. Biophys J 1998; 74:879–891.

Mitchell DC, Litman BJ. Effect of cholesterol on molecular order and dynamics in highly polyunsaturated phospholipid bilayers. Biophys J 1998; 75:896–908.

Muldoon MF, Manuck SB, Matthews KA. Lowering cholesterol concentration and mortality: a quantitative review of primary prevention trials. Br Med J 1990; 301:309–314.

Nabekura J, Noguchi K, Witt MR, Nielsen M., Akaike N. Functional modulation of human recombinant gamma-aminobutyric acid type A receptor by docosahexaenoic acid. J Biol Chem 1998; 273:11,056–11,061.

Needham D, Nunn RS. Elastic deformation and failure of lipid bilayer membranes containing cholesterol. Biophys J 1990; 58:997–1009.

Neuringer M, Connor WE, Van Petten C, Barstad L. Dietary omega-3 fatty acid deficiency and visual loss in infant rhesus monkeys. J Clin Invest 1984; 73:272–276.

Niebylski CD, Salem N Jr. A calorimetric investigation of a series of mixed-chain polyunsaturated phosphatidylcholines: effect of sn-2 chain length and degree of unsaturation. Biophys J 1994; 67:2387–2393.

O'Brien DF, Costa LF, Ott RA. Photochemical functionality of rhodopsin-phospholipid recombinant membranes. Biochemistry 1977; 16:1295–1303.

Olbrich K, Rawicz W, Needham D, Evans E. Water permeability and mechanical strength of polyunsaturated lipid bilayers. Biophys J 2000; 79:321–327.

Palczewski K, Kumasaka T, Hori T, Behnke CA, Motoshima H, Fox BA, et al. Crystal structure of rhodopsin: A G protein-coupled receptor. Science 2000; 289:739–745.

Poling JS, Vicini S, Rogawski MA, Salem N Jr. Docosahexaenoic acid block of neuronal voltage-gated K^+ channels: subunit selective antagonism by zinc. Neuropharmacology 1996; 35:969–982.

Polozova A, Litman BJ. Cholesterol dependent recruitment of di22:6 PC by a G protein-coupled receptor into lateral domains. Biophys J 2000; 79(5):2632–2643.

Rawicz W, Olbrich KC, McIntosh T, Needham D, Evans E. Effect of chain length and unsaturation on elasticity of lipid bilayers. Biophys J 2000; 79:328–339.

Salem N Jr. Omega-3 fatty acids: molecular and biochemical aspects. In: Spiller GA, Scala J, eds., New Protective Roles for Selected Nutrients. Alan R. Liss, New York, 1989, pp. 109–228.

Schroeder F, Woodford JK, Kavecansky J, Wood WG, Joiner, C. Cholesterol domains in biological membranes. Mol Membr Biol 1995; 12:113–119.

Separovic F, Gawrisch K. Effect of unsaturation on the chain order of phosphatidylcholines in a dioleoylphosphatidylethanolamine matrix. Biophys J 1996; 71:274–282.

Skuladottir GV, Schioth HB, Gudbjarnason, S. Polyunsaturated fatty acids in heart muscle and alpha 1-adrenoceptor binding properties. Biochim Biophys Acta 1993; 1178:49–54.

Slater SJ, Kelly MB, Taddeo FJ, Ho C, Rubin E, Stubbs CD. The modulation of protein kinase C activity by membrane lipid bilayer structure. J Biol Chem 1994; 269:4866–4871.

Smaby JM, Momsen MM, Brockman HL, Brown RE. Phosphatidylcholine acyl unsaturation modulates the decrease in interfacial elasticity induced by cholesterol. Biophys J 1997; 73:1492–1505.

Stillwell W, Ehringer WD, Dumaual AC, Wassall SR. Cholesterol condensation of alpha-linolenic and gamma-linolenic acid-containing phosphatidylcholine monolayers and bilayers. Biochim Biophys Acta 1994; 1214:131–136.

Stinson AM, Wiegand RD, Anderson RE. Fatty acid and molecular species compositions of phospholipids and diacylglycerols. Exp Eye Res 1991; 52:213–218.

Stinson AM, Wiegand RD, and Anderson RE. Recycling of docosahexaenoic acid in rat retinas during n-3 fatty acid deficiency. J Lipid Res 1991; 32:2009–2017.

Straume M, Litman BJ. Influence of cholesterol on equilibrium and dynamic bilayer structure of unsaturated acyl chain phosphatidylcholine vesicles as determined from higher order analysis of fluorescence anisotropy decay. Biochemistry 1987; 26:5121–5126.

Straume M, Litman BJ. Equilibrium and dynamic bilayer structural properties of unsaturated acyl chain phosphatidylcholine–cholesterol–rhodopsin recombinant vesicles and rod outer segment disk membranes as determined from higher order analysis of fluorescence anisotropy decay. Biochemistry 1988; 27:7723–7733.

Tanskanen A, Vartianinen E, Tuomilehto J, Viinamaki H, Lehtonen J, Puska P. High serum cholesterol and risk of suicide. Am J Psychiatry 2000; 157:648–650.

van Ginkel G, van Langen H, Levine YK. The membrane fluidity concept revisited by polarized fluorescence spectroscopy on different model membranes containing unsaturated lipids and sterols. Biochimie 1989; 71:23–32.

Wiedmann TS, Pates RD, Beach JM, Salmon A, Brown MF. Lipid–protein interactions mediate the photochemical function of rhodopsin. Biochemistry 1988; 27:6469–6474.

Zerouga M, Jenski LJ, Stillwell W. Comparison of phosphatidylcholines containing one or two docosahexaenoic acyl chains on properties of phospholipid monolayers and bilayers. Biochim Biophys Acta 1995; 1236:266–272.

3
Role of Docosahexaenoic Acid in Determining Membrane Structure and Function

Lessons Learned from Normal and Neoplastic Leukocytes

Laura J. Jenski and William Stillwell

1. WHY IS DOCOSAHEXAENOIC ACID OF INTEREST?

Understanding the basis for lipid diversity remains one of the major unsolved problems in membrane biology. A plausible explanation of why biological membranes are composed of hundreds to perhaps thousands of different lipid species, when it initially appears that only a few would suffice, has not been forthcoming. Clearly, unique functions for many of the different lipids have yet to be uncovered. All phospholipids have a role in forming the lipid bilayer, providing proper membrane surface charge and supplying the appropriate hydrophobic environment in which membrane proteins flourish. Each lipid species' unique role, if one exists, must reside with some other aspect of the membrane, perhaps a role in maintaining unusual structure (e.g., nonbilayer phase, regions of high curvature, etc.), supporting specific enzyme regulation, or furnishing a specialized role (e.g., platelet-activating factor). These functions may be accomplished through the existence of unique lipid domains (i.e., specialized and diverse membrane areas created by differential affinities of phospholipids for other membrane components) (Edidin, 1993).

Of particular interest to us is understanding the membrane role for the omega-3 fatty acid, docosahexaenoic acid (DHA, $22:6^{\Delta 4,7,10,13,16,19}$), the longest chain and most unsaturated fatty acid commonly found in biological membranes (Salem et al., 1986). It represents the extreme example of an omega-3 fatty acid. Once incorporated into a membrane, DHA will become an important component of the hydrophobic interior where it will affect membrane order or fluidity, thickness, domain size and stability, permeability, lipid phase, and, through interaction with membrane proteins, even biochemical activity (Salem et al., 1986; Stubbs & Smith, 1984). The presence of DHA in membranes is necessary for normal neurologic development (Menkes et al., 1962) and vision (Neuringer et al., 1988), as well as for benefits in various diseases, including atherosclerosis

From: *Fatty Acids: Physiological and Behavioral Functions*
Edited by: D. Mostofsky, S. Yehuda, and N. Salem Jr. © Humana Press Inc., Totowa, NJ

(Dyerberg & Jorgensen, 1982), autoimmune inflammatory diseases such as rheumatoid arthritis (James et al., 2000) (Calder, 1997) and multiple sclerosis (Bernsohn & Stephanides, 1967), and cancer (Bougnoux, 1999). How DHA exerts these health benefits is unclear. This fatty acid may act by altering the production of eicosanoid hormones from arachidonic acid, undergoing peroxidation, and affecting cell signaling and gene expression. The diverse array of conditions modulated by DHA implies that DHA's action is fundamental to various cell types and processes. We propose that one fundamental action is in determining membrane (domain) structure and function.

2. WHERE IS DHA FOUND?

In mammals, DHA is found in high amounts in three types of cellular membranes, those of the rod outer segment (ROS) (Wiegand & Anderson, 1983), synaptosomes (Breckenridge et al., 1972) and sperm (Neill & Masters, 1973). In these cells, DHA can represent more than 50% of the total membrane phospholipid acyl chains (Salem et al., 1986). At such high levels, much of the DHA exists as di-DHA phospholipid species (Miljanich et al., 1979). In these tissues, DHA is tenaciously retained at the expense of other fatty acids (Salem et al., 1986) and so levels are kept constantly high and are relatively unresponsive to dietary fluctuations. In all other tissues, DHA is found at much lower levels. In these cells, DHA exists primarily in hetero acid phospholipids with the *sn*-1 chain composed mainly of the saturated fatty acids palmitic or stearic acid and the *sn*-2 chain DHA (Anderson & Sperling, 1971). In leukocytes, DHA represents about 2–5% of total fatty acids (Fritsche & Johnston, 1990; Hinds & Sanders, 1993; Yaqoob et al., 1995). This value is dependent on diet (Fritsche & Johnston, 1990; Hinds & Sanders, 1993; Yaqoob et al., 1995) and increases with age (Huber et al., 1991). DHA is found predominantly in phosphatidylethanolamine (PE) and phosphatidylserine (PS), but it is also present in other phospholipid classes, including phosphatidylcholine (PC) and phosphatidylinositol (PI) (Robinson et al., 1993; Salem et al., 1986; Tiwari et al., 1988; Zerouga et al., 1996). Omega-3 fatty-acid-rich diets can increase the amount of DHA two to three fold in all phospholipid species of normal lymphocytes (Robinson et al., 1993), and can significantly increase the DHA content of phospholipids in leukemia and lymphoma cells grown in vivo as ascites (El Ayachi et al., 1990; Jenski et al., 1993). Culture of leukemia cells in DHA-enriched medium can increase the DHA content of phospholipids severalfold (Chow et al., 1991; Zerouga et al., 1996). For example, in our laboratory, DHA was incubated with T27A leukemia cells in culture, where it was initially incorporated into PE. DHA levels in PE of the cultured T27A cells increased from 3.3% of the total fatty acids in control cells to 21% after 12 h of incubation in DHA and 29% after 48 h of incubation (Zerouga et al., 1996). By comparison, PC–DHA rose from undetectable levels to 5.9% at 12 h and 9.6% at 24 h. After 48 h, T27A cells exhibited new DHA incorporation into PE about 2.7 times greater than into PC. When the 2.1-fold excess of PC over PE is factored in, it was estimated that a new incoming DHA prefers PE over PC by about 5.7 times. In a variety of cell types, DHA has been shown to be rapidly incorporated primarily into the plasma membrane of synaptosomes and mitochondria (Suzuki et al., 1997). Once incorporated from the diet into cellular membranes, DHA is tenaciously retained at the expense of other fatty acids (Salem et al., 1986).

3. WHAT IS THE STRUCTURE OF DHA?

With 22 carbons and 6 double bonds, DHA is the longest and most unsaturated fatty acid commonly found in membranes (Salem et al., 1986; Whiting et al., 1961). It would seem logical that when incorporated into phospholipids, this unusual fatty would support membranes that would be thicker than most and exceptionally fluid. Perhaps these two properties might be a good place to begin to search for DHA's health role.

3.1. Length of the DHA Chain

Docosahexaenoic acid's conformation remains a matter of conjecture. An early computer simulation by Applegate and Glomset (Applegate & Glomset, 1991) predicted that DHA would assume a rigid, extended "angle iron" shape. In fact, the "angle iron" conformation does predict better packing efficiency by forming a DHA, sn-2 groove into which the saturated sn-1 chain fits. However, this static model would likely pertain only to the gel state. The "angle iron" conformation is also at odds with experimental observations demonstrating that ROS membranes with very high levels of DHA are quite thin (27–28 Å) (Dratz et al., 1985). In comparison, bilayers made from phospholipids with much shorter 18:0 saturated chains are about 29–30 Å (Lewis & Engelman, 1983). In fact, nuclear magnetic resonance (NMR) studies predict that DHA's chain length is similar to that of palmitic acid (Mitchell et al., 1998).

These observations indicate that whatever DHA's conformation is, it must be compact. A second model, based on the ROS membrane, predicts a much different structure for DHA. A "molecular spring model" predicts a helical structure where DHA lengthens and shortens to accommodate conformation changes in rhodopsin (Dratz & Holte, 1992). Conformational energy calculations suggest DHA can lengthen or shorten over a range of 3–4 Å with a small input of energy (Dratz et al., 1985). NMR order parameters and spin lattice relaxation times support the idea that DHA performs rapid structural transitions between extended and looped conformations (Holte et al., 1998; Koenig et al., 1997; Mitchell et al., 1998). It is therefore likely that whatever DHA's structural role in membranes is, it does not support a thick membrane; in fact, DHA's conformation is quite compact.

3.2. DHA and Membrane Fluidity

It is often erroneously assumed that more fatty acyl double bonds automatically means a more fluid membrane. By differential scanning calorimetry, however, it has been reported that for a series of PCs where the sn-1 chain is stearic acid and the sn-2 chain has zero, one, two, three, four, or six (DHA) double bonds, the phase transition temperature (T_m) decreases upon the addition of a first and second double bond (Coolbear et al., 1983; Dratz & Deese, 1986). However, further double bonds actually *increases* T_m so that 18:0,22:6 PC has a slightly higher T_m than does 18:0,18:3 PC or 18:0,20:4 PC. There are many reports in the literature indicating that the addition of DHA to already fluid biological membranes does not further enhance their fluidity as measured by steady-state polarization of fluorescent membrane probes. Dietary fish-oil studies on hepatocytes (Clamp et al., 1997), intestinal microvillus (Wahnon et al., 1992), erythrocytes (Popp-Snigders et al., 1986; Popp-Snijders et al., 1987), platelets (Gibney & Bolton-Smith, 1988), and mitochondria (Stillwell et al., 1997) failed to find DHA-associated changes in membrane fluidity. With Y-79 retinoblastoma cells cultured in DHA-enriched media, Treen et al.

(Treen et al., 1992) were not able to detect any fluidity changes using the membrane probes DPH and TMA–DPH despite a fourfold increase in DHA. With phosphatidylcholine bilayers, Stubbs et al. (Stubbs et al., 1980), Wassall et al. (Wassall et al., 1992) and Ehringer et al. (Ehringer et al., 1990) could ascribe no unusual fluidity to phospholipid bilayers containing polyunsaturated chains. Membrane fluidity experiments, however, are not all in agreement. Numerous reports of DHA-induced increases in fluidity exist (Salem et al., 1986). Even in cases where no DHA-induced fluidity could be detected by steady-state polarization probes, fluidity changes were noted by lateral mobility probes. Salem and Niebylski (Salem & Niebylski, 1995) used steady-state fluorescence polarization and time-resolved correlation times to monitor fluidity in a series of PC lipid vesicles containing from zero to six double bonds. The steady-state measurements showed a large increase in fluidity upon the addition of the first double bond, a smaller increase with the second, and further but smaller fluidity increases up to four double bonds. No further increase could be measured for DHA with six double bonds. In contrast, with time-resolved anisotropy, there was a progressive graded decrease in the correlation time for each additional double bond, perhaps indicating an increase in fluidity. The relationship between number of double bonds and fluidity is a complex one, very dependent on the technique employed (Stubbs & Smith, 1984). We contend that whatever DHA's major role in membranes is, the simple concept of enhanced fluidization is inadequate.

4. MEMBRANE PROPERTIES OF DHA

If DHA's major effect on membranes is controlling neither membrane thickness nor global fluidity, what is it doing in membranes?

4.1. Membrane Order

^{2}H-Nuclear membrane resonance was used to determine the orientational order parameter of a series of PCs with perdeuterated stearic acid in the *sn*-1 chain and 18:1, 18:2, 18:3, 20:4, 20:5, and 22:6 in the *sn*-2 position (Holte et al., 1995). The profiles showed that unsaturation causes inhomogeneous disordering along the *sn*-1 chain. Going from one to six double bonds in the *sn*-2 chain results in a 0.01 reduction in S_{CD} (order parameter) in the plateau region and a maximum difference of 0.03 at C_{14} of the sn-1 chain. From these studies, it was determined that the molecular area of the *sn*-1 chain increases three to four times more around carbon-14 than it does for carbons in the top part of the chain. This suggests a wedge shape for 18:0,22:6 PC (a slightly wider acyl chain base than polar head) that loosens packing at the interface but is not so pronounced as to cause formation of a nonlamellar inverted H_{II} phase, Consequently, a "free volume" exists at the lipid–water interface. This space would be more available for small solutes such as water (Holte et al., 1995). These studies could detect only a slight change in membrane thickness.

4.2. Hydration

The wedge shape of DHA-containing PCs, proposed by Holte et al. (Holte et al., 1995) from ^{2}H-NMR, predicts that looser lipid packing at the aqueous interface would result in deeper penetration of water and other solutes into the bilayers. As a result, DHA favors increased hydration of the headgroup and interchain region. The looser surface packing would also favor insertion of proteins into DHA-rich portions of the membrane (Mitchell et al., 1998). Fluorescence measurements on polyunsaturated lipids also confirms that water content in the bilayer hydrocarbon region increases with increasing chain unsaturation (Mitchell & Litman, 1998).

4.3. Permeability

The increased hydration of DHA-containing membranes implies that these membranes should display enhanced water permeability. This prediction was confirmed by Huester et al. using ^{17}O-NMR (Huster et al., 1997). Diffusion-controlled water permeability increased from 155 μm/s for 18:0,18:1 PC to 330 μm/s for 18:0,18:3 PC and 412 μm/s for 18:0,22:6 PC. DHA-enhanced membrane permeability has been reported for several other solutes as well. Despite exhibiting the same bulk membrane fluidity, Ehringer et al. (Ehringer et al., 1990) reported that DHA as either the free fatty acid or as part of a mixed-chain PC had a much larger effect on erythritol permeability than did linolenic acid (α-18:3). Stillwell et al. (Stillwell et al., 1993a) reported that DHA enhanced bilayer permeability to carboxyfluorescein twofold to three fold more than oleic acid (18:1). Vesicles made from lipid extracts of rod outer segment membranes (DHA rich) were shown by Hendriks et al. (Henricks et al., 1976) to be 10 times more permeable to ^{22}Na$^+$ than vesicles made from lipids extracted from other (DHA poor) retinal lipids. Brand et al. (Brand et al., 1994) noted a relationship between H$^+$ permeability of liposomes made from mitochondrial lipid extracts and DHA content. Stillwell et al. (Stillwell et al., 1997) incorporated 18:0,22:6 PC into vesicles made from rat liver mitochondrial extracts and found a linear relationship between H$^+$ permeability and bilayer DHA content. Finally, Demel et al. (Demel et al., 1972) showed that DHA incorporated into the *sn*-2 position of PC increased permeability to glucose, erythritol, and glycerol.

Docosahexaenoic acid has also been linked to increases in membrane permeability in intact organelles and living cells. Stillwell et al. (Stillwell et al., 1997) showed that an increase in membrane DHA either through diet or by fusing the mitochondria with 18:0,22:6 PC large unilamellar vesicles (LUV) generated mitochondria with substantially enhanced H$^+$ permeability. Crespo-Armas et al. (Crespo-Armas et al., 1994) reported a *decrease* in H$^+$ permeability related to a decrease in rat mitochondrial membrane DHA content. In a series of reports, Jenski and co-workers (Jenski et al., 1991; Stillwell et al., 1993a) demonstrated that enhanced DHA membrane content in T27A leukemia cells resulted in increased leakage to ^{51}Cr and erythritol. Burns and Spector (Burns & Spector, 1990) reported that dietary DHA makes L1210 murine ascites lymphoblastic leukemia cells more permeable to the anti-cancer drugs mitoxantrone and doxorubicin. Although there are dissenters, there is substantial evidence that DHA enhances membrane permeability.

4.4. Packing

As with fluidity, it is often assumed that there must be an increase in molecular area with increasing number of acyl chain double bonds. Although this is true upon the addition of the first and second double bond, the molecular area is about the same for a third, fourth, and sixth double bond (Demel et al., 1972). Stillwell and co-workers have used the lipid-packing-sensitive probe MC540 to show that 18:0,22:6 PC bilayers are more loosely packed than 18:0,18:1 PC bilayers (Stillwell et al., 1993b) and that incorporation of DHA into membranes of mitochondria (Stillwell et al., 1997) and T27A tumor cells decreases lipid packing (Stillwell et al., 1993b). The various fatty acids exhibit different compressibility as a function of lateral pressure measured on a Langmuir film balance. Although DHA is much more expanded than oleic acid at low lateral pressures, at higher, more biologically relevant lateral pressures (> 30 mN/m) both fatty acids occupy about the same space (Urquhart et al., 1992). In fact, lateral compressibility moduli derived from pressure-area isotherms indicate that monolayers are more compressible (lower modulus) with increasing double bonds (Smaby et al., 1997). DHA is therefore more compressible than oleic acid. The large compressibility measured for DHA implies that less energy would be required to deform DHA-rich membranes (Smaby et al., 1997). Koenig et al. (Koenig et al., 1997) noted that for mixed-chain 18:0,22:6 PC, the stearic acid chain is far less compressible than is DHA. Lateral area compressibility under osmotic stress reveals that 75% of the net area change in 18:0,22:6 PC results from area change of DHA (Koenig et al., 1997).

4.5. Phase Transitions

The DHA-containing phospholipids exhibit very low phase transition temperatures. For example, 18:0,22:6 PC has its T_m at about –9°C (Stillwell et al., 2000), whereas 22:6, 22:6 PC's transition is at –68.4°C (Kariel et al., 1991) and the transition enthalpy · H for homo acid is extremely low (· H = 0.5 kcal/mol) (Kariel et al., 1991). The low T_m's are the result of packing restrictions resulting from steric effects caused by DHA's multiple rigid double bonds. The restrictions result in reductions in intermolecular and intramolecular van der Waal's interactions. The broad, low · H transitions are consistent with the notion that interaction between saturated *sn*-1 chains stabilize the gel state in hetero acids (Kariel et al., 1991). Furthermore, differential scanning calorimetry (DSC) isotherms are often multi-component, suggesting microclustering and domain formation (Niebylski & Salem, 1994).

4.6. Fusion

The membranes naturally enriched in DHA (ROS, synaptosomes, and sperm) are partly characterized by their predisposition to undergo membrane vesicle formation and fusion and, perhaps, this is related to DHA's location in these membranes. Years ago, polyunsaturated fatty acid-induced membrane fusion was reported by Ahkong et al. (Ahkong et al., 1973) and Meers et al. (Meers et al., 1988). Later, Lavoie et al. (Lavoie et al., 1991) proposed that polyunsaturated fatty acids may be involved in GTP-dependent membrane-fusion events. More recently, Talbot et al. (Talbot et al., 1997) proposed that acyl chain unsaturation and membrane curvature may be important participants in PEG-induced fusion of lipid vesicle model membranes. Negative curvature stress induced by lipids have also been linked to viral fusion (Epand, 1998). In a series of studies, Stillwell and Jenski have fused DHA-containing phospholipid vesicles with several types of membranes, including other types of phospholipid vesicles (Ehringer et al., 1990), mitochondria (Stillwell et al., 1997), and T27A tumor cells (Jenski et al., 1995b; Kafawy et al., 1998). DHA as either the free fatty acid or as part of a mixed chain PC (18:0,22:6 PC) enhanced lipid vesicle fusion to a much larger extent than did either stearic acid or α-linolenic acid (Ehringer et al., 1990).

4.7. Nonlamellar Phase and Curvature Stress

From pressure–area isotherms on a Langmuir film balance, Smaby et al. (Smaby et al., 1997) reported the average area/molecule at 30 mN/m for a series of PCs decreased from

70 Å² for 16:0,22:6 PC to 63 Å² for 16:0,18:1 PC. The diunsaturated series were 3–5 Å²/molecule larger than the saturated/unsaturated series. Holte et al. (Holte et al., 1995) using ²H-NMR, found an increase in area three to four times larger toward the methyl end of the acyl chains than near the lipid–water interface, suggesting that DHA-containing PC occupies a wedge shape, becoming slightly wider at the acyl chain center. Phospholipids with this shape are known to demonstrate a greater propensity to form an inverted H_{II} phase (Cullis & de Kruijff, 1979). Using ³¹P-NMR, Brzustowicz et al. (Brzustowicz et al., 2000) have found evidence that 18:0,22:6 PC can indeed form the H_{II} phase. Because of their much smaller headgroup size, it would be predicted that DHA–PEs should exhibit even more curvature strain than would DHA–PCs (Giorgione et al., 1995). Also, it is often in PEs that DHA accumulates in biological membranes (Zerouga et al., 1996). When lipids that are prone to form a nonlamellar phase are constrained in a planar lamellar phase, it results in bending or curvature energy that can be released by the interaction with some proteins such as rhodopsin and protein kinase C (Epand, 1998; Giorgione et al., 1995). Therefore, it appears that in excess lipid, DHA-containing phospholipids could drive inverted H_{II}-phase patches in membranes. However, in biological membranes, a large excess of these lipids seems unlikely and the role of DHA-containing phospholipids would be in creating negative curvature strain, altering the activity of a variety of proteins.

5. INTERACTION OF DHA WITH OTHER LIPIDS

5.1. Cholesterol

It has been proposed that membrane structure can be determined in part by the strength of its lipid–lipid interactions, driving the formation of lipid microdomains. A major structural lipid in plasma membranes is cholesterol, where it may exceed 50 mol% of the total lipids. How cholesterol interacts with DHA in membranes is now starting to emerge. It is well established that cholesterol is not randomly distributed in either model or biological membranes, but, instead, exists in transbilayer and lateral domains (Hui, 1988; Schroeder et al., 1991). Cholesterol has been shown to induce lateral phase separation, producing liquid-ordered (cholesterol rich) and liquid-disordered (cholesterol poor) domains (Almeida et al., 1992). Cholesterol has been shown to exhibit preferable interactions with a wide variety of membrane lipids. By DSC, Demel et al. (Demel et al., 1977) reported that cholesterol associates with phospholipids in the following sequence: sphingomyelin (SM) > PC >> PE. Cholesterol has even been shown to be excluded from membranes rich in PE (Bruckdorfer & Sherry, 1984). This is of particular interest because, in many membranes, PE is the phospholipid species that is most enriched in DHA. A weak association of cholesterol with DHA was reported by Demel et al. (Demel et al., 1972) and Ghosh et al. (Ghosh et al., 1973) using pressure–area isotherms on a Langmuir film balance and by Kariel et al. (Kariel et al., 1991) using DSC. More recently, Brzustowicz et al. (Brzustowicz et al., 1999) by solid-state NMR and x-ray diffraction suggested that cholesterol may be excluded from DHA-enriched lipid domains. NMR experiments suggest that when forced into the same locale, cholesterol interacts preferentially with the saturated *sn*-1 chain in mixed-chain phospholipids (Brzustowicz et al., 1999; Huster et al., 1998). In the presence of DHA, steric constraints imposed by the multiple rigid double bonds deter close contact with cholesterol's rigid sterol moiety, enhancing the sterol's interaction with the saturated acyl chain. As a result, cholesterol is excluded from dipolyunsaturated membranes (Brzustowicz et al., 1999). By DSC, the addition of up to 50 mol% cholesterol to 22:6,22:6 PC has little impact on T_m or · H of the phase transition (Kariel et al., 1991), whereas the transition of 16:0,22:6 PC is completely eliminated by only 35 mol% cholesterol (Hernandez-Borrell & Keough, 1993) Also, as measured by its elasticity moduli, 30 mol% cholesterol diminishes the in-plane elasticity of 16:0,22:6 PC but not 22:6,22:6 PC (Smaby et al., 1997). These and other experiments indicate that cholesterol-rich domains would likely be devoid of DHA.

5.2. α-Tocopherol

With six double bonds, DHA is the most oxidizable of all fatty acids and so must be stringently protected from oxidation. For this reason, Maggio et al. (Maggio et al., 1977) proposed that the antioxidant α-tocopherol could be particularly important in membranes (or domains) that are enriched in polyunsaturated fatty acids such as DHA, resulting in its being nonhomogeneously distributed. Stillwell et al. (Stillwell et al., 1992) demonstrated that in free ethanolic solution and in PC liposomes, α-tocopherol interacts better with PCs having more double bonds. More recently, these workers used DSC, fluorescence of MC540, fluorescence polarization of 1,6-diphenyl-2,3,5-hexatriene (DPH), proton permeability, and lipid peroxidation to show that α-tocopherol exerts a much larger effect than cholesterol on DHA-rich membranes (Stillwell et al., 1996). Their multicomponent DSC curves revealed the likely existence of DHA–α-tocopherol membrane domains. These experiments predict that DHA would partition into cholesterol-poor, α-tocopherol-rich lipid domains.

6. DHA AND THE INDUCTION OF LIPID MICRODOMAINS

6.1. Detergent-Resistant Membranes

At present, detergent-resistant membranes (DRMs) represent the best studied example of a lipid microdomain isolated from a biological membrane (Brown & London, 1997; Brown & London, 1998). Part of a cellular membrane is soluble in cold nonionic detergent, usually Triton X-100, and is in the liquid-disordered (l_d) phase (Brown & London, 1998). The remaining insoluble fraction, known as DRMs, are enriched in cholesterol and sphingolipids and are found in the liquid-ordered (l_o) phase. It has been reported, for example, that 41% of the erythrocyte plasma membrane (Yu et al., 1973) and 35% of the mastocytoma plasma membrane (Mescher & Apgar, 1986) lipids are insoluble in Triton. It has been demonstrated that cholesterol and sphingolipids promote the l_o state (Sankaram & Thompson, 1990) and induce detergent resistance and the glycerol lipids (phospholipids) comprise the bulk of the detergent-soluble l_d phase (Brown & London, 1998; Schroeder et al., 1998). It is in this detergent-soluble, cholesterol-poor phase that one would expect to find most of the DHA–phospholipids. Unfortunately, the majority of studies have concentrated on the detergent-insoluble phase and so have missed the significance of DHA's possible role in forming membrane microdomains. Perhaps much of the attention placed on DRMs involves their undisputed link to related signaling events. If a related family of DHA-altered activities could be linked to the detergent-soluble fraction, it might stimulate more interest in the other portion of cell membranes.

6.2. Model Membranes

Several studies linking DHA to lipid microdomains in model phospholipid membranes have been reported. The most thoroughly characterized of these models is that of Litman and Mitchell for the ROS membrane. These studies are discussed in Chapter 2. Two simple liquid crystalline–gel state phase-separated models have been reported by Niebylski and Litman (Niebylski & Litman, 1996; Niebylski & Littman, 1997) and by Dumaual et al. (Dumaual et al., 2000). Niebylski and Litman employed a novel set of fluorescence probes, 16:0,16:0 PE–pyrene and 22:6,22:6 PE–pyrene, to demonstrate that bilayers composed of 16:0,16:0 PC and 22:6,22:6 PC can phase separate (Niebylski & Littman, 1997). These investigators also confirmed phase separation in their model membranes by DSC (Niebylski & Litman, 1996). Dumaual et al. (Dumaual et al., 2000) examined phase separation in 16:0,16:0 PC/18:0,22:6 PC by DSC on bilayers and pressure–area (Π–A) isotherms and fluorescence digital imaging microscopy on monolayers. From DSC studies, it was concluded that dipalmitoyl–phosphatidylcholine (DPPC)/18:0,22:6 PC phase separates into DPPC-rich and 18:0,22:6 PC-rich phases. In monolayers, phase separation was indicated by changes in Π–A isotherms, implying phase separation where 18:0,22:6 PC is "squeezed out" of the remaining DPPC monolayer. Although both of these reports demonstrate the formation of DHA-enriched domains in model membranes, each suffers from the same shortcoming. They are examples of liquid crystalline–gel separations and, as such, have limited biological relevance. More complex model membranes composed of 18:0,22:6 PC or PE/sphingomyelin/cholesterol (mimicking the plasma membrane) are currently under investigation in our laboratories.

6.3. Trans-Membrane Domains

An asymmetric lipid distribution between the exoplasmic and cytoplasmic leaflets of plasma membranes is typical (Devaux, 1991). For example, it is well documented that highly unsaturated species of PE and PS are found primarily on the inner leaflet of many membranes. This has been reported for human erythrocytes (Knapp et al., 1994), murine synaptosomal plasma membranes (Fontaine et al., 1980), and human lymphocytes (Bougnoux et al., 1985), among others. It has also been shown that polyunsaturated fatty acids, including DHA, are found in higher concentrations in the aminophospholipids in the inner, cytoplasmic leaflet (Crinier et al., 1990; Hullin et al., 1991). As a result, the cytoplasmic leaflet of erythrocytes is more fluid than the exoplasmic leaflet (Morrot et al., 1986). The addition of polyunsaturated fatty acids to membranes have been shown to translocate cholesterol to the outer leaflet, where its efflux from membranes is enhanced (Dusserre et al., 1995).

7. DHA, MEMBRANE PROTEINS, AND LEUKOCYTE FUNCTION

7.1. Introduction to DHA and Proteins

Docosahexaenoic acid has the potential to affect cell function by modulating two aspects of membrane proteins: synthesis (particularly via transcriptional control) (Sellmayer et al., 1997) and structure (within the lipid milieu of the membrane). Although understanding the effect of DHA on proteins is still in its infancy, two examples, rhodopsin and protein kinase C (PKC), have been particularly well studied. The effect of DHA on rhodopsin is reviewed in detail in Chapter 2.

It is well documented that nonlamellar-preferring lipids such as those with DHA acyl chains can profoundly influence the activity of proteins (Epand, 1998). Lipids that are cylindrical in shape will have a uniform lateral pressure throughout the bilayer and will have little free volume or negative curvature strain (Epand, 1998). If lipids are wedge shaped, they will have increased lateral pressure toward the bilayer center with a negative curvature stress and a larger free volume near the aqueous interface. Protein kinase C is an amphitropic enzyme (it can exist in both a water-soluble and membrane-bound form). Its activity can be increased by lipids with negative curvature stress that is caused by changes in the physical property of the bilayer and not by the formation of small domains of inverted phases (Mosior et al., 1996). With protein kinase C, increased hexagonal phase propensity results in both an increased partitioning of the enzyme to the membrane as well as an increase in the activity of the membrane-bound form of the enzyme. Because protein kinase C may only penetrate the membrane a little, it is influenced only by the lateral pressure near the interface. A different physical feature of the membrane related to interfacial properties and not directly to curvature stress is responsible for the correlation between the presence of nonlamellar-forming lipids and protein kinase C activity (Giorgione et al., 1995; Mosior et al., 1996). It is the release of negative curvature stress that is believed to augment the activity of rhodopsin (Brown, 1994). Therefore, there is a correlation between protein function and nonlamellar-forming lipids for both protein kinase C and rhodopsin, but the molecular correlation is different (Epand, 1998). Also for protein kinase C, increasing unsaturation of PS decreases enzyme activity, indicating that PS's role is the result of specific protein–lipid interactions (Slater et al., 1996). Furthermore, fluidity is not a factor because cholesterol, a molecule known to reduce fluidity, has no effect on PKC activity (Slater et al., 1996). PKC activity was shown to be biphasic with increasing unsaturated PE (Slater et al., 1994; Souvignet et al., 1991). The initial increase in PKC activity is believed to be the result of curvature stress, whereas the subsequent decrease in activity is the result of changes in surface pressure (headgroup spacing). This suggests that a possible function of DHA is to enhance partitioning of PKC into the membrane.

In addition to rhodopsin and protein kinase C, the activity of numerous other proteins has been shown to be altered by nonlamellar phase-preferring lipids, including DHA. Included in the long and growing list of likely targets for DHA–enzyme interactions are the phospholipases A_2 (Huang et al., 1996) and C (Sanderson & Calder, 1998), CTP : phosphocholine cytidyltransferase (Arnold & Cornell, 1996), mitochondrial ubiquinone–cytochrome-c reductase (Li et al., 1995), Ca^{2+}-ATPase (Hui et al., 1981), GDP mannose–dolicolphosphate mannosyl transferase (Jensen & Schutzbach, 1988), cytochrome P450SCC (Schwarz et al., 1997), adenine nucleotide translocator (Streicher-Scott et al., 1994), the insulin receptor (McCallum & Epand, 1995), Na,K-ATPase (Mayol et al., 1999), the γ-aminobutyric acid type A receptor (Nabekura et al., 1998), lecithin : cholesterol acyltransferase (Parks et al., 1992), K^+ (Poling et al., 1995), Na^+ (Kang & Leaf, 1996) and Ca^{2+} (Pepe et al., 1994) channels, gap junctions (Hasler et al., 1991) and even the ionophores alamethicin and gramicidin A (Epand, 1998).

7.2. Introduction to Leukocytes

Leukocytes or white blood cells are a heterogeneous group of cells that aid in our body's defense against invading organisms and substances. Leukocytes include lymphocytes, neutrophils, monocytes/macrophages, dendritic cells, mast cells, basophils, eosinophils, and megakaryocytes. Diverse in their functions and characteristics, leukocytes do have two properties in common that utilize membrane proteins: the need to adhere to other cells and the engagement of signaling pathways to carry out effector activities, including, in some cases, proliferation. Adhesion is carried out by an assortment of cell-surface proteins and carbohydrates, whose select distribution among various tissues and at different stages of cell development and differentiation regulate the migration and ultimate cellular interactions of leukocytes. Signaling is a profoundly complex process utilizing seemingly countless proteins, including those at the plasma membrane that initiate the pathway that extends through the cytoplasm to the nucleus. Lymphocytes (T-cells, B-cells) undergo a special recognition process wherein interaction of their cell-surface antigen receptors with antigen results in a signal requisite for proliferation and differentiation. In the case of T-cells, antigen alone is not recognized, rather, peptides derived from foreign materials are bound to normal cell-surface glycoproteins (major histocompatibility complex [MHC] molecules) on "antigen-presenting cells," and the antigen receptors on T-cells interact with the peptide–MHC complex; if the interaction is sufficiently avid and other necessary costimulatory signals are received, the T-cell becomes activated. Large granular lymphocytes with natural-killing activity also interact with MHC molecules, but in this case, the interaction produces an inhibitory instead of an activating signal. Neoplasms arising from leukocytes may not necessarily retain antigen reactivity but do practice adhesion during metastasis, and are prime examples of dysregulated cell signaling.

Thus, to explore the role of DHA in modulating leukocyte membrane function, we will focus on three categories of membrane proteins: adhesion proteins, MHC molecules, and membrane-associated signaling proteins. In the case of adhesion proteins and signaling proteins, a few examples will be used to illustrate the point.

8. DHA AND ADHESION MOLECULES

Communication between leukocytes and the rest of the body that they patrol requires the intimate association of cells and, thus, the reciprocal expression of receptor and ligand on the respective cell types. Molecules involved in leukocyte adhesion and costimulation are exemplified by the integrins, CD11a/CD18 (LFA-1), CD11b/CD18 (Mac-1), CD11c/CD18 (CR4), members of the immunoglobulin supergene family, CD54 (ICAM-1), CD4, CD8, CD106 (VCAM-1), and C-type lectins (e.g., CD62E [E-selectin]). Leukocyte adhesion is regulated to allow the cells to bind first with low avidity and then, following stimulation, with high avidity, and, finally, to detach for continued migration and surveillance. Cytokines dramatically affect the surface expression of adhesion molecules by regulating gene transcription; by contrast, fatty acids such as DHA have the potential to affect adhesion by modifying the membrane lipid environment and, thus, the conformation and activity of adhesion proteins. Fatty acids are also implicated in regulation of gene expression by direct or indirect actions on nuclear factors (Sellmayer et al., 1997).

The structure of the plasma membrane plays a distinct role in modulating the adhesion of leukocytes to other cells. T-Lymphocytes with more loosely packed membrane lipids, as monitored by fluorescence of the membrane probe MC540, displayed greater adherence to macrophages than T-lymphocytes having a tighter lipid packing (Del Buono et al., 1989). MC540 also detected loosely packed lipids in membranes of viable cells enriched in DHA-containing phospholipids (Stillwell et al., 1993b), suggesting that DHA-rich membranes may facilitate cell–cell adhesion. However, when incorporated biosynthetically into cells, DHA was found to reduce the adherence of monocytes to (cytokine-stimulated) endothelial cells, in parallel with reduced endothelial expression of adhesion molecules VCAM-1, E-selectin, and ICAM-1 (De Caterina et al., 1995; De Caterina et al., 2000; De Caterina & Libby, 1996). DHA also suppressed ICAM-1 expression on activated monocytes (Hughes et al., 1996b). Fatty acids, in the order arachidonic acid (AA)>DHA>eicosapentaenoic acid (EPA), increased neutrophil adherence in vitro concomitantly with upregulated expression of Mac-1 (but not the other integrins LFA-1 and CR4) (Bates et al., 1993). DHA decreased adhesion of human peripheral blood lymphocytes to activated endothelial cells and reduced endothelial VCAM-1 expression and L-selectin as well as LFA-1 expression on lymphocytes, but did not influence endothelial ICAM-1 or E-selectin expression (Khalfoun et al., 1996a). Some of the disparities reported for DHA and EPA relate to the stimulus used to induce adhesion molecules on cells (Weber et al., 1995). Both DHA and EPA are reported to reduce the production of mRNA for various adhesion molecules (Collie-Duguid & Wahle, 1996; Wahle & Rotondo, 1999), suggesting that DHA may affect adhesion through transcriptional control of adhesion proteins as well as through direct effects on membrane structure.

CD8 and CD4, present on mature T-cell subsets, are not adhesion molecules *per se* but, rather, provide additional signals during antigen recognition, thereby enhancing the likelihood of T-cell activation. Oth et al. (Oth et al., 1990) observed that a fish-oil diet reduced CD4 expression on lymphoma cells (grown as ascites) without affecting expression of CD8, CD11a/CD18, and MHC I. The dietary fatty acids may have affected CD4 gene expression or modified the membrane so as to deform or mask the epitope detected by the anti-CD4 monoclonal antibody. With regard to the latter, we demonstrated that cells modified with DHA presented as a fatty acid in culture medium or in phospholipids fused into the plasma membrane of lymphocytes displayed altered expression of two of three CD8 epitopes (one epitope decreased, another increased, the last was unchanged) (Jenski et al., 1995a), arguing for a direct effect of DHA on CD8 conformation or the lateral distribution of CD8 and its interaction with other "masking" proteins perhaps in specific lipid microdomains.

9. DHA AND MAJOR HISTOCOMPATIBILITY COMPLEX PROTEINS

Fatty acids are reported to affect the expression of MHC molecules. In some cases, the protein's synthesis may be affected, whereas other experimental designs point to a direct effect of the lipid on plasma membrane-bound MHC. The function of MHC molecules is to bind peptides and present these peptides to specific T-cells, thereby clonally activating T-cells reactive to the peptide. The two key classes of MHC molecules functioning in this fashion are class I (MHC I) and class II (MHC II). MHC I molecules are present on all nucleated cells in the body and predominantly bind peptides generated by proteosome-mediated cleavage of cytosolic proteins. The tissue distribution of MHC II is limited to B-cells, dendritic cells, and macrophages, although expression may be induced on various other cell types, primarily by cytokines. MHC II molecules bind peptides produced in endolysosomes by proteolytic cleavage of exogenously derived proteins. These two types of molecules share an overall general structure, that of four extracellular protein domains, of which the two distal to the plasma membrane directly participate in peptide binding, transmembrane region(s), and cytoplasmic tail. MHC I molecules have three extracellular domains provided by an α-chain and the fourth by an associated non-MHC protein, $β_2$-microglobulin. Two MHC II polypeptides, the α-chain and the β-chain, each provide two extracellular domains.

The expression of MHC I molecules may be modified by DHA. We have demonstrated altered expression of MHC I, CD8, and CD90 (Thy-1) on murine lymphocytes and leukemia cells enriched in DHA through diet or cell culture (Jenski et al., 1995a; Jenski et al., 1993). In addition, we have shown that MHC I expression can be modulated by the direct insertion of DHA-containing phospholipids into the plasma membrane of viable cells; because one MHC I epitope increased concurrently with the decrease of another epitope, the effect of DHA was not a global loss of the MHC protein (Pascale et al., 1993). This is an important observation, as it implies that the role of the DHA-containing phospholipids is to induce a physical change in MHC I that pre-exists at the cell surface, rather than exclusively through modification of MHC I biosynthesis. This conclusion was drawn more directly from the observation that more monoclonal antibodies against an MHC I conformation-dependent epitope bound to purified MHC I reconstituted into DHA-containing phosphatidylcholine liposomes than liposomes composed of other fatty acids (Jenski et al., 2000). Finally, catalytic hydrogenation of viable murine leukemia cells reduced the membrane content of linoleic acid (LA), AA, and DHA but did not affect the expression of MHC I, although another protein displayed increased surface expression (Benko et al., 1987).

That plasma membrane structure is important for the expression of MHC I molecules is also suggested by experiments with other lipids. Cholesterol plays a role in modifying MHC surface expression, and, in general, cholesterol's effects oppose those of phospholipids (phosphatidylcholine). Incubation of mouse splenocytes with cholesteryl hemisuccinate for 1–2 h in vitro decreased the expression of MHC I and increased lipid packing as monitored by DPH fluorescence, whereas a PC lipid mixture from hen egg yolk increased the expression of MHC I and fluidized the membrane (Muller et al., 1983). Cholesterol enrichment of cultured human B-lymphoblasts led to increased MHC I clustering (Bodnar et al., 1996), possibly as a result of conformational changes in the MHC protein. Bene et al. (1997) suggested that membrane depolarization may induce physical changes in the lipid bilayer and thereby produce conformational changes in human MHC I molecules.

In general, DHA or mixtures of omega-3 fatty acids decrease the expression of MHC II molecules. As designed, most experiments detect changes in MHC II production rather than direct modification of plasma membrane structure. For example, human volunteers fed a fish-oil supplement displayed reduced expression of MHC II on peripheral blood monocytes before and after treatment with the cytokine γ-interferon (Hughes et al., 1996a), and dietary fish oil decreased MHC II expression on rat lymphatic dendritic cells (Sanderson et al., 1997) and murine peritoneal exudate cells (primarily B-cells and macrophages) (Huang et al., 1992). The nature of the stimulus used to induce MHC II expression is an important consideration in evaluating DHA's effects. Macrophages (thioglycollate-elicited peritoneal exudate cells) from n3-rich fish-oil-fed mice expressed more MHC II after a brief treatment with platelet-activating factor contrast than did macrophages from n6-rich safflower oil (Erickson et al., 1997), and similar results were obtained with EPA or DHA added in vitro. When EPA and DHA, combined in ratios commonly found in fish oil, were added to cultures of unstimulated human monocytes, MHC II expression was unaffected; however, these n3 combinations did inhibit MHC II expression on γ-interferon-stimulated monocytes (Hughes & Pinder, 1997). When these n3 fatty acids were added individually to unstimulated human monocyte cultures, EPA inhibited but DHA enhanced MHC II expression (both inhibited expression on stimulated monocytes) (Hughes et al., 1996b). In vitro, DHA inhibited MHC II expression on murine macrophages (thioglycollate-elicited peritoneal exudate cells) treated with γ-interferon in a fashion not reversed with leukotrienes or 5-HETE, implying an action for DHA independent of the lipoxygenase pathway (Khair-El-Din et al., 1996). With regard to function, peptides were shown to bind with greater affinity to purified murine MHC II in the presence of PC, PS, PI, and cardiolipin (but not PE, sphingomyelin, or cholesterol), although the role of very long-chain polyunsaturated fatty acids (e.g., DHA) was not tested (Roof et al., 1990).

10. DHA AND CELL SIGNALING

The influence of DHA on leukocyte signaling has focused on several membrane proteins, including surface receptors, ion channels, kinases, phosphatases, and phospholipases. Here, we examine three examples germane to leukocytes: the interleukin-2 receptor (IL2R), protein kinase C, and calcium channels (i.e., calcium mobilization).

10.1. IL2R

Interleukin-2 signals through a trimeric plasma membrane receptor composed of α, β, and γ polypeptide chains, thereby promoting cell cycling. T-Lymphocytes are a primary, although not an exclusive, target for IL2. Mitogen-stimulated peripheral blood lymphocytes from human volunteers consuming EPA and DHA ethyl esters displayed a reduced surface density of CD25, the IL2R α-chain induced by cell activation, implying reduced immunological responsiveness (Soyland et al., 1994). DHA acts, at least in part, at the level of transcription, dramatically decreasing IL2R mRNA in concanavalin A-stimulated murine splenic lymphocytes (Jolly et al., 1998). In this study, DHA- and EPA-enriched diets were similar in their inhibitory effects on IL2R mRNA synthesis, showing approximately twice the inhibition produced by AA supplementation relative to the LA-rich safflower-oil control diet. DHA or EPA added to phytohemagglutinin-stimulated human peripheral blood mononuclear cells in culture-inhibited lymphocyte proliferation independently of oxidation and eicosanoid hormones, but flow cytometry suggested that these fatty acids, in this case, increased the surface expression of CD25 (Khalfoun et al., 1996b).

10.2. Protein Kinase C

Protein kinase C, which exists in various isoforms, is activated when it translocates to the inner leaflet of the plasma membrane and interacts with the lipid bilayer including diacylglycerol (DAG). Omega-3 fatty acids induce or enhance PKC activation or translocation, often more so than other polyunsaturated fatty acids, and these actions have been reported in various cells including leukocytes. DHA, EPA, and AA induced the translocation of PKCα, PKC-βI, PKC-βII, and PKC-ε isozymes to a particulate fraction in parallel with enhanced respiratory burst in stimulated macrophages (Huang et al., 1997). DHA also stimulated respiratory burst in neutrophils, however, in a fashion independent of PKC but involving calmodulin (Poulos et al., 1991). Under certain conditions, DHA displays extraordinary ability to activate PKC; Marignani et al. (Marignani et al.,

1996) used an in vitro system of PKC in lipid vesicles to demonstrate that 18:0,22:6-*sn*-glycerol was more effective than 18:0,20:4-*sn*-glycerol, 18:0,20:5-*sn*-glycerol, or dioleoylglycerol in stimulating PKC activity.

10.3. Calcium Mobilization

Within seconds after receiving a signal, many cells display increased levels of cytosolic free calcium. The sources of the cytosolic free calcium are intracellular stores, such as the endoplasmic reticulum, and the extracellular milieu. IP_3, generated in concert with DAG from phosphatidylinositol 4,5-biphosphate (PIP_2) cleavage by phospholipase C, stimulates mobilization of stored calcium. Depletion of these stores stimulates calcium influx from the extracellular milieu. Fatty acids affect calcium mobilization and are thus assumed to affect cell activation. Chow et al. (Chow et al., 1990) stimulated the human T leukemia cell line Jurkat with antibodies to the CD3 component of the T-cell antigen receptor in the presence of various unsaturated free fatty acids. DHA, as well as EPA, α-linolenic acid (ALA), AA, LA, and oleic acid, did not affect the initial rise in cytosolic free calcium but appeared to prevent the extracellular calcium influx required for sustained cytosolic-free-calcium levels. Fatty acids did not affect CD3 expression as measured by flow cytometry, nor involve PKC in their mode of action, and thus the initial interpretation was a direct effect of fatty acids on the receptor-operated calcium channels. It was shown subsequently that the free fatty acids inhibited sustained cytosolic free calcium levels in anti-CD3-stimulated Jurkat cells by increasing calcium extrusion, presumably through activation of the plasma membrane calcium ATPase (Breittmayer et al., 1993). In untreated Jurkat cells (i.e., without additional stimulation), free polyunsaturated fatty acids directly mobilized intracellular calcium pools (these pools were also sensitive to anti-CD3 and IP_3) (Chow & Jondal, 1990); DHA appeared to be more effective than the other n3 (EPA, ALA) and n6 fatty acids (AA, LA). When various n3 and n6, but not n9, fatty acids were esterified into membrane phospholipids of Jurkat cells, the anti-CD3-induced rise in cytosolic free calcium was dampened, presumably because the influx of extracellular calcium, rather than release from intracellular stores, was impaired by the fatty acids (Chow et al., 1991).

11. CONCLUDING REMARKS

Omega-3 fatty acids have long been recognized as beneficial foodstuff, but their actions are complex and thus poorly understood. There is now a resurgence of interest in omega-3 fatty acids, particularly DHA, among basic scientists and clinicians. DHA is being used in a variety of forms, from dietary supplements to novel drug conjugates, to benefit an extensive and varied series of conditions, including normal vision and neurologic development, heart disease, autoimmune disease, and cancer. The emerging literature suggests a diversity of mechanisms of action for DHA: modulation of eicosanoid hormone production, generation of free radicals, regulation of gene expression, and the fundamental actions of DHA on membrane structure and function. It is the current state of the art, however, that the physiological and even cellular processes affected by DHA are not clearly connected to a single action of DHA, and thus there is urgent need for the interdisciplinary research that will link the "what," "how," "when," and "for whom" at the molecular, cellular, and organismal levels.

REFERENCES

Ahkong QF, Fisher D, Tampion W, Lucy JA. Fusion of erythrocytes by fatty acids, esters, retinol, and alpha-tocopherol. Biochem J 1973; 136:147–155.

Almeida PFF, Vaz WLC, Thompson TE. Lateral diffusion in the liquid phases of dimyristoylphosphatidylcholine/cholesterol lipid bilayers. A free volume analysis. Biochemistry 1992; 31:6739–6747.

Anderson RE, Sperling L. Lipids of the ocular tissue. VII. Positional distribution of the fatty acids in the phospholipids of bovine rod outer segments. Arch Biochem Biophy 1971; 144:673–677.

Applegate KR, Glomset JA. Computer-based modeling of the conformation and packing properties of docosahexaenoic acid. J Lipid Res 1991; 27:658–680.

Arnold RS, Cornell RB. Lipid regulation of CTP : phosphocholine cytidyltransferase: electrostatic, hydrophobic and synergistic interactions of anionic phospholipids and diacylglycerol. Biochemistry 1996; 35:9917–9924.

Bates EJ, Ferrante A, Harvey DP, Poulos A. Polyunsaturated fatty acids increase neutrophil adherence and integrin receptor expression. J Leukocyte Biol 1993; 53:420–426.

Bene L, Szollosi J, Balazs M, Matyus L, Gaspar R, Ameloot M, et al. Major histocompatibility complex class I protein conformation altered by transmembrane potential changes. Cytometry 1997; 27:353–357.

Benko S, Hilkmann H, Vigh L, van Blitterswijk WJ. Catalytic hydrogenation of fatty acyl chains in plasma membranes; effect on membrane lipid fluidity and expression of cell surface antigens. Biochim Biophys Acta 1987; 896:129–135.

Bernsohn J, Stephanides LM. Aetiology of multiple sclerosis. Nature 1967; 215:821–823.

Bodnar A, Jenei A, Bene L, Damjanovich S, Matko J. Modification of membrane cholesterol level affects expression and clustering of class I HLA molecules at the surface of JY human lymphoblasts. Immunol Lett 1996; 54:221–226.

Bougnoux P. n-3 Polyunsaturated fatty acids and cancer. Curr Opin Clin Nutr Metab Care 1999; 2:121–126.

Bougnoux P, Salem NJ, Lyons C, Hoffman T. Alterations in themembrane fatty acid composition of human lymphocytes and cultured transformed cells induced by interferon. Mol Immunol 1985; 22:1107–1113.

Brand MD, Couture P, Hulbert AJ. Liposomes made from mammalian liver mitochondria are more polyunsaturated and leakier to protons than those from reptiles. Comp Biochem Physiol 1994; 108:181–188.

Breckenridge WC, Gombos G, Morgan IG. The lipid composition of adult rat brain synaptosomal membranes. Biochim Biophys Acta 1972; 266:695–707.

Breittmayer J-P, Pelassy C, Cousin J-L, Bernard A, Aussel C. The inhibition by fatty acids of receptor-mediated calcium movements in Jurkat T-cells is due to increased calcium extrusion. J Biol Chem 1993; 268:20,812–20,817.

Brown DA, London E. Structure of detergent-resistant membrane domains: Does phase separation occur in biological membranes? Biochem Biophys Res Commun 1997; 240:1–7.

Brown DA, London E. Structure and origin of ordered lipid domains in biological membranes. J Membr Biol 1998; 164:103–114.

Brown MF. Modulation of rhodopsin function by properties of the membrane bilayer. Chem Phys Lipids 1994; 73:159–180.

Bruckdorfer KR, Sherry MK. The solubility of cholesterol and its exchange between membranes. Biochim Biophys Acta 1984; 769:187–196.

Brzustowicz MR, Stillwell W, Wassall SR. Molecular organization of cholesterol in polyunsaturated membranes: a solid state ^{2}H NMR investigation. FEBS Lett 1999; 451 197–202.

Brzustowicz MR, Zerouga M, Cherezov V, Caffrey M, Stillwell W, Wassall SR. Solid state NMR and X-ray diffraction studies of cholesterol molecular organization in polyunsaturated phospholipid membranes. Biophys J 2000; 78:184A.

Burns PC, Spector AA. Effects of lipids on cancer therapy. Nutr Rev 1990; 48:223–240.

Calder PC. n-3 Polyunsaturated fatty acids and cytokine production in health and disease. Ann Nutr Metab 1997; 41:203–234.

Chow SC, Ansotegui IJ, Jondal M. Inhibition of receptor-mediated calcium influx in T cells by unsaturated non-esterified fatty acids. Biochem J 1990; 267:727–732.

Chow SC, Jondal M. Polyunsaturated free fatty acids stimulate an increase in cytosolic Ca^{2+} by mobilizing the inositol 1,4,5-trisphosphate-sensitive Ca^{2+} pool in T cells through a mechanism independent of phosphoinositide turnover. J Biol Chem 1990; 265:902–907.

Chow SC, Sisfontes L, Jondal M, Bjorkhem I. Modification of membrane phospholipid fatty acyl composition in a leukemic T cell line: effects on receptor mediated intracellular Ca^{2+} increase. Biochim Biophys Acta 1991; 1092:358–366.

Clamp AG, Ladha S, Clark DC, Grimble RF, Lund EK. The influence of dietary lipids on the composition and membrane fluidity of rat hepatocyte plasma membrane. Lipids 1997; 32:179–184.

Collie-Duguid ESR, Wahle KWJ. Inhibitory effect of fish oil n-3 polyunsaturated fatty acids on the expression of endothelial cell adhesion molecules. Biochem Biophys Res Commun 1996; 220:969–974.

Coolbear KP, Berde CB, Keough KMW. Gel to liquid-crystalline transitions of aqueous dispersions of polyunsaturated mixed-acid phosphatidylcholines. Biochemistry 1983; 22:1466–1473.

Crespo-Armas A, Azavache V, Torres SH, Anchustegui B, Cordero, Z. Changes produced by experimental hypothyriodism in fibre type composition and mitochondrial properties of rat slow and fast twitch muscles. Acta Cientifica Venezolana 1994; 45:42–44.

Crinier S, Morrot G, Neumann J-M, Devaux PF. Lateral diffusion of erythrocyte phospholipids in model membranes-comparison between inner and outer leaflet components. Eur Biophys J 1990; 18:33–41.

Cullis PR, de Kruijff B. Lipid polymorphism and the functional roles of lipids in biological membranes. Biochim Biophys Acta 1979; 559:399–420.

De Caterina R, Cybulsky MA, Clinton SK, Gimbrone MA Jr, Libby P. Omega-3 fatty acids and endothelial leukocyte adhesion molecules. Prostaglandins Leukotrienes Essential Fatty Acids 1995; 52:191–195.

De Caterina R, Liao JK, Libby P. Fatty acid modulation of endothelial activation. Am J Clin Nutr 2000; 71:213S–223S.

De Caterina R, Libby P. Control of endothelial leukocyte adhesion molecules by fatty acids. Lipids 1996; 31:S57–S63.

Del Buono BJ, White SM, Williamson PL, Schlegel RA. Plasma membrane lipid organization an dthe adherence of differentiating lymphocytes to macrophages. J Cell Physiol 1989; 138:61–69.

Demel RA, Geuts van Kessel WS. M., Van Deenen LL. M. The properties of polyunsaturated lecithins in monolayers and liposomes and the interactions of these lecithins with cholesterol. Biochim Biophys Acta 1972; 266:26–40.

Demel RA, Jansen JWCM, Van Dijck PWM, Van Deenen LLM. The preferential interaction of cholesterol with different classes of phospholipids. Biochim Biophys Acta 1977; 465:1–10.

Devaux PF. Static and dynamic lipid asymmetry in cell membranes. Biochemistry 1991; 30:1163–1173.

Dratz EA, Deese AJ. The role of docosahexaenoic acid (22:6w3) in biological membranes: examples from photoreceptors and model membrane bilayers. In: Simopoulos AP, et al., eds. Health Effects of Polyunsaturated Fatty Acids in Sea Foods. Academic, New York, 1986, pp. 353–379.

Dratz EA, Holte LL. The molecular spring model for the function of docosahexaenoic acid (22:6w-3) in biological membranes. In: Sinclair A, Gibson R, eds. Essential Fatty Acids and Eicosanoids. American Oil Chemists Society, Champaign, IL, 1992, pp. 122–127.

Dratz EA, Van Breeman JF, Kamps KM, Keegstra W, Van Bruggen EF. Two-dimensional crystallizationof bovine rhodopsin. Biochim Biophys Acta 1985; 832:337–342.

Dumaual AC, Jenski LJ, Stillwell W. Liquid crystalline/gel state phase separation in docosahexaenoic acid-containing bilayers and monolayers. Biochim Biophys Acta 2000; 1463:395–406.

Dusserre E, Pulcini T, Bourdillon, M-C, Ciavatti M, Berthezene F. w-3 Fatty acids in smooth muscle cell phospholipids increase membrane cholesterol efflux. Lipids 1995; 30:34–41.

Dyerberg J, Jorgensen KA. Marine oils and thrombosis. Prog Lipid Res 1982; 21:255–269.

Edidin M. Patches and fences: probing for plasma membrane domains. J Cell Sci 1993; 17(Supp):165–169.

Ehringer W, Belcher D, Wassall S, Stillwell W. A comparison of the effects of linolenic (18:3Ω3) and docosahexaenoic (22:6Ω3) acids on phospholipid bilayers. Chem Phys Lipids 1990; 54:79–88.

El Ayachi N, Begin M, Mercier D, Ells G, Oth D. Susceptibility of RDM4 lymphoma cells to LAK-mediated lysis is decreased in tumor bearers fed fish oil high fat regimen. Cancer Lett 1990; 49:217–224.

Epand RM. Lipid polymorphism and protein-lipid interactions. Biochim Biophys Acta 1998; 1376:353–368.

Erickson KL, Howard AD, Chakrabarti R, Hubbard NE. Alteration of platelet activating factor-induced macrophage tumoricidal response, IA expression, and signal transduction by n-3 fatty acids. Adv Exp Med Biol 1997; 407:371–378.

Fontaine RN, Harris RA, Schroeder F. Aminophospholipid asymmetry in murine synaptosomal plasma membrane. J Neurochem 1980; 34:269–277.

Fritsche KL, Johnston PV. Effect of dietary omega-3 fatty acids on cell-mediated cytotoxic activity on BALB/c mice. Nutr Res 1990; 10:577–588.

Ghosh D, Williams MA, Tinoco J. The influence of lecithin structure on their monolayer behavior and interactions with cholesterol. Biochim Biophys Acta 1973; 291:351–362.

Gibney MJ, Bolton-Smith C. The effect of a dietary supplement of n-3 polyunsaturated fat on platelet lipid composition, platelet function, and platelet plasma membrane fluidity in healthy volunteers. Br J Nutr 1988; 60:5–12.

Giorgione J, Epand RM, Buda C, Farkas T. Role of phospholipids containing docosahexaenoyl chains in modulating the activity of protein kinase C. Proc Natl Acad Sci USA 1995; 92:9767–9770.

Hasler CM, Trosko JE, Bennink MR. Incorporation of n-3 fatty acids into WB-F344 cell phospholipids inhibits gap junctional intercellular communication. Lipids 1991; 26:788–792.

Henricks TH, Klompmakers AA, Daemen FJM, Bonting SL. Biochemical aspects of the visual process. XXXII. Movement of sodium ions through bilayers composed of retinal and rod outer segment lipids. Biochim Biophys Acta 1976; 433:271–281.

Hernandez-Borrell J, Keough KMW. Heteroacid phosphatidylcholines with different amounts of unsaturation respond differently to cholesterol. Biochim Biophys Acta 1993; 1153:277–282.

Hinds A, Sanders TAB. The effect of increasing levels of dietary fish oil rich in eicosapentaenoic and docosahexaenoic acids on lymphocyte phospholipid fatty acid composition and cell-mediated immunity in the mouse. Br J Nutr 1993; 69:423–429.

Holte LL, Koenig BW, Strey HH, Gawrisch KBJ. Structure and dynamics of the docosahexaenoic acid chain in bilayers studied by NMR and X-ray diffraction. Biophys J 1998; 74:A371.

Holte LL, Senaka AP, Sunwell TM, Gawrisch K. 2H Nuclear magnetic resonance order parameter profiles suggest a change of molecular shape for phosphatidylcholines containing a polyunsaturated acyl chain. Biophys J 1995; 68:2396–2403.

Huang H-W, Goldberg EM, Zidovetzki R. Ceramide induces structural defects into phosphatidylcholine bilayers and activates phospholipase A2. Biochem Biophys Res Commun 1996; 220:834–838.

Huang S-C, Misfeldt ML, Fritsche KL. Dietary fat influences Ia antigen expression and immune cell populations in the murine peritoneum and spleen. J Nutr 1992; 122:1219–1231.

Huang ZH, Hii CS. T., Rathjen DA, Poulos A, Murray AW, Ferrante A. n-6 and n-3 Polyunsaturated fatty acids stimulate translocation of protein kinase Cα, -βI, -βII, and -ε and enhance agonist-induced NADPH oxidase in macrophages. Biochem J 1997; 325:553–557.

Huber LA, Xu QB, Jurgens G, Bock G, Buhler E, Gey KF, et al. Correlation of lymphocyte lipid composition, membrane microviscosity and mitogen response in the aged. Eur J Immunol 1991; 21:2761–2765.

Hughes DA, Pinder AC. N-3 polyunsaturated fatty acids modulate the expression of functionally associated molecules on human monocytes and inhibit antigen presentation *in vitro*. Clin Exp Immunol 1997; 110:516–523.

Hughes DA, Pinder AC, Piper Z, Johnson IT, Lund EK. Fish oil supplementation inhibits the expression of major histocompatibility complex class II molecules and adhesion molecules on human monocytes. Am J Clin Nutr 1996; 63:267–272.

Hughes DA, Southon S, Pinder AC. (n-3) Polyunsaturated fatty acids modulate the expression of functionally associated molecules on human monocytes *in vitro*. J Nutr 1996; 126:603–610.

Hui SW. The spatial distribution of cholesterol in membranes. In: Yeagle PL, ed. The Biology of Cholesterol. CRC, Boca Raton, FL, 1988, pp. 213–232.

Hui SW, Stewart TP, Yeagle PL, Albert AD. Bilayer to non-bilayer transition in mixtures of phosphatidylethanolamine and phosphatidylcholine: implications for membrane properties. Arch Biochem Biophy 1981; 207:227–240.

Hullin F, Bossant MJ, Salem NJ. Aminophospholipid molecular species asymmetry in the human erythrocyte plasma membrane. Biochim Biophys Acta 1991; 1023:335–340.

Huster D, Albert JJ, Arnold K, Gawrisch K. Water permeability of polyunsaturated lipid membranes measured by ^{17}O NMR. Biophys J 1997; 73:856–864.

Huster D, Arnold K, Gawrisch K. Influence of docosahexaenoic acid and cholesterol on lateral lipid organization in phospholipid mixtures. Biochemistry 1998; 37:17,299–17,308.

James MJ, Gibson RA, Cleland LG. Dietary polyunsaturated fatty acids and inflammatory mediator production. Am J Clin Nutr 2000; 71:343S–348S.

Jensen JW, Schutzbach JS. Modulation of dolichyl-phosphomannose synthase activity by changes in the lipid environment of the enzyme. Biochemistry 1988; 27:6315–6320.

Jenski LJ, Bowker GM, Johnson MA, Ehringer WD, Fetterhoff T, Stillwell W. Docosahexaenoic acid-induced alteration of Thy-1 and CD-8 expression on murine splenocytes. Biochim Biophys Acta 1995; 1236:39–50.

Jenski LJ, Nanda PK, Jiricko P, Stillwell W. Docosahexaenoic acid-containing phosphatidylcholine affects the binding of monoclonal antibodies to purified K^b reconstituted into liposomes. Biochim Biophys Acta, 2000; 1467(2):293–306.

Jenski LJ, Sturdevant LK, Ehringer WD, Stillwell W. ω-3 Fatty acids increase spontaneous release of cytosolic components from tumor cells. Lipids 1991; 26:353–358.

Jenski LJ, Sturdevant LK, Ehringer WD, Stillwell W. Omega-3 fatty acid modification of membrane structure and function: I. Dietary manipulation of tumor cell susceptibility to cell and complement-mediated lysis. Nutr Cancer 1993; 19:135–146.

Jenski LJ, Zerouga M, Stillwell W. ω-3 Fatty acid-containing liposomes in cancer therapy. Proc Soc Exp Biol Med 1995; 210:227–233.

Jolly CA, McMurray DN, Chapkin RS. Effect of dietary n-3 fatty acids on interleukin-2 and interleukin-2 receptor a expression in activated murine lymphocytes. Prostaglandins Leukotrienes Essential Fatty Acids 1998; 58:289–293.

Kafawy O, Zerouga M, Stillwell W, Jenski LJ. Docosahexaenoic acid in phosphatidylcholine mediates cytotoxicity more effectively than other omega-3 and omega-6 fatty acids. Cancer Lett 1998; 132:23–29.

Kang JX, Leaf A. Evidence that polyunsaturated fatty acids modify Na^+ channels by directly binding to the channel protein. Proc Natl Acad Sci USA 1996; 93:3542–3546.

Kariel N, Davidson E, Keough KM. W. Cholesterol does not remove the gel–liquid crystalline phase transition of phosphatidylcholines containing two polyenoic acyl chains. Biochim Biophys Acta 1991; 1062:70–76.

Khair-El-Din TA, Sicher SC, Vazquez MA, Lu CY. Inhibition of macrophage nitric-oxide production and Ia-expression by docosahexaenoic acid, a constituent of fetal and neonatal serum. Am J Reprod Immunol 1996; 36:1–10.

Khalfoun B, Thibault G, Bardos P, Lebranchu Y. Docosahexaenoic and eicosapentaenoic acids inhibit in vitro human lymphocyte–endothelial cell adhesion. Transplantation 1996; 62:1649–1657.

Khalfoun B, Thibault G, Lacord M, Gruel Y, Bardos P, Lebranchu Y. Docosahexaenoic and eicosapentaenoic acids inhibit lymphoproliferative responses in vitro but not the expression of T cell surface activation markers. Scand J Immunol 1996; 43:248–256.

Knapp HR, Hullin F, Salem NJ. Asymmetric incorporation of dietary n-3 fatty acids into membrane aminophospholipids of human erythrocytes. J Lipid Res 1994; 35:1283–1291.

Koenig BW, Strey HH, Gawrisch K. Membrane lateral compressibility determined by NMR and x-ray diffraction: effect of acyl chain polyunsaturation. Biophys J 1997; 73:1954–1966.

Lavoie C, Jolicoeur M, Paiement J. Accumulation of polyunsaturated free fatty acid coincident with the fusion of rough endoplasmic reticulum membranes. Biochim Biophys Acta 1991; 1070:274–278.

Lewis BA, Engelman DM. Lipid bilayer thickness varies linearly with acyl chain length in fluid phosphatidylcholine vesicles. J Mol Biol 1983; 166:211–217.

Li L, Zheng LX, Yang FY. Effect of propensity of hexagonal II phase formation on the activity of mitochondrial ubiquinol–cytochrome c reductase and H(+)-ATPase. Chem Phys Lipids 1995; 76:135–144.

Maggio B, Diplock AT, Lucy JA. Interactions of tocopherols and ubiquinones with monolayers of phospholipids. Biochem J 1977; 161:111–121.

Marignani PA, Epand RM, Sebaldt RJ. Acyl chain dependence of diacylglycerol activation of protein kinase C activity in vitro. Biochem Biophys Res Commun 1996; 225:469–473.

Mayol V, Duran MJ, Gerbi A, Dignat-George F, Levy S, Sampol J, et al. Cholesterol and omega-3 fatty acids inhibit Na, K ATPase activity in human endothelial cells. Atherosclerosis 1999; 142:327–333.

McCallum CD, Epand RM. Insulin receptor autophosphorylation and signaling is altered by modulation of membrane physical properties. Biochemistry 1995; 34:1815–1824.

Meers P, Hong K, Papahadjopoulos D. Free fatty acid enhancement of cation-induced fusion of liposomes; symergism with synexin and other promoters of vesicle aggregation. Biochemistry 1988; 27:6784–6794.

Menkes JH, Alter M, Steigleder GK, Weakley DR, Sung JH. A sex-linked recessive disorder with retardation of growth, peculiar hair and focal cerebral and cerebellar degeneration. Pediatrics 1962; 29:764–779.

Mescher MF, Apgar JR. The plasma membrane "skeleton" of tumor and lymphoid cells: a role in cell lysis? Adv Exp Med Biol 1986; 184:387–400.

Miljanich GP, Sklar LA, White DL, Dratz EA. Disaturated and dipolyunsaturated phospholipids in the bovine retinal rod outer segment disk membrane. Biochim Biophys Acta 1979; 552:294–306.

Mitchell DC, Gawrisch K, Litman BJ, Salem NJ. Why is docosahexaenoic acid essential for nervous system function? Biochem Soc Trans 1998; 26:365–370.

Mitchell DC, Litman BJ. Molecular order and dynamics in bilayers consisting of highly polyunsaturated phospholipids. Biophys J 1998; 74:879–891.

Morrot G, Cribier S, Devaux PF, Geldwerth D, Davoust J, Bureau JF, et al. Asymmetric lateral movement of phospholipids in the human erythrocyte membrane. Proc Natl Acad Sci USA 1986; 83:6863–6867.

Mosior M, Golini ES, Epand RM. Chemical specificity and physical properties of the lipid bilayer in the regulation of protein kinase C by anionic phospholipids: evidence for the lack of a specific binding site for phosphatidylserine. Proc Natl Acad Sci USA 1996; 93:1907–1912.

Muller CP, Stephany DA, Shinitzky M, Wunderlich JR. Changes in cell-surface expression of MHC and Thy1.2 determinants following treatment with lipid modulating agents. J Immunol 1983; 131:1356–1362.

Nabekura J, Noguchi K, Witt MR, Nielsen M, Akaike N. Functional modulation of human recombinant gamma-aminobutyric acid type A receptor by docosahexaenoic acid. J Biol Chem 1998; 273:11,056–11,061.

Neill AR, Masters CJ. Metabolism of fatty acids by ovine spermatozoa. J Reprod Fertil 1973; 34:279–287.

Neuringer N, Anderson GJ, Conner WE. The essentiality of n-3 fatty acids for the development and function of the retina and brain. Ann Rev Nutr 1988; 8:517–541.

Niebylski CD, Litman BJ. Phase separation in bilayers containing disaturated and dipolyunsaturated acyl chains. Biophys J 1996; 70:A418.

Niebylski CD, Littman BJ. Lateral phase heterogeneity in dipolyunsaturated and disaturated phospholipid bilayers monitored with novel pyrene-labeled phospholipids. Biophys J 1997; 72:A305.

Niebylski CD, Salem NJ. A calorimetric investigation of a series of mixed-chain polyunsaturated phosphatidylcholines: effect of sn-2 chain length and degree of unsaturation. Biophys J 1994; 67:2387–2393.

Oth D, Mercier G, Tremblay P, Therien HM, Begin ME, Ells G, et al. Modulation of CD4 expression on lymphoma cells transplanted to mice fed (n-3) polyunsaturated fatty acids. Biochim Biophys Acta 1990; 1027:47–52.

Parks JS, Thuren TY, Schmitt JD. Inhibition of lecithin : cholesterol acyltransferase activity by synthetic phosphatidylcholine species containing eicosapentaenoic acid or docosahexaenoic acid in the sn-2 position. J Lipid Res 1992; 33:879–887.

Pascale AW, Ehringer WD, Stillwell W, Sturdevant LK, Jenski LJ. Omega-3 fatty acid modification of membrane structure and function. II. Alteration by docosahexaenoic acid of tumor cell sensitivity to immune cytolysis. Nutr Cancer 1993; 19:147–158.

Pepe S, Bogdanov K, Hallaq H, Spurgeon H, Leaf A, Lakatta E. w3 Polyunsaturated fatty acid modulates dihydropyridine effects on L-type Ca^{2+} channels, cytosolic Ca^{2+}, and contraction in adult rat cardiac myocytes. Proc Natl Acad Sci USA 1994; 91:8832–8836.

Poling JS, Karanian JW, Salem N Jr, Vicini S. Time- and voltage-dependent block of delayed rectifier potassium channels by docosahexaenoic acid. Mol Pharmacol 1995; 47:381–390.

Popp-Snigders C, Shouten JA, Van Blitterswijck WJ, Van der Veen EA. Changes in membrane lipid composition of human erythrocytes after dietary supplementation of (n-3) polyunsaturated fatty acids. Biochim Biophys Acta 1986; 854:31–37.

Popp-Snijders C, Schouten JA, Heine RJ, van der Meer J, van der Veen EA. Dietary supplementation of omega-3 polyunsaturated fatty acids improves insulin sensitivity in non-insulin-dependent diabetes. Diabetes Res 1987; 4:141–147.

Poulos A, Robinson BS, Ferrante A, Harvey DP, Hardy SJ, Murray AW. Effect of 22–32 carbon n-3 polyunsaturated fatty acids on superoxide production in human neutrophils: synergism of docosahexaenoic acid with f-met-leu-phe and phorbol ester. Immunology 1991; 73:102–108.

Robinson DR, Xu L-L, Knoell CT, Tateno S, Olesiak W. Modification of spleen phospholipid fatty acid composition by dietary fish oil and by n-3 fatty acid ethyl esters. J Lipid Res 1993; 34:1423–1434.

Roof RW, Luescher IF, Unanue ER. Phospholipids enhance the binding of peptides to class II major histocompatibility molecules. Proc Natl Acad Sci USA 1990; 87:1735–1739.

Salem N Jr, Niebylski CD. The nervous system has an absolute molecular species requirement for proper function. Mol Membr Biol 1995; 12:131–134.

Salem NJ, Kim H-Y, Yergey JA. Docosahexaenoic acid: membrane function and metabolism. In Simopoulos AP, Kifer RR, and Martin RE, eds. Health Effects of Polyunsaturated Fatty Acids in Seafoods. Academic, New York, 1986, pp. 319–351.

Sanderson P, Calder PC. Dietary fish oil appears to prevent the activation of phospholipase C-γ in lymphocytes. Biochim Biophys Acta 1998; 1392:300–308.

Sanderson P, MacPherson GG, Jenkins CH, Calder PC. Dietary fish oil diminshes the antigen presentation activity of rat dendritic cells. J Leukocyte Biol 1997; 62:771–777.

Sankaram MB, Thompson TE. Interaction of cholesterol with various glycerophospholipids and sphingomyelin. Biochemistry 1990; 29:10,670–10,675.

Schroeder F, Jefferson JR, Kier AB, Knittel J, Scallen TJ, Wood WG, et al. Membrane cholesterol dynamics: cholesterol domains and kinetic pools. Proc Soc Exp Biol Med 1991; 196:235–252.

Schroeder RJ, Ahmed SN, Zhu Y, London E, Brown DA. Cholesterol and sphingolipid enhance the Triton X-100 insolubility of glycosylphosphatidylinositol-anchored proteins by promoting the formation of detergent-insoluble ordered membrane domains. J Biol Chem 1998; 273:1150–1157.

Schwarz D, Kisselev P, Wessel R, Pisch S, Bornscheur U, Schmid RD. Possible involvement of nonbilayer lipids in the stimulation of the activity of cytochrome P450SCC (CYP11A1) and its propensity to induce vesicle aggregation. Chem Phys Lipids 1997; 85:91–99.

Sellmayer A, Danesh U, Weber PC. Modulation of the expression of early genes by polyunsaturated fatty acids. Prostagladins Leukotrienes Essential Fatty Acids 1997; 57:353–357.

Slater SJ, Kelley MB, Taddeo FJ, Ho C, Rubin E, Stubbs CD. The modulation of protein kinase C activity by membrane lipid bilayer structure. J Biol Chem 1994; 269:4866–4871.

Slater SJ, Kelly MB, Yeager MD, Larkin J, Ho C, Stubbs CD. Polyunsaturation in cell membranes and lipid bilayers and its effects on membrane proteins. Lipids 1996; 31:S189–S192.

Smaby JM, Momsen MM, Brockman HL, Brown RE. Phosphatidylcholine acyl chain unsaturation modulates the decrease in interfacial elasticity induced by cholesterol. Biophys J 1997; 73:1492–1505.

Souvignet C, Pelosin JM, Daniel S, Chambaz EM, Ransac S, Verger R. Activation of protein kinase C in lipid monolayers. J Biol Chem 1991; 266:40–44.

Soyland E, Lea T, Sandstad B, Drevon A. Dietary supplmentation with very long -chain n-3 fatty acids in man decreases expression of the interleukin-2 receptor (CD25) on mitogen-stimulated lymphocytes from patients with inflammatory diseases. Eur J Clin Invest 1994; 24:236–242.

Stillwell W, Dallman T, Dumaual AC, Crump FT, Jenski LJ. Cholesterol versus α-tocopherol: effects on properties of bilayers made from heteroacid phosphatidylcholines. Biochemistry 1996; 35:13,353–13,362.

Stillwell W, Ehringer W, Jenski LJ. Docosahexaenoic acid increases permeability of lipid vesicles and tumor cells. Lipids 1993; 28:103–108.

Stillwell W, Ehringer WD, Wassall SR. Interaction of a-tocopherol with fatty acids in membranes and ethanol. Biochim Biophys Acta 1992; 1105:237–244.

Stillwell W, Jenski LJ, Crump FT, Ehringer W. Effect of docosahexaenoic acid on mouse mitochrondrial membrane properties. Lipids 1997; 32:497–506.

Stillwell W, Jenski LJ, Zerouga M, Dumaual AC. Detection of lipid domains in docosahexaenoic acid-rich bilayers by acyl chain-specific FRET probes. Chem Phys Lipids 2000; 104:113–132.

Stillwell W, Wassall SR, Dumaual AC, Ehringer W, Browning CW, Jenski LJ. Use of merocyanine (MC540) in quantifying lipid domains and packing in phospholipid vesicles and tumor cells. Biochim Biophys Acta 1993; 1146:136–144.

Streicher-Scott J, Lapidus R, Sokolove PM. The reconstituted mitochondrial adenine nucleotide translocator: effects of lipid polymorphism. Arch Biochem Biophy 1994; 315:548–554.

Stubbs CD, Smith AD. The modification of mammalian membrane polyunsaturated fatty acid composition in relation to membrane fluidity and function. Biochim Biophys Acta 1984; 779:89–137.

Stubbs CD, Tsang WM, Belin J, Smith D, Johnson SM. Incubation of exogenous fatty acids with lymphocytes. Changes in fatty acid composition and effects on rotational relaxation time of 1,6-diphenyl-1,3,5-hexatriene. Biochemistry 1980; 19:2756–2762.

Suzuki H, Manabe S, Wada O, Crawford MA. Rapid incorporation of docosahexaenoic acid from dietary sources into brain microsomal, synaptosomal and mitochondrial membranes in adult mice. Int J Vitam Nutr Res 1997; 67:272–278.

Talbot WA, Zheng L, Lentz BR. Acyl chain unsaturation and vesicle curvature alter outer leaflet packing and promote poly(ehtylene glycol)-mediated membrane fusion. Biochemistry 1997; 36:5827–5836.

Tiwari RK, Venkatraman JT, Cinader B, Flory J, Wierzbicki A, Goh YK, et al. Influence of genotype on the phospholipid fatty acid composition of splenic T and B lymphocytes in MRL/MpJ-lpr/lpr mice. Immunol Lett 1988; 17:151–158.

Treen M, Uauay RD, Jameson DM, Thomas VL, Hoffman RR. Effect of docosahexaenoic acid on membrane fluidity and function in intact cultured Y-79 retinoblastoma cells. Arch Biochem Biophy 1992; 294:564–570.

Urquhart R, Chan RY, Li OT, Tilley L, Grieser F, Sawyer WH. Omega-6 and omega-3 fatty acids: monolayer packing and effects on bilayer permeability and cholesterol exchange. Biochem Int 1992; 26:831–841.

Wahle KW, Rotondo D. Fatty acids and enothelial cell function: regulation of adhesion molecule and redox enzyme expression. Curr Opin Clin Nutr Metab Care 1999; 2:109–115.

Wahnon R, Cogan U, Mokady S. Dietary fish oil modulates the alkaline phosphatase activity and not the fluidity of rat hepatocyte plasma membrane. J Nutr Metab 1992; 29:279–288.

Wassall SR, Yang McCabe RC, Ehringer WD, Stillwell, W. Effects of dietary fish oil on plasma high density lipoproteins: electron spin resonance and fluorescence polarization studies of ordering and dynamics. J Biol Chem 1992; 267:8168–8174.

Weber C, Erl W, Pietsch A, Danesch U, Weber PC. Docosahexaenoic acid selectively attenuates induction of vascular cell adhesion molecule-1 and subsequent monocytic cell adhesion to human endothelial cells stimulated by tumor necrosis factor-alpha. Arteriosclerosis, Thromb Vascular Biol 1995; 15:622–628.

Whiting LA, Harvey CC, Century B, Worwitt MK. Dietary alterations of fatty acids of erythrocytes and mitochondria of brain and liver. J Lipid Res 1961; 2:412–418.

Wiegand RD, Anderson RE. Phospholipid molecular species of frog rod outer segment membranes. Exp Eye Res 1983; 37:159–173.

Yaqoob P, Newsholme EA, Calder PC. Influence of cell culture conditions on diet-induced changes in lymphocyte fatty acid composition. Biochim Biophys Acta 1995; 1255:333–340.

Yu J, Fischman DA, Steck TL. Selective solubilization of proteins and phospholipids from red blood cell membranes by nonionic detergents. Journal of Supramol Struct 1973; 3:233–248.

Zerouga M, Stillwell W, Stone J, Powner A, Jenski LJ. Phospholipid class as a determinant in docosahexaenoic acid's effect on tumor cell viability. Anticancer Res 1996; 16:2863–2868.

4

Effects of Essential Fatty Acids on Voltage-Regulated Ionic Channels and Seizure Thresholds in Animals

Robert A. Voskuyl and Martin Vreugdenhil

1. INTRODUCTION

If polyunsaturated fatty acids (PUFAs) have an antiarrhythmic action on the heart, do they also have a suppressant action on other excitable cells, such as neuronal tissue? The answer to this question is important, because there are several conditions where lowering of neuronal excitability obviously has a beneficial effect and where there is still a need for new and better treatments. Neuropathic pain and neurological disorders such as epilepsy are two examples. The work of Leaf and co-workers (Leaf, Kang, Xiao, Billman & Voskuyl, 1999a; Leaf, Kang, Xiao, Billman & Voskuyl, 1999b) has convincingly demonstrated that the stabilizing effect of PUFAs on the electrical activity of isolated cardiac myocytes stems mainly from inhibition of voltage-regulated sodium and calcium currents. Some studies in neuronal preparations have pointed in the same direction (Park & Ahmed, 1992; Takenaka, Horie & Hori, 1987; Takenaka, Horie, Hori & Kawakami, 1988). In this respect, PUFAs bear a striking resemblance to local anesthetics and a number of antiepileptic drugs. Furthermore, one of the antiepileptic drugs that interact with sodium channels, valproic acid, is actually a short-chain fatty acid. This remarkable coincidence prompted us to investigate the possible antiepileptic action of a number of PUFAs on isolated hippocampal neurons and in an experimental epilepsy model in vivo. The second reason why PUFAs deserve interest in the context of epilepsy is the recent re-emergence of the ketogenic diet as an alternative for antiepileptic drug treatment. The diet has a high fat content and is low in carbohydrates. Although it does not seem to work in all patients, it has been remarkably successful in treating, in particular, the most difficult types of epilepsy in children. At present, it is unclear why and how the diet works, but it is not unreasonable to suppose that the composition of the diet and the different fat components should have an influence. This subject is discussed in detail in Chapter 17.

2. EPILEPSY, SYNDROMES, SEIZURES, AND ANTIEPILEPTIC DRUGS

Epilepsy is one of the most common neurological disorders, with a prevalence of approximately 0.5–1% of the world population. It is not a homogenous disease entity, but a collection of diverse syndromes. These are broadly divided in localization-related (focal, local, partial) epilepsies and generalized epilepsies, which involve the whole

From: *Fatty Acids: Physiological and Behavioral Functions*
Edited by: D. Mostofsky, S. Yehuda, and N. Salem Jr. © Humana Press Inc., Totowa, NJ

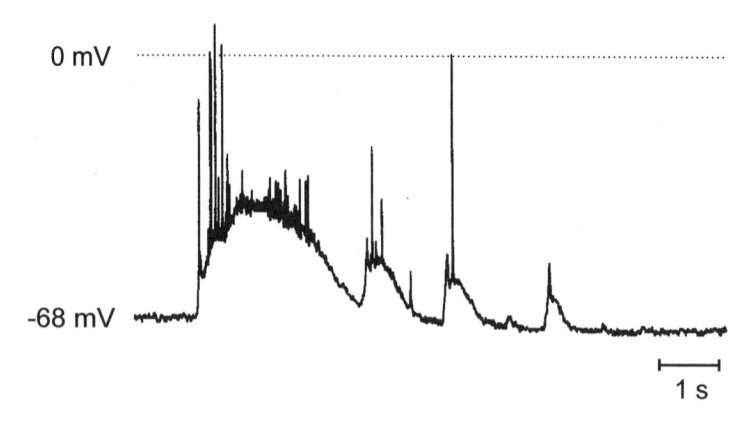

Fig. 1. Epileptiform discharges. Spontaneous epileptiform bursts recorded intracellularly in a CA1 pyramidal cell in a hippocampal slice of a young rat. The epileptiform activity was induced by perfusing the slice with 50 µ*M* 4-aminopyridine. Most of the action potentials are clipped by the low sampling rate.

brain. However, the complete official classification of epileptic syndromes and epilepsies (1989) comprises almost 40 different types. Whatever the syndrome or type of epilepsy, they all have in common the sudden and unpredictable, repeated occurrence of epileptic seizures, which, again, may take a large variety of appearances. Epileptic seizures are characterized by abnormal, excessive synchronous discharges in neuronal cerebral networks, accompanied by behavioral manifestations, ranging from a hardly noticeable arrest of activity that lasts only a few seconds, to the dramatic tonic–clonic convulsions that may last several minutes.

In the face of this bewildering complexity, it will be no surprise that different types of epilepsy and seizures ask for different drugs. Fortunately, in the majority of patients, seizures can be effectively controlled by one or more of the presently available drugs, allowing them a reasonable normal life. Nevertheless, there remains a considerable group of about 25% in which seizures cannot be suppressed by any drug or only at dosages that cause unacceptable side effects. Therefore, there is a continuing need for new drugs and new therapies. From a mechanistic point of view, the three major targets for antiepileptic drugs are enhancement of synaptic inhibition, depression of synaptic excitation, and containment of excessive action potential generation (e.g., Löscher, 1998; White, 1999). The importance of the latter is illustrated in Fig. 1, which shows a typical epileptiform burst of action potentials riding on top of a strong depolarization, induced in isolated brain tissue by one of the many available convulsants. As there is a remarkable similarity in how PUFAs decrease the excitability of cardiac myocytes and how local anesthetics, phenytoin, and other antiepileptic drugs do the same job for neurons, we will first describe the action potential mechanism at some length. Excellent, more detailed reviews on the molecular properties of sodium channels and actions of anticonvulsants have been published recently (Catterall, 1999; Ragsdale & Avoli, 1998; Taylor & Narasimhan, 1997).

3. VOLTAGE-REGULATED SODIUM CHANNELS

Figure 2 shows how conformational changes of the sodium channels are involved in the generation of action potentials. Sodium channels are large glycoproteins that form a

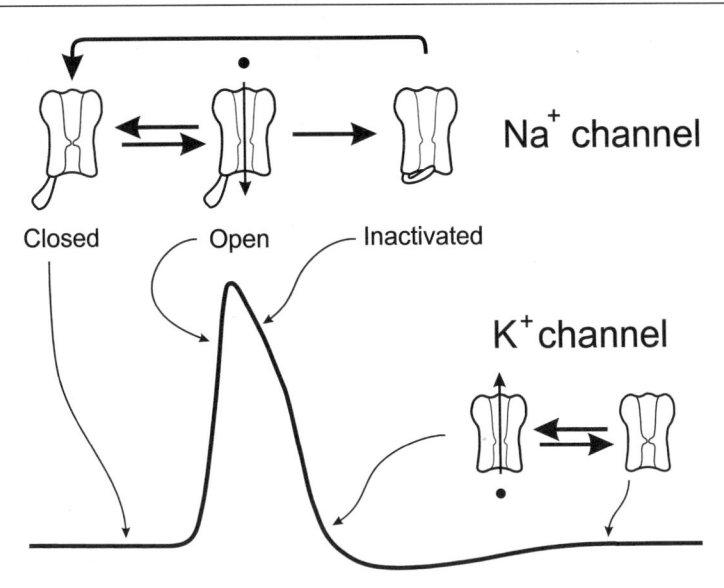

Fig. 2. Action potential generation. Opening and closing kinetics of sodium and potassium channels during an action potential. The curved arrows indicate the predominant conformation of the channels during the different phases of an action potential.

pore through the membrane. The opening and closing of this channel is regulated by the membrane potential. In neurons that are not active, the resting potential is about −70 mV (inside negative with respect to the outside), and at this potential, the large majority of the channels is in a nonconducting or closed conformation. This is indicated here schematically by a constriction in the pore. When the neuron is depolarized (e.g., by an excitatory postsynaptic potential), sensors in the sodium channel detect the voltage change and this is the signal for the channel to open. This allows sodium ions, for which the pore is selective, to rush into the cell and depolarize the cell further. As a result of this positive feedback mechanism, the membrane potential will change from −70 mV to approx +20 mV within a fraction of a millisecond. Depolarization not only causes opening or *activation* of the channel but, at a slightly slower time-scale, also closing or *inactivation* of the channel. The latter has been envisioned as a "hinged-lid" mechanism (the lid being an intracellular loop of the channel protein), where the lid physically plugs the pore. The inactivation of the channel blocks the influx of sodium, allowing the neuron to restore the resting potential, a process which is helped by opening of voltage-sensitive potassium channels. At the time-scale of the action potential (i.e., about a millisecond), the potassium channels simply open upon depolarization and close when the membrane potential is restored, but more slowly than the sodium channels.

It is important to realize that the conformational change from the "Closed" to the "Open" state is a reversible step, but the conversion from the "Open" to the "Inactivated" state is irreversible. Thus, upon continuing depolarization, all sodium channels eventually become trapped in the inactivated (nonconducting) state, from which they can only escape to the (also nonconducting) closed state by hyperpolarization (i.e., by restoration of the resting potential). This removal of inactivation is a relatively slow process and determines how quick a neuron can fire a new action potential. Properties of the activation and inactivation processes can be conveniently studied by clamping the membrane potential to

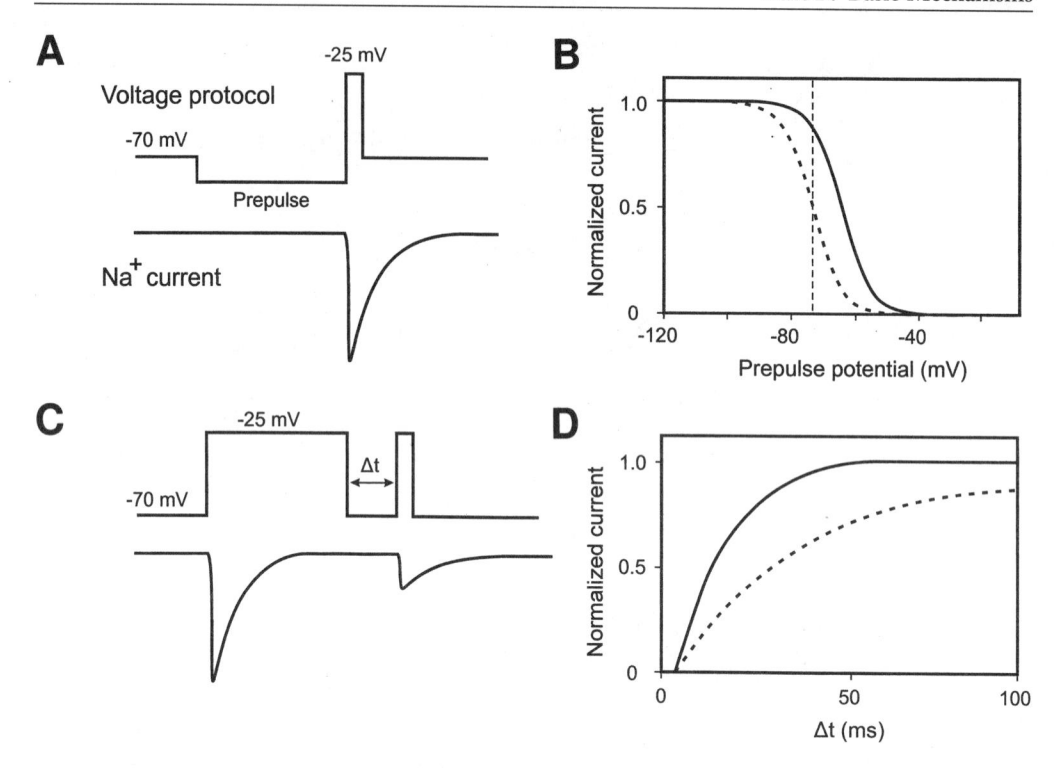

Fig. 3. Voltage-clamp protocols. Typical voltage-clamp protocols to assess the inactivation curve (**A** and **B**) and the recovery from inactivation (**C** and **D**). (**A**) The voltage protocol is shown in the top trace, the transient sodium current in the bottom trace (by convention an inward current is shown as a downward deflection). From rest, the membrane potential is first "stepped" to a conditioning potential (prepulse), and from there to a value near 0 mV for maximal activation of the sodium current. In this case, the prepulse is hyperpolarizing to remove sodium inactivation. (**B**) Inactivation curve (i.e., peak current as a function of the conditioning prepulse). The dashed line illustrates the typical leftward shift induced by phenytoin or carbamazepine. The dotted vertical line at about –75 mV illustrates that after a leftward shift, the maximal sodium current that can be achieved from that potential is reduced, and by that, the chance of generating an action potential. (**C**) In this protocol, all sodium channels are first brought into the inactivated state by activating them with a long depolarizing voltage step. After a recovery at the original resting potential for a certain time Δt, the sodium current is activated again. (**D**) Rate of recovery. The dashed line illustrates the delay of recovery by, for example, phenytoin.

different values in a stepwise fashion. The current needed to keep the potential at that level then provides information on the kinetics of the channel. Figure 3 shows some typical voltage-clamp protocols.

It should also be mentioned that the sodium channel contains a number of intracellular phosphorylation sites. The degree of phosphorylation of these sites is under the control of cyclic AMP or protein kinase C. This provides further opportunities for modulation of sodium currents, which may modify high-frequency firing.

Local anesthetics like lidocaine and anticonvulsants like phenytoin do not interfere with the process of activating sodium channels itself and thus leave the action-potential-generating mechanism intact. However, they cause a voltage- and use-dependent block of sodium current that reduces the number of channels available for action potential generation (Schwartz & Grigat, 1989; Taylor & Narasimhan, 1997; Vreugdenhil & Wadman, 1999), as do PUFAs in cardiac myocytes (Xiao, Kang, Morgan & Leaf, 1995; Xiao, Wright, Wang, Morgan & Leaf, 1998). The block is poor when the cell is activated from a strongly negative membrane potential, but it becomes progressively more effective when starting from more depolarized levels. This is illustrated by the typical leftward shift of the inactivation curve in Fig. 3B. Activating the neuron from a potential of about –75 mV (dotted vertical line), the peak sodium current is about 90% of the maximal achievable current, as indicated by the solid curve. In other words, at rest, 90% of the sodium channels are available for activation. If the curve is shifted to the left, as indicated by the dashed line, the peak sodium current will be only 50% of the maximum if activation starts from –75 mV. Only 50% of the channels are available for opening (i.e., 50% are in the inactivated state), and the chance of reaching the threshold for action potential generation is reduced.

Furthermore, the block increases with repeated activation of the sodium channel. This can be explained by assuming preferential, but slow, binding to the inactivated channel. Binding is therefore incomplete during the short time window after a single activation, but will accumulate with repeated activation. Furthermore, most anticonvulsants also delay the recovery of inactivation. This combined action will keep more and more channels in the inactivated state and fewer channels are available for opening. The various anticonvulsants, local anesthetics, and PUFAs differ only in minor details in their actions. The voltage- and use-dependent block of sodium channels is a very desirable property for anticonvulsants, because it means that the block becomes effective only when the neuron is depolarized and firing at high frequency, which is the case during epileptic activity. This limitation of the frequency at which a neuron can fire action potentials is very characteristic and often the first indication of an effect on sodium channels (McLean & Macdonald, 1986a; McLean & Macdonald, 1986b). In cardiac myocytes, PUFAs have an analogous action (Leifert, McMurchie & Saint, 1999; Xiao et al., 1995), providing a low-pass filter for repetitive contraction, in effect preventing cardiac arrhythmias in rat (Hock, Beck, Bodine & Reibel, 1990) and man (de Logeril et al., 1994). Under normal conditions when no high-frequency firing is needed, the functioning of the channel is hardly changed. Because activation is not affected, the shape of the action potential is otherwise unaffected.

Kang and Leaf provided evidence that PUFAs block sodium currents only in the free form (Kang & Leaf, 1994). Incorporated in the membrane, bound to albumin, or esterified, they are inactive. They also showed that the efficacy depends on the number of double bonds, the n-3 fatty acids docosahexaenoic acid (DHA) and eicosapentaenoic acid (EPA) being the most effective (Kang & Leaf, 1994). Finally, they showed that PUFAs have a similar action on calcium channels (Xiao, Gomez, Morgan, Lederer & Leaf, 1997), which behave in much the same way as sodium channels.

4. EFFECTS OF FATTY ACIDS ON SODIUM CURRENTS IN HIPPOCAMPAL NEURONS

Reduction in sodium currents in squid giant axons (Takenaka et al., 1987; Takenaka et al., 1988) and dorsal root ganglion cells cultured in PUFA containing media (Park & Ahmed, 1992) has been reported earlier. We have tested whether PUFAs have a similar action on sodium currents in pyramidal neurons from the hippocampal CA1 area. Our choice of neuron type was inspired by the crucial role this hippocampal output structure plays in temporal lobe epilepsy (Meier & Dudek, 1996).

Neurons were acutely isolated from enzyme-treated CA1 tissue pieces, cut from the hippocampus removed from the brain of adult Wistar rats. Detailed methods are given in two earlier articles (Vreugdenhil et al., 1996; Vreugdenhil & Wadman, 1999). The neurons selected for recording had a truncated apical dendrite and allowed adequate voltage control in the whole-cell patch configuration. Voltage-dependent calcium currents and potassium currents were pharmacologically blocked.

4.1. Voltage Dependence of Sodium Current Inactivation Is Affected

To assess the voltage dependence of the inactivation process, we determined the voltage dependence of steady-state inactivation, using a double-pulse voltage protocol. During the first conditioning pulse of 0.5 s, the fraction of channels in the inactivated state was set by stepping to different potentials. The result was tested by a second step to −25 mV (voltage protocol is given as an inset in Fig. 4A). Figure 4A gives a typical series of current traces recorded using the double-pulse protocol. After a hyperpolarizing conditioning pulse, a maximal inward current was recorded that decayed exponentially. Following a depolarizing conditioning pulse, only a small fraction of the current was left. The normalized peak current amplitude as a function of the conditioning voltage was fit with a Boltzmann equation (Fig. 4B) that describes the voltage dependence with a potential of half-maximal inactivation (V_h; −60.6 ± 0.3 mV, n = 128) and a factor V_c proportional to the slope at V_h (5.6 ± 0.1 mV). After two measurements in virtually lipid-free control solution containing 1 mg of delipidated bovine serum albumin (BSA), the voltage dependence of steady-state inactivation was determined in the presence of fatty acids in the BSA-free perfusate. We first tested the effect of 16 μM of the PUFA cis-4,7,10,13,16,19 docosahexaenoic acid (DHA: C22:6n-3). Sixteen micromolar DHA shifted the voltage dependence of steady-state inactivation by 13 mV to more hyperpolarized levels, with a small increase of V_c. We quantified the effect of different fatty acids on the shift in voltage dependence of steady-state inactivation. Figure 4C gives the negative shift in V_h (ΔV_h) for 16 μM of the PUFAs DHA, cis-5,8,11,14,17-eicosapentaenoic acid (EPA: C20:5n-3), and cis-9,12-octadecadienoic acid (linolenic acid; LA: C18:2n-6), the monounsaturated fatty acid cis-9-octadecaenoic acid (oleic acid; OA: C18:1n-9) and the unsaturated fatty acid hexadecanoic acid (palmitic acid; PA: C16:0) and, as a control, continued perfusion with BSA. There was a clear relationship between the amount of unsaturated bonds and the potency of the fatty acid to shift V_h, with the monounsaturated OA and the saturated PA not different from the BSA control. The same relationship was found in cardiac myocytes, where the potency was found to be related to membrane fluidity (Leifert et al., 1999). In correlation with the shift in the potential of half-maximal activation ΔV_h, the slope factor V_c increased significantly for the PUFAs (0.8 ± 0.2 mV for DHA, 1.3 ± 0.2 mV for EPA, and 0.6 ± 0.2 mV for LA).

As the prediction was that the refractory period would increase as a result of a slower recovery from inactivation, we tested the effect of DHA on the time-course of the recovery from inactivation. Using a double-pulse protocol, the current was first completely inactivated by a 20-ms conditioning pulse to −25 mV. After a recovery period of increasing duration at either −70 mV or −80 mV, the fraction of channels that were recovered from inactivation was assessed with a second pulse to −25 mV (Fig. 4D). The recovery from inactivation as a function of interval was fit by an exponential function with a time

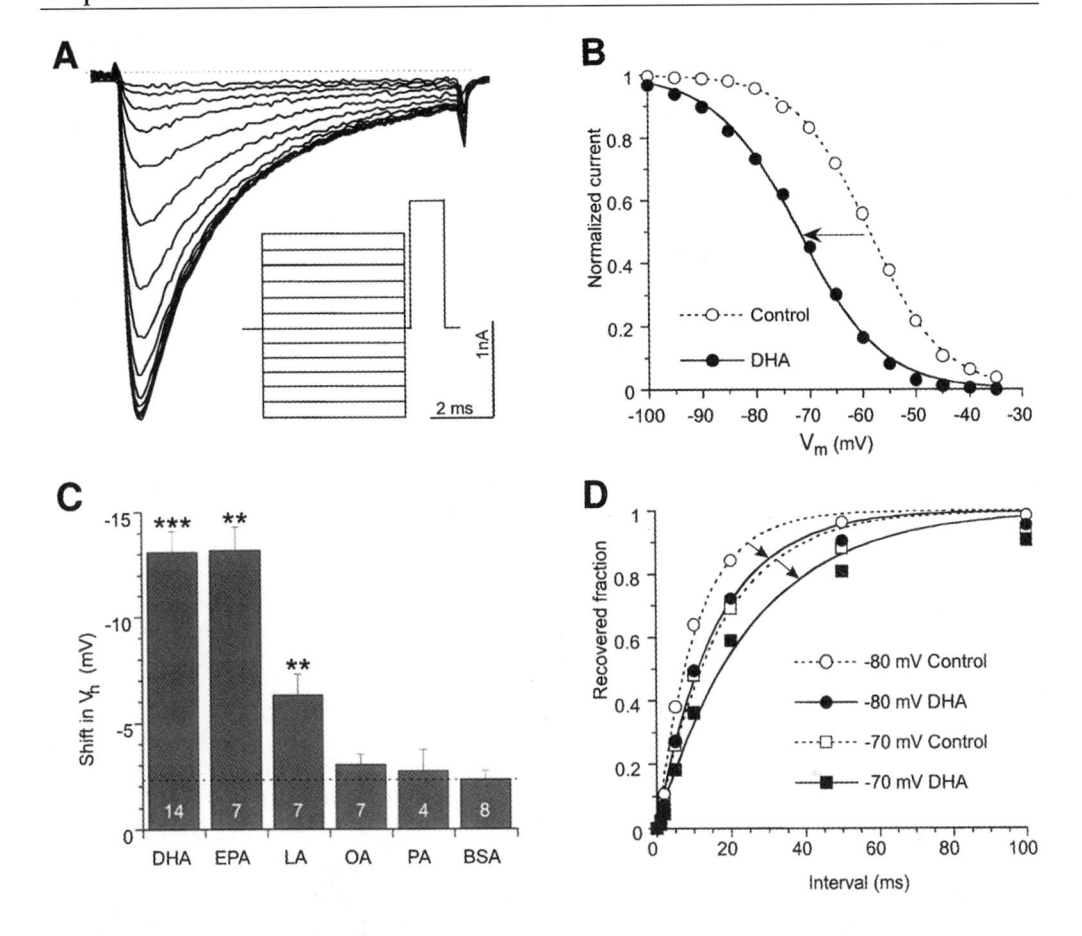

Fig. 4. PUFAs shift voltage dependence of sodium current inactivation. (**A**) Sodium current inactivation. A set of sodium currents recorded in an isolated CA1 neuron by a 10-ms depolarizing potential step to –25 mV, following different 0.5-s conditioning steps to potentials ranging from –140 mV to –35 mV. (**B**) Steady-state inactivation of the sodium current in the cell in Fig. 1A. Peak sodium current amplitude is normalized to the maximal current given as a function of conditioning step potential (*V*) in fat-free control solution (open symbols) and after perfusion with 16 μ*M* DHA (filled symbols). Data are fitted with a Boltzmann equation of the form $I/I_{max} = 1/(1 + e((V - V_h)/V_c))$, indicating a shift in the potential of half-maximal inactivation V_h of 13 mV and an increase of the slope factor V_c by 1.1 mV. (**C**) Potency of different fatty acids. The shift in V_h induced by 16 μ*M* of the PUFAs DHA (*n* = 14), EPA (*n* = 7), and LA (*n* = 7), the mono-unsaturated OA (*n* = 7), and the saturated PA (*n* = 4) are given and compared to the unspecific time-dependent shift with continued perfusion with BSA-containing control solution (*n* = 8). Error bars indicate SEM. Unpaired Student's *t*-test significance is indicated as $^{**}p < 0.01$; $^{***}p < 0.001$. (**D**) Recovery from inactivation. The time-course of the recovery from inactivation was assessed using two 20-ms steps to –25 mV with an interval of varying duration. After complete inactivation by the first step, the sodium current was allowed to recover from inactivation during the interval. The current peak amplitude, normalized to the unconditioned amplitude, is given for the same cell as in Fig. 1A,B as a function of interval duration, for an interval potential of –80 mV (circles) and –70 mV (squares), in control solution (open symbols), and after perfusion with 16 μ*M* DHA. Data are fit with a monoexponential function. The rate of recovery from inactivation is increased with more hyperpolarized interval potentials. DHA shifts this voltage dependence by about 10 mV in this neuron.

constant τ of 11.6 ± 0.4 ms at -80 mV and of 21.4 ± 0.4 ms at -70 mV ($n = 115$). Sixteen micromolar DHA slowed down the rate of recovery from inactivation at -80 mV to the same level as at -70 mV in the control solution, indicating a shift in the voltage dependence of recovery from an inactivation of about 10 mV to more hyperpolarized levels. The increase in τ at -80 mV induced by 16 μM of the fatty acids was 4.9 ± 0.9 ms for DHA, 2.7 ± 0.8 ms for EPA, 2.1 ± 0.9 ms for LA, -0.2 ± 0.8 ms for OA, and 0.1 ± 0.8 ms for PA. When compared to the small change with continued perfusion with BSA (-1.3 ± 0.7 ms), only the PUFAs DHA, EPA, and LA reduced the rate of recovery from inactivation significantly.

4.2. Sodium Current Activation Is Not Affected

The voltage dependence of activation was assessed by determining the sodium conductance at different depolarizing steps, from -120 mV, and fitting the voltage–conductance relationship with a Boltzmann equation (*see* Vreugdenhil et al., 1996). The potential of half-maximal activation was -31.1 ± 0.4 mV ($n = 86$). At a concentration of 16 μM, not one of the tested PUFAs affected the voltage dependence of activation. DHA and EPA suppressed the sodium conductance significantly (by $35 \pm 7\%$ and $10 \pm 6\%$ respectively). This suppressive effect was absent at lower concentrations and was more pronounced in cardiac myocytes at higher concentrations (Leifert et al., 1999).

4.3. Concentration–Effect Relationship

The PUFA-induced ΔV_h was determined for different concentrations of DHA or EPA. The concentration–effect relationship for DHA (Fig. 5A) and EPA (Fig. 5B) was fit with a Hill equation. With a Hill coefficient of 2, the maximal effect was a shift of -11.2 ± 0.6 mV for DHA and -11.4 ± 0.2 mV for EPA. The concentration of half-maximal effect (EC_{50}) was 2.1 ± 0.3 μM for DHA and 3.7 ± 0.2 μM for EPA. DHA and EPA are much more effective than the anticonvulsant carbamazepine. The ΔV_h induced by 15 μM carbamazepine under identical conditions ($\Delta V_h = -4.3$ mV) (Vreugdenhil & Wadman, 1999) can already be achieved by 1.6 μM DHA or 2.9 μM EPA.

5. EFFECTS OF PUFA ON CALCIUM CURRENTS

In addition to an effect on sodium currents, many known anticonvulsant drugs reduce calcium currents as well (Elliott, 1990; Rogawski & Porter, 1990). PUFAs were shown to suppress calcium currents in cardiac myocytes (Xiao et al., 1997). In a parallel study, we tested the effect of DHA and EPA on voltage-dependent calcium currents in acutely isolated CA1 neurons and with similar voltage protocols to assess the voltage dependence of activation and steady-state inactivation of the high-voltage activated calcium current as for the sodium current. For details, *see* Vreugdenhil et al., 1996; Vreugdenhil & Wadman, 1994. As for the sodium current, the calcium current activation characteristics were not affected by PUFAs, but the calcium current inactivation accelerated and the voltage dependence of the steady-state inactivation shifted to more hyperpolarized levels. The concentration–effect relationship for DHA and EPA are shown in Fig. 5A, B. With a Hill coefficient of 2, the maximal effect of DHA was -6.7 ± 0.4 mV with an EC_{50} of 2.2 ± 0.4 μM. EPA was clearly less effective and had an $EC_{50} > 15$ μM. The monounsaturated fatty acid OA had no significant effect at 20 μM.

Fig. 5. Concentration–effect relationships. **(A)** The concentration–effect relation ship of DHA. The shift in V_h is given as a function of DHA concentration for sodium currents (I_{Na}, squares) and calcium currents (I_{Ca}, circles) after subtraction of the unspecific time-dependent shift with continued perfusion with BSA-containing control solution. Data are fit with a Hill equation of the form $\Delta_{max}/(1 + (c/EC_{50})^h)$, where Δ_{max} is the maximal effect, c is the concentration of the drug, EC_{50} is the concentration of half maximal effect and h is the Hill coefficient. The EC_{50} was approx 2 μM for both the sodium current and the calcium current. **(B)** Concentration–effect relationship for EPA. The EC_{50} for the sodium current was lower than that for the calcium current.

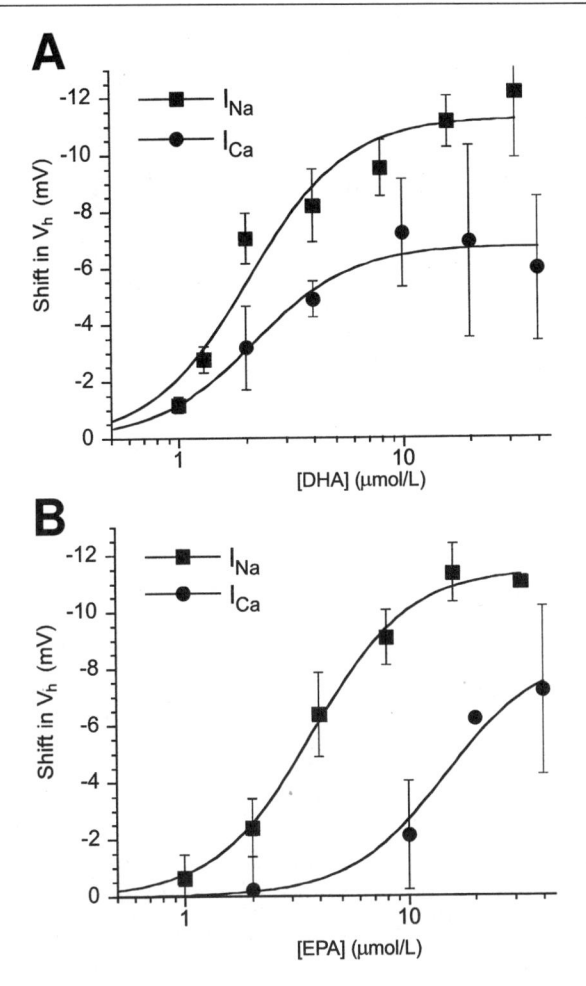

6. EFFECTS OF FATTY ACIDS IN VIVO

Yehuda and co-workers were the first to demonstrate that PUFAs can have an anticonvulsant effect in vivo (Yehuda, Carasso & Mostofsky, 1994). They administered a mixture of α-linolenic/linoleic acid in a ratio of 1 : 4 to rats for 3 wk and assessed the protection against acute convulsant doses of pentylenetetrazole (PTZ), repeated subconvulsive doses of PTZ (chemical kindling), in rats made epileptic by a $FeCl_3$ injection in the amygdala and in rats made seizure-prone to acoustic stimulation by repeated injection with p-cresol. In all epilepsy models, the treatment either prevented the occurrence of seizures or increased the threshold for convulsions and diminished the severity and duration.

We investigated whether an anticonvulsant effect could be demonstrated after acute doses (Voskuyl, Vreugdenhil, Kang & Leaf, 1998). The cortical stimulation model was used for this purpose, as it allows frequent testing of the threshold for convulsive activity in individual animals and can, therefore, provide the time-course of effect in a single experiment (Voskuyl, Dingemanse & Danhof, 1989; Voskuyl, Hoogerkamp & Danhof, 1992).

In this experimental epilepsy model, convulsive activity is induced by electrical stimulation of the motor area of the cortex. Because the bilateral, cortical electrodes are chronically implanted, the stimulation can take place via a cable in otherwise freely moving animals. A current pulse train is used that slowly increases in amplitude with each pulse. This results in a progressive pattern of behavioral, convulsive activity that can be seen during stimulation. Within this pattern two thresholds have been defined, the threshold for localized seizure activity (TLS) and, with continued stimulation, the threshold for generalized seizure activity (TGS). If stimulation is stopped before the TLS is reached (typically the start of clonic activity of the forelimbs), convulsive activity is immediately aborted. Crossing of the TGS is characterized by self-sustained seizure activity, continuing for 10–40 s after stimulation has stopped. "Postictal" (i.e., postseizure) threshold increases are avoided if the TGS is not crossed. Testing a great number of anticonvulsant drugs, it was found that drugs could increase the TLS, the difference between the TLS and TGS, or both (Della Paschoa, Hoogerkamp, Edelbroek, Voskuyl & Danhof, 2000; Hoogerkamp et al., 1996; Hoogerkamp, Vis, Danhof & Voskuyl, 1994). (It should be noted that if a drug increases the TLS, the TGS rises by necessity, because it is reached only when the stimulus intensity has passed the TLS.) Phenytoin and carbamazepine preferentially affect the TGS, but valproate increases the TLS primarily. Fresh emulsions of fatty acids made by mixing a small volume of concentrated fatty acid in alcohol, with 2 mL of physiological saline containing BSA and subsequent sonication, were administered via a previously implanted jugular vein canula. As pilot experiments had shown that bolus injections were ineffective, the emulsion was infused over a period of 30 min. We tested DHA, EPA, LA, OA, and vehicle. DHA and EPA moderately, but highly significantly, increased the TLS and TGS (Table 1). Both thresholds started to rise at the end of the infusion and reached a maximum after 7 h. The next day, the thresholds had partly returned to baseline. LA and OA had similar but smaller effects, but OA did not change the TLS. The vehicle had no effect at all. The time-courses of the threshold changes are illustrated in Fig. 6. Although it was less conspicuous than in cardiomyocytes and hippocampal neurons, the efficacy of the fatty acids appeared to correlate with the number of double bonds in these experiments as well.

In comparison with phenytoin and carbamazepine, which induce changes in TGS of several hundred microampere, the effects of the fatty acids were rather small. However, a more conspicuous difference was the slow increase in threshold over a period of several hours. With intravenous administration of phenytoin, carbamazepine, or valproate, the maximal effect is reached almost immediately after injection, after which the threshold returns to baseline in 4–6 h. The pharmacokinetics were not investigated in detail, but the time-course of effect certainly did not follow the plasma concentration. Blood samples taken from some rats that were treated with DHA or EPA indicated that the plasma concentration was maximal at the end of the infusion and dropped to undetectable levels after 6 h (probably already after 3 h).

Table 1
Increase in the Thresholds for Localized (TLS) and for
Generalized Seizure Activity (TGS), 6 h After the Start of Infusion

	TLS (µA)	TGS (µA)	n
DHA	77 ± 17	130 ± 19	7
EPA	73 ± 13	125 ± 20	7
LA	49 ± 10	75 ± 11	9
OA	16 ± 7	54 ± 9	9
Vehicle	15 ± 15	8 ± 11	6

Note: Values are expressed as mean ± sem. One-way analysis of variance indicated a highly significant difference between groups, both for the TLS and TGS ($p < 0.001$). Subsequent multiple range tests, using the least-squares difference procedure, demonstrated that, with respect to the TLS, EPA, DHA, and LA differed from OA and control. With respect to the TGS, EPA and DHA differed from LA and OA and the latter two, in turn, from the control.

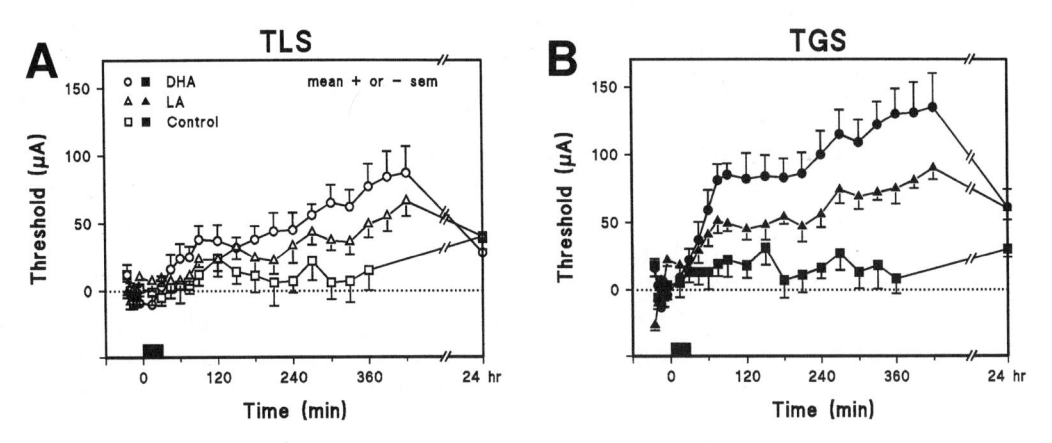

Fig. 6. Threshold increase in the cortical stimulation model. **(A)** Change in TLS induced by 30-min iv infusion of 40 µmol fatty acid, shown for DHA ($n = 7$), LA ($n = 9$), or vehicle ($n = 6$). The TLS is expressed as the increase with respect to the mean value during the preinfusion period. The infusion is indicated by black bar. Data are mean ± sem. **(B)** Change in TGS, measured simultaneously with the TLS in the same animals.

7. POTENTIAL ROLE OF PUFAS IN THE TREATMENT OF EPILEPSY

In conclusion, in vitro studies on hippocampal neurons and in vivo studies in experimental epilepsy models have shown that PUFAs, especially DHA, in principle can have an anticonvulsant action. At low-micromolar concentrations, they exhibit the same effects as well-known antiepileptic drugs like carbamazepine, phenytoin, and valproic acid, in that they reduce sodium currents by shifting the voltage dependence of the inactivation in the hyperpolarizing direction and delaying the recovery of inactivation, without effect on activation. The manner in which the sodium current is reduced should increase the threshold for action potential generation and limit the firing rate, which has, indeed, been shown recently for CA1 and CA3 hippocampal neurons (Xiao & Li, 1999). The selective effect appears to be positively correlated to the number of unsaturated bonds and to depend on the presence of a free carboxylic acid group. Furthermore, calcium currents are modified in the same way. This may be expected to reduce calcium-dependent neurotransmission and postsynaptic calcium influx in neurons. The combined action on sodium and calcium currents is characteristic for drugs effective against partial seizures (Elliott, 1990; Rogawski & Porter, 1990) and is already achieved by low-micromolar concentrations.

Although it is tempting to attribute the anticonvulsant effect in vivo to the direct effects of PUFAs on sodium and calcium channels, there is, as yet, no proof that this is actually the case. The in vivo studies reveal some complicating factors. The two classical screening models for anticonvulsant action are the maximal electroshock (MES) test and the subcutaneous pentylenetetrazole (PTZ) test. The majority of the presently available anticonvulsants have been initially identified with one of these tests and often they show efficacy in only one of the tests. Phenytoin and carbamazepine, which presumably act primarily by suppressing sodium currents, are the two prototype compounds that show selective activity in the MES test, but not in the PTZ test. However, so far, the anticonvulsant action of PUFAs has not yet been demonstrated in the MES test, but only in the PTZ test. The selective effect of carbamazepine and phenytoin in the MES test is paralleled in the cortical stimulation model by a strong and immediate effect on the TGS, whereas the effect of PUFAs was moderate and reached a peak only after several hours. Furthermore, in the studies by Yehuda and co-workers, the fatty acids appeared to become effective only after the animals had been administered a fatty acid mixture for 3 wk. Although the authors did not discuss this point, it seems that this time-consuming design was chosen because acute doses were not effective. This raises the question of whether the administered PUFAs actually reach their targets and whether reduction of sodium and calcium currents indeed underlies the anticonvulsant action or that other mechanisms are involved. For example, a new class of "background" potassium channels has recently been discovered that is found only in the central nervous system and is exclusively activated by PUFAs (Fink et al., 1998). Activation of these channels would be indistinguishable from the aforementioned effects on Na$^+$ channels. A role for these K$^+$ channels in the protective effects of PUFAs against kainic acid-induced epileptic seizures has been suggested (Lauritzen et al., 2000).

It is unlikely that PUFAs do not reach the brain, because they can pass the blood-brain barrier rapidly. Thus, it seems more likely that they are quickly stored somewhere, so that the free concentration in the extracellular space remains too low, at least initially, to interact with ion channels. From these stores, they can be released more slowly, accounting for a delayed effect. One possibility is that they are incorporated in neuronal membranes and later liberated by activity-dependent lipase (Dumuis, Sebben, Haynes, Pin & Bockaert, 1988). Alternatively, astroglial cells may buffer the rapid rise in fatty acid concentration and later release them at a much slower rate to the immediate vicinity of the neurons.

The possibility of other mechanisms involved in the action of PUFAs should be considered. Some studies suggest that a reduction in sodium current may also be accomplished by activation of protein kinase C (Godoy & Cukierman, 1994; Linden & Routtenberg, 1989), possibly by modulating the degree of phosphorylation of the sodium channel (*see* Catterall, 1999). As an example of the complexity of correlating mechanisms observed at the molecular and cellular levels to the in vivo anticonvulsant action,

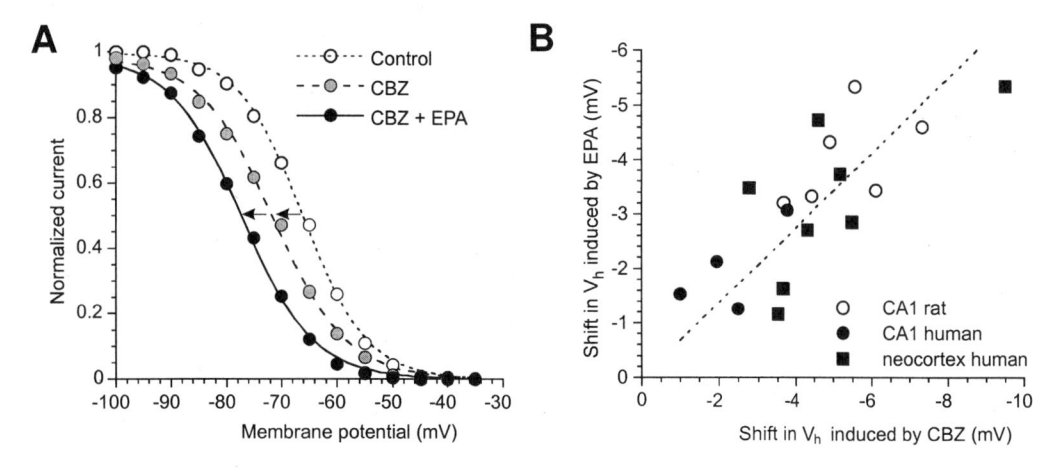

Fig. 7. Synergistic action of PUFAs. **(A)** Voltage dependence of steady-state inactivation of the sodium current in the control solution (open symbols) after perfusion with 15 μM carbamazepine (CBZ, gray symbols) and after the addition of 0.5 μM EPA (black symbols). Data are fit with a Boltzmann equation. Like PUFAs, CBZ induces a shift in V_h and increases the slope factor V_c. The addition of 0.5 μM EPA induces an extra shift in V_h, but without affecting the V_c, whereas 0.5 μM EPA alone has no discernible effect on V_h. **(B)** The shift in V_h induced by addition of 0.5 μM EPA on top of 15 μM CBZ is given against the shift induced by 15 μM CBZ alone for CA1 neurons from healthy rats (open circles), for CA1 neurons from patients with pharmaco-resistant temporal lobe epilepsy (filled circles), and for neocortical neurons from epilepsy patients (squares). The potentiating effect of subthreshold concentrations of EPA is similar for all groups.

valproic acid can serve as an example. Although the typical effects on sodium channel inactivation and lowering of repetitive firing have been described in cultured neurons (McLean & Macdonald, 1986b; Van den Berg, Kok & Voskuyl, 1993), the expected changes in firing behavior could not be confirmed in hippocampal slices (Albus & Williamson, 1998). Thus, it remains an open question as to what extent the effects of valproate on sodium channels contribute to the antiepileptic action. Other proposed mechanisms are increased GABA-mediated inhibition via stimulation of GABA synthesis and release, reduction of the release of the proconvulsant amino acid γ-hydroxybutyric acid, and attenuation of synaptic excitation mediated by *N*-methyl-D-aspartate (NMDA) receptors (*see* Löscher, 1999 for a review). DHA has also been reported to affect GABA-mediated synaptic inhibition (Hamano, Nabekura, Nishikawa & Ogawa, 1996) and NMDA responses (Nishikawa, Kimura & Akaike, 1994). However, the actions were in the opposite direction and would be expected to counteract an anticonvulsant effect via sodium and calcium channels (this could, in part, be responsible for the moderate effect in vivo). It illustrates that much work still needs to be done to correlate molecular mechanisms of PUFAs to responses of the intact organism and differences between different members of this class of compounds should be considered.

8. SYNERGISTIC ACTION OF PUFA

Nevertheless, there are some exciting avenues to explore, of which one is the already mentioned possible role of PUFAs in the ketogenic diet, as is covered in Chapter 17. Another aspect is the enhancement of the action of conventional drugs by nanomolar concentrations of PUFAs. This interesting phenomenon was observed at very low doses of PUFAs. Figure 7A gives an example where 15 µM carbamazepine shifted the voltage dependence of the sodium current steady-state inactivation by 5.4 mV to more hyperpolarized levels. The addition of 0.5 µM EPA almost doubled this shift, without affecting the slope of the curve. When tested in isolation, this dose of EPA was without discernible effect (*see* Fig. 5B). This overadditive or synergistic effect was also found for DHA (not shown). In a similar fashion, subthreshold levels of the two-branched fatty acid valproic acid (100 µM) boosted the effect of 15 µM carbamazepine or 10 µM phenytoin on the V_h of sodium currents (M. Vreugdenhil, unpublished data). This synergistic action of valproic acid in vitro was confirmed in vivo in the cortical stimulation model for anticonvulsant action of drugs: subthreshold doses of valproic acid significantly boosted the anticonvulsant effect of phenytoin (Della Paschoa, Kruk, Hamstra, Voskuyl & Danhof, 1998).

We determined the synergistic effect of 0.5 µM EPA on 15 µM carbamazepine in CA1 neurons from healthy rats, in CA1 neurons isolated from the hippocampus surgically removed from patients with intractable temporal lobe epilepsy, and in neocortical pyramidal neurons from the same patients (for details on methods, *see* Vreugdenhil & Wadman, 1999). Figure 7B shows a clear relation between the shift in the inactivation curve induced by 15 µM carbamazepine and the additional shift induced on top of 0.5 µM EPA. This suggests that the effect of a therapeutically relevant dose of carbamazepine can be boosted by subthreshold levels of PUFAs.

This knowledge might be relevant for those epilepsy patients who need intolerably high doses of antiepileptic drugs to control their seizures. The shift in inactivation curve induced by 15 µM carbamazepine in CA1 neurons from the hippocampus of epileptic patients with hippocampal sclerosis (Vreugdenhil et al., 2000), as well as in CA1 neurons from chronic epileptic (kindled) rats (Vreugdenhil & Wadman, 1999), was smaller than that in CA1 cells from healthy rats (Fig. 7B). Instead of doubling the dose of carbamazepine, very low concentrations of PUFAs might provide the shift required for seizure control. The other way around, the success of treatment with antiepileptic drugs like carbamazepine might depend on PUFA levels in the brain, which can vary considerably, according to diet (de Logeril et al., 1994) and metabolism. This synergistic action may provide a basis for developing rational polytherapy of PUFAs and carbamazepine or phenytoin.

REFERENCES

Commission on Classification and Terminology of the International League Against Epilepsy. Proposal for revised classification of epilepsies and epileptic syndromes. Epilepsia 1989;30:389–399.

Albus H, Williamson R. Electrophysiologic analysis of the actions of valproate on pyramidal neurons in the rat hippocampal slice. Epilepsia 1998; 39(2):124–139.

Catterall WA. Molecular properties of brain sodium channels: an important target for anticonvulsant drugs. In: Delgado-Escueta AV, Wilson WA, Olsen, RW, Porter RA, eds. Jasper's Masic Mechanisms of the Epilepsies, 3rd ed. Advances in Neurology, Vol. 79. Lippincott Williams & Wilkins, Philadelphia, 1999, pp. 441–456.

de Logeril M, Renaud S, Mamelle N, Salen P, Martin J-L, Monjaud I, et al. Mediterranean alpha-linolenic acid-rich diet in secondary prevention of coronary heart disease. Lancet 1994; 143:1454–1459.

Della Paschoa OE, Hoogerkamp A, Edelbroek PM, Voskuyl RA, Danhof M. Pharmacokinetic-pharmacodynamic correlation of lamotrigine, flunarizine, loreclezole, CGP40116 and CGP39551 in the cortical stimulation model. Epilepsy Res 2000; 40:41–52.

Della Paschoa OE, Kruk MR, Hamstra R, Voskuyl RA, Danhof M. Pharmacodynamic interaction between phenytoin and sodium valproate changes seizure thresholds and pattern. Br J Pharmacol 1998; 125(5):997–1004.

Dumuis A, Sebben M, Haynes L, Pin JP, Bockaert J. NMDA receptors activate the arachidonic acid cascade in striatal neurons. Nature 1988; 336:68–70.

Elliott P. Action of antiepileptic and anaesthetic drugs on Na- and Ca-spikes in mammalian non-myelinated axons. Eur J Pharmacol 1990; 175:155–163.

Fink M, Lesage F, Duprat F, Heurteaux C, Reyes R, Fosset M, Lazdunski M. A neuronal two P domain K$^+$ channel stimulated by arachidonic acid and polyunsaturated fatty acids. The EMBO Journal 1998; 17:3297–3308.

Godoy CM, Cukierman S. Multiple effects of protein kinase C activators on Na+ currents in mouse neuroblastoma cells. J Membr Biol 1994; 140(2):101–110.

Hamano H, Nabekura J, Nishikawa M, Ogawa T. Docosahexaenoic acid reduces GABA response in substantia nigra neuron of rat. J Neurophysiol 1996; 75:1264–1270.

Hock CE, Beck LD, Bodine LC, Reibel DK. Influence of dietary n-3 fatty acids on myocardial ischemia and reperfusion. Am J Physiol 1990; 259:H1518–H1526.

Hoogerkamp A, Arends RHGP, Bomers AM, Mandema JW, Voskuyl RA, Danhof M. Pharmacokinetic/pharmacodynamic relationship of benzodiazepines in the direct cortical stimulation model of anticonvulsant effect. J Pharmacol Exp Ther 1996; 279(2):803–812.

Hoogerkamp A, Vis PW, Danhof M, Voskuyl RA. Characterization of the pharmacodynamics of several antiepileptic drugs in a direct cortical stimulation model of anticonvulsant effect in the rat. J Pharmacol Exp Ther 1994; 269:521–528.

Kang JX, Leaf A. Effects of long-chain polyunsaturated fatty acids on the contraction of neonatal rat cardiac myocytes. Proc Natl Acad Sci USA 1994; 91(21):9886–9890.

Lauritzen I, Blondeau N. Heurteaux C, Widmann C, Romey G, Lazdunski M. Polyunsaturated fatty acids are potent neuroprotectors. The EMBO Journal 2000; 19:1784–1793.

Leaf A, Kang JX, Xiao Y-F, Billman GE, Voskuyl RA. The antiarrhythmic and anticonvulsant effects of dietary N-3 fatty acids. J Membr Biol 1999; 172:1–11.

Leaf A, Kang JX, Xiao Y-F, Billman GE, Voskuyl RA. Experimental studies on antiarrhythmic and antiseizure effects of polyunsaturated fatty acids in excitable tissue. J Nutr Biochem 1999; 10:440–448.

Leifert WR, McMurchie EJ, Saint DA. Inhibition of cardiac sodium currents in adult rat myocytes by n-3 polyunsaturated fatty acids. J Physiol 1999; 520:671–679.

Linden DJ, Routtenberg A. cis-Fatty acids, which activate protein kinase C, attenuate Na$^+$ and Ca^{2+} currents in mouse neuroblastoma cells. J Physiol 1989; 419:95–119.

Löscher W. New visions in the pharmacology of anticonvulsion. Eur J Pharmacol 1998; 342:1–13.

Löscher W. Valproate: a reappraisal of its pharmacodynamic properties and mechanisms of action. Prog Neurobiol 1999; 58:31–59.

McLean MJ, Macdonald RL. Carbamazepine and 10,11-epoxycarbamazepine produce use and voltage-dependent limitation of rapidly firing action potentials of mouse central neurons in cell culture. J Pharmacol Exp Ther 1986; 238:727–738.

McLean MJ, Macdonald RL. Sodium valproate, but not ethosuximide, produces use- and voltage-dependent limitation of high frequency repetitive firing of action potentials of mouse central neurons in cell culture. J Pharmacol Exp Ther 1986; 237(3):1001–1011.

Meier CL, Dudek FE. Spontaneous and stimulation-induced synchronized burst afterdischarges in the isolated CA1 of kainate-treated rats. J Neurophysiol 1996; 76:2231–2239.

Nishikawa M, Kimura S, Akaike N. Facilitatory effect of docosahexaenoic acid on N-methyl-D-aspartate response in pyramidal neurones of rat cerebral cortex. J Physiol 1994; 475:83–93.

Park CC, Ahmed Z. Alterations of plasma membrane fatty acid composition modify the kinetics of Na$^+$ current in cultured rat diencephalic neurons. Brain Res 1992; 570:75–84.

Ragsdale DS, Avoli M. Sodium channels as molecular targets for antiepileptic drugs. Brain Res Rev 1998; 26:16–28.

Rogawski MA, Porter RJ. Antiepileptic drugs: pharmacological mechanisms and clinical efficacy with consideration of promising developmental stage compounds. Pharmacol Rev 1990; 42(3):223–286.

Schwartz J, Grigat G. Phenytoin and carbamazepine: potential- and frequency-dependent block of sodium currents in mammalian myelinated nerve fibres. Epilepsia 1989; 30:286–294.

Takenaka T, Horie H, Hori H. Effects of fatty acids on membrane currents in the squid giant axon. J Membr Biol 1987; 95:113–120.

Takenaka T, Horie H, Hori H, Kawakami T. Effects of arachidonic acid and the other long-chain fatty acids on the membrane currents in the squid giant axon. J Membr Biol 1988; 106:141–147.

Taylor CP, Narasimhan LS. Sodium channels and therapy of central nervous system diseases. Adv Pharmacol 1997; 39:47–98.

Van den Berg RJ, Kok P, Voskuyl RA. Valproate and sodium currents in cultured hippocampal neurons. Exp Brain Res 1993; 93:279–287.

Voskuyl RA, Dingemanse J, Danhof M. Determination of the threshold for convulsions by direct cortical stimulation. Epilepsy Res 1989; 3:120–129.

Voskuyl RA, Hoogerkamp A, Danhof M. Properties of the convulsive threshold determined by direct cortical stimulation in rats. Epilepsy Res 1992; 12:111–120.

Voskuyl RA, Vreugdenhil M, Kang JX, Leaf A. Anticonvulsant effect of polyunsaturated fatty acids in rats, using the cortical stimulation model. Eur J Pharmacol 1998; 341(2–3):145–152.

Vreugdenhil M, Bruehl C, Voskuyl RA, Kang JX, Leaf A, Wadman WJ. Polyunsaturated fatty acids modulate sodium and calcium currents in CA1 neurons. Proc Natl Acad Sci USA 1996; 93:12,559–12,563.

Vreugdenhil M, Wadman WJ. Kindling-induced long-lasting enhancement of calcium current in hippocampal CA1 area of the rat: relation to calcium-dependent inactivation. Neuroscience 1994; 59(1):105–114.

Vreugdenhil M, Wadman WJ. Modulation of sodium currents in rat CA1 neurons by carbamazepine and valproate after kindling epileptogenesis. Epilepsia 1999; 40:1512–1522.

White HS. Comparative anticonvulsant and mechanistic profile of the established and newer antiepileptic drugs. Epilepsia 1999; 40(Suppl. 5):S2–S10.

Xiao YF, Gomez AM, Morgan JP, Lederer WJ, Leaf A. Suppression of voltage-gated L-type Ca^{2+} currents by polyunsaturated fatty acids in adult and neonatal rat ventricular myocytes. Proc Natl Acad Sci USA 1997; 94:4182–4187.

Xiao YF, Kang JX, Morgan JP, Leaf A. Blocking effects of polyunsaturated fatty acids on Na^+ channels of neonatal rat ventricular myocytes. Proc Natl Acad Sci USA 1995; 92(November), 11,000–11,004.

Xiao YF, Li X. Polyunsaturated fatty acids modify mouse hippocampal neuronal excitability during excitotoxic or convulsant stimulation. Brain Res 1999; 846:112–121.

Xiao, Y.-F., Wright SN, Wang GK, Morgan JO, Leaf A. N-3 fatty acids suppress voltage-gated Na^+ currents in HEK293t cells transfected with the α-subunit of the human cardiac Na^+ channel. Proc Natl Acad Sci USA 1998; 95:2680–2685.

Yehuda S, Carasso RL, Mostofsky DI. Essential fatty acid preparation (SR-3) raises the seizure threshold in rats. Eur J Pharmacol 1994; 254:193–198.

5 Role of Dietary Fats and Exercise in Immune Functions and Aging

Jaya T. Venkatraman and David Pendergast

1. INTRODUCTION

Immunosenescence is a complex remodeling of the immune system that may contribute significantly to morbidity and mortality in the elderly. Much evidence suggests an association between immune function and well-being during aging and longevity. Despite several studies on the immune system in the elderly, little is known of the biological basis of immunosenescence in humans. Undoubtedly, some diseases to which the elderly are particularly susceptible, such as infections and autoimmune and neoplastic pathologies, include dysregulation of several immune functions in their pathogenesis. On the other hand, recent studies in healthy centenarians suggest that the immunological changes observed during aging are consistent with a reshaping, rather than a generalized deterioration, of the main immune functions. The infection rate and severity increase with aging as a result of decrease in immune function with aging. The primary changes of the aged host are in the T-lymphocytes, perhaps because of the involution of the thymus. Secondary changes (environmental) may be the result of changes in diet, drug intake, physical activity, and so forth or, alternatively, the result of underlying diseases (Wick and Grubeck-Loebenstein, 1997). There is a paradoxical increase in autoimmunity, and the responsiveness to exogenous antigens and tumors are reduced and may play a proinflammatory role in the development of many pathological conditions (Weyand et al., 1998). Recurrent stress, infections, and inflammation over the life-span play a significant role in producing age-associated changes in all systems. Genetic selection for infections has been implicated in coronary heart disease (McCann et al., 1998), early onset of Parkinson's following influenza encephalitis, and Alzheimer's disease (Stoessl, 1999).

The restoration of immune functions of the aged individuals is possible and might be beneficial for them to cope with various diseases associated with aging (Hirokawa, 1997). Physiological thymic atrophy is controlled by both extrathymic and intrathymic factors and is not a totally irreversible process. The process of thymic atrophy might be explained by a further understanding of the relationship between the neuroendocrine and the immune systems (Hirokawa et al., 1994). Although the most obvious age-related structural alteration of the immune system occurs in the thymus, the role of thymic involution in immunosenescence is still not well understood.

From: *Fatty Acids: Physiological and Behavioral Functions*
Edited by: D. Mostofsky, S. Yehuda, and N. Salem Jr. © Humana Press Inc., Totowa, NJ

1.1. Immune Theory of Aging

Original immunologic theory of aging suggests that aging in mammals is a self-destructive process leading to a decline in immune response. The failure in immune homeostasis is associated with a consequent rise in autoimmunity (Walford, 1987). Age-related phenomena such as increased prevalence of autoantibodies and monoclonal immunoglobulins reflect dysregulation of the senescent immune system rather than a simple decline in responsiveness. The age-related changes primarily occur in the T-cell-dependent immune system and are associated with increased susceptibility to infections and incidence of autoimmune phenomena in the elderly. One of the characteristics of all somatic cells is a finite life-span. Cells may proliferate until they reach a point, after which they can no longer produce daughter cells. This observation is central to the clonal exhaustion hypothesis, a mechanism cited to explain age-associated immune dysfunction. In this hypothesis, repeated division of lymphocytes leads to a replicative limit, after which the cells enter the senescent phase but are not lost from the pool of T-cells. Advancing age would then be associated with an increase in the number of T-cells that are unable to proliferate to a stimulus that induces a proliferative response in T-cells from younger individuals.

2. IMMUNE SYSTEM CHANGES WITH AGING

2.1. Immune Function

2.1.1. AGING AND IMMUNE CELL SUBSETS

The innate immunity is preserved over the life-span. Immune changes, dysregulations, mainly affect acquired immunity and lead to a gradual increase in T-helper (Th) 2/T-helper 1 cells. This change is the result initially of decreased thymic function, and later of accumulative antigen pressure over the life-span. T-Cell subset distribution in peripheral lymphocytes changes with age to a higher memory/naive cell ratio, accompanied by changes in cytokine production patterns (Fernandes and Venkatraman, 1993; Fernandes et al., 1997; Thoman and Weigle, 1989; Venkatraman and Fernandes, 1994; Venkatraman et al., 1994). Proliferative capacity of T-cells declines with age (Venkatraman & Fernandes, 1997; Makinodan, 1998; Makinodan et al., 1987). The composition of T-cell subsets in the periphery gradually change with age, resulting in the alteration of T-cell functions in the elderly. There appears to be a macrophage–lymphocyte disequilibrium in aged persons (Lesourd, 1999). There is a shift in the equilibrium of peripheral T- and B-lymphocyte subsets, T helper 1 subset (Th1) to Th2 and $CD5^-$ to $CD5^+$ cells with aging. T-Cell responses are more affected than B-cell responses that may result in T-cell-mediated dysregulation of antibody responses and low affinity and self-reactive antibodies (Doria and Frasca 1997). T-Cell activity may be restricted to immune surveillance of neoplastic transformation. Even the B-cells exhibit intrinsic defects and natural-killer (NK) cells have a profound loss of activity. Aging leads to replacement of virgin cells by memory cells and to the accumulation of cells with defects in signal transduction.

The aging immune system is characterized by a progressive decline in the responsiveness to exogenous antigens and tumors in combination with a paradoxical increase in autoimmunity. From a clinical viewpoint, deficiencies in antibody responses to exogenous antigens, such as vaccines, have a major impact and may reflect intrinsic B-cell defects or altered performance of helper T-cells. Aging is associated with the emergence of an unusual CD4 T-cell subset characterized by the loss of CD28 expression (Weyand et al., 1998). CD28 is the major costimulatory molecule required to complement signaling through the antigen receptor for complete T-cell activation. CD4$^+$ CD28$^-$ T-cells are long-lived, typically undergo clonal expansion in vivo, and react to autoantigens in vitro. Despite the deficiency of CD28, these unusual T-cells remain functionally active and produce high concentrations of interferon-γ (IFN-γ) and interleukin-2 (IL-2). The loss of CD28 expression is correlated with a lack of CD40 ligand expression, rendering these CD4 T-cells incapable of promoting B-cell differentiation and immunoglobulin secretion, thus aberrations in immune responsiveness (Weyand et al., 1998). Aging-related accumulation of CD4$^+$ CD28$^-$ T-cells should result in an immune compartment skewed toward autoreactive responses and away from the generation of high-affinity B-cell responses against exogenous antigens. We propose that the emergence of CD28-deficient CD4 T-cells in the elderly can partially explain age-specific aberrations in immune responsiveness.

In a cohort of apparently healthy and well-nourished elderly women, total T (CD3$^+$), T-helper (CD4$^+$), or T-cytotoxic (CD8$^+$) cell number, NK cell number, cytotoxicity, phagocytosis, and subsequent oxidative burst were similar to values in young women. However, they had lower T-cell proliferation responses and significantly reduced response to phytohemagglutinin (Krause et al., 1999). T-cells show the largest age-related differences in distribution and function with aging with thymus involution as the apparent underlying cause (Shinkai et al., 1998).

2.1.2. AGING AND CYTOKINES

Complex remodeling of cytokine production is likely to be a characteristic of immunosenescence. Cytokines are produced by various immunocompetent cells in response to appropriate stimuli, are able to mediate many immune functions, orchestrate the immune system continuously, and act as molecular signals between immunocompetent cells. An excessive or insufficient production of cytokines may contribute to infectious, immunological, and inflammatory diseases. They act via specific receptor sites on cytokine-secreting cells (autocrine action) or immediately adjacent immune cells (paracrine action). The production of IL-2 is decreased, with a decrease of total T-cell count, and often with changes in T-cell subsets and proliferative response to mitogens during aging. T-cell-dependent functions are most dramatically compromised, which is most likely a result of age-related involution of the thymus gland. Defects in T-cell proliferative capacity/responsiveness, IL-2 production and receptor expression, signal transduction, and cytotoxicity are frequently cited problems associated with immunosenescence (Cinader et al., 1993). As a consequence of their differentiated state, memory T-cells can express higher levels and a greater variety of lymphokines than naive T-cells; the former more efficiently generate and regulate humoral and cell-mediated immune responses to recall antigens than naive T-cells (Ernst et al., 1990). Transforming growth factor-β (TGF-β) for instance, is known to have immunosuppressive properties. A balance between proflammatory and anti-inflammatory cytokines derived from Th1 or Th2 cells are very essential for the maintenance of a sound immune system as they stimulate growth, differentiation and functional development of immune cells.

Following a primary encounter with an antigen, naive T-cells express early-response genes, including those that encode the T-cell growth factor, IL-2 and IL-2 receptor chains. IL-2 functions through high-affinity receptors to drive activated cells to proliferate and differentiate into effector T-cells that mediate primary humoral or cell-mediated immune responses (Ernst et al., 1995). In the process, a large number of differentiated, antigen-specific T-cells (memory cells) is generated. The memory cells are in the differentiated state and can express cytokines that naive T-cells cannot. The naive-to-memory cell conversion is accompanied by quantitative changes in the expressed levels of several membrane molecules that can alter cellular function (Ernst et al., 1995). The age-dependent increase in memory T-cells, which are well differentiated and unable to mutate their TCR genes somatically, could result in a peripheral pool of blood cells that respond well to previously encountered antigens but cannot recognize and respond to new antigens. The increased pool of memory T-cells could represent a greater risk for dysregulation by perhaps overproducing certain cytokines, which may suppress appropriate immune responses or amplify inappropriate ones (Ernst et al., 1995). Lymphocytes of the elderly show decreased proliferation after induction with mitogens, decreased release of IL-2, soluble IL-2R, IFN-γ, and other Th1 cytokines and increase in Th2 cytokines such as IL-4 and IL-10, suggesting a dysregulation of the Th1/Th2 system in the elderly (Rink et al., 1998). Cytokines and their antagonists are significant factors in host responses to infections and inflammatory stimuli and the elderly have higher levels of cytokine antagonists IL-1RA and sTNF-R, and neopterin (produced by activated macrophages and monocytes) in their plasma and IL-2 production by PBMN cells is decreased (Catania et al., 1997).

2.1.3. Neuroendocrine System

Numerous interactions exist among the nervous, endocrine, and immune systems and are mediated by neurotransmitters, hormones, and cytokines. Aging is associated with enhanced responsiveness of the T-cell compartment and alterations in temporal architecture of the neuroendocrine-immune system (Mazzoccoli et al., 1997). There are relevant integrations between pituitary–thyroid hormones and immune factors favoring the development and maintenance of both thymic and peripheral immune efficiency. The GH-insulin-like growth factor (IGF)-I axis is dysregulated in aging, in catabolic states, and in critical illness. The pineal gland seems to regulate, via circadian secretion of melatonin, all basic hormonal functions and immunity (Pierpaoli, 1998). Studies with in vivo models have shown that this fundamental role of the pineal gland decays during aging. Melatonin is a ubiquitous molecule and can be found in a large variety of cells and tissues. Binding sites and "receptors" have been identified in many tissues and cells of the neuroendocrine and immune system.

2.1.4. Apoptosis and Aging

The progressive decline in immune response and increased incidence of autoimmune phenomenoa might be the result of modified cellular mechanisms affecting the immune system in the course of aging. The apoptotic deletion of activated T-cells has been proposed as the key mechanism to maintain T-cell homeostasis, and in this respect, CD95 (Fas antigen) seems to play a major role in this course of events (Potestio et al., 1998). Immune senescence is a sum of dysregulations of the immune system and its interaction with other systems. Two predominent features of immune senescence are altered T-cell phenotype and reduced T-cell response (Mountz et al., 1997). Cell lines derived from human premature aging diseases have a higher sensitivity to Fas-mediated apoptosis. Data indicate that, in vivo, there is a gradual decrease in apoptosis with aging, resulting in the accumulation of senescent T-cells with increased DNA damage. Lymphocytes from elderly have an increased expression of CD95 and enhanced apoptosis (Potestio et al., 1998). Recent evidence suggests that apoptotic deletion of activated mature lymphocytes is an essential physiological process implicated in both the regulation of the immune response and the control of the overall number of immunocompetent cells. During the course of aging, numerous alterations of these signaling pathways may shift the balance toward cell death (Phelouzat et al., 1996).

2.1.5. Aging and Transmembrane Signaling

Aging is also associated with alterations in transmembrane signaling and decreased GTPase activity in lymphocytes and granulocytes. A defective protein kinase C (PKC)-α translocation and a decreased tyrosine kinase activity following TCR or CD3 stimulation in T-cells from the elderly might also contribute to alteration in immune responses in the elderly. The activation of other protein kinases after T-cell activation with PHA (p56 lck, JAK kinases) was also affected by aging, suggesting a slower activation and a lower degree of phosphorylation. The src family of protein tyrosine kinase p56 lck is expressed in T-cells, NK cells, and lymphoid cell lines (Guidi et al., 1998). P56 co-precipitates with CD4 and CD8 molecules, CD2 and CD28, and the β-chain of IL-2 receptors. These kinases have a crucial role in T-cell activation. The protein expression and degree of phosphorylation are reduced in elderly subjects compared to adult controls, suggesting that aging may lead to a drastic impairment in the ability of PBL to be efficiently activated by mitogens or costimuli (both proliferation and IL-2 production) in the elderly (Guidi et al., 1998). Early events associated with transmembrane signaling, such as PKC translocation, IP3 generation, and Ca^{2+} mobilization, are impaired by aging, whereas PLC-γ1 activity seems to be unaffected. This suggests that the signaling steps upstream of PLC-γ1 activation, namely thyrosine phosphorylation, may be defective in the elderly.

2.1.6. Aging and Telomerases

Normal human cells undergo a finite number of cell divisions and ultimately enter a nondividing state called replicate senescence. Telomerase shortening has been proposed as the molecular clock that triggers senescence. A causal relationship seems to exist between telomerase shortening and in vitro cellular senescence (Bodnar et al., 1998). The reduction of proliferative capacity of cells from elderly donors and patients with premature aging syndromes and the accumulation in vivo of senescent cells with altered patterns of gene expression implicate cellular senescence in aging and age-related pathologies. A common finding in different cells suffering replicative senescence is the shortening of the telomers (Solana and Pawelec, 1998). Telomers are essential genetic elements that stabilize chromosome ends. The synthesis of these elements is controlled by telomerase, an enzyme that synthesizes new telomeric DNA, thus compensating for the loss that occurs as a result of the "end-replication problem" inherent in DNA replication. Some tumor-suppressor genes (p21, p53, Rb) have been reported to accumulate in senescent cells. Escape from senescence is greatly enhanced in cells that have lost Rb or p16, and both alterations behave redundantly with respect to each other.

2.1.7. Aging and Thymus

Thymic regrowth and reactivation of thymic endocrine activity may occur in older animals by different endocrinological or nutritional manipulations than in young animals. Intrathymic transplantation of pineal gland or treatment with melatonin, implantation of a growth hormone (GH) secreting tumor cell line or treatment with exogenous

GH, exogenous luteinizing hormone-releasing hormone (LH-RH), exogenous thyroxine or triiodothyronine, and nutritional interventions such as arginine or zinc supplementation have shown potential (Fabris et al., 1997). These data strongly suggest that thymic involution is a phenomenon secondary to age-related alterations in neuroendocrine–thymus interactions and it is the disruption of such interactions in old age that is responsible for age-associated dysfunction. The effect of GH, thyroid hormones, and LH-RH may be the result of the presence of the specific hormone receptors on thymic epithelial cells supposed to produce thymic peptides. Melatonin or other pineal factors may also act through specific receptors, but experimental evidence is still lacking. The role of zinc, whose turnover is usually reduced in old age, may range from reactivation of zinc-dependent enzymes, required for both cell proliferation and apoptosis, to the reactivation of thymulin, a zinc-dependent thymic hormone. Recent preliminary data obtained both in animal and human studies suggest that the endocrinological manipulations are capable of restoring thymic activity in old age and may act also by normalizing the altered zinc pool (Fabris et al., 1997).

A significant age-related decrease in RAG-1 and RAG-2 expression has been reported in thymocytes of mice aged-12 mo and over compared to young mice (Yehuda et al., 1998). Cells derived from immature thymocytes of the old donors fail to express RAG-1 and RAG-2, the bone-marrow-derived cells did not exhibit this trend, and there was no difference in Vβ rearrangement of the TCR. T-cell progenitors have a potential to give rise to T-cells with TCR rearrangements, and the expression is determined by the thymic stroma.

2.2. Exercise and Aging

Exercise poses a stress on the immune system; however, moderate chronic exercise may enhance positive immune functions. Severe or intense long-term exercise may compromise the immune system by suppressing concentrations of lymphocytes, natural-killer cell activity, lymphocyte proliferation, and secretory IgA. During the period of immune impairment ("open window"), microbial agents may invade the host and cause infections. The changes in immune functions, lower exercise capacity, and nutritional factors in the elderly may lead to a greater and longer "open window" period, increased risk of infections which may contribute to morbidity and mortality.

The specific impact of intrinsic aging has not yet been clearly dissociated from genetic traits and age-related differences in nutritional status, habitual physical activity, and exposure to psychological stressors (Rumyantsev 1998). Age-related reduction in muscle is a direct cause of the age-related decrease in muscle strength. There has been a growing belief that an appropriate regular dose of endurance exercise might slow the age-related decline in immune function by enhancing or suppressing various immunological stimuli. A decline in lean body mass, referred to as sarcopenia, and an accompanying increase in fat are known to occur during aging. The consequences of these physiological changes may include decreased physical activity, altered energy metabolism, and impaired resistance to infection. The mechanisms behind these age-related events remain unknown, but they may include changes in some of the humoral and cytokine mediators that seem to regulate body composition. Age-related loss in skeletal muscle mass has been and is a direct cause of the age-related decrease in muscle strength.

There is an average loss of 10% or more per decade in maximal aerobic power with aging (Kasch et al., 1999; Ceddia et al 1999). The reduced $\dot{V}_{O_2max}$ is associated with a reduced proportion of the cardiac output going to exercising skeletal muscle and a reduced cardiac output reserve (Beere et al., 1999). The elderly have higher ventilation during exercise both below and above the anaerobic threshold, which is related to higher lactic acid levels (Prioux et al., 2000). The anaerobic power of the elderly is inevitability reduced in both legs and arms (Marsh et al 1999). There is a reduction in the responsiveness of the sympathoadrenergic to exercise with aging, resulting in a reduced secretion of adrenaline for the medulla (Zouhal et al., 1999). Thus, a given activity or exercise would impose a relatively greater stress to the elderly.

2.3. Exercise and Immune Function

Aging leads to a diminution of resting immune function, increasing the risk of infection, tumor development, and autoimmune diseases (Shephard and Shek, 1995). The production of IL-2 is decreased, sometimes with a decrease of total T-cell count, and often with changes in T-cell subsets and proliferative responses to mitogens. However, NK cell activity remains unchanged. In theory, moderate exercise training should help to reverse the adverse effects of aging upon the immune system. However, there have been relatively few studies comparing the immune responses of young and older individuals to acute exercise and to training. A single bout of moderate exercise seems to be well tolerated by the elderly. The NK cell response is as much as in younger individuals, but perhaps because of a low initial proliferative capacity, older subjects show less stimulation of lymphocyte proliferation by moderate activity and less suppression with exhausting exercise. Perhaps because resting immune function is less than in the young, moderate training programs seem to stimulate immune function to a greater extent in the elderly than in young subjects (Shephard et al., 1994). The proliferative response of the T-cells is enhanced in aging rodents, whereas in young animals, it is suppressed. Moreover, the resting NK cell activity of elderly human subjects seems to be increased by training. Nevertheless, the therapeutic use of exercise must be cautious in the elderly, because aging also enhances susceptibility to overtraining.

Aging affects the muscle precursor cells (satellite cells or myoblasts) and their regeneration after exercise (Grounds 1998). Aging may also affect proliferation and fusion of myoblasts in response to injury; signaling molecules that stimulate satellite cells with aging; host environment, inflammatory cells, growth factors and their receptors, and the extracellular matrix. There is a reduction in growth hormone, total and free testosterone, and cortisol, both at baseline and after exercise. The decreased anabolic effects on muscles may explain the loss of muscle mass and strength with aging (Hakkinen et al., 1998). The more primitive components of immune defense, including natural killer cells, phagocytes, acute-phase proteins, and regulatory cytokines, may be altered in elderly. Several investigators have found that the proportion of neutrophils will be increased in the circulation following physical exercise. An increase in neutrophils has been correlated with an increase in plasma cortisol. The number of circulating monocytes and NK cells change as a result of exercise. Nonspecific immunity may be further stressed by maximal physical exertion that may contribute to an increased susceptibility of the elite sports person to infection.

At rest, the total number of lymphocytes and $CD4^+$ and $CD8^+$ T-cell subset populations was lower in the elderly than the young (Mazzeo et al., 1998). There is significant exercise-induced leukocytosis, primarily lymphocytosis and neutrophilia, in both young and elderly; however, the magnitude in the elderly is reduced (Ceddia et al., 1999). The elderly have higher memory and lower CD4 and CD8 T-cells than the young. Acute maximal exercise increases CD8⁺ and CD4⁺ in the elderly and young. The aged recruited fewer numbers of CD4⁺ naive and transitional cells than the young (Ceddia et al., 1999). Proliferative responses are lower in the elderly than the young. Proliferative response does not increase above rest in elderly subjects as it does in young subjects. Middle-aged subjects had a blunted growth hormone response to exercise than the young subjects, and this was not altered by training (Zaccaria et al., 1999) and which could not be explained by augmentation of somatostatin tone and/or diminished GH-release hormone (Marcell et al., 1999).

The immune system is loosely linked to the neuroendorcine system, which responds to stress. Increased cortisol, catecholamines, glucocorticoids, adrenaline, β-endorphin, growth hormones, other stress hormones, and insulin are decreased during aging. There is an increase in soluble IL-2 receptors (sIL-2R), soluble intercellular adhesion molecule-1 (sICAM-1), soluble TNF-β receptors (sTNF-R), and neopterin with exercise. These insoluble matter increases suggest that immune activation may be involved in the pathogenesis of impaired immune function after exercise. The natural-killer cell response to a single bout of exercise is normal in older individuals, but immediately after exercise, the elderly subjects manifest less suppression of phytohemagglutinin (PHA)-induced lymphocyte proliferation than younger individuals (Shinkai et al 1998). Strenuous exercise seems to induce a more sustained post-exercise suppression of cellular immunity in the elderly than in the young. The magnitude of the increase in cortisol is substantially lower in trained subjects who exercise at the same level or intensity as untrained subjects. Habitual physical activity enhances NK cell activity through mitogenesis and reduced cytokine production (Shinkai et al., 1998).

2.4. Aging, Exercise, and Antioxidant Defense System

The oxidant/antioxidant balance is an important determinant of immune cell function, including maintaining integrity and functionality of membrane lipids, cellular proteins, nucleic acids, and for control of signal transduction and gene expression in immune cells. Enzymatic and nonenzymatic antioxidants play a vital role in protecting tissues from excessive oxidative damage during exercise. Antioxidant enzymes play an important role in defending the cells against free-radical-mediated oxidative damage. In rats, hepatic and myocardial antioxidant enzymes are declined at older age, whereas activity of glutathione-related enzymes in the liver and mitochondrial enzymes in the heart are increased significantly. Skeletal muscle antioxidant enzymes are uniformly elevated during aging. The generation of oxygen free radicals and other reactive oxygen species may be the underlying mechanism for exercise-induced oxidative damage, but a causal relationship remains to be established. Depletion of each of the antioxidant systems increases the vulnerability of various tissues and cellular components to reactive oxygen species.

As acute strenuous exercise and chronic exercise training increase the requirement for various antioxidants, it is conceivable that dietary supplementation of specific antioxidants would be beneficial. Older subjects may be more susceptible to oxidative stress and may benefit from the antioxidant protection provided by vitamin E. During severe oxidative stress such as strenuous exercise, the enzymatic and nonenzymatic antioxidant systems of skeletal muscle are not able to cope with the massive free-radical formation, which results in an increase in lipid peroxidation. Vitamin E decreases exercise-induced lipid peroxidation. The exercise may increase superoxide anion generation in the heart, and the increase in the activity of superoxide dismutase (SOD) in skeletal muscle may be indirect evidence for exercise-induced superoxide formation. Therefore administration of SOD may prevent exercise-induced oxidative stress.

An acute bout of exercise can increase the activity of certain antioxidant enzymes in various tissues. The mechanism of this activation is unclear. Exercise training has little effect on hepatic or myocardial enzyme systems, but can cause adaptive responses in skeletal muscle antioxidant enzymes, particularly glutathione peroxidase. These findings suggest that both aging and exercise may impose an oxidative stress to the body. Optimal levels of antioxidants are required for maintenance of the immune response across all age groups. The genetic material of our cells is susceptible to damage by a wide variety of extrinsic and intrinsic entities. The amount of genetic material damage accumulated in vivo will depend on an individual's ability to defend against and/or repair DNA damage (Barnett and Barnett, 1998). T-cells in vivo have been shown to accumulate DNA damage and mutations over time, occurring mainly in naive and memory phenotypes. Dietary supplementation with thiolic antioxidants prevents mitochondrial degeneration in the aged and preserves immune function in aged mice. Treatment with α-tocopherol and antioxidants eliminated abnormal nuclear factor kappa B (NFκB) activity, reduced peroxide levels in membrane lipids, and corrected the dysregulated expression of cytokines. This suggested that abnormal activation of NFκB may be due to the activation of intermediate kappa B (IκB) kinase by excesses of reactive oxygen species and that it may be responsible for the dysregulated expression of certain cytokines observed in aging.

3. THERAPEUTIC APPROACHES

Aging is associated with the decline in multiple areas of immune function, but, to date, no single mechanism has emerged as responsible for all of the observed changes. It is being increasingly acknowledged that autoimmune processes play a proinflammatory role in the development of many pathological conditions, such as atherosclerosis. It is likely that the mechanisms underlying age-related changes in immunity are multifactorial, with both genetic and environmental factors playing a significant role (Burns et al., 1997). The progression of and recovery from infectious diseases depends, at least in part, on immune responses and nutritional status in the elderly. Nutrition therapy may improve the immune responses of elderly people, particularly those with total-, protein-, fat-, or micronutrient-energy balances on prescribed low protein or fat diets (Lesourd, 1997). Micronutrient supplementation improves the immune status in the aged individuals. Episodes of disease in the aged leads to the depletion of the body's nutritional stores and causes protein-energy malnutrition, undernutrition associated with immunodeficiency, and, finally cachexia (Lesourd, 1999). It has been suggested that interventions include grafting cells and tissues, dietary manipulation, free-radical scavengers or antioxidants, thymic peptides, endocrine manipulation, and physical exercise (Hirokawa, 1997). There are a number of factors that can modify immunity. These include age, genetic, metabolic, environmental, anatomical, physiological, and microbial factors.

3.1. Nutrition Intervention

Undernutrition is a common factor underlying many pathological conditions in the elderly and a common symptom in over 50% of hospitalized elderly patients. Macronutrient and micronutrient deficits are though to be partially responsible for the depressed immune system in the elderly, even in the apparently healthy independent-living elderly (Lesourd, 1999). Protein-energy malnutrition is associated with decreased lymphocyte proliferation, reduced cytokine release, and decreased antibody responses to vaccines. Deficits in zinc, selenium, and vitamin B_6 in the elderly compound the depression of the immune system. The elderly do not have an adequate supply of fats for proliferation of lymphocyte and macrophage populations, antibody production, and hepatic synthesis of acute-phase proteins. Fat oxidation is reduced in the elderly at rest, during exercise, and in response to meal ingestion, as a result, in part, to the loss of fat-free mass, decreased intrinsic fat oxidation in muscle, and hormonal factors. These changes in fat oxidation are only partially corrected with training (Calles-Escandon and Poehlman, 1997). Lipids are known to have modulatory effects on the cellular immune system at the biochemical and molecular levels, including the production and expression of cytokins. Dietary ω-6 lipids (linoleic acid) generally increase the levels of proinflammatory cytokines and inflammatory prostaglandins (PGs), while ω-3 lipids (eicosapentaenic acid [EPA] and docosahexaenoic acid [DHA]) may decrease the levels of these cytokines and inflammatory PGs. Destruction of polyunsaturated fatty acids (PUFAs) is an important reaction to the synthesize the very long-chain fatty acids that are necessary for membrane function and formation of eicosanoids. Reduced fat intake may modulate lymphoid cell subsets, suppression of pokeweed mitogen, $CD4^+/CD8^+$ ratio, proliferative response to mitogens, cytokine production, and so forth, thus increasing negative or inflammatory effects of exercise. Dietary lipids have an effect on cytokine balance (IL-2 and TGF-β) and immune function.

Diets that are low in protein, zinc, selenium, vitamin B_6, and fat may collectively depress immune function. This type of diet may be associated with either a low-caloric-intake diet or low-fat, low-meat-products diet. Zinc is an essential trace element for many biological functions, including immune functions. Indeed, zinc is required for the biological activity of a thymic hormone, called thymulin in its zinc-bound form, and is important for the maturation and differentiation of T-cells. With advancing age, zinc, thymic functions, and peripheral immune efficiency show a progressive decline. Supplementing zinc in old age restores immune efficiency.

Nutrients have a profound effect upon the production and actions of cytokines. Protein-energy malnutrition, dietary ω-3 polyunsaturated fatty acids, and vitamin E suppress the production of specific cytokines. The synthesis of acute-phase proteins and glutathione is dependent on the adequacy of dietary sulfur-containing amino acids. The consequences of the modulatory effects of previous and concurrent nutrient intake on cytokine biology are the depletion of resources and damage to the host, which ranges from mild and temporary to severe, chronic, or lethal.

It has been suggested that high pharmacological doses of zinc or vitamin E may improve immune function in the elderly (Lesourd, 1997). Vitamin E supplementation has been show to improve some aspects of immune function in aging animal and humans. Vitamin E supplementation in old mice exhibited a high significant reduction in lung viral titer, which is associated with antioxidant effects (Han and Meydani, 1999). Conjugated linoleic acid (CLA) has been suggested to have immuno-enhancing properties. Aged mice that consumed diets containing CLA (1 g CLA /100 g for 8 wk) (Hayek et al., 1999) significantly increased all CLA isomers measured in hepatic neutral lipids and phospholipids. Old mice fed 1 g CLA/100 g had significantly higher proliferative response to Con A, whereas CLA had no effect on NK cell activity, prostaglandin E_2 (PGE_2) production or delayed-type hypersensitivity (DTH) in young or old mice. The observation that altering lipid substrates effects immunosuppressive eicosanoid production in the macrophages has generated many questions regarding the importance of specific PUFAs upon host immune response (Chaet et al., 1994).

Fish oils are rich in the long-chain ω-3 polyunsaturated fatty acids (PUFAs), eicosapentanoic (20:5ω-3) and docosahexaenoic (22:6ω-3) acids. Linseed oil and green plant tissues are rich in the precursor fatty acid, α-linolenic acid (18:3ω-3). Most vegetable oils are rich in the ω-6 PUFA linoleic acid (18:2ω-6), the precursor of arachidonic acid (20:4ω-6). Arachidonic-acid-derived eicosanoids such as prostaglandin E_2 are proinflammatory and regulate the functions of cells of the immune system. Consumption of fish oils leads to the replacement of arachidonic acid in cell membranes, diminishes lymphocyte proliferation, T-cell-mediated cytotoxicity, NK cell activity, macrophage-mediated cytotoxicity, monocyte and neutrophil chemotaxis, major histocompatibility class II expression, and antigen presentation, production of proinflammatory cytokines (IL-1, IL-6, TNF), and adhesion molecule expression. Studies on animal models indicate that fish oil reduces acute and chronic inflammatory responses, improves survival to endotoxin, and prolongs the survival of grafted organs in models of autoimmunity. Feeding fish oil reduces cell-mediated immune responses. Fish-oil supplementation may be clinically useful in acute and chronic inflammatory conditions and following transplantation. ω-3 PUFAs may exert their effects by modulating signal transduction and/or gene expression within inflammatory and immune cells (Calder, 1998).

Dietary long-chain (ω-3) fatty acids from fish oil and low-intensity exercise have been reported, independently, to inhibit tumor growth in rats and is perhaps related to alterations in immune function. Individually, but not in combination, long-chain (ω-3) fatty acids and low-intensity exercise may be advantageous by augmenting cell-mediated immune function and NK cell cytotoxicity in healthy rats (Robinson and Field, 1998). Lipid sources could alter the development of autoimmune diseases and the life-span of short-lived mice (Fernandes, 1995; Venkatraman and Fernandes, 1993). The decline in autoimmune disease found with fish oil (FO) is linked to the decrease in ω-6 PUFAs, such as 18:2,20:4, that serve as precursors for proinflammatory prostaglandins of the E series. FO is also reported to decrease the levels of c-*myc* and c-*Ha-ras* oncogenes expression in the spleens of autoimmune mice and increase the levels of TGF-β1 (Chandrasekar et al., 1995; Fernandes et al., 1994; Venkatraman and Chu, 1999; Venkatraman and Chu, 1999a). The increase of TGF-β1 in the spleens may be anti-inflammatory and immunosuppressive (Fernandes et al., 1994). ω-3 lipids are reported to reduce the activation of macrophages, expression of class II antigens, production of IL-1, TNF mRNA, LTB_4 and TXB_2 levels, and platelet aggregation. It is possible that ω-3 lipids would maintain normal immune function by preventing the loss of IL-2 and IFN-γ and lowering TGFβ levels in T- and B-cells. Several age-associated diseases, particularly autoimmune diseases with a viral etiology, appear to be exacerbated by high-fat diets with a large proportion of vegetable oils high in ω-6 fatty acids. ω-3 lipids containing FO supplemented with adequate levels of vitamin E enhanced the activity and mRNA levels of hepatic antioxidant enzymes in autoimmune mice (Venkatraman et al., 1994a). These oils could increase autoimmune disease by increasing free-radical formation and decreasing the levels of antioxidant enzymes, thus further decreasing immune function by inhibiting the development of anti-inflammatory cytokines such as IL-2 and TGF-β. In contrast, ω-3 lipids could protect against autoimmunity by enhancing TGF-β mRNA levels and preventing an increase in oncogene expression (Fernandes etal, 1994; Venkatraman and Chu, 1999). Immune cell reactivity is directly influenced by the enrichment of lymphoid cell membranes with particular fatty acids (Venkatraman and Fernandes, 1994). Specifically, membrane enrichment in linoleic acid was associated with immunosuppression. Immunosuppression may occur through subsequent increase in the content of membrane arachidonic acid (a precursor for PGE_2) which results from high 18:2ω6 supplementation. PGs of the E_2 series suppress blastogenesis, lymphokine production, and cytotoxicity. In aging *ad libitum* (AL)-fed rats, increased membrane levels of 20:4ω6 may increase precursors for PGE_2 and thereby immune responses. It has been indicated that aging can alter the fatty acid composition of immune cells, and this change may alter the fluidity of microenvironment in the plasma membranes. Membrane receptors can be modulated by lipid composition, microenvironment, and physicochemical characteristics of membranes, which can be altered by different dietary treatments. The percent composition of 18:2, 20:4, 22:4, and 22:5 were found to be significantly affected by age in the nonadherent spleen cells for total lipids, phosphatidyl choline, and phosphatidyl ethanolamine fractions of aged AL-fed rats. Subcellular membrane fatty acids changes and membrane rigidity may result in impairment of immune regulation along with increasing free-radical production and/or other immunosuppressive products such as PGs and certain cytokines such as IL-4, IL-5, IL-6, and IL-10 and lowering levels of IL-2 and IFN-γ production. Diets containing fish oil are known to extend the life-span and inhibit free-radical formation by modulating fatty acid composition of microsomal and mitochondrial membranes of livers and spleens of animals of various ages (Byun et al., 1995; Fernandes et al., 1990; Laganiere et al., 1990). Dietary ω-3 lipids induce apoptosis and apoptosis mediators in splenic lymphocytes of mice and Fas apoptotic gene expression (Fernandes et al., 1995; Fernandes et al., 1998; Reddy Avula et al., 1999).

Although supplementation with ω-3 fatty acids did not significantly alter the humoral immune response to keyhole limpet hemocyanin (KLH) in geriatric beagles (Wander et al., 1997), it significantly suppressed the cell-mediated immune response based on results of a delayed-type hypersensitivity (DTH) skin test. After consumption of the 1.4:1 diet, stimulated mononuclear cells produced 52% less PGE_2 than those from dogs fed the 31:1 diet.

3.2. Exercise Training

Few studies have followed the impact of long-term training on the immune systems of elderly people. Given that a number of age-related changes occur in many systems (e.g., neuroendocrine) known to alter immune function both at rest and during exercise, it would be of value to learn the extent to which both acute and chronic exercise influence immune function in the elderly. In older humans, aerobic exercise training lowers the heart rate at rest, reduces levels of the heart rate and plasma catecholamines at the same absolute submaximal workload, and improves left ventricular performance during peak exercise, but it does not reduce, and may even increase, basal sympathetic nerve activity. With age, there is a slow but significant reduction in muscle mass and ability to perform certain physical activities. This may be the result of changes with the age of muscle composition and protein turnover, as well as the decrease of trophic influence in neural control of muscles of the elderly. Maintenance of aerobic fitness in the older group of exercisers partially prevented the age-associated decline in platelet protein kinase C activity and in stimuli-induced enzyme redistribution.

Endurance training improves the physical capacity and body composition of elderly persons; however this effect is lost after 1 yr, unless the program is continued (Morio 2000; Tsuji et al., 2000; Kasch et al., 1999; Woods et al., 1999). $\dot{V}_{O_{2max}}$ and its trainability includes a significant genetic component (Bouchard et al., 1999). Exercise training can increase cardiac output and skeletal muscle blood flow and thus $\dot{V}_{O_{2max}}$ (Beere et al., 1999). Exercise training increases oxidative metabolism, thus decreasing anaerobic metabolism, which leads to lower ventilation (Prioux et al., 2000). Exercise training does not reverse the reduced responsiveness of the adrenal medulla to sympathetic nervous activity with aging, thus lower adrenaline (Zouhal et al., 1999).

The balance between pro-oxidants and antioxidants is altered with aging (Ji et al., 1998). Exercise increases the disturbances of intracellular pro-oxidant–antioxidant homeostasis that poses a serious stress threat to the cellular antioxidant defense system (Ji et al., 1998). The elderly demonstrate a less resilient leukocytosis and a different lymphoproliferative response following acute maximal exercise (Ceddia et al., 1999). Insulin-like growth factor-I (IGF-1) declines with age, as does aerobic power, but they are not related (Haydar et al., 2000). Acute exercise may stress the immune system, whereas chronic exercise training may enhance it. Exercise-induced changes in the immune system include the release of inflammatory mediators, the activation of various white blood cells, and the complement and induction of acute-phase proteins (Venkatraman et al., 2000; Venkatraman et al., 2000a). There are signs of immunosuppression, such as decreased T- and B-cell function or impaired cytotoxic or phagoycytic activity. Because of the reduced exercise capacity, the relative stress of exercise on the immune system may be exaggerated, leading to vulnerability to acute and chronic inflammation and reduced post-exercise recovery. Increasing total caloric intake by 25% to match energy expenditure and the dietary fat intake to 32% in athletes appears to reverse the negative effects on immune function and lipoprotein levels reported on a low-fat diet (Pendergast et al., 1997; Venkatraman et al., 1997). Increasing the dietary fat intake of athletes to 42%, while maintaining caloric intake equal to expenditure, does not negatively affect immune competency or blood lipoproteins, but it improves endurance exercise performance at 60–80% of $\dot{V}_{O_{2max}}$ in runners (Pendergast et al., 1997; Venkatraman et al., 1997). Both exercise and a high-fat diet increase the level of IL-2 and lowered the level of IL-6, suggesting that it is possible to modulate the level of specific proinflammatory cytokines through increased fat intake, thus offsetting the proinflammatory effects of exercise (Venkatraman et al., 1997; Venkatraman and Pendergast, 1998).

Aging is characterized by shrinking of muscle fibers and protein loss from these muscle fibers. Bone is also lost, and matrix and mineral levels are also lost equally. The predominant breakdown of synthesis is probably the fundamental cause of both muscle and bone loss. Little can be done to prevent this by dietary means, but physical activity is of vital importance in helping to maintain the integrity of both muscle and bone. Resistance training is an effective means of preserving or increasing skeletal muscle mass and functional status in the elderly. In addition, resistance training has been demonstrated to increase energy requirements, protein retention, bone mass, and levels of physical activity in the healthy elderly as well as the very old and frail. The influence of 4 wk of anaerobic training program with 30-min sessions of weight lifting per week in middle-aged, moderately trained men (40–50 yr) was studied, and significant increases of the mean arm muscle force by 7% was found (Weiss et al., 1995). Highly conditioned elderly women retained greater lymphocyte proliferative response to PHA, IL-2 production, and greater NK cell activity than sedentary peers, and two groups had similar leukocyte and lymphocyte subset compositions. These observations indicate higher per cell functional activities in the well-trained group, suggesting that regular habitual physical activity may help to encounter the age-related decline in the potential of the T-cells (Nieman et al., 1993). IFN-γ and IL-4 production were also higher in the elderly runners. Although the mechanism by which exercise training enhances the immune function is unclear, one possibility might be a persisting input of IL-1 from macrophages active in repairing muscle microtraumata.

A single bout of moderate exercise seems to be well tolerated by the elderly. The NK cell response is found to be equal to that of younger individuals. However, older subjects show less stimulation of lymphocyte proliferation by moderate activity and also show less suppression with exhausting exercise. This depressed response may occur because of the depressed resting immune function or as the elderly have markedly less activity than the young, and thus a moderate training program seems to stimulate immune function to a greater extent in older than in young subjects. During exercise, the release of catecholamines and cortisol may be suppressive to NK activity, whereas the release of IL-1, IFN, and β-endorphin may be stimulating. Although NK activity in the healthy elderly may be similar to the young at baseline, they would show a diminution of their response to acute stressors (i.e., exercise or IL-2), as has been commonly observed in other systems with aging. The preserved responsiveness to NK cells in the older groups is even more remarkable because catecholamine secretion is increased with exercise in the elderly, compared to the young, which would presumably mediate more suppressive influences on NK activity with age.

4. SUMMARY AND CONCLUSION

It would seem that modulation of lipid intake and exercise training may mediate and reverse some of the age-associated depression in the immune system. As human aging generally is accompanied by a reduction in the level of physical activity and impaired responsiveness of the immune system, there may be a relationship between these concurrent changes and the increased incidence of, and mortality from, cancer, autoimmune disorders, and chronic infectious diseases with age. The studies on the immune system of centenarians suggest that the immune system in this population may not be declining with age, instead it is being constantly remodeled and reshaped as required. Appropriate fat and protein intake may improve immune function. Lipids modulate immune function by several factors and mediators. Both the quantity and type of dietary lipids are known to have modulatory effects on the cellular immune system at biochemical and molecular levels. The mechanisms by which lipids modulate the immune function may involve several factors, including the production and expression of cytokines. Dietary ω-6 lipids generally increase the levels of proinflammatory cytokines and inflammatory PGs, whereas ω-3 lipids may decrease the levels of these cytokines and inflammatory PGs. Moderate exercise in the elderly may further help them to lead a disease-free life by preserving their immune system. Further cross-sectional and longitudinal studies are required that would examine the type, intensity, and total quantity of exercise needed to optimize the immune function and the way in which this optimum dose may also change with aging. It is important that the dose of physical activity needed to optimize the immune function be defined more clearly at various points during the aging process, and that its preventive efficacy must be explored extensively. Despite the relative normality of the response to moderate exercise in the elderly, they have a reduced tolerance of free radicals and greater vulnerability to micro-injuries and an acute-phase response from overexertion. It might thus be anticipated that they would more easily reach the point where natural immunity is suppressed and vulnerability to infection is enhanced. Older adults may have to adopt a more cautious approach and follow a moderate exercise regimen along with a nutritionally well-balanced diet. Training in theory has a number of actions which could help to reverse the impact of aging upon the immune system, including a direct modulation of sympathetic activity in the neurohypophysis, a reduction of cross-linkages and a diminution of free-radical formation.

REFERENCES

Barnett YA, Barnett CR. DNA damage and mutation: contributors to the age-related alterations in T cell-mediated immune responses? Mech Ageing Dev 1998; 102:165–175.

Beere PA, Russell SD, Morey MC, Kitzman DW, Higginbotham MB. Aerobic exercise training can reverse age-related peripheral circulatory changes in health older men. Circulation 1999; 100:1085–1094.

Bodnar, AG, Ouellette M, Frolkis, M, Holt SE, Chiu CP, Morin GB, et al. Extension of life span by introduction of telomerase into normal human cells. Science 1998; 279:349–352.

Bouchard C, An P, Rice T, Skinner JS, Wilmore JH, Gagnon J, et al. Familial aggregation of $V_{O_{2(max)}}$ response to exercise training: results from the HERITAGE Family study. J Appl Physiol 1999; 87:1003–1008.

Burns EA, Goodwin JS. Immunodeficiency of aging. Drugs Aging 1997; 11:374–397.

Byun DS, Venkatraman JT, Yu BP, Fernandes, G. Modulation of antioxidant activities and immune response by food restriction in aging Fischer-344 rats. Aging: Clin Exp Res 1995; 7:40–48.

Calder PC. Immunoregulatory and anti-inflammatory effects of n-3 polyunsaturated fatty acids. Braz J Med Biol Res 1998; 31:467–490.

Calles-Escandon J, Poehlman ET. Aging, fat oxidation and exercise. Aging 1997; 9:57–63.

Catania A, Airaghi L, Motta P, Manfredi MG, Annoni G, Pettenati C, et al. Cytokine antagonists in aged subjects and their relation with cellular immunity. J Gerontol: Biol Sci 1997; 52A:B93–B97.

Ceddia MA, Price EA, Kohlmeier CK, Evans JK, Lu Q, McAuley E, et al. Differential leukocytosis and lymphocyte mitogenic response to acute maximal exercise in the young and old. Med Sci Sports Exerc 1999; 31:829–836.

Chaet MS, Garcia VF, Arya G, Ziegler MM. Dietary fish oil enhances macrophage production of nitric oxide. J Surg Res 1994; 57:65–68.

Chandrasekar B, Troyer DA, Venkatraman JT, Fernandes G. Dietary ω-3 lipids delay the onset and progression of autoimmune lupus nephritis by inhibiting transforming growth factor β mRNA and protein expression. J Autoimmun 1995; 8:381–393.

Cinader B, Doria G, Facchini A, Kurashima C, Thorbecke GJ, Weksler ME. Aging and the immune system: Workshop #4 of 8th International Congress of Immunology, Budapest, Hungary. Aging: Immunol Infec Dis 1993; 4:47–53.

Doria G, Frasca D. Genes, immunity, and senescence: looking for a link. Immunol Rev 1997; 160:159–170.

Ernst DN, Hobbs MV, Torbett BE, Glasebrook AL, Rehse MA, McQuitty DN, et al. Differences in the subset composition of CD4+ T cell population from young and old mice. Aging: Immunol Infec Dis 1990; 2:105–109.

Ernst DN, Weigle O, Hobbs MV. Aging and lymphokine gene expression by T cell subsets. Nutr Rev 1995; 53:S18–S26.

Fabris N, Mocchegiani E, Provinciali M. Plasticity of neuroendocrine–thymus interactions during aging. Exp Gerontol 1997; 32:415–429.

Fernandes, G. Effects of calorie restriction and omega-3 fatty acids on autoimmunity and aging. Nutr Rev 1995; 53:S72–S79.

Fernandes A, Venkatraman JT, Tomar V, Fernandes G. Effect of treadmill exercise and food restriction on immunity and endocrine hormone levels in rats. FASEB J 1990; 4:751A.

Fernandes G, Venkatraman JT. Effect of food restriction on immunoregulation and aging. In R. R. Watson (Ed.), Handbook of Nutrition in the Aged. 2nd ed. CRC, Boca Raton, FL, 1993, pp. 331–346.

Fernandes G, Bysani C, Venkatraman JT, Tomar V, Zhao W. Increased TGFβ and decreased oncogene expression by ω-3 lipids in the spleen delays autoimmune disease in B/W mice. J Immunol 1994; 152:5979–5987.

Fernandes G, Chandrasekar B, Mountz JD, Zhao W. Modulation of Fas apoptotic gene expression in spleens of B/W mice by the source of dietary lipids with and without calorie restriction. FASEB J 1995; 9:4559.

Fernandes G, Troyer DA, Jolly CA. The effects of dietary lipids on gene expression and apoptosis. Proc Nutr Soc 1998; 57:543–550.

Fernandes G, Venkatraman JT, Turturro A, Attwood VG, Hart RW. Effect of food restriction on life span and immune functions in long-lived Fisher-344 × Brown Norway F1 rats. J Clin Immunol 1997; 17:85–95.

Grounds MD. Age-associated changes in the response of skeletal muscle cells to exercise and regeneration. Ann NY Acad Sci 1998; 854:78–91.

Guidi, L.,Tricerri A, Frasca D, Vangeli M, Errani AR, Bartoloni C. Psychoneuroimmunology and aging. Gerontology 1998; 44:247–261.

Hakkinen K, Pakarinen A, Newton RU, Kraemer WJ. Acute hormone responses to heavy resistance lower and upper extremity exercise in young versus old men. Eur J Appl Physiol Occup Physiol 1998; 77:312–319.

Han SN, Meydani SN. Vitamin E and infectious diseases in the aged. Proc Nutr Soc 1999; 58:697–705.

Haydar ZR, Blackman MR, Tobin JD, Wright JG, Fleg JL. The relationship between aerobic exercise capacity and circulating IGF-1 levels in healthy men and women. J Am Geriatr Soc 2000, 48:139–145.

Hayek MG, Han SN, Wu D, Watkins BA, Meydani M, Dorsey JL, et al. Dietary conjugated linoleic acid influences the immune response of young and old C57BL/6NCrlBR mice. J Nutr 1999; 129:32–38.

Hirokawa K. Reversing and restoring immune functions. Mech Ageing Dev 1997; 93:119–124.

Hirokawa K, Utsuyama M, Kasai M, Kurashima C, Ishijima S, Zeng YX. Understanding the mechanism of the age-change of thymic function to promote T cell differentiation. Immunol Lett 1994; 40:269–277.

Ji LL, Leeuwenburgh C, Leichtweis S, Gore M, Hollander J, Bejm J. Oxidative stress and aging. Role of exercise and its influences on antioxidant systems. Ann NY Acad Sci 1998; 854:102–117.

Kasch FW, Boyer JL, Schmidt PK, Wells RH, Wallace JP, Verity LS, et al. Aging of the cardiovascular system during 33 years of aerobic exercise. Age Ageing 1999; 28:531–536.

Krause D, Mastro AM, Handte G, Smiciklas-Wright H, Miles MP, Ahluwalia, N. Immune function did not decline with aging in apparently healthy, well-nourished women. Mech Ageing Dev 1999; 112:43–57.

Laganiere S, Yu BP, Fernandes G. Studies on membrane lipid peroxidation in omega-3 fatty acid-fed autoimmune mice: effect of vitamin E supplementation. In: Bendich A, Phillips M, Tangerdy RB, eds. Adv Exp Med Biol, Vol. 262, Plenum Press, New York, 1990, pp. 95–103.

Lesourd BM. Nutrition and immune in the elderly: modification of immune response with nutritional treatments. Am J Clin Nutr 1997; 66:478S–484S.

Lesourd BM. Immune response during disease and recovery in the elderly. Proc Nutr Soc 1999; 58:85–98.

Makinodan T. Studies on the influence of age on immune response to understand the biology of immunosenescence. Exp Gerontol 1998; 33:27–38.

Makinodan T, Chang MP, Norman DC, Li SC. Vulnerability of T cell lineage to aging. In: Goidl E, ed. Aging and Immune Response: Cellular and Humoral Aspects. Marcel Dekker, New York, 1987, pp. 27–39.

Marcell TJ, Wiswell RA, Hawkins SA, Tarpenning KM. Age-related blunting of growth hormone secretion during exercise may not be due to increased somatosatatin tone. Metab: Clin Exp 1999; 48:665–670.

Marsh GD, Paterson DH, Govindasamy D, Cunningham DA. Anaerobic power of the arms and legs of young and older men. Exp Physiol 1999; 84:589–597.

Mazzeo RS, Rajkumar C, Rolland J, Blaher B, Jennings G, Eshler M. Immune response to a single bout of exercise in young and elderly subjects. Mech Ageing Dev 1998; 100:121–132.

Mazzoccoli G, Correra M, Bianco G, De Cata A, Balzanelli M, Giuliani A, et al. Age-related changes of neuro-endocrine-immune interactions in healthy humans. J Biol Regul Homeostat Agents 1997; 11:143–147.

McCann SM, Licinio J, Wong ML, Yu WH, Karanth S, Rettorri V. The nitric oxide hypothesis of aging. Exp Gerontol 1998; 33:813–326.

Morio B, Barra V, Ritz P, Fellmann N, Bonny JM, Beaufrere B, et al. Benefit of endurance training in elderly people over a short period is reversible. Eur J Appl Physiol Occup Physiol 2000; 81:329–336.

Mountz JD, Wu J, Zhou T, Hsu H-C. Cell death and longevity: implications of Fas-mediated apoptosis in T-cell senescence. Immunol Rev 1997; 160:19–30.

Nieman DC, Henson DA, Gusewitch G, Warren BJ, Dotson RC, Butterworth DE, Nehlsen-Cannarella SL. Physical activity and immune function in elderly women. Med Sci Sports Exerc 1993; 25:823–831.

Pendergast DR, Horvath PJ, Leddy JJ, Venkatraman JT. The role of dietary fat on performance metabolism and health. Am J Sports Med 1997; 24:S53–S58.

Phelouzat MA, Arbogast A, Laforge T, Quadri RA, Proust JJ. Excessive apoptosis of mature T lymphocytes is a characteristic feature of human immune senescence. Mech Ageing Dev 1996; 88:25–38.

Pierpaoli W. Neuroimmunomodulation of aging. A program in the pineal gland. Ann NY Acad Sci 1998; 840:491–497.

Potestio M, Caruso C, Gervasi F, Scialabba G, D'Anna C, Di Lorenzo G, et al. Apoptosis and ageing. Mech Ageing Dev 1998; 102:221–237.

Prioux J, Ramonatxo M, Hayot M, Mucci P, Prefaut C. Effects of ageing on the ventilatory response and lactate kinetics during incremental exercise in man. Eur J Appl Physiol Occup Physiol 2000; 81:100–107.

Reddy Avula CP, Zaman AK, Lawrence R, Fernandes G. Induction of apoptosis and apoptotic mediators in Balb/c splenic lymphocytes by dietary n-3 and n-6 fatty acids. Lipids 1999; 34:921–927.

Rink L, Cakman I, Kirchner H. Altered cytokine production in the elderly. Mech Ageing Dev 1998; 102:199–209.

Robinson LE, Field CJ. Dietary long-chain (n-3) fatty acids facilitate immune cell activation in sedentary, but not exercise-trained rats. J Nutr 1998; 128:498–504.

Rumyantsev SN. Constitutional and non-specific immunity to infection. Rev Sci Tech 1998; 17:26–42.

Shephard PJ, Rhind S, Shek PN. Exercise and the immune system. Natural killer cells, interleukins and related responses. Sports Med 1994; 18:340–369.

Shephard RJ, Shek PN. Exercise, aging and immune function. Intl J Sports Med 1995; 16:1–6.

Shinkai S, Konishi M, Shephard RJ. Aging and immune response to exercise. Can J Physiol Pharmacol 1998; 76:562–572.

Solana R, Pawelec G. Molecular and cellular basis of immunosenescence. Mech Ageing Dev 1998; 102:115–129.

Stoessl AJ. Etiology of Parkinson's disease. Can J Neuro Sci 1999; 26:S5–S12.

Thoman ML, Weigle WO. The cellular and sub-cellular bases of immunosenescene. Adv Immunol 1989; 46:221–261.

Tsuji I, Tamagawa A, Nagatomi R, Irie N, Ohkubo T, Saito M, et al. Randomized controlled trial of exercise training for older people (sendai Silver Center Trial; SSCT): study design and primary outcome. J Epidemiol 2000; 10:55–64.

Venkatraman JT, Chu W. Effects of dietary ω-3 and ω-6 lipids and vitamin E and on proliferative response, lymphoid cell subsets, production of cytokines by spleen cells and splenic protein levels for cytokines and oncogenes in MRL/MpJ-*lpr* mice. J Nutr Biochem 1999; 10:582–597.

Venkatraman JT, Chu W. Effects of dietary ω-3 and ω-6 lipids and vitamin E on serum cytokines, lipid mediators and anti-DNA antibodies in a mouse model for rheumatoid arthritis. J Am Coll Nutr 1999; 18:602–612.

Venkatraman JT, Fernandes G. Modulation of immune function during aging by dietary restriction. In: Yu BP, ed. Modulation of Aging Processes by Dietary Restriction. CRC, Boca Raton, 1994, pp. 193–219.

Venkatraman JT, Fernandes G. Exercise, immunity and aging. Aging: Clin Exp Res 1997; 9:42–55

Venkatraman JT, Attwood VG, Turturro A, Hart RW, Fernandes G. Maintenance of virgin T-cells and immune functions by food restriction during aging in long-lived B6D2F1 female mice. Aging: Immunol Infect Dis 1994; 5:13–25.

Venkatraman JT, Bysani C, Kim JD, Fernandes G. Effect of n-3 and n-6 fatty acids on activities and expression of hepatic antioxidant enzymes in autoimmune-prone NZB/NZWF1 mice. Lipids 1994; 29:561–568.

Venkatraman JT, Fernandes G. ω-3 lipids in health and disease. Nutr Res 1993; 13:19S–45S.

Venkatraman JT, Pendergast DR. Effects of the level of dietary fat intake and endurance exercise on plasma cytokines in runners. Med Sci Sports Exerc 1998; 30:1198–1204.

Venkatraman JT, Horvath PJ, Pendergast DR. Lipids and exercise immunology. In: Nieman DC, Pedersen B, eds. Nutrition and Exercise Immunology. CRC, Boca Raton, 2000, pp. 43–74.

Venkatraman JT, Leddy JJ, Pendergast DR. Dietary fats and immune status in athletes: clinical implications. Med Sci Sports Exerc 2000; 32:S1–S7.

Venkatraman JT, Rowland JA, DeNardin E, Horvath PJ, Pendergast DR. Influence of the level of dietary lipid intake and maximal exercise on immune status in runners. Med Sci Sports Exerc 1997; 29:333–344.

Walford RL. MHC regulation of aging: an extension of the immunologic theory of aging. In: Warner HR, Butler RN, Sprott RL, Schneider EL, eds. Modern Biological Theories of Aging. Raven, New York, 1987, pp. 243–255.

Wander RC, Hall JA, Gradin JL, Du SH, Jewell DE. The ratio of dietary (n-6) to (n-3) fatty acids influences immune system function, eicosanoid metabolism, lipid peroxidation and vitamin E status in aged dogs. J Nutr. 1997; 127:1198–205.

Weiss C, Kinscherf R, Roth S, Friedmann B, Fischbach T, Reus J, et al. Lymphocyte subpopulations and concentrations of soluble CD8 and CD4 antigen after anaerobic training. Intl J Sports Med 1995; 16:117–121.

Weyand CM, Brandes JC, Schmidt D, Fulbright JW, Goronzy JJ. Functional properties of CD4+ CD28− T-cells in the aging immune system. Mech Ageing Dev 1998; 102:131–147.

Wick G, Grubeck-Loebenstein B. The aging immune system: primary and secondary alterations of immune reactivity in the elderly. Exp Gerontol 1997; 32:401–13.

Woods JA, Ceddia MA, Wolters BW, Evans JK, Lu Q, McAuley E. Effects of 6 months moderate aerobic exercise training on immune function in the elderly. Mech Ageing Dev 1999; 109:1–19.

Yehuda AB, Friedman G, Wirtheim E, Abel L, Globerson A. Checkpoints in thymocytopoiesis in aging: expression of the recombination activating genes RAG-1 and RAG-2. Mech Ageing Dev 1998; 102:239–247.

Zaccaria M, Varnier M, Piazza P, Noventa D, Ermolao A. Blunted growth hormone response to maximal exercise in middle-aged versus young subjects and no effect of endurance training. J Clin Endocrinol Metab 1999; 84:2303–2307.

Zouhal H, Gratas-Delamarche A, Rannou F, Granier P, Bentue-Ferrer D, Delamarche P. Between 21 and 34 years of age, aging alters the cateholamine responses to supramaximal exercise in endurance trained athletes. Intl J Sports Med 1999; 20:343–348.

II Phospholipid and Fatty Acid Composition and Metabolism

6 On the Relative Efficacy of α-Linolenic Acid and Preformed Docosahexaenoic Acid as Substrates for Tissue Docosahexaenoate During Perinatal Development

Meng-Chuan Huang and J. Thomas Brenna

1. PERINATAL N-3 AND N-6 FATTY ACID METABOLISM AND NEURAL DEVELOPMENT

The polyunsaturated fatty acids (PUFA), α-linolenic acid (LNA, 18:3n-3) and linoleic acid (LA, 18:2n-6) are indispensable dietary components in mammals. It is thought that the primary metabolic role of LA and LNA are as essential precursors for conversion to their respective long-chain metabolites. Of these long-chain PUFA (LCP, C>20) docosahexaenoic acid (DHA, 22:6n-3) and arachidonic acid (AA, 20:4n-6), have received the most attention, because of their distinct functional roles in fetal and infant health. They are present in human breast milk at small but significant amounts, ranging from 0.1% to 0.6% and 0.5% to 1%, respectively (Jensen, 1996). DHA and AA are the most abundant n-3 and n-6 LCP found in the lipids of humans and are concentrated in the brain and central nervous system (CNS), which is second only to adipose tissue in containing the greatest percentage of lipid. The brain is comprised of 50–60% lipids (dry wt) serving mostly structural roles, as opposed to storage of energy of triacylglycerol in adipose tissues. A related organ, the retina, is also rich in DHA, which comprises approximately 50% of phospholipid acyl groups in retinal photoreceptor membranes (Fliesler & Anderson, 1983; Sastry, 1985). The functional importance of DHA is related to its functional significance in supporting perinatal neurological and visual development. Depletion of DHA in the brains of monkeys during fetal and early postnatal development results in permanent deficits of visual function despite later biochemical repletion of brain and retinal DHA (Neuringer, Connor, Lin, Barsted, & Luck, 1986; Neuringer, Connor, Petten, & Barstad, 1984). AA is a structural component of membrane phospholipids and is known to exert its biological actions by serving as the precursor of eicosanoids, which exercise ubiquitous physiological influences.

Docosahexaenoic acid and AA can be obtained either directly from the diet or synthesized from their C_{18} precursors LNA or LA, respectively, via a series of equivalent

From: *Fatty Acids: Physiological and Behavioral Functions*
Edited by: D. Mostofsky, S. Yehuda, and N. Salem Jr. © Humana Press Inc., Totowa, NJ

desaturation and elongation reactions. The synthetic pathway of n-3 and n-6 fatty acids have long been known to interact (Mohrhauser & Holman, 1963); the dietary level of one series of fatty acids can interfere with the metabolism of the other (Lands, 1995). For instance, a dietary excess of n-6 fatty acids can reduce the tissue concentration of n-3 fatty acids, leading to concerns about deficits of n-3 LCP. Similarly, an excess of n-3 fatty acids can hinder conversion of LA to n-6 LCP. Thus, moderate ratios of dietary n-3 and n-6 C_{18} precursors are required to facilitate balanced LCP biosynthesis for fast-growing nervous tissue to achieve optimal function, especially when dietary preformed DHA and AA are absent.

The roles of n-3 and n-6 LCP in infant nutrition have been extensively reviewed by several authors (Connor, Neuringer, & Reisbick, 1992; Innis, 1991; Jensen, 1996), and it is agreed that DHA is essential to perinatal neurological development. In the early 1970s, Benolken et al. reported that the amplitude of electroretinogram a-waves was lower in rats fed a fat-free diet than in those raised on a fat sufficient diet (Benolken, Anderson, & Wheeler, 1973). The key components causing electroretinogram changes were later identified as dietary n-3 fatty acids (Wheeler & Benolken, 1975). Further, the association between n-3 fatty acid deficiency and subsequent impaired neural outcomes was confirmed in rodents by a number of investigators (Bourre, Durand, Pascal, & Youyou, 1989; Yamamoto, Saitoh, Moriuchi, Nomura, & Okuyama, 1987).

Strong evidence that formula-fed human infants may be at risk of n-3 fatty acid or DHA deficiency arises from infant rhesus monkey studies. Neuringer et al. first reported that dietary n-3 fatty acid deprivation depressed retinal and other visual functions in primates. Infant monkeys fed an n-3 fatty-acid-depleted diet (with 18:2n-6/18:3n-3 ratio = 255) had impaired visual acuity development as assessed by electroretinography and the preferential looking method compared to those fed n-3-adequate diets (Neuringer et al., 1986; Neuringer et al., 1984). This functional deficit was ascribed to DHA deficiency because the neural deficits were positively correlated with loss of DHA from the cerebral cortex and retina. In a related human study, Farquharson et al. reported on the cerebral cortex fatty acid composition of term and preterm infants expired as a result of sudden infant death syndrome. They showed that formula-fed infants had poorer DHA levels than breast-fed infants, with the average brain DHA 22–23% lower for formula-fed relative to breast-fed (Farquharson, Cockburn, Patrick, Jamieson, & Logan, 1992; Farquharson et al., 1995).

Until 10 yr ago, most commercial infant formulas had high LA/LNA ratios, ranging from 10 to 50 (Jensen, Hagerty, & McMahon, 1978) and did not contain preformed LCP. Infants fed such formulas would have to derive all their LCP from C_{18} precursors and might be expected to have impaired conversion of LNA to DHA because of high dietary LA. These findings sparked numerous clinical trials in preterm/term infants to establish PUFA compositions for optimal neural development during infancy (Crawford, Costeloe, Ghebremeskel, & Phylactos, 1998; Heird, Prager, & Anderson, 1997). These studies generally show that preterm infant visual and cognitive function improves with addition of LCP, particularly DHA, to formula, whereas functional data in term infants sometimes shows improvement and sometimes not, leading to a consensus that both DHA and AA should be added to preterm formula, and although no universal consensus has emerged for term formula as yet. In most of these studies, changes in fatty acids in blood compartments (plasma, red cells) were considered as indices of the respective fatty acid status in neural tissues such as brain and retinas, because ethical constraints limit sampling to blood even though the target organs are the brain and retina. However, caution is warranted because data derived from animal studies (Hrboticky, Mackinnon, & Innis, 1990; Pawlosky, Ward, & Salem, 1996; Rioux, Innis, Dyer, & MacKinnon, 1997; Su et al., 1999a) and human studies (Makrides, Neumann, Byard, Simmer, & Gibson, 1994) show that blood-compartment fatty acid profiles are sensitive to short-term dietary changes and do not necessarily mirror that in neural tissues. In addition, practical constraints limit the number of dietary treatments that can be investigated. The net outcome of all these trials is confidence in the safety and efficacy of LCP supplementation in a limited number of dietary PUFA levels. In clinical studies of growing human infants, it is impossible to determine the relationship of dietary PUFA levels and tissue n-3/n-6 fatty acid accretion, information required to establish optimal dietary LCP levels, because of the inability to sample target tissue. For this, animal studies are required.

2. CONSIDERATIONS FOR ANIMAL STUDIES

It has been noted that the timing of brain growth varies among species and that studies aimed at investigating brain development must make observations relevant to the period when the brain is growing. The period of brain development during which brain growth rate exceeds that of other organs is known as the brain growth spurt (Dobbing & Sands, 1979). Animals born relatively mature, such as the guinea pig, called "precocial," generally grow their brains *in utero* and, consequently, have their highest demands for DHA prenatally. Animals born immature such as rats ("altricial"), grow their brains postnatally. The human brain growth spurt is rare in that it is perinatal. The infant brain accumulates DHA from the beginning of the third trimester *in utero* and continues up to the first 24 mo of early neonatal growth (Martinez, Ballabriga, & Gil-Gibernau, 1988). It is this period when human neural development is most dependent on adequate supplies of nutrients.

Equally important to brain nutrition is the relative brain-weight percentage. For nonprimates, the percentage of total body weight represented by the brain is less than 5% at birth (Brody, 1945). For instance, the rat brain is about 2% of body weight at birth. In contrast, the brain weight of rhesus monkeys, baboons, and other primates is greater than 10% at birth, whereas the brain weight of term human infants is about 14% at birth. Because nutrient requirements are best expressed in terms of a fraction of dietary energy, it stands to reason that requirements for an animal supporting the growth of a brain of 10% body weight are more applicable to humans than an animal with a brain of 2% body weight.

The desaturase activities required for conversion of LNA and LA to DHA and AA, respectively, have been detected in perinatal animals and humans. Δ-6 and Δ-5 desaturase activities have been found in prenatal animal tissues, including rat liver and brain (Bourre, Piciotti, & Dumont, 1990). In humans, expression of these enzyme activities were found in neonates (Poisson et al., 1993) and in fetuses as early as 17–18 wk of gestation and are stable throughout the third trimester (Chambaz et al., 1985; Rodriguez et al., 1998). Using stable-isotope-tracer methodology, human term and preterm infants were shown, in vivo, to be capable of converting the C_{18} precursor to the long-chain metabolites, DHA and AA, subsequent to oral administration of ^{13}C-labeled LNA and LA, respectively (Carnielli et al., 1996; Salem, Wegher, Mena, & Uauy, 1996; Sauerwald et al., 1997). In these and subsequent studies, a wide range of variability was observed in the appearance of labeled

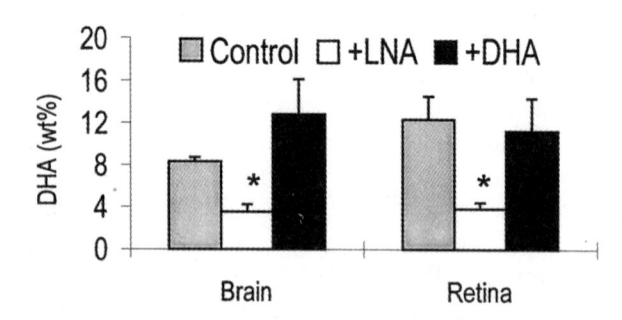

Fig. 1. Graphical representation of study by Anderson and co-workers. After laying hens were fed an n-3-deficient diet for 2 mo, their chicks were fed one of three diets for three wk: control and n-3-deficient supplemented with equivalent amounts of LNA (+LNA) or DHA (+DHA). Brain DHA levels (plotted on the ordinate) for the +LNA are 25% of those for the +DHA group. (Based on data from Anderson, Connor, & Corliss, 1990).

LCP in the plasma, indicating a wide range of DHA synthetic competence (Uauy, Mena, Wegher, Nieto, & Salem, 2000). All evidence supports the ability of human fetuses/infants to convert LNA to DHA in vivo. However, none of these studies addresses the quantitative efficacy of this process because, as with all clinical studies, sampling from blood compartments cannot be used to reliably compute the adequacy of LCP biosynthesis and incorporation for neural tissues.

3. DIETARY LNA AND DHA AS SUBSTRATES FOR BRAIN AND RETINA DHA: COMPOSITIONAL STUDIES

In nontracer experiments, relative contribution of dietary LNA and DHA to CNS DHA accretion have been examined in various species, including chicks, rat pups, newborn piglets, and guinea pigs. Anderson and colleagues studied the relative efficacy of LNA and DHA in restoring neural DHA levels in newly hatched chicks, as presented in Fig. 1 (Anderson, Connor, & Corliss, 1990). Laying hens were fed a n-3-deficient diet for 2 mo, and their hatched chicks were then fed a control diet or n-3-deficient diets supplemented with LNA (+LNA) or DHA (+DHA) at 3.6 wt% (0.44% kcal) for 3 wk. After 3 wk, the DHA group showed brain DHA levels similar to that of controls (12.3% vs 8.3%), whereas dietary LNA alone brought the level to only 25% of the controls. Similar results were observed for retinal DHA accretion. It was concluded that dietary DHA exerts a fourfold greater potency compared to LNA as a substrates for brain and retina DHA accretion.

Based on this finding, Arbuckle et al. postulated that neonatal pigs fed formula with an LNA concentration four times greater than that used in the chick study (1.7% kcal vs 0.44% kcal) would support retinal DHA at levels similar to a sow-fed group (Arbuckle & Innis, 1992). Four test formulas were used: high LNA (1.7% kcal), low LNA (0.3% kcal), low LNA plus fish oil ($C_{20} + C_{22}$ = 0.4% kcal) with n-3 LCP concentration used comparable to those used in human clinical trials (Carlson, Rhoides, Rao, & Goldgar, 1987; Uauy, Birch, Birch, Tyson, & Hoffman, 1990) and sow-reared with milk containing 1.1 wt% as LNA and 0.1 wt% as DHA. In retina phosphatidylethanolamine and in synaptic membranes, piglets fed the high-LNA diets, but not the low-LNA group, exhibited DHA comparable to that in the sow-reared group and the low-LNA plus fish oil group. It was thus concluded that dietary LNA is 24% as effective as C_{20} and C_{22} n-3 fatty acids as a source of membrane DHA, yielding, again, a bioequivalence of about 4 : 1.

Current LCP-free infant formulas contain an LA/LNA ratio of between 7 and 10. One strategy to induce greater conversion of LNA to DHA is to reduce this ratio substantially, thereby reducing competition for LCP biosynthesis from LA. Woods and co-workers tested this hypothesis in an artificial feeding study in rat pups (Woods, Ward, & Salem, 1996). DHA accretion in the brain and retina was compared among artificially reared rat pups fed diets containing LA/LNA ratios of 10 : 1, 1 : 1, and 1 : 12 (LNA = 3%, 15%, and 25% of total fatty acids, respectively) and compared to dam-reared rat pups. They found that brain DHA was comparable to that of dam-reared pups consuming milk containing 1.1% DHA only in the 1 : 12 group LA/LNA ratio. Retinal DHA accretion, however, was still 10% lower in this treatment compared to the dam-reared group. They conclude that efficacy of DHA accretion from dietary DHA is at least 20-fold greater than LNA.

It is notable that in this study, as in all dietary fatty acid studies, the LCP content of reference-group diets is critical in assessing adequate tissue levels. Dietary LCP are known to produce higher tissue LCP levels than their corresponding C_{18} precursors, possibly because the storage of dietary LCP in their nascent state is energetically more efficient than extensive metabolism prior to storage. By the same reasoning, dietary LNA and LA may be converted to LCP only to the extent that LCP is required, with increasing levels of dietary LNA and LA shunted to storage or other metabolic fates. Thus, higher levels of LCP in reference groups consuming LCP may reflect excess rather than active metabolic requirements. In the present study, the estimated conversion ratio depends on the dietary fatty acid composition of the dams' feed, which contained 6 wt% of n-3 LCP, and induced relatively high milk LCP concentrations, and there is no evidence of any functional significance for these higher DHA levels.

In a human term infant study, Jensen et al. investigated the effects of formulas with differing LA/LNA ratios on plasma phospholipid fatty acids and transient visual evoked responses up to 240 d of life (Jensen et al., 1997). Infants who consumed formula for 4 mo with an LA/LNA ratio of 4.8 had higher plasma phospholipid DHA but lower AA compared to those who consumed formulas with LA/LNA ratios of 9.7 or higher, showed no improvement in visual responses, and were significantly lower in body weight than the highest LA/LNA group (44 : 1). From these data, it was concluded that low LA/LNA ratios may not be an appropriate means to normalize plasma fatty acid content because of adverse effects on growth, and that there was no major improvement in the functional outcome measured.

Very recently, Abedin et al. reported a study on weanling guinea pigs in which LNA and LA levels were studied along with LCP treatments. Brain and retina DHA levels were analyzed in 15-wk-old guinea pigs after 12 wk of feeding with one of five diets: low LNA ("Lo LNA"), medium LNA ("Med LNA"), high LNA (LNA: 7%; "Hi LNA"), and two diets constructed to resemble DHA and AA concentration comparable to human breast milk (Lo LCP: 1%, 0.6% DHA) or three times more human breast milk (Hi LCP), as shown in Fig. 2 (Abedin, Lien, Vingrys, & Sinclair, 1999). It was found that brain and retina DHA levels were similar between animals fed the high-LNA diet and animals fed Lo LCP after 12 wk of feeding. This indicates a straightforward bioequivalence between dietary LNA and DHA of about 10 : 1 for brain and retina phospholipid DHA accretion. It should be noted that the weanling guinea pigs are well past their brain growth spurt,

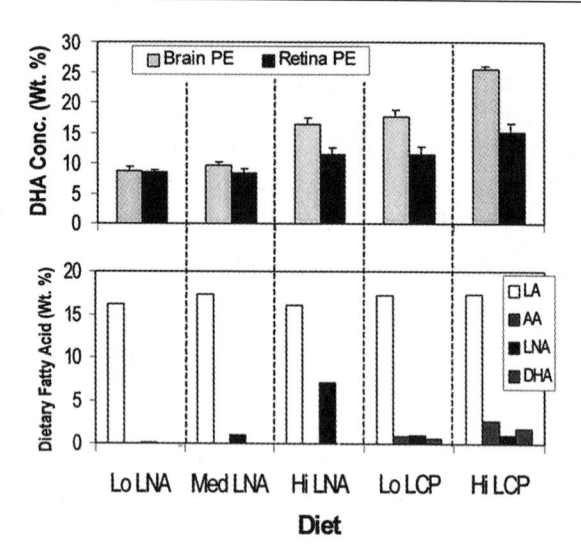

Fig. 2. Graphical representation of study by Abedin and co-workers. The lower panel presents the dietary fatty acid concentrations fed for 12- to 15-wk-old guinea pigs. The Lo LCP diet is intermediate in LCP content for human breast milks; Hi LCP has threefold higher LCP concentrations. The upper panel shows the outcome variable, DHA concentration in brain or retina phosphatidylethanolamine (PE). Brain and retina DHA were similar between animals fed the Hi LNA diet or Lo LCP. (Based on data from Abedin, Lien, Vingrys, & Sinclair, 1999).

which may explain why no differences in brain or retina DHA were found in the Lo LNA and Med LNA diet groups.

Even when there are acute demands for brain DHA, LNA appears to serve as a relatively poor substrate for its synthesis and accretion. Bourre and colleagues (Bourre et al., 1989; Bourre, Youyou, Durand, & Pascal, 1987) found that complete recovery of neural DHA took 2–3 mo in rats initially fed an n-3-deficient diet that were refed with LNA as the sole n-3 fatty acid. These compositional studies all confirm that dietary DHA is considerably more effective than LNA for DHA accretion.

4. DIETARY LNA AND DHA AS SUBSTRATES FOR BRAIN AND RETINA DHA: TRACER STUDIES

Data generated from compositional studies cannot distinguish endogenous and exogenous sources of n-3 fatty acids or the metabolic origin of precursors. Therefore, precise conversion ratios can be best established with tracer studies. Early studies were performed with radiotracers in rodents, were mostly short term in nature, and established important aspects of basic physiological handling of LNA and DHA in target organs. A series of recent and ongoing studies were conducted in our laboratory have established quantitative aspects of brain DHA accretion in primates. We first review selected studies in rodents, then discuss our primate results in detail.

4.1. Neonatal Conversion in Rodents

Several studies have examined the question of the uptake of unsaturates in the developing brain (Anderson & Connor, 1988; Dhopeshwarkar, Subramanian, & Mead, 1971; Hassma & Crawford, 1976; Sinclair, 1975; Sinclair & Crawford, 1972), with delivery of the radioisotope either orally or by injection. Among them, only two have addressed the relative efficacy of LNA and DHA as precursors for brain DHA accumulation in newborns. It is well established that rats lay down most of their brain DHA postnatally, ending around 30 days of life (Sinclair & Crawford, 1972). In order to isolate the ability of the brain, apart from the liver, as a synthetic site for DHA from LNA, Anderson and co-workers (Anderson & Connor, 1988) compared brain uptake of DHA and LNA by injecting the 5 µCi of each labeled fatty acid intravenously in suckling rats that had been functionally hepatectomized to minimize liver contributions. At 30 min postdose, it was found that brain lipid radioactivity steadily increased with increasing degree of unsaturation (16:0<18:2n-6<18:3n-3<22:6n-3). Specifically, ^{14}C-DHA radioactivity at 30 min was 1.5-fold greater ^{14}C-LNA. In this experiment, as with many radiotracer experiments, the tracer was detected as total radioactivity in a mixture, in this case organ extracts, with no chemical speciation; that is, it is not possible to determine whether the detected radioactivity was in the form of LNA or DHA. Because it is unlikely that much LNA is converted to DHA at 30 min postdose, it is possible to conclude that total uptake of ^{14}C-DHA or ^{14}C-LNA by suckling rats is greater for ^{14}C-DHA. It is not possible to estimate the conversion of LNA to DHA in the brain from this study design.

Sinclair's results in suckling rats were consistent with these findings (Sinclair, 1975). He provided an oral dose of ^{14}C-LNA or ^{14}C-DHA at 2 wk of life and measured radioactivity in specific fatty acids at 22 and 48 h postdose. As a percent of dose, ^{14}C-LNA uptake was relatively low compared to ^{14}C-DHA (brain: 0.29% vs 2.71%; liver: 3.29% vs 19.8%). Here, the relative efficacy of LNA and DHA for brain DHA could be estimated, and at 22 h postdose was measured to be 59:1 in favor of DHA. A similar trend was found for ^{14}C-LA and ^{14}C-AA at 22 and 48 hr postdose, consistent with later data Hassam and Crawford obtained for orally fed labeled AA versus LA (Hassma & Crawford, 1976). Diet control was not well described in these two tracer studies (Anderson & Connor, 1988; Sinclair, 1975), and data reported in primates, to be discussed below, indicates that the timing of sampling is not sufficiently long for all brain DHA derived from labeled LNA or DHA to reach a constant level. Thus, no overall conversion efficacy for dietary LNA versus DHA can be inferred from these two studies. They nevertheless unequivocally show that DHA is preferred dietary form for brain DHA accretion at short times postdose.

4.2. Perinatal Conversion in Primates

Several considerations suggest that quantitative requirements for brain growth should be modeled in primates. The human brain constitutes about 14% of body weight at birth, and the brains of most primates constitute more than 10% of body weight at birth. In contrast, most non-primate brains are less than 4% of body weight at birth. The relative anatomy of the brains and neural tissues of primates is much more similar to one another to that of non-primates. Dietary recommendations for essential fatty acids are usually cast in terms of the relative fraction of dietary energy required to support tissue growth and maintenance. Because the fraction of body weight occupied by the primate brain is twofold to fourfold greater than that of common laboratory species, it is *a priori* expected that dietary requirements for laboratory species cannot be directly translated into recommendations for humans. For this reason, nonhuman primates represent the only realistic model of quantitative human brain requirements. The baboon is a particularly attractive model because, like humans, it is an omnivorous, social animal found over a wide range in the wild (Su et al., 1999b). It is also a moderate-size animal that can be jacketed and catheterized and is increasingly used for studies of metabolism in pregnancy (Giussani et al., 2000; Ma et al., 2000; Tame et al., 1998).

We have used the baboon in a series of studies designed to investigate the relative efficacy of a ^{13}C-uniformally labeled LNA (LNA*) and DHA (DHA*), provided either in the diet or in a postprandial form, as substrates for brain DHA* accretion. The studies are designed to permit direct, quantitative comparison between the two n-3 tracers, identified as DHA* in specific tissue. Analysis is accomplished with high-precision gas chromatography–combustion–mass spectrometry, which yields high sensitivity and permits the use of milligrams per kilogram doses to minimize physiological perturbations (Brenna, 1994; Goodman & Brenna, 1992). Here, we review data concerning the three major perinatal states: doses to mothers, to fetuses, and to neonates.

4.3. Materno-fetal Conversion in Primates

The fetal supply of n-3 fatty acids is ultimately from the placenta. The delivery of fatty acids from the maternal circulation to the developing fetus has been reported by others in several studies. Although the human placenta lacks both Δ-6 and Δ-5 desaturase activities (Chambaz et al., 1985; Haggarty, Page, Abramovich, Ashton, & Brown, 1997; Innis, 1991), it does preferentially sequester DHA and AA (Ruyle, Connor, Anderson, & Lowensohn, 1990) and assists in concentrating LCP in the fetal circulation during the last intrauterine trimester.

To quantitatively estimate the relative contributions of the mother and fetus to supplying fetal brain DHA requirements, we studied the metabolism of LNA* and DHA* in pregnant animals, administering doses to the maternal or fetal bloodstream. Pregnant baboons consumed a LCP-free diet for approximately 8 wk prior to ^{13}C tracer dose, to minimize effects of uncontrolled intake of LCP incidentally included in fishmeal of commercial primate diets. At the beginning of the third trimester (140 d of gestation age [dGA], term = 182 d), a [U-^{13}C]-LNA or [U-^{13}C]-DHA dose as nonesterified fatty acid in a soy oil emulsion was given intravenously to the pregnant animals or the fetuses. When fats are consumed, lipoprotein lipase located at the maternal side of the placenta cleaves LCP esterified in the triglycerides or phospholipids of lipoproteins into free fatty acids (Dutta-Roy, 2000; Stammers, Stephenson, Colley, & Hull, 1995), which can then bind to fetal plasma proteins and be transported to fetal tissues (Knipp, Audus, & Soares, 1999; Naval et al., 1992). Thus, intravenous administration of nonesterified fatty acid is a prominent physiological form of nascent dietary polyunsaturates.

Figure 3a shows brain DHA* kinetic data for LNA* and DHA* doses. Each point represents a single animal, and the solid lines are plots of least squares exponential fits, both with correlation coefficients $r^2 > 0.90$. The data for LNA*-derived DHA* is multiplied by 5 to bring it on scale. The ratio of these fitted curves in the plateau region is plotted as a dashed line against the right ordinate. Brain DHA* resulting from either dose plateaus by about 15 d, and the ratio at long times settles out at about 20 : 1. We take this figure to be the bioequivalence of the two fatty acids as sources of brain DHA.

Results on accretion and bioequivalence of preformed DHA-derived DHA and LNA-derived DHA in the primate brain is compiled in Table 1. For pregnant animal dosing (Greiner et al., 1997), baboon fetal brain DHA* plateaued by 15 d postdose with 1.6% Dose of the preformed DHA* dose recovered in the brain and 0.075% of the LNA* dose recovered as brain DHA*, for a ratio of 20 : 1 in favor of DHA. When the dose was given

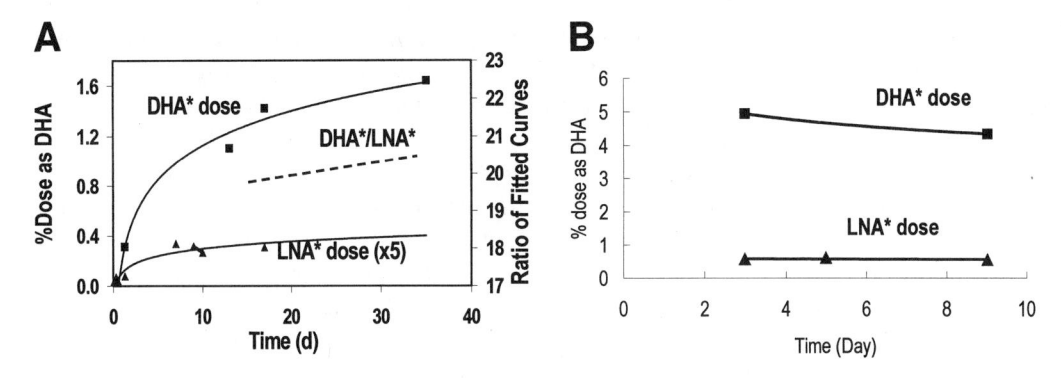

Chapter 6 / α-LNA and DHA During Perinatal Development

Fig. 3. (A) Kinetics of brain DHA* accretion in fetal baboons following an intravenous dose of either LNA* or DHA* to the mother. Each point represents a single animal. Plotted lines are exponents with $r^2 > 0.90$, and the data from LNA*-dosed animals are multiplied by 5 to bring them on scale. The plateau of the ratios of fitted curves, plotted against the right axis, shows that preformed DHA is about 20-fold more efficacious than LNA* at supplying brain DHA. Data are expressed as %Dose as DHA. Each point represents one single animal; error within a data point is within 10%. **(B)** Fetal brain DHA* accretion resulting from DHA* or LNA* intravenous dose directly to the fetal. At the 3-d postdose, brain DHA* accretion plateaued in both groups. The pooled DHA* accretion for the DHA dose and LNA* dose groups are 4.64±0.43% and 0.57±0.03% respectively, yielding an estimated bioequivalence for LNA-derived DHA* versus DHA-derived DHA* of 8 : 1.

Table 1
Bioequivalence of Preformed DHA*-Derived DHA* Versus LNA*-Derived DHA* in the Baboon Brain During the Perinatal Period

	Fetal dose[a]	Maternal dose[b]	Neonatal dose[c]
LNA*%Dose as DHA*	0.6%	0.075%	0.23%
DHA*%Dose as DHA	4.6%	1.6%	1.7%
Bioequilvalence[d]	8	20	12

Note: Data are expressed as percentage of total dose (%Dose) administered.

[a]Dose to pregnant baboons consuming an LCP-free diet for 8 wk prior to ^{13}C tracer administration. (Data from Sheaff Greiner et al., 1999).

[b]Dose to fetuses of pregnant baboons consuming an LCP-free diet for 8 wk prior to ^{13}C tracer administration. (Data from Su et al., 1998).

[c]Dose to neonatal baboons consuming commercial human infant formula containing 18:2n-6/18:3n-3 = 10 and no LCP for 6 wk of life. At wk 4, animals were given a [U-^{13}C]-LNA or [U-^{13}C]-DHA dose orally. (Data from Su et al., 1999).

[d]Definition: preformed DHA*-derived DHA*/LNA*-derived DHA*.

directly to the fetal bloodstream via an indwelling catheter at 140 dGA, brain DHA* was constant by 3 d postdose originating from either a LNA* dose or a DHA* dose, as shown in the kinetic plot in Fig. 3b. Fetal doses were recovered as brain DHA* at 4.6% Dose and 0.6% Dose respectively for DHA*-dosed and LNA*-dosed groups (Su, Corso, Nathanielsz, & Brenna, 1998), yielding a bioequivalence of 8 : 1. These are also the first data demonstrating in vivo that the fetal primate is capable of converting LNA to DHA. In the retina, DHA* was 23-fold greater in the maternally dosed study, compared to 13-fold greater for the fetal dose study.

These data permit several conclusions to be drawn. In all of our studies investigating kinetics of brain DHA* accretion, later-time points postdose always result in greater accumulation; we have been unable to find a time, even at 35 d postdose, at which the amount of label from DHA*, administered in any form, has decreased. This observation is consistent with the long-held opinion that DHA is avidly retained by the growing brain. Because brain loss of DHA is negligible compared to uptake, we can use the ratio of plateau levels in the brain to estimate the relative efficacy of dietary LNA and DHA without resorting to kinetic models requiring many time-points or to area under the curve calculations, which may not produce true kinetic constants at long times. The bioequivalence measures presented in Table 1 are based on these plateau levels and show, as expected, that LNA or DHA in the maternal bloodstream are not as efficiently transferred to the fetal brain as those to the fetal bloodstream. It is important to note that these values apply only to animals consuming diets that are LCP-free and have an n-6/n-3 ratio of about 10 : 1. It is expected that the inclusion of LCP in the diets of an alteration of the n-6/n-3 ratio would alter the expression of genes in the two competing pathway and would also alter the relative efficacies of DHA and LNA.

4.4. Neonatal Conversion in Primates

It has been known for several years through in vivo tracer studies that human term and preterm infants are capable of synthesizing C_{20} and C_{22} LCP from the C_{18} precursors (Carnielli et al., 1996; Salem et al., 1996; Sauerwald et al., 1997). These and more recent studies suggest that this capability is highly variable from individual to individual (Uauy et al., 2000). However, all studies to date have sampled blood compartments, and estimates of relative conversion based on such measurements are tenuous at best. We recently reported the bioequivalence of dietary LNA and DHA as precursors for primate neonate brain DHA accretion based on direct measurements of brain DHA accretion (Su et al., 1999a). Neonate baboons were fed a commercial infant formula with 18% of total fatty acids by weight as LA and 1.8% as LNA, which gives an LA/LNA ratio of 10 for 6 wk. Doses of LNA* or DHA* were administered orally at 4 wk and animals were sacrificed at 6 wk.

In the brain, 1.7% of the preformed DHA* dose and 0.23% of the LNA* dose was detected as DHA*, showing that preformed DHA is seven times more efficient than LNA as a source for brain DHA. The correspondence of this neonatal bioequivalence with that found in our fetal study of 4.6%/0.56% = 8, presented in Table 1, is remarkable. These neonates were, on average, about 224 d conceptual age, whereas the fetuses were about 140 dGA. This 12-wk advantage in maturity for the neonates did not translate into a significantly greater conversion/accretion efficiency. The difference in absolute accretion between neonates and fetuses (4.6% Dose vs 1.7% Dose for DHA) is large enough to indicate that the fetus is more efficient in handling n-3 fatty acids than the neonates. The dose administered to neonates was oral, not intravenous as with fetuses, and no absorption estimates were reported; thus, caution is required before drawing this conclusion. Brain growth in the neonatal baboon, although still rapid, is not as high on a relative basis as for the 140 dGA fetus, which might partly explain this result.

Studies have shown that relative plasma concentrations of DHA and AA exceed those of their C_{18} precursors, LNA and LA in newborns, whereas the opposite situation is found in maternal plasma (Al, Hornstra, van der Schouw, Bulstra-Ramakers, & Huisjes, 1990; Ruyle et al., 1990). The data from our maternal dose study (Greiner et al., 1997) show a similar pattern of C_{18} precursors versus C_{20} and C_{22} LCP between the fetus and the mother. A mechanism driving this concentration difference is selective transfer of PUFA by the placenta. Campbell et al. (Campbell, Gordon, & Dutta-Roy, 1996) reported that human placenta membranes preferentially bind to fatty acids in the order, AA>LA>LNA>>>18 : 1. In whole human perfused placenta, the order of selectively for transfer from the maternal to the fetal circulation was reported as DHA>LNA>LA>18 : 1>AA (Haggarty et al., 1997). Other perfusion studies indicate that the human placenta preferentially incorporates PUFA into phospholipids on the fetal side, thereby eliminating the possibility of back transport (Kuhn & Crawford, 1986). Overall, these studies consistently show that the placenta assists in essential fatty acid transport to the fetus.

In our maternal-dose study, DHA* plateaued at 15–35 d postdose in both DHA* and LNA* dose groups. This observation prompted us to choose 2 wk postdose as the time for collection of tissues in the neonatal study. By that time, DHA* accretion would have stabilized and would reflect overall, integrated levels of brain DHA (Su et al., 1999a). We computed that a minimum of 92% of the LNA*-derived DHA* was present in the brain at 2 wk postdose, indicating that, at most, a modest 4% per week of brain DHA turnovers during the dose period. In addition to being a strong confirmation of the slow turnover hypothesis, this estimate also shows that the measurement reflects the actual integrated bioequivalence.

The term "bioequivalence," as used here, is chosen to imply a relative efficacy in accretion between two sources of brain DHA, in analogy to the use of the term in reference, for instance, to retinol and β-carotene. The crucial clinical issue for infant formulations is to establish the amount of DHA to be added to LCP-free formulas as a precursor for neonate brain development. In our neonate study, the commercial formula contained 1.8% of calories as LNA, and the only dietary DHA that these animals consumed was from the dose. Thus, the bioequivalence of 7 : 1 applies directly to the addition of small amounts of DHA to formula, meaning that the addition of DHA at 0.26% of calories may provide an equal amount of brain DHA as the entire 1.8% calories as LNA. Factors driving the addition of less DHA include possible interference with AA metabolism, the possibility of contaminants added incidentally in DHA oils, and expense. The potency of DHA relative to LNA suggests that the addition of amounts as small as 0.1% of calories would support brain growth, a figure similar to the lowest levels of DHA found in human breast milk. Finally, we note that the purely biochemical nature of our studies to date cannot establish whether LNA can completely substitute for DHA. Studies in human preterms suggest that it cannot, whereas those in term infants remain controversial (Cunnane, Francescutti, Brenna, & Crawford, 2000).

5. DIETARY LNA METABOLISM OTHER THAN FOR DHA

Many studies including our own show that very little LNA is detected in neural tissues, including the brain, retina, and retinal pigment epithelium (RPE). It has long been known that these organs are very low in LNA, despite measurements showing that LNA traverses the blood-brain barrier (Edmond, Higa, Korsak, Bergner, & Lee, 1998). One possible interpretation of the ineffectiveness of LNA as a precursor for brain DHA is the major diversion of LNA to other metabolic pathways. Studies investigating metabolic routes of LNA* administered to animals suggest that very little LNA is sequestered in the brain for making DHA.

In a feto-maternal rhesus monkey experiment conducted in our laboratory, Scheaff Greiner et al. found that only 0.24% of LNA* appeared in fetal brain DHA after a LNA* dose to the mother, with recycling of carbon from LNA into saturates and monosaturates being the predominant metabolic pathway (Sheaff Greiner et al., 1996). Studies in artificially reared rats show that all brain palmitate is made in the brain from acetate derived from PUFA when the route of consumption is oral (Marbois, Ajie, Korsak, Senshsarma, & Edmond, 1992). Other small-animal studies show that 0.1% and 0.02% of labeled-LNA (^{14}C-LNA) is found in the brains of guinea pigs (Fu & Sinclair, 2000a) and developing rats, the latter showing preference for LNA use for cholesterol and palmitate of 16 times and 30 times, respectively, over DHA (Menard, Goodman, Corso, Brenna, & Cunnane, 1998). Preferential labeling of palmitate was also found in brains of suckling rats fed orally (Dhopeshwarkar et al., 1971; Sinclair, 1975) or injected intraperitoneally with ^{14}C-LNA (Dhopeshwarkar et al., 1971). Total β-oxidation represents a substantial source of LNA loss from the body. Leyton et al. (Leyton, Drury, & Crawford, 1987) reported that the rate of ^{14}C-LNA oxidation measured as CO_2 recovered in reference to administered dose was the fastest among unsaturates, including LA, 18:3n-6, and AA, and at a rate similar to 18 : 1 and 12 : 0, serving as the most efficient energy substrate. A recent tracer study has identified the skin as a previously unidentified route of major loss for LNA in the guinea pig. The 46% of dose found in the nonesterified fatty acid fraction of skin and fur lipids suggests that this use may, in part, account for the poor conversion efficiency to DHA and awaits confirmation by measurements in humans (Fu & Sinclair, 2000b).

These data clearly identify significant metabolic roles for LNA other than as the precursor of DHA. Based on the experimental results from many laboratories, we can conclude that LNA may have essential roles other than as a DHA precursor and that consumption of LCP-free diets in the perinatal period puts infants at risk of inadequate DHA for the rapid-growing neural tissue.

6. CONCLUSION

Data from compositional studies and tracer studies unequivocally show that DHA is a better substrate for brain and retina DHA accretion compared to LNA. In nontracer studies, the conversion ratio between LNA and DHA to neural DHA reported was over a range of 4–20, depending on outcome variable. In radiotracer studies, DHA* is incorporated into neural tissues at a faster rate compared to the labeled C_{18} precursor. Although, it has been shown that human infants are capable of converting LNA to DHA in vivo, the relative efficacy of LNA and DHA as the substrate for neural DHA cannot be obtained quantitatively from blood compartments. Nonhuman primate studies have shown that fetal brain DHA accretion reached plateau levels following intravenous doses of either ^{13}C-LNA or ^{13}C-DHA to the pregnant mother or fetus, suggesting brain DHA turnover rate during perinatal is slow and bioequivalence of LNA- or DHA-derived DHA can be computed only after the DHA plateau is reached. Brain bioequivalence obtained from nonhuman primates using stable-isotope methodology is 7 : 1 or 8 : 1 when administered directly to the developing animal, and 20 : 1 when administered to the pregnant female. These data suggest that the addition of modest amounts of DHA to infant formula, as low as 0.1% of calories, should measurably improve DHA status. No definitive evidence yet exists for an essential role for LNA. However, several metabolic roles for LNA have been identified, for which LNA is shunted at much higher rates that to DHA synthesis. The weight of evidence suggests that at least a modest supply of DHA in formula would improve the health of preterm and term infants.

REFERENCES

Abedin L, Lien EL, Vingrys AJ, Sinclair AJ. The effects of dietary alpha-linolenic acid compared with docosahexaenoic acid on brain, retina, liver, and heart in the guinea pig. Lipids 1999; 34(5): 475–482.
Al MD, Hornstra G, van der Schouw YT, Bulstra-Ramakers MT, Huisjes HJ. Biochemical EFA status of mothers and their neonates after normal pregnancy. Early Hum Dev 1990; 24(3):239–248.
Anderson G, Connor WE. Uptake of fatty acids by developing rat brain. Lipids 1988; 23:286–290.
Anderson GJ, Connor WE, Corliss JD. Docosahexaenoic acid is the preferred dietary n-3 fatty-acid for the development of the brain and retina. Pediatr Res 1990; 27(1):89–97.
Arbuckle LD, Innis SM. Docosahexaenoic acid in developing brain and retina of piglets fed high or low alpha-linolenate formula with and without fish oil. Lipids 1992; 27(2):89–93.
Benolken RM, Anderson RE, Wheeler TG. Membrane fatty acids associated with the electrical response in visual excitation. Science 1973; 182:1253–1254.
Bourre JM, Durand G, Pascal G, Youyou A. Brain-cell and tissue recovery in rats made deficient in N-3 fatty-acids by alteration of dietary-fat. J Nutr 1989; 119(1):15–22.
Bourre JM, Piciotti M, Dumont O. Delta-6 desaturase in brain and liver during development and aging. Lipids 1990; 25:354–356.
Bourre JM, Youyou A, Durand G, Pascal G. Slow recovery of the fatty acid composition of sciatic nerve in rats fed a diet initially low in n-3 fatty acids. Lipids 1987; 22(7):535–538.
Brenna JT. High-precision gas isotope ratio mass spectrometry: recent advances in instrumentation and biomedical applications. Acc Chem Res 1994; 27:340–346.
Brody, S. Bioenergetics and Growth, with Special Reference to the Efficiency Complex in Domestic Animals. Reinhold, New York, 1945.
Campbell FM, Gordon MJ, Dutta-Roy AK. Preferential uptake of long chain polyunsaturated fatty acids by isolated human placental membranes. Mol Cell Biochem 1996; 155(1):77–83.
Carlson SE, Rhoides PG, Rao VS, Goldgar DE. Effect of fish oil supplementation on the n-3 fatty acid content of red blood cell membranes in preterm infants. Pediatr Res 1987; 21:507–510.
Carnielli VP, Wattimena DJ. L., Luijendijk IH. T., Boerlage A, Degenhart HJ, Sauer PJJ. The very low birth weight premature infant is capable of synthesizing aracidonic and docosahexaenoic acids from linoleic and linolenic acids. Pediatr Res 1996; 40(1):169–174.
Chambaz J, Ravel D, Manier M-C, Pepin D, Mulliez N, Bereziat G. Essential fatty acids interconversion in the human fetal liver. Biol Neonate 1985; 47:136–140.
Connor WE, Neuringer M, Reisbick S. Essential fatty acids: the importance of n-3 fatty acids in the retina and brain. Nutr Rev 1992; 50(4):21–29.
Crawford MA, Costeloe K, Ghebremeskel K, Phylactos A. The inadequacy of the essential fatty acid content of present preterm feeds [Erratum: Eur J Pediatr 1998; 157(2):160]. Eur J Pediatr 1998; 157(Suppl 1):S23–S27.
Cunnane SC, Francescutti V, Brenna JT, Crawford MA. Breast-fed infants achieve a higher rate of brain and whole body docosahexaenoate accumulation than formula-fed infants not consuming dietary docosahexaenoate. Lipids 2000; 35(1):105–111.
Dhopeshwarkar GA, Subramanian C, Mead JF. Fatty acid uptake by the brain. V. Incorporation of (I-^{14}C)linolenic acid into adult rat brain. Biochim Biophy Acta 1971; 239(2):162–167.
Dobbing J, Sands J. Comparative aspects of the brain growth. Early Hum Dev 1979; 3(1):79–83.
Dutta-Roy AK. Transport mechanisms for long-chain polyunsaturated fatty acids in the human placenta. Am J Clin Nutr 2000; 71(Suppl 1):315S-322S.
Edmond J, Higa TA, Korsak RA, Bergner EA, Lee WN. Fatty acid transport and utilization for the developing brain. J Neurochem 1998; 70(3):1227–1234.
Farquharson J, Cockburn F, Patrick WA, Jamieson EC, Logan RW. Infant cerebral cortex phospholipid fatty-acid coposition and diet. Lancet 1992; 340:810–813.
Farquharson J, Jamieson EC, Abbasi KA, Patrick WJA, Logan RW, Cockburn F. Effect of diet on the fatty acid compostition of the major phospholipids of infant cerebral cortex. Arch Dis Child 1995; 72:198–203.

Fliesler SJ, Anderson RE. Chemisty and metabolism of lipids in the vertebrate retina. Prog Lipid Res 1983; 22:79–131.

Fu Z, Sinclair AJ. Increased alpha-linolenic acid intake increases tissue alpha-linolenic acid content and apparent oxidation with little effect on tissue docosahexaenoic acid in the guinea pig. Lipids 2000; 35(4):395–400.

Fu Z, Sinclair AJ. Novel pathway of metabolism of alpha-linolenic acid in the guinea pig. Pediatr Res 2000; 47(3):414–417.

Giussani DA, Farber DM, Jenkins SL, Yen A, Winter JA, Tame JD, Net al. Opposing effects of androgen and estrogen on pituitary–adrenal function in nonpregnant primates. Biol Reprod 2000; 62(5):1445–1451.

Goodman KJ, Brenna JT. High sensitivity tracer detection using high precision gas chromatography combustion isotope ratio mass spectrometry and highly enriched [U-13C]-labeled precursors. Anal Chem 1992; 64(10), 1088–1095.

Greiner RC, Winter J, Nathanielsz PW, Brenna JT. Brain docosahexaenoate accretion in fetal baboons: bioequivalence of dietary alpha-linolenic and docosahexaenoic acids. Pediatr Res 1997; 42(6):826–834.

Haggarty P, Page K, Abramovich DR, Ashton J, Brown D. Long-chain polyunsaturated fatty acid transport across the perfused human placenta. Placenta 1997; 18(8):635–642.

Hassma AG, Crawford MA. The differential incorporation of labelled linoleic, gamma-linolenic, dihomo-gamma-linolenic and arachidonic acids into the developing rat brain. J Neurochem 1976; 27(4):967–968.

Heird WC, Prager TC, Anderson RE. Docosahexaenoic acid and the development and function of the infant retina. Curr Opin Lipidol 1997; 8(1):12–16.

Hrboticky N, Mackinnon MJ, Innis SM. Effect of a vegetable oil formula rich in linoleic-acid on tissue fatty-acid accretion in the brain, liver, plasma, and erythrocytes of infant piglets. Am J Clin Nutr 1990; 51(2):173–182.

Innis SM. Essential fatty acids in growth and development. Prog Lipid Res 1991; 30(1):39–103.

Jensen CL, Prager TC, Fraley JK, Chen H, Anderson RE, Heird WC. Effect of dietary linoleic/alpha-linolenic acid ratio on growth and visual function of term infants. J Pediatr 1997; 131(2):200–209.

Jensen RG. The lipids in human milk. Prog Lipid Res 1996; 35(1):53–92.

Jensen RG, Hagerty MM, McMahon KE. Lipids of human milk and infant formulas: a review. Am J Clin Nutr 1978; 31(6):990–1016.

Knipp GT, Audus KL, Soares MJ. Nutrient transport across the placenta. Adv Drug Deliv Rev 1999; 38(1):41–58.

Kuhn DC, Crawford MC. Placental essential fatty acid transport and prostaglandin synthesis. Prog Lipid Res 1986; 25:345–353.

Lands WE. Long-term fat intake and biomarkers. Am J Clin Nutr 1995; 61(3 Suppl), 721S–725S.

Leyton J, Drury PJ, Crawford MA. In vivo incorporation of labeled fatty acids in rat liver lipids after oral administration. Lipids 1987; 22(8):553–558.

Ma Y, Lockwood CJ, Bunim AL, Giussani DA, Nathanielsz PW, Guller S. Cell type-specific regulation of fetal fibronectin expression in amnion: conservation of glucocorticoid responsiveness in human and nonhuman primates. Biol Reprod 2000; 62(6):1812–1817.

Makrides M, Neumann MA, Byard RW, Simmer K, Gibson RA. Fatty acid composition of brain, retina, and erythrocytes in breast- and formula-fed infants. Am J Clin Nutr 1994; 60(2):189–194.

Marbois BN, Ajie HO, Korsak RA, Sensharma DK, Edmond J. The origin of palmitic acid in brain of the developing rat. Lipids 1992; 27(8):587–592.

Martinez M, Ballabriga A, Gil-Gibernau JJ. Lipids of the developing human retina: I. Total fatty acids, plasmologens, and fatty acid compostition of ethanolamine and choline phosphoglycerides. J Neurosci Res 1988; 20:484–490.

Menard CR, Goodman KJ, Corso TN, Brenna JT, Cunnane SC. Recycling of carbon into lipids synthesized de novo is a quantitatively important pathway of alpha-[U-13C]linolenate utilization in the developing rat brain. J Neurochem 1998; 71(5):2151–2158.

Mohrhauser H, Holman RT. The effect of dose level of essential fatty acids upon fatty acid composition of the rat liver. J Lipid Res 1963; 6:494–497.

Naval J, Calvo M, Laborda J, Dubouch P, Frain M, Sala-Trepat JM, et al. Expression of mRNAs for alpha-fetoprotein (AFP) and albumin and incorporation of AFP and docosahexaenoic acid in baboon fetuses. J Biochem (Tokyo) 1992; 111(5):649–654.

Neuringer M, Connor W, Lin D, Barsted L, Luck S. Biochemical and functional effects of prenatal and postnatal omega 3 fatty acid deficiency on retina and brain in rhesus monkeys. Proc Natl Acad of Sci USA 1986; 83:4021–4025.

Neuringer M, Connor WE, Petten CV, Barstad L. Dietary omega-3-fatty acid deficiency and visual loss in infant rhesus monkeys. J Clin Invest 1984; 73:272–276.

Pawlosky RJ, Ward G, Salem N Jr. Essential fatty acid uptake and metabolism in the developing rodent brain. Lipids 1996; 31(Suppl):S103–S107.

Poisson JP, Dupuy RP, Sarda P, Descomps B, Narce M, Rieu D, et al. Evidence that liver-microsomes of human neonates desaturate essential fatty-acids. Biochim Biophy Acta 1993; 1167(2):109–113.

Rioux FM, Innis SM, Dyer R, MacKinnon M. Diet-induced changes in liver and bile but not brain fatty acids can be predicted from differences in plasma phospholipid fatty acids in formula- and milk-fed piglets. J Nutr 1997; 127(2):370–377.

Rodriguez A, Sarda P, Nessmann C, Boulot P, Leger CL, Descomps, B. Delta6- and delta5-desaturase activities in the human fetal liver: kinetic aspects. J Lipid Res 1998; 39(9):1825–1832.

Ruyle M, Connor WE, Anderson GJ, Lowensohn RI. Placental-transfer of essential fatty-acids in humans— venous arterial difference for docosahexaenoic acid in fetal umbilical erythrocytes. Proc Natl Acad Sci USA 1990; 87(20):7902–7906.

Salem JN, Wegher B, Mena P, Uauy R. Arachidonic and docosahexaenoic acids are biosynthesized from their 18-carbon precursors in human infants. Proc Natl Acad Of Sci USA 1996; 93:49–54.

Sastry PS. Lipids of nervous tissue: composition and metabolism. Prog Lipid Res 1985; 24:69–176.

Sauerwald TU, Hachey DL, Jensen CL, Chen H, Anderson RE, Heird WC. Intermediates in endogenous synthesis of C22:6 omega 3 and C20:4 omega 6 by term and preterm infants. Pediatr Res 1997; 41(2):183–187.

Sheaff Greiner RC, Zhang Q, Goodman KJ, Giussani DA, Nathanielsz PW, Brenna JT. Linoleate, alpha-linolenate, and docosahexaenoate recycling into saturated and monounsaturated fatty acids is a major pathway in pregnant or lactating adults and fetal or infant rhesus monkeys. J Lipid Res 1996; 37(12):2675–2686.

Sinclair AJ. Incorporation of radioactive polyunsaturated fatty acids into liver and brian of developing rat. Lipids 1975; 10(3):175–184.

Sinclair AJ, Crawford MA. The incorporation of linolenic acid and docosahexaenoic acid into liver and brain lipids of developing rats. Fed Exp Biol Soc Lett 1972; 26:127–129.

Stammers J, Stephenson T, Colley J, Hull D. Effect on placental transfer of exogenous lipid administered to the pregnant rabbit. Pediatr Res 1995; 38(6):1026–1031.

Su HM, Bernardo L, Mirmiran M, Ma XH, Corso TN, Nathanielsz PW, et al. Bioequivalence of dietary alpha-linolenic and docosahexaenoic acids as sources of docosahexaenoate accretion in brain and associated organs of neonatal baboons. Pediatr Res 1999; 45(1):87–93.

Su HM, Bernardo L, Mirmiran M, Ma XH, Nathanielsz PW, Brenna JT. Dietary 18:3n-3 and 22:6n-3 as sources of 22:6n-3 accretion in neonatal baboon brain and associated organs. Lipids 1999; 34(Suppl):S347–S350.

Su H-M, Corso TN, Nathanielsz PW, Brenna JT. n-3 Fatty acid accretion after a 13C-18:3n-3 dose to fetal baboons. Third Congress of the International Society for the Study of Fatty Acids and Lipids, Lyon, 1998.

Tame JD, Winter JA, Li C, Jenkins S, Giussani DA, Nathanielsz PW. Fetal growth in the baboon during the second half of pregnancy. J Med Primatol 1998; 27(5):234–239.

Uauy R, Mena P, Wegher B, Nieto S, Salem N Jr. Long chain polyunsaturated fatty acid formation in neonates: effect of gestational age and intrauterine growth. Pediatr Res 2000; 47(1):127–135.

Uauy RD, Birch DG, Birch EE, Tyson JE, Hoffman DR. Effect of dietary omega-3–fatty-acids on retinal function of very-low-birth-weight neonates. Pediatr Res 1990; 28(5):485–492.

Wheeler TG, Benolken RM. Visual membranes: specificity of fatty acid precursors of the electrical response to illumination. Science 1975; 188(4195):1312–1314.

Woods J, Ward G, Salem NJ. Is docosahexaenoic acid necessary in infant formula? Evaluation of high linolenate diets in the neonatal rat. Pediatr Res 1996; 40:687–694.

Yamamoto N, Saitoh M, Moriuchi A, Nomura M, Okuyama H. Effect of dietary alpha-linolenate/linoleate balance on brain lipid compostions and learning ability of rats. J Lipid Res 1987; 28:144–151.

7

Recent Advances in the Supply of Docosahexaenoic Acid to the Nervous System

Robert J. Pawlosky and Norman Salem Jr.

1. DHA IN THE MEMBRANE ENVIRONMENT OF THE NEURON

Docosahexaenoic acid (DHA, 22:6n-3) is the most abundant polyunsaturated fatty acid (PUFA) acylated to the aminophospholipids phosphatidylethanolamine (PE) and phosphatidylserine (PS) in membranes of neurons within the central nervous system (CNS) (Naughton, 1981; Salem et al., 1986). It can occur in concentrations exceeding 30-mol% of the fatty acids (Salem, 1989). The high enrichment of DHA in synaptosomes is especially striking and suggests that DHA has unique properties that are required for optimal neuronal function. This concentration in the CNS is even more remarkable when one considers that sources of n-3 fatty acids are disproportionately limited in the terrestrial food chain compared to the much more abundant n-6 fatty acids.

Although the precise function of DHA in the membrane has yet to be determined, the biophysical properties of membranes that are enriched with this fatty acid appear to be optimized for signal transduction through neuronal pathways. An example illustrating this apparent optimization is the visual transduction process through the photoreceptor cells and neurons within the retina. DHA is more highly enriched in the outer segment disks of rod and cone cells than in any compartment of any other cell type (Salem, 1989). The high concentration of DHA in these membranes relative to other fatty acids appears to be important for facilitating the formation of the photon-initiated transition state of rhodopsin (Litman & Mitchell, 1996).

Considering that DHA may impart unique properties to membrane domains nuclear magnetic resonance (NMR) investigations have been used to determine changes in the order parameters of the carbon atoms of the fatty acid acylated at the *sn-1* position of phospholipids in relation to the degree of unsaturation of the fatty acid at the *sn-2* position (Holte et al., 1995; Holte et al., 1996; Mitchell & Litman, 1998). The results have suggested that phospholipids with fatty acids having a high degree of unsaturation acylated at the *sn-2* position can give rise to highly transitional molecular shapes, which may impart unique packing properties within the matrix of the neuronal membrane. It is reasonable to infer from the biophysical data that rapid alterations in the protein conformation of rhodopsin may require a membrane environment capable of transitioning through highly alterable dynamic states (Holte et al., 1996). The high concentration of

DHA in the rod outer segments may impart unique membrane characteristics adapted to facilitating rhodopsin activation and association with G-proteins.

With regard to the physiological consequences of the fatty acid composition of disk membranes, several studies have documented the effects that a low intake of dietary n-3 fatty acids has on the electroretinograms (ERG) of young animals that may be directly attributed to the low content of DHA in the disks (Neuringer et al., 1986; Weisinger et al., 1996; Pawlosky et. al., 1997). In most species (Neuringer et al., 1986; Connor & Neuringer, 1988; Pawlosky et. al., 1997), the n-6 fatty acid homolog of DHA, DPAn-6 (22:5n6), increases in concentration in the disks of animals fed the low n-3 fatty acid diet. The preponderance of evidence from these studies demonstrate convincingly that low amounts of DHA in the retina result in a diminished rod cell response to light stimulus as observed in the altered voltage potentials in the electroretinograms. It is theoretically plausible, then, to presume that "enhanced" signal transduction through other neuronal pathways that are responsive to ligand-mediated G-protein activation (analogous to light-activated rhodopsin stimulation) may be related to the concentration of DHA in synaptic membranes. Conversely, it may be proposed that dietary deficiencies, disease states, genetic abnormalities, or environmental conditions that diminish the supply of DHA to neurons will adversely affect cell signaling in the central nervous system.

2. GENETIC AND DIETARY FACTORS WHICH INFLUENCE N-3 FATTY ACID METABOLISM

Because of the inability to synthesize n-3 fatty acids *de novo*, all animals require these fatty acids in their diet to meet their demand for maintaining a high concentration of DHA in the brain. Although little direct evidence exists in any species concerning the quantitative conversion of n-3 fatty acid precursors to DHA, it has been estimated based on rodent studies that an n-3 fatty acid intake of 0.5% of energy as α-linolenic acid (LNA) is needed in order to maintain an adequate level of DHA in the brain (Bourre et al., 1989). However, it must be recognized that the ability to biosynthesize DHA from LNA or other n-3 fatty acids varies among different animal species (Rivers et al., 1975; Hassam et al., 1977; Sinclair et al., 1979; Clandinin et al., 1985; Scott & Bazan, 1989: Salem & Pawlosky, 1994; Pawlosky et al., 1994: Fu & Sinclair, 2000). Moreover, the composition of fat in the diet has a significant influence on the liver production of long-chain PUFAs (Salem & Pawlosky, 1994; Pawlosky et al., 1994). For instance, it was observed that when nonhuman primates were fed a diet that contained relatively low levels of long-chain PUFAs (where eicosapentaenoic acid [EPA] and DHA were present at a level of 0.54% and 0.64% of the total dietary fat, respectively) the formation of labeled-DHA from labeled-LNA was inhibited in the liver (Pawlosky & Salem, 1993). However, both arachidonic acid (AA) (from labeled-LA) and docosapentaenoic acid (DPAn-3) were synthesized in the liver and detected in the blood of the same animals on this diet. When animals were then placed on a diet devoid of long-chain PUFAs, the synthesis of DHA was observed in the liver, and labeled-DHA was detected in the blood after 3 wk. This strongly suggests that the conversion of DPAn-3 to DHA in the liver is partly controlled by the concentration of DHA in the diet. It is interesting to theorize whether the regulation of the biosynthesis of DHA from DPAn-3 is maintained at the level of transcription of a Δ-6 desaturase which is needed to catalyze the conversion of 24:5n3 to 24:6n-3 (Marzo et al., 1996). If so, this form of regulation would have the advantage of selectively controlling DHA production, yet it would not necessarily inhibit the synthesis of other long-chain PUFAs (e.g., arachidonic acid).

Because of the genetic and dietary factors, which are capable of controlling and influencing the production of long-chain PUFAs, different species have developed various independent strategies for obtaining DHA. In the cat family, for instance, preformed DHA in the diet appears to be necessary to maintain a high concentration of DHA in the CNS (Pawlosky et al., 1997). The need for preformed DHA in the diet is probably caused by a low PUFA biosynthetic capability of this species (Hassam, 1977; Rivers, 1975; Sinclair, 1979) as well as an inherent inability to produce DHA in the feline liver (Pawlosky et al., 1994). However, there is increasing evidence that suggests that the production of DHA from LNA may be a highly inefficient process in other species, as well (Menard et al., 1998; Su et al., 1999). Nevertheless, it appears that the majority of species (other than members of the cat family) have some capacity to biosynthesize DHA from LNA in their livers.

Although, the liver has long been recognized as an important site of PUFA biosynthesis (Buzzi et al., 1997; Clandinin et al., 1985; Pawlosky et al., 1992; Schenck et al, 1996), a number of animal studies in various species have shown that long-chain PUFAs (in particular, 22:6n-3) can be synthesized by different components of the nervous system (Dhopeswarkar et al., 1974; Clandinin et al., 1985; Delton-Vandenbrouke et. al, 1997; Chen et. al, 1999; Moore, 1993; Moore et al., 1991; Pawlosky et al., 1994; Pawlosky et al., 1996; Protstein, 1996). The cells of the nervous system, like other cells of the body, take up DHA and other n-3 fatty acids from lipoproteins that are carried in the blood. There is evidence that the preferred form of DHA for uptake into the brain is as a lysophospholipid rather than as a free fatty acid (Bernoud et al., 1999). This route may offer an efficient transfer of DHA into the neuron for phospholipid synthesis and membrane biogenesis. Although felines are the only species in which it has been demonstrated that the entire brain accretion of biosynthesized DHA is the result of production that occurs within the CNS (Pawlosky et al., 1994) other species carry on similar intra-CNS processes to obtain at least part of their DHA (Pawlosky et al., 1996). Figure 1 depicts a representation of the feline strategy for the accretion of DHA in brain. From dietary sources, LNA or EPA is taken up into the liver where the fatty acids are converted into DPAn-3. DPAn-3 is released from the liver and carried on lipoproteins to the CNS, where it is converted to DHA. There is similar evidence from other species that have shown that brain cells (Moore et al., 1991) or cells isolated from the cerebral vasculature (Delton-Vandenbroucke et al., 1997) are capable of biosynthesizing DHA from n-3 fatty acid precursors.

Several investigators have described plausible mechanisms for the production of DHA in the CNS. Moore and co-workers described the biosynthesis of DHA and transport of fatty acids through microcapillary cerebral endothelial cells, astrocytes, and neurons (Moore et al., 1991; Moore, 1993). In this model, microcapillary endothelial cells produce DPAn-3 from LNA, which is turned over to the astrocytes to be synthesized into DHA. The astrocytes then release DHA, which is taken up by neurons. In contrast, Delton-Vandenbroucke and co-workers found that cerebral vascular cells produced appreciable amounts of labeled-DHA from DPAn-3 (Delton-Vandenbroucke et al., 1997). They theorize that cerebral endothelial cells will convert circulating DPAn-3 into DHA, which is then taken up into the brain. These reports suggest that in the CNS, unlike the liver in which biosynthesis of DHA from LNA takes place entirely within the hepatocyte,

DHA Accretion in the CNS
Feline Metabolism

Nervous System

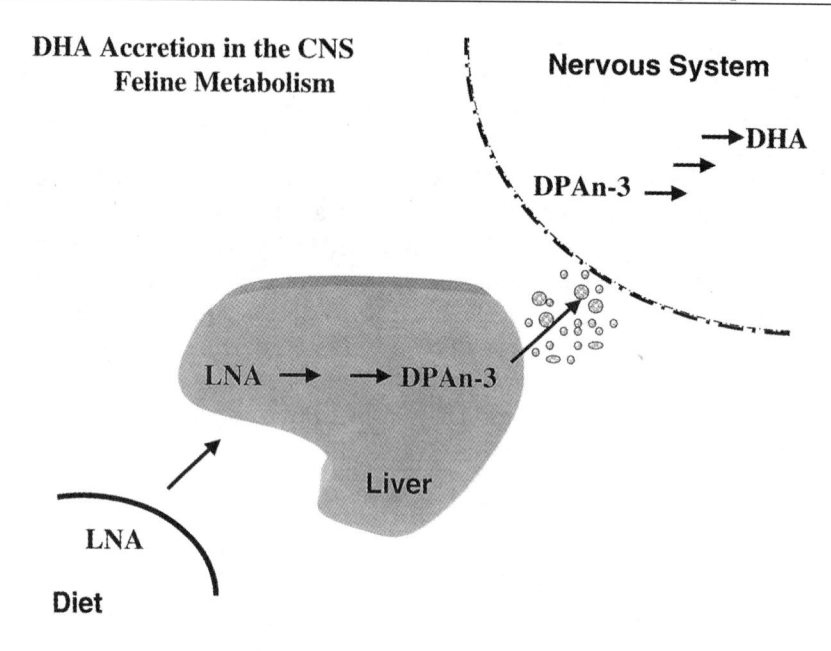

→ DHA

DPAn-3 →

LNA → → DPAn-3

Liver

LNA

Diet

Fig. 1. A schematic representation of the biosynthesis and accretion of DHA into the CNS of felines. Dietary n-3 fatty acids (LNA) are taken up into the liver and synthesized into DPAn-3 (22:5n-3). DPAn-3 is released from the liver and carried on lipoproteins in the blood to the nervous system. The synthesis of DHA is completed in the brain from DPAn-3.

more than a single cell type (either an astrocyte or endothelial cell) may act in a synergistic fashion to contribute to the synthesis and accretion of DHA.

3. DHA BIOSYNTHESIS AND EARLY DEVELOPMENT

The accumulation of DHA in the brain is especially important during brain growth periods (Green & Yavin, 1998), and although no known systematic study has been undertaken which attempts to compare rates of DHA biosynthesis to the development of the CNS or brain growth in any species, there is evidence to suggest that the capacity to synthesize DHA may be correlated with early brain development in some species (Rodriguez et al., 1998; Pawlosky et. al., 1996; Salem et al., 1996; Greiner et al., 1997; Su et al., 1999). Using stable isotopically labeled fatty acids and mass spectrometry, the biosynthesis of DHA has been demonstrated in both human infants (Salem et al., 1996; Carnielli et al., 1996; Sauerwald et al., 1996) and in fetal baboons (Greiner et al., 1997; Su et al., 1999). An example that illustrates the inherent capacity to synthesize long-chain PUFAs during the early developmental period was provided by felines (Pawlosky & Salem, 1996). Adult felines do not actively synthesize long-chain PUFAs from the 18-carbon precursors (Sinclair et al., 1979; Pawlosky et al., 1994). However, juvenile felines before weaning were capable of synthesizing labeled long-chain PUFAs in their livers, which could then be detected in the blood using mass spectrometry. When the mothers were given an oral dose of labeled-18-carbon essential fatty acids, neither the maternal blood nor milk contained any of the biosynthesized long-chain PUFAs. Apparently, the demands of pregnancy and lactation did not provide sufficient stimulus to activate the biosynthetic pathway in adult cats. As the young animals developed (and begin accepting meat in their diet), the ability to synthesize long-chain PUFAs diminished. This active capacity to synthesize long-chain PUFAs in juvenile felines is intriguing and suggests that development may be a significant factor, which triggers an enhanced biosynthesis of long-chain PUFAs during nervous system formation. Notably, the capacity to synthesize DHA in immature felines may not be sufficient to meet their brain requirement for DHA. It was observed that animals reared on any of several corn-oil-based diets had very low amounts of either DPAn-3 or DHA in their brains at 8 wk of age (Pawlosky et al., 1997). It may be inferred from this study that throughout feline development, the maternal diet must contain some amount of preformed DHA (EPA or DPAn-3 may partially substitute for DHA) for adequate accumulation of brain DHA.

4. SUPPLYING DHA TO THE BRAIN: RODENT AND FELINE MODELS OF DHA SYNTHESIS

Although much of our understanding concerning the regulation of DHA biosynthesis has been ascertained from studies in rodents, it does not appear that either rats or mice are the most appropriate model for understanding n-3 metabolism in humans. Rodents seem to be more adept at synthesizing long-chain PUFAs than either felines or rhesus monkeys and are less influenced by dietary alterations (Salem & Pawlosky, 1994b). Rodents maintain a capacity to synthesize DHA in their livers (Scott & Bazan, 1989) as well as in the CNS (Pawlosky et al., 1996b) and have more active desaturases than several other species when measured in vitro (Willis, 1981). Using stable isotopically labeled substrates, Pawlosky and Salem demonstrated that DHA precursors could be taken up into the brain of developing rats and mice at a time when the brain is rapidly growing (Pawlosky et al, 1996b). The uptake of labeled-DPAn-3 into the brain appeared to be appreciable, as there was a ratio of approximately 1:5 of labeled-DPAn-3 to that of labeled-DHA in whole-brain preparations. Both of these labeled-fatty acids were synthesized from labeled-LNA in the liver and carried in the blood to the brain. Gradually, there was a disappearance of labeled-DPAn-3 from the brain. Over the same period (about 7 d), the labeled-DHA continued to increase. The quantitative importance of the brain biosynthetic pathway for supplying part of the CNS with DHA during development is unknown, but based on differential uptake of labeled-LNA into the rapidly growing cerebellum compared to that of the frontal cortex region, it is clear that these precursors are indeed taken into the brain parenchyma during CNS development.

Early studies in domestic felines demonstrated that cats had a low Δ-6 desaturase activity, which severely limited their capacity to synthesize arachidonic acid from linoleic acid (Hassam et al., 1977; Sinclair et al., 1979). Owing to a low desaturase activity, it may be assumed that synthesis of long-chain n-3 PUFAs arising from LNA would also be very limited. It was later shown that a low essential fatty acid diet could stimulate the synthesis of both long-chain n-6 and n-3 PUFAs via a Δ-6 desaturase (Pawlosky et al., 1994). In the liver, the route for the biosynthesis of DHA from LNA is initiated on smooth endoplasmic reticulum. Through a series of alternating enzymatic processes that desaturate and elongate LNA, DPAn-3 is produced. It is believed that DPAn-3 is then elongated to 24:5n3 and desaturated (by a Δ-6 desaturase) to 24:6n-3 (Luthria et al., 1996). This fatty acid is transferred to peroxisomes, where it is partially oxidized to form DHA, which is then reincorporated into phospholipids in the microsomal membranes. Felines provide an interesting model for studying the accretion of DHA in the CNS because they lack the capacity to produce any DPAn-6 or DHA in their livers. The ability to synthesize DPAn-3 but not DHA in their livers indicates that the second desaturation by Δ-6 desaturase may not be active. Consequently, in order to maintain a high level of DHA in the brain, felines must carry out the conversion of DPAn-3 to DHA within the CNS. The Δ-6 desaturase in the brain may thus be specific for catalyzing the conversion of 24:5n3 to 24:6n-3. The compartmental separation of the different desaturation steps lends support to the suggestion that distinct forms of Δ-6 desaturase are needed in the production of DHA (Marzo et al., 1996).

5. ACCRETION OF DHA INTO THE CNS OF YOUNG BABOONS

The synthesis and accretion of DHA in the CNS has been studied in fetal and neonatal baboons (Greiner et al., 1997; Su et al, 1999). Unlike LA, which may be extensively esterified to membrane phospholipids, LNA is present only in small amounts in membrane complex lipids. Presumably most of the available plasma LNA would be used for synthesis of long-chain PUFAs. However, when 4-wk-old baboons were given oral doses of either ^{13}C-labeled LNA or ^{13}C-labeled-DHA, only about 0.2% of labeled-LNA was found in the brain as ^{13}C-labeled DHA after 2 wk (Su et al., 1999). Such low utilization of LNA for brain accretion of DHA suggests a highly inefficient process in the biosynthesis of DHA during a period of rapid brain growth. Utilization of labeled-DHA for brain accretion of DHA was nearly seven times more efficient than that of LNA. The amount of labeled-DHA in the retina was between 12-fold and 15-fold greater in animals receiving labeled-DHA compared to those receiving the labeled-LNA. Part of the inefficient use of LNA for DHA synthesis may be explained by the apparent recycling of the carbon atoms of LNA into other pathways. Cunnane and co-workers demonstrated that a large proportion of LNA is partially oxidized and returned to the acetate pool for synthesis into nonessential fatty acids and cholesterol in neonatal rats (Menard et al, 1998). Also, Fu and Sinclair recently found that much of the radiolabeled LNA fed to young guinea pigs was presumed oxidized (39%) or found as labeled-LNA in the fur and skin (46%) (Fu & Sinclair, 2000). In humans, a high oxidation rate of LNA has also been observed when it was found that as much as 20% of the ^{13}C-lableled LNA was oxidized and expired as CO_2 within the first 12 h after receiving an oral dose (Vermunt et al., 2000). The inefficient conversion of LNA to DHA and the large loss and recycling of LNA into other metabolic pools is puzzling and requires further investigation to be fully explained.

6. CONSERVATION OF DHA IN THE CNS

Conservation of DHA in the CNS may be an especially important determinant in the maintenance of DHA in the brain and retina. Brenna and co-workers estimated that the DHA turnover in the brain of neonatal baboons was low, only 4% per week based on stable isotope analysis (Brenna et al., 1999). The epithelial cells that line the retina and form a barrier between the photoreceptor cells and the blood appear to be specialized for the uptake of n-3 fatty acids (Rodriguez de Turco et al., 1991; Wang & Anderson, 1993). In the retina, a series of isotope studies have demonstrated that DHA may be conserved through a recycling process that involves the transfer of DHA between the pigment epithelial tissue and the photoreceptor cells (Stinson et al., 1991; Gordon et al., 1992; Rodriguez et al., 1999). It appears that DHA can be recycled to the rod cells from sloughed off outer disk membranes after phagocytosis by the retinal pigment epithelial cells (Gordon et al., 1992). This process appears effective in maintaining the high concentration of DHA in photoreceptors as the disks undergo constant membrane biogenesis and renewal.

7. SUMMARY

The development of a fully quantitative procedure to determine mechanisms for accretion of DHA in the brain of any species is needed. Species differences, aged-related determinants, and diet appear to be the major factors controlling and influencing the biosynthesis and accretion of DHA in the brain. The use of stable-isotope-labeled substrates in animal and human studies has provided a powerful tool for determining the metabolic fate of the fatty acids in the synthesis of DHA (Brenna, 1994; Brenna et al., 1997, Pawlosky et al., 1992; Sauerwald et al., 1997). These studies have revealed important new information in the area of metabolism of fatty acids in general (Salem et al., 1996; Salem et al., 1999; Greiner et al., 1997; Su et al, 1999) and specifically within the brain (Pawlosky et al., 1994; Menard et al., 1998; Pawlosky et al., 1996). It appears from several lines of evidence in animals that much of the ingested LNA is not available for synthesis of DHA. It may be recycled into cholesterol or nonessential fatty acids, lost to oxidation (Menard et al., 1998) or taken up by other tissue compartments (Fu & Sinclair, 2000). In studying in vivo metabolism in humans, it appears that the biosynthesis of DHA in the liver is similar to other species in that most of the available LNA is not converted to DHA (Salem et al., 1999; Vermunt et al., 2000). The complexity in determining the in vivo metabolism of essential fatty acids and the influences of diet, genetics, age, gender, and disease requires the use of more sophisticated analytical tools to comprehend the various interactions of these parameters. Mathematical approaches utilizing physiologic compartmental models are now available to researchers to be used for fuller descriptive analysis. Such modeling programs can assess the complex interactions of the kinetics of metabolism (isotope data), dietary conditions (specific dietary intake values), tissue steady-state determinants (the homeostasis within compartments), and population statistics (subject variability) through a diverse set of experimental and clinical conditions. Mathematical approaches can be employed to describe quantitative differences among animal species, the effects of development and aging on EFA metabolism, and alterations in dietary habits in a individual or throughout a given population. It is expected that as such new approaches are applied to questions of brain accretion of DHA, a well-formed biochemical understanding and better nutritional guidelines will become available, leading to the adoption of sound nutritional policies. The use of such mathematical approaches to describe EFA metabolism in humans is in the near future. It is expected that one of the immediate applications of these new quantitative approaches will be the application to questions concerning the adequacy of infant formulas in supplying both long-chain n-6 and n-3 PUFAs to the brain.

REFERENCES

Bernoud N, Fenart L, Moliere P, Dehouck MP, Lagarde M, Cecchelli R, et al. Preferential transfer of 2-docosahexaenoyl-1-lysophosphatidylcholine through an in vitro blood-brain barrier over unesterified docosahexaenoic acid. J Neurochem 1999; 72(1):338–345.

Bourre JM, Durand G, Pascal G, Youyou A. Brain cell and tissue recovery in rats made deficient in n-3 fatty acids by alterations of dietary fat. J Neurochem 1989; 119:12–22.

Brenna JT. Use of stable isotopes to study fatty acid and lipoprotein metabolism in man. Prostaglandins Leukotrienes Essential Fatty Acids 1997; 57(4&5):467–472.

Brenna JT. High-precision gas isotope ratio mass spectrometry: recent advances in instrumentation and biomedical applications. Arch Chem Res 1994; 27:340–346.

Buzzi M, Henderson RJ, Sargent JR. Biosynthesis of docosahexaenoic acid in trout hepatocytes proceeds via 24-carbon intermediates. Biochem Physiol 1997; 116B(2):263–267.

Carnielli VP, Wattimena DJ, Luijendijk IH, Boerlage A, Degenhart HJ, Sauer JP. The very low birth weight premature infant is capable of synthesizing arachidonic and docosahexaenoic acids from linoleic and linolenic acids. Pediatr Res 1996; 40(1):169–174.

Chen H, Ray J, Scarpino V, Acland GM, Aguirre GD, Anderson RE. Synthesis and release of docosahexaenoic acid by the RPE cells of prcd-affected dogs. Invest Ophthalmol Vision Sc 1999; 40:2418–2422.

Clandinin MT, Wong K, Hacker RR. Synthesis of chain elongation–desaturation products of linoleic acid by liver and brain microsomes during development of the pig. Biochem J 1985; 226:305–309.

Connor WE, Neuringer M. The effects of n-3 fatty acid deficiency and repletion upon the fatty acid composition and function of the brain and retina. In: Karnovsky ML, Leaf A, Bolis LC, eds. Biological Membranes: Aberrations in Membrane Structure and Function. Alan R Liss, New York, 1988, pp. 275–294.

Delton-Vandenbrouke I, Grammas P, Anderson RE. Polyunsaturated fatty acid metabolism in retinal and cerebral microvascular endothelial cells. J Lipid Res 1997; 38:147–159.

Dhopeshwarkar GA, Subranamanian C. Metabolism of linolenic acid in developing brain: I. Incorporation of radioactivity from 1-14C-linolenic acid into brain fatty acids. Lipids 1974; 10:238–241.

Fu Z, Sinclair AJ. Novel pathway of the metabolism of alpha-linolenic acid in the Guinea pig. Pediatr Res 2000; 47(3):414–417.

Gordon WC, Rodriguez de Turco EB, Bazan NG. Retinal pigment epithelial cells play a central role in the conservation of docosahexaenoic acid by photoreceptor cells after shedding and phagocytosis. Curr Eye Res 1992; 11(1):73–83.

Green P, Yavin E. Mechanisms of docosahexaenoic acid accretion in the fetal brain. J Neurol Res 1998; 52:129–136.

Greiner RC, Winter J, Nathanielsz PW, Brenna JT. Brain docosahexaenoate accretion in fetal baboons: Bioequivalence of dietary alpha-linolenic and docosahexaenoic acids. Pediatr Res 1997; 42(6):826–834.

Hassam AG, Rivers JPW, Crawford MA. The failure of the cat to desaturate linoleic acid: its nutrient implications. Br J Nutr 1977; 39:227–231.

Holte LL, Peter SA, Sinnwell TM, Gawrisch K. ^{2}H nuclear magnetic resonance order parameter profiles suggest a change of molecular shape for phosphatidylcholines containing a polyunsaturated acyl chain. Biophys J 1995; 68(6):2396–2403.

Holte LL, Separovic F, Gawrisch K. Nuclear magnetic resonance investigation of hydrocarbon chain packing in bilayers of polyunsaturated phospholipids. Lipids 1996; 31:S199–S203

Litman J, Mitchell DC. A role for phospholipid polyunsaturation in modulating membrane protein function. Lipids 1996; 31S:S193–S197.

Luthria DL, Mohammed BS, Sprecher WH. Regulation of the biosynthesis of 4,5,10,13,16,19-docosahexaenoic acid. J Biol Chem 1996; 271(27):16,020–16,025.

Marzo I, Alva AA, Pineiro A, Naval J. Biosynthesis of docosahexaenoic acid in human cells: evidence that two different delta-6 desaturase activities may exist. Biochim Biophys Acta 1996; 1301:263–272.

Menard CR, Goodman KJ, Corso TN, Brenna JT, Cunnane SC. Recycling of carbon into lipids synthesized de novo is a quantitatively important pathway of alpha-[U-13C] linolenate utilization in the developing rat brain. J Neurochem 1998; 71:2151–2158.

Mitchell DC, Litman BJ. Molecular order and dynamics in bilayers consisting of highly polyunsaturated phospholipids. Biophys J 1998; 74:879–891.

Moore SA. Cerebral endothelium and astrocytes cooperate in supplying docosahexaenoic acid to neurons. Adv Exp Med Biol 1993; 331:229–233.

Moore SA, Yoder E, Murphy S, Dutton GR, Spector AA. Astrocytes not neurons produce docosahexaenoic acid (22:6n-3) and arachidonic acid (20:4n-6). J Neurochem 1991; 56:518–524.

Naughton JM. Supply of polyenoic fatty acids to the mammalian brain: the ease of conversion of the short chain essential fatty acids to their longer chain polyunsaturated metabolites in liver, brain, placenta and blood. Intl J Biochem 1980; 13:21–32.

Neuringer M, Connor WE, Lin DS, Barstad L, Luck S. Biochemical and functional effects of prenatal and postnatal n-3 fatty acid deficiency on retina and brain in rhesus monkeys. Proceedings of the National. Academy of Science USA 1986; 83:4021–4025.

Pawlosky RJ, Sprecher HW, Salem N Jr. High sensitivity negative ion GC–MS method for the detection of desaturated and chain elongated products of deuterium-labeled linoleic and linolenic acids. J Lipid Res 1992; 33:1711–1717.

Pawlosky RJ, Salem N Jr. The metabolism of essential fatty acids in mammals. In: Sinclair A, Gibson R, eds. The 3rd International Conference on Eicosanoids and Essential Fatty Acids. American Oil Chemists' Society, Champaign, IL, 1993, pp. 26–30.

Pawlosky R, Barnes A, Salem N Jr. Essential fatty acid metabolism in the feline: relationship between liver and brain production of long-chain polyunsaturated fatty acids. J Lipid Res 1994; 35(11), 2032–2040.

Pawlosky R, Denkins Y, Ward G, Salem N Jr. Retinal and brain accretion of long-chain polyunsaturated fatty acids in developing felines: the effects of corn oil-based maternal diets. Am J Clin Nutr 1997; 65:465–472.

Pawlosky R, Salem N Jr. Is dietary arachidonic acid necessary for feline reproduction? J Nutr 1996; 126:1081S–1085S.

Pawlosky RJ, Ward G, Salem N Jr. Essential fatty acid uptake and metabolism in the developing rodent brain. Lipids 1996; 31S:S103–S107.

Protstein NP, Pennacchiotti GL, Sprecher H, Aveldano MI. Active synthesis of C24:5n-3 fatty acid in retina. Biochem J 1996; 316:859–864.

Rivers JPW, Sinclair AJ, Crawford MA. Inability of the cat to desaturate essential fatty acids. Nature 1975; 258:171–173.

Rodriguez A, Sarda P, Nessmann C, Boulot P, Poisson J-P, Leger CL, et al. Fatty acid desaturase activities and polyunsaturated fatty acid composition in human liver between the seventeenth and thirty-sixth gestational weeks. Am J Obstetr Gynecol 1998; 179(4):1063–1070.

Rodriguez de Turco EB, Gordon WC, Bazan NG. Rapid and selective uptake, metabolism and cellular distribution of docosahexaenoic acid among rod and cone photoreceptor cells in the frog retina. J Neurosci 1991; 111(1):3667–3678.

Rodriguez de Turco EB, Parkins N, Ershov AV, Bazan NG. Selective retinal pigment epithelial cell lipid metabolism and remodeling conserves photoreceptor docosahexaenoic acid following phagocytosis. J Neurosci Res 1999; 57:479–486.

Salem N Jr, Kim H-Y, Yergey JA. Docosahexaenoic acid: membrane function and metabolism. In: Simopoulos AP, Kifer RR, Martin R, eds. The Health Effects of Polyunsaturates in Seafoods. Academic, New York, 1986, pp. 263–317.

Salem N Jr, Omega-3 fatty acids: molecular and biochemical aspects. In: Spiller GA, Scala J, eds. New Protective Roles for Selected Nutrients Alan R. Liss, New York, 1989, pp. 109–228.

Salem N Jr, Pawlosky RJ. Health Policy Aspects of Lipid Nutrition and Early Development. In: Galli C, Simopoulos AP, Tremoli E, eds. Fatty Acids and Lipids: Biological Aspects. World Review of Nutrition and Diet Vol. 75. Karger, Basel, 1994, pp. 46–51.

Salem N Jr, Pawlosky RJ. Arachidonate and docosahexaenoate biosynthesis in various species and compartments in vivo. In: Galli C, Simopoulos AP, Tremoli E, eds. Fatty Acids and Lipids: Biological Aspects World. Review of Nutrition and Diet Vol. 75. Karger, Basel, 1994, pp. 114–119.

Salem N Jr, Wegher B, Mena P, Uauy R. Arachidonic and docosahexaenoic acids are biosynthesized from their 18-carbon precursors in human infants. Proc Natl Acad of Sci USA 1996; 93:49–54.

Salem N Jr, Pawlosky R, Wegher B, Hibbeln J. In vivo conversion of linoleic acid to arachidonic acid in human adults. Prostaglandins Leukotrienes Essential Fatty Acids 1999; 60(5–6):407–410.

Sauerwald TU, Hachey DL, Jensen CL, Chen H, Anderson RE, Heird WC. Intermediates in endogenous synthesis of C22:6n-3 and C20:4n6 by term and preterm infants. Pediatr Res 1991; 41(2):183–187.

Schenck PA, Rakoff H, Emken EA. Delta-8 desaturation in vivo of deuterated eicosatrienoic acid by mouse liver. Lipids 1996; 31(6):593–600.

Scott BL, Bazan NG. Membrane docosahexaenoate is supplied to the developing brain by the liver. Proc Natl Acad Sci USA 1989; 86:2903–2907.

Sinclair AJ, McLean JG, Monger EA. Metabolism of linoleic acid in the cat. Lipids 1979; 14:932–936.

Stinson AM, Wiegand RD, Anderson RE. Recycling of docosahexaenoic acid in rat retinas during n-3 fatty acid deficiency J Lipid Res 1991; 32:2009–2017.

Su H-M, Bernardo L, Mirmiran M, Ma X-H, Nathanielsz PW, Brenna JT. Dietary 18:3n3 and 22:6n-3 as sources of 22:6n-3 accretion in neonatal baboon brain and associated organs. Lipids 1999; 34(3):S347–S350.

Vermunt SH, Mensink RP, Simonis MM, Hornstra G. Effects of dietary alpha-linolenic acid on the conversion and oxidation of 13C-alpha-linolenic acid. Lipids 2000; 35(2):137–142.

Wang N, Anderson RE. Synthesis of docosahexaenoic acid by the retina and retinal pigment epithelium. Biochemistry 1993; 32:13,703–13,709.

Weisinger HS, Vingrys AJ, Sinclair AJ. Effect of dietary n-3 deficiency on the electroretinogram in the guinea pig. Ann Nutr Metab 1996; 40(2):91–98.

Willis AL. Unanswered questions in EFA and PG research. Prog Lipid Res 1981; 20:839–850.

8   Quantifying and Imaging Brain Phospholipid Metabolism In Vivo Using Radiolabeled Long Chain Fatty Acids

Stanley I. Rapoport

1. INTRODUCTION

Phospholipids are major components of neuronal and glial membranes and participate in membrane remodeling and signal transduction (Axelrod, Burch, & Jelsema, 1988; Fisher & Agranoff, 1987; Porcellati, Goracci, & Arienti, 1983; Stephenson et al., 1994). Many of their functions involve the release of the essential polyunsaturated fatty acids (FAs), arachidonate (20:4n-6) and docosahexaenoate (22:6n-3) in signaling processes. Docosahexaenoate can modulate membrane fluidity and neuronal recovery following injury and participate in signal transduction and synaptic plasticity, whereas arachidonate and its bioactive metabolites (eicosanoids, leukotrienes, and monohydroxyeicosatetraenoic acids) are important second messengers (Axelrod, 1995; Horrocks & Yeo, 1999; Wolfe & Horrocks, 1994). However, in pathological conditions, such as inflammation, ischemia, and trauma, large quantities of FAs are liberated from phospholipids and contribute to cell dysfunction or death (Bazán & Rodriguez de Turco, 1980; Rabin et al., 1997). Abnormal brain phospholipid metabolism also occurs in essential FA deficiency (Bourre et al., 1989; Contreras et al., 1999b), Alzheimer disease (Farooqui, Rapoport, & Horrocks, 1997a; Ginsberg, Rafique, Xuereb, Rapoport, & Gershfeld, 1995; Pettegrew, Moossy, Withers, McKeag, & Panchalingam, 1988), chronic alcohol exposure (Pawlosky & Salem Jr., 1995), and possibly in human depression and bipolar disorder (Hibbeln, 1998; Stoll et al., 1999).

For these many reasons, it would be of interest to quantify and image in vivo FA kinetics in brain phospholipids, in animals and in humans. Our laboratory has elaborated a method and model to do this (Rapoport, In press; Rapoport et al., 1997; Robinson et al., 1992).

2. FATTY ACID MODEL

2.1. Experimental Method

The FA method involves the intravenous injection or infusion of a radiolabeled albumin-bound FA, then measuring plasma and brain radioactivities. In awake rodents, the brain can be prepared for biochemical analysis or quantitative autoradiography using a [^{3}H]FA or [^{14}C]FA. In humans or higher primates, positron emission tomography (PET) can be used to noninvasively image regional incorporation of [^{11}C]FA (Arai et al., 1995; Chang et al., 1997a; Rapoport, Chang, Connolly, Carson, & Eckelman, 1999; Rapoport et al., 2000; Robinson et al., 1992).

To determine regional incorporation coefficients k^* (Eq. 3) in awake rats, following tracer injection labeled and cold FA concentrations in arterial plasma are measured until the animal is killed after 10–15 min. After removing the brain, frozen sections are cut on a cryostat and prepared for quantitative autoradiography using radioactive standards. For this, [^{3}H]FAs are preferred to minimize nonvolatile aqueous background radioactivity because of β-oxidation. Values for k^* are calculated by dividing net brain radioactivity by integrated plasma radioactivity.

To determine FA turnover rates and half-lives in brain phospholipids, the labeled FA is infused intravenously at a programmed rate for 5 or 10 min to establish constant plasma radioactivity. Arterial blood is withdrawn at regular intervals to monitor plasma concentrations of labeled and unlabeled FAs. After infusion, the animal is rapidly anesthetized and its brain is subjected to focused-beam microwave irradiation to stop metabolism. One half is used to quantify acyl-CoA, the other to quantify phospholipids and the distribution of the tracer by established analytical methods. A correction is made for radioactivity in blood (Chang et al., 1997a; Chang, Bell, Purdon, Chikhale, & Grange, 1999; Deutsch, Grange, Rapoport, & Purdon, 1994; Deutsch, Rapoport, & Purdon, 1996; Grange et al., 1995; Rapoport, In press; Washizaki, Smith, Rapoport, & Purdon, 1994).

2.2. Metabolic Pathways

Figure 1 illustrates the diffusional and metabolic pathways that can be taken by a FA entering the brain from blood. In blood, the FA may exist as the unesterified species in plasma, esterified within phospholipids, cholesterol, and triglycerides of lipoproteins, or covalently bound within erythrocytes and platelets (Bazán, 1990; Staufenbiel, 1988). Unesterified FAs in plasma with chain length ≤ 22 carbons are largely bound to circulating albumin (> 99% binding), whereas longer-chain unesterified FAs are preferentially bound to circulating lipoproteins (Shafrir, Gatt, & Khasis, 1965; Wosilait & Soler-Argilaga, 1975).

The demonstration of lipoprotein receptors on the luminal surface of the cerebrovascular endothelium suggested that they mediate brain uptake of FAs esterified within lipoproteins (Dehouck et al., 1997; Méresse, Delbart, Fruchart, & Cecchelli, 1989). However, in awake adult rats, such receptor-mediated entry is unimportant and its flux term, J_2 (see Fig. 2), can be neglected. Thus, circulating FAs cross the blood-brain barrier essentially in only the unbound unesterified form, after being hydrolyzed from lipoproteins by lipoprotein lipase within blood or at the cerebral capillary bed (Fig. 1) (Brecher & Kuan, 1979; Purdon, Arai, & Rapoport, 1997; Spector, In press). The process involves simple diffusion and not additionally, as in the heart, facilitated diffusion by a translocase (a translocase is not found in brain) (Glatz, In press; Luiken et al., 1999). Incorporation from plasma into brain phospholipids is proportional to the FA plasma concentration and is determined by brain metabolic demand. This is because the K_m for acyl-CoA synthetase (the enzyme which converts the FA to acyl-CoA) is 180 μmol, twice the total brain unesterified FA concentration (Deutsch, Rapoport, & Purdon, 1997; Rabin et al., 1997; Rapoport & Spector, Submitted; Sigiura et al., 1995).

Chapter 8 / In Vivo Brain Phospholipid Kinetics

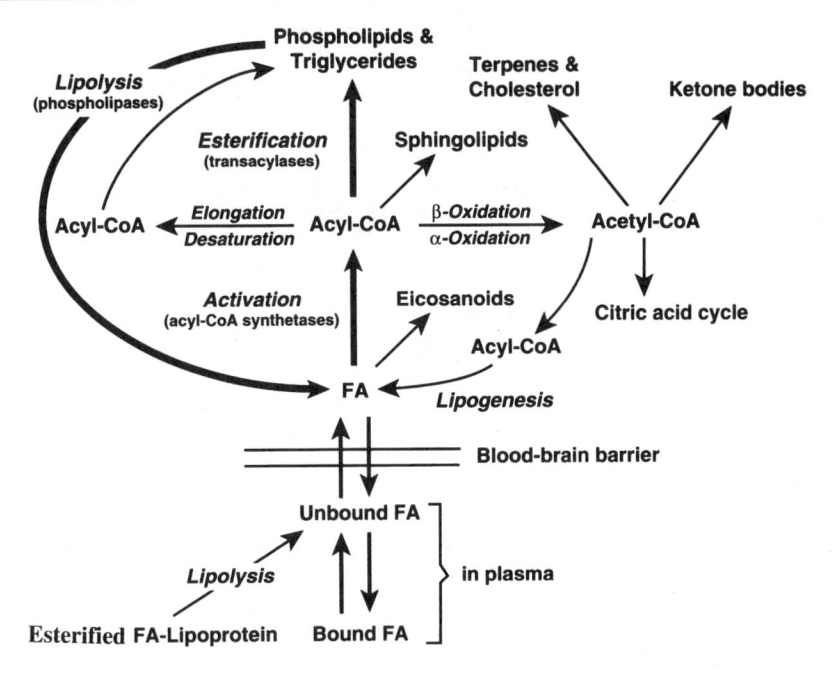

Fig. 1. Brain metabolic pathways for fatty acids. Thick arrows indicate major pathways followed rapidly by a FA after its entry into brain from plasma, and the enzymes involved in the pathways. Thin arrows indicate alternative pathways that are followed to a lesser extent within the 5- to 20-min experimental period of the FA method, the major one of concern from the acyl-CoA pool involving β-oxidation within mitochondria (*see* text).

About 5% of unesterified FA is extracted as blood passes through the rat brain, and extraction is independent of blood flow (Chang et al., 1997a; Pardridge & Mietus, 1980; Robinson et al., 1992; Yamazaki, DeGeorge, Bell, & Rapoport, 1994). After entering the brain, the FA rapidly and preferentially follows the pathway indicated by the thick arrows in Fig. 1. From the FA pool, it is activated to acyl-CoA (FA-CoA) by an acyl-CoA synthetase (Semenkovich, 1997; Watkins, 1997), then enzymatically incorporated into stable lipids (predominantly phospholipids) by an acyltransferase (Yamashita, Sugiura, & Waku, 1997). At steady state, the rate of incorporation into phospholipids equals the rate of release back into the unesterified FA pool via the brain free FA pool, to complete a "cycle." Release is catalyzed by a phospholipase A_1 (PLA_1) acting at the stereospecifically numbered (*sn*)-*1* site for saturated FAs, or by a phospholipase A_2 (PLA_2) acting at the *sn-2* site for polyunsaturated FAs (Dennis, 1994; Pete, Ross, & Exton, 1994).

Thin arrows in Figure 1 indicate alternative pathways that may be taken by a FA once within the brain. Those that are quantitatively significant within the few minutes of our pulse-labeling studies are the conversion of polyunsaturated FAs to bioactive metabolites (*see* Sec. I), and β-oxidation within mitochondria. The extent of β-oxidation, thus the rate of formation of nonvolatile aqueous metabolites, depends on the FA and how it is labeled.

The saturated [1-^{14}C]palmitate tracer is not ideally suited to estimate brain incorporation of palmitate into phospholipids with quantitative autoradiography, because about half of the tracer is converted via β-oxidation to background aqueous nonvolatile labeled metabolites, mainly glutamate and aspartate (Miller, Gnaedinger, & Rapoport, 1987; Noronha, Larson, & Rapoport, 1989). Using [9,10-^{3}H]palmitate can overcome this limi-

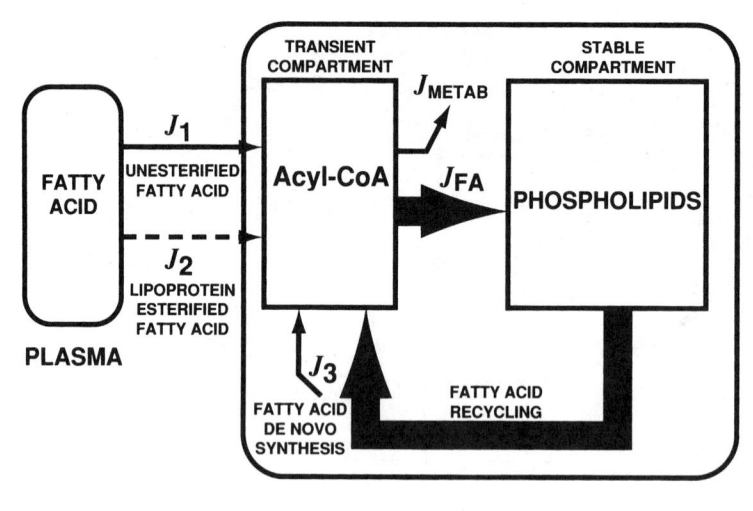

Fig. 2. Model for relation between net FA flux from final precursor pool (acyl-CoA) into brain lipids (J_{FA}) and the net flux rates from plasma unesterified FA (J_1) and from plasma esterified FA (J_2) into acyl-CoA. J_2 is negligible in the adult brain.

tation, as 75% of the tritium is converted to [³H]H$_2$O during β-oxidation (Greville & Tubbs, 1968), and can be removed during drying. For PET scanning with [1-¹¹C]palmitate, β-oxidation of carbon-labeled palmitate can be reduced by pretreatment with an inhibitor of its entry into mitochondria, methylpalmoxirate (methyl-2-tetradecylglycidate) (Chang et al., 1998; Chang, Wakabayashi, & Bell, 1994; Rapoport, 1999; Rapoport et al., 1999; Tutwiler, Ho, & Mohrbacher, 1981).

Unlike saturated FAs, carbon-labeled polyunsaturated arachidonate and docosahexaenoate can be used with autoradiography or PET scanning without the inhibitor, because they are minimally oxidized (Osmundsen, Cervenka, & Bremer, 1982); only 15% of their label is found in the brain nonvolatile aqueous compartment 20 min after an intravenous injection. [³H]Arachidonate or [³H]docosahexaenoate produce only 10% nonvolatile aqueous background activity.

2.3. Brain Compartments

The complex representation of Fig. 1 can be simplified to Fig. 2, which identifies three compartments that have to be assessed experimentally to apply the FA model. These compartments are (1) plasma unesterified FA, (2) the precursor brain FA-CoA pool, and (3) the "stable" brain phospholipid compartment (Rapoport et al., 1997; Robinson et al., 1992). Fluxes between them—J_1, J_2, J_3, J_{FA}—are defined in the legend to Fig. 2. The simplified Fig. 2 can be used because: (1) the half-life of the FA tracer in plasma is less than 1 min, (2) FA uptake into brain from blood is independent of cerebral blood flow over a wide range of unlabeled FA plasma concentrations (Chang et al., 1997a; Yamazaki et al., 1994), (3) labeled FA in plasma rapidly equilibrates (1 min or less) with label in FA-CoA, the precursor pool for FA incorporation into brain phospholipids, and (4) the FA tracer in brain is rapidly incorporated into brain phospholipids (80–90% within 1 min) (Rapoport et al., 1997; Robinson et al., 1992).

Rapid entry of a FA from plasma into the brain FA-CoA pool allows increased neuronal demand for the FA to be easily met by the large reservoir of unesterified FA in plasma. One to two minutes after a step elevation in plasma [9,10-^{3}H]palmitate or labeled arachidonate, specific activity of the respective brain FA-CoA pool has reached a steady state (Grange et al., 1995; Washizaki et al., 1994). At this time, the ratio of FA-CoA specific activity to plasma FA specific activity (λ in Eq. 4) is 0.02–0.04, attesting to marked dilution of plasma-derived FA-CoA by FA released from brain phospholipids (Fig. 2).

2.4. Operational Equations

We have derived and validated operational equations to quantify FA fluxes from plasma into the brain FA-CoA pool and from the FA-CoA pool into individual brain phospholipids, and turnover rates and half-lives within these phospholipids (Rapoport et al., 1997; Robinson et al., 1992). The incorporation rate of a FA radiotracer from plasma into a stable brain lipid compartment i is given by,

$$dt/dc^*_{br,i} = k^*_i c^*_{pl} \tag{1}$$

where k^*_i (mL/s/g, or s^{-1}) is the incorporation coefficient, c^*_{pl} is the plasma concentration of labeled unesterified FA, and $c^*_{br,i}$ is the brain concentration of label in i. Integration of Eq. 1 to time T of sampling gives k^*_i,

$$k^*_i = c^*_{br,i} / \int_0^T c^*_{pl} dt \tag{2}$$

For whole brain or for a brain region in which analysis is by quantitative autoradiography or PET, we have an overall k^* equal to

$$k^* = \sum_i k^*_i = c^*_{br}(T) / \int_0^T c^*_{pl} dt \tag{3}$$

where $c^*_{br}(T)$ equals the brain radioactivity at time T.

Dilution factor λ represents the extent to which brain FA-CoA is derived from unesterified FA in plasma (J_1), compared with esterified plasma FA (J_2) and FA produced by recycling plus *de novo* synthesis within brain (J_3) [*de novo* synthesis via phosphatidic acid is less than 1% of the recycling contribution (Murphy, 1998)]. As $J_2 << J_1$ in adult rodents (Purdon et al., 1997), λ can be simplified to

$$\lambda = J_1/(J_1 + J_2 + J_3) \approx J_1/(J_1 + J_3) \tag{4}$$

λ can be determined experimentally as the ratio of acyl-CoA specific activity to plasma specific activity after programmed tracer infusion establishes the latter to be constant:

$$\lambda = \frac{c^*_{FA-CoA} / c_{FA-CoA}}{c^*_{pl} / c_{pl}} \tag{5}$$

where c_{pl} is the plasma concentration of unlabeled unesterified FA, and c^*_{FA-CoA} and c_{FA-CoA} are the labeled and unlabeled acyl-CoA concentrations.

The rate of $J_{FA,i}$ incorporation of the *unlabeled* FA from the FA-CoA pool into brain compartment i (Fig. 2) equals (Robinson et al., 1992):

$$J_{FA,i} = k^*_i c_{pl} / \lambda \tag{6}$$

Thus, the turnover rate (percent per unit time) of the unlabeled FA in brain phospholipid *i*, reflecting de-esterification followed by re-esterification, is given as

$$F_{FA,i} = J_{FA,i} / c_{br,i} \qquad (7)$$

and the half-life resulting from de-esterification–re-esterification equals

$$t_{1/2,i} = F_{FA,i} / 0.693 \qquad (8)$$

3. EXPERIMENTAL APPLICATIONS OF THE FATTY ACID MODEL

3.1. Labeling and De-esterification–Re-esterification Half-Lives of Fatty Acids in Brain Phospholipids

3.1.1. SPECIFICITY OF PHOSPHOLIPID LABELING

A radiolabeled long-chain FA is injected intravenously in an awake rat, disappears from plasma with a half-life of less than 1 min. The brain is maximally labeled within 10 min, and the label is mainly in phospholipids. Regional differences in FA incorporation can be distinguished on quantitative autoradiographs (Fig. 5). Labeling is specific with regard to the FA tracer, the phospholipid, and the *sn* sites on the glycerol backbone (Tables 1 and 2). Specificity is conferred by substrate- and FA-specific phospholipases, acyl-CoA synthetases, and acyltransferases and by intracellular FA and acyl-CoA binding proteins (Fig. 1) (Gossett et al., 1996; Mikkelsen & Knudsen, 1987; Sellner, Chu, Glatz, & Berman, 1995). Labeled palmitate enters the *sn-1* position mainly of phosphatidylcholine (PC), labeled arachidonate the *sn-2* position mainly of phosphatidylinositol (PI) and PC, and labeled docosahexaenoate the *sn-2* position mainly of phosphatidylethanolamine (PE) and PC (DeGeorge, Nariai, Yamazaki, Williams, & Rapoport, 1991; DeGeorge, Noronha, Bell, Robinson, & Rapoport, 1989; Nariai et al., 1994; Nariai, DeGeorge, Greig, & Rapoport, 1991a; Noronha, Bell, & Rapoport, 1990). Using a combination of FA tracers can elucidate the kinetics of FAs in different *sn* positions of different phospholipids.

3.1.2. HALF-LIVES OF FATTY ACIDS IN INDIVIDUAL PHOSPHOLIPIDS ARE MINUTES TO HOURS

Table 3 summarizes recycling half-lives due to deesterification-reesterification of each of three different brain phospholipids of awake rats (Chang et al., 1999; Grange et al., 1995). The half-lives were calculated by equation 8 using experimental values for λ (Table 3), the steady-state ratio of acyl-CoA specific activity to plasma specific activity (Eq. 5). λ ranged from 0.02 to 0.04 and half-lives in some phospholipid classes were just a few hours. In another study, half-lives of arachidonate were found to range from 10 min to 6.6 h among molecular species of PC and from 1.6 h to 4.1 h among molecular species of PI (Shetty, Smith, Washizaki, Rapoport, & Purdon, 1996). Short half-lives and low values for λ imply active recycling of FAs within brain phospholipids, consistent with their role in signaling and membrane remodeling. Indeed, when these processes are blocked by pentobarbital anesthesia, the half-life of palmitate is prolonged 10-fold (Contreras, Chang, Kirkby, Bell, & Rapoport, 1999a). As two molecules of ATP are required to activate a FA to FA-CoA, the short half-lives (equivalent to high turnover rates) in Table 3 lead to the calculation that FA recycling in phospholipids consumes

Table 1
Distribution of Radioactivity Within Brain Lipid Fractions in Awake Rats, 15 min After a 5-min Intravenous Infusion of Each of Three Labeled Fatty Acids

Fraction	[9,10-^{3}H] Palmitate	[1-^{14}C] Arachidonate	[1-^{14}C] Docosahexaenoate
	Percent Total Brain Radioactivity		
Lipid fraction	89.2 ± 1.1	89.4 ± 0.5	87.1 ± 1.0
Phospholipids	51.7 ± 0.6	72.9 ± 0.8	64.7 ± 0.9
Sphingomyelin	1.21 ± 0.18	—	—
Phosphatidylcholine	31.7 ± 1.1	27.4 ± 0.2	17.2 ± 0.8
Phosphatidylserine	0.92 ± 0.20	2.09 ± 0.3	2.59 ± 0.30
Phosphatidylinositol	7.34 ± 0.20	35.6 ± 0.4	6.92 ± 0.68
Phosphatidylethanolamine	10.5 ± 0.9	7.14 ± 0.22	37.6 ± 0.8

Note: Values are means ± SEM for three to seven rats.
Source: Data from Nariai et al., 1994; Noronha et al., 1990.

Table 2
Distribution of Radiolabeled Fatty Acids Within *sn-1* and *sn-2* Sites of Brain Phospholipids That Receive Most of the Label, 15 min After Intravenous Injection in Awake Rats

Fatty Acid Tracer	Phospholipid	sn-1 Position	sn-2 Position
[9,10-^{3}H]Palmitate	PC	82.5 ± 1.9	17.5 ± 1.9
	PE	90.7 ± 0.5	9.3 ± 0.5
[1-^{14}C]Arachidonate	PC	7.8 ± 1.9	92.2 ± 1.9
	PI	0.0 ± 0.0	100 ± 0
[1-^{14}C]Docosahexaenoate	PC	22.9 ± 5.2	77.1 ± 5.2
	PE	0 ± 0	100 ± 0

Note: Values are means ± SEM for three to seven rats.
Source: From Nariai et al., 1994.

Table 3
Estimated Acyl-CoA Dilution Coefficients λ and Half-Lives of Fatty Acids in Brain Phospholipids of Awake Rats

Fatty acid	λ	PI	PC	PE	PS
		Half-life (h)			
Arachidonate	0.03 ± 0.00[a]	3.4	2.9	47	4.7
Docosahexaenoate	0.03 ± 0.00	3.9	22	58	347
Palmitate	0.02 ± 0.00	2.4	9.9	17	7.4

[a]Mean ± SEM (*n* = 5).
Source: Data from Chang & Jones, 1998; Contreras et al., 1999a; Grange et al., 1995; Washizaki et al., 1994.

some 5% of the ATP consumed by brain (Purdon & Rapoport, 1998). The net quantities of fatty acids which turn over per day in rat brain phospholipids are quite large, of the order of the net fatty acid concentrations in the phospholipids (Table 4).

3.2. The Fatty Acid Method Can Identify Molecular Targets for Centrally Acting Drugs

3.2.1. PHOSPHOLIPASE A_2 TARGET FOR LITHIUM

Fatty acid recycling (de-esterification–re-esterification (Sun & MacQuarrie, 1989)) within brain phospholipids is regulated by enzymes belonging to large families of phospholipases, acyl-CoA synthetases, and acyltransferases (Fig. 1) (Dennis, 1994; Farooqui, Yang, Rosenberger, & Horrocks, 1997b; Pete et al., 1994; Semenkovich, 1997; Watkins, 1997; Yamashita et al., 1997). Thus, a change in a FA half-life or turnover rate in response to a drug might help to identify which of these enzymes is the drug target and help to develop better drugs. This approach has proven effective with regard to lithium, used to treat bipolar disorder (Barchas, Hamblin, & Malenka, 1994).

We found that λ for arachidonoyl-CoA was increased 4.5-fold and arachidonate turnover was reduced by 80% in rats fed lithium for 6 wk to produce a "therapeutic" brain level of 0.7 mM (Table 4), whereas λ and turnover rates of docosahexaenoate and palmitate were unaffected (Chang et al., 1997a; Chang et al., 1999). The addition of 1.0–10.0 mM LiCl to the postnuclear supernatant from brains of control and lithium-treated rats did not inhibit PLA_2 activity (Chang & Jones, 1998). These results suggested that lithium interfered with transcriptional or posttranslational regulation of an arachidonate-selective PLA_2. Indeed, chronic lithium treatment of rats reduced both the mRNA and protein levels of a cytosolic 85-kDa $cPLA_2$ (type IV) without affecting the levels of another arachidonate-specific PLA_2, $iPLA_2$ (Fig. 3A,B) (Rintala et al., 1999). $cPLA_2$ releases arachidonate at the normal low level of cytoplasmic Ca^{2+} (Kramer & Sharp, 1997), and is located in neurons (Kishimoto, Matsumura, Kataoka, Morii, & Watanabe, 1999).

The 80% reduction in arachidonate turnover in the lithium-treated rats and the specificity of the lithium effect on $cPLA_2$ suggest that $cPLA_2$ is the major determinant of arachidonate release during receptor-mediated signaling in normal brain (Axelrod et al., 1988). In contrast, phospholipase C (PLC) activation in the phosphoinositide cycle has been argued to be the major source of brain arachidonate, through the formation of inositol-1,4,5-trisphosphate (IP_3) and diacylglycerol. IP_3 can increase free cytosolic Ca^{2+}, thereby activating calcium-dependent PLA_2 to liberate arachidonate (Dennis, 1994), and diacylglycerol may be acted on by diacylglycerol lipase to liberate arachidonate (Agranoff & Fisher, 1994; Cooper, Bloom, & Roth, 1996; Fisher & Agranoff, 1987). A major PLC source is unlikely, however. In stimulated platelets, a model for brain PLC signaling, the diacylglycerol lipase pathway contributes less than 20% to arachidonate release compared with direct activation of PLA_2, and IP_3 does not increase cytosolic Ca^{2+} sufficiently to activate PLA_2 (Mahadevappa & Holub, 1986; Moriyama, Urade, & Kito, 1999; Reddy, Rao, & Murthy, 1994). Furthermore, diacylglycerol lipase activity in the brain is low and may be absent in astrocytes (Murphy, Jeremy, Pearce, & Dandona, 1985). PLC induced release of brain arachidonate can occur, however, in pathological conditions like ischemia, trauma, and excitotoxicity (Bazán & Rodriguez de Turco, 1980; Deutsch et al., 1997; Reddy & Bazan, 1987).

Table 4
Fatty Acid Concentrations and Their Turnover Rates in Phospholipids of Rat Whole Brain

	Total brain	PC	PS	PI	PE	SM	PA	PC+PS +PI+PE
Phospholipid concentration (µmol/g brain)[a]								
Phospholipid	64.4	25.0	8.5	2.2	23.7	3.7	1.3	
Fatty acid concentration in individual phospholipids (µmol/g brain)[b]								
Palmitate	29.3	17.7	0.32	0.42	2.91	0.37[c]		21.35
Arachidonate	13.0	2.42	0.58	1.65	6.92	Tr		11.57
Docosahexaenoate	15.27	1.69	3.72	0.14	11.08	Tr		16.63
Fatty acid turnover in individual phospholipids (µmol per g brain/d)[d]								
Palmitate		28.6	0.72	2.93	2.86			35.1
Arachidonate		10.62	1.64	6.06	1.80			20.1
Docosahexaenoate		1.26	0.18	0.59	3.19			5.2

[a] Data from Wells & Dittmer, 1967.
[b] Data from Chang et al., 1999.
[c] Data from Marshall, Fumagalli, Niemiro, & Paoletti, 1966.
[d] Product of daily turnover rate and FA concentration in phospholipid.

Table 5
Selective Reduction by Chronic Lithium of Arachidonate Turnover in Rat Brain Phospholipids

		Turnover (% per hour)			
Fatty Acid	λ	PI	PC	PE	PS
Arachidonate					
Control	0.04 ± 0.00	15.3 ± 0.4	18.3 ± 0.6	1.1 ± 0.1	11.8 ± 0.6
Lithium	0.18 ± 0.02**	2.6 ± 0.1**	5.0 ± 0.3**	0.2 ± 0.0**	2.2 ± 0.2**
Docosahexaenoate					
Control	0.03 ± 0.00	17.7 ± 1.7	3.1 ± 0.4	1.2 ± 0.1	0.2 ± 0.0
Lithium	0.03 ± 0.00	31.0 ± 9.0	4.5 ± 1.2	1.6 ± 0.4	0.3 ± 0.1
Palmitate					
Control	0.02 ± 0.00	29.1 ± 2.6	7.0 ± 0.4	4.1 ± 0.4	9.4 ± 1.1
Lithium	0.02 ± 0.00	26.0 ± 1.2	5.1 ± 0.3*	3.5 ± 0.1	9.4 ± 0.9

Note: Data are means ± SEM ($n = 5$). Brain concentration of lithium is 0.7 mM. Statistically different from control mean, *$p < 0.05$, **$p < 0.01$.
Source: Data from Chang et al., 1997a; Chang et al., 1999.

3.3. Imaging Brain Signal Transduction and Neuroplasticity In Vivo

3.3.1. PLA2 ACTIVATION BY ACUTE CHOLINERGIC STIMULATION

Receptor-mediated, guanosine 5'-triphosphate (GTP) protein-dependent signal transduction, initiated by stimulation of cholinergic M_1 receptors or dopaminergic D_2 receptors, can activate PLA_2 to release arachidonate and/or docosahexaenoate from the *sn*-2 position of phospholipids (Axelrod et al., 1988; DeGeorge et al., 1991; Jones, Arai, Bell, & Rapoport, 1996; Jones, Arai, & Rapoport, 1997). This is illustrated for arachidonate in Fig. 4, which also shows that unlabeled arachidonic acid that is liberated by PLA_2 into the brain unesterified pool can be rapidly labeled by tracer arachidonate injected into plasma (Grange et al., 1995; Washizaki et al., 1994). A fraction of the unesterified

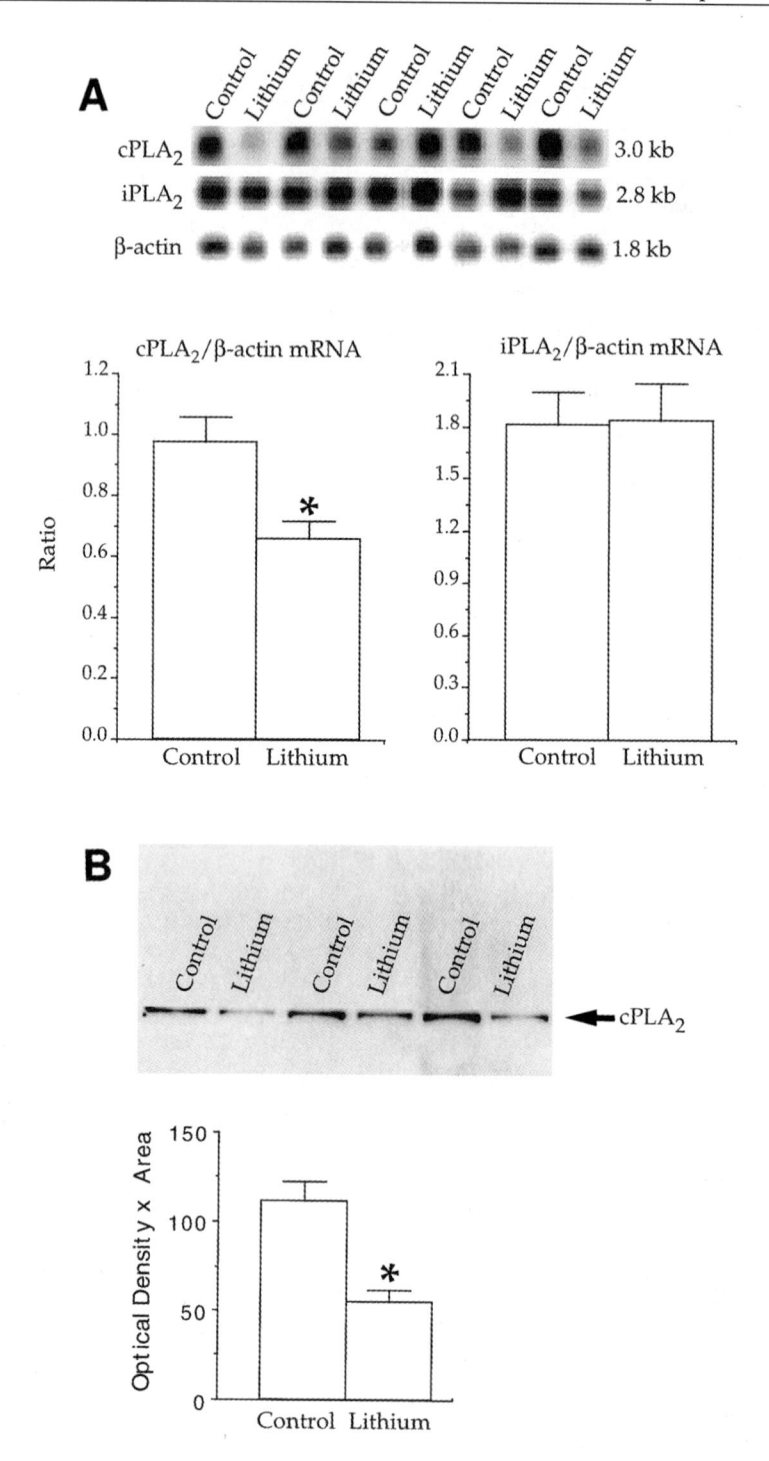

Fig. 3. (A) Autoradiograms of Northern blots (top), and mRNA levels normalized to mRNA for β-actin (bottom) for cPLA$_2$ and iPLA$_2$ in brains of control rats and rats treated with lithium for 6 wk. Means ± SEM ($n = 5$). *$p < 0.001$. **(B)** Autoradiograms of Western blots (top) and levels of cPLA$_2$ protein (bottom) in brains from control and chronic lithium-treated rats. Means ± SEM ($n = 8$). *$p < 0.001$. (From Rintala et al., 1999).

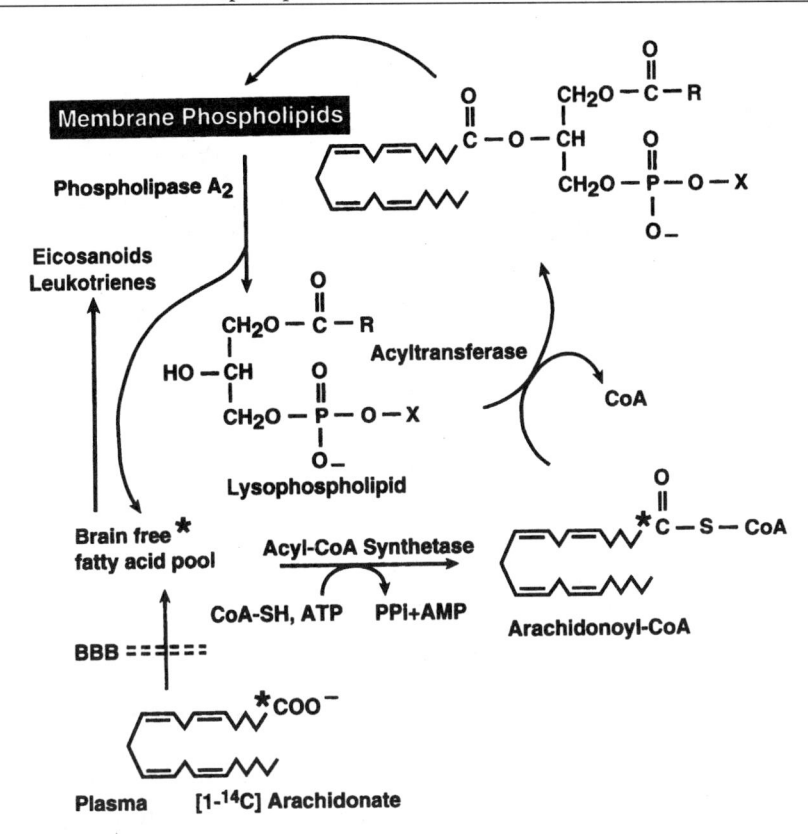

Fig. 4. Cycling of arachidonate in a brain phospholipid following its release by receptor-initiated activation of PLA$_2$. Following activation, unlabeled arachidonate is released from phospholipid (upper right) into the unesterified FA pool, which has been labeled by an intravenous injection of tracer. A fraction of the unesterified FA is converted to bioactive leukotrienes and eicosanoids. The remainder, with labeled and unlabeled FA taken up from blood into which label has been injected, is re-esterified into the brain arachidonoyl-CoA pool, from which it is incorporated into a lysophospholipid to reconstitute a "labeled" phospholipid at the site of PLA$_2$ activation.

arachidonate will be converted to bioactive eicosanoids, whereas the larger remainder, now labeled, will be re-esterified almost immediately into the lysophospholipid through the serial actions of acyl-CoA synthetase and acyltransferase. The extent of "reincorporation" will be proportional to the degree of initial PLA$_2$ activation and can be imaged in vivo using quantitative autoradiography or PET.

The cholinergic agonist, arecoline, was administered L.P. to awake rats to stimulate PLA$_2$ via brain muscarinic M$_1$ receptors (DeGeorge, Ousley, McCarthy, Lapetina, & Morell, 1987). Incorporation of intravenously injected [1-^{14}C]arachidonate was increased largely into the *sn-2* position of brain PI and PC, whereas incorporation of [1-^{14}C]docosahexaenoate was increased into the *sn-2* position of PE and PC (DeGeorge et al., 1991). These increases occurred specifically at synaptic membranes (Table 6) (Jones et al., 1996; Jones et al., 1997) and could be blocked by the muscarinic antagonist atropine or reduced by the PLA$_2$ inhibitor manoalide (Grange, Rabin, Bell, Rapoport, & Chang, 1998). Arecoline, which also increases cerebral blood flow (Maiese, Holloway,

Table 6
Incorporation Following Arecoline of [³H]arachidonate
But Not [³H]palmitate into Synaptic Membrane Fractions Rat Brain

Subcellular Fraction	Control	Arecoline	Percent Change[a]
Incorporation coefficient, $k^* \times 10^5$ (mL/s/g brain)			
Myelin			
Arachidonate	0.24 ± 0.01	0.41 ± 0.10	
Palmitate	0.37 ± 0.05	0.38 ± 0.06	
Synaptosomal plasma membranes			
Arachidonate	0.52 ± 0.06	1.39 ± 0.11[b]	+267%
Palmitate	0.67 ± 0.05	0.78 ± 0.06	
Synaptosomes with mitochondria			
Arachidonate	1.14 ± 0.12	5.09 ± 1.04[b]	+446%
Palmitate	2.69 ± 0.34	3.34 ± 0.56	
Somal mitochondria			
Arachidonate	0.53 ± 0.12	0.66 ± 0.15	
Palmitate	0.25 ± 0.06	0.30 ± 0.06	
Microsomes			
Arachidonate	0.94 ± 0.11	3.40 ± 0.28[b]	+361%
Palmitate	2.43 ± 0.18	3.00 ± 0.40	

Note: Italicized fractions showed synaptophysin immunoreactivity.
[a]Only significant percent changes are calculated
[a]Mean ± SD ($n = 4$) differs from control, $p < 0.001$.
Rats given 15 mg/kg ip arecoline and 4 mg/kg sc methylatropine before iv [³H]arachidonate.
Source: from Jones et al., 1996.

Larson, & Soncrant, 1994), did not stimulate PLA$_1$-related uptake of [9,10-³H]palmitate. Autoradiography confirmed that increased incorporation occurred at sites of muscarinic M$_1$ receptors, particularly within the neocortex (Fig. 5). Taken together, the data indicate that arecoline activated PLA$_2$-mediated release of arachidonate at M$_1$ synapses, independently of changes in cerebral blood flow.

INCREASED PLA2-MEDIATED CHOLINERGIC SIGNALING IN ANIMAL MODEL OF ALZHEIMER DISEASE

The nucleus basalis of Meynert, which provides muscarinic M$_2$ cholinergic input to the ipsilateral neocortex and is known to lose neurons in Alzheimer disease (Whitehouse et al., 1982), was unilaterally lesioned in rats (Nariai, DeGeorge, Lamour, & Rapoport, 1991b). After 2 wk of such a lesion, levels of presynaptic M$_2$ and postsynaptic M$_1$ receptors are reduced in the ipsilateral neocortex (Bogdanovic et al., 1993). However, when arecoline was administered 2 wk after unilateral lesioning, incorporation of [³H]arachidonate was greater in ipsilateral than contralateral cortex (Fig. 6). The results suggest an increased ability of postsynaptic nerve terminals to activate PLA$_2$ despite fewer M$_1$ receptors, thus suggesting upregulation at the level of G protein or PLA$_2$. They provide a new direction for examining cholinergic signaling with PET in Alzheimers disease.

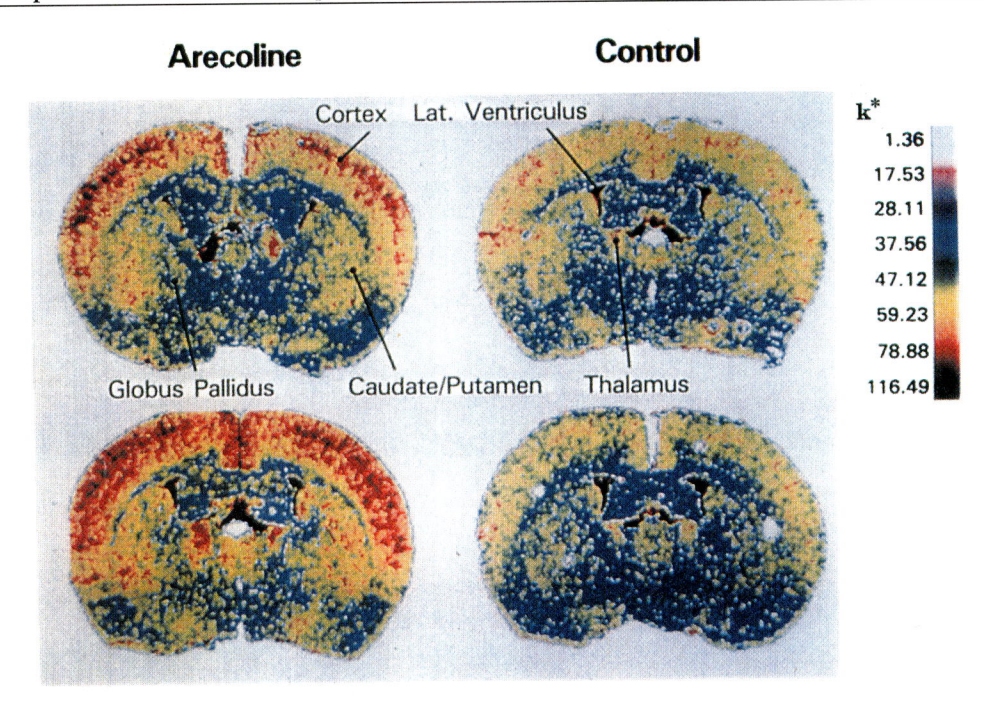

Fig. 5. Autoradiographs of coronal sections of brain from rat administered intraperitoneal saline and intravenous [1–^{14}C]arachidonate (right) or intraperitoneal arecoline and intravenous [1–^{14}C]arachidonate (left). Highlighted areas (gp = globus pallidus, hp = CA2 region hippocampus, IV and VB = layers IV and VB of cortex, f = mammalothalamic tract) are known to have high densities of M_1 cholinergic receptors.

INCREASED MEMBRANE REMODELING IN THE NEOCORTEX WITH A CHRONIC NUCLEUS BASALIS LESION

Phosphatidylcholine and PE are the major phospholipids within neuronal membranes (Table 4); their *sn-1* site is preferentially labeled by intravenously injected radioactive palmitate (Table 2). Using labeled palmitate in rats, it has been possible to image regional membrane changes involving these lipids in experimental Wallerian regeneration, chronic unilateral enucleation of the eye, experimental brain tumor, chronic auditory deprivation, brain development, and the chronic unilateral nucleus basalis lesion (Nariai et al., 1991b; Robinson et al., 1992; Yamazaki, Noronha, Bell, & Rapoport, 1989). Two weeks after a unilateral lesion of the nucleus basalis in the rat, baseline incorporation of tracer palmitate into ipsilateral neocortex was elevated, suggesting increased PC- and PE-dependent membrane remodeling (De Micheli, Chang, & Rapoport, 1996). This suggestion is supported by evidence of increased synthesis of neocortical phospholipid at 2 wk postlesion (Holmes, Nitsch, Erfurth, & Wurtman, 1993).

3.3.2. DOPAMINERGIC SYSTEM

INCREASED DOPAMINERGIC SIGNALING IN AN ANIMAL MODEL OF PARKINSON DISEASE

PLA_2 is involved in dopaminergic signaling via the D_2 receptor. The density of this receptor is increased in basal ganglia circuitry in Parkinsons disease (Cooper et al., 1996; Guttman, 1992; Scherman et al., 1989 Graham, 1990 #3706). A chronic unilateral lesion

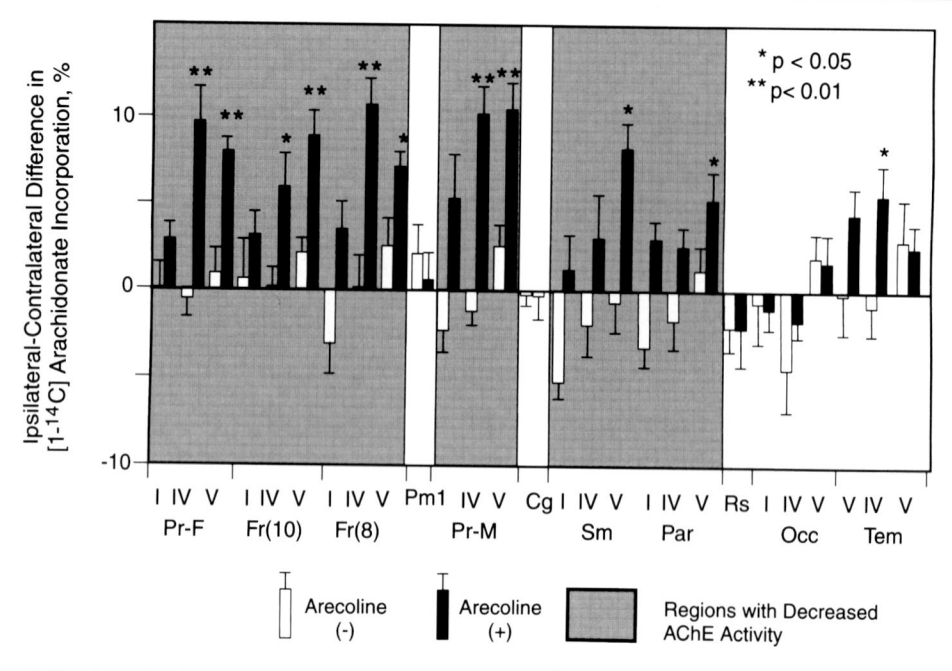

Fig. 6. Regional brain incorporation coefficient k^* of [1-^{14}C]arachidonate, as percent contralateral value, at 2 wk after a unilateral ablation of the nucleus basalis in rats, with or without prior administration of arecoline. The gray background identifies regions with decreased acetylcholinesterase (AChE) activity. Lined bars are means ± SEM. I, IV, V: layers of neocortex; Pr-F, prefrontal; Fr, frontal; Sm, sensorimotor; M, motor; Par, parietal; Occ, occipital, Tem, temporal. (From Nariai et al., 1991b).

of the substantia nigra in rats, produced by injecting 6-hydroxydopamine, increased baseline incorporation of [^{3}H]arachidonate throughout the ipsilateral basal ganglia and frontal cortex consistent with the normal inhibitory role of the substantia nigra (Hayakawa et al., 1998). Additionally, administration of the D_2 agonist quinpirol to these rats increased [^{3}H]arachidonate incorporation more into the ipsilateral than contralateral regions, consistent with higher ipsilateral D_2 receptors densities (Graham, Crossman, & Woodruff, 1990; Hayakawa et al., 1997).

3.4. PET Imaging of Labeled Fatty Acid Incorporation into the Human Brain

Positron emission tomography (PET) has been used to quantify local glucose metabolism and blood flow in the human brain and to image brain receptor densities (Rapoport, 1995). However, to date, PET has not been employed successfully in humans to image signal transduction "beyond the receptor," the "downstream" process by which neurotransmitters and drugs are closely linked to cognition and behavior (Cooper et al., 1996). In view of our results on dopaminergic and cholinergic signaling in normal and lesioned rodents (*see* Section 3.3.), a PET method for in vivo imaging of FA incorporation into the human brain might be of use for examining disrupted signaling in Alzheimer and Parkinson disease.

We have employed PET to image regional brain incorporation coefficients k^* (Eq. 3) of [1-^{11}C]arachidonate and [1-^{11}C]palmitate in anesthetized macaques. With the latter tracer, methylpalmoxirate was administered to decrease its β-oxidation (Arai et al., 1995; Chang et al., 1997a; Chang et al., 1998). Ongoing research suggests that PET imaging with [1-^{11}C]FAs can be accomplished in humans while remaining within radiation limits for clinical research (Rapoport, 1999; Rapoport et al., 1999; Rapoport et al., 2000). The neocortex, basal ganglia and thalamus are well outlined by incorporated [1-^{11}C]arachidonate, and incorporation is increased 30% in visual cortical and thalamic sensory areas by visual stimulation. Regional values for k^* for both tracers are less in white than gray matter and are about one-fifth of what they are in the rat brain.

4. DISCUSSION

This critical review describes an experimental method and mathematical model to quantify in vivo incorporation rates, half-lives, and turnover rates of FAs into brain phospholipids. FA incorporation is independent of cerebral blood flow and is thus a direct measure of brain phospholipid metabolism. Because specific FAs enter specific phospholipids at stereospecific positions, a combination of saturated and polyunsaturated FA labels can be used to investigate the active participation of phospholipids in brain signal transduction, membrane remodeling, and neuroplasticity.

Operational equations from our FA model have been used to estimate FA fluxes, turnover rates, and half-lives within brain phospholipids in vivo, after values for the dilution factor λ of the acyl-CoA pool were experimentally determined. This can be accomplished by a rapid method involving oligonucleotide solid-phase extraction (Deutsch et al., 1997). Low experimental values for λ in the awake rat, between 0.02 and 0.04, are consistent with very short half-lives resulting from de-esterification–re-esterification of FAs within some phospholipids (minutes to hours), very rapid turnover rates and high ATP consumption (Purdon & Rapoport, 1998).

The FA model can help to identify enzymatic targets of centrally acting drugs, thus providing a "reverse engineering" approach for further drug discovery (Liang, Furhman, & Somogyi, 1998). Studies employing the model suggest that lithium may act against bipolar disorder by reducing arachidonate turnover within the brain by targeting transcription of an arachidonic-specific cPLA$_2$.

The FA model provides a theoretical and experimentally proven way to image in vivo brain signal transduction involving PLA$_2$. Recent success in imaging [1-^{11}C]arachidonate incorporation into the macaque and human brain with PET, as well as incorporation of [1-^{11}C]palmitate when using methylpalmoxirate to prevent its brain β-oxidation (Arai et al., 1995; Chang et al., 1998; Chang, Grange, Rabin, & Bell, 1997b; Rapoport et al., 1999; Rapoport et al., 2000), suggests that the FA method can be exploited with PET to image brain signaling and neuroplasticity in humans in health and disease. Indeed, our observations on rat models of Alzheimer and Parkinson disease suggest that PET and [1-^{11}C]arachidonate might be used with drug activation to image abnormal, receptor-specific signal transduction preclinically and in relation to pharmacotherapy.

The FA model is sufficiently general for future use with appropriate modifications to quantify FA metabolism in organs other than brain. In the heart, for example, [1-^{11}C]palmitate has been used for some time with PET to image oxidative metabolism in mitochondria, where this FA largely undergoes β-oxidation (Schelbert et al., 1986). However, in heart as in brain, labeled arachidonate is preferentially incorporated into phospholipids instead of undergoing β-oxidation or being incorporated into neutral lipids (Murphy, Rosenberger, & Rapoport, In press), and thus might be used with appropriate modeling to study heart phospholipids under normal and disease conditions.

REFERENCES

Agranoff BW, Fisher SK. Phosphoinositides. In: Siegel GJ, Agranoff BW, Albers RW, Molinoff PB, eds. Basic Neurochemistry. 5th ed. Raven, New York, 1994, pp. 417–448.

Arai T, Wakabayashi S, Channing MA, Dunn BB, Der MG, Bell JM, Herscovitch P, Eckelman WC, Rapoport SI, Chang MC. Incorporation of [1-carbon-11]palmitate in monkey brain using PET. J Nucl Med 1995; 36:2261–2267.

Axelrod J. Phospholipase A2 and G proteins. Trends Neurosci 1995; 18:64–65.

Axelrod J, Burch RM, Jelsema CL. Receptor-mediated activation of phospholipase A_2 via GTP-binding proteins: arachidonic acid and its metabolites as second messengers. Trends Neurosci 1988; 11:117–123.

Barchas JD, Hamblin MW, Malenka RC. Biochemical hypotheses of mood and anxiety disorders. In: Siegel GJ, Agranoff BW, Albers RW, Molinoff PB, eds. Basic Neurochemistry. 5th ed. Raven, New York, 1994, pp. 979–1001.

Bazán NG. Supply of n-3 polyunsaturated fatty acids and their significance in the central nervous system. In: Wurtman RJ, Wurtman JJ, eds. Nutrition and the Brain Raven, New York, 1990, Vol. 8, pp. 1–24.

Bazán NG, Rodriguez de Turco E. Membrane lipids in the pathogenesis of brain edema: phospholipids and arachidonic acid, the earliest membrane components changed at the onset of ischemia. Adv Neurol 1980; 28:197–205.

Bogdanovic N, Islam A, Nilsson L, Bergström L, Winblad B, Adem Å. Effects of nucleus basalis lesion on muscarinic receptor subtypes. Exp Brain Res 1993; 97:225–232.

Bourre J-M, Francois M, Youyou A, Dumont O, Piciotti M, Pascal G, et al. The effects of dietary a-linolenic acid on the composition of nerve membranes, enzymatic activity, amplitude of electrophysiological parameters, resistance to poisons and performance of learning tasks in rats. J Nutr 1989; 119:1880–1892.

Brecher P, Kuan HT. Lipoprotein lipase and acid lipase activity in rabbit brain microvessels. J Lipid Res 1979; 20:464–471.

Chang MC. J., Arai T, Freed LM, Wakabayashi S, Channing MA, Dunn BB, et al. Brain incorporation of [1-^{11}C]-arachidonate in normocapnic and hypercapnic monkeys, measured with positron emission tomography. Brain Res 1997; 755:74–83.

Chang MC. J., Bell JM, Purdon AD, Chikhale EG, Grange E. Dynamics of docosahexaenoic acid metabolism in the central Neurochem Res 1999; 24:399–406.

Chang MCJ, Connolly C, Hill D, Purdon AD, Hayakawa T, Grimes G, et al. Pharmacokinetics of methyl palmoxirate, an inhibitor of β-oxidation, in rats and humans. Life Sci 1998; 63:PL297–PL302.

Chang MCJ, Grange E, Rabin O, Bell JM. Incorporation of [U-^{14}C]palmitate into rat brain: effect of an inhibitor of β-oxidation. J Lipid Res 1997; 38:295–300.

Chang MCJ, Jones CR. Chronic lithium treatment decreases brain phospholipase A_2 activity. Neurochem Res 1998; 23:887–892.

Chang MCJ, Wakabayashi S, Bell JM. The effect of methyl palmoxirate on incorporation of [U-^{14}C]palmitate into rat brain. Neurochem Res 1994; 19:1217–1223.

Contreras MA, Chang MC. J., Kirkby D, Bell JM, Rapoport SI. Reduced palmitate turnover in brain phospholipids of pentobarbital-anesthetized rats. Neurochem Res 1999; 24:833–841.

Contreras MA, Greiner RS, Chang MC. J., Myers CS, Bell JM, Salem Jr N, et al. Alpha-linolenic acid deprivation alters the in vivo turnover of docosahexaenoic acid in rat brain phospholipids. J Neurochem 1999; 72:S65D.

Cooper JR, Bloom FE, Roth RH. The Biochemical Basis of Neuropharmacology. 7th ed. Oxford University Press, Oxford, 1996.

De Micheli E, Chang MCJ, Rapoport SI. In vivo imaging of cortical membrane remodeling in rats with chronic unilateral ablation of nucleus basalis magnocellularis: use of radiolabeled palmitic acid. Brain Res 1996; 735:36–41.

DeGeorge JJ, Nariai T, Yamazaki S, Williams WM, Rapoport SI. Arecoline-stimulated brain incorporation of intravenously administered fatty acids in unanesthetized rats. J Neurochem 1991; 56:352–355.

DeGeorge JJ, Noronha JG, Bell JM, Robinson P, Rapoport SI. Intravenous injection of [1-^{14}C]arachidonate to examine regional brain lipid metabolism in unanesthetized rats. J Neurosci Res 1989; 24:413–423.

DeGeorge JJ, Ousley AH, McCarthy KD, Lapetina EG, Morell P. Acetylcholine stimulates selective liberation and re-esterification of arachidonate and accumulation of inositol phosphates and glycerophosphoinositol in C62B glioma cells. J Biol Chem 1987; 262:8077–8083.

Dehouck B, Fenart L, Dehouck MP, Pierce A, Torpier G, Cecchelli R. A new function for the LDL receptor: transcytosis of LDL across the blood-brain barrier. J Cell Biol 1997; 138:877–889.

Dennis EA. Diversity of group types, regulation, and function of phospholipase A2. J Biol Chem 1994; 269:13,057–13,060.

Deutsch J, Grange E, Rapoport SI, Purdon AD. Isolation and quantitation of long chain acyl-coenzyme A esters in brain tissue by a solid phase extraction. Anal Biochem 1994; 220:321–323.

Deutsch J, Rapoport SI, Purdon AD. Isolation and HPLC separation of polyunsaturated species of rat brain acyl-CoA produced during decapitation-ischemia. Phosphorus, Sulfur Silicon 1996; 109–110:389–392.

Deutsch J, Rapoport SI, Purdon AD. Relation between free fatty acid and acyl-CoA concentrations in rat brain following decapitation. Neurochem Res 1997; 22:759–765.

Farooqui AA, Rapoport SI, Horrocks LA. Membrane phospholipid alterations in Alzheimer's disease: deficiency of ethanolamine plasmalogens. Neurochem Res 1997; 22:523–527.

Farooqui AA, Yang H-C, Rosenberger TH, Horrocks LA. Phospholipase A2 and its role in brain tissue. J Neurochem 1997; 69:889–901.

Fisher SK, Agranoff BW. Receptor activation and inositol lipid hydrolysis in neural tissues. J Neurochem 1987; 48:999–1017.

Ginsberg L, Rafique S, Xuereb JH, Rapoport SI, Gershfeld NL. Disease and anatomic specificity of ethanolamine plasmalogen deficiency in Alzheimer's disease brain. Brain Res 1995; 698:223–226.

Glatz JF. The role of membrane-associated proteins in cellular fatty acid uptake. J Mol Neurosci, in press.

Gossett RE, Frolov AA, Roths JB, Behnke WD, Kier AB, Schroeder F. Acyl-CoA binding proteins: multiplicity and function. Lipids 1996; 31:895–918.

Graham WC, Crossman AR, Woodruff GN. Autoradiographic studies in animal models of hemi-parkisonism reveal dopamine D2 but not D1 receptor supersensitivity. I. 6-OHDA lesions of ascending mesencephalic dopaminergic pathways in the rat. Brain Res 1990; 514:93–102.

Grange E, Deutsch J, Smith QR, Chang M, Rapoport SI, Purdon AD. Specific activity of brain palmitoyl-CoA pool provides rates of incorporation of palmitate in brain phospholipids in awake rats. J Neurochem 1995; 65:2290–2298.

Grange E, Rabin O, Bell JM, Rapoport SI, Chang MCJ. Manoalide, a phospholipase A_2 inhibitor, inhibits arachidonate incorporation and turnover in brain phospholipids of the awake rat. Neurochem Res 1998; 23:1251–1257.

Greville GD, Tubbs PK. The catabolism of long-chain fatty acids in mammalian tissues. Essays Biochem 1968; 4:155–212.

Guttman M. Dopamine receptors in Parkinson's disease. Neurol Clin 1992; 10:377–386.

Hayakawa T, Chang M, Bell J, Seemann R, Rapoport SI, Appel NM. Effect of D2 dopamine receptor activation on [^{3}H]arachidonic acid incorporation in rats with unilateral 6-hydroxydopamine lesions. Soc Neurosci Abstr 1997; 23:2432.

Hayakawa T, Chang MCJ, Bell JM, Seemann R, Rapoport SI, Appel NM. Fatty acid incorporation depicts brain activity in a rat model of Parkinson's disease. Brain Res 1998; 807:177–181.

Hibbeln J. Fish consumption and major depression. Lancet 1998; 351:1213.

Holmes TC, Nitsch RM, Erfurth A, Wurtman RJ. Phospholipid and phospholipid metabolites in rat frontal cortex are decreased following nucleus basalis lesions. Ann NY Acad Sci 1993; 695:241–243.

Horrocks LA, Yeo YK. Health benefits of docosahexaenoic acid. Pharmacol Res 1999; 40:211–225.

Jones CR, Arai T, Bell JM, Rapoport SI. Preferential in vivo incorporation of [^{3}H]arachidonic acid from blood into rat brain synaptosomal fractions before and after cholinergic stimulation. J Neurochem 1996; 67:822–829.

Jones CR, Arai T, Rapoport SI. Evidence for the involvement of docosahexaenoic acid in cholinergic stimulated signal transduction at the synapse. Neurochem Res 1997; 22:663–670.

Kishimoto K, Matsumura K, Kataoka Y, Morii H, Watanabe Y. Localization of cytosolic phospholipase A2 messenger RNA mainly in neurons in the rat brain. Neuroscience 1999; 92:1061–1077.

Kramer RM, Sharp JD. Structure, function and regulation of Ca^{2+}-sensitive cytosolic phospholipase A2 (cPLA2). FEBS Lett 1997; 410:49–53.

Liang S, Furhman S, Somogyi R. Reveal, a general reverse engineering algorithm for inference of genetic network architectures. Pacific Symposium on Biocomputers 1998, pp. 18–29.

Luiken JJ, Schaap FG, van Nieuwenhoven FA, van der Vusse GJ, Bonen A, Glatz JF. Cellular fatty acid transport in heart and skeletal muscle as facilitated by proteins. Lipids 1999; 34:S169–S175.

Mahadevappa VG, Holub BJ. Diacylglycerol lipase pathway is a minor source of released arachidonic acid in thrombin-stimulated human platelets. Biochem Biophys Res Commun 1986; 134:1327–1333.

Maiese K, Holloway HH, Larson DM, Soncrant TT. Effect of acute and chronic arecoline treatment on cerebral metabolism and blood flow in the conscious rat. Brain Res 1994; 28(641), 65–75.

Marshall E, Fumagalli R, Niemiro R, Paoletti R. The change in fatty acid composition of rat brain phospholipids during development. J Neurochem 1966; 13:857–862.

Méresse S, Delbart C, Fruchart J-C, Cecchelli R. Low-density lipoprotein receptor on endothelium of brain capillaries. J Neurochem 1989; 53:340–345.

Mikkelsen J, Knudsen J. Acyl-CoA-binding protein from cow. Binding characteristics and cellular and tissue distribution. Biochem J 1987; 248:709–714.

Miller JC, Gnaedinger JM, Rapoport SI. Utilization of plasma fatty acid in rat brain: distribution of [14-C]palmitate between oxidative and synthetic pathways. J Neurochem 1987; 49:1507–1514.

Moriyama T, Urade R, Kito M. Purification and characterization of diacylglycerol lipase from human platelets. J Biochem 1999; 125:1077–1085.

Murphy E. Personal communication, 1998

Murphy EJ, Rosenberger TA, Patrick CB, Rapoport SI. Intravenously injected [1-^{14}C]arachidonic acid targets phospholipids and [1-^{14}C]palmitic acid targets neutral lipids in heart of awake rats. Lipids 2000; 35(8):891–898.

Murphy S, Jeremy J, Pearce B, Dandona P. Eicosanoid synthesis and release from primary cultures of rat central nervous system astrocytes and meningeal cells. Neurosci Lett 1985; 61:61–65.

Nariai T, DeGeorge JJ, Greig NH, Genka S, Rapoport SI, Purdon AD. Differences in rates of incorporation of intravenously injected radiolabeled fatty acids into phospholipids of intracerebrally implanted tumor and brain in awake rats. Clin Exp Metastasis 1994; 12:213–225.

Nariai T, DeGeorge JJ, Greig NH, Rapoport SI. In vivo incorporation of [9,10-^{3}H]palmitate into a rat metastatic brain-tumor model. J Neurosurg 1991; 74:643–649.

Nariai T, DeGeorge JJ, Lamour Y, Rapoport SI. In vivo brain incorporation of [1-^{14}C]arachidonate in awake rats, with or without cholinergic stimulation, following unilateral lesioning of nucleus basalis magnocellularis. Brain Res 1991; 559:1–9.

Noronha JG, Bell JM, Rapoport SI. Quantitative brain autoradiography of [9,10-^{3}H]palmitic acid incorporation into brain lipids. J Neurosci Res 1990; 26:196–208.

Noronha JG, Larson DM, Rapoport SI. Regional cerebral incorporation of plasma [14-C]palmitate, and cerebral glucose utilization in water-deprived Long–Evans and Brattleboro rats. Exp Neurol 1989; 103:267–276.

Osmundsen H, Cervenka J, Bremer J. A role for 2,4-enoyl-CoA reductase in mitochondrial β-oxidation of polyunsaturated fatty acids. Effects of treatment with clofibrate on oxidation of polyunsaturated acylcarnitines by isolated rat liver. Biochem J 1982; 208:749–757.

Pardridge WM, Mietus LJ. Palmitate and cholesterol transport through the blood-brain barrier. J Neurochem 1980; 34:463–466.

Pawlosky RJ, Salem N Jr. Ethanol exposure causes a decrease in docosahexaenoic acid and an increase in docosapentaenoic acid in feline brains and retinas. Am J Clin Nutr 1995; 61:1284–1289.

Pete MJ, Ross AH, Exton JH. Purification and properties of phospholipase A1 from bovine brain. J Biol Chem 1994; 269:19,494–19,500.

Pettegrew JW, Moossy J, Withers G, McKeag D, Panchalingam K. ^{31}P nuclear magnetic resonance study of the brain in Alzheimer's disease. J Neuropathol 1988; 47:235–248.

Porcellati G, Goracci G, Arienti G. Lipid turnover. In: Lajtha A, ed. Handbook of Neurochemistry. Plenum, New York, 1983, Vol. 5, pp. 277–294.

Purdon AD, Rapoport SI. Energy requirements for two aspects of phospholipid metabolism in mammalian brain. Biochem J 1998; 335:313–318.

Purdon D, Arai T, Rapoport SI. No evidence for direct incorporation of esterified palmitic acid from plasma into brain lipids of awake adult rat. J Lipid Res 1997; 38:526–530.

Rabin O, Deutsch J, Grange E, Pettigrew KD, Chang MC. J., Rapoport SI, et al. Changes in cerebral acyl-CoA concentrations following ischemia–reperfusion in awake gerbils. J Neurochem 1997; 68: 2111–2118.

Rapoport SI. Anatomic and functional brain imaging in Alzheimer's disease. In: Bloom FE, Kupfer DJ, eds. Psychopharmacology: The Fourth Generation of Progress. Raven, New York, 1995, pp. 1401–1415.

Rapoport SI. Investigative New Drug Application #46,139: Methyl palmoxirate. Food and Drug Administration, Department of Health and Human Services, Rockville, MD, 1999.

Rapoport SI. In vivo fatty acid incorporation into brain phospholipids in relation to plasma availability, signal transduction and membrane remodeling. J Mol Neurosci, in press.

Rapoport SI, Chang MC, Connolly K, Carson R, Eckelman WC. In vivo brain imaging of signal transduction using [^{11}C]arachidonic acid and positron emission tomography. Soc Neurosci Abstr 1999; 25(Part 1):1145.

Rapoport SI, Chang MC, Connolly K, Kessler D, Bokde A, Carson RE, et al. In vivo imaging of phospholipase A2-mediated signaling in human brain using [11C]arachidonic acid and positron emission tomography (PET). J Neurochem 2000; 74:S21.

Rapoport SI, Purdon D, Shetty HU, Grange E, Smith Q, Jones C, et al. In vivo imaging of fatty acid incorporation into brain to examine signal transduction and neuroplasticity involving phospholipids. Ann NY Acad Sci 1997; 820:56–74.

Rapoport SI, Spector AA. Delivery and turnover of plasma-derived essential polyunsaturated fatty acids in mammalian brain, submitted.

Reddy S, Rao GH, Murthy M. Differential effects of phorbol 12-myristate 13-acetate and diacylglycerols on thromboxane A_2-independent phospholipase A_2 activation in collagen-stimulated human platelets. Biochem Med Metab Biol 1994; 51:118–128.

Reddy TS, Bazan NG. Arachidonic acid, stearic acid, and diacylglycerol accumulation correlates with the loss of phosphatidylinositol 4,5-bisphosphate in cerebrum 2 seconds after electroconvulsive shock: complete reversion of changes 5 minutes after stimulation. J Neurosci Res 1987; 18:449–455.

Rintala J, Seemann R, Chandrasekaran K, Rosenberger TA, Chang L, Contreras MA, et al. An arachidonic acid-specific 85 kDa cytosolic phospholipase A2 is a target for chronic lithium in rat brain. Neuroreport 1999; 10:3887–3890.

Robinson PJ, Noronha J, DeGeorge JJ, Freed LM, Nariai T, Rapoport SI. A quantitative method for measuring regional in vivo fatty-acid incorporation into and turnover within brain phospholipids: review and critical analysis. Brain Res Reviews 1992; 17:187–214.

Schelbert HR, Henze E, Sochor H, Grossman RG, Huang SC, Barrio JR, et al. Effects of substrate availability on myocardial C-11 palmitate kinetics by positron emission tomography in normal subjects and patients with ventricular dysfunction. Am Heart J 1986; 111:1055–1064.

Scherman D, Desnos C, Darchen F, Pollak P, Javoy-Agid F, Agid Y. Striatal dopamine deficiency in Parkinson's disease: role of aging. Ann Neurol 1989; 26:551–557.

Sellner P, Chu W, Glatz J, Berman NE. Developmental role of fatty acid-binding proteins in mouse brain. Brain Res Developmental Brain Res 1995; 89:33–46.

Semenkovich CF. Regulation of fatty acid synthase (FAS). Prog Brain Res 1997; 36:43–53.

Shafrir E, Gatt S, Khasis S. Partition of fatty acids of 20–24 carbon atoms between serum albumin and lipoproteins. Biochim Biophys Acta 1965; 98:365–371.

Shetty HU, Smith QR, Washizaki K, Rapoport SI, Purdon AD. Identification of two molecular species of rat brain phosphatidylcholine that rapidly incorporate and turn over arachidonic acid in vivo. J Neurochem 1996; 67:1702–1710.

Sigiura T, Kudo N, Ojima T, Mabuchi-Itoh K, Yamashita A, Waku K. Coenzyme A-dependent cleavage of membrane phospholipids in several rat tissues: ATP-independent acyl-CoA synthesis and the generation of lysophospholipids. Biochim Biophys Acta 1995; 1255:167–176.

Spector AA. Plasma free fatty acid and lipoproteins as sources of polyunsaturated fatty acid in brain. J Mol Neurosci, in press.

Staufenbiel, M. (1988). Fatty acids covalently bound to erythrocyte proteins undergo a differential turnover in vivo. J Biol Chem, 263:13,615–13,622.

Stephenson DT, Manetta JV, White DL, Chiou XG, Cox L, Gitter B, et al. Calcium-sensitive cytosolic phospholipase A_2 (cPLA$_2$) is expressed in human brain astrocytes. Brain Res 1994; 637:97–105.

Stoll AL, Severus WE, Freeman MP, Rueter S, Zboyan HA, Diamond E, et al. Omega 3 fatty acids in bipolar disorder: a preliminary double-blind, placebo-controlled trial. Arch Genl Psychiatry 1999; 56:407–412.

Sun GY, MacQuarrie RA. Deacylation–reacylation of arachidonoyl groups in cerebral phospholipids. Ann NY Acad Sci 1989; 559:37–55.

Tutwiler GF, Ho W, Mohrbacher RJ. 2-Tetradecylglycidic acid. Methods Enzymol 1981; 72:533–551.

Washizaki K, Smith QR, Rapoport SI, Purdon AD. Brain arachidonic acid incorporation and precursor pool specific activity during intravenous infusion of unesterified [^{3}H]arachidonate in the anesthetized rat. J Neurochem 1994; 63:727–736.

Watkins PA. Fatty acid activation. Prog Lipid Res 1997; 36:55–83.

Wells MA, Dittmer JC. A comprehensive study of the postnatal changes in the concentration of the lipids of developing rat brain. Biochemistry 1967; 6:3169–3175.

Whitehouse PJ, Price DL, Struble RG, Clark AW, Coyle JT, DeLong MR. Alzheimer's disease and senile dementia: loss of neurons in the basal forebrain. Science 1982; 215:1237–1239.

Wolfe LS, Horrocks LA. Eicosanoids. In: Siegel GJ, Agranoff BW, Albers RW, Molinoff PB, eds. Basic Neurochemistry, 5th ed. Raven, New York, 1994, pp. 475–490.

Wosilait WD, Soler-Argilaga C. A theoretical analysis of multiple binding of palmitate by bovine serum albumin: the relationship to uptake of free fatty acids by tissues. Life Sci 1975; 17:159–166.

Yamashita A, Sugiura T, Waku K. Acyltransferases and transacylases involved in fatty acid remodeling of phospholipids and metabolism of bioactive lipids in mammalian cells. J Biochem (Tokyo) 1997; 122:1–16.

Yamazaki S, DeGeorge JJ, Bell JM, Rapoport SI. Effect of pentobarbital on incorporation of plasma palmitate into rat brain. Anesthesiology 1994; 80:151–158.

Yamazaki S, Noronha JG, Bell JM, Rapoport SI. Incorporation of plasma [^{14}C]palmitate into the hypoglossal nucleus following unilateral axotomy of the hypoglossal nerve in adult rat, with and without regeneration. Brain Res 1989; 477:19–28.

9 Carbon Recycling
An Important Pathway in α-Linolenate Metabolism in Fetuses and Neonates

Stephen C. Cunnane

1. INTRODUCTION

The aim of this chapter is to review the evidence for carbon recycling as a significant pathway in α-linolenate (18:3n-3) metabolism in the fetal and neonatal period. β-Oxidation is generally thought of as the complete utilization of an organic compound as a fuel. However, β-oxidation also provides carbon for acetyl-CoA units that are recovered in other newly synthesized metabolites. In the case of α-linolenate, this "carbon recycling" appears to be directed primarily toward *de novo* lipogenesis which is of interest from two points of view.

First, several reports over the last 25 yr show that carbon recycling from α-linolenate in neonatal rats and near-term primate fetuses consumes the majority of α-linolenate that is not used as a fuel. Second, α-linolenate cannot be synthesized *de novo* by mammals and is the main dietary precursor to the n-3 long-chain polyunsaturated fatty acids (LC-PUFAs), eicosapentaenoate (20:5n-3) and docosahexaenoate (22:6n-3), which have important regulatory and membrane functions, respectively. Thus, α-linolenate is a vitamin-like nutrient that would not be expected to be easily β-oxidized or extensively carbon recycled into other lipids when it already has an important precursor role for membrane components needed for normal development of the visual and nervous systems.

There are numerous other substrates for lipid synthesis; therefore, as a whole, the importance of β-oxidation and carbon recycling of α-linolenate lies in the significant negative effect these processes have on the availability of α-linolenate for esterification or conversion to n-3 LC-PUFA. The health attributes of α-linolenate in young mammals are most commonly linked to its desaturation–chain elongation to n-3 LC-PUFA, principally docosahexaenoate. In adults, α-linolenate may have biological activity without conversion to n-3 LC-PUFA. Such activity appears to include reducing the risk of cancer, attenuating some cardiovascular risk factors and modulating the immune system (Cunnane, 1995). However, the majority of α-linolenate's effects on the early development of the visual and nervous systems appear to occur through conversion to docosahexaenoate. Most infants consuming a milk formula still receive α-linolenate as the only n-3 PUFA. Thus, beyond what they are born with, they are required to synthesize all the n-3 LC-PUFA that will be incorporated into the developing tissues until they start to consume foods containing docosahexaenoate.

From: *Fatty Acids: Physiological and Behavioral Functions*
Edited by: D. Mostofsky, S. Yehuda, and N. Salem Jr. © Humana Press Inc., Totowa, NJ

Therefore, it is important to know the bioavailability of α-linolenate (i.e., the portion of consumed α-linolenate that is actually available for desaturation–chain elongation). It is also important to understand the factors that are most likely to influence the partitioning of dietary α-linolenate toward β-oxidation or carbon recycling compared to retention in the body as an n-3 PUFA. This chapter will provide evidence that β-oxidation and carbon recycling consume the majority of dietary α-linolenate in two models of mammalian fetal/neonatal development.

2. BIOAVAILABILITY OF α-LINOLENATE

Primarily through enzyme assays, dietary supplementation trials and in vivo tracer studies, much has been learned about the metabolic steps involved in the desaturation–chain elongation pathway converting α-linolenate through the main intermediates, eicosapentaenoate and docosapentaenoate (22:5n-3), to docosahexaenoate. The desaturation–chain elongation steps occur in the endoplasmic reticulum and are part of the electron transport chain. In addition to desaturation–chain elongation, it now appears that peroxisomal chain shortening of 24 carbon n-3 LC-PUFA is required to form docosahexaenoate in mammalian cells. The desaturases, in particular, are dependent on several mineral cofactors (iron, zinc, copper), whereas chain elongation appears to depend on pyridoxine. Hence, beyond the need for an adequate dietary supply of α-linolenate, normal fat absorption, and competition from n-6 PUFA at many levels, adequate dietary availability of all these cofactor nutrients plays an important role in sustaining adequate synthesis of n-3 LC-PUFA (Ackman & Cunnane, 1991).

The capacity to convert α-linolenate to docosahexaenoate also varies with age, between species, during pregnancy and lactation, nutritional status, and some degenerative as well as inherited diseases. Recent studies suggest that pressure from β-oxidation during energy deficit or undernutrition has at least as great a negative effect on the bioavailability of α-linolenate as all of these other variables (*see* Section 3). Synthesis from the small amounts of dietary hexadecatrienoate has a modest positive effect on availability of α-linolenate (Cunnane et al, 1995). However, β-oxidation probably plays a greater role than almost any other factor in determining α-linolenate availability for n-3 LC-PUFA synthesis during early development because β-oxidation followed by carbon recycling into *de novo* lipid synthesis consumes significant amounts of dietary α-linolenate in neonatal rats and fetal primates. Nutritional conditions that promote fatty acid β-oxidation for energy are usually considered to exclude simultaneous fatty acid synthesis, but this is not the case with α-linolenate, at least in the neonatal period.

3. β-OXIDATION UNDER NORMAL CONDITIONS

Carbon recycling depends on the availability of acetyl-CoA derived from the β-oxidized precursor—in this case, α-linolenate. Thus, for carbon recycling to be a significant sink for the carbon skeleton of α-linolenate, β-oxidation has to be an important route of α-linolenate metabolism. In support of this claim, many different experimental models using both infants and adults demonstrate that α-linolenate is readily β-oxidized (Tables 1–3). These models either depend on a tracer (typically ^{14}C, ^{13}C) to measure the degree of oxidation in vivo or in vitro (Leyton et al. 1987; Clouet et al. 1989; Gavino & Gavino, 1991; Emmison et al. 1995; McCloy et al. 2000), or they employ whole body fatty acid balance methodology (Chen & Cunnane, 1993; Cunnane et al. 1993; Chen et al. 1995; Cunnane & Anderson, 1997a).

Table 1
Comparison of Whole Body (Leyton et al., 1987) and Mitochondrial
(Gavino & Gavino [1991], Clouet et al [1989]) β-Oxidation of Long-Chain Fatty Acids

	Gavino and Gavino[a] (1991)	Clouet et al.[b] (19889)	Leyton et al.[c] (1987)
Palmitate	4.6	—	32
Stearate	1.9	—	23
Oleate	4.4	4.3	59
Linoleate	5.0	5.9	50
α-Linolenate	7.1	6.7	65

[a] Acylcarnitine production (nmol/mg/min).
[b] β-Oxidation to acid soluble products (nmol/mg/min).
[c] Percent dose β-oxidized in 24 h.

3.1. Tracer Methodology

Tracer methodology exploits the ability to monitor the metabolism of individual precursors passing through intermediate stages and products in specific pathways. By comparing different labeled precursors, one can evaluate the relative rates of β-oxidation of different fatty acids, often in specific cell types or subcellular compartments. The advantage is that the analysis is usually specific to the labeled compounds under study. Two general approaches to comparatively assessing α-linolenate using tracer methodology have been used; in vivo measurement of CO_2 production and monitoring of reaction products, either acyl carnitines or acid-soluble products such as ketones.

Leyton et al. (1987) used young rats (23–26 d old) and recovered $^{14}CO_2$ from breath in a metabolic chamber at various time-points up to 24 h after oral dosing of 5–6 µCi of tracer in 0.2 mL olive oil. They found that 65% of the α-linolenate dose was completely oxidized within 24 h, an amount two to three times greater than for palmitate or stearate (Table 1). The same amount of tracer was given regardless of the amount typically found in the diet or in body lipids. This inequivalent dilution of the tracer potentially confounded the interpretation of the data. However, oxidation of oleate was closest to that of α-linolenate, yet there is 10- to 20-fold more oleate in tissue lipids. In agreement with Leyton et al. (1987), we have also found that ^{13}C-labeled α-linolenate and oleate are similarly β-oxidized in healthy women (McCloy et al. 2000). Thus, the effects of differential tracer dilution do not necessarily affect the rank order of β-oxidation of these long-chain fatty acids.

Recovery of radiolabeled acylcarnitines after incubation of isolated rat liver mitochondria was greatest for α-linolenoyl CoA than for various other long-chain acyl-CoAs (Gavino & Gavino, 1991; Table 1). Their data are in agreement with those of Clouet et al. (1989) and Emmison et al. (1995), both of whom evaluated recovery of β-oxidized tracer on the basis of its appearance in acid-extractable products, mainly ketones (Tables 1 and 2).

These in vitro and in vivo tracer studies all demonstrate relatively high β-oxidation of α-linolenate compared to other long-chain fatty acids commonly present in the diet. They therefore suggest a potentially high availability of carbon from α-linolenate for recycling into lipid synthesis. In the study by Emmison et al. (1995), β-oxidation of 18 carbon fatty acids was threefold to fourfold higher in the early postnatal period compared to that in adults, pointing to the neonatal period as time during which carbon recycling of α-linolenate could be more active than in adults (Table 2).

Table 2
Recovery of β-Oxidized Carbon in Acid-Soluble β-Oxidation Products, Primarily Ketones, from α-Linolenate Compared to Linoleate and Oleate

	Day 5	Day 11	Day 21	Adult
Oleate	100	72	52	23
Linoleate	172	119	82	34
α-Linolenate	251	158	114	80

Note: The data are from primary cultures of rat hepatocytes and use the values for oleate oxidation in 5-d-old rats as a reference (100%).
Source: Adapted from Emmison et al (1995).

3.2. Whole-Body Fatty-Acid-Balance Methodology

Whole-body fatty-acid-balance methodology exploits the fact that in the absence of other n-3 PUFA, the diet is the only source of α-linolenate besides what is already present in the body. By comparing α-linolenate intake to accumulation of the sum of all n-3 PUFA determined over a balance period of several days to several weeks (deducting excretion and conversion to n-3 LC-PUFA), one can calculate α-linolenate disappearance or β-oxidation by difference. The advantages of the whole-body balance method are that (1) it reflects metabolism in the whole body rather than in an isolated tissue or subcellular compartment and (2) during energy deficit, this method alone can provide an estimate of α-linolenate oxidation derived from existing body stores as well as from the diet.

We developed the whole-body fatty-acid-balance method specifically as an alternative to tracer methodology so as to confirm the net availability of α-linolenate for conversion to docosahexaenoate. Most of our work with this method has been applied to establishing how nutritional perturbations affect α-linolenate β-oxidation during periods of increased nutrient demand such as pregnancy (see Section 4). However, two of our studies reported data for young, healthy, growing rats during balance period of 16–26 d. In the first study, rats consuming 0.3% energy as α-linolenate β-oxidized 78% of the α-linolenate intake (Chen et al. 1996). In the second study, rats consuming 1.8% energy as α-linolenate oxidized 84% of this higher level of α-linolenate intake (Cunnane & Anderson, 1997a).

These tracer and whole-body balance β-oxidation studies demonstrate three important points. First, several very different models show that relative to the other 16–20 carbon fatty acids commonly present in the human diet, α-linolenate appears to be the most easily β-oxidized. Most of these tracer studies were qualitative and did not clarify the proportion of normal dietary fatty acid or tracer intake that was β-oxidized. Hence, they did not establish the bioavailability of α-linolenate. Second, in contrast to the tracer data, the balance data are useful to quantify bioavailability of α-linolenate. There is broad agreement between tracer and balance data that upward of 65% of dietary α-linolenate is β-oxidized under normal conditions of dietary energy sufficiency, growth, health, and with a usual range of α-linolenate intake (Leyton et al, 1987; Chen e al, 1995; Cunnane and Anderson, 1997a). β-Oxidation of α-linolenate maybe somewhat less in suckling rats (50–55%; Sinclair, 1975) and in pregnancy (Table 3). The clear conclusion is that, despite its apparent nutritional importance, most α-linolenate is normally completely β-oxidized for energy even in rapidly growing young animals with a relatively high requirement for all nutrients.

Table 3
Adverse Influence of Acute (48 h Fasting During Pregnancy) and Chronic
Undernutrition (Pregnant and Zinc Deficient, or Weight Cycled) on β-Oxidation
of α-Linolenate in Rats as Determined by the Whole-Body Fatty-Acid-Balance Method

	Intake of α-linolenate (g)	Accumulation of n-3 PUFAs (g)	Disappearance of α-linolenate[a] (g)	(% intake)
Pregnant rats[b]:				
Control	308	100	208	68
48 h Fasted	225	−95	315	140
Pregnant rats[c]:				
Control	1173	597	576	49
Zinc deficient	715	−265	980	137
Weight Cycled[d]:				
Control	483	108	375	78
Weight-cycled	351	40	311	89

[a]Disappearance excludes losses via excretion
[b]Data from Chen and Cunnane (1993)
[c]Data from Cunnane et al (1993)
[d]Data from Chen et al (1995)

4. β-OXIDATION DURING ENERGY DEFICIT

Large proportions of the world's population do not have sufficient energy intake. Among those that usually have access to adequate food, many transiently but repetitively choose to restrict food intake through dieting. Our whole-body fatty-acid-balance studies in rats show that both short- and long-term dietary energy deficits rapidly and significantly increase β-oxidation of α-linolenate (Table 3).

We first reported whole-body balance data for α-linolenate oxidation using pregnant rats undergoing a 48-h fast at d 13–15 of pregnancy (Chen & Cunnane, 1993). This acute period of fasting increased α-linolenate oxidation from 68% to 142% of intake, resulting in a net whole-body loss of α-linolenate, which was not recovered during 6 d refeeding. A similar outcome including whole-body loss of α-linolenate was observed with more chronic undernutrition caused by zinc deficiency during pregnancy (Cunnane et al. 1993; Table 3). Hence, nutritional insults during a period of increased need for nutrients (pregnancy) are capable of preventing α-linolenate accumulation and actually depleting whole-body stores of α-linolenate and n-3 LC-PUFA. Weight cycling in young rats impairs α-linolenate accumulation and doubles its β-oxidation, this despite adequate α-linolenate intake (Chen et al. 1995).

Collectively, these studies demonstrate several important points about the β-oxidation of α-linolenate. First, like other long-chain dietary fatty acids, α-linolenate is readily used as a fuel. Second, acute or chronic undernutrition depletes body stores of α-linolenate regardless of the adequacy of α-linolenate intake. Third, when fasting or chronic undernutrition accompanies a condition of higher energy and nutrient demand such as pregnancy, β-oxidation of α-linolenate can exceed intake by twofold to fourfold in rats, thereby eliminating all α-linolenate consumed and also significantly depleting α-linolenate body stores (Cunnane et al, 1993; Table 3). Fourth, in these short-term studies, reduced accu-

Table 4
Desaturation–Chain Elongation Compared to Carbon Recycling of Labeled
α-Linolenate into Fatty Acids Synthesized *De Novo* in 16- to 17-d-Old
(Sinclair, 1975) or 8-d-Old (Menard et al. 1998) Suckling Rat Pups

		Tracer recovered as		
	α-Linolenate	n-3 LC-PUFAs	Sat/Mono[a]	Cholesterol
Sinclair (1975)	2	16	50	26
Menard et al. (1998)	nd[b]	12	39	49

Note: Data are given as percent administered dose recovered 48 h later in the main lipid components of brain total lipids.

[a]Sat/Mono: Tracer recovered in palmitate, stearate, and oleate combined (Sinclair, 1975) or in palmitate only (Menard et al. 1998).
[b]Not detected.

mulation and/or the frank depletion of α-linolenate from the body during energy deficit appears to exceed that which can be achieved by dietary n-3 PUFA deficiency alone, implying that there is normally no obvious means of preserving body stores of α-linolenate. Finally, although this experiment has not yet been done in α-linolenate deficiency, severe linoleate deficiency does not prevent linoleate β-oxidation (Cunnane & Anderson 1997b), suggesting that this process is probably an obligatory part of 18-carbon PUFA metabolism.

In summary, these studies show that β-oxidation of α-linolenate for energy can take precedence over all other known pathways of its metabolism, including the increased need for docosahexaenoate during pregnancy. Consequently, energy sufficiency is a probably the main prerequisite for normal rates of desaturation–chain elongation of α-linolenate to docosahexaenoate. However, energy sufficiency alone does not prevent the normal and, apparently, obligatory β-oxidation of α-linolenate. Furthermore, β-oxidation of α-linolenate is easily capable of supplying sufficient carbon to account for the amount of carbon recycling that has been reported from α-linolenate.

5. CARBON RECYCLING OF α-LINOLENATE

5.1. Radiotracer Studies in the Suckling Rat

Sinclair (1975) appears to have been the first to report the extensive carbon recycling from α-linolenate into lipids synthesized *de novo* in suckling rats (Table 4). In that study, various radiotracer-labeled PUFA were separately given orally to 16- to 17-d-old suckling rat pups. The total lipids of brain and liver were subsequently extracted and the tracer incorporation into PUFA versus incorporation into cholesterol and saturated and monounsaturated fatty acids was determined for these two tissues. Twenty-two hours after dosing with ^{14}C-α-linolenate, 26% of the label in brain lipids was in cholesterol. Of the label recovered in brain fatty acids, 65–70% was in saturated and monounsaturated fatty acids within 48 h (Sinclair, 1975). In total, however, only 0.3% of the labeled α-linolenate was found in brain lipids 48 h after dosing, whereas 49% was completely β-oxidized. At about the same time, a similar result showing substantial amounts of carbon recycling from α-linolenate was obtained in a similar experiment in suckling rats (Dhopeswarkar & Subramanian, 1975).

These radiotracer studies used a very low mass of tracer; therefore, it is likely that the tracer represented the normal metabolism of α-linolenate and would probably not have overloaded any steps leading to docosahexaenoate synthesis, thereby potentially skewing the tracer's metabolism toward β-oxidation or carbon recycling. Furthermore, these were healthy, suckling rat pups; thus, nutritional circumstances were also likely to reflect normal α-linolenate metabolism. The influence of dietary docosahexaenoate (present in the milk) in these studies may well have increased carbon recycling from α-linolenate, but this was not assessed in these studies. Thus, under normal circumstances, carbon recycling appears to consume the majority of α-linolenate appearing in lipids of the suckling rodent brain. Apart from these two reports, this topic received little or no attention until the 1990s.

5.2. NMR Spectroscopy Studies in the Suckling Rat

We have used in vivo ^{13}C-NMR nuclear magnetic resonance spectroscopy to trace the metabolism of PUFA in several different models. In an in vivo ^{13}C-NMR spectroscopy experiment designed to noninvasively evaluate the metabolism of uniformly carbon-13 [U-^{13}C]-α-linolenate, we observed that NMR signal enrichment in the brain occurred not only in the double bonds of PUFA (as expected if docosahexaenoate were being synthesized) but also in areas of the spectrum occupied by signals from saturated acyl and cholesterol carbons (Cunnane et al, 1994). Closer inspection by in vitro NMR spectroscopy of the extracted brain lipids showed a specific pattern of ^{13}C enrichment of apparently unbroken carbon pairs derived directly from α-linolenate in brain cholesterol of suckling rats given an oral dose of a mixture of [U-^{13}C]-α-linolenate, linoleate, and their respective 16-carbon PUFA precursors. These NMR data were confirmed and quantified by isotope-ratio mass spectrometry. They showed that, in suckling rats, 16- and 18-carbon n-6 and n-3 PUFAs are readily β-oxidized to acetate, which condenses to form acetoacetate, then mevalonate, and, finally, brain cholesterol (Likhodii & Cunnane 1995). At that time, we did not establish the contribution of α-linolenate itself to carbon recycling and *de novo* lipogenesis; this was subsequently done by gas chromatography–isotope-ratio mass spectrometry (GC–IRMS).

5.3. GC–IRMS Studies in the Suckling Rat

Our NMR studies confirmed earlier radiotracer reports (Sinclair, 1975; Dhopeswarkar & Subramanian, 1975) that α-linolenate could be readily used by suckling rats for brain cholesterol synthesis, but they did not establish the quantitative extent to which this occurred for α-linolenate itself. We selected GC–IRMS for this purpose primarily because a quantitative comparison could be made for incorporation of ^{13}C derived from α-linolenate into docosahexaenoate and individual lipids synthesized *de novo*.

Suckling rats were given an oral dose of 3.3 mg [U-^{13}C]-α-linolenate at 10 d old and were killed 3 h to 30 d later. The ^{13}C enrichment was quantitatively determined in palmitate, docosahexaenoate, and cholesterol of liver, gut, and brain (Menard et al., 1998). The tracer dose was higher than for a radiotracer but was within physiological limits, thereby avoiding the possibility of overloading any of the pathways under investigation. No [U-^{13}C]-α-linolenate itself was found in the brain at any time-point, but enrichment in several n-3 PUFA, especially docosahexaenoate, was detected. Many saturated fatty acids in the brain were also labeled as was brain cholesterol (Table 4).

"Atom % excess" is the commonest notation for expressing ^{13}C enrichment. Similar atom % excess values were seen in liver and brain docosahexaenoate and cholesterol, qualitatively demonstrating substantial ^{13}C incorporation into lipids synthesized *de novo*. Atom % excess data do not take into account the actual concentrations of these lipids in an organ nor do they account for the size of each lipid pool under study, both of which had to be considered to determine the percentage of the administered tracer dose that was used in these pathways. When expressed as microgram of ^{13}C in each lipid pool (docosahexaenoate, palmitate, or cholesterol)/whole brain or per milligram of lipid in the brain, peak carbon recycling of α-linolenate into brain *de novo* lipogenesis as represented by palmitate and cholesterol labeling accounted for 20–40 times more carbon from ^{13}C-α-linolenate than did brain conversion to docosahexaenoate. Combined with the data of Sinclair (1975), these studies show that 48 h after the tracer was given, its recovery in newly synthesized lipids exceeded its conversion to docosahexaenoate by fivefold to sevenfold (Table 4).

We have recently extended the observations on the dominance of carbon recycling as a route of α-linolenate metabolism in the brain to the suckling rat lung in which the data are similar to those obtained for the brain (Cunnane, unpublished data). Our report (Menard et al. 1998) extends those of Sinclair (1975) and Dhopeswarkar and Subramanian (1975) showing that under physiological conditions in the suckling rat, carbon recycling appears to be quantitatively much more important as a route of α-linolenate utilization than desaturation–chain elongation to docosahexaenoate. When the proportion of α-linolenate that is β-oxidized to CO_2 is taken into account (51%; Sinclair, 1975), the loss of α-linolenate to β-oxidation and carbon recycling normally exceeds that going to docosahexaenoate by about 11-fold.

These data do not prove that sufficient docosahexaenoate cannot be made from α-linolenate. Rather, they establish that <10% of dietary α-linolenate is *normally* available for docosahexaenoate synthesis in the suckling rat. Results obtained using whole-body-balance methodology (*see* Section 2) suggest that this 10% availability of α-linolenate for synthesis of n-3 LC-PUFA decreases substantially during undernutrition.

5.4. IRMS Studies in the Near-Term Fetal Primate

The suckling rat has been the predominant model used to assess carbon recycling from α-linolenate. However, a recent study using the near-term fetal as well as infant rhesus monkeys reported similar data as in the suckling rat (Sheaff-Greiner et al 1996; Table 5). This study made an important contribution for three reasons. First, it demonstrated that the process of carbon recycling from α-linolenate is not a metabolic idiosyncracy of neonatal rats but also occurs in primates. Second, it showed that carbon recycling from α-linolenate is an important route of its metabolism in the fetal state as well as in neonates. Third, this study reported a comparison of carbon recycling among α-linolenate, linoleate, and docosahexaenoate, showing that carbon recycling consumed more α-linolenate than these other PUFAs but that a significant proportion of docosahexaenoate was also recycled into *de novo* fatty acid synthesis (Table 5).

6. FUTURE DIRECTIONS

It is clear that β-oxidation and carbon recycling consume the majority of α-linolenate in the fetal/neonatal period in rats and primates. Several issues need to be resolved in

Table 5
Desaturation–Chain Elongation Compared to Carbon Recycling of [^{13}C]-α-Linolenate and Docosahexaenoate into Fatty Acids Synthesized *De Novo* in Near-Term Rhesus Monkey Fetuses

	Precursor	n-3 LC-PUFAs[a]	Sat/Mono
^{13}C-α-Linolenate			
Liver	19	39	40
Brain	nd[b]	20	78
Retina	nd	45	52
^{13}C-Docosahexaenoate			
Liver	93	94	4
Brain	69	76	16
Retina	87	90	8

[a]Eicosapentaenoate, n-3 docosapentaenoate, and docosahexaenoate combined.
[b]Not detected.
Note: Data are given as percent of administered dose recovered 5 d later in the main lipid components of total lipids of each tissue shown. Note that recycling into cholesterol was not determined.
Source: Adapted from Sheaff-Greiner et al (1996).

order to establish the relevance of these studies to human infant development and health. First, does β-oxidation and carbon recycling of α-linolenate (and other PUFAs) occur to a similar extent in human infants as in the models described here? Although several stable-isotope studies have been done in human infants, to my knowledge, data on carbon recycling of PUFA or even on their β-oxidation have not yet been reported. Second, what role does the intake of n-3 LC-PUFA play in modulating the carbon recycling from α-linolenate. Do n-3 LC-PUFAs, in fact, tend to make α-linolenate redundant in the neonatal period and, therefore, available to be used in other pathways? Third, if carbon recycling of PUFAs occurs because of redundancy, one would expect greater carbon recycling of PUFAs in adult animals, which should have a proportionally lower PUFA requirement, but this is not the case for linoleate (Cunnane and Anderson, 1997). Fourth, why is accumulation of lipid products of carbon recycling from α-linolenate greatest in the brain? Fifth, are saturated and monounsaturated fatty acids similarly carbon recycled into newly synthesized brain lipids or is the recycling of α-linolenate notable or potentially important for any reason?

Although one is reluctant to ascribe any special significance to a single fatty acid being used in a pathway that probably accepts many different substrates, it is also peculiar that a vitamin-like nutrient would be such a good substrate in the carbon recycling pathway leading to *de novo* lipogenesis for the brain. Two potentially relevant observations in this context are (1) that the brain appears to synthesize the overwhelming majority of the lipids present in it (Edmond et al. 1991; Edmond et al. 1998), and (2) ketones derived from fat oxidation are very important as brain lipid substrates during early postnatal development (Cunnane et al. 1999).

7. CONCLUSIONS

The aim of this chapter was to provide evidence that β-oxidation and carbon recycling compete sufficiently effectively with the desaturation–chain elongation pathway for dietary α-linolenate that, in some circumstances, conversion of α-linolenate to docosahexaenoate is

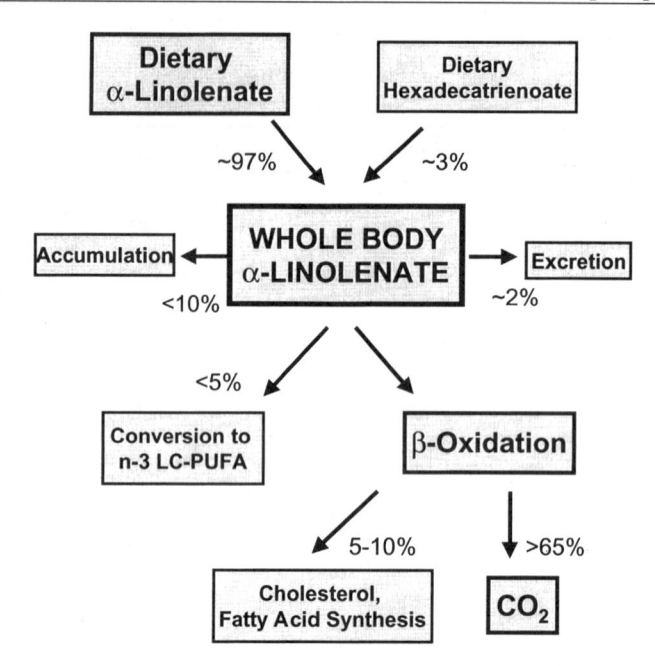

Fig. 1. Overall scheme showing the predominant directions in α-linolenate metabolism. The main routes affecting bioavailability of α-linolenate are shown in larger, bold lettering. Most incoming α-linolenate is directly from the diet, but 3–4% could be derived from dietary hexadecatrienoate present in green plant material. Most losses of α-linolenate are via β-oxidation, which is usually >65% but can increase to over 100% during energy deficit. Small amounts of α-linolenate intake (usually <3%) are excreted. Accumulation of α-linolenate itself is usually no more than 10% of intake but can be negative (body losses) with energy deficit. Conversion to n-3 LC-PUFA generally appears to be <5% of α-linolenate intake in young rats but may be as high as 10% during pregnancy. The quantitative extent of recycling into *de novo* lipogenesis appears to be at least 10% of α-linolenate intake in neonates and is easily equivalent to conversion to n-3 LC-PUFA.

reduced or even prevented. Carbon recycling into lipids synthesized *de novo* clearly captures a large proportion of α-linolenate consumed by suckling rats (i.e., during a period in which demand for docosahexaenoate is probably highest). β-Oxidation captures about 50% of α-linolenate intake during early postnatal development (Sinclair, 1975), a value that rises to at least 65% in mature animals and can exceed 100% of α-linolenate intake during energy deficit in pregnancy (Fig. 1). Hence, β-oxidation can deplete body stores as well as eliminate all dietary α-linolenate, thereby curtailing docosahexaenoate synthesis.

One assumption behind the apparent paradox of extensive carbon recycling of a vitamin-like nutrient is that α-linolenate is an important nutrient during late fetal and early postnatal development. Perhaps this assumption is wrong. Perhaps α-linolenate is not very important as long as docosahexaenoate is present in the diet. The studies reported here used fetal or suckling animals that were receiving docosahexaenoate from the mother's circulation or milk. Perhaps α-linolenate is largely redundant for neonates when docosahexaenoate is also present. This warrants further investigation.

The studies described here suggest that only about 10% of α-linolenate consumed by suckling rats is available for esterification or conversion to n-3 LC-PUFA. Although this is much less than is actually consumed, perhaps it is enough. Further work comparing functional outcomes to the ratio of recycling versus conversion to docosahexaenoate should clarify whether α-linolenate can be extensively carbon recycled but still meet the requirements for n-3 LC-PUFA synthesis. My main hope is that the important, perhaps dominant, role of the "degradative" pathways of β-oxidation and carbon recycling will be carefully considered in assessing the availability of α-linolenate for conversion to docosahexaenoate during early development.

ACKNOWLEDGMENTS

NSERC, MRC, Milupa AG, Roche, Martek, and Unilever are thanked for financial support of the author's research on this topic. External collaborators and trainees who have contributed to method development, data, or ideas described here include Matthew Anderson, Krystian Belza, Tom Brenna, Zhen-yu Chen, Michael Crawford, Sergei Likhodii, Ursula McCloy, Rory McDonagh, Chantale Menard, and Jilin Yang.

REFERENCES

Ackman RG, Cunnane SC. Long chain polyunsaturated fatty acids: sources, biochemistry and nutritional/clinical implications. Adv App Lipid Res 1991; 1:161.
Chen Z-Y, Cunnane SC. Refeeding after fasting increases apparent oxidation of n-3 and n-6 fatty acids in pregnant rats. Metabolism 1993; 42:1206.
Chen Z-Y, Menard CR, Cunnane SC. Weight cycling progressively depletes carcass and adipose tissue linoleic acid and α-linolenic acid in young rats. Br J Nutr 1996; 75:583.
Clouet P, Niot I, Bezard J. Pathway of α-linolenic acid through the mitochondrial outer membrane in the rat liver and influence on the rate of oxidation. Biochem J 1989; 263:867.
Cunnane SC. Metabolism and function of α-linolenic acid in humans. In: Cunnane SC, Thompson, LU, eds. Flaxseed in Human Nutrition. American Oil Chemists' Society, Champaign, IL, 1995, p. 99.
Cunnane SC, Anderson MJ. The majority of linoleate is partitioned towards oxidation or storage in the growing rat. J Nutr 1997a; 127:146.
Cunnane SC, Anderson MJ. Pure linoleate deficiency in the rat. Influence on growth, accumulation of n-6 polyunsaturates and oxidation of ^{14}C-linoleate. J Lipid Res 1997; 38:805.
Cunnane SC, Craig K, Brookes S, Koletzko B, Demmelmair H, Singer J, et al. Synthesis of linoleate and α-linolenate by chain elongation in the rat. Lipids 1995; 30:781.
Cunnane SC, Menard CR, Likhodii SS, Brenna JT, Crawford MA. Carbon recycling into de novo lipogenesis is a major pathway in neonatal metabolism of linoleate and α-linolenate. Prostaglandins Leukotrienes Essential Fatty Acids 1999; 60:387.
Cunnane SC, Yang J, Chen Z-Y. Low zinc intake increases apparent oxidation of linoleic and α-linolenic acids in the pregnant rat. Can J Physiol Pharmacol 1993; 71:205.
Cunnane, SC, Williams SCR, Bell JD, Brookes S, Craig K, Iles RA, et al. Utilization of [U-^{13}C]-polyunsaturated fatty acids in the synthesis of long chain fatty acids and cholesterol accumulating in the neonatal rat brain. J Neurochem 1994; 62:2429.
Dhopeswarkar GA, Subramanian C. Metabolism of linolenic acid in the developing brain. Incorporation of radioactivity from [1-^{14}C]-linolenic acid into brain fatty acids. Lipids 1975; 10:230.
Edmond J, Korsak RA., Morrow JW, Torok-Booth G, Catlin DH. Dietary cholesterol and the origin of cholesterol in the brain of the developing rat. J Nutr 1991; 121:1323.
Edmond J, Higa TA, Korsak RA., Bergner EA, Lee WNP. Fatty acid transport and utilization for the developing brain. J Neurochem 1998; 70:1227.
Emmison N, Gallagher PA, Coleman RA. Linoleic and linolenic acids are selectively secreted in triacylglycerol by hepatocytes from neonatal rats. Am J Physiol 1995; 269:R80.
Gavino GR, Gavino VC. Rat liver outer mitochondrial carnitine palmitoyltransferase activity towards long-chain polyunsaturated fatty acids and the CoA esters. Lipids 1991; 26:266–270.
Leyton J, Drury PJ, Crawford MA. Differential oxidation of saturated and unsaturated fatty acids in vivo in the rat. Br J Nutr 1987; 57:383.

Likhodii SS, Cunnane SC. Utilization of carbon from ^{13}C-polyunsaturates in the synthesis of brain cholesterol in the neonatal rat: a ^{13}C-NMR study. Magn Reson Med 1995; 34:803.

McCloy U, Cunnane SC, Pencharz PB. Metabolism of ^{13}C unsaturated fatty acids by healthy women. FASEB, 2000, in press.

Menard CR, Goodman K, Corso T, Brenna JT, Cunnane SC. Recycling of carbon into lipids synthesized de novo is a quantitatively important pathway of [U-^{13}C]-α-linolenate utilization in the developing rat brain. J Neurochem 1998; 71:2151.

Sheaff-Greiner RC, Zhang Q, Goodman KJ, Gissani DA, Nathanielsz P, Brenna JT. Linoleate, α-linolenate and docosahexaenoate recycling into saturated and monounsaturated fatty acids is a major pathway in pregnant or lactating adults and fetal or infant rhesus monkeys. J Lipid Res 1996; 37:998.

Sinclair AJ. Incorporation of radioactive polyunsaturated fatty acids into liver and brain of the developing rat. Lipids 1975; 10:175.

III DHA AND CNS DEVELOPMENT

10 Impact of Dietary Essential Fatty Acids on Neuronal Structure and Function

M. Thomas Clandinin, R. A. R. Bowen, and Miyoung Suh

1. BRAIN DEVELOPMENT: THE FORMATIVE STAGES

The form of nervous system brain development at early stages is remarkably similar in adult vertebrates. Distinct characteristics of the neural plate and neural tube, from which the nervous system originates in all vertebrates, suggests that development of the central nervous system occurs through similar overall mechanisms. Cells, regions, and various brain structures do not develop uniformly as in other tissues and organs (Dobbing, 1990; Dobbing & Sands, 1979; Herschkowitz, 1989). Well-defined stages of growth occur anatomically and biochemically (Albers, 1985; Gottlieb et al., 1977) and result in significant growth spurts of critical periods during fetal and neonatal life. Brain development occurs in stages: induction of the neural plate, localized proliferation of cells in different regions, migration of cells, formation of identifiable parts of the brain by cell aggregation, differentiation of immature neurons, formation of connections, selective cell death, and modification of connections (Cowan, 1979). These changes within the developing nervous system proceed in a caudal (brainstem) to rostral sequence (Jacobson, 1970). Caudal brain structures include phylogenetically older brain structures, whereas rostral structures are phylogenetically more recent (McLean, 1970). Brain development results in increase in weight and size that is not necessarily parallel: The greatest growth in size occurs prior to the greatest gain in weight (Marshall, 1968). Different parts of the brain grow at different speeds and not all regions reach their fastest rates at the same time. In this respect, the "growth spurt" and velocity curves as defined by Dobbing (Dobbing & Sands, 1979; Dobbing, 1968, 1971) represent rates of change in total brain weight over time and not individual regions of the brain. The early concepts developed by Dobbing did not encompass other developmental processes that are now also understood to be critical periods of growth and highly susceptible to nutritional insult or reflect interrelationships of growth occurring in subregions of the developing brain (Morgane et al., 1993).

1.1. Neurogenesis

Neurons are generated close to the ventricular zone of the neural tube. A complex set of factors including neuron type, position in mitotic gradients, and phylogenetic status, determines the time of neuronal origin (Jacobson, 1970). The number of neurons initially formed in any brain region is determined by the duration of the proliferative period (a few days to several weeks), the duration of the cell cycle (which in a young embryo is a few hours and increases to 4 or 5 d as development progresses), and the number of precursor cells (Cowan, 1979).

Production of neurons occurs over varying lengths of time with different time schedules for formation. For example, neurons originating first in a region of the mature nervous system are phylogenetically older and larger neurons (Jacobson, 1970). Large neurons apparently become postmitotic early because they have long axons that reach their targets while the embryo is still small. Neurogenesis peaks around the 14th wk of gestation and is completed by the 25th wk when the adult number of neurons is present (Lou, 1982). In rats, these developmental events occur at about 18 and 20 d of gestation, respectively. Neurons of the hippocampus and the cerebellum continue to proliferate postnatally. The timetable of neuronal differentiation cannot be simply deduced from its time of origin. The type of neuron, its regional situation, and the timing of arrival of axons with which synapses will form are factors that also affect neuronal differentiation. Thus, the period when neurons are most susceptible to change in essential fatty acid supply for membrane synthesis is variable.

1.2. Gliogenesis

Glial cells tend to originate after neurons in any particular region of the brain. Formation of glia differs from neuronal formation. Many proliferative cells that generate glia lie outside the neuroepithelium, at or near the site where they will be located in the adult (Bayer, 1985). Production of glial cells continues throughout adult life (Jacobson, 1970). Thus, it is logical to expect that the dietary essential fatty acid supply will have impact on glial cells throughout life. Gliogenesis is primarily a postnatal event (Das, 1977); however, in some brain regions, it is detected before birth (Das, 1977; Rodier, 1980). Early gliogenesis is completed by the 15th wk of gestation in humans and the 16th d of gestation in rats (Morgane et al., 1993). Damage to the glial population is rarely permanent (Bayer & Altman, 1975; Bayer, 1977; Sjostrand, 1965). However, a qualitative imbalance or quantitative deficit of food intake produces alterations in the ontogeny and function of the nervous system (Dobbing, 1970; Greenwood & Craig, 1987; Vitiello & Gombos, 1987; Winick, 1969).

1.3. Connections to Function

Nerve cells generated at or close to the inner, proliferative surface of the neural tube subsequently migrate past other cells to final locations. The temporal origins of neurons may be related to later establishment of anatomical connections (Bayer, 1985). There are two spatiotemporal gradients for neuron migration. In nuclear regions of the brain (such as the thalamus and hypothalamus), the oldest neurons produced by neuroepithelium are pushed farther out as younger neurons are generated. In these regions, cells accumulate in an "outside-in" or "pushing" gradient (Rakic, 1977). The "inside-out" or "passing" gradient occurs in regions of the brain that have a laminar structure (such as cerebral and cerebellar cortices). Cell accumulation in these regions occurs as younger neurons migrate past older neurons, which remain closest to the neuroepithelium (Angevine & Sidman, 1961). Despite the heterogeneity of the cell population, development of the nervous system's complex form is reliable. Development must contain mechanisms for neurons to migrate to their proper destination. This arrangement of neurons must be completed successfully for normal functioning to result.

Nervous system functional connections are established through growth of neurites (axons and dendrites) and formation of synapses. Synaptogenesis is the contact between axons and target cells that starts before neurogenesis is completed (Jacobson, 1978) and follows a cell-specific, region-specific timetable. Most neurons generate and receive many processes, producing many more synaptic connections. These membrane connections are high in essential fatty acids and the composition of the lipids can be altered by changes in diet fat intake (Foot et al., 1982). Periods of programmed cell death and synapse reorganization follow in the final stages of brain morphogenesis. In each brain region, brain morphogenesis ensues in a specifically timed series of events. Each period becomes a critical period for development of the next (Bayer, 1989). Time-scales are further complicated by migration of cells between regions. Because of the brain's general lack of regeneration potential and dependence on specialized interactions, any misdirected, mistimed, or absent developmental cues can lead to structural aberrations (Heuther, 1990). These structural changes are irreversible and result in functional deficits if future developmental changes are unable to compensate for these structural changes. Thus, nutritional insults that alter brain development in the fetus or neonate may permanently alter a myriad of complicated functional connections in later stages of infant development.

2. NEURONAL TISSUE LIPID

Second only to adipose tissue, the brain is the most lipid-concentrated organ in the body. Nervous tissue contains 50% lipid on a dry-weight basis, or 10% on a wet-weight basis (Sastry, 1985). Many of the complex lipids present in brain are found in other tissues, but the molecular species present may also differ. This lipid plays a role in modifying the structure, fluidity, and function of brain membranes (Brenner, 1984; Bourre et al., 1993; Dyer & Greenwood, 1991; Foot et al., 1982, 1983; Stubbs & Smith, 1984). A variety of complex lipids exists in the brain and the composition and metabolism of these lipids change with development and age (Rouser et al., 1971; Svennerholm & Stallberg-Stenhagen, 1968; Wykle, 1977).

2.1. Accretion of Essential Fatty Acid in the Brain

Formation of neuronal membranes requires synthesis and assembly of membrane phospholipids containing significant amounts of essential fatty acid, primarily arachidonic acid (20:4n-6) and docosahexaenoic acid (22:6n-3). Little quantitative evidence exists prior to 24 wk of gestation in the human. Accretion of essential fatty acids during the last trimester of intrauterine development has been estimated (Clandinin et al., 1981a).

Analyses of whole-body fat content (Widdowson, 1968; Widdowson et al., 1979) indicates that preterm infants, with appropriate weight for gestational age of 1300 g at birth, have a total-body fat content of about 30 g compared with the term infant of 3500 g with a total-body fat content of 240 g. Clandinin et al. (1981b) estimated that approximately 2783 mg of n-6 fatty acids and 387 mg n-3 fatty acids accrue in adipose tissue each week *in utero*. For premature infants, birth after only a few more weeks of intrauterine development would dramatically increase the potential reserve of fatty acids in adipose tissue both for total fatty acids used for energy production and for essential fatty acids used for synthesis of structural tissues. These estimates also support the body of research by Van Houwelingen et al. (1992), suggesting that the growing fetus represents a large draw upon maternal essential fatty acid stores and perhaps that a limitation in the size of the maternal essential fatty acid stores may impact on fetal growth and development, particularly brain growth and development.

During the third trimester of human development, n-3 and n-6 fatty acids accrue in fetal tissues as an essential component of structural lipids and rapid synthesis of brain tissues occurs. This rapid synthesis causes increases in cell size, cell type, and cell number (Clandinin et al., 1980a). Brain lipid levels increase rapidly during this period. Levels of 18:2n-6 and 18:3n-3 are consistently low in the brain during the last trimester of pregnancy (Clandinin et al., 1980a). However, accretion of long-chain essential fatty acid desaturation products 20:4n-6 and 22:6n-3 occur and the absolute accretion rates of the n-3 fatty acids, specifically 22:6n-3, are greater in the prenatal period compared with the postnatal period (Clandinin et al., 1980a, b). It is apparently critical that the developing fetus obtain the correct types and amounts of fatty acids to ensure complete and proper development of brain membranes. Timing of the availability of these fatty acids is also a factor. Collectively, this quantitative information indicates that large amounts of docosahexaenoic acid (22:6n-3) and arachidonic acid (20:4n-6) are required during development of neural tissue when cellular differentiation and active synaptogenesis are taking place.

2.2. Visual System Development

The entire visual system spans caudal to rostral regions of the brain and includes some lateral areas of the brain. The visual pathway involves a chain of visual processing events that begin with the retina in the eye. During visual processing, incoming light strikes the photoreceptors and generates electrical signals, which are sent to bipolar cells and ganglion cells. These networks send visual information along the optic nerve to the visual cortex. "Funneling" of information within the eye is the result of an individual neuron receiving, converging, and combining impulses from several incoming nerve fibers. Thus, the separate signals of each nerve fiber are integrated into an entirely new message based on all the inputs. The retina content of 22:6n-3 is very high and normally increases during development. This change in lipid composition is apparently important to visual system function although it cannot be easily distinguished functionally from the simultaneous increase in 22:6n-3 in the entire visual pathway.

2.3. Sources of Essential Fatty Acids for the Developing Brain

The developing fetal brain can synthesize saturated and monounsaturated fatty acids (Cook, 1978; Menon & Dhopeshwarkar, 1982). High rates of lipogenesis in fetal liver (Smith & Abraham, 1970) may act as a source of fatty acids for the fetus. Initially the placenta, by controlling passage of 20:4n-6 and 22:6n-3, determines the amount of these fatty acids available to the fetus, but then fetal metabolism may synthesize 20:4n-6 and 22:6n-3 (Crastes de Paulet et al., 1992). The age at which this progression begins and is completely achieved is not known. As synthesis of long-chain polyunsaturated fatty acids by human fetal liver has not been clearly demonstrated under physiological conditions, this progression remains hypothetical (Crastes de Paulet et al., 1992). Thus, the contribution of fatty acid synthesis to 20:4n-6 and 22:6n-3 accumulation by the neonate would be affected by the age of the infant at birth. *In utero,* it is clear that the fetus relies on the mother for a supply of many fatty acids, particularly the essential fatty acids.

Crawford et al. (1976) describe a process in which, compared with parent essential fatty acids, the relative percentage of 20:4n-6 and 22:6n-3 increases in phosphoglycerides progressively from maternal blood to placental and to fetal blood, liver, and brain. By this process, termed biomagnification, specific mechanisms within the placenta hypothetically result in sequestration and release of specific fatty acids to fetal circulation. Neuringer et al. (1984) also report that in the monkey and human fetus the levels of 22:6n-3 and 20:4n-6 are higher in fetal blood compared with maternal blood, whereas the opposite is true for their precursors. Thus, the importance of these long-chain polyunsaturated fatty acids is evident in their preferential, active transfer across the placenta to the fetus in a lipid form normally impermeable to the placental barrier. Biomagnification by selective sequestering of long-chain polyunsaturated fatty acids on the fetal side of the placenta is likely (Uauy et al., 1989). King et al. (1971) found that, when comparing adipose tissue triglyceride between infants and mothers, infants have greater levels of palmitic and palmitoleic acids. The predominance of these two fatty acids in newborns indicates that glucose plays an important role in fetal fat synthesis. Embryonic and fetal lipids in early gestation are derived from maternal fatty acids that cross the placenta, but with advancing gestational age, there is a gradual shift to *de novo* synthesis from glucose in fetal tissue (Poissennet et al., 1988).

Some fatty acids are transferred to the fetus across the placenta. It also appears that the degree of placental and fetal synthesis of fatty acids varies with gestational age (Poissennet et al., 1988; Robertson & Sprecher, 1968). Clandinin et al., (1980a) reported that 80% of human fetal brain 22:6n-3 accrues between 26 and 40 wk of gestation. Infants born prior to 32 wk gestation have low concentrations of brain 22:6n-3 (Clandinin et al., 1980a). Early in pregnancy, there is apparently a great dependence on maternal fatty acids to provide the fetus with lipids. This may have important implications for the low-birthweight, premature infant and for the shift of essential fatty acid from maternal stores to fetal tissues during fetal growth.

2.4. Essential Fatty Acid Synthesis and Metabolism in the Brain

Many questions remain about the role of brain in synthesizing 20:4n-6 and 22:6n-3 from dietary precursors. Animal studies with radiolabeled 18:2n-6 and 18:3n-3 indicate that the brain is capable of desaturating and elongating 18:2n-6 and 18:3n-3 to 20:4n-6 and 22:6n-3, respectively (Anderson and Connor, 1988; Clandinin et al., 1985a; Cohen & Bernsohn, 1978; Cook, 1978; Dhopeshwarkar et al., 1971a, b; Dhopeshwarkar and Subramanian, 1976; Sinclair & Crawford, 1972). These studies of the brain fail to clarify which cell types within this tissue can provide 20:4n-6 and 22:6n-3. Studies with isolated brain cells provide evidence that both neuronal and glial cells may desaturate and elongate 18:2n-6 and 18:3n-3 to 20:4n-6 and 22:6n-3, respectively (Anderson & Connor, 1988; Clandinin et al., 1985a; Cohen & Bernsohn, 1978; Dhopeshwarkar & Subramanian, 1976; Yavin & Menkes, 1974).

Microvessels and plasma contain amounts of 18:3n-3 and 20:5n-3; thus, it is possible that these fatty acids could supply 22:6n-3 to the brain (Clandinin et al., 1997; Edelstein, 1986). Moore et al. (1990) investigated whether cerebroendothelial cells could desaturate and elongate 18:2n-6 and 18:3n-3 to 20:4n-6 and 22:6n-3, respectively. Isolated cerebroendothelial cells were incubated with radiolabeled 18:2n-6 or 18:3n-3 and were found to take up 18:2n-6 and 18:3n-3 equally and desaturate and elongate 18:2n-6 to 22:4n-6 and 18:3n-3 to 22:5n-3 (Moore et al., 1990). The major metabolite of 18:2n-6 was

20:4n-6 and that of 18:3n-3 was 20:5n-3. Desaturation and elongation of 18:3n-3 by cerebroendothelial cells far exceeded that of 18:2n-6, suggesting some specificity for n-3 fatty acids. Delton-Vandenbroucke et al. (1997) demonstrated that cerebroendothelial cells can metabolize 22:5n-3 to 20:5n-3, 22:6n-3, 24:5n-3, and 24:6n-3. The presence of 24 carbon n-3 metabolites in cerebroendothelial cells lipids and culture media suggests that metabolism of n-3 fatty acids in cerebroendothelial cells were using a delta-4 desaturase-independent pathway similar to that shown in rat liver (Voss et al., 1991). Thus, cerebroendothelial cells can desaturate and elongate 18:2n-6 and 18:3n-3 to 20:4n-6 and produce a small amount of 22:6n-3, respectively.

Moore et al. (1991) determined that astrocytes from rat brain can desaturate and elongate 18:2n-6 and 18:3n-3 to 20:4n-6 and 22:6n-3, respectively. Rat type I astrocytes from either cerebrum or cerebellum cultured for 12 d were incubated with radiolabeled 18:2n-6 or 18:3n-3 formed 20:4n-6 and 22:6n-3, respectively. In contrast, cultures of rat cerebral or cerebellum neuronal cells did not desaturate fatty acid to produce 20:4n-6 and 22:6n-3. Instead, the neuronal cells appear to only elongate 18:2n-6 and 18:3n-3. Hence, astrocytes, not neuronal cells, appear to synthesize 20:4n-6 and 22:6n-3 in brain. Cerebroendothelial cells and astrocytes together supply 22:6n-3 to neuronal cells (Moore, 1993). In cocultures, astrocytes synthesize and release large amounts of 22:6n-3 from 20:5n-3 made by the cerebroendothelial cells (Moore, 1993). Neuronal cells then take up 22:6n-3 released from astrocytes and incorporate fatty acid into neuronal cell plasma membranes (Moore, 1993). Thus, cerebroendothelial cells and astrocytes seem to be needed together to synthesize 22:6n-3 for uptake by the neuronal cell during growth.

These observations on the cellular partitioning of synthesis are based primarily on the rat and are largely undetermined for the human brain.

3. IMPACT OF DIETARY ESSENTIAL FATTY ACIDS ON NEURONAL CELL COMPOSITION AND FUNCTION

Brain membranes were generally viewed as resistant to structural change by both endogenous and exogenous factors. Data have shown that brain membranes are much more sensitive to changes in composition induced by dietary fat than previously thought (Bourre et al., 1989a; Foot et al., 1982; Jope & Jenden, 1979; Lee, 1985; Wurtman et al., 1981). Moreover, the extent of the changes in brain membrane composition by dietary fat varies among brain regions, cell types, and organelles (reviewed by Clandinin et al., 1997; Clandinin et al., 1991; Hargreaves & Clandinin, 1990).

Earlier studies examining the role of dietary fat on brain membrane composition have used rodents that were fed 18:2n-6-deficient diets for several weeks to a few generations. Results from these studies demonstrate qualitative changes in brain membrane fatty acid composition associated with essential fatty acid deficiency (i.e., increase in 20:3n-9 and decrease in 20:4n-6; Koblin et al., 1980; Paoletti & Galli, 1972; Sun & Sun, 1974).

By feeding nutritionally adequate diets, dietary intake of 18:2n-6, 18:3n-3, or the proportion of 18:2n-6 to 18:3n-3, particularly during development, has been shown to influence the content of long-chain polyunsaturated fatty acids in membrane lipids by changing the composition of the whole brain, oligodendrocytes, myelin, astrocytes; mitochondrial, microsomal, and synaptosomal membrane (Bourre et al., 1984; Foot et al., 1982; Lamptey & Walker, 1976; Tahin et al., 1981). Feeding diets with a 18:2n-6 to 18:3n-3 fatty acid ratio between 4 : 1 and 7 : 1 to rats from birth to 1, 2, 3, and 6 wk of age indicates that diet alters neuronal and glial cell membrane fatty acid composition differently, and in a region- and time-specific manner (Jumpsen et al., 1997a, b).

Bowen et al. (1999) studied the effects of increasing dietary 18:3n-3 content by decreasing the dietary 18:2n-6 to 18:3n-3 fatty acid ratio on increasing the 22:6n-3 content of neuronal cells in 2-wk-old rat pups. The findings demonstrated that increasing dietary 18:3n-3 content from 1.6% to 17.5% of the total fatty acids in the diet did not significantly increase the 22:6n-3 content in phosphatidylcholine, phosphatidylethanolamine, or phosphatidylserine in neuronal cells of rat pups at 2 wk of age (Bowen et al., 1999). Lack of a significant increase in 22:6n-3 content in neuronal cells after feeding a high dietary 18:3n-3 intake may be the result of limited desaturase activity at 2 wk of age (Bourre et al., 1990). Thus, dietary fat, in the absence of essential fatty acid deficiency, can alter the content of long-chain polyunsaturated fatty acid in brain membranes, and high dietary intake of 18:3n-3 significantly increases the 18:3n-3 content in other tissues (Bowen & Clandinin, 2000).

The functional effect of diet-induced changes in the fatty acid composition of membrane phospholipids in brain has been the subject of some research. Dietary-fat-induced change in membrane fatty acid composition can affect membrane function by modifying membrane fluidity and thickness, lipid-phase properties, polar lipid composition, or specific interactions with membrane proteins (Clandinin et al., 1985b). Previous studies showed that deficiency of n-3 fatty acid in the diet changed membrane physical properties, membrane-bound enzymes, receptor activity, and signal transduction (Lee et al., 1986; Salem et al., 1988; Stubbs & Smith, 1984). However, these changes induced by dietary n-3 deficiency do not reflect an overall change in membrane fluidity but, rather, indicate selective changes in the microenvironment of the membrane-bound proteins (Bazan, 1990; Salem et al., 1988). Many membrane-bound proteins have a specific requirement for the annular lipids surrounding them (Yeagle, 1989). The annular lipids allow the membrane protein to achieve an active conformation for function in the brain (Capaldi, 1977; Marinetti & Cattieu, 1982; Tanford, 1978; Yeagle, 1989). Hence, any changes in these annular lipids by dietary fat may influence the function of proteins in the membrane. Several studies have shown that diets, in the absence of essential fatty acid deficiency, containing various dietary fats change the activities of membrane-bound enzymes, receptors, and carrier-mediated transport (reviewed by Clandinin et al., 1997). It is conceivable that these changes in membrane function by dietary fat can have an impact on brain function.

Animal studies have shown that when rodents or monkeys were maintained on an n-3-deficient diet, electroretinogram abnormalities (Benolken et al., 1973; Bourre et al., 1989b; Weisinger et al., 1996a), reduced visual acuity (Neuringer et al., 1984, 1986), altered stereotyped behavior (Reisbick et al., 1994), and decreased level of learning and memory occur (Bourre et al., 1989; Mills et al., 1988; Lamptey & Walker, 1976; Yamamoto et al., 1987, 1988). Dietary n-3 fatty acid deficiency affects brain functions of preterm infants as measured by cortical visual evoked potential, electroretinograms, and behavioral testing of visual acuity (Birch et al., 1992; Carlson et al., 1993a; Uauy et al., 1990). Human preterm infants fed infant formulas without 22:6n-3 were also shown to have abnormal electroretinograms, as well as decreased visual acuity compared to infants fed formulas containing 22:6n-3 (Carlson et al., 1993a; Uauy et al., 1990). Furthermore, infants fed vegetable oil based formulas with no 22:6n-3 had lower cognitive scores compared to infants fed formulas with 22:6n-3 (Carlson et al., 1994; Lucas et al.,

1992; Makrides et al., 1995). The mechanism for these changes in membrane proteins and brain function induced by diet fat is not known.

3.1. Docosahexaenoic Acid Supply and Conservation

It is of interest how 22:6n-3 is supplied to the developing retina to meet both its requirement and to be conserved in enriched amounts in the membrane. The intestine (Li et al., 1992) and liver (Scott & Bazan, 1989) appear to play an important role. Oral doses of [C^{14}]18:3n-3 and [C^{14}]22:6n-3 are esterified to triacylglycerols and phospholipids by the intestinal absorptive cells and transported in chylomicrons to the liver (Li et al., 1992). The liver synthesizes 22:6n-3 from 18:3n-3 and can deliver 22:6n-3 to the retina and brain via lipoproteins or albumin (Scott & Bazan, 1989). Phospholipid, triglyceride, and cholesterol esterified with 22:6n-3 are carriers of 22:6n-3 to nervous tissues (Li et al., 1992; Martin et al., 1994; Wang & Anderson, 1993b). Retinal pigment epithelium is actively involved in the preferential uptake of 22:6n-3 from the circulation and subsequently transports this fatty acid to the rod outer segment (Wang & Anderson, 1993a,b). Wang et al. (1992) confirmed that the retina and brain selectively takes up C_{22} fatty acid from the circulation because rats fed rapeseed oil (22:1n-9, 43%) for 4 mo from birth did not incorporate 22:1n-9 into the rod outer segment or brain but incorporated this fatty acid into other organs (Poulos et al., 1994). It is not understood how the neural cells take up specific fatty acids. A fatty-acid-binding protein may play a role in uptake of 22:6n-3 in the retina (Lee et al., 1995). Recently, a gene has been identified that is responsible for genetically determined degeneration of the macula, resulting in permanent blindness in humans (Zhang et al., 2001). This gene, *ELOVL4*, is homologous with an essential fatty acid elongase, thus providing a clear link between abnormality of essential fatty acid metabolism and loss of retinal function.

During essential fatty acid deficiency, 22:6n-3 content in the retina and rod outer segment, like the brain, are only slightly reduced (Anderson & Maude, 1971; Anderson et al., 1974; Wiegand et al., 1991). This indicates that there must be specific mechanisms to conserve 22:6n-3. Wiegand et al. (1991) suggested the conservation mechanisms in the retina are through recycling within the retina or between the pigment and the retina, or selective sequestration of 22:6n-3 from the blood. Since then, many studies have confirmed the intimate involvement of retinal pigment epithelium by this mechanism (Chen et al., 1992; Gordon & Bazan, 1993; Stinson et al., 1991). For example, 22:6n-3 increases in frog pigment epithelium following rod photoreceptor shedding and returns to the normal level within 8 h following the shedding events (Chen et al., 1992). The phagosomal fatty acid labeling matches the fatty acid composition of the rod outer segment tip (Gordon & Bazan, 1993) and 22:6n-3 containing lipids are released, removed, and recycled back to the photoreceptors (Chen & Anderson 1993a; Gordon & Bazan, 1993). More recently, Chen and Anderson (1993b) found that 22:6n-3 is incorporated into triglyceride in the retinal pigment epithelial cell, suggesting that this neutral lipid participates in selective enrichment of 22:6n-3. Retinal pigment epithelium can be a source of 22:6n-3 for the retina because it is able to synthesize 22:6n-3 from precursors in vitro (Wang & Anderson, 1993a). Interphotoreceptor retinoid-binding protein (IRBP) located in the interphotoreceptor matrix may also be involved in the recycling of 22:6n-3 because this fatty acid shows the highest affinity for IRBP among all the fatty acids tested (Chen et al., 1993). There is a steep transient concentration of 22:6n-3 between photoreceptor and pigment epithelium cells (Chen et al., 1996).

3.2. Dietary Modulation of Retinal Fatty Acid Composition and Function

The effect of maternal diet on the modulation of n-3 fatty acid in the retina of offspring has been studied in newly hatched chicks (Anderson et al., 1989), juvenile felines (Pawlosky et al., 1997) and piglets (Arbuckle & Innis, 1993). Docosahexaenoic acid appears to be the preferred fatty acid for raising the level of 22:6n-3 in the retina and brain among different sources of n-3 fatty acids (Anderson et al., 1990). Feeding corn oil supplemented with 22:6n-3 is able to restore the 22:6n-3 level in n-3-deficient chicks (Anderson & Connor, 1994) and felines (Pawlosky et al., 1997) probably through replacement of dipolyunsaturated molecular species (22:6n-3 - 22:6n-3) (Lin et al., 1994). These observations imply that 18:3n-3 alone in the diet as the n-3 fatty acid source may not be adequate for meeting the 22:6n-3 requirement for brain and retinal development, because 18:3n-3 is a less efficient precursor for 22:6n-3 when there is low Δ^6 desaturase activity (Anderson et al., 1990; Kohn et al., 1994).

Although retina and rod outer segment tenaciously retain 22:6n-3 during essential fatty acid deficiency (Connor et al., 1990, 1991; Wiegand et al., 1991), severe unbalanced n-6/n-3 diets or depleted n-3 fatty acid levels in membrane can cause abnormal change in biochemical and physiological membrane function. The level of 22:6n-3 in n-3 fatty acid-deficient chick brain and retina is restored by a diet containing 22:6n-3 (Anderson & Conner, 1994) and also after n-3 deficiency in the rhesus monkey (Neuringer et al., 1986; Neuringer & Connor, 1986). Functionally, n-3 fatty acid-deficient monkeys show delayed recovery of the dark adapted electroretinogram and impaired visual acuity at an early age (Neuringer et al., 1986), suggesting that n-6 fatty acids are not interchangeable with n-3 fatty acid in maintaining normal retinal function. After repletion with fatty acids from fish oil, the 22:6n-3 level increased rapidly after feeding, but no improvement in the electroretinogram occurred (Neuringer & Conner, 1989). Developing felines fed corn oil from gestation to 8 wk of age are devoid of a dietary source of 22:6n-3 and display increased a- and b-wave implicit time compared to diets containing long-chain polyunsaturated fatty acids (Pawlosky, 1997). Guinea pigs fed safflower oil through three generations also exhibit significantly decreased levels of 22:6n-3 and reduced both peak to peak and a-wave in the electroretinogram (Weisinger et al., 1996a, b, 1999). Rats raised for several generations with a fat-free diet also show decreased amplitude of the a-wave, which reflects the electrical potential of the photoreceptor membrane and altered rod outer segment disk renewal (Anderson et al., 1974). These studies indicate that n-3 fatty acids exert a key role in normal visual function that is dependent on the n-6/n-3 balance in the diet. It also indicates that the most critical time for providing an adequate diet would be during pregnancy and early lactation.

In term and preterm human infants, studies usually compare feeding breast milk versus infant formula with or without long-chain polyunsaturated fatty acids. Leaf et al. (1996) found correlation between percent intake of breast milk (>50% vs <50%) and 22:6n-3 in both plasma and red blood cells. Uauy and co-workers (1990, 1992) have reported that very low-birthweight neonates fed a soybean-oil-based infant formula have poorer early electroretinogram response, visual evoked potential (VEP) and forced choice preferential looking (FPL) compared to neonates fed human milk or marine-oil-containing formula. Preterm infants fed formula with marine oil (0.3% 20:5n-3 and 0.2% 22:6n-3) appear to have enhanced visual acuity measured by the Teller Acuity Card procedure until 4 mo of age (Carlson et al., 1993b). With a similar formula but using different tools

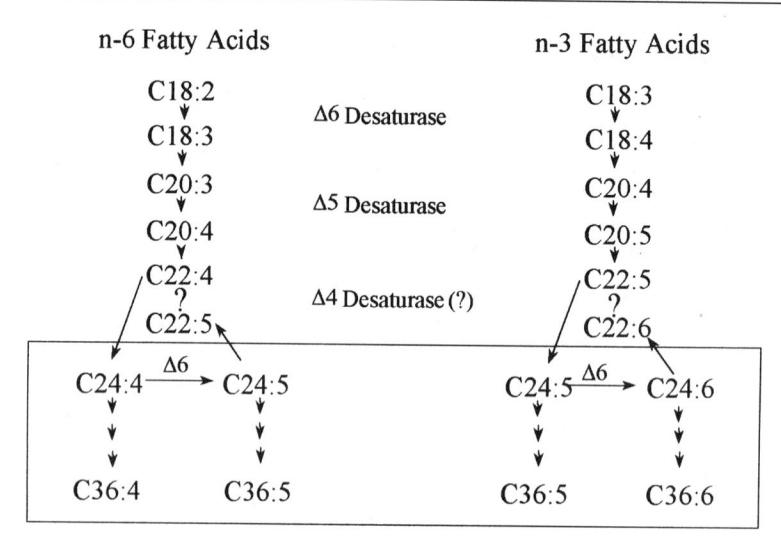

Fig. 1. Metabolic pathway of essential fatty acids. Recent evidence indicated that 22:6n-3 are produced by β-oxidation of 24:6n-3, which is desaturated from 24:5n-3, after elongation from 22:5n-3. Very long-chain fatty acids in the box are found in the retina; however, the metabolism and function is not known.

of visual function measurement, Werkman and Carlson (1996) have demonstrated the beneficial effect of 22:6n-3 on discrete looks to both novel and familiar stimuli and short overlook duration for novelty preference (visual recognition memory). The preterm infant measured at 52 wk of postconceptual age after feeding long-chain polyunsaturated fatty acid (0.3%, 22:6n-3) show the maturation pattern of visual evoked potential similar to breast-fed infants (Faldella et al., 1996). Jongmans et al. (1996) found that infants born prematurely with the absence of other major neurological symptoms are at risk for abnormal visual function and perceptual-motor difficulties at earlier school life. From the above studies, it is clear that an appropriately balanced diet can enhance visual function in infants.

Term infants fed breast milk for the first 4 mo of life increase visual acuity, measured by the Teller Acuity Card, more rapidly compared to formula-fed infants (Jorgensen et al., 1996). Carlson et al. (1996) have reported that term infants fed formula with 2% 18:3n-3 and 0.1% 22:6n-3 show better 2-mo visual acuity than infants fed formula. Makrides et al. (1995) also have found that visual evoked potential and visual acuity of breast-fed neonates and infants fed a formula supplemented with 22:6n-3 are better than those of placebo-formula-fed infants at 16 and 30 wk of age. Although these results show an effect of diet on the development of the visual function in different groups of infants, other studies have provided comparable results (Jorgensen et al., 1998).

3.3. Effect of Dietary Fat on Very Long-Chain Fatty Acids and Rhodopsin Content

Retina membrane phospholipids, particularly phosphatidylethanolamine, contain a high level of 22:6n-3 (Birch et al., 1992; Suh et al., 1994). In the rod outer segment of the retina, significant amounts of 22:6n-3 in phosphatidylserine and phosphatidylcholine also occur (Suh et al., 1994). Increased dietary intake of n-3 fatty acids increases the n-3

Chapter 10 / Dietary EFA and Brain Development

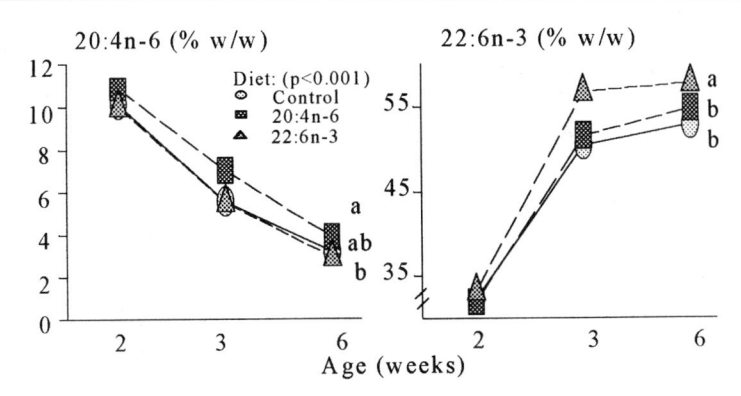

Fig. 2. Developmental profiles and effect of dietary 20:4n-6 and 22:6n-3 on 20:4n-6 or 22:6n-3 in phosphatidylethanolamine of photoreceptors.

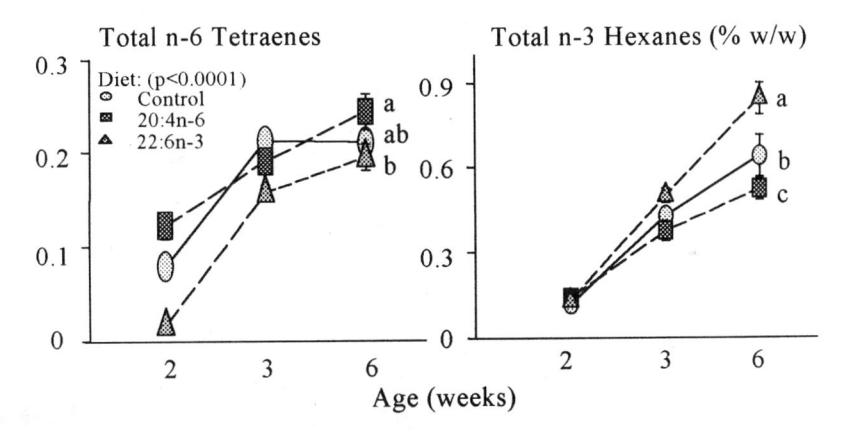

Fig. 3. Developmental profiles and effect of dietary 20:4n-6 and 22:6n-3 on very long chain fatty acids in phosphatidylcholine of photoreceptors.

fatty acid content of the rod outer segment (Suh et al., 1994, 1996; Lin et al., 1994). In addition to 22:6n-3, phosphatidylcholine contains VLCFA from C_{24}–C_{36} carbons in the chain length of n-6 and n-3 types (Fig. 1). Feeding 20:4n-6 and 22:6n-3 increased 20:4n-6 and 22:6n-3 levels in phospholipids (Fig. 2) and increased total n-6 tetraenoic VLCFA and total n-3 hexaenoic VLCFA (Fig. 3). Both of these VLCFA increase to 6 wk of age (Fig. 3). The rhodopsin content is also affected by dietary treatment. The highest rhodopsin content occurs in the retinas of animals fed diets containing 20:4n-6 or 22:6n-3 compared to diets without 20:4n-6 or 22:6n-3 (Fig. 4). The kinetics of rhodopsin disappearance after light exposure is also greatest in animals fed 22:6n-3. Small manipulations of dietary level of 20:4n-6 and 22:6n-3 are important determinants of fatty acid composition of membrane lipid, visual pigment content, and rhodopsin kinetics in the developing photoreceptor cell (Suh et al., 2000). The rod outer segment is the major site of photon absorption. Changes in VLCFA that are induced by dietary fats may influence different properties related with rhodopsin function.

This area of research has considerable potential to reveal important mechanisms through which highly unsaturated fatty acids play functional roles in excitable mem-

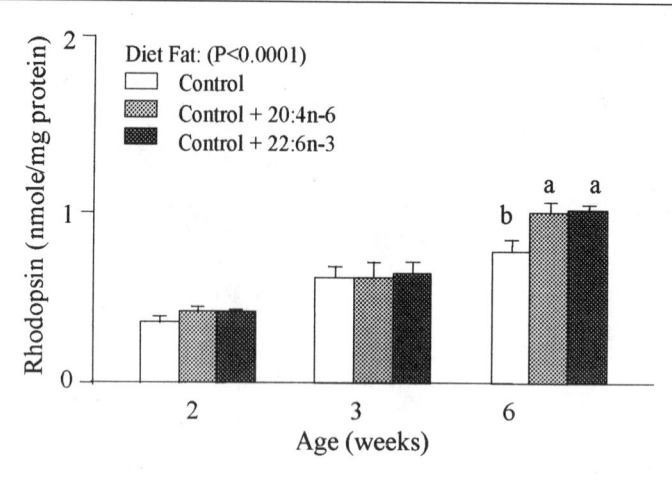

Fig. 4. Developmental profiles and effect of dietary 20:4n-6 and 22:6n-3 on rhodopsin content in the retina.

branes that relate to the diet of the infants and a variety of developmental functions (Aveldano, 1987; Suh et al., 1994, 1996). The biological function and role of these VLCFA composition is not known. It is generally believed that these functional changes in the retina are caused in some way by change occurring in the fatty acid constituents of phospholipids present in the retinal system associated with visual function.

CONCLUSION

Future directions should focus on establishing functional and/or behavioral implications caused by varying the n-6 to n-3 fatty acid ratio within the ranges recommended for infant feeding by experts and by including or omitting long-chain polyunsaturated fatty acids, namely docosahexaenoic and arachidonic acids. A discrepancy remains regarding the degree to which a newborn human infant is capable of desaturating and elongating 18:2n-6 and 18:3n-3 to their longer-chain homologs. Immediately after birth, there is no significant accretion of chain elongation–desaturation products, suggesting that a limiting factor may initially exist in the synthesis of these products or that mobilization of these long-chain polyunsaturated fatty acids from the liver exceeds the capacity of the liver to synthesize them from dietary precursors (Clandinin et al., 1989). The capability for desaturation exists, but is it sufficient to support the requirements of the developing nervous system? If it is sufficient, in what instances, diseases, or dietary circumstances might it be compromised? Premature or very low-birthweight infants are likely candidates. Is the full-term infant also a likely candidate in certain circumstances? What factors are likely to affect this activity? Maternal nutritional status during pregnancy might be one example. Another example is the occurrence of fatty acids in the diet. Fatty acids do not occur individually but with other fatty acids and are associated with vitamins and other nutrients. These clusters of nutrients likely affect how individual fatty acids are metabolized.

Little quantitative evidence is available to indicate the neuronal source of 20:4n-6, 22:4n-6, 22:5n-3, 22:6n-3, and n-6 and n-3 VLCFA in the fetus. During intrauterine development, does the fetus rely entirely on placental synthesis and a transfer mechanism to obtain these essential structural long-chain polyunsaturated fatty acids or are they synthesized by the fetus or in specific tissues? Because development of the brain is not uniform and many vulnerable periods exist prenatally and postnatally, the use of whole-brain analyses is conceptually quite limiting. In the same regard, because phospholipids vary in fatty acid composition, analyses of total brain phospholipids are also limiting. In terms of correlating functional deficits with morphological alterations, Vitiello and Gombos (1987) noted that these could occur only in some cases. These authors point out that relationships between structure and function in the brain are not always well defined. Thus, a single morphological structure cannot always be identified as the structure essential for a certain function (Vitiello & Gombos, 1987). The need exists to focus our efforts on developing innovative tests to dissect out functions of essential fatty acid in neuronal cells that are interpretable in terms of brain regions, groups of neurons, or neurotransmitter activity and function. These tests would need to be applicable in some form for use in infants and children and would then lead to the means to develop meaningful markers of essential fatty acid status during fetal and infant brain development and also during the process of aging in the brain.

REFERENCES

Albers RW. Coordination of brain metabolism with function during development. In: Wiggins RC, McCandles, DW, Enna, SJ, eds. Developmental Neurochemistry. University of Texas, Austin, 1985, pp. 180–192.

Anderson GJ, Connor WE. Uptake of fatty acids by the developing rat brain. Lipids 1988; 23:286–290.

Anderson GJ, Connor WE, Corliss JD. Docosahexaenoic acid is the preferred dietary ω-3 fatty acid for the development of the brain and retina. Pediatr Res 1990; 27:81–87.

Anderson GJ, Connor WE, Corliss JD, Lin DS. Rapid modulation of the ω-3 docosahexaenoic acid levels in the brain and retina of the newly hatched chick. J Lipid Res 1989; 30:433–441.

Anderson GJ, Connor WE. Accretion of n-3 fatty acids in the brain and retina of chicks fed a low-linolenic acid diet supplemented with docosahexaenoic acid. Am J Clin Nutr 1994; 59:1338–1346.

Anderson RE, Maude MB. Lipids of ocular tissues VIII. The effects of essential fatty acid deficiency on the phospholipids of the photoreceptor membrane of rat retina. Arch Biochem Biophys 1971; 151:270–276.

Anderson RE, Benolken RM, Dudley PA, Landis DJ, Wheeler TG. Polyunsaturated fatty acids of photoreceptor membranes. Exp Eye Res 1974; 18:205–213.

Angevine JB, Sidman RL. Autoradiographic study of cell migration during histogenesis of cerebral cortex in the mouse. Nature 1961; 192:766–768.

Arbuckle LD, Innis SM. Docosahexaenoic acid is transferred through maternal diet to milk and to tissues of natural milk-fed piglets. J Nutr 1993; 123:1668–1675.

Aveldano MI. A novel group of very long chain polyenoic fatty acids in dipolyunsaturated phosphatidylcholines from vertebrate retina. J Biol Chem 1987; 262:1172–1179.

Bayer SA, Altman J. Radiation-induced interference with postnatal hippocampal cytogenesis in rats and its long-term effects on the acquisition of neurons and glia. J Compar Neurol 1975; 163:1–20.

Bayer SA. Glial recovery patterns in rat corpus collosum after x-irradiation during infancy. Exp Neurol 1977; 54:209–216.

Bayer SA. The development of the CNS. In Wiggins DW, McCandles DW, Enna SJ, eds. Developmental Neurochemistry. University of Texas Press, Austin, 1985, pp. 8–46.

Bayer SA. Cellular aspects of brain development. Neurotoxicology 1989; 10:307–320.

Bazan NG. Supply of n-3 polyunsaturated fatty acids and their significance in the central nervous system. In Wurtman KJ and Wurtman JJ, eds. Nutrition and the Brain. Raven Press Ltd, New York, NY.

Benolken RM, Anderson RE, Wheeler TG. Membrane fatty acids associated with the electrical response in visual excitation. Science 1973; 182:1253–1254.

Birch EE, Birch DG, Hoffman DR, Uauy R. Dietary essential fatty acid supply and visual acuity development. Invest Ophthalmol Vis Sci 1992; 33:3242–3253.

Bourre JM, Bonneil M, Clement M, Dumont O, Durand G, Nalbone G, Lafont H. Function of dietary polyunsaturated fatty acids in the nervous system. Prostaglandins Leukotrienes Essential Fatty Acids 1993; 48:5–15.

Bourre JM, Durand G, Pascal G, Youyou A. Brain cell and tissue recovery in rats made deficient in n-3 fatty acids by alteration of dietary fat. J Nutr 1989; 119:15–22.

Bourre JM, Francis M, Youyou A, Dumont O, Piciotti M, Pascal G, et al. The effects of dietary α-linolenic acid on the composition of nerve membranes, enzymatic activity, amplitude of electrophysiological parameters, resistance to poisons and performance of learning tasks in rats. J Nutr 1989; 119:1880–1892.

Bourre JM, Pascal G, Durand G, Masson M, Dumont O, Piciotti M. Alterations in the fatty acid composition of rat brain cells (neurons, astrocytes and oligodendrocytes) and of subcellular fractions (myelin and synaptosomes) induced by a diet devoid of n-3 fatty acids. J Neurochem 1984; 43:342–348.

Bourre JM, Piciotti M, Dumont O. Δ-Desaturase in brain and liver during development and aging. Lipids 1990; 25:354–356.

Bowen RAR, Clandinin MT. High dietary 18:3n-3 increases the 18:3n-3 but not the 22:6n-3 content in the whole body, brain, skin, epididymal fat pads, and muscles of suckling rat pups. Lipids 2000; 35:389–394.

Bowen RAR, Wierzbicki AA, Clandinin MT. Does increasing dietary linolenic acid content increase the docosahexaenoic acid content of phospholipids in neuronal cells of neonatal rats? Pediatr Res 1999; 45:815–819.

Brenner RR. Effect of unsaturated acids on membrane structure and enzyme kinetics. Prog Lipid Res 1984; 23:69–96.

Capaldi RA., ed. Membrane Proteins and Their Interactions with Lipids. Marcel Dekker, New York, 1977, p. 1.

Carlson SE, Ford AJ, Werkman SH, Peeples JM, Koo WWK. Visual acuity and fatty acid status of term infants fed human milk and formula with and without docosahexaenoate and arachidonate from egg yolk lecithin. Pediatr Res 1996; 39:882–888.

Carlson SE, Werkman SH, Peeples JM, Cooke RJ, Tolley EA. Arachidonic acid status correlates with first year growth in preterm infants. Proc Natl Acad of Sci USA 1993; 90:1073–1077.

Carlson SE, Werkman SH, Peeples JM, Wilson WM III. Growth and development of premature infants in relation to n-3 and n-6 fatty acid status. World Rev Nutr Diet 1994; 75:63–69.

Carlson SE, Werkman SH, Rhodes PG, Tolley EA. Visual-acuity development in healthy preterm infants: effect of marine-oil supplementation. Am J Clin Nutr 1993; 58:35–42.

Chen H, Anderson RE. Differential incorporation of docosahexaenoic and arachidonic acids in frog retinal pigment epithelium. J Lipid Res 1993; 34:1943–1955.

Chen H, Anderson RE. Metabolism in frog retinal pigment epithelium of docosahexaenoic and arachidonic acids derived from rod outer segment membrane. Exp Eye Res 1993; 57:369–377.

Chen H, Wiegand RD, Koutz CA, Anderson RE. Docosahexaenoic acid increases in frog retinal pigment epithelium following rod photoreceptor shedding. Exp Eye Res 1992; 55:93–100.

Chen Y, Houghton LA, Brenna JT, Noy N. Docosahexaenoic acid modulates the interactions of the interphotoreceptor retinoid-binding protein with 11-*cis*-retinal. J Biol Chem 1996; 271:20,507–20,515.

Chen Y, Saari JC, Noy N. Interactions of all-*trans*-retinol and long-chain fatty acids with interphotoreceptor retinoid-binding protein. Biochemistry 1993; 32:11,311–11,318.

Clandinin MT. Influence of diet fat on membranes. In: Bittar E, Bittar N, eds. Membrane and Cell Signaling. Principles of Medical Biology. JAI, Greenwich, CT, 1997, Vol. 7A, pp. 93–119.

Clandinin MT, Chappell JE, Van Aerde JEE. Requirements of newborn infants for long-chain polyunsaturated fatty acids. Acta Paediatr Scand 1989; 351(Suppl):63–71.

Clandinin MT, Chappell JE, Heim T, Swyer PR, Chance GW. Fatty acid utilization in perinatal *de novo* synthesis of tissues. Early Hum Dev 1981; 5:355–366.

Clandinin MT, Chappell JE, Heim T, Swyer PR, Chance GW. Fatty acid accretion in fetal and neonatal liver: implications for fatty acid requirements. Early Hum Dev 1981; 5:7–14.

Clandinin MT, Chappell JE, Leong S, Heim T, Swyer PR, Chance GW. Intrauterine fatty acid accretion rates in human brain: implications for fatty acid requirements. Early Hum Dev 1980; 4:121–129.

Clandinin MT, Chappell JE, Leong S, Heim T, Swyer PR, Chance GW. Extrauterine fatty acid accretion rates in human brain: implications for fatty acid requirements. Early Hum Dev 1980; 4:130–138.

Clandinin MT, Cheema S, Field CJ, Garg ML, Venkatraman JT, Clandinin TR. Dietary fat: exogenous determination of membrane structure of cell function. Fed Am Soc Exp Biol J 1991; 5:2761–2769.

Clandinin MT, Field CJ, Hargreaves K, Morson L, Zsigmond E. Role of diet fat in subcellular structure and function. Can J Physiol Pharmacol 1985; 63:546–556.

Clandinin MT, Van Aerde JE, Parrott A, Field CJ, Euler A. R, Lien EL. Assessment of the efficacious dose of arachidonic and docosahexaenoic acids in preterm infant formulas: fatty acid composition of erythrocyte membrane lipids. Pediatr Res 1997; 42:819–825.

Clandinin MT, Wong K, Hacker RR. Synthesis of chain elongation–desaturation products of linoleic acid by liver and brain microsomes during development of the pig. Biochem J 1985; 226:305–309.

Cohen SR, Bernsohn J. The in vivo incorporation of linolenic acid into neuronal and glial cells and myelin. J Neurochem 1978; 30:661–669.

Connor WE, Neuringer M, and Lin DS. Dietary Effects on brain fatty acid composition: The reversibility of n-3 fatty acid deficiency turnover of docosahexaenoic acid in the brain erythrocytes and plasma of rhesus monkeys. J Lipid Res 1990; 31:237–247.

Connor WE, Neuringer M, and Reisbick S. Essentiality of ω3 fatty acids: Evidence from the primate model and implications for human nutrition. In: Simopoulos AP, Kifer RR, Martin RE, and Barlow SM, eds. Health effects of ω3 polyunsaturated fatty acids in seafoods. 1991; 66:118–132, Karger Press, Basel, Switzerland.

Cook HW. *In vitro* formation of polyunsaturated fatty acids by desaturation in rat brain: some properties of the enzymes in developing brain and comparisons with liver. J Neurochem 1978; 30:1327–1334.

Cowan WM. The development of the brain. In: The Brain. New York: Scientific American 241(3):113–133.

Crastes de Paulet P, Sarda P, Boulot P, Crastes de Paulet A. Fatty acids blood composition in fetal and maternal plasma. In: Ghisolfi J, Putet G, eds. Essential Fatty Acids in Infant Nutrition. John Libbey Eurotext, Montrouge, France, 1992.

Crawford MA, Hall B, Laurance BM, Munhambo A. Milk lipids and their variability. Curr Med Res Opin 1976; 4(1):33–42.

Das GD. Gliogenesis during early embryonic development in the rat. Experientia 1977; 33:1648–1649.

Delton-Vandenbroucke I, Grammas P, Anderson RE. Polyunsaturated fatty acid metabolism in retinal and cerebral microvascular endothelial cells. J Lipid Res 1997; 38:147–159.

Dhopeshwarkar GA, Subramanian C. Biosynthesis of polyunsaturated fatty acids in the developing brain: I. Metabolic transformations of intracranially administered 1-^{14}C linolenic acid. Lipids 1976; 11:67–71.

Dhopeshwarkar GA, Subramanian C, Mead JF. Fatty acid uptake by the brain. IV. Incorporation of [1-^{14}C] linolenic acid into the adult rat brain. Biochim Biophys Acta 1971; 231:8–14.

Dhopeshwarkar GA, Subramanian C, Mead JF. Fatty acid uptake by the brain. IV. Incorporation of [1-^{14}C] linolenic acid into the adult rat brain. Biochim Biophys Acta 1971; 231:162–167.

Dobbing J, Sands J. Comparative aspects of the brain growth spurt. Early Hum Dev 1979; 3:79–83.

Dobbing J. Vulnerable periods in developing brain. In: Davison AN, Dobbing J, eds. Applied Neurochemistry. Charles C. Thomas, Springfield, IL, 1968, pp. 241–261.

Dobbing J. Undernutrition and the developing brain: the use of animal models to elucidate the human problem. Adv Exp Med Biol 1971; 13:399–412.

Dobbing J. Early nutrition and later achievement. Proc Nutr Soc 1990; 49:103–118.

Dyer JR, Greenwood CE. Neural 22-Carbon fatty acids in the weanling rat respond rapidly and specifically to a range of dietary linoleic acid to alpha-linolenic fatty acid rations. J Neurochem 1991; 56:1921–1931.

Edelstein C. Biochemistry and biology of the plasma lipoproteins. In: Scanu AM, Spector AA, eds. Marcel Dekker, New York, 1986, pp. 495–505.

Faldella G, Govoni M, Alessandroni R, Marchiani E, Paolo S, Biagi PL, Spano C. Visual evoked potentials and dietary long chain polyunsaturated fatty acids in preterm infants. Arch Dis Child 1996; 75:F108–F112.

Foot M, Cruz TF, Clandinin MT. Effect of dietary lipids on synaptosomal acetylcholinesterase activity. Biochem J 1983; 211:507–509.

Foot M, Cruz TF, Clandinin MT. Influence of dietary fat on the lipid composition of rat synaptosomal and microsomal membranes. Biochem J 1982; 208:631–640.

Gordon WC, Bazan NG. Visualization of [^{3}H]Docosahexaenoic acid trafficking through photoreceptors and retinal pigment epithelium by electron microscopic autoradiography. Invest Ophthalmol Vis Sci 1993; 34:2402–2411.

Gottlieb A, Keyder I, Epstein HT. Rodent brain growth stages: an analytical review. Biol Neonate 1977; 32:166–176.

Greenwood CE, Craig REA. Dietary influences on brain function: implications during periods of neuronal maturation. In: Rassin DK, Haber B, Drujan B, eds. Current Yopics in Nutrition and Diseases. Alan R. Liss, New York, 1987, pp. 159–216.

Hargreaves KM, Clandinin MT. Dietary lipids in relation to postnatal development of the brain. Upsala J Med Sci 1990; 48(Suppl):79–95.

Herschkowitz N. Brain development and nutrition. In: Evrard P, Minkowski A, eds. Developmental Neurobiology. Raven Press, New York, 1989, p. 297.

Heuther G. Malnutrition and developing synaptic transmitter systems: lasting effects, functional implications. In: van Gelder NM, Butterworth RF, Drujan DB, eds. (Mal) Nutrition and the Infant Brain. Wiley-Liss, New York, 1990, pp. 141–156.

Jacobson M. Histogenesis and morphogenesis of the nervous system. In: Jacobson M, ed. Developmental Neurobiology. Holt, Rinehart & Winston, New York, 1970, pp. 1–74.

Jacobson M. Developmental Neurobiology. Plenum, New York, 1978, pp. 115–210.

Jhang K, Kniazeva M, Han M, Li W, Yu Z, Yang Z, et al. A 5-bp deletion in *ELOVL4* is associated with two related forms of autosomal dominant macular dystrophy. Nature Genet 2001; 27:89–93.

Jongmans M, Mercuri E, Henderson S, De vries L, Sonksen P, Dubowitz L. Visual function of prematurely born children with and without perceptual-motor difficulties. Early Hum Dev 1996; 45:73–82.

Jope RS, Jenden DJ. Choline and phospholipid metabolism and the synthesis of acetylcholine in rat brain. J Neurosci Res 1979; 4:69–82.

Jorgensen MH, Hernell O, Lund P, Holmer G, Michaelsen KF. Visual acuity and erythrocyte docosahexaenoic acid status in breast-fed and formula-fed term infants during the first four months of life. Lipids 1996; 31:99–105.

Jorgenson MH, Holmer G, Lund P, Hernell O, Fleischer Michaelsen K. Effect of formula supplemented with docosahexaenoic acid and γ-linolenic acid on fatty acid status and visual acuity in term infants. J Pediatr Gastroenterol Nutr 1998; 26:412–421.

Jumpsen JA, Lien EL, Goh YK, Clandinin MT. Small changes of dietary (n-6) and (n-3) fatty acid content alter phosphatidylethanolamine and phosphatidylcholine fatty acid composition during development of neuronal and glial cells in rats. J Nutr 1997; 127(5):724–731.

Jumpsen JA, Lien E, Goh YK, Clandinin MT. During neuronal and glial cell development diet n-6 to n-3 fatty acid ratio alters the fatty acid composition of phosphatidylinositol and phosphatidylserine. Biochim Biophys Acta 1997; 1347:40–50.

King KC, Adam PAJ, Laskowski DE, Schwartz R. Sources of fatty acids in the newborn. Pediatrics 1971; 47:192–198.

Koblin DD, Dong DE, Deady JE, Eger EI. Alteration of synaptic membrane fatty acid composition and anesthetic requirement. J Pharm Exp Ther 1980; 212:546–552.

Kohn G, Sawatzki G, Van Biervliet JP, Rosseneu M. Diet and the essential fatty acid status of term infants. Acta Pediatr Scand 1994; 402(Suppl):69–74.

Koletzko B, Braun M. Arachidonic acid and early human growth: is there a relations? Ann Nutr Metab 1991; 35:128–131.

Lamptey MS, Walker BL. A possible role for dietary linoleic acid in the development of the young rat. J Nutr 1976; 106:86–93.

Leaf A, Gosbell A, McKenzie L, Sinclair A, Favilla I. Long chain polyunsaturated fatty acids and visual function in preterm infants. Early Hum Dev 1996; 45:35–53.

Lee AG, East JM, Froud RJ. Are essential fatty acids essential for membrane function? Prog Lipid Res 1986; 25:41–46.

Lee J, Jiao XP, Gentleman S, Wetzel MG, O'Brien P, Chader GJ. Soluble binding proteins for docosahexaenoic acid are present in neural retina. Invest Ophthalmol Vis Sci 1995; 36:2032–2039.

Lee RE. Membrane engineering to rejuvenate the aging brain. Can Med Assoc J 1985; 132:325–327.

Li J, Wetzel MG, O'Brien PJ. Transport of n-3 fatty acids from the intestine to the retina in rats. J Lipid Res 1992; 33:539–548.

Lin DS, Anderson GJ, Connor WE, Neuringer M. Effect of dietary n-3 fatty acids upon the phospholipid molecular species of the monkey retina. Invest Ophthalmol Vis Sci 1994; 35:794–803.

Lou HC. Developmental Neurology. Raven, New York, 1982, pp. 1–132

Lucas A, Morley R, Cole TJ, Lister G, Leeson-Payne C. Breast milk and subsequent intelligence quotient in children born preterm. Lancet 1992; 339:261–264.

Makrides M, Neuman M, Simmer K, Pater J, Gibson R. Are long-chain polyunsaturated fatty acids essential nutrients in infancy? Lancet 1995; 345:1463–1468.

Marinetti GV, Cattieu K. Tightly (covalently) bound fatty acids in cell membrane proteins. Biochim Biophys Acta 1982; 685:109–116.

Marshall WA. The growth of the brain. In: Development of the Brain. Marshall WA, ed. Oliver and Boyd, Edinburgh, 1968, pp. 1–14.

Martin RE, De Turco EBR, Bazan N. Developmental maturation of hepatic n-3 polyunsaturated fatty acid metabolism: supply of docosahexaenoic acid to retina and brain. J Nutral Biochemistry 1994; 5:151–160.

McLean PD. The triune brain, emotion and scientific bias. In: Schmitt, FO, ed. The Neurosciences: Second Study Program. Rockefeller University, New York, 1970, pp. 336–348.

Menon NK, Dhopeshwarkar GA. Essential fatty acid deficiency and brain development. Prog Lipid Res 1982; 21:309–326.

Mills DE, Young RP, Young C. Effects of prenatal and early postnatal fatty acid supplementation on behavior. Nutr Res 1988; 8:273.

Moore SA. Cerebral endothelium and astrocytes cooperate in supplying docosahexaenoic acid to neurons. Adv Exp Med Biol 1993; 331:229–233.

Moore SA, Yoder E, Spector AA. Role of the blood-brain barrier in the formation of long-chain n-3 and n-6 fatty acids from essential fatty acid precursors. J Neurochem 1990; 55:391–402.

Moore SA, Yoder E, Murphy S, Dutton GR, Spector AA. Astrocytes, not neurons, produce docosahexaenoic acid (22:6n-3) and arachidonic acid (20:4n-6). J Neurochem 1991; 56,518–524.

Morgane PJ, Austin-Lafrance R, Bronzino J, Tonkiss J, Diaz-Cintra S, Cintra L, et al. Prenatal malnutrition and development of the brain. Neurosci Behav Rev 1993; 17:91–128.

Neuringer M, Connor WE. N-3 Fatty acids in the brain and retina: evidence for their essentiality. Nutr Rev 1986; 44:285–294.

Neuringer M, Connor WE. Omega-3 fatty acids in the retina. In: Galli C, Simopolos AP, eds. Dietary Omega-3 and Omega-6 Fatty Acids. Plenum, New York, 1989, pp. 177–180.

Neuringer M, Connor WE, Lin DS, Barstad L, Luck S. Biochemical and functional effects of prenatal and postnatal omega-3 fatty acid deficiency on retina and brain of rhesus monkeys. Proc Natl Acad Sci USA 1986; 83:4021–4025.

Neuringer M, Connor WE, Van Patten C, Barstad L. Dietary omega-3 fatty acid deficiency and visual loss in infant rhesus monkeys. J Clin Invest 1984; 73:272–276.

Paoletti R, Galli C. Lipid Malnutrition and the Developing Brain. Ciba Foundation Symposium. Elsevier Biomedical, Amsterdam, 1972, pp. 121–140.

Pawlosky RJ, Denkins Y, Ward G, Salem N Jr. Retinal and brain accretion of long-chain polyunsaturated fatty acids in developing felines: the effects of corn oil-based maternal diets. Am J Clin Nutr 1997; 65:465–472.

Poissennet CM, LaVelle M, Burdi AR. Growth and development of adipose tissue. J Pediatr 1988; 113:1–9.

Poulos A, Gibson R, Sharp P, Beckman K, Tech M, Grattan-Smith P. Very long chain fatty acids in X-linked adrenoleukodystrophy brain after treatment with Lorenzo's oil. Neurology 1994; 36:741–746.

Rakic PJ. Genesis of the dorsal lateral geniculate nucleus in the Rhesus monkey: site and time origin, kinetics and proliferation, routes of migration and pattern distribution of neurons. Compar Neurol 1977; 176:23–52.

Reisbick S, Neuringer M, Hasnain R, Connor WE. Home cage behavior of Rhesus monkeys with long-term deficiency of omega-3 fatty acids. Physiol Behav 1994; 55:231–239.

Robertson AF, Sprecher H. A review of human placental lipid metabolism and transport. Acta Paediatr Scand 1968; 183(Suppl):2–18.

Rodier PM. Chronology of neuron development: animal studies and their clinical implications. Dev Med Child Neurol 1980; 22:525–545.

Rouser G, Yamamoto A, Kritchevsky G. Cellular membranes. Structure and regulation of lipid class composition species differences, changes with age and variations in some pathological states. Arch Intern Med 1971; 127:1105–1121.

Salem N Jr, Shingu T, Kim HY, Hullin F, Bougnoux P, Karanian JW. Specialization in membrane structure and metabolism with respect to polyunsaturated lipids. In: Karnovsky ML, Leaf A, Bollis LC, eds. Biological Membranes: Aberrations in Nembrane Structure and Function. Alan R. Liss, New York, 1988, pp. 319–333.

Sastry PS. Lipids of nervous tissue: composition and metabolism. Prog Lipid Res 1985; 24:69–176.

Scott BL, Bazan NG. Membrane docosahexaenoate is supplied to the developing brain and retina by the liver. Proc Natl Acad Sci USA 1989; 86:2903–2907.

Sinclair AJ, Crawford MA. The accumulation of arachidonate and docosahexaenoate in developing rat brain. Neurochemistry 1972; 19:1753–1758.

Sjostrand J. Proliferative changes in glial cells during nerve regeneration. Zeitschr Fur Zellforsch Mikroskop Anat 1965; 68:481–493.

Smith S, Abraham S. Fatty acid synthesis in developing mouse liver. Arch Biochem Biophys 1970; 136:112–121.

Stinson AM, Wiegand RD, Anderson RE. Recycling of docosahexaenoic acid in rat retinas during n-3 fatty acid deficiency. J Lipid Res 1991; 32:2009–2017.

Stubbs CD, Smith AD. The modification of mammalian membrane polyunsaturated fatty acid composition in relation to membrane fluidity and function. Biochim Biophys Acta 1984; 779:89–137.

Suh M, Wierzbicki AA, Clandinin MT. Dietary fat alters membrane composition in rod outer segments in normal and diabetic rats: impact on content of very-long-chain (C$\geq$24) polyenoic fatty acids. Biochim Biophys Acta 1994; 1214:54–62.

Suh M, Wierzbicki AA, Clandinin MT. Dietary 20:4n-6 and 22:6n-3 modulates the profile of long- and very-long-chain fatty acids, rhodopsin content and kinetics in developing photoreceptor cells. Pediatr Res 2000; 48(4):524–530.

Suh M, Wierzbicki AA, Clandinin MT. Relationship between dietary supply of long-chain fatty acids and membrane composition of long- and very long chain essential fatty acids in developing rat photoreceptors. Lipids 1996; 31:61–64.

Sun GT, Sun AY. Synaptosomal plasma membranes: acyl group composition of phosphoglycerides and (Na$^+$ + K$^+$)-ATPase activity during essential fatty acid deficiency. J Neurochem 1974; 22:15–18.

Svennerholm L, Stallberg-Stenhagen S. Changes in the fatty acid composition of cerebrosides and sulfatides of human nervous tissue with age. J Lipid Res 1968; 9:215–225.

Tahin QS, Blum M, Carafoli E. The fatty acid composition of subcellular membranes of rat liver, heart and brain: diet-induced modifications. Eur J Biochem 1981; 121:5–13.

Tanford C. The hydrophobic effect and the organization of living matter. Science 1978; 200:1012–1018.

Uauy R, Birch DG, Birch EE, Tyson JE, Hoffman DR. Effect of dietary omega-3 fatty acids on retinal function of very-low-birth-weight neonates. Pediatr Res 1990; 28:485–492.

Uauy R, Birch EE, Birch DG, Peirano P. Visual and brain function measurements in studies of n-3 fatty acid requirements of infants. J Pediatr 1992; 120:S168–S180.

Uauy RM, Treen M, Hoffman DR. Essential fatty acid metabolism and requirements during development. Semin Perinatol 1989; 13:118–130.

Van Houwelingen AC, Puls J, Hornstra G. Essential fatty acid status during early human development. Early Hum Dev 1992; 31:97–111.

Vitiello F, Gombos G. Cerebellar development and nutrition. In: Rassin DK, Haber B, Drujan B, eds. Basic and Clinical Aspects of Nutrition and Brain Development. Alan R. Liss, New York, 1987, pp. 99–130.

Voss A, Reinhart M, Sankarappa S, Sprecher H. The metabolism of 7,10,13,16,19-docosapentaenoic acid to 4,7,10,13,16,19-docosahexaenoic acid in rat liver is independent of a 4-desaturase. J Biol Chem 1991; 266:19,995–20,000.

Wang N, Anderson RE. Synthesis of docosahexaenoic acid by retina and retinal pigment epithelium. Biochemistry 1993; 32:13,703–13,709.

Wang N, Anderson RE. Transport of 22:6n-3 in the plasma and uptake into retinal pigment epithelium and retina. Exp Eye Res 1993; 57:225–233.

Wang N, Wiegand RE, Anderson RE. Uptake of 22-carbon fatty acids into rat retina and brain. Exp Eye Res 1992; 54:933–939.

Weisinger HS, Vingrys AJ, Sinclair AJ. The effect of docosahexaenoic acid on the electoretinogram of the guinea pig. Lipids 1996; 31:65–70.

Weisinger HS, Vingrys AJ, Sinclair AJ. Effect of dietary n-3 deficiency on the electroretinogram in the guinea pig. Ann Nutr Metab 1996; 40:91–98.

Weisinger HS, Vingrys AJ, Bui BV, Sinclair AJ. Effects of dietary n-3 fatty acid deficiency and repletion in the guinea pig retina. Invest Ophthalmol Vis Sci 1999; 40:327–338.

Werkman SH, Carlson SE. A randomized trial of visual attention of preterm infants fed docosahexaenoic acid until nine months. Lipids 1996; 31:91–97.

Widdowson EM. Growth and composition of the human fetus and newborn. In: Assali NS, ed. Biology of Gestation. Academic, New York, 1968, pp. 1–49.

Widdowson EM, Southgate DAT, Hay EN. In: Visser HKA, ed. Proceedings of the Fifth Nutricia Symposium on Nutrition and Metabolism of the Fetus and Infant. Martinus Nijhoff, The Hague, 1979, pp. 169–177.

Winick M. Malnutrition and brain development. J Pediatr 1969; 74:667–679.

Wiegand RD, Koutz A, Stinson AM, Anderson RE. Conservation of docosahexaenoic acid in rod outer segments of rat retina during n-3 and n-6 fatty acid deficiency. J Neurochem 1991; 57:1690–1699.

Wurtman RJ, Hefti F, Melamed E. Precursor control of neurotransmitter synthesis. Pharmacology Review 1981; 32:315–335.

Wykle RL. Brain. In: Snyder F, ed. Lipid Metabolism in Mammals. Plenum, New York, 1977, pp. 317–366.

Yamamoto N, Saitoh M, Moriuchi A, Nomura M, Okuyama H. Effect of dietary linolenate/linoleate balance on brain lipid compositions and learning ability of rats. J Lipid Res 1987; 28:144–151.

Yamamoto N, Hashimoto A, Takemoto Y, Okuyama H, Nomura M, Kitjaima R, et al. Effect of the dietary alpha-linolenate/linoleate balance on lipid compositions and learning ability of rats. II. Discrimination process, extinction process, and glycolipid composition. J Lipid Res 1988; 29:1013–1021.

Yavin E, Menkes JH. Polyenoic acid metabolism in cultured dissociated brain cells. J Lipid Res 1974; 15:152–157.

Yeagle PI. Lipid regulation of cell membrane structure and function. Fed Am Soc Exp Biol J 1989; 3:1833–1842.

11
Dietary N-3 Fatty Acid Deficiency and Its Reversibility
Effects Upon Brain Phospholipids and the Turnover of Docosahexaenoic Acid in the Brain and Blood

William E. Connor, Gregory J. Anderson, and Don S. Lin

1. INTRODUCTION

Humans need fat in the diet in order to survive, grow, and prosper. Over 60 yr ago, Burr and Burr demonstrated that fat was an essential component of the diet (1929). Rats reared on a fat-free diet failed to grow and reproduce and also developed renal disease, fatty liver, dermatitis, and necrosis of the tail. Later studies identified the deficient components of the diet as the polyunsaturated fatty acids with two or more double bonds (Holman, 1968). Until recently, linoleic acid, an 18:2\n-6 fatty acid commonly found in vegetable oils and many other foods, was deemed the primary essential fatty acid, together with its derivative n-6 fatty acids, of which arachidonic acid (20:4) was the most important. The second series of highly polyunsaturated fatty acids includes the n-3 fatty acids, α-linolenic acid (18:3) and its longer-chained, more polyunsaturated derivative, docosahexaenoic acid (22:6). n-3 Fatty acids are also contained in a wide variety of natural foodstuffs, including some but not all liquid vegetable oils, particularly soybean and rapeseed oils, and green leafy vegetables. The higher n-3 fatty acid derivative, docosahexaenoic acid, is found particularly in plants of the sea, phytoplankton, or, as one ascends the food chain, in shellfish, fish, and sea mammals and in the tissues of terrestrial animals but not in land plants. Docosahexaenoic acid is especially predominant in the brain, retina, and spermatozoa but is also found in phospholipid membranes throughout the body, including the myocardium (Lin et al., 1993). It now appears that both n-6 and n-3 fatty acids are essential for health.

Aside from essential amino acids, the n-3 and n-6 fatty acids constitute the largest chemical component of the cerebral cortex and retina that can be obtained only from the diet. The body cannot synthesize either the n-3 or the n-6 structure. Once the basic n-3 structure is consumed in the form of linolenic acid (18:3), the body can synthesize the longer-chained and highly polyunsaturated fatty acid docosahexaenoic acid (22:6), which is the n-3 fatty acid so predominant in the nervous system. Similarly, in the n-6 series, linoleic acid (18:2) is converted to arachidonic acid (20:4), the predominant n-6 fatty acid

From: *Fatty Acids: Physiological and Behavioral Functions*
Edited by: D. Mostofsky, S. Yehuda, and N. Salem Jr. © Humana Press Inc., Totowa, NJ

in most tissues, and also to 22:4 and 22:5. Longer-chained fatty acids of the n-6 series (22:5) have become of interest because they increase in the n-3 fatty acid dietary deficiency syndrome, which will be described subsequently. n-3 and n-6 fatty acids compete for the same desaturase enzymes; high levels of linoleic acid inhibit the conversion of linolenic acid in the n-3 series to the next form, 18:4. Thus, a high ratio of linoleic acid to linolenic acid is likely to produce the greatest depletion of longer-chained n-3 fatty acids such as docosahexaenoic acid (DHA) 22:6. It is DHA with its extreme degree of unsaturation that is present in particularly high concentrations in the synaptic membranes of the brain and in the outer segment membranes of the rods and cones of the retina

Membrane lipids constitute 50–60% of the solid matter in the brain (O'Brien, 1986), and phospholipids are quantitatively the most significant component of membrane lipids (Crawford and Sinclair, 1972). A major proportion of brain phospholipids contain long-chain polyunsaturated fatty acids of the two essential fatty acid classes, n-6 and n-3 (Crawford et al., 1976, O'Brien et al., 1964). These fatty acids usually occupy the *sn*-2 position of the brain phospholipid molecules. Normally, docosahexaenoic acid (22:6n-3, DHA) is the predominant polyunsaturated fatty acid of the phospholipids of the cerebral cortex and retina. The primate brain gradually accumulates its full complement of DHA during intrauterine life and during the first year after birth (Clandinin et al., 1980a, Clandinin et al., 1980b).

In our previous reports, we have shown that infant rhesus monkeys born from mothers fed an n-3 fatty acid-deficient diet and then also fed a deficient diet after birth developed low levels of n-3 fatty acids in the brain and retina and impairment in visual function (Neuringer et al., 1984, Connor et al., 1984, Neuringer et al., 1986). The specific biochemical markers of the n-3-deficient state were a marked decline in the DHA of the cerebral cortex and a compensatory increase in n-6 fatty acids, especially docosapentaenoic acid (22:5n-6). Thus, the sum total of the n-3 and n-6 fatty acids remained similar, about 50% of the fatty acids in phosphatidylethanolamine and phosphatidylserine, indicating the existence of mechanisms in the brain to conserve polyunsaturation of membrane phospholipids as much as possible despite the n-3-deficient state.

In this chapter, our intent is to review the rapid changes that take place in the cerebral cortex of n-3 fatty acid-deficient monkeys when their diet is subsequently supplied with ample dietary n-3 fatty acids. Juvenile rhesus monkeys who had developed n-3 fatty acids deficiency since intrauterine life were repleted with a fish-oil diet rich in n-3 fatty acids, DHA, and 20:5n-3 (eicosapentaenoic acid, EPA). The fatty acid composition was determined for the lipid classes of plasma and erythrocytes and for the phospholipid classes and molecular species of frontal cortex samples obtained from serial biopsies and at the time of autopsy. From these analyses, the half-lives of DHA and EPA in the phospholipids of plasma, erythrocytes, and cerebral cortex were estimated. The deficient brain rapidly regained a normal or even supernormal content of DHA with a reciprocal decline in n-6 fatty acids, demonstrating that the fatty acids of the gray matter of the brain turn over with relative rapidity under the circumstances of these experiments.

The establishment of the n-3 fatty acid deficiency state was documented by the biochemical changes, electroretinographic abnormalities, and visual acuity loss, as described previously in these monkeys before repletion (Neuringer et al., 1984, Connor et al., 1984, Neuringer et al., 1986). Beginning at 10–24 mo of age, five juvenile monkeys were then given the same semipurified diet with fish oil replacing 80% of the safflower oil as the fat source (Connor et al., 1990a). The remaining 20% safflower oil provided ample amounts of n-6 fatty acids as linoleic (18:2n-6) at 4.5% of calories (Table 1).

Table 1
Fatty Acid Composition of the Experimental Diets (Percent of Total Fatty Acids)

Fatty acid	Deficient diet (safflower oil)	Repletion diet (fish oil)[a]	Control diet (soy oil)
16:0	7.1	15.8	10.7
18:0	2.5	3.1	4.2
Saturated	10.0	27.7	16.4
18:1(n-9)	13.3	13.0	23.7
Monounsaturated	13.8	28.4	24.2
18:2(n-6)	76.0	1.7	53.1
20:4 (n-6)	—	1.1	—
Total (n-6)	76.5	3.6	53.4
18:3(n-3)	0.3	0.7	7.7
20:5 (n-3)	—	16.4	—
22:6 (n-3)	—	11.0	—
Total (n-3)	0.3	34.2	7.7
n-6/n-3	255.0	0.1	7.0

[a]Consists of 4 : 1 mixture of fish oil (San Omega® or MaxEPA®) and safflower oil.

Changing from the n-3 fatty acid-deficient diet (safflower oil) to the n-3-rich diet (fish oil) increased the total plasma n-3 fatty acids greatly, from 0.1% to 33.6% of total fatty acids (Fig. 1). EPA, which was especially high in the fish oil, contributed the major increase, from 0% to 22.1%, and represented 66% of the total n-3 fatty acid increase. DHA increased from 0.1% to 8.3% and 22:5n-3 from 0% to 2.1%. A major reciprocal decrease occurred in the n-6 fatty acid linoleic acid, which was reduced from 54.3% to 9.2% of total fatty acids and total n-6 fatty acids fell from 65.4% to 15.5%. The change in arachidonic acid, however, was relatively small, from 6.4% to 5.5%. These changes were certainly manifest as early as 2 wk after repletion and were completed by 6–8 wk. The restored fatty acid values then remained constant until autopsy.

The four major plasma lipid fractions (phospholipids, cholesteryl esters, triglycerides, and free fatty acids) exhibited similar changes in response to the fish-oil diet. In the phospholipid fraction, the n-3 fatty acids increased from 0.4% to 34% of total fatty aicds in the fish oil diet. EPA increased from 0% to 19%, accounting for 55 percent of the total increase. Linoleic acid reciprocally decreased from 36% to 5% and total n-6 fatty acids from 48% to 12% of total fatty acids. In cholesteryl esters, n-3 fatty acids increased from 0.2% to 36%. An increase in EPA from 0% to 31% accounted for 86% of the increase, a much greater proportion than in phospholipids, whereas DHA only increased from 0% to 4%. The decline in n-6 fatty acids from 77% to 23% was largely accounted for by a decrease in linoleic acid from 73% to 17%. Similar changes were seen in the triglycerides and free fatty acid fractions.

The major unsaturated fatty acids of the phospholipids of erythrocytes from before fish oil feeding to the time of the last cortical biopsy (after 12–28 wk of fish oil feeding) are displayed in Fig. 2. The total n-3 fatty acids increased from 1.3% to 31% of total fatty acids after the fish-oil diet. EPA and DHA had similar increases, from 0.2% to 14% for EPA and from 0.2% to 13% for DHA. In erythrocytes, the n-6 fatty acids showed a significant decrease, which occurred reciprocally. Arachidonic acid fell from 18% to 6% and linoleic acid declined from 24% to 7%. The n-3 fatty acids (EPA, DHA) after the fish-

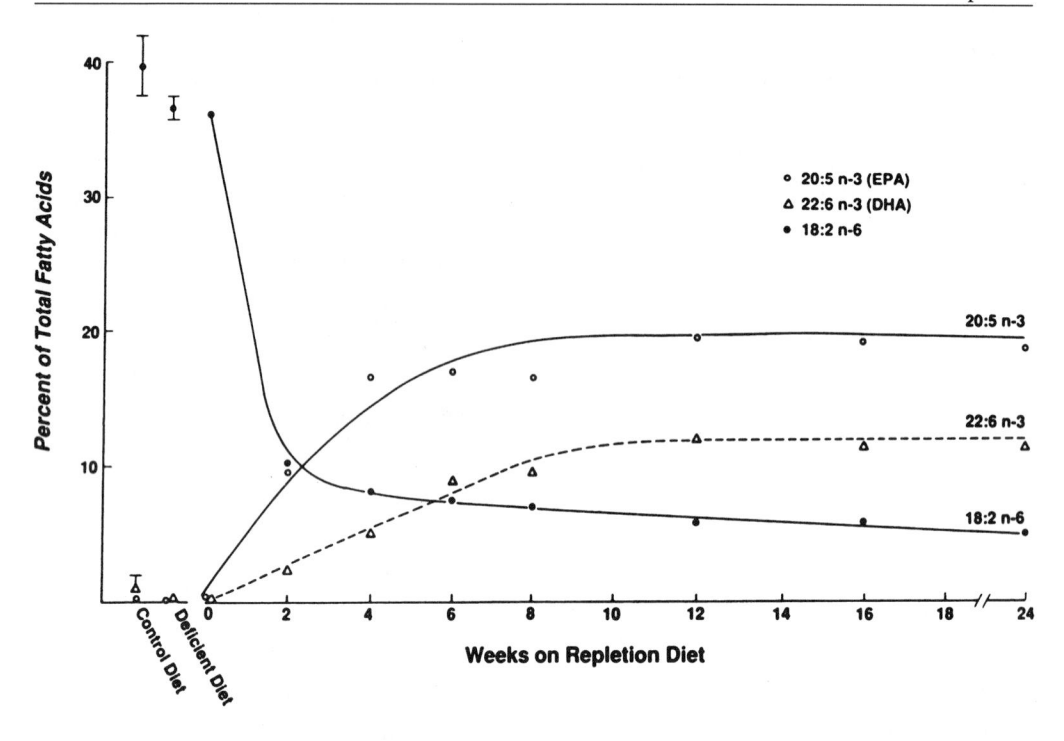

Fig. 1. The time-course of mean fatty acid changes in plasma phospholipids after the feeding of fish oil. Note that reciprocal changes of the two major n-3 (EPA and DHA) and the major n-6 (18:2) polyunsaturated fatty acids occurred as n-3 fatty acids increased and n-6 fatty acids decreased. The concentrations of these fatty acids in the plasma phospholipids of monkeys fed the control soybean oil and safflower oil diet from our previous study (Neuringer et al, 1986) are given for comparison. Expressed as percentage of total fatty acids, DHA in control monkeys was 1.1 ± 0.7%; EPA 0.2 ± 0.1%; 18:2n-6, 39.6 ± 2.3%. In deficient monkeys, DHA was 0%; 18:2n-6 was 36.7 ± 0.7%.

oil feeding increased greatly from almost 0% to about 14% of total fatty acids. The fatty acid composition of erythrocytes reached a new steady state by 12 wk after fish-oil feeding and remained constant until autopsy, but marked changes in fatty acid composition were already extensive after 2 wk of the repletion diet.

Dramatic changes in the fatty acids of the frontal cortex were detected within 1 wk after the fish-oil diet was given as demonstrated in the individual data in the frontal lobe biopsy specimens from five juvenile monkeys. All four major phospholipid classes of the brain underwent extensive remodeling of their constituent fatty acids. The data in Fig. 3 is for the fatty acids of phosphatidylethanolamine from each of the five experimental monkeys. By 12–28 wk, the total n-3 fatty acids increased from 4% to 36% of total fatty acids (Connor et al., 1990b). The major increase was in DHA, from 4% to 29%, whereas EPA and 22:5n-3, another n-3 fatty acid found in fish oil, each increased from 0% to almost 3%. To be emphasized, as will be discussed later, is the apparent conversion of EPA to DHA in the brain. The total n-6 fatty acids reciprocally decreased from 44% to 16% of the total fatty acids, with the major reduction occurring in 22:5n-6, from 18% to 2%, and 22:4n-6, from 12% to 4%. There was also a moderate decrease of arachidonic acid from 12.8% to 8.9% of total fatty acids. Again, a major remodeling of the phospholipid fatty acids from n-6 to n-3 fatty acids was evident.

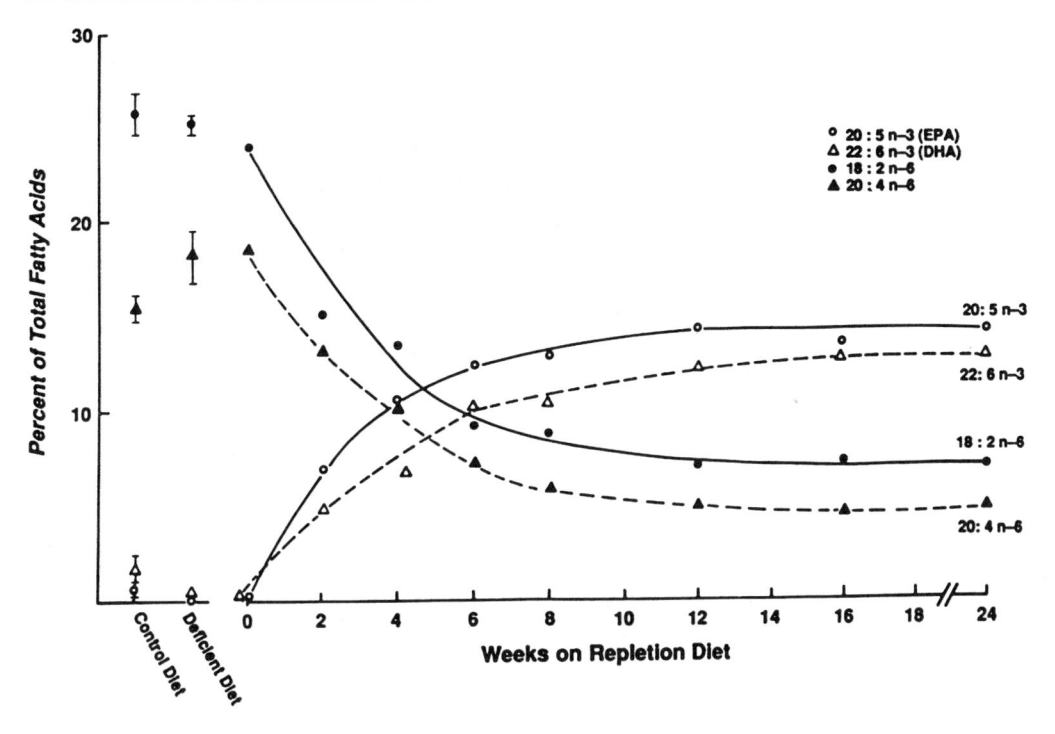

Fig. 2. The time-course of mean fatty acid changes in erythrocyte phospholipids after the feeding of fish oil. Note that reciprocal changes of the two major n-3 (EPA and DHA) and the two major n-6 (18:2 and 20:4) polyunsaturated fatty acids occurred as n-3 fatty acids increased and n-6 fatty acids decreased. The concentrations of these four fatty acids in the erythrocyte phospholipids of monkeys fed the control soybean-oil and the deficient safflower-oil diet from our previous study (Neuringer et al, 1986) are given for comparison. Expressed as percentage of total fatty acids, DHA in control monkeys was 1.7 ± 0.9%; EPA, 0.5 ± 0.1%; 20:4 n-6, 15.4 ± 0.5%; 18:2 n-6, 25.7 ± 1.0%. In deficient monkeys, DHA was 0.2 ± 0%; EPA was 0%; 20:4 n-6, 18:2 ± 1.3; 18:2 n-6 was 25.2 ± 0.4%.

In the phosphatidylserine (PS) fraction, the total n-3 fatty acids also increased greatly from 4.8% to 36% of total fatty acids with DHA increasing from 5% to 32%. Total n-6 fatty acids decreased from 38% to 11%. Again, the major and reciprocal decrease was in 22:5n-6, from 20% to 4%. In the phosphatidylinositol (PI) fraction, total n-3 fatty acids increased from 3% to 18%, whereas n-6 fatty acids decreased from 43% to 25% and 22:5n-6 decreased from 11% to 0.7%. Although the content of n-3 fatty acids in phosphatidylcholine (PC) is relatively small even in the normal brain, this fraction also showed an increase in the n-3 fatty acids from 0.4% to 5.1% after fish-oil feeding. Unlike PE and PS, the fish-oil feeding did not decrease the arachidonic acid levels in PI and PC.

In summary, fish-oil feeding to n-3 fatty acid-deficient monkeys resulted in reciprocal changes in the levels of n-3 and n-6 fatty acids in the phospholipids of cerebral cortex. The two major 22-carbon n-3 and n-6 fatty acids, DHA and 22:5n-6, were responsible for the greatest changes. Figure 3 plots the changes in these two fatty acids in phosphatidylethanolamine, plus the analogous values for juvenile monkeys fed a control (soybean oil) diet and deficient (safflower oil) diet from our previous study (Connor et al., 1990b). In control monkeys, the DHA and 22:5n-6 contents of frontal cortex were 22.3 ± 0.3 and 1.4 ± 0.3 percent of total fatty acids, respectively, whereas in deficient monkeys they were

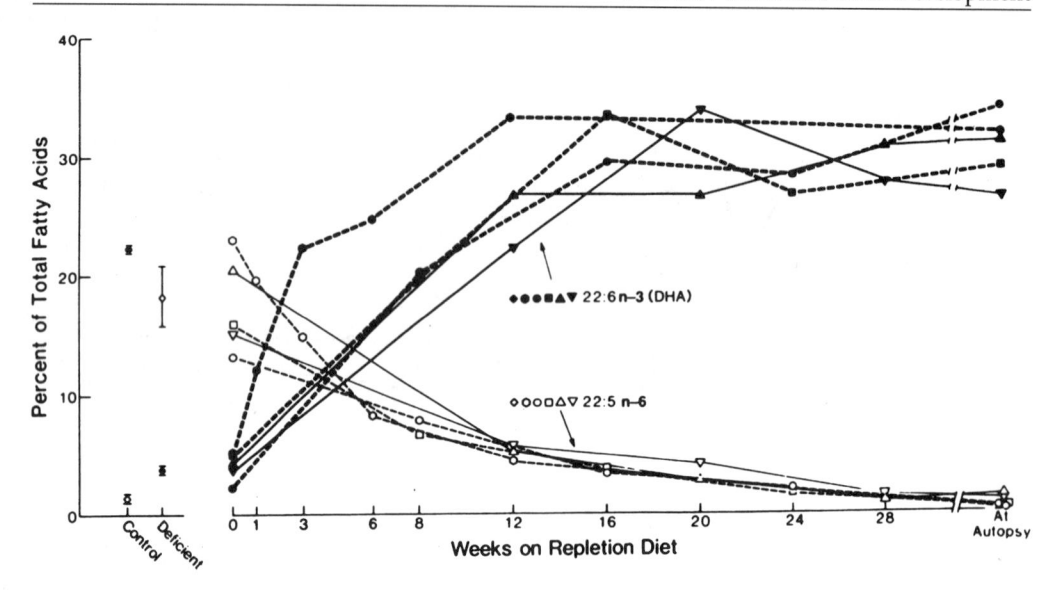

Fig. 3. The time course of fatty acid changes in phosphatidylethanolamine of the cerebral cortex of five juvenile monkeys fed fish oil for 43–129 wk. As DHA increased, 22:5n-6 decreased reciprocally. Levels of DHA and 22:5n-6 in phosphatidylethanolamine of the frontal cortex of monkeys fed control (soybean oil) and deficient diets from a previous study (Neuringer et al, 1986) are given for comparison. DHA and 22:5n-6 in control monkeys were 22.3 ± 0.3% and 1.4 ± 0.3% of total fatty acids, respectively. DHA and 22:5n-6 in the deficient monkeys were 3.8 ± 0.4% and 18.3 ± 2.5%, respectively.

nearly the reverse, 3.8 ± 0.4% DHA and 18.3 ± 2.5% 22:5n-6 (Neuringer et al., 1986). Reflecting the high content of the long-chain n-3 fatty acids of fish oil, the DHA content of cerebral cortex in the fish-oil-fed monkeys became even higher than in the soybean-oil-fed control monkeys (29.3 ± 2.6 vs 22.3 ± 0.3% $p < 0.025$).

2. TURNOVER OF FATTY ACIDS IN VARIOUS TISSUES

Using the serial data for the cerebral cortex, plasma, and erythrocytes, we constructed accumulation and decay curves for several key fatty acids in these tissues, which provided gross estimates of their turnover times after fish-oil feeding to n-3 fatty acid-deficient monkeys (Table 2). For cerebral cortex, a steady state was reached after 12 wk of fish-oil feeding for DHA, but 22:5n-6 took longer to decline to the low levels found in the cortex of control animals. The half-lives of DHA in cerebral phospholipids ranged from 17 to 21 d: 21 d for phosphatidylethanolamine, 21 d for phosphatidylserine, 18 d for phosphatidylinositol, and 17 d for phosphatidylcholine. The corresponding values for 22:5n-6 in these same phospholipids were 32, 49, 14, and 28 d, respectively. The half-lives of linoleic acid, EPA, and DHA in plasma phospholipids were estimated to be 8, 18, and 29 d, respectively. In the phospholipids of erythrocytes, linoleic acid, arachidonic acids, EPA, and DHA had half-lives of 28, 32, 14, and 21 d, respectively.

3. EFFECTS OF DIETARY n-3 FATTY ACIDS ON THE PHOSPHOLIPID MOLECULAR SPECIES OF THE MONKEY BRAIN

Phospholipids form an important component of the lipid biolayer of cell membranes. For any naturally occurring phospholipid, a large number of molecular species may exist.

Table 2
Turnover of Fatty Acids in Various Tissues (Half-Life In Days)

	DHA	EPA	22:5n-6
Plasma phospholipids	29	18	
Erythrocyte phospholipids	21	14	
Cerebral cortex			
Phosphatidylethanolamine	21		32
Phosphatidylserine	21		49
Phosphatidylinositol	18		14
Phosphatidylcholine	17		28

Each molecular species is defined by the chemical nature of the polar headgroup, the type of linkage to glycerol, and the aliphatic chains at both the *sn-1* and *sn-2* positions. Different molecular species have different metabolic and physical properties. The molecular species composition of membrane phospholipids is associated with membrane fluidity and the function and activity of membrane-bound enzymes (Lynch and Thompson, 1984, Lin et al., 1990).

In the present study, we compared the molecular species composition of ethanolamine glycerophospholipids of the frontal cortex of monkeys fed three different diets; a control soy-oil (n-3 fatty acid sufficient) diet, a safflower-oil (n-3 fatty acid deficient) diet, or a fish-oil diet rich in several n-3 fatty acids (Table 1) (Hargreaves and Clandinin, 1988). Our results demonstrated the decisive effect of different dietary fatty acids upon the composition of the molecular species of brain phospholipids. These 80 odd different molecular species provide insights into the widespread biochemical diversity imposed by the different diets.

The safflower- and fish-oil diets had a great effect upon the proportions of molecular-species containing 22:6n-3 within the diacyl subclass of the ethanolamine glycerophospholipids of the cerebral cortex (Table 3). Fish oil, which was fed to monkeys made deficient in n-3 fatty acids by prior feeding of the safflower-oil diet, raised the proportion of the species 18:0–22:6 to over 42% of total species (Fig. 4). This change was accompanied by a reciprocal decrease in species containing 22:5n-6, which rose in safflower-oil-fed animals (Fig. 5).

When moneys were fed the n-3 fatty acid-deficient diet, the species containing n-3 fatty acids decreased to 9% of total, compared to 41% in the control (soy oil) monkeys. The amount of decrease was similar for all species containing an n-3 fatty acid. Conversely, the n-6 series increased to 77% with the species containing 22:4n-6 and 22:5n-6 being responsible for most of the increase.

Fish-oil feeding to the n-3-deficient monkeys increased the proportion of species containing an n-3 fatty acid from 9% to 61% (Table 3). This n-3 enrichment of the brain was much higher than in control monkeys (41%). Because 20:5n-3 (EPA) was present in the fish oil (Table 3), fish-oil-fed monkeys synthesized a new molecular species not found in the control brain, namely 18:0–20:5n-3. The increase in n-3 fatty acids, however, did not change the ratio of *sn-1* of 16:0 to 18:0 to 18:1 for a given *sn-2* n-3 fatty acid. The total species of the n-6 series were depressed to 25% of total, with 22:5 and 22:4 decreasing to very low levels. In addition, species with 20:4 in the *sn-2* position were decreased somewhat.

Table 3
Dietary-Induced Changes in the Major Molecular Species
of Diacyl Ethanolamine Phosphoglyceride of Monkey Brain (mol%)

	Soy oil	Safflower oil	Fish oil	p
n-3				
18:0–20:5	ND^a	ND^a	2.5 ± 0.4^b	<0.001
16:0–22:5	Trace	ND	0.8 ± 0.2	
18:0–22:5	1.1 ± 0.2	0.7 ± 0.4	4.0 ± 0.2	
16:0–22:6	6.9 ± 0.9^a	1.6 ± 0.5^b	9.1 ± 0.7^c	0.003
18:0–22:6	30.9 ± 1.7^a	6.2 ± 0.6^b	41.8 ± 3.6^c	<0.001
18:1–22:6	2.1 ± 0.5^a	0.8 ± 0.6^b	3.1 ± 0.2^c	0.013
n-6				
16:0–18:2	0.5 ± 0.1	ND	1.0 ± 0.1	
18:0–18:2	0.2 ± 0.1	ND	Trace	
16:0–20:3	0.3 ± 0.1	0.8 ± 0.1	0.6 ± 0.1	
18:0–20:3	1.5 ± 0.1	0.5 ± 0.0	1.8 ± 0.2	
16:0–20:4	2.3 ± 0.5	1.9 ± 0.0	1.6 ± 0.3	
18:0–20:4	18.9 ± 1.6^a	17.8 ± 1.6^a	13.9 ± 2.2	0.046
18:1–20:4	4.6 ± 0.7	3.6 ± 0.1	3.6 ± 1.4	
16:0–22:4	1.9 ± 0.2^a	2.8 ± 0.4^b	ND^c	<0.001
18:0–22:4	8.9 ± 0.2^a	14.6 ± 0.4^b	1.4 ± 0.2^c	<0.001
16:0–22:5	1.1 ± 0.5^a	5.6 ± 0.8^b	0.3 ± 0.1^c	0.027
18:0–22:5	4.8 ± 2.5^a	28.0 ± 4.7^b	1.3 ± 0.6^c	0.053
18:1–22:5	ND^a	1.3 ± 0.0^b	ND^a	<0.001
16:0–18:0	0.6 ± 0.2	0.3 ± 0.4	0.2 ± 0.3	
16:0–18:1	3.1 ± 0.1^a	ND^b	4.8 ± 0.7^c	0.001
18:0–18:1	4.8 ± 0.4	3.2 ± 1.3	5.1 ± 0.9	
18:1–18:1	1.8 ± 0.4	1.5 ± 0.4	2.5 ± 0.4	
18:1–20:1	0.6 ± 0.7	ND	Trace	
18:0–20:3n-9	0.2 ± 0.2	3.0 ± 1.6	0.5 ± 0.2	
sn-1 16:0	16.9 ± 1.6^a	13.2 ± 0.6^b	18.1 ± 0.7^a	0.007
sn-1 18:0	71.1 ± 1.9	73.9 ± 2.8	72.5 ± 2.8	
sn-1 18:1	9.7 ± 1.4	6.9 ± 0.0	9.2 ± 1.7	
sn-2 n-3	41.0 ± 2.9^a	9.2 ± 0.8^b	61.3 ± 4.1^c	<0.001
sn-2 n-6	44.5 ± 1.6^a	76.6 ± 6.8^b	25.4 ± 3.2^c	<0.001
sn-2 n-9	10.6 ± 0.8^{a,b}	7.7 ± 3.5^a	13.5 ± 2.1^b	0.013

Note: Data are shown as means ± SD. ND = not detected. Values with unlike superscripts within a given row are different at the indicated *p*-value.

As emphasized, dietary fatty acids produced drastic modification of the molecular species of brain ethanolamine phospholipids. Hargreaves and Clandinin reported similar findings in the rat (Connor et al., 1997). Using argentation thin-layer chromatography (TLC), which is unable to resolve individual molecular species, they fed fish-oil or linseed-oil diets to rats, which resulted in an increased microsomal and synaptic membrane content of phosphatidylethanolamine species containing six double bonds, and a decrease in species containing five double bonds, compared with animals fed soy or safflower oil.

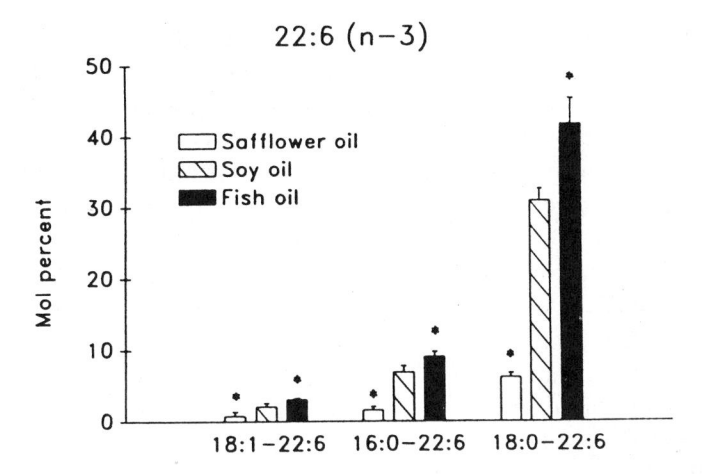

Fig. 4. Diet-induced changes in the proportion of diacyl ethanolamine glycerophospholipids molecular species containing 22:6n-3. *Monkeys fed the safflower-oil and fish-oil diets had values significantly different from the soy-oil-fed monkeys and from each other (*see* Table 3).

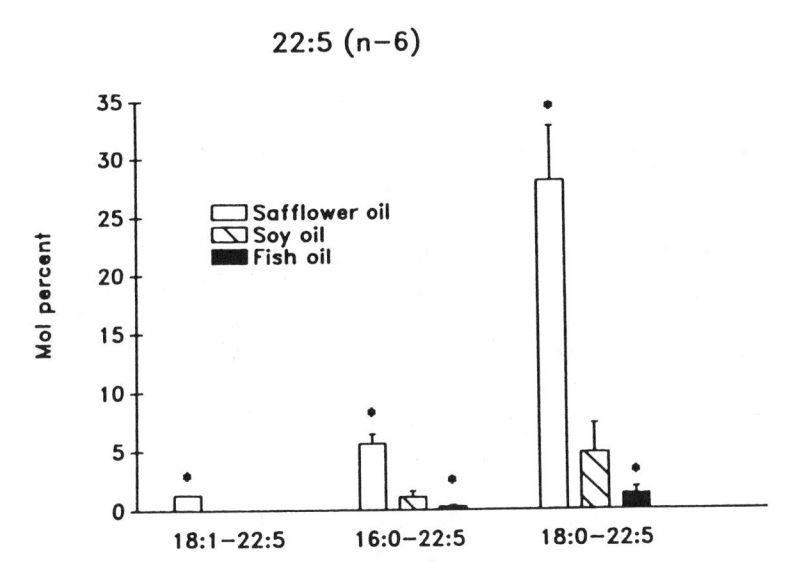

Fig. 5. Diet-induced changes in proportion of diacyl ethanolamine glycerophospholipids molecular species containing 22:5n-6. *Monkeys fed the safflower-oil and fish-oil diets had values significantly different from the soy-oil-fed monkeys (*see* Table 3). Note that for the soy-oil- and fish-oil-fed monkeys, the species 18:1–22:5 was completely absent.

Because the function of individual phospholipid molecular species is unclear at this time, the precise implication of our findings is not known. However, phospholipids are an important part of the lipid bilayer of the cell membrane. Different phospholipid molecular species would be expected to have different metabolic and physical properties. Our recent observation seems to support this assumption. When we examined the fatty acid composition of the testes of monkeys with different ages, we found that there was an increase in DHA and a decrease in arachidonic acid in the monkey testes during puberty (Reisbick et al., 1990). Interestingly, these changes were mainly due to the increase of a single molecular species, 16:0–22:6 and the decrease of 18:0–20:4 (Connor et al., unpublished data). The other molecular species with 22:6 and 20:4 at *sn-2* were little changed. This selective change of individual molecular species suggested that each molecular species may have a specific role in cellular function. Furthermore, the distribution of molecular species was found to be closely associated with cell membrane fluidity, function, and the activity of membrane-bound enzymes (Lynch and Thompson, 1984, Lin et al., 1990, Reisbick, et al., 1990). It is therefore, conceivable that the large changes in molecular species composition seen in the present study could have a significant effect on cellular metabolism and function. In fact, as demonstrated in our studies, yet already one important behavior, water ingestion, has been altered greatly by the deficient diet (Dawson, 1985). These monkeys displayed polydipsia, presumably of central origin. As already mentioned, deficient monkeys also had impaired vision and abnormal electroretinograms (Connor et al., 1984, Neuringer et al., 1986).

The substantial diet-induced changes in the phospholipid molecular species composition seen in the present study indicate a significant turnover of brain phospholipids. This turnover must have resulted from either degradation and resynthesis or deacylation–reactylation of phospholipids (Van Golde et al., 1969). Attempting to ascertain the metabolic pathway responsive for the observed changes, we calculated the ratios of 18:0–22:6 to 16:0–22:6 and 18:0–22:5 to 16:0–22:5 for the various diets. In the control (soy oil) group, the 18:0–22:6/16:0–22:6 ratios were 4.5, 3.6, and 0.9 in the diacyl, alkenylacyl, and alkylacyl subclasses, respectively. These ratios were unchanged in the safflower- and fish-oil groups. The analogous ratios for species containing 22:5n-6, a fatty acid that rises in concentration when animals are fed a diet deficient in n-3 fatty acids, were 4.5, 3.4, and 1.7, respectively, in the control diet and were unchanged on the safflower-oil diet.

From this comparison, it seems possible that the decrease of the 22:6n-3 species and the corresponding increase of the 22:5n-6 species seen in animals on the deficient diet may be the result of simple replacement of 22:6 with 22:5 through deacylation–reacylation. This is supported by the results of dual-labeling experiments in rat liver (Careaga-Houck and Sprecher, 1989), which suggested that *de novo* synthesis via phosphatidic acid is highly operative for the formation of monoenoic and dienoic molecular species of phosphatidylcholine, but that tetraenoic molecules are synthesized mainly by acylation of lysophosphatidylcholines.

On the other hand, the relative amounts of species containing 22:6 and 22:5 paired with *sn-1* 18:1, as opposed to *sn-1* 18:0 and 16:0, were quite different. This result suggested that if deacylation–reacylation was responsible for the observed remodeling of brain molecular species, it did not affect the three species containing 22:6 (i.e., 18:0–22:6, 16:0–22:6, and 18:1–22:6) in an equal manner. Furthermore, the responses of the three species containing 20:4 toward dietary change were also quite different. Comparing the control to the fish-oil diets, 18:1–20:4 in all three subclasses either remained unchanged or increased, analogous to the response of 18:1–20:4 in ethanolamine glycerophospholipids from neutrophils of rats fed fish oil (Philbrick et al., 1987). At the same time, 16:0–20:4 and 18:0–20:4 and 18:0–22:5 in the fish-oil-fed monkeys significantly decreased, as expected from the fatty acid composition data in this and other experiments (Yeo et al., 1989, Dawson and Richter, 1950, Ansell and Dohmen, 1957, Crokin and Sun, 1978).

In conclusion, the phospholipid molecular species composition of the primate brain was changed by dietary fat in a way that did not simply mirror changes in the fatty acid composition of the brain. These effects on molecular species composition included disparate responses to dietary fish oil for the various molecular species containing a given *sn*-2 fatty acid, as well as an asymmetrical replacement of 22:6n-3 by 22:5n-6 in the brains of n-3 fatty acid-deficient monkeys.

4. DISCUSSION: GENERAL

Before 1940 it was generally considered that phospholipids, once laid down in the nervous system of mammals during growth and development, were comparatively static entities. However, later studies using (^{32}P)orthophosphate showed that brain phospholipids as a whole are metabolically active in vivo (Ansell and Dohmen, 1957, Crokin and Sun, 1978). In the present study, by following the changes in phospholipid fatty acid composition, we have demonstrated that an n-3 fatty acid-enriched diet can rapidly reverse a severe n-3 fatty acid deficiency in the brains of primates. The phospholipid fatty acids of the cerebral cortex of juvenile monkeys are in a dynamic state and are subject to continuous turnover under certain defined conditions.

The observed changes in fatty acid composition could be the result of a complete breakdown and resynthesis of cortical phospholipids or a turnover of only the fatty acids in the *sn*-2 position, which has the higher proportion of polyunsaturated acyl groups. Turnover of fatty acids in the *sn*-2 position is well known and is commonly referred to as deacylation–reacylation (Fisher and Rowe, 1980, Anderson and Connor, 1988). This process could have an important role in maintaining optimal membrane composition without the high-energy cost associated with *de novo* phospholipid synthesis.

The reversibility of n-3 fatty acid deficiency in the monkey cerebral cortex was relatively rapid in our study. Effects of fish-oil feeding were seen within 1 wk after its initiation. By that time, DHA in the phosphatidylethanolamine of the cerebral cortex had more than doubled. The DHA concentration in phosphatidylethanolamine reached the control value of 22% in 6–12 wk after fish-oil feeding. We and others have demonstrated that the uptake of DHA and other fatty acids occurs within minutes after their intravenous injection bound to albumin (Pitkin et al., 1972). Furthermore, DHA is taken up by the brain in preference to other fatty acids (Pitkin et al., 1972). In contrast to the rapid incorporation of DHA from 4% to 13% in the first week, 22:5n-6 only decreased form 23% to 20% during the first week (Holub and Kuksis, 1978). This asymmetry may indicate that DHA did not exclusively displace 22:5n-6 from the *sn*-2 position of brain phospholipids.

In this study, the half-life of DHA of phosphatidylethanolamine in the cerebral cortex was similar to the half-lives of DHA in plasma and erythrocyte phospholipids, roughly 21 d. These data suggest that the blood-brain barrier present for cholesterol (Olendorf, 1975, Trapp and Bernson, 1977) and other substances may not exist for the fatty acids of the plasma phospholipids because of the relatively rapid uptake of plasma DHA into the brain. The mechanisms of transport of these fatty acids remain to be investigated.

Similar reversals of biochemical deficiencies of n-3 fatty acids or of total essential fatty acids have been studied in the rodent brain under somewhat different experimental conditions (Walker, 1967, Sanders et al., 1984, Youyou et al., 1986, Homomayoun et al., 1988, Leyton et al., 1987). In the experiments by Youyou et al., complete recovery from the n-3 fatty acid deficiency, as measured by an increase of DHA and a decrease of 22:5n-6, required 13 wk as compared to 6–12 wk in our monkeys. There were major differences between this study and ours, which may be responsible for the different recovery rates observed. Most importantly, we fed monkeys fish oil, which is high in

DHA and EPA, whereas they fed young rats soy oil, which is lower in total n-3 fatty acids and contains only the precursor of DHA, linolenic acid (18:3n-3). The *in situ* biosynthesis of DHA from 18:3n-3 may be a rate-limiting factor because of low desaturase activity and because the precursor 18:3n-3 is oxidized much more readily than DHA (Anderson et al., 1989). Dietary 18:3n-3 is also less effective in promoting biochemical recovery in n-3 fatty acid-deficient chicks than dietary DHA or EPA (Anderson et al., manuscript in preparation). Indeed, in our repletion studies (Lynch and Thompson, 1984, Connor et al., manuscript in preparation), dietary 18:3 (n-3) was also much slower than dietary EPA/DHA in achieving repletion. By 4 wk after repletion with EPA/DHA, the erythrocyte and brain already showed substantial recovery. On the other hand, monkeys repleted with 18:3n-3 showed no change in erythrocyte n-3 fatty acids. Because erythrocyte and brain DHA levels in repleting monkeys are closely correlated (Cocchi et al., 1984), we can infer that brain n-3 fatty acids had also not yet begun to recover at this time-point. Ultimately, however, dietary 18:3n-3 was effective at reversing the n-3 fatty acid deficiency. The effects of dietary n-3 fatty acids from fish oil, linseed oil, and soy oil upon the lipid composition of the rat brain have also been reported by several groups of investigators (Lin et al., 1990, Bourre et al., 1988, Tarozzi et al., 1984, Carlson 1986, Op den Kamp et al., 1985).

Because of the reciprocal changes of n-3 and n-6 fatty acids with fish-oil feeding, the sum total of n-3 and n-6 polyunsaturation of the brain of animals fed n-3 deficient and fish-oil diets remained very similar. However, the unsaturation index of phospholipids of the frontal cortex was higher in the fish-oil-fed monkeys than in n-3 deficient monkeys. The functional significance of this difference in the unsaturation index is not known. Because phospholipids rich in polyunsaturated fatty acids constitute an integral part of brain and retinal membranes, the degree of unsaturation of these fatty acids may have an important influence on the structure of the membranes and their functions, via changes in biophysical properties and/or the activities of membrane-bound proteins including enzymes, receptors, or transport systems (Connor and Neuringer, 1988).

In our n-3-deficient monkeys, the electroretinograms showed several abnormalities before fish oil was fed (Connor et al., 1990b). After repletion, when the concentration of DHA had been restored to above normal levels in the brain and retina, the electroretinograms remained abnormal (Moore et al., 1991). The reason for the failure of the electroretinogram to improve is unknown but may relate to the time of repletion in the animal's development or to the use of fish oil containing EPA as well as DHA. The use of purified DHA might be a more physiologic way of repleting n-3 fatty acid-deficient monkeys. As demonstrated in the present experiment, fish-oil feeding was able not only to reverse the n-3 fatty acid-deficient state in the brain but also to increase the n-3 acid content in brain phospholipids above control levels. Furthermore, EPA increased from zero in control monkeys to 3.1% after fish-oil feeding and 22:5n-3 increased from 0.1% to 3.5%. At the same time, arachidonic acid and 22:4n-6 decreased to below control levels. Whether this "overload" of n-3 fatty acids and perhaps unphysiologic reduction of n-6 fatty acids in brain phospholipids was advantageous or detrimental in terms of membrane function is uncertain. Similar considerations would apply to the retina and its functioning.

An interesting question relates to the blood and brain ratios of EPA and DHA. The concentrations of EPA were high in the plasma and low in the brain, with DHA high in both brain and blood. For the erythrocytes, the EPA/DHA was 1.0; for the brain, the ratio was 0.12. It is clear that EPA was metabolized, in contrast to DHA, between the blood compartment and the brain. EPA could have entered the brain and there been converted to DHA by astrocytes, which do have that capacity (Yerram et al., 1989). Less likely, it could have been catabolized with one pathway leading to the synthesis of the series 3 prostaglandins (Reisbick et al., 1994).

5. BEHAVIORAL CHANGES IN N-3 FATTY ACID DEFICIENCY

Two behavioral changes were noted in rhesus monkeys fed an n-3-deficient diet and monitored remotely (television) over 24-h periods of time. These were polydipsia and stereotypical activity. The n-3 fatty acid-deficient monkeys were found to visit the waterspout much more often and drank more water than control. The total 24-h water intake for the deficient group was 365 ± 39 mL/kg compared to 158 ± 50 mL/kg for the control group ($p<0.05$). Both groups consumed the liquid-formula diet in the same amounts. However, as expected, the fluid output in the urine plus the stool was considerably more in the n-3-deficient group. This amounted to 268 g/kg in the deficient group and 121 g/kg in the control group. These changes were not related to abnormalities of the serum electrolytes or blood sugar. The authors hypothesized that the increased fluid intake and higher fluid output may be related to direct or indirect effects on the central or peripheral control mechanisms for drinking or urinary excretion which may be mediated by altered composition of neural or other membranes or changes in eicosanoids metabolism (Neuringer et al., 1984).

An observational study with the blind observers found that rhesus monkeys whose diet was deficient in omega-3 fatty acid initiated more bouts of stereotyped behavior in their cages than monkeys who had received a control diet abundant in omega-3 fatty acids. These stereotyped behaviors associated with the omega-3 fatty acid deficiency were typical of rhesus monkeys raised as partial social isolates or those whose surroundings had been disrupted. The control monkeys had much less stereotypical behavior. It was noted also that stereotypic behavior decreased after meals in males, but not in females. Again, these seemed to be related to central effects on the behavior of n-3-deficient monkeys (Connor et al., 1969).

6. APPLICABILITY TO HUMANS

What are the implications of these studies of n-3 fatty acid deficiency and subsequent repletion in rhesus monkeys to human beings? First, these data strongly suggest that any n-3 fatty acid-deficient state will be corrected by fish oil containing EPA and DHA or in other studies, a diet containing n-3 fatty acids from soy oil (18:3n-3 linolenic acid). The brain phospholipids will readily assemble the correct amounts of DHA in the *sn*-2 position of the phospholipid molecular species. Furthermore, other fatty acids of the n-6 series, which occupy that position in the deficient state, will ultimately be removed and replaced by DHA. It is not certain yet if functional abnormalities would likewise be corrected because the appropriate function may need to take place at a certain stage of development or it may not occur at all or to a lesser degree when there has been biochemical correction.

A second point is that the diagnosis of an n-3 fatty acid deficiency state in the tissues, especially the brain and retina, can well be approximated by analysis of plasma and red blood cells. This is particularly true if the deficient state occurs during pregnancy and from birth onwards, as has already been demonstrated (Cocchi et al., 1984). DHA in erythrocytes correlated well with the DHA of the cerebral cortex.

A third point is that the brain presumably has the capacity to synthesize DHA from precursor forms such as EPA. The presumption of this statement is based on the fact that high plasma and red blood cell concentrations of EPA are not mirrored by similar concentrations in the brain; instead, it is DHA that predominates so greatly. The alternative explanation is that EPA was rapidly metabolized once it enters the brain, but there is no evidence to support this particular point. Finally, normal monkeys fed 18:3n-3 from soy oil have a pronounced fall in DHA concentrations from birth onward and deficient monkeys repleted with soy oil have increased DHA concentration but not more than the normal soy-oil-fed monkeys. This same situation prevails in humans in that soy-oil formula-fed human infants likewise have a fall in blood DHA concentrations that does not occur in breast-fed infants or infants fed formulas supplemented with DHA. These data suggest that both the shorter chain n-3 fatty acids (e.g., 18:3 n-3), and the longer-chain n-3 fatty acids (e.g., DHA and EPA), will correct the n-3 fatty acid deficiency. In all probability, however, the EPA and DHA from fish oil applies a more suitable correction and is perhaps more physiological. Pure DHA might be even better. Notably, human milk supplies a variety of n-3 fatty acids to the infant, including EPA and DHA (Lin et al., 1990).

Several questions are raised by the rapid incorporation of dietary DHA and other n-3 fatty acids from fish oil into phospholipid membranes of the cerebral cortex of juvenile rhesus monkeys. Would primate brains of "normal" fatty acid composition incorporate dietary DHA just as avidly as the brains of n-3-deficient monkeys? This situation is analogous to humans consuming large quantities of fish oil and has been tested in adult rats (Neuringer et al., 1988), where a 30% rise in brain DHA was observed. Would such a change in primate brain be deleterious or advantageous? Quantities of EPA in the erythrocytes and cerebral cortex of the fish-oil-supplemented monkeys were much higher than is normally the case. These abnormal levels might lead to functional disturbances, but no information is available about this point. Future studies of fish-oil feeding to normal adult monkeys may provide answers to these questions, especially if molecular species of fatty acids of the phospholipid classes are determined and if the function of the "changed" organs is measured. For example, when we analyzed individual phospholipid molecular species of the brains of monkeys fed different diets, we observed highly significant dietary effects (Neuringer et al., 1988).

If DHA turns over as rapidly in the adult normal brain as in the deficient monkey brain, then perhaps the brain should be provided with a constant supply of DHA or other n-3 fatty acids. Ultimately, dietary sources of n-3 fatty acids would be desirable in both adults and infants (Neuringer et al., 1988). Whether the n-3 fatty acid supplied from the diet should be as 18:3n-3 or preformed DHA or both is not completely known. It is possible that ample amounts of DHA could be synthesized from 18:3n-3 via the desaturation and elongation pathways. However, the active uptake of DHA by the infant rat brain over other fatty acids suggests preference for acquiring preformed DHA directly from the blood (Connor et al., 1969). In view of the significant impact of diet on brain composition, it will be important in the future to address the question of the appropriate amount and type of dietary n-3 fatty acids for optimal brain development during infancy and for maintenance during adult life (Neuringer et al., 1986).

REFERENCES

Anderson G, Connor WE. Uptake of fatty acids by developing rat brain. Lipids 1988; 23:286–290.
Anderson GJ, Connor WE, Corliss JD. Docosahexaenoic acid is the preferred dietary n-3 fatty acid for the development of the brain and retina. Pediatr Res 1989; 27:89–97.

Anderson GJ, Lin DS, Neuringer M, Connor WE. The reversal of prenatal n-3 fatty acid deficiency: effect upon the blood, retina and brain, in preparation.

Ansell, GB, Dohmen H. The metabolism of individual phospholipids in rat brain during hypoglycemia, anesthesia and convulsions. J Neurochem 1957; 2:1–10.

Bourre JM, Bonneil M, Dumont O, Piciotti M, Nalbone G, Lafont H. High dietary fish oil alters brain polyunsaturated fatty acid composition. Biochim Biophys Acta 1988; 960:458–461.

Burr GO, Burr MM. A new deficiency disease produced by the rigid exclusion of fat from the diet. J Biol Chem 1929; 82:345–367.

Careaga-Houck M, Sprecher H. Effect of a fish oil diet on the composition of rat neutrophils lipids and the molecular species of choline and ethanolamine glycerophospholipids. J Lipid Res 1989; 30:237–247.

Carlson SE, Carver JD, House SG. High fat diets varying in ratios of polyunsaturated fatty acid and linoleic to linolenic acid: a comparison of rat neural and red cell membrane phospholipids. J Nutr 1986; 11:718–725.

Clandinin MT, Chappel JE, Leong S, Heim T, Swyer PR, Chance GW. Intrauterine fatty acid accretion rates in human brain: implications for fatty acid requirements. Early Hum Dev 1980a; 4:121–129.

Clandinin MT, Chappel JE, Leong S, Heim T, Swyer PR, Chance GW. Extrauterine fatty acid accretion rates in infant brain: implications for fatty acid requirements. Early Hum Dev 1980b; 4:131–138.

Cocchi M, Pignatti C, Carpigiani M, Tarozzi G, Turchetto E. Effect of C 18:3 (n-3) dietary supplementation on the fatty acid composition of the rat brain. Acta Vitamonol Enzymol 1984; 6:151–156.

Connor WE, Hoak JC, Warner ED. Plasma free fatty acids, hypercoagulability and thrombosis. In: Sherry S, Brinkhous KM, Genton ED, and Stengle JM, eds. Thrombosis. National Academy of Sciences, Washington DC, 1969, pp. 355–373.

Connor, W.E, Lin DS, Neuringer M. Biochemical markers for puberty in the monkey testes: desmosterol and docosahexaenoic acid. J of Clin Endo Metab 1997; 82:1911–1916.

Connor, WE, Lin, DS, et al. Will the docosahexaenoic acid content in the erythrocytes predict brain docosahexaenoic acid? manuscript in preparation.

Connor WE, Neuringer M. The effects of n-3 fatty acid deficiency and repletion upon the fatty acid composition and function of the bran and retina. In: Karnowsky ML, Leaf A, Bolis LC, eds. Biological Membranes: Aberrations in Membrane Structure and Function. Alan R. Liss, New York, 1988, pp. 275–294.

Connor WE, Neuringer M, Barstad L, Lin DS. Dietary deprivation of linolenic acid in rhesus monkeys: effects on plasma and tissue fatty acid composition and on visual function. Trans Assoc Am Physicians 1984; 97:1–9.

Connor WE, Neuringer MA, Lin DS. Dietary effects on brain fatty acid composition: the reversibility of n-3 fatty acid deficiency and turnover of docosahexaenoic acid in the brain, erythrocytes, and plasma of rhesus monkeys. J Lipid Res 1990; 31:237–247.

Crawford MA, Casperd NM, Sinclair AJ. The long-chain metabolites of linoleic and linolenic acids in liver and brain in herbivores and carnivores. Comp Biochem Physiol 1976; 54B:395–401.

Crawford MA, Sinclair AJ. Nutritional influences in evolution of mammalian brain. In: Lipids, Malnutrition and the Developing Brain. Ciba Foundation Symposium. Associated Scientific, Amsterdam 1972, pp. 267–287.

Crokin, DR, Sun GY. Characterization of the enyzmic transfer of arachidonly groups to 1 Acyl-phosphoglycerise in mouse synaptosome fraction. J Neurochem 1978; 30:77–82.

Dawson RMC. Enzymatic pathways of phospholipid metabolism in the nervous system. In: Eichberg J., ed. Phospholipids in the Nervous System. Wiley, New York, 1985, pp 45–73.

Dawson, RMC, Richter D. The phosphorous metabolism of the brain. Proc R Soc London 1950; B137:252–267.

Fisher SK, Rowe CE. The acylation of lysophosphatidylcholine by subcellar fractions of guinea pig cerebral cortex. Biochim Biophys Acta 1980; 618:231–241.

Hargreaves KM, Clandinin MT. Dietary control of diacylphosphatidylethanolamine species in brain. Biochim Biophys Acta 1988; 962:98–104.

Holman RT. Essential fatty acid deficiency. Prog Chem Fats Other Lipids 1968; 9:275–348.

Holub BJ, Kuksis A. Metabolism of molecular species of diacylglycerophospholipids. Adv Lipid Res 1978; 16:1–125.

Homomayoun P, Durand G, Pascal G, Bourre JM. Alteration in fatty acid composition of adult rat brain capillaries and choroid plexus induced by a diet deficient in n-3 fatty acids: slow recovery after substitution with a non-deficient diet. J Neurochem 1988; 51:45–48.

Leyton J, Drury PJ, Crawford MA. Differential oxidation of saturated and unsaturated fatty acids in vivo in the rat. Br. J. Nutr 1987; 57:383–393.

Lin DS, Connor WE, Anderson GJ, Neuringer M. The effect of diet upon the phospholipid molecular species composition of monkey brain. J Neurochem 1990; 55:1200–1207.

Lin DS, Connor WE, Wolf DP, Neuringer M, Hachey DL. Unique lipids of primate spermatozoa: desmosterol and docosahexaenoic acid. J Lipid Res 1993; 34:491–499.

Lynch DV, Thompson GA. Retailored lipid molecular species: a tactical mechanism for modulating membrane properties. Trans Biochem Sci 1984; 9:442–445.

Moore SA, Yoder E, Murphy S, Dutton GR, Spector AA. Astrocytes, not neurons, produce docosahexaenoic acid (22:6 omega-3) and arachidonic acid (20:4 omega-6). J Neurochem 1991; 56:518–524.

Neuringer M, Anderson GJ, Connor WE. The essentiality of n-3 fatty acids for development and function of retina and brain. Annu Rev Nutr 1988; 8:517–541.

Neuringer M, Connor WE. Omega-3 fatty acids in the brain and retina: evidence for their essentiality. Nutr Rev 1986; 44:285–294.

Neuringer M, Connor WE, Lin DS, Barstad L, Luck SJ. Biochemical and functional effects of prenatal and postnatal omega-3 fatty acid deficiency on retina and brain in rhesus monkeys. Proc Natl Acad Sci USA 1986; 83:4021–4025.

Neuringer M, Connor WE, Van Patten C, Barstad L. Dietary omega-3 fatty acid deficiency and visual loss in infant rhesus monkeys. J Clin Invest 1984; 73:272–276.

O'Brien JS. Stability of the myelin membrane. Science 1986; 147:1099–1107.

O'Brien JS, Fillerup DL, Mead JF. Quantification and fatty acid and fatty aldehyde composition of ethanolamine, choline, and serine glycerophosphatides in human cerebral gray and white matter. J Lipid Res 1964; 5:329–338.

Olendorf WH. Permeability of the blood-brain barrier. In: Tower DB, ed. The Nervous System. Raven, New York, 1975, 279–289.

Op den Kamp JAF, Roelofsen B, Van Deenan LLM. Structural and dynamic aspects of phosphatidyl choline in human erythrocyte membrane. Trends Biochem Sci 1985; 10:320–323.

Philbrick DJ, Mahadevappa VG, Ackman RG, Holub BJ. Ingestion of fish oil or a derived n-3 fatty acid concentrate containing eicosapentaenoic acid affects fatty acid compositions of individual phospholipids of rat brain, sciatic nerve, and retina. J Nutr 1987; 17:1663–1670.

Pitkin RM, Connor WE, Lin DS. Cholesterol metabolism and placental transfer in pregnant rhesus monkey. J Clin Invest 1972; 15:2582–2592.

Reisbick S, Neuringer M, Hassan R, Connor WE. Polydipsia in rhesus monkeys deficient in omega-3 fatty acids. Physiol Behav 1990; 47:315–323.

Reisbick S, Neuringer M, Hasnain R, Connor WE. Home cage behavior of Rhesus monkeys with long-term deficiency of omega-3 fatty acids. Physiol Behav 1994; 55:231–239.

Salem N, Kim H-Y, Yergey JA. Docosahexaenoic acid: membrane function and metabolism. In: Simopoulos AP, ed. Health Effects of Polyunsaturated Fatty Acids in Seafoods. Academic, New York, 1986, pp. 263–317.

Sanders TAB, Mistry M, Naismith DJ. The influence of a maternal diet rich in linoleic acid on bran and retinal docosahexaenoic acid in the rat. Br J Nutr 1984; 51:57–66.

Swanson JE, Black JM, Kinsella JE. Dietary menhaden oil: effects on the rate and magnitude of modification of phospholipid fatty acid composition of mouse heart and brain. Br J Nutr 1988; 59:535–545.

Tarozzi G, Barzanti V, Biagi PL, Coochi M, Lodi R, Maranesi M, Pignatti C, et al. Fatty acid composition of single brain structures following different alpha-linolenate dietary supplementations. Acta Vitamonol Enzymol 1984; 6:157–163.

Trapp BD, Bernson J. Changes in phosphoglyceride fatty acids of rat brain induced by linoleic and linolenic acids after pre and postnatal fat deprivation. J Neurochem 1977; 28:1009–1013.

Van Golde LMG, Scherphof GL, Van Deenen LLM. Biosynthetic pathways in the formation of individual molecular species of rat liver phospholipids. Biochem Biophys Acta 1969; 176:635–637.

Walker BL. Maternal diet and brain fatty acids in young rats. Lipids 1967; 2:497–500.

Yeo YK, Philbrick DJ, Holub BJ. Altered acyl chain compositions of alkylacyl, alkenylacyl, and diacyl subclasses of choline and ethanolamine glycerophospholipids in rat heart by dietary fish oil. Biochem Biophys Acta 1989; 1001:25–30.

Yerram NR, Moore SA, Spector AA. Eicosapentaenoic acid metabolism in brain microvessel endothelium: effect on prostaglandin formation. J Lipid Res 1989; 30:1747–1757.

Youyou A, Durand G, Pascal G, Piccotti M, Dumont O, Bourre JM. Recovery of altered fatty acid composition induced by a diet devoid of n-3 fatty acids in myelin, synaptosomes, mitochondria and microsomes of developing rat brain. J Neurochem 1986; 46:224–227.

12 The Role of Omega-3 Polyunsaturated Fatty Acids in Retinal Function

Algis J. Vingrys, James A. Armitage, Harrison S. Weisinger, B. V. Bui, Andrew J. Sinclair, and Richard S. Weisinger

1. POLYUNSATURATED FATTY ACIDS

The precursors of both n-6 and n-3 polyunsaturated fatty acids (PUFAs), linoleic acid and α-linolenic acid, respectively, are essential for mammals as they are required for normal physiological function and cannot be synthesized *de novo* (Holman, 1968). They can only be accumulated by placental transfer or by dietary intake. Once accretion of these fatty acids has occurred, metabolic, conservation and recycling pathways sustain them (Bazan et al., 1994). Unlike mammals, plants can synthesize these precursor PUFAs (linoleic and α-linolenic acids) so they are found in abundance in the chloroplast membranes of plants, in certain vegetable oils, and in the tissues of plant-eating animals (Nettleton, 1991). The best sources of α-linolenic acid are vegetable oils, such as perilla (Yoshida et al., 1993) rapeseed (canola), linseed, walnut, and soybean (Nettleton, 1991). They are also abundant in shellfish, fish, and fish products and can be found in low amounts in green, leafy vegetables and baked beans (Nettleton, 1991; Sinclair, 1993).

Once ingested, these essential fatty acids can be metabolized into longer, more unsaturated products (Holman, 1968). This process involves sequential desaturation (adding double bonds) and chain elongation (adding carbon atoms), as shown in Fig. 1. The important aspect of Fig. 1 is that the n-6 and n-3 families compete for the enzymes responsible for desaturation (Sinclair, 1993). The main metabolite of the n-6 series is arachidonic acid (20:4n-6, AA), whereas eicosapentaenoic acid (20:5n-3, EPA) and docosahexaenoic acid (22:6n-3, DHA) are the main metabolites of the n-3 series (Holman, 1968). The metabolic pathways leading to DHA are complicated by involving retroconversion from 24:6n-3 to 22:6n-3 (DHA) (Voss et al., 1991).

The enzymatic competition between n-6 and n-3 families requires a dietary balance in the precursor fatty acids, as an overabundance in one will lead to a relative deficiency in the metabolites of the other. Conversely, a deficiency in one series leads to increased synthesis of the other series (Galli et al., 1971).

Following ingestion, fatty acids are esterified to phospholipids and triglycerides by the intestine and transported to the liver within chylomicrons and very low density lipoproteins

From: *Fatty Acids: Physiological and Behavioral Functions*
Edited by: D. Mostofsky, S. Yehuda, and N. Salem Jr. © Humana Press Inc., Totowa, NJ

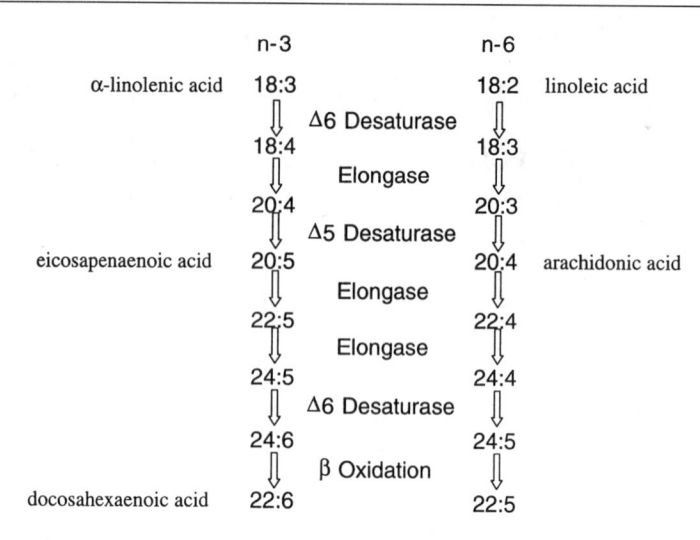

```
                      n-3                      n-6
      α-linolenic acid   18:3                  18:2   linoleic acid
                          ⇓    Δ6 Desaturase    ⇓
                         18:4                  18:3
                          ⇓      Elongase       ⇓
                         20:4                  20:3
                          ⇓    Δ5 Desaturase    ⇓
    eicosapenaenoic acid 20:5                  20:4   arachidonic acid
                          ⇓      Elongase       ⇓
                         22:5                  22:4
                          ⇓      Elongase       ⇓
                         24:5                  24:4
                          ⇓    Δ6 Desaturase    ⇓
                         24:6                  24:5
                          ⇓     β Oxidation     ⇓
   docosahexaenoic acid  22:6                  22:5
```

Fig. 1. The n-3/n-6 metabolic pathways. Precursors of the n-3 (18:3n-3, linolenic acid) and n-6 (18:2n-6, α-linoleic acid) are converted by a series of desaturation and (adding double bonds) and elongation (adding carbon atoms to the hydrocarbon backbone) reactions. Note that the same enzymes catalyze n-3 and n-6 desaturation and elongation reactions. Major metabolites are indicated. PUFAs with 20-carbon backbones (20:4n-6, arachidonic acid, and 20:5n-3, eicosapentaenoic acid) are precursors to the eicosanoids (prostaglandins, leukotrienes, thromboxanes). Docosahexaenoic acid (22:6n-3) is also indicated. Note that only a limited part of the metabolic pathway is shown in this figure.

(Bazan et al., 1994). The liver is thought to be the major supplier of DHA to neural tissues following synthesis from precursor fatty acids (Bazan et al., 1992, 1994; Sinclair, 1975). The pigmented epithelium of the retina (RPE) has also been shown to produce DHA (Rotstein et al., 1996; Wang & Anderson, 1993).

1.1. The Retinal PUFA Profile

The retina is highly enriched with DHA, particularly in the membranes of the photoreceptor outer segments (Anderson & Maude, 1970; Fliesler & Anderson, 1983; Weisinger et al., 1995). Because all outer segments are renewed over a 9- to 10-day period (Anderson, 1983; Cibis et al., 1995; Landis et al., 1973; Young, 1967; Young & Bok, 1969) the retina requires a reliable and continuous supply of DHA. This is achieved by recycling of DHA by the RPE (Bazan et al., 1994); active synthesis and transport from the liver by circulating plasma lipoproteins (Aveldano de Caldironi & Bazan, 1980; Bazan et al., 1994); and selective uptake by the RPE (Anderson et al., 1994). Fatty acids are important for the genesis of outer segments and synaptic membranes of the photoreceptors (Fliesler & Anderson, 1983).

The fact that such an elaborate system of DHA maintenance is in place suggests that its presence serves some important function in the retina. Otherwise, why not let the DHA levels fluctuate as a function of dietary intake? On a similar issue Crawford and Sinclair (1972) found that proportions of arachidonic, docosatetraenoic, and docosahexaenoic acids in brain gray matter lipids were constant in over 30 species. In contrast, other tissues such as liver and muscle reflect species dietary intake (Crawford & Sinclair, 1972). The fact that PUFAs comprise approx 7% (dry weight) of brain gray matter (Sinclair, 1975), are in high proportions in critical tissues such as the central nervous system (Bell et al., 1997; Salem Jr., 1989) and show little difference among many species of mammal, including human (Anderson & Maude, 1970; Anderson & Risk, 1974), implies a critical role in mammalian physiology.

The fatty acid composition of the retina is similar in profile to that of brain gray matter, although the retina has a greater content of DHA. The whole retina is composed of approx 20% lipid, with the lipid content of the photoreceptor layer being estimated at 25% (dry weight) (Fliesler & Anderson, 1983). Of this lipid, approx 50–60% is phospholipid (Fliesler & Anderson, 1983). It has been reported that long-chain (20–22 carbon) PUFAs constitute 37% of all fatty acids in the rat retina (Futterman et al., 1971), and that DHA makes up 21% of fatty acids in the human retina (Futterman & Stevens Andrews, 1964). This is consistent with assays of primate photoreceptor outer segments where the DHA content was reported to be in excess of 40 mol%. The rod outer segment membranes are reported to be composed of approx 50% lipid and 50% protein by weight and of this lipid, phospholipids constitute 90–95% (i.e., 80–85 mol%) (Connor & Neuringer, 1988).

It is apparent that the central nervous system, and retina in particular, appears to go to great lengths to sustain its levels of DHA. The following discussion will consider biological membranes and retinal function and discuss the possible role that DHA has in its physiology.

2. PHOSPHOLIPIDS AND BIOLOGICAL MEMBRANES

Phospholipids and cholesterol are abundant in all biological membranes (Gurr & James, 1980) and phospholipid composition has been found to be relatively constant in any one tissue, between different species, across a wide range of vertebrates (Lin et al., 1994). In general, the phospholipid composition of a typical membrane is as follows: phosphatidylethanolamine (PE), 30–35%; phosphatidylcholine (PC), 40–50%; phosphatidylserine (PS), 5–10%; others, 5–25% (Gurr & James, 1980). Phospholipids can vary in the fatty acid chain that is attached at both the *sn*-1 and *sn*-2 positions of their backbone. The *sn*-1 position is generally occupied by a saturated fatty acyl chain, whereas the fatty acyl chain at the *sn*-2 position is usually unsaturated (Gurr & James, 1980). These chains contain an even number of carbon atoms, usually between 14 and 24, and it is this variation that provides most of the permutations of phospholipid structure (Gurr & James, 1980).

Fleischer and Rouser (1965) proposed that biological membranes can be classified in two categories that differ in their biochemical and functional characteristics. The first describes stable, rigid membranes (e.g., myelin) that contain high proportions of cholesterol and sphingomyelin, as well as saturated and monoenoic fatty acids. The second are found in tissues that require greater fluidity or metabolic activity, such as in cell organelles and synapses, and photoreceptor outer segments. These membranes typically contain very high levels of PUFAs, particularly DHA (Fleischer & Rouser, 1965). Indeed, PUFAs (in phospholipids) have been associated with cellular activities such as ion transport (Gerbi et al., 1993), activation of membrane bound enzymes (Bernsohn & Spitz, 1974; Bourre et al., 1989), signal transduction via effects on inositol phosphate, diacylglycerol, and protein kinase C (Kurlack & Stephenson, 1999), and gene transcription (Clarke & Jump, 1994).

Membrane fluidity (the reciprocal of membrane microviscosity) is a vital parameter in terms of cell function. The fluidity of a membrane reflects the stability and permeability of a cell (McMurchie, 1988). Some regard fluidity as a state of disorder (Huster et al., 1998). It also determines the positioning and motility of membrane-bound proteins, thus modulating their function and, with that, cellular function (McMurchie, 1988). Membrane fluidity can be affected by temperature, pressure, membrane potential, pH and the presence of divalent cations, such as Ca^{2+} (McMurchie, 1988; Papahadjopoulos, 1978). There are many intrinsic factors that can also alter membrane fluidity, and an excellent review of these factors is given by McMurchie (1988). However, of importance for this chapter, is that variations in fatty acid acyl chain configuration can influence fluidity. The most significant sources of variation include chain length and the degree of saturation. Increased chain lengths discourage close packing of molecules, hence increasing fluidity. Likewise, the degree of unsaturation also increases fluidity. Quinn (1981) has shown membrane fluidity to be inversely proportional to the ratio of saturated fatty acids (usually 16:0 and 18:0) to *cis* unsaturates. Huster *et al* (1988) propose that cholesterol forms PC-enriched microdomains in membranes containing PC, PE, and PS with DHA in the *sn*-2 position where the saturated fatty acids in the *sn*-1 position are preferentially oriented toward the cholesterol molecules.

One index of membrane fluidity is the packing free volume f_v. This is defined as the difference between the measured molecular volume (from density measurements) and the molecular volume determined for the sum of the atomic volumes making up the molecule (Littman & Mitchell, 1996). As a result of disorder effects in acyl chain packing, this property is always positive. Cholesterol acts to decrease f_v by promoting acyl chain interactions typically between saturated molecules at the *sn*-1 position (Littman & Mitchell, 1996).

2.1. The Retina and Its Activation by Light

Because the retina contains such high levels of DHA, it should be a good tissue in which to consider the effects of DHA deprivation. In order to do so, we need to understand retinal structure and function and the methods that can be used to study them. The following subsections will develop these issues.

The retina is a complex of cells (Rodieck, 1988) that can be identified by their anatomical, physiological and molecular properties as photoreceptors, bipolar cells, horizontal cells, amacrine cells, and ganglion cells (Dowling, 1987). The other major cells of the retina are the glia, of which the Müller cells (Ripps & Witlovsky, 1985) and the retinal pigment epithelium (RPE) (Steinberg et al., 1970) are important for our discussion. As will be described later, the Müller cell is an important player in generating the electroretinogram (ERG) because of its potassium buffering capacity, whereas the RPE sustains photoreceptor function by removing waste (Young, 1967) and recycling key molecules [e.g., opsin, docosahexaenoic acid (Bazan et al., 1994)]. The RPE is the primary site of reconformation of the photopigment chromophore (Anderson, 1983; Cibis et al., 1995; Young & Bok, 1969), prior to transport and reincorporation into the photoreceptor inner segment and has a regulatory role in maintaining nutrient supply and ionic homeostasis (Miller & Steinberg, 1977).

The process of vision begins in the photoreceptor outer segment, where light is absorbed by the photopigment (rhodopsin) to produce a cascade of reactions within the receptor (Stryer, 1986). Rhodopsin contains a membrane-bound opsin (protein) and a light-

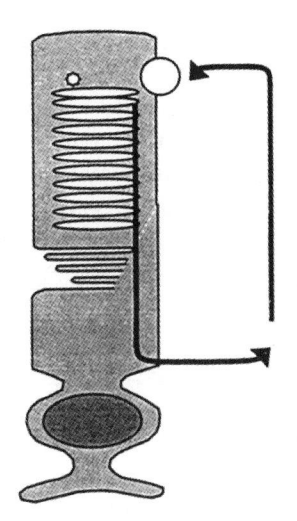

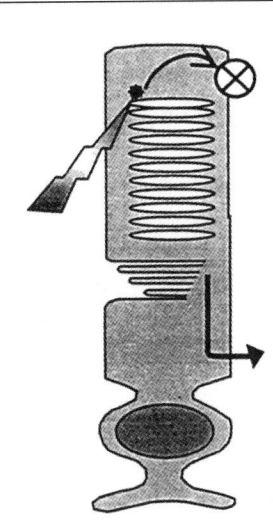

Fig. 2. Schematic of the physiological process and potassium fluxes that underlie the ERG. In the dark (left panel), the rod photoreceptor sustains a current circulating from the outer to inner segment via a gated cationic channel. Stacked membrane discs hold the light labile photopigment. After light activates the photopigment (right panel), a series of cascade events take place to close the cationic channels (*see* text). This reduces the dark current and results in the fast PIII component of the ERG that produces the leading edge of the a-wave.

sensitive chromophore. It is one of the superfamily of membrane-bound receptors that include the angiotensin II (AT1) receptor, the serotonin (5-HT) receptor, as well as many others (Stryer, 1986). These receptors all have seven transmembrane domains arranged in an α-helix. One consequence of their transmembrane structure is that the characteristics of the membrane can impact on their function. Moreover, the similarities in their structure may mean that the effect that DHA has on this entire family of receptors may be elucidated by considering the rhodopsin light response. The next subsection considers the light response of rhodopsin and later we discuss mechanisms by which DHA can modify this response.

2.2. The Light Response

The chromophore of rhodopsin is activated by light to produce a conformational change in the molecule and to yield its active state called meta-rhodopsin II (Liebman et al., 1987). This active state interacts with other membrane-bound proteins and, via a G-protein cascade, leads to closure of ionic channels in the outer segment (Fig. 2) that produces a change in the ionic fluxes within the surrounding extracellular fluid (Faber, 1969; Karwoski & Proenze, 1978). The current changes can be detected and form components of the ERG. We will first discuss the physiological processes involved in vision in order to understand better the various ERG components.

2.3. Phototransduction

Rod photoreceptors are neurons specialized for the capture of light (Dowling, 1987; Schnapf & Baylor, 1987). These cells are long and thin, with their outer portions, termed the rod outer segment (ROS), being dedicated to the capture of light (Rodieck, 1988). Rod outer segments contain many stacked organelles, called discs, that are composed of a lipid bilayer and that carry the visual pigment molecule, rhodopsin, in high concentration (Dowling, 1987; Fulton et al., 1995a; Fulton et al., 1991; Stryer, 1981). The fatty acids of the ROS lipids are dominated by polyunsaturated fatty acids (Anderson & Risk, 1974; Fliesler & Anderson, 1983) and especially docosahexaenoic acid, which is believed to have a functional role in the activation of rhodopsin (Brown, 1994; Bush et al., 1994). This issue will be considered in greater detail later.

The outer segment membranes contain nonselective cyclic guanosine monophosphate (cGMP), gated cationic channels (Lamb, 1986; Nicholls et al., 1992; Yau, 1994), that are open in the dark. This allows a flux of current into the outer segment (Fig. 2), which is called the *dark current,* down the concentration gradient maintained by a Na^+K^+-ATPase pump (Lamb, 1986). Phototransduction is initiated with the light-evoked conformational change to the chromophore from an 11-*cis*-retinaldehyde to the all-*trans* conformation (Anderson, 1983; Liebman et al., 1987; Stryer, 1981). This process goes through several intermediary states (Anderson, 1983; Cibis et al., 1995), which are identifiable by their spectral absorption characteristics (Kraft et al., 1993). The reaction that is actually responsible for the initiation of visual excitation occurs after the formation of the intermediate, meta-rhodopsin II (MII) (Liebman et al., 1987).

Considerable amplification follows MII formation, with each MII molecule reacting with up to 100 molecules of the disc membrane-bound G-coupled protein, transducin (Lamb & Pugh Jr, 1992; Liebman et al., 1987; Stryer, 1986). Transducin (T) actually has three polypeptides (T_α, T_β, and T_γ) bound with guanosine diphosphate (GDP) in the dark (Stryer, 1986). Reaction with MII results in the cleavage of T_α.GDP from $T_{\beta\gamma}$. The subcomponent T_α.GDP binds with cGMP phosphodiesterase (PDE), which is also composed of P_α and P_β subcomponents, and an inhibitory P_γ subcomponent (Cibis et al., 1995; Lamb & Pugh Jr, 1992; Stryer, 1986). Release of P_γ results in activation of PDE, which hydrolyzes cGMP to 5'-cGMP. Because one activated PDE molecule can hydrolyze thousands of cGMP molecules per second (Cibis et al., 1995) the bleaching of one chromophore molecule results in amplification to produce 10^5 hydrolyzations (Cibis et al., 1995; Stryer, 1986). The nonselective cationic channels of the outer segment are ligand-activated using the cyclic nucleotide, cGMP (Liebman et al., 1987; Lipton et al., 1977) and it is these that close after light hydrolyzes the cGMP (Fig. 2). This results in a reduced cationic influx into the outer segment (Lamb, 1986) that is detected as the ERG a-wave (*see* Subheading 2.5.).

2.4. Inactivation of the Transduction Process

It is important for the visual system to reset itself after light capture (Lyubarski & Pugh, 1996; Pepperberg et al., 1996); it does this by a process of inactivation. Inactivation begins with the active opsin (MII) being phosphorylated by rhodopsin kinase and bound by arrestin (Liebman et al., 1987); this inhibits further binding to T_α.GTP. The inhibitory component of PDE, P_γ, binds to T_α.GTP, to prevent that molecule from cleaving other P_γ from the PDE complex (Cibis et al., 1995). P_γ, itself, reassociates with the $P_{\alpha\beta}$ in a process that also involves the inactivation of T_α.GTP (to T_α.GDP and then $T_{\alpha\beta\gamma}$.GDP) (Cibis et al., 1995).

The opsin is rejoined with 11-*cis*-retinaldehyde, following regeneration of the chromophore, via the RPE (Anderson, 1983; Cibis et al., 1995). Recent work by Rodriguez de Turco and colleagues (1997) has shown that rhodopsin is cotransported with

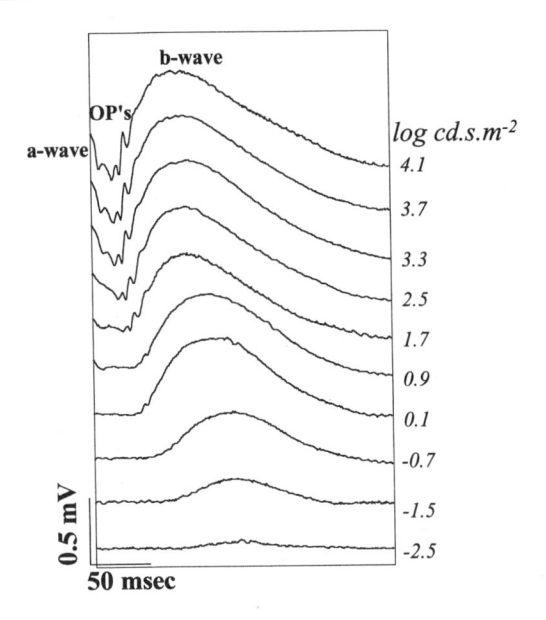

Fig. 3. Typical ERG waveforms obtained from the rat. These signals were collected for a range of light intensities (shown on the right-hand side). Note how at low light levels, a (positive) slow, rod driven b-wave is seen. This grows and speeds up as a result of cone contribution and develops oscillatory potentials (OPs). The negative trough is called the a-wave and reflects photoreceptor activation (*see* Fig. 2).

docosahexaenoate–phospholipids within post-Golgi vesicles in the photoreceptor inner segments for reincorporation into new outer segments.

2.5. The Electroretinogram

The ERG is a measure of the current flowing within and across the retina (Fig. 2) in its response to incident light (Rodieck, 1972). To maximize the response, the entire retina needs to be stimulated using a Ganzfeld bowl as the light source. The voltage change detected by the ERG is a composite of subcomponent potentials that are summed to give the total waveform (Granit, 1933). The following discussion provides a brief description of the main current generators and subcomponents of the ERG waveform (Fig. 3). For a more complete description of the ERG and some of its applications, *see* Weisinger et al. (1996c).

The morphology of the ERG depends on the level of retinal adaptation (Fig. 3) and stimulus variables such as, light intensity, wavelength, duration, and interstimulus interval (Bui, 1998; Weisinger, 1995; Weisinger et al., 1996b). Although there are many forms of electroretinogram, the most commonly recorded is that derived from a dark adapted retina following a single bright flash of up to 1 ms in duration. A bright, fast flash is needed to ensure that the kinetics of the transduction cascade are resolved. The International Society of Clinical Electrophysiology in Vision provide useful guidelines on testing and extraction of ERG signals (Marmor et al., 1999). In our experience, minor adaptations to these protocols are required when working with small animals.

The chronological order of the oscillations that appear following a light flash are the Early Receptor Potential (ERP), the negative a-wave, the positive b-wave, and the positive

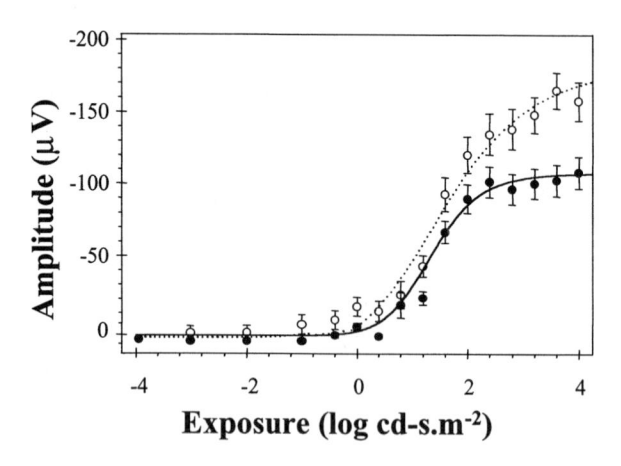

Fig. 4. Average a-wave amplitude (±SEM) as a function of intensity from n-3-deficient (filled circles, $n = 12$) and control rats (unfilled circles, $n = 12$). The lines show the best-fitting Naka–Rushton function, which indicate that n-3-deprived animals have a loss of approx 35% amplitude and a sensitivity shift of 0.3 log units compared to controls. Note that below -2.0 log $cd/s/m^2$ the a-wave vanishes into noise.

c-wave (Weisinger et al., 1996b). Figure 3 shows the appearance of the ERG as a function of light intensity. The main features are the development of a slow, positive-going rod b-wave under low light levels, the subsequent genesis of a negative trough or a-wave at brighter light levels, and the superimposition of a faster cone-driven b-wave on which can be seen four to six oscillations called the oscillatory potentials. It should be noted that although the faster peak is sometimes called the cone response, it, in fact, represents a mixed cone and rod contribution. The rod and cone contributions can be identified by digitally subtracting the isolated cone response (*see* Subheading 2.9.) from the mixed signal. A positive potential referred to as the d-wave can be found at light offset when using flashes of longer duration (Armington, 1974; Bui, 1998). This is sometimes referred to as the OFF-response (Frishman & Karwoski, 1991) which, for short duration stimuli, is thought to summate with the ON-response to give the b-wave (Armington, 1974). As the d-wave is only isolated effectively with long-duration stimuli, it will not be considered further.

There are several ways of describing the various ERG components for comparative purposes. In some cases, investigators may simply wish to consider the dose-related response of an ERG component as a function of light intensity. This has been described with a three-parameter Hill plot, termed the Naka–Rushton relationship (Fulton & Rushton, 1978; Naka & Rushton, 1966), and is adequate for comparative purposes, as shown in Fig. 4. It will provide a more robust and accurate estimate of the amplitude than will a measure derived from a single intensity. Amplitudes are measured for the a-wave from baseline to the maximum negative trough, with the time from the flash taken to reach this trough being its implicit time. For the b-wave, the amplitude is taken as the maximum peak-to-peak value from the trough of the a-wave to the peak of the b-wave. Again, the corresponding time of this b-wave peak is defined as the implicit time. As the Naka–Rushton response of the b-wave shows a plateau at moderate intensities, it is customary to fit the lower limb of this response with the three-parameter Hill plot (Fulton & Rushton, 1978; Naka & Rushton, 1966). Although physiological sub-

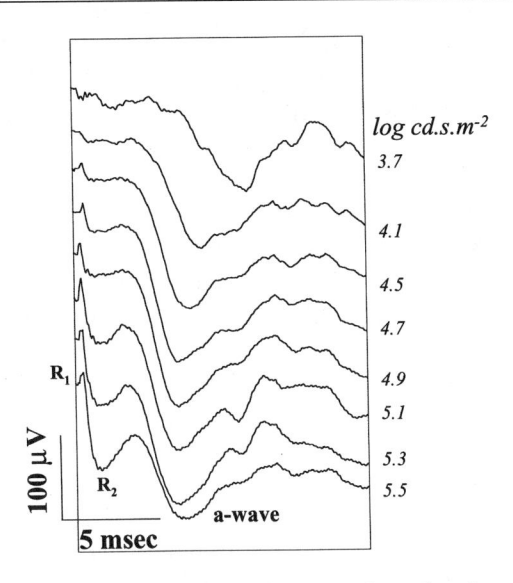

Fig. 5. Representative ERP waveforms extracted from raw data of rat (*see* text for details) showing both R1 and R2 components. The ERP has been isolated using a fast sampling rate (20 kHz) and metallic electrodes. Note that the R1 component is not well nor reliably represented (*see* the text for details).

strates have been proposed to underlie aspects of the intensity–response relationship in the past (Fulton & Rushton, 1978), more recent modeling techniques (Cideciyan & Jacobson, 1996; Hood & Birch, 1994; Lamb & Pugh Jr, 1992) can be used to better define the generation of these waveforms.

The underlying potentials of the dark-adapted ERG to bright flash stimuli have separate origins within the retina (Frishman & Steinberg, 1990). These generators were first described by Granit (1933) and are the fast PIII, generated by the photoreceptors (Penn & Hagins, 1969; Bush & Sieving, 1994), the Müller cell-generated postreceptoral potentials, the slow PIII (Witkovsky et al., 1975) and PII (Ripps & Witlovsky, 1985), and the Trans-Epithelial Potential (TEP, PI) (Steinberg et al., 1985). Interactions among these generators give rise to the a-, b-, and c-waves. As this chapter specifically considers the receptoral response, we will not dwell on responses other than the ERP and fast PIII. Readers interested in these other components should consider the review published by Weisinger et al. (1996c).

2.6. The Early Receptor Potential

The ERP is the earliest component of the ERG (Fig. 5) that can be detected following light capture (Brown, 1968; Cone, 1964; Cone & Cobbs, 1969). The signal appears to originate from altered electron flows about the rhodopsin molecule, following its conformational change to meta-rhodopsin. It is likely that if DHA were important in facilitating this conformational change, then DHA deficiency should alter the ERP. Bush et al. (1994) report that the rate of active meta-rhodopsin formation was slower in vitro in the absence of DHA implying that ERP changes should be found in vivo. To our knowledge, such in vivo work has not been performed, but the work of Bush et al. suggest that it might be a useful marker of DHA deficiency (*see* Section 6; Dratz et al., 1993; Dratz & Holte, 1993;

Littman & Mitchell, 1996). Our preliminary findings support this possibility (*see* Fig. 7) by indicating a loss of ERP in n-3-deficient animals. However, as these data were collected using a 2-kHz sampling rate, it needs to be replicated with higher sampling rates in future studies.

The ERP waveform has two components: a fast positive R1 and a slower negative R2 (Fig. 5). The R1 is usually smaller than the R2 (35–50% R2 amplitude) and both are only seen at high light levels (>2.5 log cd.s/m^2), although isolating the R1 is not easily achieved.

One of the reasons that the ERP is not studied often is that it is hard to isolate reliably. ERP isolation requires fast sampling, bright-light stimuli, nonmetallic electrodes, and the removal of artifactual responses (melanin response). As the ERP commences immediately after light stimulation (no delay), it requires high sampling rates for proper isolation (>20 kHz; Fig. 5). A small R2 can sometimes be found superimposed on the leading edge of the a-wave using slower sampling rates (2–4 kHz), but these do not give reliable nor repeatable signals because of undersampling.

If metallic electrodes are used they will give rise to a photovoltaic artifact (electron release from metal molecules by light photons). This can interfere with the leading edge of the ERP (R1) and can only be overcome by using nonmetallic electrodes (Dawson & Galloway, 1991). A metallic electrode with high sampling rates can be used to isolate the R2 (Fig. 5), as the photovoltaic artifact has a very fast time-course. In many cases, this approach may be adequate, but it needs to be mentioned that the ratio R1/R2 is known to provide useful additional information (Dawson & Galloway, 1991). The problem is that the extraction of the R1 is further complicated by the temporal characteristics of the light source and the melanin response. Any pigment molecule can interact with light to generate a potential, as does melanin within the eye. This produces a fast electronegative potential (<500 µs) that interferes with the R1 component of the ERP frustrating resolution of the R1. Moreover, both the melanin response and R1 will be smeared if a fast flash is not used (buildup <100 µs). The usual technique for R1 isolation is to remove the photovoltaic and melanin responses from traces by digital subtraction of signals collected at low light levels (Fig. 6, >2.5 log cd.s/m^2) where the ERP is not visualized, but these other signals have already saturated. When this is done, good R2 waveforms will be obtained with a less robust R1 (Fig. 5). One interesting aspect of the ERP is that, contrary to the other ERG components, this signal does not saturate with increasing exposure (Fig. 6).

2.7. Fast PIII: The Photoreceptor Response

The light-activated amplification cascade described earlier results in a cornea-negative potential arising from the photoreceptors at high intensities, known as the fast PIII (Penn & Hagins, 1969; Bush & Sieving, 1994). The fast PIII forms the leading edge of the conventionally measured a-wave (Hood & Birch, 1990a; Lamb & Pugh Jr, 1992) at moderate light levels. Its generation requires the interaction of several membrane-bound proteins and these are likely to be affected by the biophysical state of the membrane and determined by DHA content (*see* later). Hyperpolarization of the photoreceptor membrane potential results in a change in the K$^+$ flux and a reduction in [K$^+$]$_0$ (Kline et al., 1985). This change in [K$^+$]$_0$ is called the distal response and underlies some of the other potentials that form the ERG (Bush & Sieving, 1994) but will not be considered in this chapter.

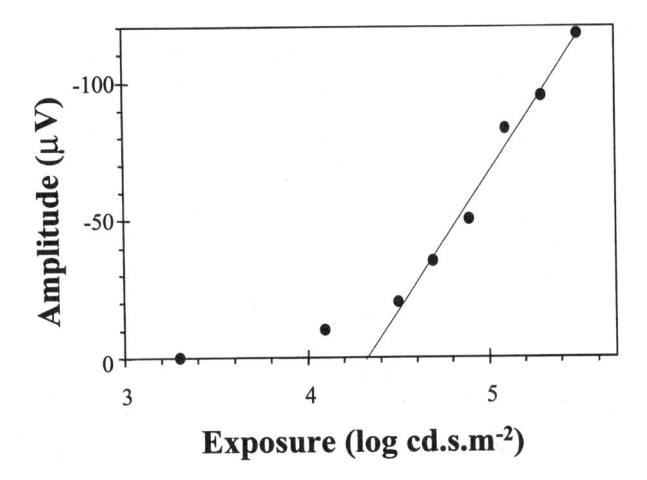

Fig. 6. Average R2 amplitude plotted as a function of flash exposure for $n = 6$ rats. Note that this shows a nonsaturating linear response, in contrast with the a-wave or other ERG components (compare with Fig. 4)

2.8. Modeling the Receptoral Photoresponse

Lamb and Pugh (1992) showed that the photoreceptoral response in a single rod outer segment (r) following a brief flash that has resulted in φ number of rhodopsin isomerizations can be described as a delayed Gaussian function of time (t) by

$$r(\phi,t) \cong [1 - \exp(-\tfrac{1}{2} \phi A[\{t - t_{eff}\}^2])] r_{max}] \qquad \text{for } t > t_{eff} \qquad (1)$$

In Eq. 1, r_{max} (μamp) is the magnitude of the saturating (maximal) photocurrent. The constant, t_{eff} (s), is the delay that arises from a number of small delays during the transduction cascade, but mostly those inherent to the recording system. The amplification, A (per isomerization per second), is the product of numerous amplifications occurring during phototransduction. Lamb and Pugh (1992) demonstrated that the model accurately described the current recorded from single rods in response to a wide range of stimulus intensities. They also noted that as intensity increased, the values of t_{eff} and A decreased (Lamb & Pugh Jr, 1992), a finding confirmed by others in single-cell and massed retinal recordings (Breton et al., 1995; Cobbs & Pugh, 1987; Lyubarski & Pugh, 1996). This is believed to arise from the saturation of PDE in the hydrolysis of cGMP (Breton et al., 1994; Lamb & Pugh Jr, 1992).

A modification of the delayed Gaussian model of Lamb and Pugh (1992) (Eq. 1), has been applied to the leading edge of the ERG (i.e., the fast PIII) (Hood & Birch, 1990a) as given by

$$\text{PIII}(i,t) \cong [1 - \exp\{-iS(t - t_d)^2\}] R_{max} \qquad (2)$$

In Eq. 2, the amplitude of the PIII (μV) represents the sum of the individual rod responses and is a function of both flash intensity, i (cd.s/m^2) and time, t (s) after the onset of the flash. S (m^2.cd^{-1}.s^{-3}) is a sensitivity parameter that scales i, R_{max} is the saturated PIII response (μV) and t_d is the delay (s). This model provides good fits to the raw waveform when fitted as an ensemble (i.e., to a group of intensities; Fig. 7) at low to medium light levels (Hood & Birch, 1990b) or when the responses to individual intensities are fitted

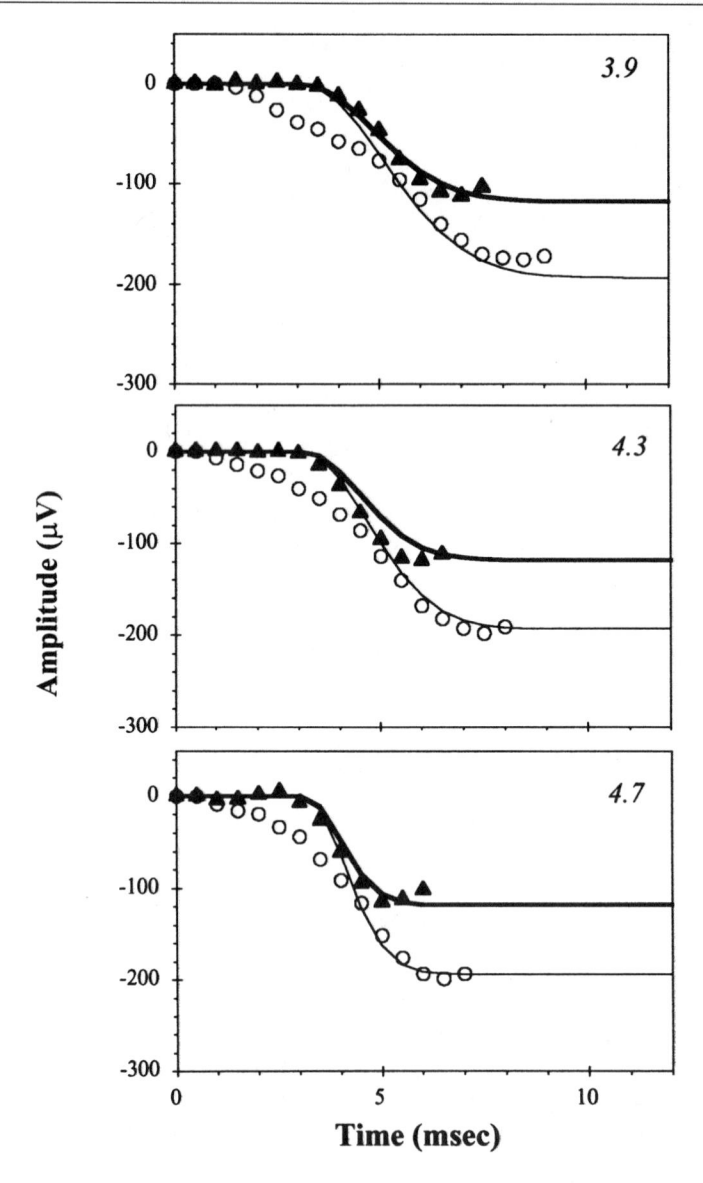

Fig. 7. Representative ensemble fitting (lines) to raw ERG data (symbols) over three exposures (shown to right) for one n-3-sufficient (unfilled symbols, thin line) and one n-3-deficient (filled symbols, thick line) rat. These data are typical of the respective diet groups and show a reduced R_{max} and lowered sensitivity (S) in deficient animals. In this figure, the sufficient animal shows a marked $R2$ response in the first 3 ms of the waveform (compare with Fig. 5), which frustrates isolation of the fast PIII.

separately (Cideciyan & Jacobson, 1996). We usually maximize the likelihood of our fit by minimizing the root-mean-square error term over an ensemble of intensities using the solver module of a Microsoft Excel spreadsheet [Fig. 7 (Bui & Vingrys, 2000)]. The maximum intensity for an ensemble fit must be chosen to show response saturation (maximal a-wave slope). More complex formulas may be applied to fit better rod or cone responses by allowing for membrane capacitance (Cideciyan & Jacobson, 1996). However, we prefer the less complicated model for rod responses because of the ill-defined aspects of membrane capacitance. When performing fits, only S and R_{max} should be floated because the other parameters (delay, capacitance, etc.) are usually fixed in physiological terms. If these variables are floated, then their interdependency may yield erroneous and physiologically irrelevant outcomes that would require single-cell studies for confirmation. In our laboratory, we usually derive t_d (and capacitance) by floating these variables in our normal controls and then refit all experimental data using the mean of this value. Sometimes, at high intensities, we find that an R2 can complicate fast P3 extraction and we allow for this by ignoring the initial 3-ms of negativity (*see* Fig. 7).

Changes in the parameters R_{max} and S can be interpreted as distinct functional alterations to either the photoreceptor or the phototransduction cascade (Birch et al., 1995; Holopigian et al., 1997; Hood & Birch, 1994a; Vingrys et al., 1998; Weisinger et al., 1999). R_{max} describes the saturated rod response and reflects the change in current flow about the outer segments following light stimulation. This saturated response has been shown to be related to the total number of cationic channels and reflects outer segment area or length (Baylor et al., 1979; Breton et al., 1994). As such, R_{max} is reduced when outer segment morphology is altered or when there is a reduction in the number of receptors, as may occur in some disease states (Hood & Birch, 1994b). However, in developing retina, R_{max} has also been shown to be related to the level of active rhodopsin available in the rod outer segments (Fulton, Hansen and Findl, 1995b) implying that the concentration of unbleached rhodopsin may be an important factor in determining the maximal rod response. The sensitivity parameter, S, can be considered as reflecting a change in flash energy. Structural factors such as rod-packing density, receptor alignment, and pigment content will all affect S. Likewise, the efficiency of the rhodopsin–transducin–phosphodiesterase cascade will affect sensitivity and will manifest as an altered slope (S) at a fixed R_{max}.

Figure 7 shows that dietary DHA deprivation, in rats, acts to reduce both R_{max} and sensitivity of the fast PIII, which is consistent with our previous reports in guinea pigs (Vingrys et al., 1998; Weisinger et al., 1999). The R_{max} loss is not great (25–50%), and in guinea pigs, it shows a dose dependency that becomes manifest when tissue DHA levels drop below 75% of their normal levels (<16% total phospholipids). On the other hand, sensitivity shows an aging component to its loss. Reductions in sensitivity were not evident in young third-generation animals, even though their tissue DHA levels were grossly abnormal (2–4% total phospholipids) and a loss of R_{max} was present (Weisinger et al., 1999). However, by 16 wk of age, a large reduction in sensitivity became manifest (75–90% loss) and this showed a dose-dependent nature (Vingrys et al., 1998; Weisinger et al., 1999). We argue that these age-related differences in R_{max} and sensitivity changes indicate for two separate lesions following DHA deprivation. Moreover, as rod morphology is normal after DHA deprivation (Bush et al., 1994; Futterman et al., 1971), we propose that our finding of a reduced R_{max} most likely reflects a lower concentration of unbleached rhodopsin in these eyes. Although not stated, our data suggest that the inactivation of rhodopsin could be affected by DHA deprivation in order to produce the altered concentration, as in vitro experiments have found normal rhodopsin concentration after DHA depletion (Bush et al., 1994). More recent experiments support this possibility (Neuringer et al., 2000). Although it is tempting to propose that the reduced pigment content would also produce the sensitivity change, our data fail to support this possibility, as normal sensitivities were found with abnormal R_{max} values. Instead, we feel that a reduced efficiency must exist in the rhodopsin–transducin–phosphodiesterase cascade in order to produce these sensitivity changes. Indeed, if the cause of this loss of efficiency involved membrane-bound processes, then our finding might imply a more widespread physiological dysfunction involving the superfamily of transmembrane proteins, of which rhodopsin is just one member (*see* Subheading 2.7.). Such a loss might be subtle initially, but show an age dependency similar to the alteration found in our visual response and deserves investigation.

2.9. Inactivation and Twin-Flash Responses

The kinetics of the inactivation process can be studied with a twin-flash procedure (Birch et al., 1995; Pepperberg et al., 1996). In this paradigm, the first, or test, flash delivers a saturating response, whereas the second flash, called the probe, is delivered at different times following the test flash in order to determine the state of recovery of the retinal response. Such methods have recently been applied to primate eyes that have been deprived of DHA (Neuringer et al., 2000) and show a deficiency in the inactivation process (30% increase in time constant) consistent with our previous suggestion. In this same study, the authors failed to find a change in R_{max} or S but did report an abnormal a-wave velocity (Neuringer et al., 2000). We interpret their finding as an indicator of altered phosphodiesterase activity consistent with an abnormal PIII sensitivity. Perhaps the sample sizes of this study limited their power to find losses in these parameters or perhaps our earlier finding for an age dependency may have frustrated its expression.

The twin-flash response can also be used to isolate cone responses from mixed rod/cone signals. It is known that the rods require a longer recovery interval between test and probe flashes (Birch et al., 1995). A twin-flash paradigm with a test/probe latency of 1–4 s would not enable the rods to recover sufficiently, meaning that the second signal would arise from cones alone. Elsewhere, using different methods, we show that the cone response has similar losses to that reported for rods following n-3 deprivation (Vingrys et al., 1998; Weisinger et al., 1999).

3. SUMMARY OF ERG FINDINGS

Studies show that moderately reduced levels of DHA (75% of normal) act to reduce the maximal a-wave response and its slope. Interestingly, gross DHA deprivation, which can be achieved by successive breeding, only results in a modest loss of function that becomes further exaggerated with age. We feel that these losses are mediated by two different physiological processes: one that involves photopigment activation kinetics and the other involves enzymatic kinetics. This former process is DHA dependent, whereas the latter process appears to show a complex age interaction. Note that whether this change is age dependent or duration of deprivation dependent is not clear from our studies and needs clarification. We submit that our findings, derived from observations of rhodopsin function, imply the potential for generalized dysfunction in many membrane-bound proteins that might manifest as more generalized systemic and central nervous system defects.

4. FUNCTIONAL AND DEVELOPMENTAL ASPECTS

4.1. Fatty Acid Deficiency

Burr and Burr (1929) identified linoleic acid as essential for growth and reproduction. In their studies, they found that rats deprived of these n-6 fatty acids developed scaly skin on their feet and tails. Growth was retarded, such that deficient animals grew to about two-thirds the size of controls. Both males and females were unable to mate successfully. If fertilization did occur, fetuses were usually reabsorbed or the pup was born deformed. Histological abnormalities were observed in both liver and kidneys so the need for dietary n-6 PUFAs became well established. However, from this time until the early 1970s, little attention was paid to the n-3 PUFAs because human diets contained relatively abundant levels of n-6 PUFAs.

Since the early 1970s, the role of n-3 fatty acids in mammalian physiology and development has been slowly elucidated, typically with the conduct of animal studies involving dietary deprivation. It is worthy of mention that greatest levels of n-3 depletion are achieved by manipulating the dietary n-6 to n-3 ratio while simultaneously decreasing the total n-3 intake (Weisinger, 1995). This accentuation of n-3 deprivation is brought about by virtue of the fact that competitive inhibition of desaturase enzymes is produced by both the n-3 and n-6 families (Fig. 1) (Gurr & James, 1980; Wainwright, 1992).

4.2. Effect of n-3 Deficiency on Retinal Function

Deficiency in n-3 fatty acids alone does not appear to result in the overt pathology found with combined n-3 and n-6 deprivation (Lamptey & Walker, 1978; Tinoco et al., 1971). Instead, n-3 deficiency has been shown to give rise to a number of subtle neural anomalies.

It is possible to partially deplete animals of n-3 fatty acids (Benolken et al., 1973; Bourre et al., 1989; Neuringer et al., 1986; Vingrys et al., 1998; Ward et al., 1996) despite recycling and conservation mechanisms (*see* Section 1) that serve to maintain the fatty acid composition of tissues during periods of dietary insufficiency. The most common modality used to achieve deprivation is feeding with semisynthetic diets, supplemented with oils low in n-3 fatty acids; the affect can be augmented by successive breeding over several generations (Weisinger et al., 1998). A recent study by Ward and co-workers (1996) employed direct gastric feeding of rats with diets controlled for both n-3 fatty acids and the ratio of n-3 to n-6. In this study, brain DHA levels were 50% of control values by 8 wk of age in the first generation, whereas the second generation showed DHA levels reduced to 10% of age-matched controls (8 wk). Such reductions in DHA are associated with a concomitant increase in 22:5n-6 (Bourre et al., 1983; Galli et al., 1971; Weisinger et al., 1995), a fatty acid differing by only one double bond from DHA, so the existence of functional loss with n-3 deprivation is somewhat surprising, as total PUFA levels are relatively constant.

4.3. Functional Deficits in n-3 Deficiency

The earliest studies investigating the effects of n-3 dietary fatty acid manipulation on vision date back to the early 1970s. Benolken and others (1973) fed rats a fat-free diet, which resulted in a 60% reduction in the retinal DHA content. This was associated with reductions in the amplitudes of the ERG a-wave (and b-wave), reflecting anomalous photoreceptor function. In a later study, the same authors (Wheeler et al., 1975) studied the effects of feeding rats diets free of fat or supplemented with either n-9, n-6, or n-3 precursors. They determined that rats showed increased ERG amplitudes when fed 2% (w/w) linoleic acid (18:2n-6, LA), but an even greater response when 2% (w/w) α-linolenic acid (18:3n-3, ALA) was used. A mixture of 1% (w/w) LA and 1% (w/w) ALA gave an intermediate response, implying that n-3 fatty acids were critical for the development of optimal retinal function.

Both Birch and co-workers (1992; 1993) and Uauy et al. (1991; 1994) assessed retinal function by electroretinography in very low birth weight infants fed differing amounts of n-3 fatty acids. They found infants fed human breast milk had the greatest amplitudes and lowest thresholds, whereas infants fed formula, low in n-3 fatty acids, showed significant reductions in all parameters of rod function at 36 wk. Cone receptor function was not affected at any age, in contrast to the studies on the guinea pig, where cone responses were also affected (Vingrys et al., 1998, Weisinger et al., 1999). These authors reported reductions in visual acuity as measured by a preferential looking technique, resulting from corn-oil-supplemented diets, compared with breast milk. The difference was small (in the order of 0.1 log units) (Birch et al., 1993). Makrides and co-workers (1994) investigated whether the disparity in neural maturation between breast-fed and formula-fed term infants could be corrected by supplementation with DHA-containing fish oil. The results of the study indicated that supplemented-formula-fed infants had better visual evoked potential acuity, in the order of 0.3 log units, than did placebo-supplemented infants. Although there were no differences in the visual performance of breast-fed and fish-oil-supplemented infants, the increased level of erythrocyte DHA in fish-oil-supplemented infants was associated with a significant decrease in the proportion of AA. All infants, regardless of diet, had normal length and weight-for-age scores.

The use of the guinea pig and successive breeding afforded a good animal model to consider the role that DHA has on retinal development (Leat et al., 1986; Weisinger et al., 1995; 1996a; 1996b; 1997; 1998). Guinea pigs are useful models of human neonatal growth, as they are one of the few mammals to show intrauterine patterns of neural development similar to primates and humans (Bui & Vingrys, 2000). Moreover, dietary deprivation and successive breeding is able to reduce the retinal DHA content to less than 3% of total phospholipid fatty acids. Although such dietary manipulation over three generations was associated with reductions in ERG amplitude (50%), the findings imply that DHA is not essential for neural function but is needed to produce optimal function. This finding was significantly different to that of Leat et al. (1986), who reported no reduction in a- or b-waves, despite very low retinal DHA levels (approx 2%). We (Weisinger et al., 1996a) argue elsewhere that the negative finding of Leat et al. (1986) resulted from the low statistical power of their experimental design, which emphasizes the importance of sample size, replication, and animal selection in such studies (*see* Ward & Wainwright (1998) for an excellent discussion).

4.4. Effect of Maternal Diet on Retinal n-3 Depletion

Because many of the studies described in this chapter employed dietary manipulation over several generations to achieve a greater effect on the fatty acid profile (e.g., (Benolken et al., 1973; Bush et al., 1994; Leat et al., 1986; Weisinger et al., 1996a)) there is a necessity to quantify the effect of the maternal dietary status, not only on the fatty acid profile but also on the rate and degree to which manipulation is possible in the infant. This issue was addressed, in part, by Connor and his colleagues (1993), who raised rhesus monkeys deprived of n-3 fatty acids during the prenatal or postnatal or combined periods. Analysis of their results indicates that there was little difference in the level of depletion achieved, when comparing plasma DHA levels of animals deprived for the gestation period compared with those commenced at 8–9 mo of postnatal life. However, as this study failed to provide neural fatty acid profiles, their finding is hard to interpret, as neural fatty acid profiles are known to differ from those of the blood components (Anderson et al., 1994). Furthermore, the study design did not enable comparison of data in age-matched animals born of mothers to whom the supply of n-3 fatty acids differed. More recently, we showed that maternal diet has a significant impact on both the rate of depletion and the final level of neural fatty acids in the neonate (Weisinger et al., 1998).

4.5. Repletion Following n-3 Fatty Acid Deficiency

One issue germane to this topic, involves the reversibility of neural deficits following an initial n-3 fatty acid deficiency. Although many studies have shown that reintroduction of α-linolenic into the diet results in the gradual restoration of neural DHA levels (Bourre et al., 1989; Galli et al., 1971; Youyou et al., 1986), there is a lack of literature regarding reversibility of function following dietary n-3 repletion. Bourre and co-workers (1989) fed rats diets containing either n-3 adequate soybean oil or n-3 deficient sunflower oil as the sole source of lipid. The sunflower-oil diet resulted in a 64% reduction in the DHA levels of second-generation animals, relative to those fed soybean oil. These fatty acid reductions were associated with lower a- and b-wave amplitudes early in life; however, these were not sustained to adulthood. A similar effect was reported by Neuringer and co-workers (1993) in nonhuman primates. Like the findings of Bourre et al. (1989), Neuringer et al. (1993) reported that some functional effects associated with DHA deficiency disappeared after 2 yr, despite continued dietary deprivation of n-3 fatty acids. However, although these studies indicate that functional losses may resolve with time, despite continued dietary n-3 deficiency, this finding has not been reproduced in guinea pigs by our laboratory (Vingrys et al., 1998). Moreover, recent studies from our laboratory show not only that n-3 deprivation induced deficits fail to resolve with time but also that rats deprived of n-3 PUFAs early in life show persistent ERG defects as adults even after n-3 supplementation has restored normal fatty acid tissue profiles. We suggest that this persistence of ERG deficits in adult life may offer evidence for a critical period for the accretion of n-3 PUFAs into the retina (Armitage et al., 2000). The reason for these different outcomes between studies is not clear and needs further clarification.

5. PROPOSED ROLE OF n-3 FATTY ACIDS

The literature poses many interesting questions regarding the necessity of long-chain n-3 fatty acids in neural function and development. The next subsection will further elucidate the role of these molecules in retinal function. Several models have been proposed on a biomolecular and biophysical level, which go some way to explaining some of the functional changes.

5.1. Role in ROS Membranes: Modulation of Protein Function

There is little doubt that the primary activating step in the phototransduction cascade is the conversion of the chromophore metarhodopsin I (MI) to metarhodopsin II (MII) (Cibis et al., 1995; Liebman et al., 1987; Yau, 1994). This reaction is in a state of equilibrium and several investigators have observed the manner with which this equilibrium (and hence transduction) can be modulated by altering the fatty acid composition of the membrane containing the rhodopsin molecule. In doing so, several models have been proposed that consider the effect that the long-chain n-3 fatty acids, such as DHA, have in modulating rhodopsin's function within the outer segment (Brown, 1994; Dratz & Holte, 1993; Littman & Mitchell, 1996).

6. BIOMECHANICAL ROLES FOR DHA

6.1. Membrane Requirements for Photoactivity

Brown (1994) provides strong evidence that the MI–MII transition is favored by the presence of relatively small phospholipid headgroups, which pack tightly to produce a condensed bilayer, together with bulky acyl chains, as found in DHA. The resultant imbalance has been shown to give rise to an unstable membrane environment, which is a necessary element in facilitating the activation of rhodopsin. In contrast, stable membranes forming regular, lamellar arrangements do not readily support activation of the rhodopsin molecule.

Brown (1994) proposes that rhodopsin's function is influenced by the mechanical properties of the bilayer carrying the molecule, rather than the specific chemical properties of the fatty acyl chains. This is best illustrated by the curious finding that native membrane behavior can be approximated in recombinant membranes composed of plant and egg phospholipids, together with phytanic acid (non-native) chains (Brown, 1994). Brown has interpreted these findings in favor of a mechanical role for DHA on the basis that mixtures of lipids are necessary, as no single lipid has been found to replicate native behavior. In particular, PC in any mixture, and even when formed as di(22:6n-3), cannot support native function implying a complex mechanical effect. Brown (1994) concluded that acyl chain length was a contributing factor, as a minimum bilayer thickness is needed to facilitate conversion of MI to MII.

Furthermore, Brown (1994) listed the following properties as those that are important in determining the activation of rhodopsin: average bilayer thickness (*see* Dratz; Subheading 6.2.); lateral compressibility (*see* Litman and Mitchell, below); curvature stresses of the lipid–water and protein–lipid interfaces; and electrostatic forces (determined by the charge of the headgroups).

6.2. Membrane Expansion

Dratz (1993) has proposed a model in which docosahexaenoic acid acts as a molecular spring within the retinal outer segment membranes. In the resting state, the chromophore of rhodopsin is enshrouded by a protein environment, which, in turn, is surrounded by a membrane composed of phospholipids. Where the membrane contains a high DHA content, its thickness, measured by x-ray diffraction, is approx 27 Å. The higher-energy state following absorption of a photon of light requires the protein to change conformation. It is argued that expansion of the membrane is necessary to facilitate this conformational change. Membranes composed of DHA have been shown to expand (thicken) by up to 7% (Dratz & Holte, 1993) with minimal energy input, implying facilitation of the process of transduction.

6.3. Membrane Fluidity

Litman and Mitchell (1996) proposed a model of altered fluidity. They suggest that when outer segment membranes are composed of highly unsaturated molecules, such as DHA, they provide an optimal domain for rhodopsin transformation after light capture. This results from the fluidity and lateral compressibility provided at the domain boundary formed between saturated *sn*-1 chains and the highly unsaturated *sn*-2 chains (typically DHA). Hence, a reduction in retinal DHA levels results in a suboptimal membrane environment, which restricts the activation of membrane-bound proteins. A reduced receptoral response is predicted in conditions of DHA deficiency, consistent with the findings of several ERG studies.

There is other evidence for increased fluidity in DHA-rich membranes. Treen and co-workers (1992) showed that cultured human retinoblastoma cells, grown in serum containing a fourfold increase of DHA, had significantly greater lateral membrane fluidity. In addition to promoting the reconformation of rhodopsin from MI to MII (Littman & Mitchell, 1996), Treen (1992) and others (Hyman & Spector, 1982; 1992) have proposed that the increase in fluidity facilitates the high-affinity uptake of choline, which has shown to be independent of both sodium and energy.

7. ROLE OF DHA IN RPE FUNCTION

Long-chain n-3 fatty acids are also believed to affect the function of the retinal pigment epithelium (RPE) and, hence, play an indirect but vital role in the physiology of the photoreceptors. Several studies demonstrating altered RPE function in conditions of n-3 fatty acid deficiency have been reported (Bush et al., 1994; Watanabe et al., 1993). An overview of these is given in Section 7.1.

7.1. Effect on RPE Phagosomes

One of the major functions of the RPE is to provide photoreceptor recycling by the phagocytosis of the shed tips of photoreceptor outer segments. It is vital that phagocytosis occurs unhindered, because the quantity of material engulfed by each RPE cell is in the order of 10 entire outer segments per day (based on 10% renewal/d; 10 rods/RPE cell) (Landis et al., 1973; Young, 1967; Young & Bok, 1969). Any defect in the normal function of the RPE is likely to promote serious photoreceptor degeneration, as in rats (Cibis et al., 1995).

Indeed, n-3 fatty acids are known to be involved in this process, because a reduction in retinal DHA levels has been associated with changes in both function (Bush et al., 1994) and size distribution (Watanabe et al., 1993) of RPE phagosomes. Hence, it is entirely possible that a deficiency in n-3 fatty acids may indirectly, through RPE dysfunction, lead to anomalous photoreceptor function.

However, other factors may also come into play. Chen and co-workers (1996) have conducted a series of elegant experiments and proposed a mechanism by which retinoids are transferred between photoreceptors and the RPE. The retinoids, 11-*cis* retinal and all-*trans* retinol, are shuttled between the photoreceptors and the RPE by a class of carrier proteins known as interphotoreceptor retinoid-binding proteins (IRBP). Chen et al. (1996) determined that retinoid binding affinities of the IRBP change, as a function of the fatty acids that associate with the protein. When IRBP is localized near the RPE, it is associated with the fatty acids most prevalent in this location, usually saturated fatty acids. Under these conditions, the hydrophilic site of the IRBP possesses a high affinity for the readily available 11-*cis* retinal, which forms a ligand at this site. As the IRBP moves toward the photoreceptors, the fatty acids associating with the IRBP are altered by the high local concentration of long-chain molecules, such as DHA, for which it has the highest affinity (Chen et al., 1993). When DHA associates with IRBP, the affinity of the hydrophilic site for 11-*cis* retinal is greatly reduced and the retinoid is released to the receptors for incorporation into rhodopsin. Because the affinity for all-*trans* retinol remains high in the presence of DHA, this now forms a ligand to the IRBP for transportation back to the RPE, where it is reisomerized. In this way, DHA acts as a switch that alters the affinity of IRBP to various conformations of the retinoid. By extension, should the DHA content of the photoreceptors be lowered, it is possible that the efficiency of IRBP in delivering 11-*cis* retinal to the receptors could be greatly reduced. This, in turn could be expected have an impact on the visual cycle and result in a lowered ERG amplitude.

However, Bush and co-workers (1994) determined the dark-adapted rhodopsin content of n-3-deficient rats was 12–15% higher than in n-3 sufficient soy-oil-supplemented rats, a finding not made in rats fed fat-free diets (Dudley et al., 1975). It is proposed that this discrepancy might reflect higher levels of outer segment membrane peroxidation, associated with the higher levels of tissue long-chain PUFAs (Penn & Anderson, 1991), after feeding with n-3-adequate diets. If the DHA deficiency were to impact directly on RPE physiology, rather than the capacity of the RPE to support photoreceptor function, this should manifest as a change in the c-wave; an issue not considered in this chapter, but one that deserves investigation.

8. SUMMARY

The body of literature indicates that n-3 fatty acids play an important role in several aspects of function of mammalian brain and retina. There is widespread agreement that dietary n-3 deficiency results in the depletion of neural phospholipid docosahexaenoic acid, despite local and systemic conservation mechanisms. In turn, depletion of tissue DHA stores appear to alter membrane characteristics. The resultant changes in membrane structure appear to affect receptoral mechanisms, which manifest as changes in function. One membrane-bound receptor to be affected is rhodopsin, and we propose that it might provide a useful model of the effects that DHA deprivation can have on other brain-dependent mechanisms. As such, DHA deprivation results in decreases in the neural activity of the retina, as measured by the ERG, reduced visual acuity (Birch et al., 1992; Makrides et al., 1994; Uauy et al., 1994), altered learning behavior (Umezawa et al., 1999), altered behavioral responses, abnormal thirst and drinking (Reisbick et al., 1992; Reisbick et al., 1990), abnormal auditory responses (Bourre et al., 1999), and abnormal olfaction (Greiner et al., 1999). We propose that these abnormalities might involve a common mechanism, as each of these processes involves rhodopsin-like (seven transmembrane α-helix) receptors, so a common dysfunction similar to that found in the visual process might underlie all of these effects.

REFERENCES

Anderson R. Biochemistry of the Eye. American Academy of Ophthalmology, San Fransisco, 1983.
Anderson R, Maude M. Lipids of ocular tissues. VIII. The effects of essential fatty acid deficiency on the phospholipdis of the photoreceptor membranes of rat retina. Arch Biochem Biophysiol 1970; 151:270–276.
Anderson R, Risk M. Lipids of ocular tissues. IX. The phospholipids of frog photoreceptor membranes. Vision Res 1974; 14:129–131.
Anderson RE, Chen H, Stinson A. The accretion of docosahexaenoic acid in the retina. World Rev Nutr Diet 1994; 75:124–127.
Armington JC. The Electroretinogram. Academic, New York, 1974.
Armitage J, Weisinger H, Vingrys A, et al. Perinatal omega-3 fatty acid deficiency alters ERG in adult rats irrespective of tissue fatty acid content. [ARVO Abstract]. Invest Ophthalmol Vis Sci 2000; 41:S245.
Aveldano de Caldironi M, Bazan N. Composition and biosynthesis of meolcular species of retina phosphoglycerides. Neurochemistry 1980; 1:381–392.
Baylor DA, Lamb TD, Yau KW. The membrane current of single rod outer segments. J Physiol 1979; 288:589–611.

Bazan NG, Gordon WC, Rodriguez de Turco EB. Docosahexaenoic acid uptake and metabolism in photoreceptors: retinal conservation by an efficient retinal pigment epithelial cell-mediated recycling process. Adv Exp Med Biol 1992; 318:295–306.

Bazan NG, Rodriguez de Turco EB, Gordon WC. Docosahexaenoic acid supply to the retina and its conservation in photoreceptor cells by active retinal pigment epithelium-mediated recycling. World Rev Nutr Diet 1994; 75:120–123.

Bell M, Dick J, Buda C. Molecular seciation of fish sperm phospohlipids: large amounts of dipolyunsaturated phosphatidylserine. Lipids 1997; 32:1085–1091.

Benolken RM, Anderson RE, Wheeler TG. Membrane fatty acids associated with the electrical response in visual excitation. Science 1973; 182:1253–1254.

Bernsohn J, Spitz F. Linoleic and linolenic acid dependency on some brain membrane-bound enzymes after lipid deprivation in rats. Biochem Biophys Res Commun 1974; 57:293–298.

Birch DG, Birch EE, Hoffman DR, et al. Retinal development in very-low-birth-weight infants fed diets differing in omega-3 fatty acids. Invest Ophthalmol Vis Sci 1992; 33:2265–2376.

Birch DG, Hood DC, Nusinowitz S, et al. Abnormal activation and inactivation mechanisms of rod transduction in patients with autosomal dominant retinitis pigmentosa and the Pro-23-his mutation. Invest Ophthalmol Vis Sci 1995; 36:1603–1614.

Birch E, Birch D, Hoffman D, et al. Breast-feeding and optimal visual development. J Pediatr Ophthalmol Strabismus 1993; 30:33–38.

Bourre J, Durand G, Erre J, et al. Changes in auditory brainstem responses in alpha-linolenic acid deficiency as a function of age in rats. Audiology 1999; 38:8–13.

Bourre J, Faivre A, Dumont O, et al. Effect of polyunsaturated fatty acids on fetal mouse brains in culture in a chemically defined medium. J Neurochem 1983; 41:1234–1242.

Bourre JM, Francois M, Youyou A, et al. The effects of dietary alpha-linolenic acid on the composition of nerve membranes, enzymatic activity, amplitude of electrophysiological parameters, resistance to poisons and performance of learning tasks in rats. J Nutr 1989; 119:1880–1892.

Breton ME, Quinn GE, Schueller AW. Development of electroretinogram and rod phototransduction response in human infants. Invest Ophthalmol Vis Sci 1995; 36:1588–1602.

Breton ME, Schueller AW, Lamb T, et al. Analysis of ERG a-wave amplification and kinetics in terms of the G-protein cascade of phototransduction. Invest Ophthalmol Vis Sci 1994; 35:295–309.

Brown KT. The electroretinogram: its components and their origins. Vision Res 1968; 8:633–677.

Brown MF. Modulation of rhodopsin function by properties of the membrane bilayer. Chem Phys Lipids 1994; 73:159–180.

Bui BV, Vingrys AJ. Development of receptoral responses in pigmented and albino guinea-pigs *(Cavea porcellus)*. Documenta Ophthalmol, 2000; 99:151–170.

Bui BV. The development of the electroretinogram in the guinea pig *(cavia porcellus)* MSc dissertation, University of Melbourne, 1998.

Burr G, Burr M. A new deficiency disease produced by the rigid exclusion of fat from the diet. J Biol Chem 1929; 82:345–367.

Bush RA, Malnoe A, Reme CE, et al. Dietary deficiency of n-3 fatty acids alters rhodopsin content and function in the rat retina. Invest Ophthalmol Vis Sci 1994; 35:91–100.

Bush RA, Sieving PA. A proximal retinal component in the primate photopic ERG a-Wave. Invest Ophthalmol Vis Sci 1994; 35(2):635–645.

Chen Y, Houghton LA, Brenna JT, et al. Docosahexaenoic acid modulates the interactions of the interphotoreceptor retinoid-binding protein with 11-*cis*-retinal. J Biol Chem 1996; 271:20,507–20,515.

Chen Y, Saari JC, Noy N. Interactions of all-*trans*-retinol and long-chain fatty acids with interphotoreceptor retinoid-binding protein. Biochemistry 1993; 32:11,311–11,318.

Cibis G, Anderson R, Chew E, et al. Fundamentals and principles of ophthalmology. In: Cibis G, Anderson R, Chew E, et al., eds. Basic and Clinical Science Course. American Academy of Ophthalmology, San Fransisco, 1995.

Cideciyan AV, Jacobson SG. An alternative phototransduction model for human rod and cone ERG a-waves: normal parameters and variation with age. Vision Res 1996; 36:2609–2621.

Clarke SD, Jump DB. Dietary polyunsaturated fatty acid regulation of gene transcription. Ann Rev Nutr 1994; 14:83–98.

Cobbs WH, Pugh EN. Kinetics and components of the flash photocurrent of isolated retinal rods of the larval salamanda, *Ambystoma tigrinum*. J Physiol 1987; 394:529–572.

Cone RA. Early receptor potential of the vertebrate retina. Nature 1964; 201:626–628.

Cone RA, Cobbs WH. Rhodopsin cycle in the living eye of the rat. Nature 1969; 221:820–822.

Connor WE, Neuringer M. The effects of n-3 fatty acid deficiency and repletion upon the fatty acid composition and function of the brain retina. In: Karnovsky ML, Leaf A, Bolls LC, eds. Biological Membranes: Aberrations in Membrane Structure and Function. Alan R. Liss, New York, 1988.

Connor WE, Lin DS, Neuringer M, et al. The comparative importance of prenatal and postnatal n-3 fatty acid deficiency: repletion at birth and later. In: Sinclair AJ, Gibson RA, eds. The Third International Congress on Essential Fatty Acids and Eicosanoids. AOCS, Champaign, IL, 1993.

Crawford MA, Sinclair AJ. Nutritional influences in the evolution of the mammalian brain, CIBA Foundation Symposium on Lipids, Malnutrition and the Developing Brain. Associated Scientific, New York, 1972.

Dawson W, Galloway N. Early receptor potential: origin and clinical applications. In: Heckenliveley J, Arden G, eds. Principles and Practice of Clinical Electrophysiollogy of Vision. Mosby Year Book, St. Louis, MO, 1991.

Dowling JE. The Retina: An Approachable Part of the Brain. Harvard University Press, Cambridge, MA, 1987.

Dratz EA, Holte LL. The molecular spring model for the function of docosahexaenoic acid (22:6n-3) in biological membranes. In: Sinclair AJ, Gibson RA, eds. The Third International Congress on Essential Fatty Acids and Eicosanoids. AOCS, Champaign, IL, 1993.

Dratz EA, Furstenau JE, Lambert CG, et al. NMR structure of a receptor-bound G-protein peptide. Nature 1993; 363:276–281.

Dudley P, Landis D, Anderson R. Further studies on the chemistry of photoreceptor membranes fed an essential fatty acid deficient diet. Exp Eye Res 1975; 21:523–530.

Faber DS. Analysis of the slow transretinal potentials in response to light. Doctoral dissertation, State University of New York, Buffalo, 1969.

Fliesler SJ, Anderson RE. Chemistry and metabolism of lipids in the vertebrate retina. Prog Lipid Res 1983; 22:79–131.

Frishman L, Karwoski C. The d-wave. In: Heckenliveley J, Arden G, eds. Principles and Practice of Clinical Electrophysiollogy of Vision. Mosby YearBook, St Louis, MO, 1991.

Frishman LJ, Steinberg RH. Origin of negative potentials in the light-adapted ERG of cat retina. J Neurophysiol 1990; 63:1333–1346.

Fulton AB, Rushton WA. The human rod ERG: correlation with psychophysical responses in light and dark adaptation. Vision Res 1978; 18(7):793–800.

Fulton AB, Dodge J, Hansen RM, et al. The quantity of rhodopsin in young human eyes. Curr Eye Res 1991; 10:977–982.

Fulton A, Hansen RM, Dorn E, et al. Development of primate rod structure and function. In: Vital-Durrand F, ed. Infant Vision. Oxford University Press, London, 1995a.

Fulton AB, Hansen RM, Findl O. The development of the rod photoresponse from dark-adapted rats. Invest Ophthalmol Vis Sci 1995b; 36:1038–1045.

Futterman S, Stevens-Andrews J. The fatty acid composition of human retinal vitamin A ester and the lipids of human retinal tissue. Invest Ophthalmol Vis Sci 1964; 3:441–444.

Futterman S, Downer JL, Hendrickson A. Effect of essential fatty acid deficiency on the fatty acid composition, morphology, and electroretinographic response of the retina. Invest Ophthalmol Vis Sci 1971; 10:151–156.

Galli C, White H, Paoletti R. Lipid alterations and their reversion in the central nervous system of growing rats deficient in eesential fatty acids. Lipids 1971; 6:378–387.

Gerbi A, Zerouga M, Debray M, et al. ffect of dietary a-linolenic acid on functional characteristic of Na$^+$/K$^+$-ATPase isoenzymes in whole brain membranes of weaned rats. Biochim Biophys Acta 1993; 1165:291–298.

Granit R. Components of the retinal action potential in mammals and their relations to the discharge in the optic nerve. J Physiol 1933; 77:207–238.

Greiner R, Moriguchi T, Hutton A, et al. Rats with low levels of brain docosahexaenoic show impaired performance in oflactory-based and spatial learning tasks. Lipids 1999; 34:S239–S243.

Gurr MI, James AT. Lipid Biochemistry. An Introduction. Chapman & Hall, London, 1980.

Holman R. Biological activities of and requirements for polyunsaturated fatty acids. Prog Chem Fats Other Lipids 1968; 9:611–680.

Holopigian K, Greenstein VC, Seiple W, et al. Evidence for photoreceptor changes in patients with diabetic retinopathy. Invest Ophthalmol Vis Sci 1997; 38:2355–2365.

Hood DC, Birch DG. A quantitative measure of the electrical activity of human rod photoreceptors using electroretinography. Vis Neurosc 1990a; 5:379–387.

Hood DC, Birch DG. The a-wave of the human ERG and rod receptor function. Invest Ophthalmol Vis Sci 1990b; 31:2070–2081.

Hood DC, Birch DG. Rod phototransduction in retinitis pigmentosa: estimation and interpretation of parameters derived from the rod a-wave. Invest Ophthalmol Vis Sci 1994; 35:2948–2961.

Huster D, Arnold K, Gawrisch K. Influence of docosahexaenoic acid and cholesterol on lateral lipid organization in phospholipid mixtures. Biochemistry 1998; 37:17299–308.

Hyman B, Spector A. Choline uptake in cultured human Y79 retinoblastoma cells: effect of polyunsaturated fatty acid compositional modifications. J Neurochem 1982; 38:650–656.

Karwoski CJ, Proenze LM. Light-evoked changes in extracellular potassium concentration in mudpuppy retina. Brain Res 1978; 142:515–530.

Kline RP, Ripps H, Dowling JE. Light induced potassium fluxes in the skate retina. J Neurosci 1985; 464:225–235.

Kraft TW, Schneeweis DM, Schnaft JL. Visual transduction in human rod photoreceptors. J Physiol 1993; 464:747–765.

Kurlack L, Stephenson T. Plausible explanations for effects of long chain polyunstaurated fatty acids on neonates. Arch Dis Child 1999; 80:148–154.

Lamb TD. Transduction in vertebrate photoreceptors: the roles of cyclic GMP and calcium. Trends Neurosci 1986; 9:224–228.

Lamb TD, Pugh EN Jr. A quantitative account of the activation steps involved in phototransduction in amphibian photoreceptors. J Physiol 1992; 449:719–758.

Lamptey M, Walker B. Learning behaviour and brain lipid composition in rats subjected to essential fatty acid deficiency during gestation, lactation and growth. J Nutr 1978; 108:358–367.

Landis DJ, Dudley PA, Anderson RE. Alteration of disc formation in photoreceptors of rat retina. Science 1973; 182:1144–1146.

Leat WMF, Curtis R, Millichamp NJ, et al. Retinal function in rats and guinea pigs reared on diets low in essential fatty acids and supplemented with linoleic or linolenic acids. Ann Nutr Metab 1986; 30:166–174.

Liebman PA, Parker KR, Dratz EA. The molecular mechanism of visual excitation and its relation to the structure and composition of the rod outer segment. Ann Rev Physiol 1987; 49:765–791.

Lin DS, Anderson GJ, Connor WE, et al. Effect of dietary n-3 fatty acids upon the phospholipid molecular species of the monkey retina. Invest Ophthalmol Vis Sci 1994; 35:794–803.

Lipton S, Rasmussen H, Dowling J. Electrical and adaptive properties of rod photoreceptors in *Bufo marinus*: II. Effects of cyclic nucleotides and prostaglandins. J Gen Physiol 1977; 70:771–791.

Littman BJ, Mitchell DC. A role for phospholipid polyunsaturation in modulating membrane protein function. Lipids 1996; 31:S193–S197.

Lyubarski A, Pugh E. Recovery phase of the murine rod photoresponse reconstructed from electroretinographic recordings. J Neurosci 1996; 16:563–571.

Makrides M, Neumann M, Simmer K, et al. Are long-chain polyunsaturated fatty acids essential nutrients in infancy? Lancet 1994; 345:1463–1468.

Marmor MF, Zrenner E. Standard for clinical electroretinography (1999 update). Documenta Ophthalmol 1999; 97:143–156.

McMurchie EJ. Dietary lipids and the regulation of membrane fluidity and function In: Aloia, RC, ed. Physiological Regulation of Membrane Fluidity. Alan R. Liss, New York, 1988.

Miller S, Steinberg R. Passive ionic properties of the frog retinal pigment epithelium. J Membr Biol 1977; 36:337–372.

Mitchell DC, Gawrisch K, Litmann BJ, Salem N Jr. Why is docosahexaenoic acid essential for nervous system function? Biochem Soc Trans 1998; 26:365–370.

Naka K, Rushton WA. H. S-potential from luminosity units in the retina of the fish (cyprinidont). J Physiol 1966; 185:587–599.

Nettleton JA. n-3 Fatty acids: comparison of plant and seafood sources in human nutrition. J Am Diet Assoc 1991; 91:331–337.

Neuringer M. The relationship of fatty acid composition to function in the retina and visual system. In: Dobbing J, ed. Lipids, Learning and the Brain: Fats in Infant Formulas, Report of the 103rd Ross Conference on Paediatric Research. Ross Laboratories, Columbus, OH, 1993.

Neuringer M, Connor WE, Lin DS, et al. Biochemical and functional effects of prenatal and postnatal omega-3 fatty acid deficiency on retina and brain in rhesus monkeys. Proc Natl Acad Sci USA 1986; 83:4021–4025.

Neuringer M, Jeffrey B, Gibson R. N-3 fatty acid deficiency alters rod phototrandsuction and recovery. [ARVO Abstract]. Invest Ophthalmol Vis Sci 2000; 41:S493.

Nicholls J, Martin A, Wallace B. eds. From Neuron to Brain. Sinauer Associates, Sunderland, MA, 1992.
Papahadjopoulos D. Calcium-induced phase changes and fusion in natural and model membranes. In: Poste G, ed. Membrane Fusion. North-Holland, Amsterdam, 1978.
Penn JS, Anderson RE. Effects of light history on the rat retina. Prog Retinal Res 1991; 11:75–98.
Penn RD, Hagins WA. Signal transmission along retinal rods and the origin of the electroretinographic a-wave. Nature 1969; 223:201–205.
Pepperberg DR, Birch DG, Hofmann KP, et al. Recovery kinetics of human rod phototransduction inferred from the two-branched a-wave saturation function. J Opt Soc Am A 1996; 13:586–600.
Quinn PJ. The fluidity of cell membranes and its regulation. Prog Biophysiol Mol Biol 1981; 38:1–104.
Reisbick S, Neuringer M, Connor WE, et al. Postnatal deficiency of omega-3 fatty acids in monkeys: fluid intake and urine concentration. Physiol Behav 1992; 51:473–479.
Reisbick S, Neuringer M, Hasnain R, et al. Polydipsia in rhesus monkeys deficient in omega-3 fatty acids. Physiol Behav 1990; 47:315–323.
Ripps H, Witlovsky P. Neuron-glia interaction in the brain and retina. Prog Retinal Res 1985; 4:181–219.
Rodieck RW. Components of the electroretinogram-a reappraisal. Vision Res 1972; 12:773–780.
Rodieck RW. The primate retina. Compar Primate Biol 1988; 4:203–278.
Rodriguez de Turco EB, Deretic D, Bazan NG, et al. Post-golgi vesicles cotransport docosahexaenoyl-phospholipids and rhodopsin during frog photoreceptor membrane biogenesis. J Biol Chem 1997; 272(16):10,491–10,497.
Rotstein NP, Pennacchiotti GL, Sprecher H, et al. Active synthesis of C24:5,n-3 fatty acid in retina. J Biochem 1996; 316:859–864.
Salem Jr., N. Omega-3 fatty acids: molecular and biochemical aspects. In: Spiller GA, Scala J, eds. New Protective Roles for Selected Nutrients. Alan R. Liss, New York, 1989.
Schnapf JL, Baylor DA. How photoreceptors respond to light. Sci Am 1987; 256:40–47.
Sinclair A. The nutritional significance of omega-3 polyunsaturated fatty acids for humans. Asean Food 1993; J8:3–13.
Sinclair AJ. Long chain polyunsaturated fatty acids in the mammalian brain. Proc Nutr Soc 1975; 34:287–291.
Steinberg RH, Linsenmeier RA, Griff ER. Retinal pigment epithelial cell contributions to the electroretingram and electrooculogram. Prog Retinal Res 1985; 4:33–66.
Steinberg RH, Schmidt R, Brown K. Intracellular responses to light from the cat retinal pigment epithelium: origin of the electroretinogram c-wave. Nature 1970; 227:728–730.
Stryer L. Biochemistry. WH Freeman, New York, 1981.
Stryer L. The cyclic GMP cascade of vision. Ann Rev Neurosci 1986; 9:87–119.
Tinoco J, Williams M, Hincenbergs I, et al. Evidence for nonessentiality of linolenic acid in the diet of the rat. J Nutr 1971; 101:937–946.
Treen M, Uauy RD, Jameson DM, et al. Effect of docosahexaenoic acid on membrane fluidity and function in intact cultured Y-79 retinoblastoma cells. Arch Biochem Biophysiol 1992; 294:564–570.
Uauy R, Hoffman DR. Essential fatty acid requirements for normal eye and brain development. Semin Perinatol 1991; 15:449–455.
Uauy RD, Birch EE, Birch DG, et al. Significance of ω3 fatty acids for retinal and brain development of preterm and term infants. World Rev Nutr Diet 1994; 75:52–62.
Umezawa M, Kogishi K, Tojo H, et al. High linoleate and high alpha linolenate diets affect learning ability and natural behaviour in SAMRI mice. J Nutr 1999; 129:431–437.
Vingrys AJ, Weisinger HS, Sinclair AJ. The effect of age and n-3 PUFA level on the ERG in the guinea pig. In: Huang Y. Sinclair A eds., Lipids and Infant Nutrition. AOCS, Champaign, IL, 1998.
Voss A, Reinhart S, Sankarappa S, et al. Metabolism of 22:5n-3 to 22:6n-3 in rat liver is independent of 4-desaturase. J Biol Chem 1991; 166:1995–2000.
Wainwright P. Do essential fatty acids play a role in brain and behavioural development? Biol Behav Rev 1992; 16:193–205.
Wang N, Anderson RE. Synthesis of docosahexaenoic acid by retina and retinal pigment epithelium. Biochemistry 1993; 32:13,703–13,709.
Ward G. Wainwright P. The contribution of animal models to understanding the role of fats in infant nutrition. In: Huang Y, Sinclair A, eds. Lipids in Nutrition. AOCS, Champaign, IL, 1998.
Ward G, Woods J, Reyzer M, et al. Artificial rearing of infant rats on milk formula deficient in n-3 essential fatty acids: a rapid method fo the production of experimental n-3 deficiency. Lipids 1996; 31:71–78.
Watanabe I, Aonuma H, Kaneko S, et al. Effect of high linoleate and high α-linoleate diets on size distribution of phagosomes in retinal pigment epithelium. In: Yasugi T, Nakamura H, Soma M, eds. Advances in Polyunsaturated Fatty Acid Research. Elsevier, Amsterdam, 1993.

Weisinger HS. The effect of docosahexanaenoic acid on the electroretinogram of the guinea pig.MSc dissertation, University of Melbourne, Melbourne, 1995.

Weisinger HS, Sinclair AJ, Vingrys AJ. Effect of dietary n-3 deficiency on the electroretinogram in the guinea pig. Ann Nutr Metab 1996; 40:91–98.

Weisinger HS, Vingrys AJ, Sinclair AJ. Dietary manipulation of long-chain polyunsaturated fatty acids in the retina and brain of guinea pigs. Lipids 1995; 30:471–473.

Weisinger HS, Vingrys AJ, Sinclair AJ. The effect of docosahexaenoic acid on the electroretinogram of the guinea pig. Lipids 1996a; 31(1):65–70.

Weisinger HS, Vingrys AJ, Sinclair AJ. Electrodiagnostic methods in vision. Parts 1–3: Clinical experimental optometry 1996b; 79:50–61; 97–105; 131–143.

Weisinger HS, Vingrys AJ, Sinclair AJ. Effect of diet on the rate of depletion of n-3 fatty acids in the retina of the guinea pig. J Lipid Res 1998; 39:1274–1279.

Weisinger HS, Vingrys AJ, Bui BV, et al. Effects of dietary n-3 fatty acid deficiency and repletion in the guinea pig retina. Invest Ophthalmol Vis Sci 1999; 40:327–338.

Wheeler TG, Benolken RM, Anderson RE. Visual membranes: specificity of fatty acid precursors for the electrical response to illumination. Science 1975; 188:1312–1314.

Witkovsky P, Dudek FE, Ripps H. Slow PIII component of the carp electroretinogram. J Gen Physiol 1975; 65:119–134.

Yau KW. Phototransduction in retinal rods and cones. Invest Ophthalmol Vis Sci 1994; 35:9–32.

Yoshida S, Yasuda A, Kawasato H, et al. Ultrastructural study of hippocampus synapse in perilla and safflower oil fed rats. In: Yasugi T, Nakamura H, Soma M, eds. Advances in Polyunsaturated Fatty Acid Research. Elsevier, Amsterdam, 1993.

Young R. The renewal of photoreceptor cell outer segments. J Cell Biol 1967; 42:392–403.

Young R, Bok D. Participation of the retinal pigment epithelium in the rod outer segment renewal process. J Cell Biol 1969; 42:392–403.

Youyou A, Durand G, Pascal G, et al. Recovery of altered fatty acid composition induced by a diet devoid of n-3 fatty acids in myelin, synaptosomes, mitochondria, and microsomes of developing rat brain. J Neurochem 1986; 46:224–228.

13 Brightness–Discrimination Learning Behavior and Retinal Function Affected by Long-Term α-Linolenic Acid Deficiency in Rat

Harumi Okuyama, Yoichi Fujii and Atsushi Ikemoto

1. INTRODUCTION

Major fatty acids in food are classified into three series depending on their metabolism in mammals (Fig. 1). Saturated fatty acids with 16- and 18-carbon chain lengths are synthesized *de novo* and are easily desaturated to monounsaturated fatty acids in mammals. Linoleic acid (18:2n-6, LA) and α-linolenic acid (18:3n-3, ALA) are synthesized in plants but not mammals. When ingested, LA is desaturated and elongated to form arachidonic acid (20:4n-6, ARA) and other n-6 fatty acids. ALA is metabolized similarly to form other n-3 fatty acids such as eicosapentaenoic acid (20:5n-3, EPA) and docosahexaenoic acid (22:6n-3, DHA). No interconversion occurs among the three series, although retroconversion occurs in mammals (e.g., from DHA to ALA) (Sprecher, 1995; Moor, 1995). Various types of food contain different proportions of these three series of fatty acids, and tissue fatty acid composition, particularly those of n-6 and n-3 fatty acids, varies depending on the choice of food. In the central nervous system, the major polyunsaturated fatty acids are ARA and DHA. In the desaturation, elongation, and chain-shortening steps, n-3 fatty acids are the preferred substrates leading to DHA. When the supply of n-3 fatty acids is limited, ARA is further metabolized to form docosapentaenoic acid (DPA, 22:5n-6). When the supply of both n-3 and n-6 fatty acids is limited, oleic acid (18:1n-9) is desaturated and elongated to form eicosatrienoic acid (20:3n-9, Mead acid), which is ineffective in suppressing the symptoms of LA deficiency (e.g., growth retardation and reproductive failure). ALA is essential for the maintenance of brain and retinal function.

Brain ARA and DHA levels are retained relatively constant in the central nervous system, but prolonged starvation of n-6 or n-3 fatty acids (e.g., through two generations) results in decreased ARA or DHA levels in brain phospholipids. LA deficiency is not common in humans because most food contains significant amounts of LA. Furthermore, the intake of LA has increased significantly in the past several decades in industrialized

From: *Fatty Acids: Physiological and Behavioral Functions*
Edited by: D. Mostofsky, S. Yehuda, and N. Salem Jr. © Humana Press Inc., Totowa, NJ

Saturated and monounsaturated fatty acid series:

$$16:0 \xrightarrow{\Delta 9} 16:1n\text{-}7 \rightarrow 18:1n\text{-}7$$
$$\downarrow$$
$$18:0 \xrightarrow{\Delta 9} 18:1n\text{-}9 \xrightarrow{\Delta 6} (18:2n\text{-}9) \rightarrow (20:2n\text{-}9) \xrightarrow{\Delta 5} 20:3n\text{-}9$$

Oleic acid   Mead acid
(animal fat)
      (olive oil & Canola oil)

Linolenic acid (n-6) series:

$$18:2n\text{-}6 \xrightarrow{\Delta 6} (18:3n\text{-}6) \rightarrow (20:3n\text{-}6) \xrightarrow{\Delta 5} 20:4n\text{-}6 \rightarrow 22:4n\text{-}6 \xrightarrow{\Delta 6} (24:4n\text{-}6) \rightarrow (24:5n\text{-}6) \rightarrow 22:5n\text{-}6$$

LA   ARA   DPA
(most cooking oils
and oil products)

α-Linolenic acid (n-3) series:

$$18:3n\text{-}3 \xrightarrow{\Delta 6} (18:4n\text{-}3) \rightarrow (20:4n\text{-}3) \xrightarrow{\Delta 5} 20:5n\text{-}3 \rightarrow 22:5n\text{-}3 \xrightarrow{\Delta 6} (24:5n\text{-}3) \rightarrow (24:6n\text{-}3) \rightarrow 22:6n\text{-}3$$

ALA   EPA   DHA
(vegetables,
perilla oil & flaxseed oil)   (seafood)   (seafood)
                                                                                                     ((brain & retina))

Fig. 1. Three biosynthetic pathways for major polyunsaturated fatty acids in mammals (desaturation, chain-elongation and chain-shortening steps). The site of desaturase action is shown by Δ9, Δ6, or Δ5. The major polyunsaturated fatty acids found in tissue lipids are linoleic (LA), arachidonic (ARA), docosapentaenoic (DPA), α-linolenic (ALA), eicosapentaenoic (EPA) and docosahexaenoic (DHA) acids. Fatty acids are designated by the carbon chains: the number of double bonds, and the position of the first double bond from the methyl terminus, as n-9, n-7, n-6, or n-3. Typical foods enriched with the indicated fatty acids are also shown.

countries as well as among urban dwellers in developing countries. Because of the competitive nature of the n-6 and n-3 fatty acid series (Fig. 1), increased amounts of LA intake competitively inhibit n-3 fatty acid metabolism, leading to relative n-3 fatty acid deficiency and vice versa (Okuyama, 1997), and the consequence of relative n-3 fatty acid deficiency in brain function is the major topic in this chapter. Because we noted different responses to n-3 fatty acid deficiency between rats and mice [e.g., water-maze learning ability was significantly different between mice fed an n-3 fatty acid-deficient diet and those fed an n-3 fatty acid-enriched diet (Nakashima, 1993) but such a difference was not reproduced in rats (Ikemoto, A. et al., unpublished observations) as reported by Wainwright (1999)], we focus here on the brightness-discrimination learning test performed in rat. Needless to say, similar consistent effects of n-3 fatty acid deficiency on

Table 1
Compositions of Major Fatty Acids in Fats and Oils Cited

Fatty acid	Lard	Olive oil	Corn oil	Soybean oil	Safflower oil	Perilla oil
16:0	26.5	9.9	11.2	10.3	7.3	5.4
18:0	12.1	3.2	2.1	3.8	2.6	1.4
18:1	42.5	75.0	34.7	24.3	13.4	15.1
18:2n-6	9.8	10.4	50.5	52.7	76.4	14.8
18:3n-3	0.7	0.8	1.5	7.9	0.2	63.0

learning behavior have been observed in other types of learning test (Bourre, 1989; Umezawa, 1995, 1999; Suzuki, 1998; Gamoh, 1999; and those described in other chapters).

The effect on central nervous system function of n-3 deficiency in the presence of LA was first examined using a combination of soybean oil and safflower oil, and it was revealed that the correct response ratio in a simple Y-maze test was greater in rats fed a soybean-oil diet through two generations than in those fed a safflower-oil diet (Lamptey, 1976). "Standing supported" behavior was also greater in the soybean-oil group (*see* Table 1 for the fatty acid composition of vegetable oils cited in this chapter). In the same year, Harman (1976) reported the results of similar experiments as follows. Comparing dietary lard, olive oil, corn oil, safflower oil, and fish oil, the number of errors in a maze test was reported to increase with the degree of unsaturation and the amount of unsaturated fatty acids in the diet; The fish-oil group had the highest number of errors. This formed one of the major bases for the so-called "free radical theory of aging." These two conclusions are apparently inconsistent with each other because the group fed soybean oil containing ALA is expected to have a higher brain DHA level than the group fed safflower oil that contains almost no ALA (Lamptey, 1976), whereas the fish-oil group is expected to have the highest brain DHA level among the groups examined (Harman, 1976). We believe that these maze tests are too simple to evaluate behavioral changes induced by n-3 fatty acid deficiency in the presence of n-6 fatty acids. In fact, a later work using a more complex X-maze test failed to reproduce the results with Y-maze test as reviewed by Bivins (1983), and we found no significant difference in the number of correct responses between rats fed a safflower-oil diet and those fed a perilla-oil diet over two generations when a radial (eight-arm) maze test was conducted, both groups of rats were clever enough to complete the radial maze test easily (unpublished observations). Instead, for more than 10 yr we have been using a computer-programmed brightness-discrimination learning apparatus (operant type), which has allowed us to evaluate ALA-deficiency-induced alterations in learning behavior. In this case, the responses of rats to diet pellets (response to reward) was not altered by n-3 fatty acid deficiency either (Yamamoto, 1987, 1988, 1991; Okaniwa, 1996).

2. ALA DEFICIENCY AFFECTS MAINLY NEGATIVE RESPONSES BUT NOT POSITIVE RESPONSES

Female rats were fed a semipurified diet supplemented with safflower oil or perilla oil (from seeds of the beefsteak plant, *Perilla frutescens*), mated at 11 wk of age, and the male progeny (F_1) were weaned at 3 wk of age to the diet of the dam. These diets brought about

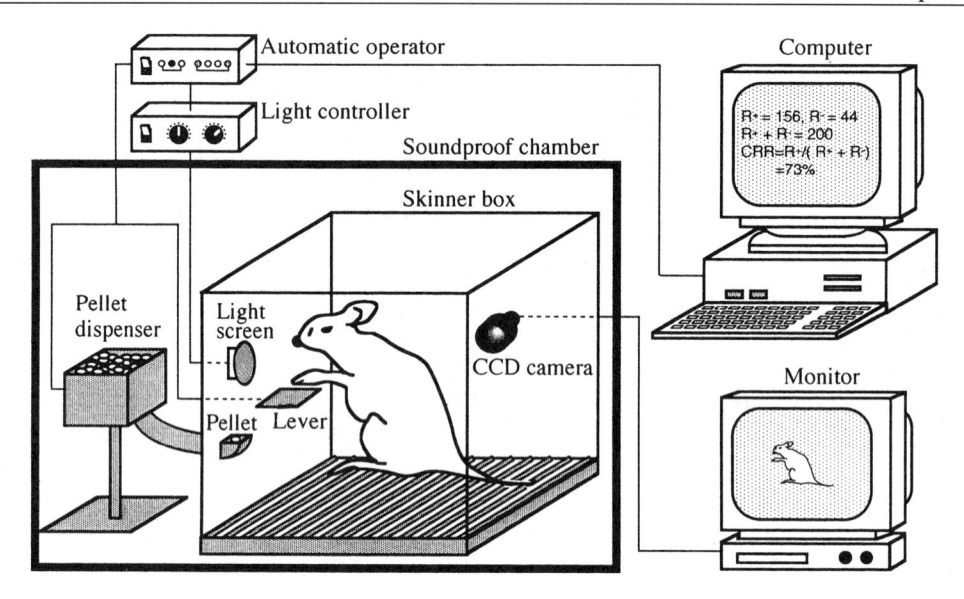

Fig. 2. Brightness-discrimination learning apparatus.

no differences in growth rate, litter size, growth of progeny, brain levels of phospholipid, glycolipid, cholesterol, and plasmalogen, and appearance. The only major difference was in the levels of n-6 and n-3 fatty acids in brain phospholipids. The DHA level in the safflower-oil group was reduced to about half the level in the perilla-oil group, which was compensated for by the increase in the levels of ARA, docosatetraenoic acid (22:4n-6) and DPA (22:5n-6).

At 11 wk of age, the body weights of F_1 rats were reduced to 85% within 1 wk and the shaping process was started while a bright light was kept on. A rat in a Skinner box was trained to press a lever to obtain diet pellets (Fig. 2), and the shaping process was judged to have reached completion when the rat obtained more than 40 pellets in 20 min.

In several sets of the test, the time required to complete the shaping process tended to be longer in the perilla-oil group than in the safflower-oil group, although the differences were not always statistically significant. The rat in the perilla-oil group tended to be more cautious and slower in starting to press the lever when it was put in the box.

After the shaping process, the brightness-discrimination learning test was started. Either a bright light or a dim light was presented on a screen randomly but at the same frequency during the 20-min test period per day. In the original schedule, the responses under the bright light (R^+) were reinforced with diet pellets, whereas no pellet was delivered on its responses under the dim light (R^-), and the responses were recorded automatically. After 25–30 d of the test, stimuli were suddenly reversed; lever-pressing responses under the dim light were reinforced with diet pellets and vice versa (reverse schedule). The results of one set of experiments are shown in Fig. 3.

Both R^+ and R^- responses increased in a similar manner during the initial several days. Then, R^+ responses kept increasing up to approx 25 d, whereas R^- responses reached plateau levels in about 10 d, and then began to decrease. The difference in the R^+ responses between the perilla-oil group and the safflower-oil group was usually not significant; in contrast, major significant differences were observed in the R^- response of the two

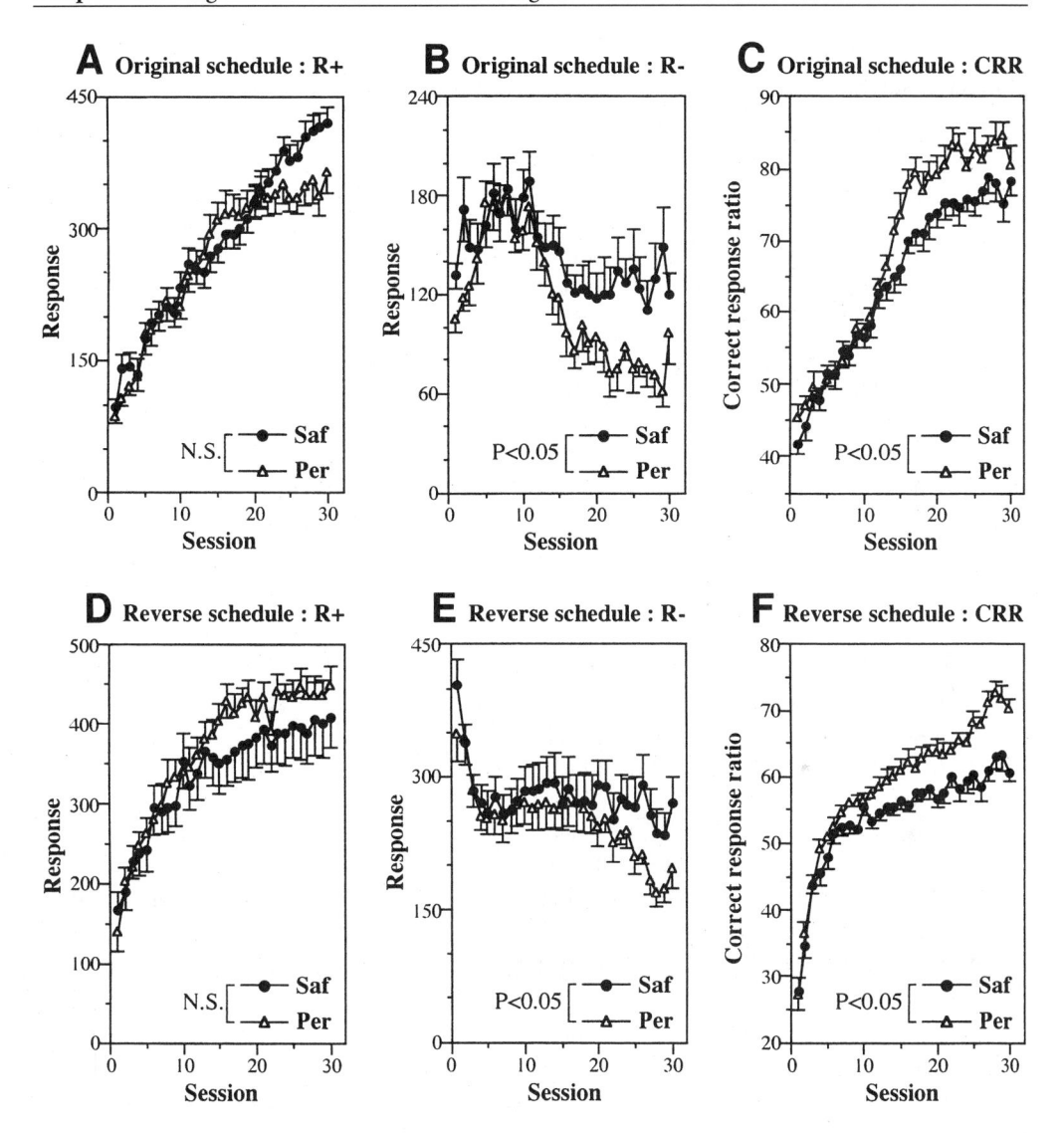

Fig. 3. Brightness-discrimination learning behavior of rats fed an ALA-restricted safflower-oil diet or ALA-sufficient perilla-oil diet over two generations. Lever-pressing responses under a bright light were reinforced with diet pellets (R^+), but no pellets were delivered on response under a dim light (R^-) in the original schedule. Similar results were obtained in the rat strains of Wistar/Kyoto, SHR, SD, and Donryu rats.

groups. The correct response ratio, calculated as $100 \times [R^+/(R^+ + R^-)]$ and taken as a measure of learning ability, was significantly higher in the perilla-oil group. In the reverse schedule, R^+ responses increased more rapidly and R^- responses decreased more rapidly in the perilla-oil group than in the safflower-oil group. Thus, the ability to discriminate set conditions was significantly higher in the perilla-oil group. It is known that brain function related to the lever-pressing response to obtain diet pellets (response to reward) is governed by relatively primitive brain regions, whereas feedback suppression of futile, vain actions (negative response) is regulated by different regions (Aosaki, 1994;

Schultz, 1993,1997). It should be noted that n-3 fatty acid deficiency in rat weakens this feedback suppression of futile (negative) responses.

3. DHA-RICH FISH OIL IS AS EFFECTIVE AS PERILLA OIL ENRICHED WITH ALA

Fish oil containing mainly DHA and a small amount of EPA was found to be as effective as perilla oil in maintaining brightness-discrimination learning ability (Okuyama, H. et al., Japan Patent Publication 1990–153629, 1990). Using a slightly different brightness-discrimination learning test, a significant difference in learning ability was also shown between the groups fed purified DHA ethyl ester and safflower oil (Fujimoto, 1993), the results of which were apparently inconsistent with those reported by Harman (1976). Thus, we interpreted that it is n-3 fatty acid, in general, that is required for the maintenance of learning ability (CRR in the brightness-discrimination learning test). As will be described later, the level of LA in the diet and, consequently, the level of n-6 fatty acids in the brain have been found to be critical factors affecting the reversibility of altered learning behavior.

4. REVERSIBILITY OF ALTERED LEARNING BEHAVIOR INDUCED BY n-3 FATTY ACID DEFICIENCY DURING GESTATION AND LACTATION: THE LEVEL OF n-6 FATTY ACIDS AS ANOTHER CRITICAL FACTOR

The effect of dual deficiency of n-6 and n-3 fatty acids on the learning behavior in the Y-maze test during gestation and lactation and after weaning was examined in rats using hydrogenated coconuts oil and corn oil (Lamptey, 1978). Gestational deprivation of n-6 and n-3 fatty acids irreversibly impaired the learning behavior of the progeny; lactational deprivation tended to impair learning; and postweaning deficiency had no effect on learning performance. Using a similar experimental design, we examined the effect of ALA (n-3) deficiency in the presence of LA (n-6) on the reversibility of altered learning behavior in rat (Okaniwa, 1996). Rats were fed a safflower-oil diet or a perilla-oil diet over the two generations (Saf-Saf and Per-Per), or the diets were shifted to different diets after weaning at 3 wk of age (Per-Saf and Saf-Per), and the learning performance was evaluated from 11 wk of age. Brain DHA level was significantly lower and learning ability was also lower only in the group fed the safflower-oil diet over two generations (Saf-Saf); supplementation of ALA after weaning was sufficient to restore learning ability and the brain DHA level. ALA deficiency after weaning decreased the brain DHA level significantly, but the level was sufficient to maintain learning ability. Thus, we concluded that n-3 fatty acid deficiency in the presence of n-6 fatty acid during gestation and lactation does not bring about irreversible change in brain function, as estimated by the brightness-discrimination learning test.

Then, we tried to estimate the efficacy of DHA (ethyl ester) on the reversibility of learning behavior. Unexpectedly, a mixture of DHA and safflower oil (1.6 en% DHA and 4.7 en% LA, Saf-DHA) failed to restore learning ability when the safflower-oil diet was shifted to the Saf-DHA diet after weaning (Table 2) (Ikemoto, A., Ph.D. Thesis, Nagoya City University, 2000). Brain DHA level of the Saf-DHA group was restored to the level of the group fed the perilla-oil diet (4.9 en% ALA and 1.2 en% LA). However, the brain

Table 2
Reversibility of ALA-Deficiency-Induced Alterations of Brightness-Discrimination Learning Behavior in the Rat

Diet after Weaning[a]	Fatty acid in the diet (energy %)			Learning ability (CRR)[b]
	LA	ALA	DHA	
Saf	5.8	<0.1	nd[c]	Low
Saf-DHA	4.7	<0.1	1.6	Low
Saf-OA-DHA	1.2	<0.1	3.1	High
Per	1.2	4.9	nd	High

[a]The safflower oil (Saf) diet was shifted to the diet containing Saf and DHA, Saf and oleic acid (OA) and DHA, or to the perilla oil (Per) diet after weaning at 3 wk of age, and the learning test was started at 11 wk of age.
[b]Correct response ratio (CRR) was taken as a measure of learning ability.
[c]Not detectable.

levels of n-6 fatty acids (ARA, 22:4n-6 and DPA) of the Saf-DHA group were between those of the safflower-oil group and the perilla-oil group. When the safflower-oil diet was shifted to a diet supplemented with a mixture of safflower oil, oleic acid (OA), and DHA (3.1 en% DHA, 1.2 en% LA) in which the LA level was adjusted to equal that of the Per diet, both learning ability (correct response ratio) and brain fatty acid profiles were restored completely to the levels of the perilla-oil group, although both R+ and R− responses of the Saf-OA-DHA group tended to be less in the original schedule compared to other groups. Thus, the level of brain n-6 fatty acids was found to be another critical factor affecting learning behavior (Table 2). Here, the importance of taking into consideration the competitive nature of n-6 and n-3 fatty acid metabolism (Fig. 1) is emphasized, the intake of n-6 should be decreased in order to improve the efficacy of supplemented n-3 fatty acids.

5. DECREASED RETINAL RESPONSE AND TURNOVER OF ROS CAUSED BY ALA DEFICIENCY

As reported by Benolken (1973), the amplitudes of both the a and b-waves of an electroretinogram decreased significantly as a result of ALA deficiency in rat (Watanabe, 1987). A diet supplemented with soybean oil (1.5 en% as ALA) did not give maximum amplitudes similar to those observed in the perilla-oil group (Fig. 4).

A part of the retinal rod outer segment (ROS) is known to undergo shedding soon after light exposure, followed by its phagosomal digestion in pigment epithelial cells and resynthesis in photoreceptor cells (LaVail, 1976; Reich-D'Almerdic, 1975). Along with these observations, the number of large phagosomes was found by image analysis to increase 1.5 h after light exposure and then decrease toward the evening (Watanabe, 1993). However, the number of large phagosomes 1.5 h after light exposure was found to be significantly less in the safflower-oil group compared with the perilla-oil group (Fig. 5). Lysosomal enzymes in retina also exhibit diurnal rhythms (Vaughan, 1990). We found that lysosomal enzyme activities were also lower in the safflower-oil group while diurnal rhythms were retained (Ikemoto, 2000c).

The contents and compositions of retinal phospholipids were essentially the same between the two dietary groups. However, the rate of phospholipid synthesis was lower

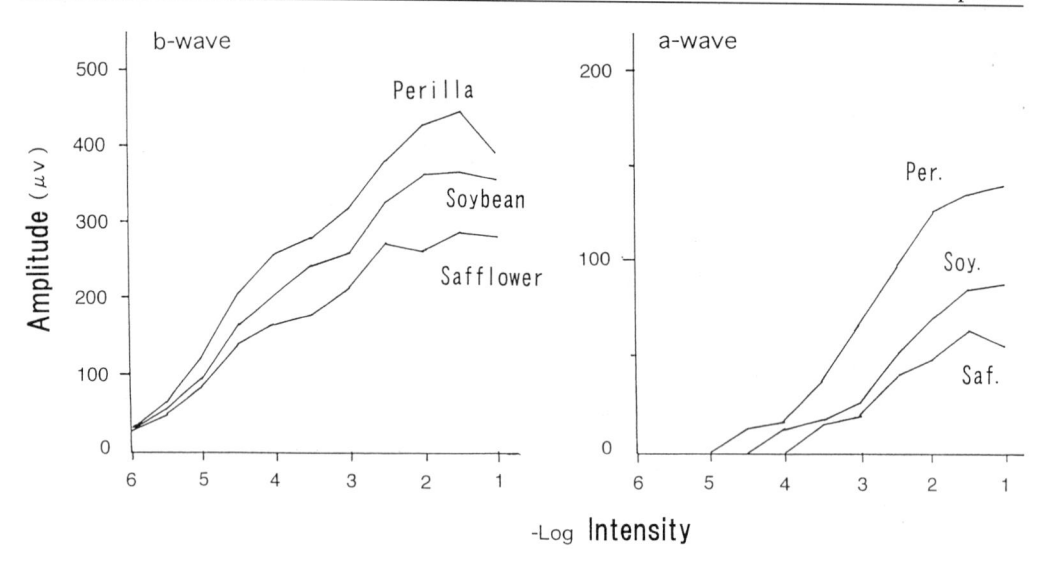

Fig. 4. Amplitudes of a- and b-waves in the electroretinogram in rats fed a semipurified diet supplemented with safflower oil or perilla oil, or a conventional diet supplemented with soybean oil. The intensity of the dim light corresponded to 3 in the abscissa, and that of the bright light was 1000 times higher.

in the safflower-oil group (Ikemoto, 2000c). These results suggest that the rate of turnover of ROS is lower under n-3 fatty acid deficiency, possibly as a result of the reduced availability of DHA for its resynthesis in photoreceptor cells (Scott,1989; Anderson,1994).

n-3 Fatty acid-deficiency-induced alterations of retinal functions have been well characterized in the monkey (Neuringer, 1986).

6. BLINDNESS TO DIM LIGHT IS NOT THE CAUSE OF INFERIOR LEARNING PERFORMANCE IN ALA-DEFICIENT RATS

As retinal response was also impaired in ALA deficient rats (Fig. 4), one can argue that the decrease in learning ability observed in the brightness-discrimination learning test is the result of retinal dysfunction but not of cognitive dysfunction. However, we interpreted that rats that were n-3 fatty acid deficient could recognize even dim light because its intensity was sufficient to evoke electroretinographic responses (Yamamoto, 1988). Moreover, the R^- responses were significantly more than R^+ responses on the first day of the test, possibly reflecting the nocturnal nature of the rat. From these observations, we interpreted that the rats under n-3 deficiency can discriminate the brightness used in the test (Yamamoto, 1988).

Recently, we found that both the perilla-oil group and the safflower-oil group exhibited a similarly high learning ability (CRR) when light intensities were set at 100 lux for bright light and 0.1 lux for dim light at the center of the Skinner box. A significantly lower learning ability in n-3 fatty acid-deficient rats was observed at the 100-lux and 3-lux setting. Rats that were n-3 fatty acid deficient appeared to be weaker in discriminating a smaller difference in brightness (100 and 3 lux vs 100 and 0.1 lux). Because n-3 fatty acid deficiency affects a variety of nervous system functions (olfactory function, pain,

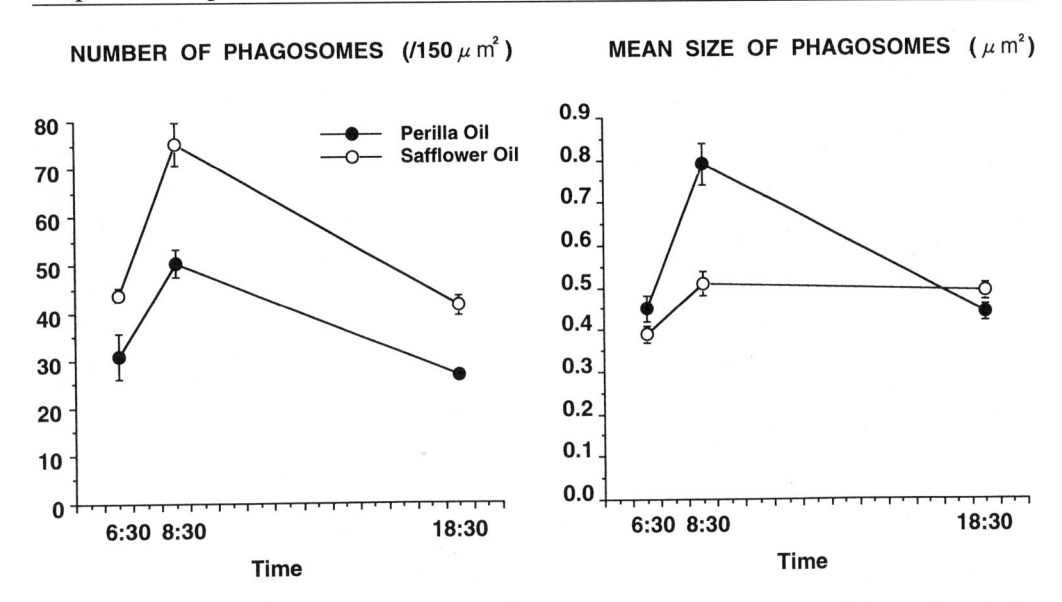

Fig. 5. Diurnal changes in the number and the mean size of phagosomes in rats fed a perilla-oil diet or a safflower-oil diet.

retinal function), it is not possible to discriminate which of the sensory and cognitive functions are affected by ALA deficiency. The brain is an organ that integrates information from the peripheral nervous system and makes a decision, and we understand that this brain function is impaired by n-3 fatty acid deficiency.

7. BIOCHEMICAL BASES FOR ALA-DEFICIENCY-INDUCED ALTERATIONS OF BEHAVIOR AND RETINAL FUNCTION

Earlier, some membrane enzymes in the brain (e.g., 5'-nucleotidase, cyclic nucleotide phosphodiesterase, acetylcholine esterase and Ca^{2+}-ATPase) were reported to be specifically associated with dietary n-3 fatty acid levels. However, we have been unable to find significant differences in the activities of these enzymes of the safflower-oil group and the perilla-oil group of rats. A possible difference in the activities of Na^+,K^+-ATPase isozymes has not been excluded (Gerbi, 1993; Tutsumi, 1995), although no significant difference in the ouavain sensitivity of this enzyme was observed between the two dietary groups. In the retina, lysosomal enzyme activities (β-glucosidase, β-glucuronidase, hexosaminidase, and acid phosphatase) as well as the levels of phospholipid *de novo* synthetic enzymes (CTP: phosphocholine cytidyltransferase; CTP: phosphoethanolamine cytidyltransferase) in the safflower-oil group were significantly lower than those in the perilla-oil group (Ikemoto, 2000c), which was suggested to be associated with the decreased turnover of ROS, as described earlier. However, the difference in lysosomal enzyme activities between the two dietary groups was not significant in the pineal glands.

The rate of phospholipid synthesis in the brain in vivo was reported to decrease in n-3 fatty acid deficient free-moving rats (Gazzah, 1995). When isolated synaptosomal fractions were examined, the rate of phospholipid synthesis from labeled ethanolamine or serine was found not to differ between safflower-oil and perilla-oil groups. However, the rate of formation of the lyso form of ethanolamine glycerophospholipids from labeled

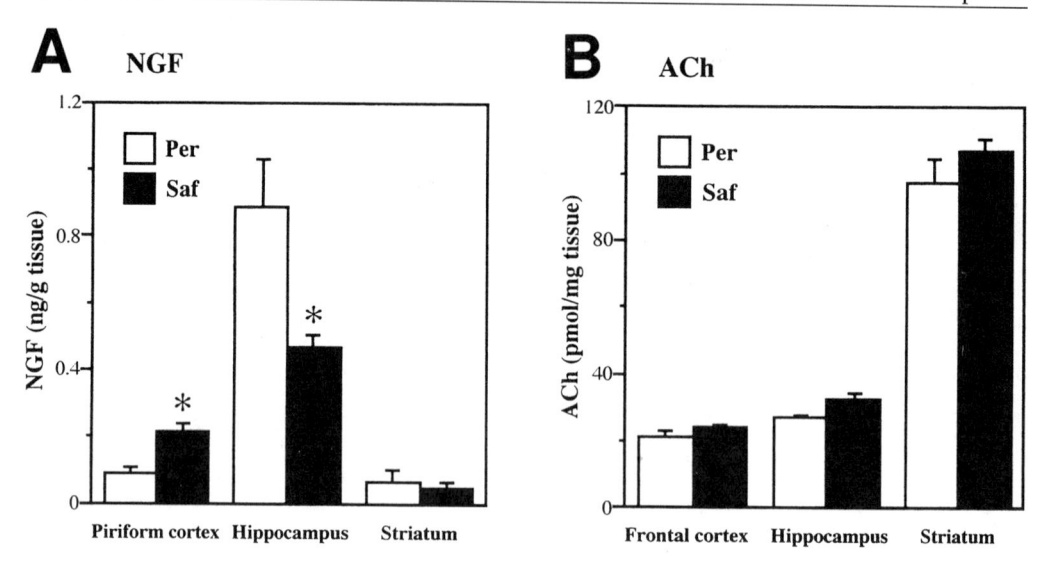

Fig. 6. Levels of nerve growth factor (NGF) and acethylcholine (Ach) in brain regions of rats fed a perilla-oil diet or a safflower-oil diet over two generations.

ethanolamine was significantly lower in the safflower-oil group (Ikemoto, 2000b). This difference in phospholipid metabolism in synaptosomes may provide clues to evaluate possible differences in the phospholipid turnover rate in the brain of rats fed n-3 fatty acid-deficient and n-3 fatty acid-sufficient diets. In contrast to phospholipid synthesis, n-3 fatty acid deficiency was reported to increase the rate of brain protein synthesis from labeled precursors in free-moving rats (Giaume, 1994). However, we found no significant difference in the rate of protein synthesis between the n-3 fatty acid-deficient and n-3 fatty acid-sufficient groups of rats (Sato, 1999). It should be noted that our data do not exclude the possibility that the rate of turnover of a limited number of proteins in brain is affected by n-3 fatty acid deficiency.

With regard to neurotransmitters, we found no significant difference in the levels of norepinephrine, 3,4-dihydrophenylacetic acid, dopamine, 5-hydroxyindoleacetic acid, homovanilinic acid, and serotonin in the frontal cortex, the hippocampus, and the striatum between conventional rats fed the safflower-oil diet and those fed the perilla-oil diet. The acetylcholine (Ach) level was very high in the striatum, but these diets had no effect on the Ach levels in brain regions (Fig. 6). In stroke-prone spontaneously hypertensive rats, DHA supplementation increased Ach levels in the frontal cortex and the hippocampus compared with those fed the safflower-oil diet (Minami, 1997a,b).

Somewhat different results were reported from other laboratories: n-3 fatty acid deficiency in rats caused a decrease in dopamine level and D2-receptor density in the frontal cortex as well as a decrease in serotonin receptor density (Delion, 1994, 1996), whereas fish oil increased the dopamine level (Chalon, 1998). A decrease in the dopamine release from the frontal cortex was also detected by the microdialysis method after thyramine stimulation but not after KCl stimulation (Zimmer, 1998, 2000). In aged rats, however, the monoamine level was not affected, whereas monoamine oxidase activity was decreased during n-3 fatty acid deficiency (Delion, 1997).

In dual deficiency of LA (n-6) and ALA (n-3) in rat, the dopamine level in the frontal cortex was reduced, which was reversed by supplementation with both ARA (n-6) and

DHA (n-3) (Owens, 1999). Thus, some inconsistent results have been obtained so far regarding the effect of n-3 fatty acid deficiency on neurotransmitter and receptor levels in different brain regions. Different experimental conditions, particularly the possible effects of minor components in some vegetable oils used (Miyazaki, 1998; Kameyama, 1996), may have obscured the results.

The neurotrophin family is involved in differentiation and survival of various neurons through p76 and trk receptors. Among them, nerve growth factor (NGF) is known to regulate certain sensory neurons in the peripheral nervous system and cholinergic neurons in the central nervous system (Thoenen, 1987). Continuous administration of NGF in vivo improved water-maze learning performance that was associated with hyperplasia of cholinergic neurons (Fisher, 1987). We measured NGF levels in brain regions and found that the NGF level in the hippocampus was roughly half, but it was doubled in the piriform cortex of rats that were n-3 fatty acid deficient (Fig. 6) (Ikemoto, 2000a).

Morphologically, Yoshida's group (Yoshida, 1997a) found decreased synaptic vesicle density in the release site in the n-3 fatty acid-deficient rat hippocampus CA1 region. Changes in hydration of the membrane surface and modification of the oligosaccharide environment of microsomes were also suggested from Fourier transform infrared spectroscopic analysis (Yoshida, 1997b). In microsomes, an increase in calcium-induced aggregation rate and an increase in reactivity to phospholipase A2 and sialidase treatment following learning task were also noted (Yoshida, 1997c). The difference between physicochemical properties of DHA (n-3) and DPA (n-6) appears to be small, but brain membranes may have strict requirements for the phospholipid molecular species containing DHA (Salem, 1995).

Thus, dietary n-3 fatty acid deficiency was shown to induce alterations in various biochemical and physicochemical parameters in rat. However, the causal relationship between these parameters and the altered learning behavior during n-3 fatty acid deficiency remains to be elucidated.

8. POSSIBLE IMPLICATIONS OF OBSERVATIONS IN RAT IN HUMAN BEHAVIOR

Rudin (1981) described a causal relationship between ALA deficiency and neurotic disorders in humans. Psychoses and neuroses of some pellagra cases unresponsive to multivitamin therapy have been treated effectively with linseed oil enriched with ALA. In 1982, Holman described numbness, paraesthesia, weakness, inability to walk, pain in the legs, and blurring of vision in a 6-yr-old girl under total parenteral nutrition taking safflower oil as the source of essential fatty acid. When the regimen was changed to an emulsion containing ALA, the neurological symptoms disappeared, and the amount of ALA required was estimated to be 0.54 en%.

In schizophrenia, the severity was reported to be less in patients taking more n-3 fatty acids, and supplementation of fish oil was effective in improving the symptoms (Peet, 1995, 1996). It was reported that the levels of polyunsaturated fatty acids such as LA (n-6), ARA (n-6) and DHA (n-3) are lower in plasma and red cell lipids in schizophrenic patients (Yao, 1994; Kaiya, 1991; Mahadik, 1996), suggesting enhanced oxidative injury, although their causal relationship has yet to be elucidated. In depression, the ARA/EPA ratio in blood showed positive correlation with clinical symptoms of depression (Adams, 1996; Maes, 1996). Hibbeln (1995a, 1995b) pointed out that DHA plays a role in mental function and depression, revealing a highly negative correlation between average fish intake and incidence of depression among countries. Based on these observations, some clinical trials have been started, aiming at suppressing the symptoms of depression. Recently, Hamazaki (1996) reported the results of a double-blind test in medical students that "aggressiveness against others" was suppressed by supplementing DHA ethyl ester. This study also revealed that the medical students used to take a much lower amount of DHA (0.2 g/d) than the average Japanese.

In classical lipid nutrition for the prevention of atherosclerosis and related diseases, hypercholesterolemia and animal fat were considered to be the major risk factors and high-LA vegetable oils were recommended. Although this recommendation was found to be ineffective (Multiple Risk Factor Intervention Trial Research Group, 1982) and even to be risky for atherosclerosis-related diseases, an increase in the incidence of violent death was observed to be associated with it (Strandberg, 1991; Muldoon, 1990). The plasma cholesterol level does not decrease significantly after prolonged dietary recommendations to raise the vegetable oil/animal fat ratio (or P/S ratio) of food; the hypocholesterolemic effect of dietary LA was found to be only transient (Okuyama, 1997, 2000). Therefore, it is not cholesterol that is associated with the increase in the incidence of violent death; rather, the increased intake of LA and the elevated n-6/n-3 ratio are likely to be the cause, in view of the results of animal experiments described earlier. Similarly, those who have experienced a heart attack have been noted to have a characteristic behavioral pattern called the Type A behavior pattern (TABP), and the causal relationship between heart attack and TABP has been studied by a group of scientists. Again, "increased intake of LA and elevated n-6/n-3 ratio," but not hypercholesterolemia, are probably the major risk factors for both heart attack and TABP (Okuyama, 1997, 2000).

Attention-deficit hyperactive disorder (ADHD) is known to be typical among atopic patients (Stevens, 1995, 1996). Atopic dermatitis is treated effectively with steroidal anti-inflammatory drugs and other antiallergic drugs that exert their effects mainly by inhibiting the cascade of LA → ARA → lipid mediators of allergic, inflammatory reactions → receptors (LA cascade). Clinically, decreasing the intake of LA and increasing the intake of n-3 fatty acids that are competitive effectors of the LA cascade and partial agonists for the lipid mediator receptors were shown to be effective for the prevention of atopic dermatitis (Kato, 2000) and other allergic hyperreactivities (Ashida, 1997). In rodents suffering from n-3 fatty acid deficiency, the observed decrease in the feedback suppression of negative responses in the brightness-discrimination learning test (Fig. 3) and the increase in anxiety in the elevated plus-maze test (Nakashima, 1993) appear to have characteristics common to ADHD in atopic children. Here, again, we would like to emphasize that both allergic hyperreactivities and behavioral anomalies such as ADHD as well as some pathological states of the brain (Yoshida, 1998) could be the result of an enhancement of the LA cascade (ie., the result of the increase in the intake of LA and the elevated n-6/n-3 ratio of dietary fat and oils.

Along with the Westernization of the dietary habits in Japan and possibly among urban dwellers in developing countries, the disease pattern has also been Westernized: an increase in the incidence of cancers of Western type, thrombotic diseases, and other inflammatory diseases. It is also a serious concern for the Japanese that changes in behavioral patterns may be occurring among young Japanese, following those of younger populations in Western industrialized countries. These changes in disease pattern and behavioral pattern could well be the result of the enhancement of the LA cascade resulting from the increase in the intake of LA and the elevated n-6/n-3 ratio of tissue phospholipids.

Simply applying the results shown in Table 2 to humans, 1.6 en% DHA in the Saf-DHA diet is not the amount easily ingested by people in industrialized countries; even the average Japanese takes less than 1 en% as EPA+DHA. Thus, it is important to reduce the intake of competing n-6 fatty acids, essentially LA that is present in most vegetable oil and oil products, and increase the intake of n-3 fatty acids such as ALA, EPA, and DHA present in seafood and vegetables. The essential amount of LA in humans is roughly 1 en%, but average people in industrialized countries are ingesting more than 6 en% LA.

9. SUMMARY

Behavioral changes induced in rat by feeding ALA (n-3)-restricted, LA-enriched diet were demonstrated reproducibly using a brightness-discrimination learning apparatus. Decreased feedback suppression of negative, futile responses was found to be characteristic to n-3-deficient rats. n-3 Deficiency in the presence of n-6 fatty acids during gestation and lactation did not bring about irreversible changes in learning behavior; feeding diets with relatively low n-6/n-3 ratios after weaning restored learning ability. Some biochemical and physicochemical parameters in the brain were found to be altered by n-3 fatty acid deficiency.

Decreased retinal responses were also noted during n-3 fatty acid deficiency, and a reduced rate of turnover of ROS (shedding, phagocytosis, and resynthesis) was suggested as one basis for the altered retinal function. Behavioral changes induced by diets with high n-6/n-3 ratios were implicated with Western-type diseases (cancers of Western type, atherosclerotic–thrombotic diseases, allergic hyperreactivity, and other inflammatory diseases) caused by an enhancement of the LA cascade (LA →ARA →lipid mediators). The importance of decreasing the n-6/n-3 ratio of fats and oils by choosing the right kind of food was emphasized, particularly for those living in industrialized countries and urban dwellers in developing countries.

ACKNOWLEDGMENTS

This work was supported in part by a Special Coordination Fund for the Promotion of Science and Technology from the Science and Technology Agency of Japan.

REFERENCES

Adams PB, Lawson S, Sanigorski A, Sinclair AJ. Arachidonic acid to eicosapentaenoic acid ratio in blood correlates positively with clinical symptoms of depression. Lipids 1996; 31:S157–S161.
Anderson RE, Chen, H, Wang N, Stinson A. The accretion of docosahexaenoic acid in the retina. World Rev Nutr Diet 1994; 75:124–127.
Aosaki T, Graybiel A, Kimura M. Effect of the nigrostriatal dopamine system on acquired neuronal responses in the striatum of behaving monkeys. Science 1994; 265:412–415.
Ashida K, Mitsunobu F, Mifune T, Hosaki Y, Yokota S, Tsugeno H, et al. A pilot study: effects of dietary supplementation with α-linolenic acid enriched perilla seed oil on bronchial asthma. Allergol Intl 1997; 46:181–185.
Benolken RM, Anderson RE, Wheeler TG. Membrane fatty acids associated with the electrical response in visual excitation. Science 1973; 182:1253–1254.
Bivins BA, Bell RM, Rapp RP, Griffen WO Jr. Linoleic acid versus linolenic acid in the development of young rat: what is essential? J Parenter Enteral Nutr 1983; 7:473–478.

Bourre JM, Francois M, Youyou A, Dumont O, Piciotti M, Pascal G, et al. The effects of dietary α-linolenic acid on the composition of nerve membranes, enzymatic activity, amplitude of electrophysiological parameters, resistance to poisons and performance of learning tasks in rats. J Nutr 1989; 119:1880–1892.

Chalon S, Delion S, Vancassel S, Belzung C, Guilloteau D, Leguiesquet AM, et al. Dietary fish oil affects monoaminergic neurotransmission and behavior in rats. J Nutr 1998; 128:2512–2519.

Delion S, Chalon S, Herault J, Guilloteau D, Besnard JC, Durand G. Chronic dietary α-linolenic acid deficiency alters dopaminergic and serotoninergic neurotransmission in rats. J Nutr 1994; 124:2466–2476.

Delion S, Chalon S, Guilloteau D, Besnard JC, Durand G. α-Linolenic acid dietary deficiency alters age-related changes of dopaminergic and serotoninergic neurotransmission in the rat frontal cortex. J Neurochem 1996; 66:1582–1591.

Delion S, Chalon S, Guilloteau D, Lejeune B, Besnard JC, Durand G. Age-related changes in phospholipid fatty acid composition and monoaminergic neurotransmission in the hippocampus of rats fed a balanced or an n-3 polyunsaturated fatty acid deficient diet. J Lipid Res 1997; 38:680–689.

Fisher W, Wictorin K, Bjorklund A, Williams LR, Varon S, Gage FH. Amelioration of cholinergic neuron atrophy and spatial memory impairment in aged rats by nerve growth factor. Nature 1987; 329:65–68.

Fujimoto K, Kanno T, Koga H, Onoderqa K, Hirano H, Nishikawa M, et al. Effects of n-3 fatty acid deficiency during pregnancy and lactation on learning ability of rats. In: Yasugi Y, Nakamura H, Soma M, eds. Advances in Polyunsaturated Fatty Acid Research Elsevier Scientific, Amsterdam, 1993, pp. 257–260.

Gamoh S, Hashimoto M, Sugioka K, Hossain MS, Hata N, Misawa Y, et al. Chronic administration of docosahexaenoic acid improves reference memory-related learning ability in young rats. Neuroscience 1999; 93:237–241.

Gazzah N, Gharib A, Croset M, Bobillier P, Lagarde M, Sarda N. Decrease of brain phospholipid synthesis in free-moving n-3 fatty acid deficient rats. J Neurochem 1995; 64:908–991.

Gerbi A, Zerouga M, Debray M, Durand G, Chanez C, Bourre JM. Effect of dietary α-linolenic acid on functional characteristic of Na$^+$/K$^+$-ATPase isozymes in whole brain membranes of weaned rats. Biochim Biophys Acta 1993; 1165:291–298.

Giaume M, Gay N, Baubet V, Gharid A, Durand G, Bobillier P, et al. N-3 fatty acid deficiency increases brain protein synthesis in the free-moving adult rat. J Neurochem 1994; 63:1995–1998.

Hamazaki T, Sawazaki, S, Itomura M, Asaoka, E, Nagao Y, Nishimura N, et al. The effect of docosahexaenoic acid on aggression in young adults. A placebo-controlled double-blind study. J Clin Invest 1996; 97:1129–1133.

Harman D, Hendricks, S, Eddy DE, Seibold J. Free radical theory of aging: effect of dietary fat on central nervous system function. J Am Geriatr Soc 1976; 24:301–307.

Hibbeln JR, Salem JN. Dietary polyunsaturated fatty acids and depression when cholesterol does not satisfy. Am J Clin Nutr 1995; 62:1–9.

Hibbeln JR, Umhau JC, Linnoila M, George DT, Ragan PW, Schoaf SE, et al. A replication study of violent and nonviolent subjects: cerebrospinal fluid metabolites of serotonin and dopamine are predicted by plasma essential fatty acids. Biol Psychiatry 1995; 44:243–249.

Holman RT, Johnson SB, Hatch TF. A case of human linolenic acid deficiency involving neurological abnormalities. Am J Clin Nutr 1982; 35:617–623.

J.Kemoto A. PhD thesis, Biochemical and behavioral evaluation of theroles of polyunsaturated fatty acids and phosphollipid metabolism in the nervous system. Nagoya City University, 2000.

Ikemoto A, Nitta A, Furukawa S, Ohishi M, Nakamura A, Fujii Y. et al. Dietary n-3 fatty acid deficiency decreases nerve growth factor content in rat hippocampus. Neurosci Lett 2000; 285:99–102.

Ikemoto A, Ohishi M, Hata N, Misawa Y, Fujii Y, Okuyama H. Effect of n-3 fatty acid deficiency on fatty acid composition and metabolism of aminophospholipids in rat brain synaptosomes. Lipids 2000; 35:1107–1115.

Ikemoto A, Fukuma A, Fujii Y, Okuyama H. Lysosomal enzyme activities are decreased in the retina but their circadian rhythms are maintained differently from those in the pineal gland in rats fed an α-linolenate-restricted diet. J Nutr 2000; 130:3059–3062.

Kaiya H, Horrobin DF, Manku MS, Fisher, N.M. Essential and other fatty acids in plasma in schizophrenics and normal individuals from Japan. Biol Psychiatry 1991; 30:357–362.

Kameyama T, Ohhara T, Nakashima Y, Naito Y, Huang M-Z, Watanabe S, et al. Effects of dietary vegetable oils on behavior and drug responses in mice. Biol Pharm Bull 1996; 19:400–404.

Kato M, Nagata Y, Tanabe A, Ikemoto A, Watanabe S, Kobayashi T, et al. Supplementary treatment of atopic dermatitis patients by choosing foods to lower the n-6/n-3 ratio of fatty acids. J Health Sci 2000; 46:1–10.

Lamptey MS, Walker BL. A possible essential role for dietary linolenic acid in the development of the young rat. J Nutr 1976; 106:86–93.

Lamptey MS, Walker BL. Learning behavior and brain lipid composition in rats subjected to essential fatty acid deficiency during gestation, lactation and growth. J Nutr 1978; 108:358–367.

LaVail MM. Rod outer segment disk shedding in rat retina: relationship to cyclic lighting. Science 1976; 194:1071–1074.

Maes M, Smith R, Christopher A, Cosyns P, Desnyder R, Meltzer H. Fatty acid composition in major depression-decreased omega 3 fractions in cholesteryl esters and increased C20:4 omega6/C20:5 omega3 ratio in cholesteryl esters and phospholipids. J Affect Disord 1996; 38:35–46.

Mahadik SP, Mukherjee S, Horrobin DF, Jenkins K, Correnti EE, Scheffer RE. Plasma membrane phospholipid fatty acid composition of cultured skin fibroblasts from schizophrenic patients—comparison with bipolar patients and normal subjects. Psychiatry Res 1996; 63:133–142.

Minami M, Kimura S, Endo T, Hamaue N, Hirafuji M, Monma Y, et al. Effects of dietary docosahexaenoic acid on survival time and stroke-related behavior in stroke-prone spontaneously hypertensive rats. Gen Pharmacol 1997; 29:401–407.

Minami M, Kimura S, Endo T, Hamaue N, Hirafuji M, Togashi H, et al. Dietary docosahexaenoic acid increases cerebral acetylcholine levels and improves passive avoidance performance in stroke-prone spontaneously hypertensive rats. Pharmacol Biochem Behav 1997; 58:1–9.

Miyazaki M, Huang M-Z, Takemura N, Watanabe S, Okuyama H. Free fatty acid fractions from some vegetable oils exhibit reduced survival time-shortening activity in stroke-prone spontaneously hypertensive rats. Lipids 1998; 33:655–661.

Moore SA, Hurt E, Yoder E, Sprecher H, Spector AA. Docosahexaenoic acid synthesis in human skin fibroblasts involves peroxisomal retroconversion of tetracosahexaenoic acid. J Lipid Res 1995; 36:2433–2443.

Muldoon MF, Manuck SB, Matthews KA. Lowering cholesterol concentrations and mortality: a quantitative review of primary prevention trials. Br Med J 1990; 301:309–314.

Multiple Risk Factor Intervention Trial Research Group Multiple risk factor intervention trial. Risk factor changes and mortality results. J Am Med Assoc 1982; 248:1465–1477.

Nakashima Y, Yuasa S, Hukamizu Y, Okuyama H, Ohhara T, Kameyama T, et al. Effect of a high linoleate and a high α-linolenate diet on general behavior and drug sensitivity in mice. J Lipid Res 1993; 34:239–247.

Neuringer M, Connor WE, Lin DS, Barstad L, Luck S. Biochemical and functional effects of prenatal and postnatal omega-3 fatty acid deficiency on retina and brain in rhesus monkeys. Proc Natl Acad Sci USA 1986; 83:4021–4025.

Okaniwa Y, Yuasa S, Yamamoto N, Watanabe S, Kobayashi T, Okuyama H, et al. A high linoleate and a high α-linolenate diet induced changes in learning behavior of rats. Effects of a shift in diets and reversal of training stimuli. Biol Pharm Bull 1996; 19:536–540.

Okuyama H, Kobayashi T, Watanabe S. Dietary fatty acids—the n-6/n-3 balance and chronic elderly diseases. Prog Lipid Res 1997; 35:409–457.

Okuyama H, Fujii Y, Ikemoto A. N-6/n-3 ratio of dietary fatty acids rather than hypercholesterolemia as the major risk factor for atherosclerosis and coronary heart disease. J Health Sci 2000; 46:157–177.

Okuyama, H. et al. Japan Patent Publication 1990–153629, 1990.

Owens SP, Innis SM. Docosahexaenoic acid and arachidonic acid prevent a decrease in dopaminergic and serotoninergic neurotransmitters in frontal cortex caused by a linoleic and α-linolenic acid deficient diet in formula-fed piglets. J Nutr 1999; 129:2088–2093.

Peet M, Laugharne J, Rangarajan N, Horrobin D, Reynolds G. Depleted red cell membrane essential fatty acids in drug-treated schizophrenic patients. J Psychiatr Res 1995; 29:227–232.

Peet M, Laugharne JD, Mellor J, Ramchand, CN. Essential fatty acid deficiency in erythrocyte membranes from chronic schizophrenic patients, and the clinical effects of dietary supplementation. Prostaglandins Leukotrienes Essential Fatty Acids 1996; 55:71–75.

Reich-D'Almeida FB, Hockley JB. In situ reactivity of the retinal pigment epithelium. I. Phagocytosis in the normal rat. Exp Eye Res 1975; 21:333–345.

Rudin DO. The major psychoses and neuroses as omega-3 essential fatty acid deficiency syndrome: substrate pellagra. Biol Psychiatry 1981; 16:837–850.

Salem N Jr, Niebyiski CD. The nervous system has absolute molecular species requirement for proper function. Mol Membr Biol 1995; 12:131–134.

Sato A, Osakabe T, Ikemoto A, Watanabe S, Kobayashi T, Okuyama H. Long-term n-3 deficiency induces no substantial change in the rate of protein synthesis in rat brain and liver. Biol Pharm Bull 1999; 22:775–779.

Schultz W, Apicella P, Ljungberg T. Responses of monkey dopamine neurons to reward and conditioned stimuli during successive steps of learning in a delayed response task. J Neurosci 1993; 13:900–913.

Schultz W. Dopamine neurons and their role in reward mechanisms. Curr Opin Neurobiol 1997; 7:191–197.

Scott BL, Bazan NG. Membrane docosahexaenoate is supplied to the developing brain and retina by the liver. Proc Natl Acad Sci USA 1989; 86:2903–2907.

Sprecher H, Luthria DL, Mohammed BS, Baykousherva SP. Reevaluation of the pathways for the biosynthesis of polyunsaturated fatty acids. J Lipid Res 1995; 36:2471–2477.

Stevens LJ, Zentall SS, Deck JL, Abate ML, Watkins BA, Lipp SR, Burgess JR. Essential fatty acid metabolism in boys with attention deficit hyperactivity disorder. Am J Clin Nutr 1995; 62:761–768.

Stevens LJ, Zentall SS, Abate ML, Kuczek T, et al. Omega-3 fatty acids in boys with behavior, learning and health problems. Physiol Behav 1996; 59:915–920.

Strandberg TE, Salomaa VV, Naukkarinen VA, Vanhanen HT, Sarna SJ, Miettinen TA. Long-term mortality after 5-year multifactorial primary prevention of cardiovascular diseases in middle-aged men. J Am Med Assoc 1991; 266:1225–1229.

Suzuki H, Park SJ, Tamura M, Ando S. Effect of the long-term feeding of dietary lipids on the learning ability, fatty acid composition of brain stem phospholipids and synaptic membrane fluidity in adult mice: a comparison of sardine oil diet with palm oil diet. Mech Ageing Develop 1998; 101:119–128.

Thoenen H, Bandtlow C, Heumann R. The physiological function of nerve growth factor in the central nervous system: comparison with the periphery. Rev Physiol Biochem Pharmacol 1987; 109:145–179.

Tsutsumi T, Yamauchi E, Suzuki E, Watanabe S, Kobayashi T, Okuyama H. Effect of a high α-linolenate and high linoleate diet on membrane-associated enzyme activities in rat brain—modulation of Na^+,K^+-ATPase activity at sub optimal concentrations of ATP. Biol Pharm Bull 1995; 18:664–670.

Umezawa M, Ohta A, Tojo H, Yagi H, Hosokawa M, Takeda T. Dietary α-linolenate/linoleate balance influences learning and memory in the senescence-accelerated mouse (SAM). Brain Res 1995; 669:225–233.

Umezawa M, Kogishi K, Tojo H, Yoshimura S, Seriu N, Ohta A, et al. High-linoleate and high-α-linolenate diets affect learning ability and natural behavior in SAMR1 mice. J Nutr 1999; 129:431–437.

Vaughan MK, Little JC, Vaughan GM, Buzzel GR, Chambers JP, Reiter RJ. Pineal lysosomal enzyme circadian rhythms in male hamsters exposed to natural decreasing photoperiod and temperature conditions. Brain Res Bull 1990; 24:561–564.

Wainwright PE, Xing HC, Ward GR, Huang, Y-S, Bobik E, Auestad N, Montalto M. Water maze performance is unaffected in artificially reared rats fed diets supplemented with arachidonic acid and docosahexaenoic acid. J Nutr 1999; 127:184–193.

Watanabe I, Kato M, Aonuma H, Hashimoto A, Naito Y, Moriuchi A, Okuyama H. Effect of dietary α-linolenate/linoleate balance on the lipid composition and electroretinographic responses in rats. Adv Biosci 1987; 62:563–570.

Watanabe I, Aonuma H, Kaneko S, Okuyama H. Effect of a high linoleate and a high α-linolenate diet on size distribution of phagosomes in retinal pigment epithelium. In: Yasugi T, Nakamura H, Soma M, eds. Advances in Polyunsaturated Fatty Acid Research. Elsevier Science, Amsterdam, 1993, pp. 269–272.

Yamamoto N, Saitoh M, Moriuchi A, Nomura M, Okuyama H. Effect of dietary α-linolenate/linoleate balance on brain lipid compositions and learning ability of rats. J Lipid Res 1987; 28:144–151.

Yamamoto N, Hashimoto A, Takemoto Y, Okuyama H, Nomura M, Kitajima R, et al. Effect of the dietary α-linolenate/linoleate balance on lipid compositions and learning ability of rats II—discrimination process, extinction process, and glycolipid compositions. J Lipid Res 1988; 29:1013–1021.

Yamamoto N, Okaniwa S, Mori S, Nomura M, Okuyama H. Effects of a high linoleate and a high α-linolenate diet on the learning ability of aged rats. Evidence against an autoxidation-related lipid peroxide theory of aging. J Gerontol 1991; 46:B17–B22.

Yao JK, van Kammen DP, Gurklis J. Red blood cell membrane dynamics in schizophrenia. III. Correlation of fatty acid abnormalities with clinical measures. Schizophr Res 1994; 13:227–232.

Yoshida S, Yasuda A, Kawazato H, Sakai K, Shimada T, Takeshita M, et al. Synaptic vesicle ultrastructural changes in the rat hippocampus induced by a combination of α-linolenate deficiency and a learning task. J Neurochem 1997; 68:1261–1268.

Yoshida S, Miyazaki M, Sakai K, Takeshita M, Yuasa S, Sato A, et al. Fourier transform infrared spectroscopic analysis of rat brain microsomal membranes modified by dietary fatty acids—Possible correlation with altered learning behavior. Biospectroscopy 1997; 3:281–290.

Yoshida S, Miyazaki M, Takeshita M, Yuasa S, Kobayashi T, Watanabe S, et al. Functional changes of rat brain microsomal membrane surface after learning task depending on dietary fatty acids. J Neurochem 1997; 68:1269–1277.

Yoshida S, Sato A, Okuyama H. Pathophysiological effects of dietary essential fatty acid balance on neural systems. Jpn J Pharmacol 1998; 77:11–22.

Zimmer L, Hembert S, Durand G, Guilloteau D, Bodard S, Besnard JC, et al. Chronic n-3 polyunsaturated fatty acid diet-deficiency acts on dopamine metabolism in the rat frontal cortex: a micro-dialysis study. Neurosci Lett 1998; 240:177–181.

Zimmer L, Delion S, Vancassel S, Durand G, Guilloteau D, Bodard S, et al. Modification of dopamine neurotransmission in the nucleus accumbens of rats deficient in n-3 polyunsaturated fatty acids. J Lipid Res 2000; 41:32–40.

IV Pathology

14 Disturbances of Essential Fatty Acid Metabolism in Neural Complications of Diabetes

Joseph Eichberg and Cristinel Mîinea

1. INTRODUCTION

In diabetic patients, the incidence of clinically manifested deficits in peripheral nervous system function increases with duration of disease and is approx 50% after 25 yr of disease. The resulting diabetic neuropathies comprise a group of distinct disorders that can affect both somatic and autonomic nerves, the most common of which is symmetric sensory polyneuropathy. Clinical signs of overt human diabetic neuropathy include decreases in nerve conduction velocity and action potential amplitude and in resistance to ischemic conduction failure. These abnormalities may be accompanied by sensory deficits and, in some cases, severe pain. In diabetes of long standing, morphological deterioration is evident, and both nerve fiber loss and segmental demyelination may occur.

The pathogenesis of diabetic neuropathy is now acknowledged to be a complex, multifactorial process, and a consideration of the role of abnormalities in essential fatty acid metabolism in the disease must be viewed in this light. Hyperglycemia is clearly a universal and important precipitating factor, but insulin deficiency may also play a role. The Diabetes Complications and Control Trial unambiguously showed that if blood sugar is maintained at as near an euglycemic level as possible, the incidence and progression of diabetic complications, including those of the peripheral nerve, is greatly reduced (The DCCT Research Group 1993). Disease onset, at least as manifested in experimentally diabetic animals, involves both relatively rapid, often reversible, metabolic aberrations and longer-term irreversible structural alterations. Moreover, at the present time, controversy exists as to what extent neuropathy is brought about by impairments associated with nerve cells (i.e., Schwann cells, axons, and nerve cell bodies in the dorsal root ganglia), as opposed to the vasculature that provides oxygen and nutrients to the nerve.

An overview of the principal overlapping and interacting abnormalities involved in the pathogenesis of diabetic neuropathy is depicted schematically in Fig. 1. Hyperglycemia brought on by insulin and/or C peptide deficiencies (Type I diabetes) or by a failure of target tissues to respond to insulin (Type II diabetes) leads to a variety of metabolic alterations. Among these is heightened polyol pathway activity, an alternative pathway

From: *Fatty Acids: Physiological and Behavioral Functions*
Edited by: D. Mostofsky, S. Yehuda, and N. Salem Jr. © Humana Press Inc., Totowa, NJ

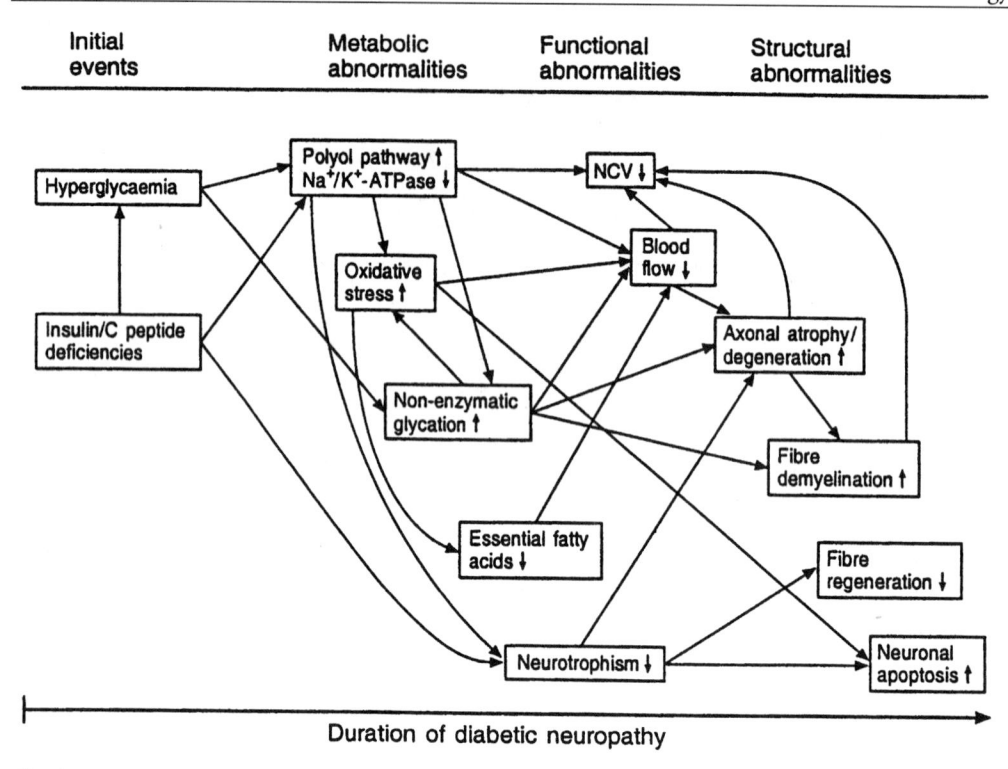

Fig. 1. Hyperglycemic conditions induce multiple abnormalities that occur generally in a temporal sequence, are frequently interactive, and are reciprocally perpetuating. (with permission of Sima, A. A. F. and Sugimoto, K., Diabetologia, (1999), 42, 773–788.)

for glucose catabolism in which glucose is first reduced to sorbitol by aldose reductase and then oxidized to fructose by sorbitol dehydrogenase:

$$\text{Glucose} + \text{NADPH} + \text{H}^+ \rightarrow \text{sorbitol} + \text{NADP}^+,$$

$$\text{Sorbitol} + \text{NAD}^+ \rightarrow \text{fructose} + \text{NADH} + \text{H}^+.$$

In addition to causing accumulation of sorbitol and fructose, increased flux through this pathway will tend to deplete NADPH and cause accumulation of NAD^+. High tissue glucose levels also leads to nonenzymatic glycation of primary amino groups proteins and nucleic acids by reducing sugars, as well as autoxidative glycosylation and lipid peroxidation. Accumulating evidence supports the existence of oxidative stress in diabetic nerve, which is brought about by enhanced production of reactive oxygen species together with developing deficits in antioxidant defense mechanisms that scavenge free radicals. The impact of these abnormalities on nerve essential fatty acid metabolism will be discussed in Section 2.

There are additional changes in diabetic nerve of experimental animals that will not be considered in detail here. These include, but are not limited to, diminished Na^+,K^+-ATPase activity, which can have adverse effects on the maintenance of the nerve resting membrane potential and perturbation of neurotrophic support. All of these pathological processes, acting in concert, lead not only to decreased nerve conduction velocity but also to reduced nerve blood flow, a condition believed to induce endoneurial hypoxia. Over the long term, compromised nerve metabolism and function result in irreversible structural alterations and gradual neuronal apoptosis (Greene et al. 1999).

2. NERVE FATTY ACID METABOLISM IN DIABETES

The impact of diabetes mellitus causes an inability of tissues to metabolize glucose properly and this, in turn, leads to accelerated triacylglycerol breakdown and enhanced β-oxidation of fatty acids. The resulting increase in fat catabolism, together with a depletion of Krebs cycle intermediates, produces a marked increase in ketone body formation. Glucose is the major metabolic fuel for the normal nerve axon and Schwann cells, although in its absence, the peripheral nerve is able to utilize ketone bodies for at least a portion of its energy needs (Winegrad and Simmons 1987). Under normal circumstances, the blood-brain barrier prevents entry of albumin-bound fatty acids into the tissue. Furthermore, nerve incubated with physiological concentrations of free fatty acid exhibits profound respiratory inhibition. Although fatty acids do not therefore constitute a significant energy source in the nerve, the tissue possess the enzymatic machinery to biosynthesize long-chain saturated and monounsaturated fatty acids *de novo* and these are chiefly found as acyl moieties of phospholipids in the myelin sheath. The nerve also contains significant, albeit relatively low, levels of polyunsaturated fatty acids (Lin 1985; Fressinaud et al. 1986; Chattopadhay et al., 1992).

3. BIOSYNTHESIS OF n-6 POLYUNSATURATED FATTY ACIDS AND EICOSANOID FORMATION

The metabolic pathways for synthesis of n-6 and n-3 families of polyunsaturated fatty acids from the essential fatty acids, linoleic acid (LA) (18:2 [n-6]) and α-linolenic acid (18:3 [n-3]), respectively, are shown in Fig. 2. Conversion of LA to arachidonic acid (AA) occurs via Δ6 desaturation to yield γ-linolenic acid (GLA), then an elongation step to produce dihomo-γ-linolenic acid (DHGLA) and Δ5 desaturation, to form AA. The Δ6 and Δ5 microsomal desaturases have been reported to utilize both NADH and NADPH as cofactors in vitro (Brenner 1977). Whether there is a more stringent pyridine nucleotide requirement in vivo is not known with certainty. Desaturase activities are especially abundant in the liver.

Progress in understanding the role of the two desaturases has been handicapped by the lack of a reliable purification protocol. Recently, both enzymes have been cloned from mouse and human and shown to share 75% nucleotide homology (Cho et al. 1999ab). Tissue distribution studies revealed that the human Δ6 desaturase is most abundant in liver and heart, but is present in many tissues, including the brain. By comparison, levels of Δ5 desaturase mRNA are considerably lower in all tissues examined. Ample evidence indicates that Δ6- and Δ5-desaturase expression is nutritionally regulated in a coordinate fashion and is inversely influenced by the quantity of essential fatty acids in the diet. At least in the liver, this occurs at the mRNA level.

Many of the biological actions of essential fatty acids are the result of their metabolic products, the eicosanoids. These are oxidized derivatives of AA and include prostaglandins and thromboxanes that are formed via the cyclo-oxygenase (COX) pathway, as well as hydroxy fatty acids and leukotrienes that arise by means of the lipoxygenase pathway. Another series of AA-derived products, the epoxy fatty acids, are produced by the cytochrome P450 epoxygenase pathway. The AA that serves as precursor for these reactions

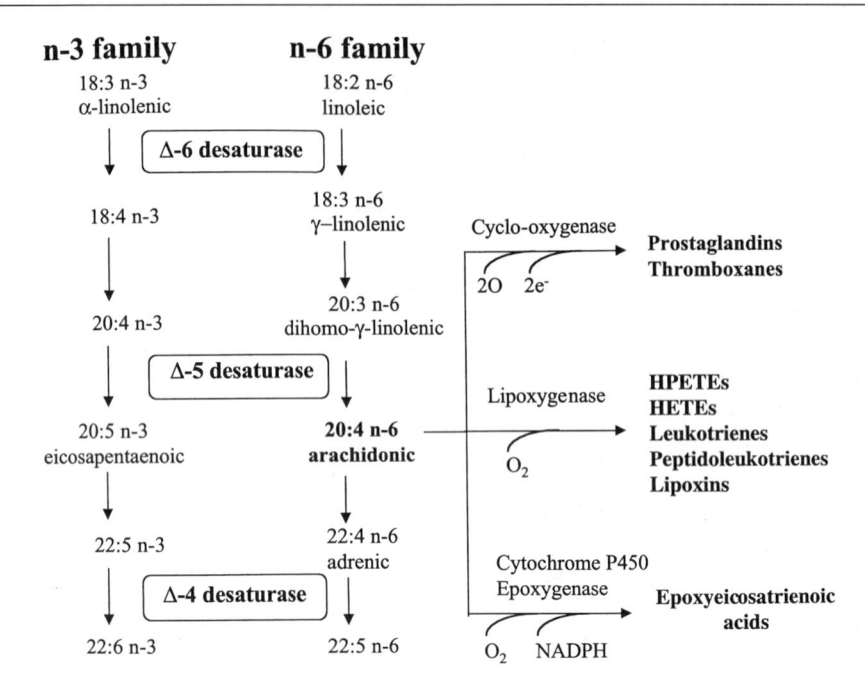

Fig. 2. Biosynthetic pathways of polyunsaturated fatty acid and eicosanoid synthesis. The terminology used indicates first the number of carbons in the acyl chain, then the number of double bonds, and then the number of carbons from the methyl terminus at which the most distal double bond is located (e.g., 20:4n-3). HPETEs: hydroperoxyeicosatetraenoic acids; HETEs: hydroxyeicosatetraenoic acids.

is released from glycerolipids by the action of phospholipase A_2. Nerve preparations are able to synthesize prostaglandins PGD_2, PGE_2, PGI_2 (prostacyclin) and thromboxane A_2 (Goswami and Gould 1985) although to what extent this occurs in the neural or vascular components of the tissue is not known. Prostacyclin and thromboxane A_2 possess well-documented vasodilator and vasoconstrictor properties, respectively. DHGLA is the precursor for an analogous series of prostaglandins, of which PGE_1, a vasodilator, is the best studied. The activity of lipoxygenases is known to give rise to a variety of hydroxy fatty acids that include 5-, 12-, and 15-hydroxyeicosatetraenoic acids, which are formed from AA, and 9- and 13-hydroxyoctadecadienoic acids, which are derived from LA.

4. ALTERED ESSENTIAL FATTY ACID METABOLISM IN DIABETIC NERVE

For many years, it has been known that the profile of glycerolipid fatty acids is altered in tissues, including brain and nerve, of humans and animals with diabetes. The most prominent changes are an increase in LA and docosahexaenoic acid (22:6 [n-3]) and a decrease in AA (Faas and Carter 1983; Holman et al. 1983; Lin et al. 1985; Chattopadhyay et al., 1992). More detailed analysis of nerve glycerolipid molecular species revealed a rather complex impact of diabetes on the distribution of esterified essential fatty acids. Most strikingly, in the nerves of rats that were diabetic for 10 wk following an injection of streptozotocin, arachidonoyl-containing molecular species (ACMS) as a fraction of total molecular species are diminished up to 50% in 1.2-diacylglycerol, phosphatidylcholine, and phosphatidylethanolamine (Zhu and Eichberg 1990, 1993). The proportions of all three major ACMS, namely 18:1/20:4, 16:0/20:4 and 18:0/20:4, are comparably reduced. A modest decline in 18:0/20:4 phosphatidylinositol, the principal molecular species of this phospholipid class, was also observed, but only after 32 wk of diabetes (Doss et al. 1997).

In general, the content of glycerolipid ACMS species drops progressively with increasing duration and severity of diabetes in the nerve and other tissues (Hu et al. 1994; Doss et al. 1997). When normal rats were maintained for 9 wk in a hypoxic environment (10% oxygen atmosphere) in an effort to induce endoneurial hypoxia, a small (approx 20%), but significant fall in the proportions of 18:1/20:4 and 16:0/20:4 occurred (Doss et al. 1997). Under the same environmental conditions, diabetic rats exhibited a substantial fall in phosphatidylcholine and phosphatidylethanolamine ACMS. However, the combination of diabetes and hypoxia did not elicit a further decrease. This result is consistent with the interpretation that sustained hypoxia and diabetes bring about the decrease in ACMS via the same mechanism.

A number of reports have established that the activities of both $\Delta 6$ and $\Delta 5$ desaturases, when assayed in vitro under optimal conditions, are deficient in non-neural tissues in both streptozotocin or alloxan diabetic rats as well as in the spontaneously diabetic BB rat strain (Mercuri et al. 1966; Eck et al. 1979; Poisson 1985; Mimouri and Poisson 1990, Ramsammy 1993). Moreover, the rate of conversion of radiolabeled LA, GLA, and DHGLA to AA in the liver of diabetic rats in vivo is markedly reduced and is restored to normal by insulin administration, as is $\Delta 6$ desaturase activity (Friedman et al. 1966, Poisson 1985, El Boustani et al. 1989). In contrast, in the diabetic kidney, the activity of the elongase required in AA synthesis was not affected, and other enzymes required for fatty acyl CoA formation and subsequent incorporation of acyl groups into glycerophospholipids showed increased activity (Ramsammy et al. 1993). Studies of the impact of diabetes on these enzyme activities in the nerve have not yet been reported.

The depletion of ACMS in glycerophospholipids that occurs in the diabetic nerve can be mimicked to a large extent in a transformed rat Schwann-cell line (initially identified as a human cell line) grown in 30 mM as compared to 5 mM glucose (Kuruvilla and Eichberg 1998a). Furthermore, the incorporation of [^{14}C]DHGLA into arachidonoyl-containing glycerophospholipids was markedly reduced when the cells were cultured at the higher glucose concentration, consistent with an impairment by elevated glucose of cellular $\Delta 5$ desaturase activity (Miinea and Eichberg 2000). However, no difference could be detected in the liberation of AA from prelabeled phospholipids as a function of glucose level, suggesting the loss of ACMS is unlikely the result of altered phospholipase A_2 activity.

In summary, the bulk of available biochemical evidence points to dysfunctional operation of the n-6 biosynthetic pathway in nerve from diabetic animals, resulting primarily from diminished activities of the two participating desaturases.

5. EFFECTS OF LA AND GLA TREATMENT ON DIABETIC NERVE

The reduced activity of the n-6 fatty acid biosynthetic pathway in the diabetic nerve could decrease the availability of AA for prostanoid formation and this could have deleterious consequences on the nerve vasculature, with resultant adverse effects on nerve function. The depletion of AA could be exacerbated by a heightened level of reactive oxygen species (ROS), which react readily with the double bonds of polyunsaturated fatty acids. A number of investigators have examined the effect of supplementing the diet of the diabetic animals with essential fatty acids, a treatment that might ameliorate the loss of AA. One approach has been to feed large amounts of LA and this has yielded variable results. In several studies, feeding a diet rich in corn oil or sunflower oil, which have a high content of LA but no GLA, failed to improve reduced nerve conduction velocity and nerve blood flow. In contrast, in one investigation corn oil administration was effective in preventing decreased nerve conduction velocity (Kuruvilla et al. 1998). In this study, the administration of LA elevated the proportion of ACMS in the nerve from normal animals, but it did not correct the fall in ACMS in diabetic nerve. Inclusion of LA in the diet was found to largely prevent diabetic cataract formation in rats, despite the persistence of high sorbitol levels in the lens (Hutton et al. 1976). Beneficial effects on diabetic complications in patients fed an increased proportion of calories in the diet as LA has been reported (Houtsmuller et al. 1982).

The presumption in these studies has been that the presence of massive amounts of LA tends to promote the diminished activity of the $\Delta 6$ desaturase and hence the overall activity of the n-6 pathway. It remains possible, however, that enhanced availability of LA might exert a beneficial action by formation of other LA metabolites with incompletely understood biological effects. Thus, 13-(S) hydroxy-octadecadienoic acid, which is synthesized from LA by 15-lipoxygenase, has been reported to enhance prostacyclin production in endothelial cells by stimulating the liberation of AA from phospholipids (Setty et al. 1987). Interestingly, there is evidence that 15-lipoxygenase pathway activity is elevated in diabetic retina and in endothelial cells cultured in elevated glucose (Brown et al. 1988; Tesfamariam et al. 1995; Ottlecz et al. 1997). Another oxidized product of LA, 9-hydroxyoctadidecanoic acid, has recently been found in increased amounts in diabetic erythrocyte membranes, perhaps as a consequence of oxidative stress-mediated peroxidation (Inouye et al. 1999).

An alternative approach to compensating for diminished n-6 pathway activity in the diabetic nerve by feeding large amounts of LA in the diet would be to eliminate the need for one of more of the steps by supplying a pathway intermediate. During the past decade, much attention has been paid to the effects of feeding diets enriched in GLA, a maneuver that should bypass the $\Delta 6$ desaturase step. For this purpose, most investigators have utilized evening primrose oil, a naturally occurring oil that contains about 70% LA and 10% GLA. Others have fed either GLA itself or triacylglycerols enriched in this fatty acid. Administration of GLA has been shown to improve diabetic nerve function in both animals and humans, as judged by correction of the nerve conduction velocity deficit, reduced blood flow, and attenuation of resistance to ischemic conduction failure, but not the buildup of polyol pathway metabolites (Julu et al. 1988; Tomlinson et al. 1989; Dines et al. 1993, Cameron and Cotter 1994, Dines et al. 1995, Kuruvilla et al. 1998). However, GLA treatment failed to prevent the depletion of AA in the diabetic nerve (Kuruvilla et al. 1998). GLA administration also amplifies the reduction in nerve Na,K-ATPase activity characteristic of diabetic animals (Lockett and Tomlinson 1992), a finding that tends to dissociate this change from a causative role in decreased nerve conduction velocity. A longer-term effect of GLA supplementation is to promote endoneurial capillary density (Cameron et al. 1991). Dietary AA also has therapeutic effects in that both nerve conduction velocity and blood flow improve when diabetic animals are fed AA-rich oils (Cotter et al. 1997).

The underlying mechanism for the beneficial action of GLA in the diabetic nerve is not fully established. An attractive possibility is that the fatty acid brings about an increase in the synthesis of vasodilatory prostanoids and thus tends to restore the perturbed balance in the synthesis of prostacyclin and thromboxane A_2, eicosanoids that have antagonistic actions on nerve vasculature. A reduction in the availability of arachidonoyl moieties may contribute to the decreased formation of prostacyclin, which occurs in diabetic nerve, but diminished activity of cyclo-oxygenase must also be taken into account (Ward et al. 1989; Stevens et al. 1993). In this regard, the level of constitutive cyclo-oxygenase (COX-1) mRNA is considerably reduced in the diabetic nerve (Fang et al. 1997). Whether an increase in thromboxane A_2 formation occurs in the nerve from diabetic animals is not known, although this is the case in non-neural tissues (Karpen et al. 1982; Peredo et al. 1994). Interestingly, prolonged dietary administration of GLA to human subjects is able to increase production of prostacyclin and decrease generation of thromboxane A_2 in plasma.

Perhaps the strongest evidence that GLA acts via stimulation of prostanoid synthesis is that flurbiprofen, a cyclo-oxygenase inhibitor, blocks improvement in nerve conduction velocity and blood flow elicited by administration of evening primrose oil (Cameron et al. 1993). Greater availability of endogenous GLA might also be expected to enhance the formation of PGE_1. This prostanoid, in addition to its vasodilatory action, tends to fluidize membranes and has been proposed to modulate nerve conduction (Horrobin et al. 1977; Horrobin 1988). It is noteworthy that several PGE_1 analogs have been shown to ameliorate decreased conduction velocity and blood flow and to increase Na^+,K^+-ATPase activity in nerves of experimentally diabetic animals (Yasuda et al. 1999). In opposition to the idea that PGE_1 is primarily involved in the therapeutic action of GLA are the results of experiments in which rats were fed a mixture of fish oil, which is rich in the n-3 fatty acid, eicosapentaenoic acid, together with either evening primrose oil or GLA. Because conversion of DHGLA, the immediate metabolic product of GLA, to AA is inhibited by eicosapentaenoic acid (Lands 1992), the predicted outcome would be enhanced formation of PGE_1 at the expense of prostacyclin synthesis. The results showed that the mixture was not as efficacious in correcting reduced conduction velocity as is GLA alone, thus arguing against a major role for PGE_1 (Cameron and Cotter 1994). Recent evidence suggests that vascular effects of GLA may also be mediated by endothelium-derived hyperpolarizing factor, which also exerts a vasodilatory action (Cotter et al. 2000).

Finally, dietary polyunsaturated fatty acids and some of their eicosanoid metabolites have recently been recognized as regulators of gene transcription, largely through their ability to serve as ligands for peroxisome proliferator-activated receptors (PPARs) (Kerston et al. 2000). These members of the steroid nuclear receptor family bind to a specific response element in the promoter of a target gene as a heterodimer with the 9-*cis*-retinoic acid receptor and bring about gene activation. The genes affected by PPARs encode proteins involved in lipid transport and metabolism. Polyunsaturated fatty acids inhibit lipogenic gene expression and activate expression of genes associated with fatty acid oxidation. Although PPARs have been primarily studied in non-neural tissues, PPAR mRNAs occur in brain, especially during development, as well as in primary neural cell cultures, and have been detected in sciatic nerve (Granneman et al. 1998; Cullingford et al. 1999). There is experimental evidence that Δ6 desaturation of LA is an essential step for the inhibition of the fatty acid synthase gene expression by polyunsaturated fatty acids (Nakamura et al. 2000). Thus, it may be speculated that perturbations in the n-6 fatty acid biosynthetic pathway that occur in diabetic complications could have profound effects on lipid metabolism.

6. ROLE OF n-3 FATTY ACIDS

In comparison to the n-6 series, much less attention has been paid to the involvement of n-3 fatty acids in diabetic neuropathy, although the beneficial effects of fish-oil supplements, a rich source of these fatty acids, in the prevention of atherosclerosis and hypertension in animal models and patients with vascular complications is well known (Lands et al. 1992). Proposals for the mechanisms by which n-3 fatty acids act include serving as precursors of vasoactive prostanoids and acting as stimulants for production of relaxing factors, such as nitric oxide (Lands et al. 1992; Boulanger 1990; McVeigh et al. 1993).

Studies of the potential roles of the n-3 series in the diabetic nerve have yielded conflicting data. Dines et al. (1993) found that fish-oil treatment was less effective than evening primrose oil or GLA in preventing the development of conduction velocity deficits in diabetic rats. Moreover, when fish oil and evening primrose oil were administered simultaneously to diabetic animals, the improvement in nerve conduction velocity and the increase in nerve capillary density elicited by evening primrose oil alone were reduced by the combination therapy. This finding may reflect the known competition between the n-3 and n-6 metabolic pathways (Lands 1992). In contrast to these findings, Gerbi et al. (1999) reported that fish-oil dietary supplementation prevented the drop in nerve conduction velocity and also corrected neuroanatomical anomalies typical of diabetic nerve.

7. INTERACTIONS BETWEEN ESSENTIAL FATTY ACID METABOLISM AND OTHER GLUCOSE-INDUCED DEFECTS

In order to comprehend fully the consequences of perturbed essential fatty acid metabolism on nerve function, it is essential to relate these changes to other metabolic alterations that occur in diabetic neuropathy. These include elevated polyol pathway flux, effects on nitric oxide synthesis, increased oxidative stress, and formation of advanced glycation products. The importance of these abnormalities is reinforced by the fact that the treatment of experimentally diabetic rats with aldose reductase inhibitors, antioxidants and aminoguanidine, an agent that inhibits nonspecific glycation events, all tend to ameliorate physiological impairments in diabetic nerve (Sima and Sugimoto 1999). In the following sections, discussion of the interrelationship of these alterations with impaired essential fatty acid metabolism will underline the complex pathology of this disorder.

8. POLYOL PATHWAY

As previously mentioned, enhanced flux through the polyol pathway that occurs in many tissues, including the nerve, of diabetic humans and animals could alter the ratio between the reduced and oxidized cofactors involved, leading to the accumulation of NADH and the depletion of NADPH. Excess cytosolic NADH may be readily shuttled into the mitochondria, thereby rendering cells "pseudohypoxic" and interfering with energy metabolism (Williamson et al. 1993). An increase in cytosolic NADH/NAD$^+$ has been documented in diabetic lens and in transformed Schwann cells cultured in elevated glucose and a rise in NADP$^+$/NADPH has also been observed in the former system (Obrosova et al. 1998; Mîinea and Eichberg, unpublished observations). The loss of NADPH may inhibit the activities of enzymes that are dependent on this cofactor, in particular glutathione reductase and NO synthase. Evidence that this may be so is provided by the findings that reduced glutathione is depleted in the nerve, as well as in the lens, and that the depletion could be corrected by administering an aldose reductase inhibitor (Lou et al. 1988; Nickander 1994).

A possible connection between polyol pathway activity and essential fatty acids is suggested by the ability of aldose reductase inhibitors to ameliorate the depletion of ACMS in cultured transformed rat Schwann cells grown in 30 mM as compared to 5 mM glucose (Kuruvilla and Eichberg 1998a). These cells exhibit a modest intracellular accumulation of sorbitol and fructose when maintained in the high-glucose medium (Yamaguchi, Miinea, and Eichberg, unpublished observations). However, as yet, no clear-cut increase in the cellular NADPH/NADP$^+$ ratio has been detected. There are indications that some aldose reductase inhibitors may possess antioxidant properties (Ou et al. 1996; Costantino et al. 1999). Taken together, these data suggest that use of polyol pathway flux blockers combined with agents able to resuscitate the diminished antioxidant defenses in cells exposed to 30 mM glucose, as well as in diabetic nerve, will aid in preservation of an active n-6 fatty acid biosynthetic pathway.

9. NITRIC OXIDE FORMATION

A role for NO in modulating neurovascular status seems likely because NO synthase inhibitors tend to blunt the corrective effect of aldose reductase inhibitors or evening primrose oil on nerve conduction velocity (Stevens et al. 1994; Cameron et al. 1996). Moreover, cotreatment of diabetic rats with an aldose reductase inhibitor and evening primrose oil restores nerve conduction velocity to normal in a strongly synergistic fashion and this effect is also antagonized by NO synthase blockers (Cameron et al. 1996). Because evening primrose oil does not affect polyol accumulation induced by diabetes, it is probable that the neurovascular effects of NO and essential fatty acids, acting via their eicosanoid metabolites, are brought about by distinct mechanisms.

10. OXIDATIVE STRESS: CHEMICAL MECHANISMS

Oxidative stress is defined as a condition in which there is an imbalance between production of reactive oxygen species (ROS) in tissues or cells as a result of pro-oxidant effects and their removal by means of antioxidant defenses that scavenge free radicals. Growing evidence indicates that the presence of excessive levels of ROS is a contributory factor in the onset of diabetic complications, as it is in the pathogenesis of atherosclerosis and several neurodegenerative diseases. Abnormally high levels of ROS may be generated in diabetic tissues as a result of several mechanisms, among which are lipid peroxidation, nonenzymatic glycation of proteins, and glucose autoxidation. These processes appear to be enhanced in the presence of transition metals (Hunt and Wolff 1991; van Dam et al. 1995). The most reactive ROS are superoxide (O_2^-) and hydroxyl radicals, both of which are free radicals with short half-lives. Superoxide is also produced as a consequence of respiratory chain electron transport and by a nonmitochondrial NADH oxidase (Mohazzab-H et al. 1997). Hydrogen peroxide is also considered to be an ROS and can serve as a source for the formation of hydroxyl radicals.

A summary of chemical reactions by which hydroxyl radicals can arise and the types of antioxidant mechanisms that destroy ROS are shown in Fig. 3. Under normal circumstances, the propagation of ROS-induced radicals is limited by several antioxidant

Radical oxygen species generation

Haber-Weiss	$H_2O_2 + O_2^{-\bullet} \longrightarrow$	$O_2 + OH^- + HO^\bullet$	(1)
Fenton	$H_2O_2 + Fe^{2+} \longrightarrow$	$OH^- + HO^\bullet + Fe^{3+}$	(2)

Scavenging mechanisms

Superoxide dismutase	$2 O_2^{-\bullet} + 2 H^+ \longrightarrow$	$H_2O_2 + O_2$	(3)
Catalase	$2 H_2O_2 \longrightarrow$	$2 H_2O_2 + O_2$	(4)
Glutathione cycle	$2 GSH + H_2O_2 \xrightarrow{GPX}$	$GSSG + 2 H_2O$	(5a)
	$GSSG + NADPH + H^+ \xrightarrow{GR}$	$2 GSH + NADP^+$	(5b)

Fig. 3. Radical oxygen species formation and antioxidant scavenging systems. Hydroxyl radicals (HO˙) are generated via either the Haber–Weiss reaction (1) in which superoxide O_2^- is reduced by hydrogen peroxide, H_2O_2, or (2) by a transition metal ion as in the Fenton reaction. Superoxide is reduced to hydrogen peroxide in the presence of superoxide dismutase (3), which is further reduced to water and molecular oxygen by catalase (4). Alternatively, hydrogen peroxide could be scavenged by reduced glutathione (GSH) (5a) in the presence of glutathione peroxidase (GPX). The oxidized glutathione (GSSG) is regenerated by glutathione reductase (GR) and NADPH (5b).

protection mechanisms. These include the operation of the glutathione cycle, which involves the concerted action of glutathione reductase and glutathione peroxidase to maintain a steady-state level of reduced glutathione (GSH), catalase and superoxide dismutase activities, ascorbic acid and vitamin E (α-tocopherol and related compounds). The formation of lipid peroxides by the reaction of polyunsaturated fatty acids with hydroxyl radicals as well as the self-propagation of this process and its termination through the oxidation of vitamin E and, as has been suggested, the subsequent reduction of vitamin E by ascorbic acid, is depicted in Fig. 4.

The importance of oxidative stress in the pathogenesis of diabetic complications, including neuropathy, has recently been strongly supported by the findings of Nishikawa et al. (2000). These investigators reported that exposure of cultured vascular endothelial cells to elevated glucose increased ROS production, as judged using a fluorescent molecular probe, and this was accompanied by increases in protein kinase C activation, formation of advanced glycation end products, and flux through the polyol pathway. Treatments that tended to normalize mitochondrial superoxide production lowered the level of ROS and prevented the other changes, all of which are associated with hyperglycemic damage. Enhanced production of ROS has also been observed in cultured Schwann cells (Mîinea and Eichberg, unpublished observations; Fig. 5).

In a somewhat different interpretation of available data, Baynes and Thorpe (1999) have stressed the importance of the formation in diabetic tissues of reactive carbonyl compounds, produced by either metabolic or nonenzymatic processes that may or may not require oxygen. The excess of reactive carbonyls is proposed to then promote glycoxidation and lipoxidation so as to cause chemical modification of proteins. These

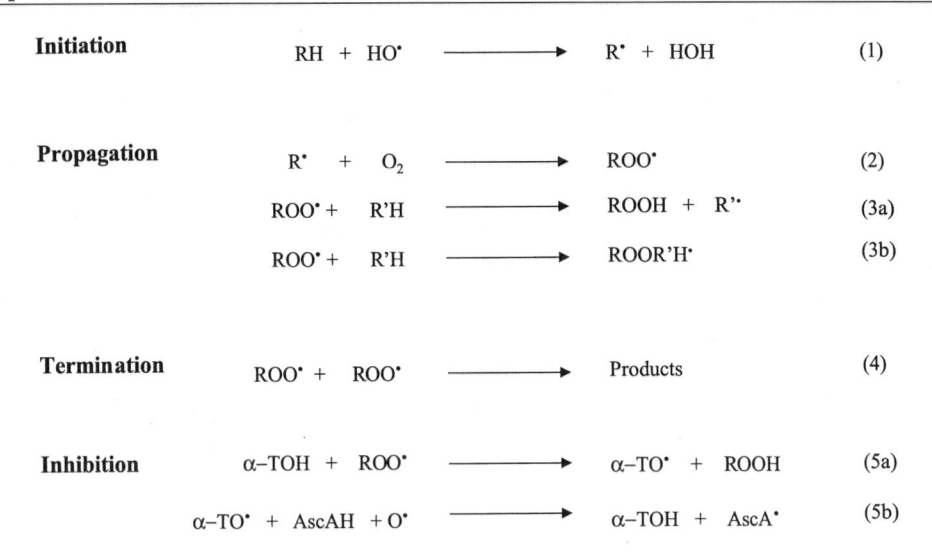

Initiation	RH + HO•	⟶	R• + HOH	(1)
Propagation	R• + O$_2$	⟶	ROO•	(2)
	ROO• + R'H	⟶	ROOH + R'•	(3a)
	ROO• + R'H	⟶	ROOR'H•	(3b)
Termination	ROO• + ROO•	⟶	Products	(4)
Inhibition	α–TOH + ROO•	⟶	α–TO• + ROOH	(5a)
	α–TO• + AscAH + O•	⟶	α–TOH + AscA•	(5b)

Fig. 4. Mechanism of lipid peroxidation and its inhibition by vitamin E. Lipid peroxidation is initiated by generating a relatively unreactive carbon-centered radical upon hydrogen abstraction by a hydroxyl radical (1). The fast formation (2) of the more reactive peroxyl radicals (ROO) ensures rapid attack of any peroxidizable substrate either by abstraction of a hydrogen atom (3a) or addition to a double bond (3b). The propagation is terminated by mutual elimination of peroxyl radicals (4) or by suppression of free-radical formation in the presence of α-tocopherol (α-TOH) (5a). The tocopheryl radical is believed to be neutralized by ascorbic acid (AscAH) (5b) and radical oxygen, and α-tocopherol then re-enters the inhibition cycle.

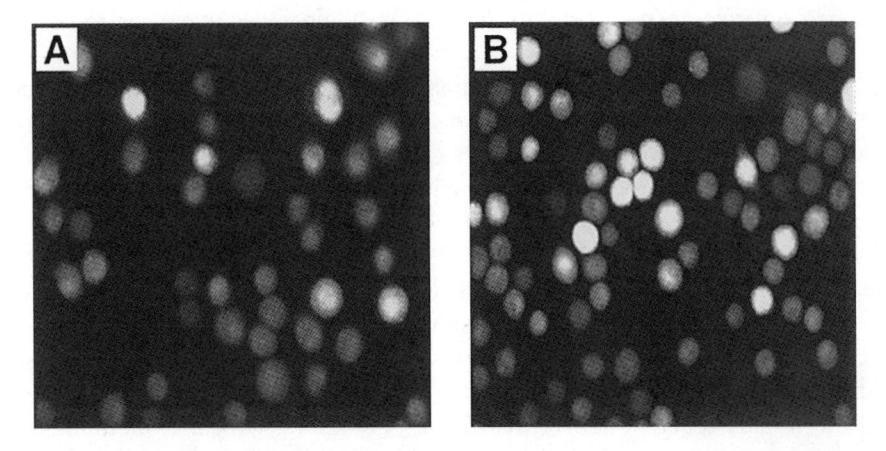

Fig. 5. Reactive oxygen species (ROS) detection in rat Schwann cells in vivo. Cells were cultured in Dulbecco's modified Eagle medium supplemented with 10% fetal bovine serum and either 5 mM (**A**) or 30 mM glucose (**B**). They were then incubated for 45 min in phosphate buffer saline, pH 7.5, containing 10 μM of the ROS-sensitive molecular probe 5- (and 6)-chloromethyl-2',7'-dichlorodihydrofluorescein diacetate (CM–H$_2$DCFDA) and viewed by fluorescence microscopy.

protein adducts may exacerbate pathological changes, leading to oxidative stress with consequent tissue damage and compromised function. These investigators point out that the carbonyl stress hypothesis has considerably broader implications than does the concept of oxidative stress for the development of diabetic complications.

11. OXIDATIVE STRESS: ANTIOXIDANTS AND LIPID PEROXIDATION

A weakening of antioxidant defenses has been convincingly demonstrated in both human diabetic patients and experimentally diabetic rodents, although there appear to be tissue-dependent differences (cf. van Dam 1995; Low et al. 1999 for references). In several non-neural tissues, GSH concentrations as well as Cu-Zn superoxide dismutase and catalase activities are decreased in diabetes. In the case of the peripheral nerve, information is relatively scant, but as compared to the brain and liver, the activities of free-radical scavenging enzymes, except for superoxide dismutase, in the normal tissue are markedly lower (Low et al. 1999). In experimental diabetic neuropathy, Cu-Zn superoxide dismutase activity is substantially reduced, whereas reports conflict regarding changes in glutathione peroxidase (Low and Nickander 1991; Hermenegildo et al. 1993; Kishi et al. 2000). GSH is also decreased in diabetic nerve, whereas catalase activity is enhanced (Nickander et al. 1994; Nagamatsu et al. 1995 Cameron et al. 1999; van Dam et al. 1999), and similar changes in antioxidant defenses are evident in diabetic retina (Obrosova et al. 2000). Within the nerve, the histochemical localization of GSH shows a heterogeneous distribution between the axon and the myelin sheath, suggesting that the peripheral nerve possesses an intricate protective mechanism against ROS which may involve Schwann-cell metabolism in controlling oxygen toxicity (Romero 1996). Cultured Schwann cells grown in elevated glucose also exhibit declines in both the GSH level and superoxide dismutase activity (Mîinea and Eichberg 2000).

Interestingly, in nerve and dorsal root ganglia of the streptozotocin-induced diabetic rat, no changes occur in the levels of antioxidant enzyme mRNAs, including glutathione peroxidase, and both Cu-Zn- and Mn-dependent superoxide dismutase (the cytosolic and mitochondrial forms, respectively, of this enzyme), whereas the expression of catalase mRNA is increased (Kishi et al. 2000). Several of the antioxidant enzyme genes undergo upregulation of expression in non-neural diabetic tissues and in endothelial cells cultured in high glucose (Ceriello et al. 1996; Khanna et al. 1996; Reddi and Bollineni 1997). This increase of gene expression could reflect an effort to mount an adaptive response to hyperglycemia-mediated oxidative stress and resultant nonenzymatic protein glycation or other free-radical-mediated damage, which causes inactivation of the existing antioxidant enzyme population.

The extent of lipid peroxidation is most commonly assessed by determining levels of one or more products derived from polyunsaturated fatty acids, namely malondialdehyde, conjugated dienes, lipid hydroperoxides, 4-hydroxyalkenals, and isoprostanes. Quantitation of malonaldehyde has often been accomplished by measurement of thiobarbituric-acid-reactive substances. Results obtained with this method should be interpreted with caution however, because thiobarbituric acid can react with many substances, especially in crude tissue preparations (Esterbauer and Cheeseman 1990). In human patients and experimental diabetes, lipid peroxidation is manifest in a variety of non-neural tissues, including the kidney, liver, and plasma, and has been observed in brain and retina as well (Low et al. 1999; Obrosova et al. 2000). The peripheral nerve from diabetic animals has been demonstrated to contain elevated conjugated dienes, malondialdehyde, and lipid hydroperoxides (Low and Nickander 1991; Nickander et al. 1994; Nagamatsu et al 1995). Evidence for the importance of lipid peroxidation in the onset of diabetic neuropathy was provided by Nickander et al. (1994), who showed that normal rats fed a vitamin E-deficient diet exhibited increased conjugated dienes and hydroperoxides and reduced GSH in the nerve, together with the development of a sensory neuropathy. These investigators also found that enhanced lipid peroxidation was associated with worsened neuropathy in diabetic rats. In another study, this group found that the antioxidant, α-lipoic acid, which improves nerve function and blood flow when administered to diabetic animals (Cameron et al. 1998), abolished lipid peroxidation induced in vitro by ascorbate–iron–EDTA or when the nerve was incubated in 20 mM glucose (Nickander et al. 1996). Van Dam et al. (1998) reported that malondialdehyde levels in diabetic rats were elevated in plasma, but not in the nerve. These investigators found, in contrast, that nerves of vitamin E-deficient rats had a markedly increased content of malondialdehyde. It has been shown that carboxymethyllysine, the major epitope recognized by antibodies against chemically modified proteins, may be formed not only via glycoxidation of glucose but also during lipid peroxidation of polyunsaturated fatty acids (Fu et al. 1996). Taken together, these findings argue for the importance of lipid peroxidation in the pathogenesis of diabetic neuropathy, but leave unsettled whether the adverse effects are localized to nerve tissue or are more systemic in nature.

12. THERAPEUTIC APPROACHES

A variety of compounds intended to combat oxidative stress have been used to treat experimental diabetic neuropathy and considerable success has been attained in correcting neural and vascular abnormalities (cf. Cameron and Cotter 1999). These substances include hydrophilic and lipophilic free-radical scavengers, as well as chelators of transition metals. A complete discussion of these approaches is beyond the scope of this chapter. However, it is pertinent to comment on several in the context of essential fatty acid metabolism. The effectiveness of the hydrophilic substance, *N*-acetylcysteine, may derive either from its antioxidant properties or its ability to prevent reduction in nerve GSH content by acting as a precursor for synthesis of the tripeptide (Sagara et al. 1996). It is of interest that exposure of transformed Schwann-cell cultures to micromolar concentrations of *N*-acetylcysteine partially prevents the depletion in ACMS (Kuruvilla and Eichberg 1998b). The lipophilic compounds include artificial antioxidants, such as butylated hydroxytoluene and probucol, as well as naturally occurring vitamin E, which was effective in preventing loss of nerve conduction velocity only at very high doses (1 g/kg body weight/d) (Cotter et al. 1995). In the same study, trials using ascorbic acid showed that this vitamin exerted only modest beneficial effects that were diminished at high concentrations, perhaps because sufficient autoxidation caused it to act as a pro-oxidant in the presence of transition metals. The effectiveness of α-lipoic acid, which has already been mentioned, may be attributed to its unique dual properties of acting as a free-radical scavenger and a transition metal chelator. Thus, it may also enhance mitochondrial function.

Certain substances administered to diabetic rats appear to exert therapeutic actions by what may be termed indirect antioxidant effects. Thus, as already noted, the effectiveness of aldose reductase inhibitors is likely, in part, based on their ability to promote GSH levels through blockade of the first step of the polyol pathway and consequent preservation of NADPH. Further, because glycoxidation and lipid peroxidation will produce ROS, agents such as aminoguanidine, which readily react with and remove reactive carbonyl intermediates, will block these chemical transformations and thus help to prevent irreversible modification of proteins (Cameron et al. 1996a). Recently, pyridoxamine has been shown to inhibit chemical modifications of proteins during lipid peroxidation in vitro. This compound acts by trapping reactive intermediates formed during this process and may prove useful in controlling the progressive tissue damage at the molecular level in diabetic neuropathy (Onorato et al. 2000).

Finally, the ability of GLA to promote essential fatty acid synthesis and prostanoid formation and the beneficial actions of selected antioxidants have prompted assessment of combining these treatments. Such an approach has been demonstrated to yield synergistic effects on improving diabetic nerve conduction velocity and blood flow. Thus, comparison of GLA and a probucol analog, each either given alone or in combination, showed a marked increase in efficacy when the two drugs were administered together (Cameron and Cotter 1996b). In analogous fashion, a combination of an aldose reductase and evening primrose oil was far more effective than either alone (Cameron et al. 1996). Treatment of diabetic rats with GLA covalently linked with ascorbic acid showed several-fold improvement over that exhibited by GLA alone on nerve function (Cameron and Cotter 1996c). When an α-lipoic acid-GLA covalent conjugate was given, a substantial synergistic effect was also observed and several physiological and neurochemical indices of experimental diabetic neuropathy were markedly improved (Cameron et al. 1998, Hounsom et al. 1998). In contrast, the effect of a docosahexaenoic acid–α-lipoic acid compound was not different from α-lipoic acid alone, again pointing to the efficacy of n-6 as compared to n-3 fatty acids.

13. CONCLUDING REMARKS

The development of abnormal n-6 essential fatty acid metabolism in the nerve and other tissues as a consequence of diabetes is well established and prevailing evidence indicates that the primary defect is diminished flux through the biosynthetic pathway, which converts LA to AA. However, it is still unknown to what extent altered coenzyme availability or free-radical-mediated damage leading to impaired desaturase gene expression or enzyme activity is responsible. Despite the reduction of both Δ6 and Δ5 desaturase activities in the diabetic nerve, the ability of dietary GLA supplementation to prevent or restore decreased nerve conduction velocity and blood flow suggests that raising the level of this intermediate in vivo sufficiently increases the rate of AA formation to enhance the supply of this precursor for prostacyclin synthesis. In addition, beneficial neurovascular effects of PGE_1 and possibly metabolites of LA formed via the lipoxygenase pathway cannot be discounted.

Regulation of factors governing blood flow in normal nerve is complex, involving synergistic interactions between several systems that determine the degree of vasoconstriction versus vasodilatation. For example, normal rats given suboptimal doses of cyclooxygenase or NO synthase inhibitors exhibit slightly slowed conduction velocity. Simultaneous administration of the inhibitors markedly reduces conduction, consistent with compensatory interactions between prostanoid and NO synthesis, which would limit adverse effects of blocking one of these pathways (Cameron et al. 1993). In this view, diabetes interferes with the integrated operation of these mechanisms, leading to diminished blood flow. It will be important to more fully appreciate the synergy existing between these modes of regulation in order to devise more effective therapies.

Successful treatment of the defects in the diabetic nerve and its vasculature that involve essential fatty acid metabolism will almost certainly require combination therapies. The coadministration of antioxidants and a source of GLA, which has shown promise thus far, can be expected to lead to the development and testing of more potent mixtures or covalent compounds.

REFERENCES

Baynes, J.W. Thorpe, S.R. Role of oxidative stress in diabetic complications. Diabetes 1999; 48:1–9.
Boulanger C, Schini V, Hendrickson H. Vanhoutte P. Chronic exposure of cultured endothelial cells to eicosapentanoic acid potentiates the release of endothelium-derived relaxing factor(s). Br J Pharmacol 1990; 99:176–180.
Brenner RB. Metabolism of endogenous substrates by microsomes. Drug Metab Rev 1977; 6:155–212.
Brown ML, Jakubowski JA, Leventis LL, Deykin D. Elevated glucose alters eicosanoid release from porcine aortic endothelial cells. J Clin Invest 1988; 82:2136–2141.
Cameron NE, Cotter MA. The relationship of vascular changes to metabolic factors in diabetes mellitus and their role in the development of peripheral nerve complications. Diabetes/Metab Rev 1994; 10:189–224.
Cameron NE, Cotter MA. Rapid reversal by aminoguanidine of the neurovascular effects of diabetes in rats: modulation by nitric oxide synthase inhibition. Metabolism 1996; 45:1147–1152.
Cameron NE, Cotter MA. Interactions between oxidative stress and gamma-linolenic acid in impaired neurovascular function of diabetic rats. Am J Physiol 1996; 271:E471–E476.
Cameron NE, Cotter MA. Comparison of the effects of ascorbyl gamma-linolenic acid and gamma-linolenic acid in the correction of neurovascular deficits in diabetic rats. Diabetologia 1996; 39:1047–1054.
Cameron NE, Cotter MA. Effects of antioxidants on nerve and vascular dysfunction in experimental diabetes. Diabetes Res Clin Pratice 1999; 45:137–146.
Cameron NE, Cotter MA. Role of linolenic acid in diabetic polyneuropathy. In: Dyck PJ, Thomas PK, eds. Diabetic Neuropathy. 2nd ed. Saunders, Philadelphia, 1999; pp. 359–367.
Cameron NE, Cotter MA, Dines KC, Robertson S, Cox D. The effects of evening primrose oil on peripheral nerve function and capillarization of in streptozotocin-diabetic rats: modulation by the cyclo-oxygenase inhibitor flurbiprofen. Br J Pharmacol 1993; 109:972–979.
Cameron NE, Cotter MA, Hohman TC. Interactions between fatty acid, prostanoid, polyol pathway and nitric oxide mechanisms in the neurovascular deficit of diabetic rats. Diabetologia 1996; 39:172–182.
Cameron NE, Cotter MA, Horrobin DH, Tritschler HJ. Effects of α-lipoic acid on neurovascular function in diabetes rats: interaction with essential fatty acids. Diabetologia 1998; 41:390–399.
Cameron NE, Cotter MA, Robertson S. Essential fatty acid supplementation. Effects on peripheral nerve and skeletal muscle function and capillarization in streptozotocin-induced diabetic rats. Diabetes 1991; 40:532–539.
Ceriello A, dello Russo P, Amstad P, Cerutti P. High glucose induces antioxidant enzymes in human endothelial cells in culture. Evidence linking hyperglycemia and oxidative stress. Diabetes 1996; 45:471–477.
Chattopadhyay J, Thompson EW, Schmid HH. Nonesterified fatty acids in normal and diabetic rat sciatic nerve. Lipids 1992; 27:513–517.
Cho HP, Nakamura MT, Clarke SD. Cloning, expression, and nutritional regulation of the mammalian Δ-6 desaturase. J Biol Chem 1999; 274:471–477.
Cho HP, Nakamura MT, Clarke SD. Cloning, expression and fatty acid regulation of the human Δ-5 desaturase. J Biol Chem 1999; 274:37,335–37,339.
Costantino L, Rastelli G, Gamberini MC, Vinson JA, Bose P, Iannone A, et al. 1-Benzopyran-4-one antioxidants as aldose reductase inhibitors. J Med Chem 1999; 42:1881–1893.
Cotter MA, Cameron NE. Effects of dietary supplementation with arachidonic acid rich oils on nerve conduction and blood flow in streptozotocin-diabetic rats. Prostaglandins Leukotrienes Essential Fatty Acids 1997; 56:337–343.
Cotter MA, Love A, Watt MJ, Cameron NE, Dines KC. Effects of natural free radical scavengers on peripheral nerve and neurovascular function in diabetic rats. Diabetologia 1995; 38:1285–1294.
Cotter MA, Jack AM, Cameron NE. Essential fatty acid effects on endothelium dependent relaxation in the mesenteric vasculature of diabetic rats. Diabetes 2000; 49(Suppl 1):222-OR.
Cullingford TE, Bhakoo K, Peuchen S, Dolphin CT, Patel R, Clark JB. Distribution of mRNAs encoding the peroxisome proliferator-activated receptor α, β, and γ and the retinoid X receptor α, β, and γ in rat central nervous system. J Neurochem 1998; 70:1366–1375.
The DCCT Research Group. The effect of intensive treatment of diabetes on the development and progression of long-term complications in insulin-dependent diabetes mellitus. N Engl J Med 1993; 329:977–986.
Dines KC, Cameron NE, Cotter MA. Comparison of the effects of evening primrose oil and triglycerides containing γ-linolenic acid on nerve conduction and blood flow in diabetic rats. J Pharmacol Exp Ther 1995; 273:49–55.

Dines KC, Cotter MA, Cameron NE. Contrasting effects of treatment with ω-3 and ω-6 essential fatty acids on peripheral nerve function and capillarization on streptozotocin-diabetic rats. Diabetologia 1993; 36:1132–1138.

Doss DJ, Kuruvilla R, Bianchi R, Peterson RG, Eichberg J. Effects of hypoxia and severity of diabetes on Na,K-ATPase activity and arachidonoyl-containing glycerophospholipid molecular species in nerve from streptozotocin diabetic rats, J Peripheral Nervous Sys 1997; 2:155–164.

Eck G, Wynn JO, Carter WJ, Faas FH. Fatty acid desaturation in experimental diabetes mellitus. Diabetes 1979; 28:479–485.

El-Boustani S, Descomps B, Monnier L, et al. In vivo conversion of dihomogammalinolenic acid into arachidonic acid in man. Prog Lipid Res 1986; 25:67–71.

Esterbauer H, Cheeseman KH. Determination of aldehydic lipid peroxidation products: malonaldehyde and 4-hydroxynonenal. Methods Enzymol 1990; 186:407–421.

Faas FH, Carter WJ. Altered microsomal phospholipid composition in the streptozotocin diabetic rat. Lipids 1983; 18:339–342.

Fang C, Jiang Z, Tomlinson DR. Expression of constitutive cyclo-oxygenase (COX-1) in rats with streptozotocin-induced diabetes; effect of treatment with evening primrose oil or an aldose reductase inhibitor on COX-1 mRNA levels. Prostaglandins Leukotrienes Essential Fatty Acids 1997; 56:157–163.

Friedman N, Gellhorn A, Benjamin W. Synthesis of arachidonic acid from linoleic acid in vivo in diabetic rats. Israel J Med Sci 1966; 2:677–682.

Fressinaud C, Rigaud M, Vallat JM. Fatty acid composition of endoneurium and perineurium from adult rat sciatic nerve. J Neurochem 1986; 46:1549–1554.

Fu MX, Requena JR, Jenkins A.J, Lyons TJ, Baynes JW, Thorpe SR. The advanced glycation end product, Nepsilon-(carboxymethyl) lysine, is a product of both lipid peroxidation and glycoxidation reactions. J Biol Chem 1996; 271:9982–9986.

Gerbi A, Maixent J-M, Ansaldi J-L, Pierlovisi M, Coste T, Pellisier J-F, et al. Fish oil supplementation prevents diabetes-induced nerve conduction velocity and neuroanatomical changes in rats. J Nutr 1999; 129:207–213.

Goswami SK. Gould RM. Prostanoid synthesis in peripheral nerve. Biochim Biophys Acta 1985; 834:263–266.

Granneman J, Skoff R, Yang X. Member of the peroxisome proliferator-activated receptor family of transcription factors is differentially expressed by oligodendrocytes. J Neurosci Res 1998; 51:563–573.

Greene DA, Stevens MJ, Obrosova I, Feldman EL. Glucose-induced oxidative stress and programmed cell death in diabetic neuropathy. Eur J Pharmacol 1999; 375:217–223.

Hermenegildo C, Raya A, Roma J, Romero FJ. Decreased glutathione peroxidase activity in sciatic nerve of alloxan-induced diabetic mice and its correlation with blood glucose levels. Neurochem Res 1993; 18:893–896.

Holman RT, Johnson SB, Gerrard JM, et al. Arachidonic acid deficiency in streptozotocin-induced diabetes. Proc Natl Acad Sci USA 1983; 80:2375–2379.

Horrobin DF, Durand LG, Manku MS. Prostaglandin E1 modifies nerve conduction and interferes with local anaesthetic action. Prostaglandins 1977; 14:103–108.

Horrobin DF. The roles of essential fatty acids in the development of diabetic neuropathy and other complications of diabetes mellitus, Prostaglandins Leukotrienes Essential Fatty Acids 1988; 44:127–131.

Hounsom L, Horrobin DF, Tritschler H, Corder R. Tomlinson DR. A lipoic acid–gamma linlenic acid conjugate is effective against multiple indices of experimental diabetic neuropathy. Diabetologia 1998; 41:839–843.

Houtsmmuller AJ, van Hal-Ferweda J, Zahn, KJ, Henkes HE. Favourable influences of linoleic acid on the progression of diabetic micro- and macro-angiopathy in adult onset diabetes mellitus. Prog Lipid Res 1982; 20:377–386.

Hu Q, Ishii E, Nakagawa Y. Differential changes in relative levels of arachidonic acid in major phospholipids from rats during the progression of diabetes. J Biochem 1994; 115:405–408.

Hunt JV, Wolff SP. The role of histidine residues in the nonenzymic covalent attachement of glucose and ascorbic acid to protein. Free Radical Res Commun 1991; 14:279–287.

Hutton JC, Scholfield PJ, Williams JF, et al. The effect of an unsaturated fat diet on cataract formation in streptozotocin-induced diabetic rat. Br J Nutr 1976; 36:161–177.

Inouye M, Mio T, Sumino K. Formation of 9-hydroxy linoleic acid as a product of phospholipid peroxidation in diabetic erythrocyte membrane. Biochim Biophys Acta 1999; 1438:204–212.

Julu PO. Essential fatty acids prevent slowed nerve conduction in streptozotocin diabetic rats. J Diabetes Complic 1988; 2:185–188.

Karpen CW, Pritchard KA Jr, Merola AJ, Pangamala RV. Alterations of the prostacyclin-thromboxane ration in streptozotocin induced diabetic rats. Prostaglandins Leukotrienes Med 1982; 8:93–103.

Kersten S, Desvergne B, Wahli W. Role of PPARs in health and disease. Nature 2000; 405:421–424.

Khana P, Wang L, Ansari NH. Semi-quantitation of mRNA by polymerase chain reaction. Levels of oxidative defense enzymes and aldose reductase in rat lenses cultured in hyperglycemic or oxidative medium. Res Commun Mol Pathol Pharmacol 1996; 92:3–18.

Kishi Y, Nickander KK, Schmelzer JD, Low PA. Gene expression of antioxidant enzymes in experimental diabetic neuropathy. J Peripheral Nervous Sys 2000; 5:11–18.

Kuruvilla R, Eichberg J. Depletion of phospholipid arachydonoyl-containing molecular species in a human Schwann cell line grown in elevated glucose and their restoration by an aldose reductase inhibitor. J Neurochem 1998a; 71:775–783.

Kuruvilla R, Eichberg J. Antioxidants enhance restorative effects of an aldose reductase inhibitor on depleted arachidonoyl moieties in Schwann cells grown in elevated glucose. Diabetes 1998b; 47(Suppl 1):A138.

Kuruvilla R, Peterson RG, Eichberg J. Evening primrose oil treatment corrects reduced conduction velocity but not depletion of arachidonic acid in nerve from streptozotocin-induced diabetic rats. Prostaglandins Leukotrienes Essential Fatty Acids 1998; 59:195–202.

Lands WEM. Biochemistry and physiology of n-3 fatty acids. FASEB Journal 1992; 6:2530–2536.

Lin C-J, Peterson R, Eichberg J. The fatty acid composition of glycerolipids in nerve, brain and other tissues of the streptozotocin diabetic rat, Neurochem Res 1985; 10:1453–1465.

Lockett MJ, Tomlinson DR. The effects of dietary treatment with essential fatty acids on sciatic nerve conduction and activity of the Na+/K+ pump in streptozotocin-diabetic rats. Br J Pharmacol 1992; 105:355–360.

Lou MF, Dickerson R, Garadi R, York BM. Glutathione depletion in the lens of galactosemic and diabetic rats. Exp Eye Res 1988; 46:517–530.

Low PA, Nickander KK. Oxygen free radical effects in sciatic nerve in experimental diabetes. Diabetes 1991; 40:873–877.

Low PA, Nickander KK, Sciotti L. Role of hypoxia, oxidative stress and excitatory neurotoxins in diabetic neuropathy. In: Dyck PJ, Thomas PK, eds. Diabetic Neuropathy. 2nd ed. Saunders, Philadelphia, 1999, pp. 317–329.

McVeigh GE, Brennan GM, Johnston GD, McDermott BJ, McGrath LT, Henry WR, et al. Dietary fish oil augments nitric oxide production or release in patients with type 2 (non-insulin-dependent) diabetes mellitus. Diabetologia 1993; 36:33–38.

Mercuri O, Pelluffo RO, Brenner RR. Depression of microsomal desaturation of linoleic to gamma-linolenic acid in the alloxan diabetic rat. Biochim Biophys Acta 1966; 116:407–411.

Mimouni V, Poisson J-P. Spontaneous diabetes in BB rats: evidence for insulin-dependent liver microsomal delta-6 and delta-5 desaturase activities. Horm Metab Res 1990; 22:405–407.

Mîinea C, Eichberg J. Altered arachidonic acid metabolism and diminished antioxidant production is induced in cultured Schwann cells by elevated glucose, Diabetes 2000; 49(Suppl 1):A166.

Mohazzab-HKM, Kaminski PM, Wolin MS. Lactate and PO2 modulate superoxide anion production in bovine cardiac myocytes: potential role for NADH oxidase. Circulation 1997; 96:614–620.

Nagamatsu M, Nickander KK, Schmelzer JD, Raya A, Wittrock DA, Tritschler H, et al. Lipoic acid improves nerve blood flow, reduces oxidative stress, and improves distal nerve conduction in experimental diabetic neuropathy. Diabetes Care 1995; 18:1160–1167.

Nakamura MT, Cho HP, Clarke SD. Regulation of hepatic delta-6 desaturase expression and its role in the polyunsaturated fatty acid inhibition of fatty acid synthase gene expression in mice. J Nutr 2000; 30:1561–1565.

Nickander KK, McPhee BR, Low PA, Tritshcler H. Alpha-lipoic acid antioxidant potency against lipid peroxidation of neural tissues in vitro and implications for diabetic neuropathy. Free Radical Biol Med 1996; 21:631–639.

Nickander KK, Schmelzer JD, Rohwer DA, Low PA. Effect of alpha-tocopherol deficiency on indices of oxidative stress in normal and diabetic peripheral nerve. J Neurol Sci 1994; 126:6–14.

Nishikawa T, Edelstein D., Du XL, Yamagishi S, Matsumura T, Kaneda Y, et al. Normalizing mitochondrial superoxide production blocks three pathways of hyperglycaemic pathway. Nature 2000; 404:787–790.

Obrosova IG, Cao X, Green DA, Stevens MJ. Diabetes-induced changes in lens antioxidant status, glucose utilization and energy metabolism: effect of DL-alpha-lipoic acid. Diabetologia 1998; 41:1442–1450.

Obrosova IG, Fathalllah L, Greene DA. Early changes in lipid peroxidation and antioxidative defense in diabetic rat retina: effect of DL-alpha-lipoic acid. Eur J Pharmacol 2000; 398:139–146.

Onorato JM, Jenkins AJ, Thorpe SR, Baynes JW. Pyridoxamine, an inhibitor of advanced glycation reactions, also inhibits advanced lipoxidation reactions. J Biol Chem 2000; 275:21,177–21,184.

Ottlecz A, Sanduja SK, Eichberg J. Altered arachidonic acid metabolism in the retina of the streptozotocin-induced diabetic rat. Invest Ophthamol Vis Sci 1997, 38(Suppl):S772.

Ou, P, Nourooz-Zadeh J, Tritschler HJ, Wolff S. Activation of aldose reductase in rat lens and metal-ion chelation by aldose reductase inhibitors and lipoic acid. Free Radical Res 1996; 25:337–346.

Peredo HA, Filinger EJ, Sanguinetti S, Lorenzo PS, Adler-Graschinsky E. Prostanoid production in hypoxic rat isolated artria: influence of acute diabetes. Prostaglandins Leukotrienes Essential Fatty Acids 1994; 51:231–234.

Poisson JP. Comparative in vivo and in vitro study of the influence of experimental diabetes on rat liver linoleic acid $\Delta 6$ and $\Delta 5$-desaturation. Enzyme 1985; 34:1–14.

Ramsammy LS, Haynes B, Josepovitz C, Kaloyanides GJ. Mechanisms of decreased arachidonic acid in the renal cortex of rats with diabetes mellitus. Lipids 1993; 28:433–439.

Reddi AS, Bollineni JS. Renal cortical expression of mRNAs for antioxidant enzymes in normal and diabetic rats. Biochem Biophys Res Commun 1997; 235:598–601.

Romero FJ. Antioxidants in peripheral nerve. Free Radical Biol Med 1996; 20:925–932.

Sagara M, Satoh J, Wada R, Yagihashi S, Takahashi K, Fukuzawa M, et al. Inhibition of development of peripheral neuropathy in streptozotocin-induced diabetic rats with N-acetylcysteine. Diabetologia 1996; 39:263–269.

Sima AAF, Sugimoto A. Experimental diabetic neuropathy: an update. Diabetologia 1999; 42:773–788.

Setty BN, Berger M, Stuart MJ. 13-Hydroxyoctadidecanoic acid (13-HODE) stimulates prostacyclin production by endothelial cells. Biochem Biophys Res Commun 1987; 146:502–509.

Stevens MJ, Dananberg J, Feldman EL, Lattimer SA, Kamijo M, Thomas TP, et al. The linked role of nitric oxide, aldose reductase and, (Na+,K+)-ATPase in the slowing of nerve conduction in the streptozotocin diabetic rat. J Clin Invests 1994; 94:853–859.

Stevens EJ, Lockett MJ, Carrington AL, Tomlinson DR. Essential fatty acid treatment prevents nerve ischaemia and associated anomalies in rats with experimental diabetes mellitus. Diabetologia 1993; 36:397–401.

Tesfamariam B, Brown ML, Cohen RA. 15-Hydroxyeicosatetraenoic acid and diabetic endothelial dysfunction in rabbit aorta. J Cardiovasc Pharmacol 1995; 25:748–755.

Tomlinson DR, Robinson JP, Compton AM, Keen P. Essential fatty acids treatment—effects on nerve conduction, polyol pathway and axonal transport in streptozotocin diabetic rats. Diabetologia 1989; 32:655–659.

van Dam PS, Sweder van Asbeck B, Erkelens DW, Marx JJM, Gispen W-H, Bravenboer B. The role of oxidative stress in neuropathy and other diabetic complications. Diabetes/Metab Rev 1995; 11:181–192.

van Dam PS, Sweder van Asbeck, B, Bravenboer B, van Oirschot JFLM, Gispen WH, Marx JJM. Nerve function and oxidative stress in diabetic and vitamin E-deficient rats. Free Radical Biol Med 1998; 24:18–26.

van Dam PS, Sweder van Asbeck B, Bravenboer B, van Oirschot JFLM, Marx JJM, Gispen WH. Nerve conduction and antioxidant levels in experimentally diabetic rats: effects of streptozotocin dose and diabetes duration. Metabolism 1999; 48:442–447.

Ward KK, Low PA, Schmelzer JD, Zochodne DW. Prostacyclin and noradradrenaline in peripheral nerve of chronic experimental diabetes in rats. Brain 1989; 112:197–208.

Winegard AI, Simmons DA. Energy metabolism in peripheral nerve. In: Dyck PJ, Thomas PK, Asburry AK, Winegard AI, Porte D Jr, eds. Diabetic Neuropathy. Saunders, Philadelphia, 1987, pp. 279–288.

Williamson JR, Chang K, Frangos M, Hasan KS, Ido Y, Kawamura T, Nyengaard JR, van der Ende M, Kilo C, Tilton RG. Hyperglycemic pseudohypoxia and diabetic complications. Diabetes 1993; 42:801–813.

Yasuda H, Kikkawa R. Role of antiprostaglandins in diabetic neuropathy. In: Dyck PJ, Thomas PK, eds. Diabetic Neuropathy. 2nd ed. Saunders, Philadelphia, 1999; pp. 368–376.

Zhu X, Eichberg J. 1,2-Diacylglycerol content and its arachidonoyl-containing molecular species are reduced in sciatic nerve from streptozotocin-induced diabetic rats. J Neurochem 1990; 55:1087–1090.

Zhu X, Eichberg J. Molecular species composition of glycerophospholipids in rat sciatic nerve and its alteration in streptozotocin-induced diabetes. Biochim Biophys Acta 1993; 1168:1–12.

15 Docosahexaenoic Acid Therapy for Disorders of Peroxisome Biogenesis

Hugo W. Moser and Gerald V. Raymond

1. INTRODUCTION

That docosahexaneoic acid (DHA) plays an important role in brain and retinal function is suggested by the fact that it is present in high concentration in both of these tissues and that its level in these tissues increases as these organs mature (Neuringer, et al., 1988). Chapters 10–13 and 17 provide support for the thesis that DHA deficiency impairs cognitive function and vision in human beings and experimental animals and that such deficiencies are significant from the perspective of human development and public health. In this chapter, we report on the effect of DHA therapy in the disorders of peroxisome biogenesis (PBD). These are relatively rare genetically determined disorders that have been recognized only recently and cause profound deficits of cognitive development, vision, hearing, as well as multiple malformations and often lead to death in infancy or childhood (Lazarow & Moser, 1994). The central feature in these disorders is the failure to form a subcellular organelle referred to as the peroxisome. This organelle normally is present in all cells other than the mature red blood cell. Its name is derived from the fact that it is a major site for the formation and degradation of hydrogen peroxide (De Duve and Baudhuin, 1966). Later, it was also shown to be essential for 50 more biochemical reactions, many of which involve the formation and degradation of complex lipids (van den Bosch, et al., 1992). Of crucial relevance to this chapter is the fact that the peroxisome is required for the synthesis of DHA (Moore, et al., 1995; Voss, et al., 1991).

Why should a full chapter in this volume be devoted to a group of disorders that are relatively rare and known mainly to physicians who specialize in genetics, neurology, ophthalmology, or liver disease? The reasons are both humanistic and biological. The severe and progressive disability associated with PBD causes severe hardship to the child and may devastate the family. DHA replacement therapy is simple to administer and provides a ray of hope for the first time. From a biologic point of view, a major point of interest is that PBD patients cannot synthesize DHA to any significant extent. This leads to DHA deficiency more profound than can be produced by any other means and thus provides a unique opportunity to examine the effects of this deficiency. DHA levels in PBD patients or animal models can be controlled and monitored precisely, because all or nearly all of their DHA is derived from what they were supplied by their mother during gestation and from the diet after birth. These levels are more difficult to control in persons

From: *Fatty Acids: Physiological and Behavioral Functions*
Edited by: D. Mostofsky, S. Yehuda, and N. Salem Jr. © Humana Press Inc., Totowa, NJ

Table 1
Major Clinical Features of Disorders of Peroxisome Assembly
and Their Occurrence in Various Peroxisomal Disorders

Feature	ZS	NALD	IRD	Bifunctional enzyme deficiency	RCDP	DHAP synthase deficiency	DHAP alkyl transferase deficiency
Average age at death or last follow-up (yr)	0.76	2.2	6.4	0.75	1.0	0.5	?
Facial dysmorphism	++	+	+	73%	++	++	++
Cataract	80%	45%	7%	0	72%	+	+
Retinopathy	71%	82%	100%	+	0	0	0
Impaired hearing	100%	100%	93%	?	71%	33%	100%
Psychomotor delay	4+	3–4+	3+	4+	4+	4+	?
Hypotonia	99%	82%	52%	4+	±	±	?
Neonatal seizures	80%	82%	20%	93%	±	?	?
Large liver	100%	79%	83%	+	0	?	
Renal cysts	93%	0	0	0	0	0	0
Rhizomelia	3%	0	0	0	93%	+	+
Chondrodysplasia punctata	69%	0	0	0	100%	+	+
Neuronal migration defect	67%	20%	±	88%	+	?	?
Coronal vertebral cleft	0	0	0	0	+	+	+
Demyelination	22%	50%	0	75%	0	0	0

Note: Percentages indicate the percentage of patients in whom the abnormality is present. 0, abnormality is absent; ± to 4+: degree to which an abnormality is present.

or animals without this gene defect, because, here, dietary DHA deficiency can be overcome to a variable and unknown extent by compensatory increases in DHA synthesis. Thus, studies in PBD patients and animal models can increase understanding of normal brain development and function and help set nutritional policies.

2. DISORDERS OF PEROXISOME BIOGENESIS

2.1. Clinical Manifestations

The disorders that are now assigned to the PBD category were described clinically and named before their relation to the peroxisome was understood. This has resulted in a clinical nomenclature that is somewhat confusing, but has been retained because of its long usage. Table 1 lists the main clinical features. Genetic analyses and detailed studies of genotype–phenotype relationships have shown that the PBD can be subdivided into two groups. The first group includes three disorders that in the past had been considered to be separate clinical entities: the Zellweger syndrome (ZS) (Bowen, et al., 1964; Wilson, et al., 1986; Zellweger, 1987); neonatal adrenoleukodystrophy (NALD) (Ulrich, et al., 1978; Kelley, et al., 1986); and infantile Refsum disease (IRD) (Scotto, et al., 1982). ZS, NALD, and IRD are now considered to represent a clinical continuum, with ZS the most severe, NALD intermediate, and IRD the mildest compared to the other forms, even though in absolute terms it still causes marked disability. "Classical ZS" is a very severe disorder, often leading to death during the first year and psychomotor development is severely compromised and sometimes absent. It is associated with a striking and characteristic defect in neuronal migration (Evrard, et al., 1978). NALD and IRD share many of the features of the features of ZS, but are somewhat milder. Patients live longer; a few have survived to the fourth or fifth decade (Moser, et al., 1995). Clinical distinction between NALD and IRD is difficult. NALD often is associated with abnormalities of the brain white matter, whereas IRD is not. The second PBD group comprises a single disorder, rhizomelic chondrodysplasia punctata (RCDP). RCDP has a distinct phenotype (Spranger, et al., 1971; Braverman, et al., 1997). The major clinical features of RCDP are shortening of the humeri and femurs, which cause the limbs to be short; cataracts, ichthyosis (dry, scaling skin), and severe psychomotor retardation. Radiographs show chondrodysplasia punctata (bony stippling of the cartilages near the joints and spine).

2.2. Genetic Bases of PBD

Advances about the biochemical and genetic bases of the PBD were achieved in close conjunction with gains in the understanding of how peroxisomes are formed normally. Lazarow and Fujiki showed in 1985 that the proteins that make up the peroxisome are formed in polyribosomes, enter the cytoplasm of the cell, and are then imported into preexisting peroxisomes. Proteins that are "destined" for the peroxisome contain targeting sequences that direct them to the organelle (Subramani, 1998; Gould, et al., 1989; Swinkels, et al., 1991). This targeting and subsequent import into the peroxisomes is a complex process, controlled by a variety of import factors, which are now referred to as peroxins (Distel, et al., 1996). More than twenty peroxins have been identified. They are numbered sequentially in accordance with the date of their publication. *All of the PBD are the result of defects of protein import into the peroxisome caused by mutations that affect one of the peroxins.* Study of the PBD and identification of the peroxins have been highly interactive. The individual peroxins were identified as a result of studies of cells from human patients or yeast in which the import of proteins into the peroxisome was impaired due to a genetic defect. The initial lead toward the identification of a specific gene defects was provided by the technique of complementation analysis (Brul, et al., 1988; Moser, et al., 1995). The underlying principle of this technique is that cells from two different patients, both deficient in a peroxisomal import, are induced to fuse. The resulting multinucleated cells are then examined for their capacity to carry out the import process. Restoration of activity can occur only if each cell line provides the gene product defective in the other, and cells that complement each other in this way thus must represent distinct genotypes. Complementation analyses demonstrated that human PBD can be caused by at least 12 different gene defects. Basic cell-biology studies to define the exact structure and role of each of the peroxins were then undertaken and have led to the definition of the gene defect in 10 of the complementation groups. Table 2 shows the

Table 2
Molecular Defects in Disorders of Peroxisone Biogenesis

Peroxin	Characteristic	Complementation group[a] KKI	Jap	Ams	Patient # Reported KKI	Ams	Other	Phenotype
1	143 kDa AAA ATPase	1	E	3	99	61		ZS, NALD, IRD
2	C3HC4 zinc-binding integral peroxisomal membrane protein 35–52 kDa	10	F		2			ZS
3	51–52 kDa integral peroxisomal membrane protein							
4	21–24 kDa peroxisomal-associated ubiquitin-conjugating enzyme							
5	PTS 1 receptor	2		4	2			ZS, NALD
6	12–127 kDa AAA ATPase	4	C	3	16	12		ZS, NALD
7	PTS 2 receptor	11		1	43			RCDP
8	71–81 kDa peroxisomal-associated protein							
9	42 kDa integral peroxisomal membrane protein							
10	C3HC4 zinc-binding integral peroxisomal membrane protein	7	B	5				ZS, NALD
11	27–32 kDa peroxisomal membrane protein involved in peroxisomal proliferation							
12	48 kDa C3HC4 zinc-binding integral peroxisomal membrane protein	3		5	6			ZS, NALD, IRD
13	SH-3 containing 40–43 kDa peroxisomal integral peroxisomal membrane protein		H		2			ZS, NALD
14	41 kDa integral membrane protein							
15	48 kDa cytosolic protein							
16	39 kDa peripheral peroxisomal membrane protein	9	D	2	1	1		ZS
17	27–30 kDa peroxisomal? intrinsic membrane protein							
18	35–39 peroxisome membrane protein zinc-finger motif							
19	Peroxisomal membrane protein, prenylated		J		1			ZS
	Unidentified	8	A		7			ZS, NALD, IRD
	Unidentified		G				2	ZS

[a]KKI = Kennedy Krieger Institute; Jap = Japan; Ams = Amsterdam.

results of these interactive studies. The knowledge gained from these basic studies has increased the understanding of the disease processes.

2.3. Genotype–Phenotype Correlations

Complementation analyses and molecular studies have had a profound effect on the classification and conceptual framework of the PBD. Table 2 lists the main clinical features and relative frequencies of the PBD. Note that the ZS, NALD, and IRD phenotypes can be associated with 11 different complementation groups and with 9 different molecular defects. This finding forms the basis for the conclusion that these three phenotypes represent a continuum, rather than distinct disease entities. In contrast, among the PBD, RCDP is associated only with a defect in *Pex7*.

2.4. Specific PBD Genotypes Often Are Associated with Widely Variant Disease Expression (Phenotype)

Table 2 shows that defects that involve the same gene often are associated with different phenotypes. For example, complementation group 1, in which the defect involves *PEX1*, may be associated with three different phenotypes (ZS, NALD, IRD). This wide range of phenotypic expression complicates the evaluation of therapeutic interventions. Although the causes of this variability of clinical expression are not fully understood, two factors that contribute to it have been identified. The first involves the nature of the mutation, as exemplified in defects that involve *PEX1*. Studies in cultured cells had shown that the c2097insT mutation of this gene abolishes peroxisome import completely, whereas the G843D mutation reduces import to 15% of control but does not abolish it completely (Collins and Gould, 1999). Seventy-five percent of patients who have an allele with the G843D mutation had the relatively mild IRD phenotype and only 9% had the severe ZS. In contrast, only 13% of patients with a c2097insT allele had the IRD phenotype, whereas the ZS phenotype was present in 37%. *Mosaicism (i.e., a variable expression of the import defect in different tissues)*, is another cause of relatively mild phenotypic expression (Roels, et al., 1996).

2.5. Single-Enzyme Defects that Mimic the Clinical Manifestations of PBD

Another factor that complicates the evaluation of therapeutic interventions is the fact that the clinical manifestations of PBD disorders may be mimicked by another category of peroxisomal disorders in which the genetic defect affects a single peroxisomal enzyme and does not alter import mechanisms or peroxisome structure. The most common example of this category is a genetic defect that involves the peroxisomal bifunctional enzyme, which is required for the degradation of very long-chain fatty acids and bile acids (van Grunsven, et al., 1998). In the Kennedy Krieger Institute series, 18% of patients whose clinical presentation resembled the ZS–NALD–IRD continuum had bifunctional enzyme deficiency and not a PBD. The RCDP phenotype may be mimicked by genetically determined enzyme defects that impair the synthesis of a group of complex lipids referred to as plasmalogens (Wanders, et al., 1994; Wanders, et al., 1992). Analysis of clinical manifestations alone does not permit a reliable distinction between single enzyme defects and the PBD. Yet, because the genetic defects are profoundly different, effects of therapeutic interventions may also differ, and we consider it desirable to specify the genetic defects in patients that are included in therapeutic trials.

2.6. Animal Models of PBD

Two mouse models of human ZS have been constructed by targeted disruption of *PEX5* (Baes, et al., 1997) or *PEX2* (Faust and Hatten, 1997). The pathological changes in the two mouse models are identical and bear a close resemblance to human ZS. The animals are hypotonic, hypoactive, feed poorly, and die in the neonatal period. Their biochemical abnormalities are similar to those in the human disease, and, most significantly, they also demonstrate the defect of neuronal migration that is characteristic of human ZS. These animal models provide a key opportunity to evaluate therapeutic interventions, particularly during fetal life. A study of the effects of maternal DHA supplementation during pregnancy has already been reported (Janssen, et al., 2000).

2.7. DHA Deficiency in Zellweger Syndrome

Martinez in Barcelona was the first to demonstrate a marked reduction of DHA levels in postmortem tissues of patients with ZS. The level in brain was reduced to 35% of control, and that in the liver and kidney to 10–20% of control (Martinez, 1990). DHA levels in plasma and red blood cells are also reduced (Moser, et al., 1999). The DHA deficiency is the result of a defect in the last step of DHA synthesis, which is catalyzed by an enzyme that is normally located in the peroxisome (Moore, et al., 1995) and fails to function normally in PBD patients. The fact that DHA levels in plasma and red blood cells can be increased by oral administration of this substance, combined with the observation that DHA in plasma is rapidly transferred into brain (Edmond, et al., 1998) led to the initiation of DHA therapy of PBD patients.

3. DHA THERAPY FOR PATIENTS WITH DISORDERS OF PEROXISOME BIOGENESIS

The rationale for DHA therapy in PBD is simple and, at first glance, compelling: DHA is important for brain and retinal function, it is deficient in PBD patients, and its level in blood and probably also in brain and retina can be increased by oral DHA administration. A serious limitation is that malformations in brain and other organs develop during fetal life and cannot be remedied by postnatal therapy. Thus, the rationale for therapy is based on the hypothesis that correction of DHA deficiency can remedy secondary ill effects that occur postnatally. Martinez has emphasized that in normal brain, DHA levels increase most rapidly during the perinatal period (Martinez, et al., 1974). Postnatally, brain DHA levels increase at a lower rate from the second month until the second year and level off thereafter (Martinez, 1992). It is postulated therefore that DHA therapy in PBD patients has the greatest chance of success if it is begun during the first 2 yr of life, and preferably shortly after birth. In this section, we will review current experience with DHA therapy in PBD patients and also in the mouse model of ZS. Normalization of DHA levels in plasma and red blood cells can be readily achieved. Martinez has also made the intriguing observation that DHA therapy may improve other peroxisomal functions. Although observational studies have reported clinical benefit particularly in patients who were treated in the early stages of the illness, evaluation of these effects is complicated by the great variability of the natural history. An argument for the desirability and feasibility of conducting a carefully designed controlled clinical trial will be presented.

4. RESULTS IN NONRANDOMIZED STUDIES

The most extensive published reports on DHA therapy of PBD patients are those of Martinez and associates (Martinez, et al., 1993; Martinez and Vazquez, 1998; Martinez, et al., 2000; Martinez, 2000).

Evaluations of nonrandomized trials of DHA in PBD disorders have also been performed with support of a grant from the European Concerted Action for Peroxisomal Leukodystrophy, BMH4-ct96-1621, by a study group chaired by Dr. M. Pineda.

4.1. Methods of DHA Administration and Dietary Regimen

Docasahexaenoic acid was administered orally either as the ethyl ester or as the triglyceride. The DHA ethyl ester was supplied by the Charleston Laboratory through a grant from the Fish Oil Test Materials Program (NIH/ADAMHA/DOC). This material is no longer available at this time. The triglyceride preparation is supplied by the Martek Bioscience Corporation, as a product referred to as DHASCO/ARASCO. The DHASCO supplies DHA, and the ARASCO provides arachidonic acid. The triglycerides contain 47% DHA and 46% arachidonic acid. They are administered as a microcapsule that can be sprinkled on food. The daily dosage of the ethyl ester varied between 100 and 500 mg, depending on the age and degree of DHA deficiency (Martinez, 2000). The initial dose of DHASCO/ARASCO was 100 mg/kg. Dosages are adjusted on the basis of plasma DHA and arachidonic acid levels, with the aim of maintaining these levels as close to normal as possible. Martinez does not recommend pharmacological supplementation with arachidonic acid, because the deficiency of arachidonic acid in PBD patients is less severe than that of DHA, and she has noted clinical regression in some patients who received arachidonic acid supplements (Martinez, 1996). She utilizes whole-milk formula enriched in DHA and arachidonic acid to supply the latter substance. Other groups have used the ARASCO formula to supply arachidonic acid, with the aim of normalizing plasma levels of this substance. A generally nutritious diet is provided, as well as supplements of vitamins A and D and also vitamin K.

4.2. Biochemical Effects

Normalization of DHA and ararchidonic acid levels in plasma and red blood cells was achieved in most patients. In addition, Martinez has reported two intriguing additional biochemical effects, namely an increase in the levels of plasmalogens in red blood cells and a reduction in the levels of very long-chain fatty acids in plasma. Abnormally low plasmalogen levels and abnormally high very long-chain fatty acid levels are hallmarks of impaired peroxisome function and are noted in most PBD patients. The partial normalization of the levels of these two substances would imply that DHA therapy has improved the general function of the peroxisome and would represent a valuable and unexpected bonus of DHA therapy. So far, we have not observed these changes in plasmalogen and very long-chain fatty acid levels in the PBD patients treated at the Kennedy Krieger Institute in whom plasma DHA levels had been normalized (Moser, unpublished observation). Studies in additional patients are in progress.

4.3. Clinical Effects and MRI Changes

Table 3 summarizes the data on DHA therapy in PBD patients that have been reported by Martinez and associates. Specific mention is in regard to patients 1, 2, 4, 6, 7, 14, 17, 18, and 20.

Table 3
Martinez DHA-Treated Patients

# and Dx	Age began	Death follow-up	Status at follow-up					Ref.
			General	Liver	Neurological	Eye	MRI	
1 N-I	2 mo	7 yr	Good	Normal	Improved		Normal	Martinez, 2000; Martinez and Vazquez, 1998
2 N-I	4 mo	2 yr	Good	Improved	Improved	Improved		Martinez, 2000
3 N-I	5 mo	4 yr						
4 ZS	6 mo	9 mo	Tone improved	Improved			Normal	Martinez et al., 2000
5 N-I	7 mo	4 yr						
6 N-I	7 mo	5 yr					Normal	Martinez, 2000
7 N-I	9 mo	5.7 yr			Improved	Improved	Normal	Martinez et al., 1996
8 N-I	9 mo	26 mo	Improved year 1					Martinez et al., 1995
9 ZS	1 yr	1.75 yr						
10 N-I	13 mo	7.1 yr						
11 N-I	14 mo	4.2 yr						
12 N-I	15 mo	6.25 yr						
13 N-I	15 mo	5.25 yr	Worse					
14 N-I	2 yr	2.67 yr						
15 N-I	3 yr	4.5 yr						
16 N-I	3 yr	3.1 yr						Martinez, 1993
17 N-I	5 yr	11 yr				Improved	Stable	Martinez and Vazquez, 1998
18 N-I	5 yr	9 yr						
19 N-I	5 yr	6.5 yr						
20 Mosaic Pex6	16 yr							Martinez, 1993; Martinez et al., 2000

Patient 1 had the NALD–IRD phenotype. DHA therapy was begun at 2 mo. At follow-up at 7 yr of age, he was found to walk normally, able to ride a bicycle, and communicate orally with elementary language, and brain magnetic resonance imaging (MRI) was normal (Martinez, et al., 2000).

Patient 2, also with the NALD–IRD phenotype, had severe steatorrhea prior to the initiation of therapy at 4 mo. The steatorrhea ceased 3 wk later; body weight, liver function, and muscle tone improved and he had good eye contact at last follow-up at age 2 yr.

Patient 4, who has the ZS phenotype, had severe liver disease and marasmus at 5 mo of age. Therapy was initiated at 6 mo; at follow-up at 9 mo of age, liver function, body weight, and muscle tone had improved.

Patient 7 appeared to be blind when she presented at 9 mo of age and was unresponsive to stimuli. After a year of therapy, she could see, sit unsupported, play, and spoke a few words (Martinez, 1996).

Patient 8 (NALD–IRD) had marked hypotonia and marasmus when therapy was started at 9 mo, and appeared blind and unresponsive. After 2 mo, the patient could follow

Table 4
Effect of DHA on Brain MRI Abnormalities in PBD Patients

	Age		MRI	
Diagnosis	Baseline	Follow-up	Baseline	Follow-up
ZS	8 mo	3 yr 8 mo	No cerebral myelin	Myelination age-appropriate
ZS	9 mo	3 yr 4 mo	Myelination equivalent to normal 4 month	Significant maturation of myelin
IRD	15 mo	3 yr	Occipital lobe and centrum ovale unmyelinated	Myelination age-appropriate
IRD	16 mo	26 mo	Frontal lobe myelin absent suggestion of demyelination I occipital lobe	More prominent myelination in frontal lobe and centrum semiovale
IRD	5 yr	7 yr	High signal intensity in central white contrast enhancement in periventricular region	Somewhat less demyelination, no contrast enhancement

objects, and after 5 mo, growth and motor control had improved. After 1 yr, she had acquired head control, smiled, played, and uttered several words (Martinez, 1995; Martinez, 1996). She died unexpectedly of septicemia at 26 mo of age (Martinez, et al., 2000).

Patient 16 had presented with West syndrome at 2 yr of age, with electroencephalographic changes typical of this disorder. Although seizures responded to clonazepam therapy, she was tetraplegic and totally unresponsive at 3 yr of age. She had repeated episodes of bronchopneumonia. DHA therapy was begun at 3 yr. She died 3 wk later during an episode of bronchopneumonia similar to those she had suffered prior to therapy.

Patient 17, with NALD–IRD, began therapy at 5 yr of age. She had psychomotor retardation, spasticity, ataxic gait, dysmetria, and poor vision. The MRI showed abnormally high signal intensity in the frontal and occipital regions, and there was bilateral signal enhancement in the frontal white matter (Table 4). Two years after initiation of therapy, vision was improved, the MRI appeared slightly improved, and contrast enhancement was no longer present.

Patient 20 has been shown to be mosaic for a *PEX6* defect (Moser, 1999). Cultured skin fibroblasts and liver biopsy samples have demonstrated a mixture of normal and abnormal cells (Pineda, et al., 1999). The patient is now 16 yr old. Developmental milestones were delayed at 2 yr. Liver biopsy showed micronodular cirrhosis. Neurologic exam at 3.5 yr showed nystagmus, head tremor, and generalized spasticity, and he was unable to walk. At 4 yr of age, he was started on a diet containing medium-chain triglycerides, pure olive oil (20% of daily fat intake), and vitamin A supplementation. This coincided with clinical improvement as evidence by the disappearance of abnormal eye movements and head tremor and reduction of spasticity. This improvement was attributed to a reduction in phytanic acid levels. Visual evoked responses improved between 4 and 6 yr of age. DHA therapy was added to the regimen at age 6.7 yr. Further improvement of the visual evoked responses occurred. At 12 yr, motor and mental improvements have been maintained. He is able to stand with support. His IQ is 60 and he attends a special school.

Table 4 lists the changes in brain magnetic resonance imaging (MRI) studies that have been observed during DHA therapy. Four of the patients whose ages ranged from

8–16 mo when therapy was begun showed increased deposition in myelin after 10–36 mo of therapy. Although these results are encouraging, it must be noted that "catch-up myelination" may occur spontaneously (van der Knaap & Valk, 1995). Of particular interest is a 5-yr-old patient (patient 17 in Table 3) who showed evidence of active demyelination (periventricular contrast enhancement), which was no longer present after 2 yr of therapy.

The clinical and neuroimaging data presented by Martinez and her collaborators show a considerable number of instances in which highly encouraging improvement was associated temporally with the administration of DHA. A cause-and-effect relationship may well exist, but it is not yet established. The natural history of PBD shows a wide range of variation (Moser, et al., 1995), and apparently spontaneous improvement in neurologic and liver function has been observed (Moser, unpublished observation). Patient 17 improved prior to the administration of DHA. Finally, DHA was not administered in isolation, it was combined with a general improvement in nutrition and, in some instances, also with other rehabilitative approaches and general stimulation, and this makes it difficult to isolate the effects of DHA administration alone. The effects of DHA therapy have also been evaluated by a study group assembled as part of the European Concerted Action for Peroxisomal Leukodystrophy. This group evaluated the results of DHA therapy in 33 multicenter observational studies of DHA therapy in 33 PBD patients. The study group concluded that nutrition improved in 6 and remained unchanged in 24. Hepatic function improved in 12 and remained unchanged in 19. Slight improvements in visual function were observed in 12, improved greatly in 2, and worsened in 5. Neurologic function was judged to have been improved in nine and to have worsened in seven. This information, which has not yet been published, was provided by Professor Frank Roels, Principal Investigator of this European collaborative study.

5. DESIGN OF A RANDOMIZED DOUBLE-BLINDED PLACEBO-CONTROLLED STUDY OF DHA THERAPY

It is our view that although the above-presented nonrandomized studies do suggest that DHA therapy can be beneficial, the results cannot be considered conclusive. We also believe that it is unlikely that additional nonrandomized studies have the capacity to resolve this question. Therefore, we have initiated a randomized prospective study that will involve 60 PBD patients. Through a fixed randomization schedule, half of the patients will be assigned to receive DHA and half will receive a placebo. The treatment assignment will be masked from the patients and their families and all evaluators who have direct patient contact. Evaluations will be performed at baseline and at 12 mo after initiation of therapy. Emphasis will be placed on the evaluation of neurological, neuropsychological, and visual functions. The neuropsychological test batteries will utilize those designed by Dr. Elsa Shapiro for the evaluation of patients with leukodystrophies or lysosomal disorders (Shapiro & Klein, 1994). The ocular evaluations are performed in collaboration with the Pediatric Ophthalmology staff at the Wilmer Eye Institute. They include assessment of visual acuity, visual fields, pupillary function, extraocular motility and muscle balance, evaluation of the media, intraocular tension and assessment of the retina, electroretinography, and flash visual evoked potentials. Fundus photography is done with a hand-held camera under adequate sedation with chloral hydrate. We have found that careful preparation has permitted performance of both the neuropsychological and ophthalmological examinations.

The code will be broken and data will be analyzed when the 12-mo follow-up has been completed for all patients. At that time, DHA therapy will be offered to all patients who desire to receive it. Clinical and laboratory studies will be monitored by an independent data and safety monitoring committee. This committee will also perform an interim analysis when 30 patients have completed assessment and has permission to terminate the study if there are significant favorable findings or unanticipated adverse outcomes. A determination of statistical power indicates that this sample size is sufficient to detect a major effect (0.862 standard deviations). With a more modest effect of 25%, however, the statistical power diminishes to 60%. The study has been approved by the Institutional Review Board at the Johns Hopkins Medical Institutions and is supported by a research grant from the Office of Orphan Products Development at the Food and Drug Administration. Fifty-two patients are enrolled in the study at this time. The code has not yet been broken. No adverse effects attributable to the medication have been observed.

6. STUDIES IN EXPERIMENTAL ANIMALS

Jannsen et al. (Janssen, et al., 2000) have conducted studies in the *PEX5 –/–* mouse model of human ZS. These animals show clinical and neuropathological abnormalities similar to those in the human disease, including the characteristic defect in neuronal migration. At birth, levels of DHA in brain were 40% less than in control littermates, but DHA levels in the liver were normal. Because the *PEX5*-deficient animals are incapable of synthesizing this substance, all of their DHA must have been supplied by the mother. The maternal supply appears sufficient to maintain normal DHA concentrations in the liver, but insufficient to maintain levels in the fetal brain. We have obtained similar results of our studies in the *PEX2 –/–* mouse model of human ZS (Su, et al., 2000). The demonstration of the fetal brain DHA deficit provides the opportunity to determine whether DHA supplementation can ameliorate the neurological deficits. To test this hypothesis the diet of the mothers' of *PEX5-1*-mutants (Janssen, et al., 2000) was supplemented with 10 mg of the DHA ethyl ester daily from embryonic d 10.5 until embryonic d 18.5 or birth. Although this therapy normalized fetal brain DHA levels, it did not alter neurological function or brain pathology. This finding suggests that the DHA deficit is not the cause of the neurological defects that are already present at birth, but does not rule out the possibility that earlier initiation of maternal therapy might be of benefit. It was not possible to evaluate the effects of postnatal DHA therapy, because the *PEX5 –/–* pups do not survive after birth. It will be possible to examine this question in a recently developed *PEX2 –/–* model that is capable of survival until postnatal day (Faust, et al., 2000). Unlike the findings reported by Martinez in human ZS patients (Martinez, 1995), DHA supplementation in the *PEX5 –/–* mouse did not lower the levels of very long-chain fatty acids and led to only a minimal increase in the levels of plasmalogens.

6. CONCLUDING REMARKS

The rationale for DHA therapy in the disorders of peroxisome biogenesis is compelling at first glance: DHA is essential for normal function of brain and retina; PBD patients cannot synthesize DHA and, thus, have a deficiency of this substance; oral administration of DHA can normalize levels of DHA in plasma and red cells and probably also in brain and retina; DHA therapy is simple and not excessively expensive and it appears to be free of significant side effects. Finally, the nonrandomized studies that have been published and have been cited in detail in this chapter suggest that DHA therapy has been of benefit. Thus, the argument can be made that this is the "end of the story" and that, in the future, every PBD patient should receive DHA as part of routine clinical management.

Although the rationale for DHA therapy in PBD patients is appealing, we recommend, nevertheless, that clinical effectiveness should be evaluated in appropriately designed randomized clinical trials before it is offered as standard clinical care. This recommendation is based on the following reasons.

1. All of the published reports are based on nonrandomized observational studies that at this time do not permit definite conclusions about clinical benefit. Although several of the reports show encouraging and highly appealing clinical progress in association with DHA therapy, interpretation is confounded by the great variability of the natural history of PBD and the fact that initiation of DHA therapy was combined (appropriately) with other rehabilitative measures. Furthermore, preliminary analysis of the evaluation of DHA therapy in 33 PBD patients which is being prepared by the European Concerted Action for Peroxisomal Leukodystrophy Consortium and has been cited in a previous section of this chapter does not provide *prima facie* evidence of clinical benefit. Although this study also has the limitation that it was observational rather than randomized, it does provide systematic multidisciplinary appraisals in a substantial number of patients. We recommend that conclusions about the results of nonrandomized studies be deferred until the analysis of this study has been completed.
2. The argument has been made that randomized studies of DHA therapy in PBD patients are unethical and that families would not agree to participate in such trials. This is a complex and emotionally charged issue. Reluctance to undertake a randomized trial of DHA therapy is based on the seriousness of the disease process, the convenience and safety of DHA administration, and the appealing rationale for its use. It is our view that these concerns are outweighed by the following considerations: (a) Even though DHA therapy has now been in use for 10 yr, its clinical effectiveness is still uncertain. (b) Randomized placebo-controlled studies provide the most rapid approach for the appraisal of clinical effectiveness. (c) Randomized trials can be designed without compromising the best interests of the patient and the family. Key to this is the supervision by an independent Data and Safety Monitoring Committee, which carries out interim analyses and has the power to terminate the study if there are significant favorable findings or unanticipated adverse outcomes. (d) Our experience with the double-blinded placebo-controlled trial of DHA therapy in PBD patients, which is now in progress at the Kennedy Krieger Institute, indicates that most families are willing to participate in a randomized study. Fifty-two patients are enrolled at this time.
3. Definitive information about the clinical effectiveness of DHA is important both for the patients and their families and for the scientific community. Provision of an unproven or ineffective therapy may provide a false sense of security and retard the introduction of new approaches and thus be counter to the patients' best interests. Furthermore, there still are unresolved disputes about the optimal modes of DHA therapy. Which DHA preparation is most effective? Should arachidonic acid supplementation be provided along with DHA? Although at this time the titration of the DHA dosage is based on total levels of this substance in plasma and red blood cells, it is possible that follow-up of the levels of this fatty acid in specific lipid moieties might be more informative. Would it be effective and safe to increase the plasma DHA levels to higher than normal? Carefully designed clinical trials are required to answer these questions.

Chapter 15 / DHA Therapy for Peroxisome Biogenesis Disorders

The question of whether DHA therapy in PBD patients improves neurological and retinal function is of critical interest to the scientific community and will influence further research directions. A point of major interest is the evaluation of the evidence whether DHA therapy leads to a general improvement in peroxisomal function, as assessed by levels of very long-chain fatty acids and plasmalogens and other measures in body fluids and accessible tissues. Prospective randomized studies are needed to assess this. Further studies in PBD animal models are required to determine whether DHA supplementation is beneficial, and they may provide leads for intrauterine therapy.

4. Evaluations of DHA therapy and other therapies that may be developed in the future will be aided by more precise definition of the genotype and phenotype of PBD patients who participate in these studies. Patients with *PEX1* deficiency who have the G843D mutation may prove of particular value. More than half of all PBD patients have *PEX1* deficiency (Gartner, et al., 1999) and 30% of patients with this gene defect have the G843D mutation. This mutation is associated with a relatively mild and homogeneous phenotype (Gartner, et al., 1999; Collins and Gould, 1999). The relatively limited range of phenotypic expression in patients with this genotype suggests that they would be particularly suitable for the evaluation of therapeutic interventions.

ACKNOWLEDGMENTS

This work was supported by grants HD 10891 and RR00052 from the National Institutes of Health and FDR 001289 from the Food and Drug Administration.

REFERENCES

Baes M, Gressens P, Baumgart E, Carmeliet P, Casteels M, Fransen M, et al. A mouse model for Zellweger syndrome. Nature Genet 1997; 17(1):49–57.

Bowen P, Lee CSN, Zellweger H, Lindenberg R. A familial syndrome of multiple cogenital defects. Bull Johns Hopkins Hosp 1964; 114:402–414.

Braverman N, Steel G, Obie C, Moser A, Moser H, Gould SJ, et al. Human PEX7 encodes the peroxisomal PTS2 receptor and is responsible for rhizomelic chondrodysplasia punctata. Nature Genet 1997; 15(4):369–376.

Brul S, Westerveld A, Strijland A, Wanders RJ, Schram AW, Heymans HS, et al. Genetic heterogeneity in the cerebrohepatorenal (Zellweger) syndrome and other inherited disorders with a generalized impairment of peroxisomal functions. A study using complementation analysis. J Clin Invest 1988; 81(6):1710–1715.

Collins CS, Gould SJ. Identification of a common PEX1 mutation in Zellweger syndrome. Hum Mutat 1999; 14(1):45–53.

De Duve C, Baudhuin P. Peroxisomes (microbodies and related particles). Physiol Rev 1966; 46(2):323–357.

Distel B, Erdmann R, Gould SJ, Blobel G, Crane DI, Cregg JM, et al. A unified nomenclature for peroxisome biogenesis factors. J Cell Biol 1996; 135(1):1–3.

Edmond J, Higa TA, Korsak RA, Bergner EA, Lee WN. Fatty acid transport and utilization for the developing brain. J Neurochem 1998; 70(3):1227–1234.

Evrard P, Caviness VS Jr, Prats-Vinas J, Lyon G. The mechanism of arrest of neuronal migration in the Zellweger malformation: an hypothesis bases upon cytoarchitectonic analysis. Acta Neuropathol (Berlin) 1978; 41(2):109–117.

Faust P, Su H-M, Moser A, Moser H. The peroxisome deficient PEX2 Zellweger mouse. J Mol Neurosci, 2000, in press.

Faust PL, Hatten ME. Targeted deletion of the PEX2 peroxisome assembly gene in mice provides a model for Zellweger syndrome, a human neuronal migration disorder. J Cell Biol 1997; 139(5):1293–1305.

Gartner J, Preuss N, Brosius U, Biermanns M. Mutations in PEX1 in peroxisome biogenesis disorders: G843D and a mild clinical phenotype. J Inherited Metab Dis 1999; 22(3):311–313.

Gould SJ, Keller GA, Hosken N, Wilkinson J, Subramani S. A conserved tripeptide sorts proteins to peroxisomes. J Cell Biol 1989; 108(5):1657–1664.

Janssen A, Baes M, Gressens P, Mannaerts GP, Declercq P, Van Veldhoven PP. Docosahexaenoic acid deficit is not a major pathogenic factor in peroxisome-deficient mice. Lab Invest 2000; 80(1):31–35.

Kelley RI, Datta NS, Dobyns WB, Hajra AK, Moser AB, Noetzel MJ, et al. Neonatal adrenoleukodystrophy: new cases, biochemical studies, and differentiation from Zellweger and related peroxisomal polydystrophy syndromes. Am J Med Genet 1986; 23(4):869–901.

Lazarow PB, Moser HW. Disorders of peroxisomal biogenesis. In: Scriver CR, Beaudel AL, Sly WS, Valle D, eds. The Metabolic and Molecular Basis of Inherited Disease. 7th ed. Chapter 71: McGraw-Hill, New York, 1994, Vol. II, pp. 2287–2324.

Martinez M. Severe deficiency of docosahexaenoic acid in peroxisomal disorders: a defect of delta 4 desaturation? Neurology 1990; 40(8):1292–1298.

Martinez M. Tissue levels of polyunsaturated fatty acids during early human development. J Pediatr 1992; 120(4 Pt 2):S129–S138.

Martinez M. Polyunsaturated fatty acids in the developing human brain, erythrocytes and plasma in peroxisomal disease: therapeutic implications. J Inherited Metab Dis 1995; 18(Suppl 1):61–75.

Martinez M. Docosahexaenoic acid therapy in docosahexaenoic acid-deficient patients with disorders of peroxisomal biogenesis. Lipids 1996; 31(Suppl):S145–S152.

Martinez M. Restoring the DHA levels in the brains of Zellweger patients. J Mol Neurosci, 2001, in press.

Martinez M, Conde C, Ballabriga A. Some chemical aspects of human brain development. II. Phosphoglyceride fatty acids. Pediatric Research 1974; 8(2):93–102.

Martinez M, Pineda M, Vidal R, Conill J, Martin B. Docosahexaenoic acid—a new therapeutic approach to peroxisomal-disorder patients: experience with two cases. Neurology 1993; 43(7):1389–1397.

Martinez M, Vazquez E. MRI evidence that docosahexaenoic acid ethyl ester improves myelination in generalized peroxisomal disorders. Neurology 1998; 51(1):26–32.

Martinez M, Vazquez E, Garcia-Silva MT, Manzanares J, Bertran JM, Castello F, et al. Therapeutic effects of docosahexaenoic acid ethyl ester in patients with generalized peroxisomal disorders. Am J Clin Nutr 2000; 71(1 Suppl):376S–385S.

Moore SA, Hurt E, Yoder E, Sprecher H, Spector AA. Docosahexaenoic acid synthesis in human skin fibroblasts involves peroxisomal retroconversion of tetracosahexaenoic acid. J Lipid Res 1995; 36(11):2433–2443.

Moser AB, Jones DS, Raymond GV, Moser HW. Plasma and red blood cell fatty acids in peroxisomal disorders. Neurochem Res 1999; 24(2):187–197.

Moser AB, Rasmussen M, Naidu S, Watkins PA, McGuinness M, Hajra AK, et al. Phenotype of patients with peroxisomal disorders subdivided into sixteen complementation groups. J Pediatr 1995; 127(1):13–22.

Moser HW. Genotype–phenotype correlations in disorders of peroxisome biogenesis. Mol Genet Metab 1999; 68(2):316–327.

Neuringer M, Anderson GJ, Connor WE. The essentiality of n-3 fatty acids for the development and function of the retina and brain. Ann Rev Nutr 1988; 8:517–541.

Pineda M, Giros M, Roels F, Espeel M, Ruiz M, Moser A, et al. Diagnosis and follow-up of a case of peroxisomal disorder with peroxisomal mosaicism. J Child Neurol 1999; 14(7):434–439.

Roels F, Tytgat T, Beken S, Giros M, Espeel M, De Prest B, et al. Peroxisome mosaics in the liver of patients and the regulation of peroxisome expression in rat hepatocyte cultures. Ann NY Acad Sci 1996; 804:502–515.

Scotto JM, Hadchouel M, Odievre M, Laudat MH, Saudubray JM, Dulac O, et al. Infantile phytanic acid storage disease, a possible variant of Refsum's disease: three cases, including ultrastructural studies of the liver. J Inherited Metab Dis 1982; 5(2):83–90.

Shapiro EG, Klein KA. Dementia in childhood: issues in neuropsychological assessment with application to the natural history and treatment of degenerative storage diseases. In: Tramontana MG, Hooer SRE, eds. Advances in Child Neuropsychology. Springer Verlag, New York, 1994, pp. 119–191.

Spranger JW, Opitz JM, Bidder U. Heterogeneity of Chondrodysplasia punctata. Humangenetik 1971; 11(3):190–212.

Su H-M, Moser AB, Moser HW, Faust PL. Low docosahexaenoic acid in brain but not in plasma and red blood cell in newborn Zellweger syndrome mouse. J Neurochem 2000; 74(Suppl):S86.

Subramani S. Components involved in peroxisome import, biogenesis, proliferation, turnover, and movement. Physiol Rev 1998; 78(1):171–188.

Swinkels BW, Gould SJ, Bodnar AG, Rachubinski RA, Subramani S. A novel, cleavable peroxisomal targeting signal at the amino-terminus of the rat 3-ketoacyl-CoA thiolase. EMBO J 1991; 10(11):3255–3262.

Ulrich J, Herschkowitz N, Heitz P, Sigrist T, Baerlocher P. Adrenoleukodystrophy. Preliminary report of a connatal case. Light- and electron microscopical, immunohistochemical and biochemical findings. Acta Neuropathol (Berlin) 1978; 43(1–2):77–83.

van den Bosch H, Schutgens RB, Wanders RJ, Tager JM. Biochemistry of peroxisomes. Ann Rev Biochem 1992; 61:157–197.

van der Knaap MS, Valk J. Magnetic Resonance of Myelin, Myelination and Myelin Disorders. 2nd ed. Springer-Verlag, Berlin, 1995, p. 38.

van Grunsven EG, van Berkel E, Ijlst L, Vreken P, de Klerk JB, Adamski J, et al. Peroxisomal D-hydroxyacyl-CoA dehydrogenase deficiency: resolution of the enzyme defect and its molecular basis in bifunctional protein deficiency. Proc Natl Acad Sci USA 1998; 95(5):2128–2133.

Voss A, Reinhart M, Sankarappa S, Sprecher H. The metabolism of 7,10,13,16,19-docosapentaenoic acid to 4,7,10,13,16,19-docosahexaenoic acid in rat liver is independent of a 4-desaturase. J Biol Chem 1991; 266(30):19,995–20,000.

Wanders RJ, Dekker C, Hovarth VA, Schutgens RB, Tager JM, Van Laer P, et al. Human alkyldihydroxy-acetonephosphate synthase deficiency: a new peroxisomal disorder. J Inherited Metab Dis 1994; 17(3):315–318.

Wanders RJ, Schumacher H, Heikoop J, Schutgens RB, Tager JM. Human dihydroxyacetonephosphate acyltransferase deficiency: a new peroxisomal disorder. J Inherited Metab Dis 1992; 15(3):389–391.

Wilson GN, Holmes RG, Custer J, Lipkowitz JL, Stover J, Datta N, Hajra A. Zellweger syndrome: diagnostic assays, syndrome delineation, and potential therapy. Am J Med Genet 1986; 24(1):69–82.

Zellweger H. The cerebro-hepato-renal (Zellweger) syndrome and other peroxisomal disorders. Dev Med Child Neurol 1987; 29(6):821–829.

16 Effects of Fatty Acids and Ketones on Neuronal Excitability
Implications for Epilepsy and Its Treatment

Carl E. Stafstrom

1. INTRODUCTION: FATTY ACIDS AND NEURONAL EXCITABILITY

The critical role of fatty acids in nervous system development have been emphasized by multiple authors in this volume and in numerous previous studies (for reviews, *see* Innis, 1991; Neuringer et al., 1994; Uauy et al., 1996; Kurlak & Stephenson, 1999; Gibson, 1999). Essential fatty acids, especially long-chain polyunsaturated fatty acids (PUFAs), are necessary for development of normal retinal and neuronal membranes (Neuringer, et al., 1988) and subsequent normal behavior and cognition (Enslen, et al., 1991; Carrie et al., 1999; McGahon et al., 1999). The importance of PUFAs for normal retinal and brain development has even led to the suggestion that infant formulas should be supplemented to achieve PUFA levels comparable to those in human breast milk, although the need for such supplementation remains controversial (Crawford, 1993; Raiten, et al., 1998; Lucas et al., 1999).

In addition to their role in brain development, fatty acids also exert important modulatory effects on neuronal excitability (Ordway, et al., 1991) and receptor-mediated signaling pathways (Hwang & Rhee, 1999). Fatty acids can either increase or decrease the firing of neurons, modulate neurotransmitter release, or alter synaptic responses (Meves, 1994; Bazan et al., 1996; McGahon et al., 1999; Leaf et al., 1999b). PUFAs have been shown to reduce neuronal sodium and calcium currents (Vreugdenhil et al., 1996) and inhibit or activate potassium channels (Poling et al., 1996; Keros & McBain, 1997; Horimoto et al., 1997). The PUFA docosahexaenoic acid (DHA) facilitates excitatory synaptic transmission mediated by *N*-methyl-D-aspartate (NMDA)-type glutamate receptors (Nishikawa et al., 1994). DHA also alters inhibitory neurotransmission in some neuron types (e.g., substantia nigra) by reducing responses to the inhibitory transmitter γ-amino-butyric acid (GABA) (Hamano et al., 1996). It is likely that PUFA effects are regionally specific and may also vary according to developmental stage. Fatty acids and their derivatives may also attenuate the response to excitotoxic injury in the neonatal period (Valencia et al., 1998).

Many of the effects of PUFAs on physiological excitability were originally shown in elegant studies on cardiac muscle cells (Kang et al., 1995). In cardiac myocytes, long-chain

From: *Fatty Acids: Physiological and Behavioral Functions*
Edited by: D. Mostofsky, S. Yehuda, and N. Salem Jr. © Humana Press Inc., Totowa, NJ

PUFAs inhibit voltage-dependent sodium channels and L-type calcium channels, thereby stabilizing the membrane, resulting in an antiarrhythmic action (Leaf et al., 1999a). Based on PUFA effectiveness in reducing excitability in the cardiac system, as well as their widespread actions on neurons, the idea arose that perhaps PUFAs could also suppress hyperexcitable neurons (i.e., those prone to epileptic firing). This notion had some precedent, as Yehuda and colleagues had already shown that a mixture of n-3 and n-6 essential fatty acids, in a specific ratio, raised seizure threshold in several experimental models (Yehuda et al., 1994). It was subsequently shown that long-chain PUFAs produce a transient elevation of seizure threshold in a cortical stimulation model (Voskuyl et al., 1998). Similar protective effects of the PUFAs DHA and eicosapentaenoic acid (EPA) against seizure activity were also described in hippocampal slices in vitro (Xiao & Li, 1999). Both compounds hyperpolarized hippocampal CA1 neurons, raising the threshold for action potential generation, and reduced baseline and convulsant-induced firing rates. Given this extensive evidence for a neuromodulatory role of fatty acids on cortical hyperexcitability, it is tempting to speculate that this class of compound might be utilized in clinical situations in which hyperexcitable neurons fire in abnormal, synchronous patterns (i.e., epilepsy). It is interesting that valproic acid, a commonly used broad spectrum antiepileptic drug, is, itself, a short-chain fatty acid.

The prospect that epilepsy might be controlled, at least partially, by dietary intake is radical but highly appealing. Although clinical dietary trials of PUFAs on reducing disorders of cardiac excitability have been promising, there have been no analogous attempts to use PUFAs for seizure control, either in patients or animal models. However, another dietary protocol already exists that is used to control intractable seizures—the ketogenic diet (KD). This high-fat, low-carbohydrate, low-protein formulation has been used for over 80 yr for seizures that are refractory to standard medications. Use of the KD declined as anticonvulsants in convenient pill form became readily available. However, since the early 1990s, there has been a resurgence in interest in the KD, mandating that its mechanism of action be addressed, with the ultimate goal of optimizing the diet's composition and effectiveness. The remainder of this chapter will discuss the KD in more detail, with emphasis on the biochemical alterations induced by the diet and possible mechanisms by which it reduces neuronal excitability. The possible role of fatty acids, especially those of the polyunsaturated type, in the KD mechanism will also be considered.

2. KETONES AND THE KETOGENIC DIET

The KD was formulated as an epilepsy treatment around 1921 (Wilder, 1921). The diet was designed to simulate the biochemical changes of fasting, which was long recognized as an efficacious, short-term way to reduce seizures in persons with epilepsy (Geyelin, 1921). The diet consists of a high percentage of fat, adequate protein for growth, and minimal carbohydrate, constituted in a ratio of 3–4 g of fat for every gram of combined protein and carbohydrate (hence the notations 3 : 1 or 4 : 1 for the classic KDs*). Fat,

---

*Ketogenic diets are described by the ratio of foods with "ketogenic potential" (K) versus "antiketogenic potential" (AK), approximated by the formula: $K/AK = [F + 0.5P]/[C + 0.1F + 0.6P]$, where F, P, and C are amounts, *by weight*, of fat, protein, and carbohydrate. Practically, KDs are usually denoted as 3 : 1 or 4 : 1. K/AK > 1.5 : 1 is considered necessary to produce ketosis. The equation adapted from Withrow, 1980.

provided mainly in the forms of heavy cream and butter, contributes about 90% of the calories. The fats in the classic KD consist of a mixture of animal and plant-derived fats; fatty acids of varying chain lengths are likely to be included, but no attempt is made to specify fat type or chain length. Therefore, the classic KD likely contains an abundance of saturated fatty acids and, possibly, a relative dearth of long-chain PUFAs; as discussed, this is of interest because of the roles of PUFAs in brain development and modulation of excitability. There have been attempts to formulate a KD using medium-chain triglyceride (MCT)-based oils; these provide a greater degree of ketosis per calorie and, hence, a higher level of ketosis with less fat in the diet (about 60% calories as fat) (Huttenlocher et al., 1971; Schwartz et al., 1989). By allowing a greater range of food choices and less restrictive fat content, it was thought that the MCT-based KD would be more palatable and tolerable for patients. Unfortunately, MCT ketogenic diets are associated with a high incidence of diarrhea and abdominal pain and are not usually used today. Formulation of KDs with a higher PUFA content might improve its effectiveness and should be a high priority for future research.

All constituents of the KD need to be measured precisely to provide the same ratio in each meal. Calories are restricted to about 75–90% of the recommended daily allowance for age. All sources of carbohydrate must be accounted for and included in the daily allotment, including calories from medications (carbohydrate-based carriers) and even toothpaste. Appropriate vitamins and minerals are added. Some protocols include restriction of fluid intake to about 60 cm^3/kg/d (Freeman et al., 1994). The goal is to produce and maintain ketosis, and even minor deviations from the proscribed regimen terminates the ketosis and the antiseizure effect of the KD (Huttenlocher, 1976). Practical details regarding clinical use of the KD may be found in Freeman et al. (1994).

The KD is certainly a radical departure from a typical dietary regimen; it must not be regarded as a fad diet but rather as a strictly controlled medical therapy. The diet is begun in the hospital. Children are ordinarily fasted for 24–36 h, followed by gradual increments in ketogenic foods over the next 3 d. This protocol allows ketosis to develop gradually, initiated by the fast and later maintained by the diet. Precautions against hypoglycemia and other acute medical complications are taken. Once a child is stable on the KD, the goal is to maintain the diet for 2 yr. The child's medical and nutritional status is monitored closely. Because the KD is a chronic therapy, justified concerns may be raised about the effects of this unusual dietary regimen on growth, metabolism, arterial atherogenesis, and renal function (Freeman et al., 1994; Couch et al., 1999). In most cases, acute side effects can be managed, although there is a 5–8% incidence of renal stones (Herzberg et al., 1990) and some serious liver and metabolic complications have been reported, especially in children on concomitant valproic acid therapy (Ballaban-Gil et al., 1998). Possible interactions with other anticonvulsant drugs need further investigation. Most other long-term changes, such as elevation of plasma lipids, resolve once the diet is discontinued (Swink et al., 1997).

Obviously, the KD is complex to formulate, administer, and monitor, demanding a high level of dedication by family members, nutritionists, and medical providers. Yet, its effectiveness in many children and adults in whom medications have afforded poor seizure control make the KD an alternative worth pursuing. The KD has proven to be a valuable adjunct to the armamentarium of anticonvulsant medications, providing excellent control in one-third to one-half of children, moderate effectiveness in at least another third, and no benefit in the remainder (Freeman et al., 1998a; Vining, 1999). Although

Table 1
Animal Models of the Ketogenic Diet: Observations and Clinical Correlates

	Observation in animal models	Clinical correlate
Age relationship	Younger animals respond better to KD	Children extract and utilize ketones from blood better than older individuals
Diet type	Classic and MCT diets both increase seizure threshold	In patients, classic and MCT KDs are equally efficacious, but the MCT has and provide seizure protection more gastrointestinal side effects effects, limiting its clinical usefulness
Latency to KD effectiveness	There is a latency of several days before a KD effect is seen in animals	Seizures may decrease during the pre-KD fast or after a latency of days to weeks
Ketosis	There is an *association* between ketosis and seizure control, but it is not clear if this is *causal* (i.e., "Ketosis is necessary but not sufficient")	There is an *association* between ketosis and seizure control, but it is not clear if this is *causal* (i.e., "Ketosis is necessary but not sufficient")
Reversal of protective effect when KD is discontinued	Rapid	Children who lose ketosis have a rapid return (within h) of seizures and EEG abnormalities
Seizure type	KD is effective in models employing a wide variety of seizure paradigms	KD is effective in a wide variety of seizure types and epilepsy syndromes

Abbreviations: KD, ketogenic diet; MCT, medium chain triglyceride; EEG, electroencephalogram

these numbers are approximate, they have held up over the 80 yr of the diet's use (Helmholz & Keith, 1930; Kinsman et al., 1992); that is, the KD is about as efficacious today as it was in the early 20th century. The beneficial response to the KD compares favorably to the newer antiepileptic drugs (Walker & Sander, 1996). The KD seems to work best in children, possibly as a result of a greater capacity of the brain to extract and utilize ketones in early life (Hawkins et al., 1971; Kraus et al., 1974). However, carefully selected adults also seem to derive benefit from the KD (Sirven et al., 1999).

Many aspects of KD therapy are based more on lore than on scientific fact. For example, the diet duration, formulation, need for initial fast and fluid restriction, and optimal patient selection (age, seizure type, seizure etiology) have not been subject to scientific scrutiny. Many of these factors can be addressed using animal models, but surprisingly few experimental studies on the KD have been carried out (Stafstrom, 1999). Earlier studies varied so much in their experimental designs (e.g., species, age, diet formulation and administration, method of seizure induction, and monitoring) that comparisons are difficult. Fortunately, the resurgence of interest in KD therapy over the past 5 yr has spawned renewed efforts to investigate the diet in experimental models (Hori et al., 1997; Bough & Eagles, 1999; Rho et al., 1999a; Thavendiranathan et al., 2000). Based on animal models, several experimental–clinical correlations can be drawn (Table 1).

3. BIOCHEMICAL ASPECTS OF KETOSIS

The KD was originally formulated to mimic the effects of fasting, as it had been known since biblical times that fasting had a beneficial effect on epilepsy (Geyelin, 1921; Wheless, 1995; Swink et al., 1997). Ever since, it has been assumed that fasting and the KD share a common mechanism of action, although this assumption has not been proven.

Dietary fats and fatty acids are ordinarily broken down by liver mitochondrial β-oxidation into two-carbon acetyl-CoA molecules. Acetyl-CoA is ordinarily funneled into the tricarboxylic acid (TCA) cycle with eventual production of energy via ATP generation, which is used for biosynthesis and other cellular functions. Ketone production under such normal circumstances is minimal. However, under conditions of fasting, starvation, or provision of high amounts of fat but little carbohydrate (i.e., the ketogenic diet), ketone production is accelerated. Acetyl-CoA cannot enter the TCA cycle because of low availability of key intermediates, the rate-limiting substrate oxaloacetate, and the rate-limiting enzyme α-ketoglutarate, which are diverted to produce glucose via gluconeogenesis. Instead, acetyl-CoA molecules are used to synthesize the four-carbon ketone bodies: β-hydroxybutyrate (β-OHB) and acetoacetate (AcAc). Therefore, both fasting and the KD produce ketosis (i.e., elevated blood levels of β-OHB and AcAc) (McGarry & Foster, 1980). The liver lacks the enzymes necessary to metabolize these compounds, so β-OHB and AcAc are exported to other body tissues for use in energy production (especially muscle and brain). Ketone bodies enter the cerebral circulation and cross the blood-brain barrier (Janigro, 1999), eventually entering neurons via monocarboxylic acid transporters (Moore et al., 1976; Pellerin et al., 1998).

Ordinarily, the brain is an obligate user of glucose as its energy source, and a minimal arterial–venous difference exists for ketone bodies. However, during ketosis, there is a metabolic shift whereby the brain oxidizes ketones derived from fats as its primary fuel source (Owen et al., 1967; Sokoloff, 1973; DeVivo et al., 1978). Brain mitochondria possess enzymes capable of breaking down β-OHB and AcAc, and, eventually, acetyl-CoA is produced that can provide energy via the TCA cycle (Sankar & Sotero de Menezes, 1999). It has been assumed that the brain itself is not ketogenic; that is, it does not possess the synthetic machinery to produce ketones. However, a recent study showed that early in development (before weaning in rodents), the brain does express ketogenic enzymes (Cullingford et al., 1998). Such established and novel metabolic pathways of ketone formation and utilization need to be studied in much greater detail in animal models. Potential mechanisms by which the KD exerts its unique anticonvulsant effect must also be explored experimentally (Schwartzkroin, 1999).

4. ANTICONVULSANT MECHANISMS OF THE KETOGENIC DIET

4.1. Mechanistic Criteria

Understanding how the KD protects against seizures would allow the design of a more effective diet or improve and simplify the current regimen. Animal models offer many advantages in exploring possible mechanisms: Diet components can be controlled precisely, including the ketogenic ratio, constituents, caloric intake, and fluid balance; biochemical and neurologic abnormalities can be quantitated; seizure type can be controlled; the effects of the experimental diet on seizure susceptibility, latency, and duration can be assessed; effects on cognition can be studied; correlations can be made at the whole

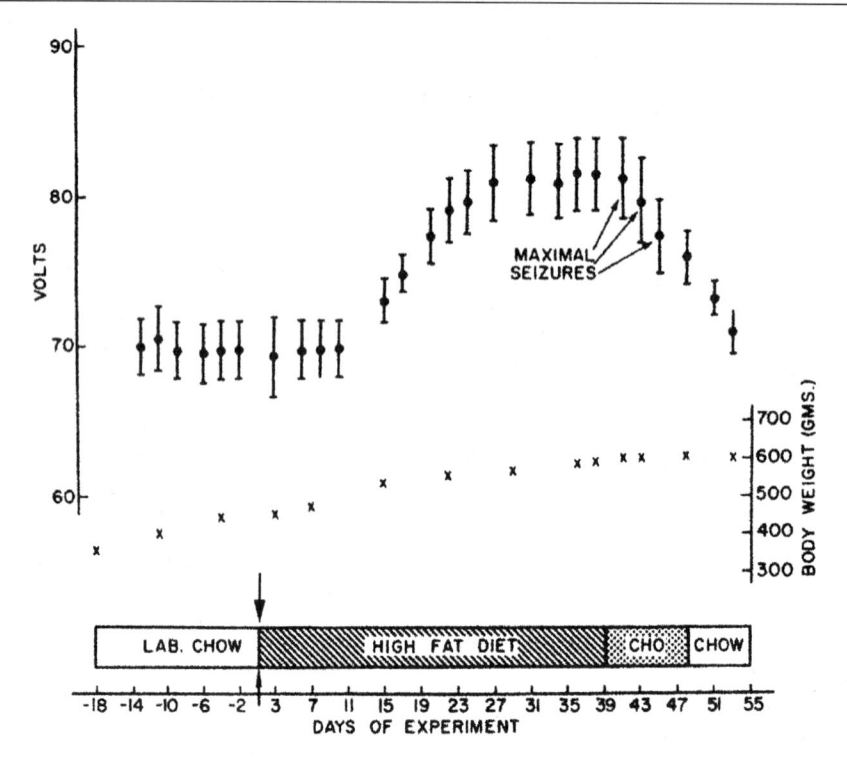

Fig. 1. Effect of dietary manipulation on electroconvulsive threshold in adult rats. Ordinate: Voltage needed to produce a "minimal convulsion" (brief, massive flexion spasm with forelimb and jaw clonus lasting 1–3 s). Abscissa: Experimental day. Animals received standard rat chow until day 0 (arrow), followed by ketogenic diet for 39 d, high carbohydrate diet for 9 d, and, finally, standard rat chow again. The threshold for a minimal convulsion remains constant at about 70 V until about 12 d on the ketogenic diet; the threshold then rises steadily to a plateau. After changing to the high carbohydrate diet, minimal seizure threshold falls and more intense ("maximal") seizures occur at the same stimulus intensity that previously produced minimal convulsions. Mean body weights are also plotted. (Reproduced with permission from Appleton & DeVivo, 1974.)

animal, local circuit, neuronal membrane, and subcellular signaling levels (Stafstrom & Spencer, 2000).

In considering mechanistic possibilities, account must be made of several clinical observations. First, the importance of ketosis must be examined. As will be discussed, ketosis appears to be *necessary but not sufficient* to explain the anticonvulsant effect of the KD.

Second, the latency to onset of the anticonvulsant action has mechanistic implications, although this factor is controversial. Based on animal studies and some clinical observations, seizure control is achieved gradually, over days to weeks. In rats, the threshold to electroshock seizures began to increase after about 12 d on a KD (Fig. 1) (Appleton & DeVivo, 1974). However, in some patients, seizures subside almost immediately, even during the fasting stage (Freeman & Vining, 1999), suggesting that a long-term metabolic adaptation is not necessary or that there may be multiple phases of biochemical adaptation, each exerting its protective function by a different mechanism. For example, during fasting, before maximal ketosis has been attained, seizure frequency may decline as a result of some unknown factor (e.g., lack of calories, lack of carbohydrate, mild acidosis,

Table 2
Possible Mechanisms of the Ketogenic Diet

Possible mechanism	Comments
Ketosis	• Necessary but not sufficient to explain anticonvulsant effect
	• Poor correlation between ketosis and seizure control
Acid–base status	• Cannot by itself explain seizure protective effect
	• Normal serum pH in patients on KD suggests that metabolic acidosis is compensated
	• Brain pH is normal
Cerebral metabolic adaptation	• Increased cerebral energy reserve (ATP/ADP)
	• Unclear how this leads to anticonvulsant effect
Acute effects on neuronal membrane	• Voltage- or ligand-gated ionic channels
	• Membrane pumps
	• Membrane composition
	• Neuromodulation
Chronic effects on epileptogenesis	• Membrane lipid alterations
	• Synaptic reorganization
	• Metabolic adaptations

early ketone production). Later, once ketosis is fully established, the protection may be due to the metabolic adaptation to persistently present ketones as the primary fuel (Nordli & DeVivo, 1997).

The third criterion, linked to the first two, is the time-course over which seizure protection disappears once the KD is terminated. Several reports document a rapid deterioration in seizure control once the KD is stopped and ketosis ceases. In two children with seizures well controlled on the KD, Huttenlocher (1976) administered glucose intravenously. Within hours, seizures returned in one child, and in the other, who was on continuous EEG monitoring, there was marked electrographic deterioration with reappearance of epileptiform discharges. Anecdotal clinical observations also stress the importance of maintaining ketosis, because eating only a few peanuts or sneaking a lollipop is sufficient to negate the beneficial effects of the KD and necessitate "starting from square one" to re-establish ketosis (Freeman et al., 1994). In animals, seizure protection also ceases when the diet is stopped. Appleton and DeVivo (1974) found a reversal of seizure threshold elevation back toward pre-KD levels when rats were switched from the high-fat diet to one enriched in carbohydrate. This effect took several days (Fig. 1). A more rapid effect was seen in experiments of Uhlemann and Neims (1972), where it took only 3 h to reverse the anticonvulsant effect.

A final criterion relates to the effect of age on KD effectiveness. As mentioned earlier, the immature brain has an enhanced ability to extract and utilize ketones, about four to five times more effectively than the adult brain (Dahlquist et al., 1972; Kraus et al., 1974; Nehlig, 1999). Each animal study that examined age as a variable verified greater benefit at younger ages (Uhlemann & Neims, 1972; Otani et al., 1984; Bough et al., 1999b).

4.2. Ketosis and Fasting

As summarized by Withrow (1980), theories of KD mechanism have evolved over time, and none is currently accepted universally (Table 2). Many of the early ideas about

KD mechanism were inferred from clinical studies (Bridge & Iob, 1931). The earliest theory held that seizure protection was a result of a sedative effect of ketones, akin to the action of phenobarbital; accordingly, the anesthetic or sedative effect engendered by ketone bodies could be overcome by glucose, which, in effect, destroyed the ketones (Helmholz & Keith, 1930). However, children maintained on the KD rarely became sedated, and the idea later arose that the beneficial effect was directly related to the ketosis produced by either fasting or the high-fat diet. It was thought that mild ketonemia was protective in mild epilepsy and that a larger degree of ketosis was necessary for seizure control in severe cases. Most subsequent reports have shown an *association* of seizure control with ketosis, although a *causal* relationship has not been proven. Ketosis occurs rapidly upon fasting or KD feeding, usually within 1 d (Keith, 1933; Dodson et al., 1976; Huttenlocher, 1976). Seizures may subside in the first few days or may be delayed several weeks (Freeman & Vining, 1999). There may be two phases or components to KD seizure control: an initial seizure reduction as a result of fasting and a later, sustained effect of the KD (Huttenlocher, 1976; Freeman, et al., 1998b). Additional uncertainty arises when the method of ketone monitoring is considered, because urine ketone levels are an inaccurate reflection of serum levels.

It is also unclear how serum ketone levels correlate with brain ketone levels. Cerebrospinal fluid ketone levels are elevated in animals (Appleton & DeVivo, 1974) and in humans fasting or on the KD (Owen et al., 1967; DeVivo, 1983), but to a lesser degree than in serum. It will be important to learn whether regional, local, or even synaptic ketone concentrations are important for an anticonvulsant effect, especially with regard to whether ketones directly alter neuronal excitability.

The relationship between ketosis and seizure threshold is now being addressed in animal studies (Bough & Eagles, 1999; Bough et al., 1999b). These investigators found a positive correlation between pentylenetetrazole (PTZ) seizure threshold and serum ketone level, suggesting that the level of ketosis could be involved in seizure protection. However, there was a marked discrepancy between the time-course of ketonemia and that of seizure threshold elevation (*see also* Appleton & DeVivo, 1974). In addition, modest seizure protection was found in a group of rats on a calorie-restricted normal diet that did not become ketonemic. Further experiments showed that when age was controlled in rats on calorie-restricted KD or normal diets, the seizure threshold bore no relationship to the level of ketonemia (Bough et al., 1999a). Although these findings suggest that the β-OHB level and seizure threshold are independent consequences of the KD, they do not exclude the possibility of a "threshold ketone level" necessary for seizure protection.

Other preliminary data reported that simply supplementing rats' water supply with ketones protects against PTZ-induced seizures (Lustig & Niesen, 1998). If confirmed, such a finding could have profound implications for treatment of refractory seizures. Work is in progress on developing forms of ketone bodies that can be administered parenterally (Birkhahn et al., 1997; Brunengraber, 1997).

4.3. Acid–Base Changes and Dehydration

Another early idea was that the acidosis produced by the KD had a salutary effect on seizures. Although it is now well known that acid–base changes modulate neuronal excitability (Deitmar & Rose, 1996; de Curtis et al., 1998; Velíšek et al., 1998; Xiong et al., 2000), and that a mild metabolic acidosis is produced early in KD treatment, several studies have shown that a nearly complete compensation occurs in children receiving the KD. Animal studies support the notion that acidosis does not have a major influence on seizure parameters. For example, acidosis did not significantly alter maximal electroshock or PTZ seizure parameters (Hendley et al., 1948). Furthermore, brain and cerebrospinal fluid pH did not change significantly in rats maintained on a KD for 5 wk (Withrow, 1980). Finally, chronic administration of the classic KD or MCT oil does not alter the intracellular pH of neurons (Davidian et al., 1978; Al-Mudallal et al., 1996).

The effects of tissue dehydration and water loss (Fay, 1930; McQuarrie & Keith, 1929; Bridge & Iob, 1931), and electrolyte depletion (Bridge & Iob, 1931; Millichap et al., 1964) were initially thought to play an important role in the antiseizure effect of the KD. Later studies failed to replicate these findings (Huttenlocher et al., 1971; Appleton & DeVivo, 1974; DeVivo et al., 1978). Although minor or transient electrolyte changes may be seen clinically in patients on the KD, a primary role is unlikely.

4.4. Effects of Lipids

Because the KD provides an inordinate amount of lipids, a natural hypothesis would be that hyperlipidemia diminishes seizure susceptibility by altering the lipid composition of neuronal or glial membranes, by affecting membrane protein mobility or function, or by some metabolic mechanism. It is uncertain which, if any, lipid components are essential or modulatory on neuronal excitability in the ketotic condition. Many studies, both human and animal, have demonstrated an elevation of serum lipids in KD-fed individuals (Appleton & DeVivo, 1974; Huttenlocher et al., 1971; Dekaban, 1966; Schwartz et al., 1989). However, the temporal correlation between the rise in serum lipids and seizure control is imprecise. The elevation of serum lipids occurs over several weeks, but a tight correlation with seizure reduction has not been established. Discontinuation of the KD quickly causes seizure recurrence while lipids remain elevated. It may be that certain types of lipids, chain lengths, or degrees of saturation could exert an important role on seizure control. In rodents, a cholesterol-rich diet protected against seizures induced by pentamethylenetetrazole and audiogenic stimuli (Alexander & Kopeloff, 1971), but in humans, there is no consistent correlation between cholesterol level and seizure control (Huttenlocher, 1976; Schwartz et al., 1989). In rats, seizure control was correlated with a higher ketogenic ratio, with maximal protection against PTZ seizures at a fat to carbohydrate plus protein ratio of 6 : 1 (Bough et al., 1999a).

As discussed, exogenous fatty acids, taken parenterally, can integrate into neuronal membranes and alter their fluidity and composition (Bourre et al., 1993). Increasing membrane fluidity (e.g., insertion of certain lipid constituents) increases membrane excitability. Therefore, a reasonable and testable hypothesis would be that lipid type exerts a differential effect on membrane excitability (Ordway et al., 1991; Jumpsen et al., 1997). Yehuda and colleagues (1994) reported that the ratio of dietary n-3 to n-6 PUFAs was the critical variable that afforded protection in several epilepsy models, including single and multiple subconvulsive doses of the convulsant PTZ. Fatty acids of the n-3 variety reduce the excitability of cardiac myocyte membranes (Kang et al., 1995), and may also exert a suppressive role on the excitability of neurons (Voskuyl et al., 1998; Vreugdenhil et al., 1996; Xiao & Li, 1999). Fatty acid levels can also rise during paroxysmal activity and, subsequently, may alter excitability by affecting ionic homeostasis (Woods & Chiu, 1991).

4.5. Cerebral Energy Metabolism

The most comprehensive biochemical theory of KD action was put forth by DeVivo and colleagues, who studied in detail the brain biochemical changes accompanying KD feeding (Appleton & DeVivo, 1974; DeVivo et al., 1978; DeVivo et al., 1975; Nordli & DeVivo, 1997). Brains of rats fed a KD underwent a metabolic adaptation in switching from carbohydrates to fats as the primary energy source. In these brains, there was an increase in "energy charge"; that is, enzymes and substrates of glycolysis and the tricarboxylic acid cycle were altered so as to increase the relative ATP/ADP ratio, resulting in increased energy reserves. The decrease in seizures resulting from this metabolic adaptation was hypothesized to be the result of greater availability of energy in the brain. Serum levels of β-OHB and AcAc were several-fold higher in KD-treated rats than in rats receiving the high-carbohydrate diet. Similarly, brain levels of β-OHB were sevenfold higher in KD-fed rats. The authors concluded that ketosis alters the cerebral glucose metabolism, resulting in an elevation of seizure threshold. Improvement of cerebral energy metabolism with the KD has been verified in humans (Pan et al., 1999).

However, the mechanism by which such an increase in energy reserves would reduce seizure propensity is not clear. To generate and sustain seizures, the brain requires sufficient energy (Siesjo et al., 1983); yet, seizures can deplete the energy supply (Schuchmann et al., 1999). While confirming that cerebral glucose levels were unaltered by the KD, Al-Mudallal and colleagues (1995) did not observe increases in cortical glycogen, glucose-6-phosphate, lactate and citrate, as did DeVivo's group (DeVivo et al., 1978). Therefore, the biochemical mechanisms of the KD are complex and remain incompletely defined.

4.6. Direct Effects on Neuronal Excitability

Neuronal excitability, and hence predisposition to seizures, is regulated by the balance between excitation and inhibition in the brain. The increased availability or function of synaptic inhibition or decreased synaptic excitation would favor decreased cortical excitability and seizure propensity. Because GABA is the primary inhibitory neurotransmitter in the central nervous system, enhanced GABA function would serve to suppress epileptic activity. The structure of GABA resembles a fatty acid (Ordway et al., 1991), raising the intriguing possibility that GABA is a primary target of KD action. Studies looking at GABA levels in animals on the KD have been variable. Total brain GABA was not increased in rats on the KD in one study (Al-Mudallal et al., 1996), but GABA was increased in synaptosomes (Erecinska et al., 1996), suggesting that small but physiologically important changes may occur at a more local or subcellular level. In mice on the KD, GABA levels in the hippocampus were increased (Rho et al., 1999b); given the pivotal role of the hippocampus in the generation and spread of epileptiform activity, even small local GABA changes in this region might exert a widespread effect.

Other recent studies have begun to examine the effects of ketones themselves on membrane excitability and synaptic function. β-OHB or AcAc were applied to in vitro hippocampal slices to test the hypothesis that ketones directly alter neuronal excitability. Niesen and Ge (1998) obtained field and whole-cell recordings from cultured hippocampal CA1 neurons of P11-25 rats. Electrographic seizures were induced by a variety of methods, including high potassium (K^+), 4-aminopyridine, bicuculline, and rapid in vitro kindling. Perfusion of either β-OHB or AcAc at physiologically relevant concentrations

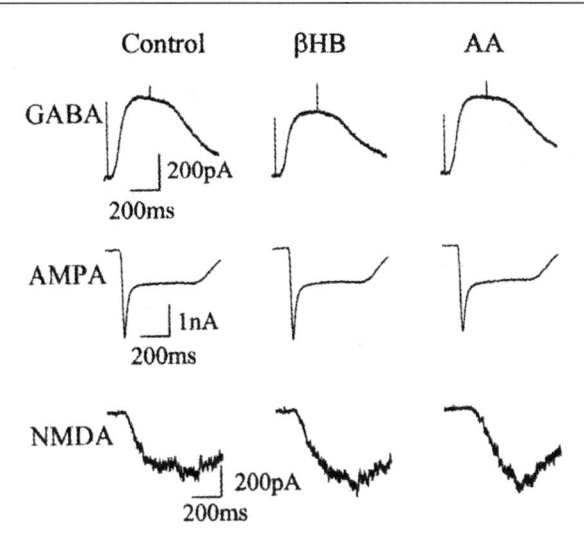

Fig. 2. Whole-cell currents in cultured rat hippocampal neurons in response to GABA (20 μM), glutamate (500 μM), and glutamate (2 μM), shown in the control condition and in the presence of 2 mM β-hydroxybutyrate (β-HB) or 1 mM acetoacetate (AA). Under the conditions employed, 500 μM glutamate reflects α-amino-3-hydroxy-5-methylisoxazole-4-proprionate (AMPA) receptor activation and the response to 2 μM glutamate reflects NMDA receptor activation. Ketone bodies had no effect on either inhibitory (GABA) or excitatory (AMPA, NMDA) currents. Reproduced with permission from Thio et al., 2000.

reduced the size and number of multiple population spikes in CA1 in the high-K$^+$ and 4-AP models. Furthermore, β-OHB eliminated spontaneous bursting and ictal-like discharges in all four models. In whole-cell recordings, β-OHB increased the amplitudes of GABA$_A$-mediated inhibitory postsynaptic potentials (IPSPs), leading the authors to conclude that GABA exerted a direct postsynaptic effect (Ge & Niesen, 1998). These findings imply that ketones may directly suppress neuronal hyperexcitability and that the level of ketosis *per se* may be the critical variable in affording the anticonvulsant benefit. However, there is a possibility that commercially available β-OHB preparations contain neuroactive impurities (Donevan et al., 1999), and these results need verification. In preliminary experiments, we confirmed that 5–10 mM β-OHB reduced spontaneous neuronal bursting in the high-K$^+$ (8 mM) model, but we were unable to demonstrate anticonvulsant effects of this compound in the bicuculline, zero-Mg^{2+} or in vitro kindling paradigms (Stafstrom et al., 1999a). It must be emphasized that each of the above results have been presented in abstract form only.

In the first full report to evaluate the direct effects of ketones on neurons, Thio and colleagues examined the effects of direct ketone application on excitatory and inhibitory synaptic transmission in hippocampus slices or cultures (Thio et al., 2000). They found no effect of either β-OHB or AcAc on any aspect of synaptic function tested, including GABA- and NMDA-evoked inhibitory or excitatory currents, respectively (Fig. 2); CA1 field excitatory postsynaptic potentials; or spontaneous epileptiform discharges elicited by 4-aminopyridine. They concluded that ketones do not directly alter membrane excitability, as neither excitation nor inhibition was altered. Clearly, additional studies into ketone modulation of cortical excitability are urgently needed.

4.7. Chronic Effects of the Ketogenic Diet

In addition to acute effects of ketones on neuronal excitability and seizure threshold, the KD may exert longer-term seizure-suppressive actions. Using the kainic acid (KA) model of chronic epilepsy, we have investigated whether the KD alters the process by which the brain becomes epileptogenic (i.e., "epileptogenesis") (Muller-Schwarze et al., 1999). KA is a glutamate analog that causes an epilepsy syndrome characterized by status epilepticus (Ben-Ari, 1985). After recovery from status epilepticus, there is a several-week "latent period," after which brief spontaneous recurrent seizures occur (Cavalheiro et al., 1982). The generation of these unprovoked spontaneous seizures suggest that the brain, which was normal before KA, has become epileptic. We tested the hypothesis that the KD, administered after status epilepticus but before the onset of spontaneous seizures, would protect against chronic epileptogenic changes (Tauck & Nadler, 1985; Cronin & Dudek, 1988; Mathern et al., 1993; Dudek & Spitz, 1997).

Two days after recovery from KA-induced status epilepticus, adult rats were divided into two diet groups: KD or standard rat chow. For the next 8 wk, spontaneous seizures were monitored using a closed-circuit video recording system (Stafstrom et al., 1992). All rats fed the KD became ketotic and remained so for the duration of dietary treatment. After 8 wk on the KD, rats were sacrificed for routine histology and Timm staining to detect mossy fiber sprouting (MFS). MFS is a form of synaptic reorganization, whereby fibers of dentate granule neurons, deprived of their synaptic targets because of seizure-induced cell death, form aberrant connections with their own dendrites. KD-fed rats had significantly fewer and briefer spontaneous seizures, and there was less sprouting in the dentate supragranular layer, suggesting that at least part of the KD's action may be to retard the process of epileptogenesis (Muller-Schwarze et al., 1999).

Kainic-acid-treated rats normally display hyperexcitable hippocampal circuitry (Tauck & Nadler, 1985; Cronin & Dudek, 1988; Mathern et al., 1997). To determine whether the KD could ameliorate KA-induced hyperexcitability changes, we examined hippocampal slices obtained from KA-treated rats on normal and ketogenic diets. Significantly fewer CA1 population spikes were evoked by Schaffer collateral stimulation in slices from KD-fed rats than from controls, suggesting that this neuronal network is less excitable after KD treatment (Stafstrom et al., 1999b). Because slices were perfused in normal artificial cerebrospinal fluid (without ketones), this reduction in excitability is independent of ketosis, reflecting a chronic stabilizing effect of the KD. It is tempting to speculate that such long-term effects may involve membrane lipid alterations.

5. PUFA: A FEASIBLE EPILEPSY TREATMENT?

The preceding discussion has established the following: (1) PUFAs are an important modulator of neuronal excitability; (2) dietary PUFA can alter several aspects of brain function; (3) the ketogenic diet is an effective therapeutic modality for some persons with epilepsy; (4) the mechanism of the ketogenic diet's seizure protective effect is unknown but may involve lipid components; (5) the ketogenic diet is composed of a high volume of fats that contain a variable proportion of PUFA. From these observations, can we make a leap to the hypothesis that PUFA may be an effective anticonvulsant? Although such a statement would be premature, there are intriguing parallels between the ketogenic diet and PUFA that warrant further scientific scrutiny. The rationale for exploring this notion further comes from previous studies in both heart and brain. Like the antiarrhythmogenic

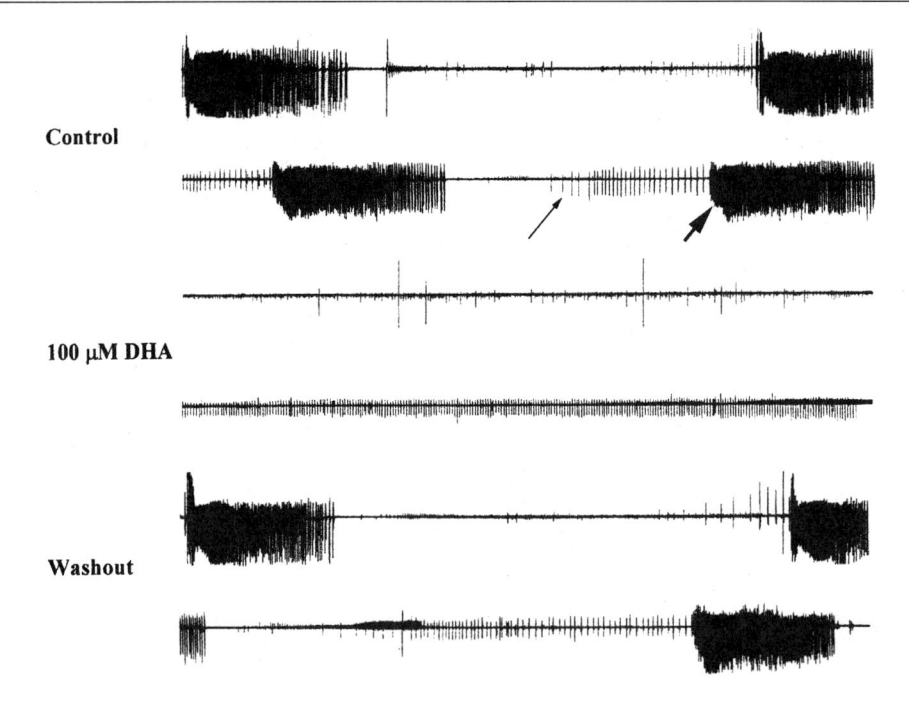

Fig. 3. Extracellular recordings from hippocampal subfield CA1 of a P14 rat, in the presence of zero external Mg^{2+}. Control and washout traces show periodic ictal epileptiform bursts (thick arrow) and interictal bursts (thin arrow). In the presence of 100 μ*M* DHA, ictal epileptiform bursts are suppressed, indicating a reduction of neuronal excitability.

effects of PUFA on cardiac myocytes, PUFAs exert a suppressive action on convulsant-evoked discharges in hippocampal slices (Xiao & Li, 1999). In vivo, PUFAs elevate the seizure threshold in a cortical stimulation model (Voskuyl et al., 1998).

For any agent to be considered a clinically useful anticonvulsant, it must be nontoxic at therapeutically effective doses, cross the blood-brain barrier, and reduce neuronal excitability, thereby suppressing seizures. To examine whether PUFAs could potentially fit those criteria, we undertook some preliminary experiments (Stafstrom & Sarkisian, 1997). Weanling rats (P21) were divided into two groups: standard rat chow and a diet enriched in n-3 PUFA (20% menhaden oil). Plasma levels of DHA and EPA were assessed chromatographically before and 2 wk into dietary therapy; rats gained weight equally and the PUFA group exhibited 7- to 15-fold increases in plasma EPA and DHA levels. After 3 wk of dietary treatment, rats were subjected to seizures induced by inhalation of the convulsant flurothyl. The latency, duration, or severity of flurothyl seizures did not differ between standard diet and PUFA diet groups. These negative results must be interpreted with caution, and further experiments are necessary. Because brain PUFA levels were not measured, we cannot assume that incorporation at the neuronal level occurred.

A more direct approach is to assess the effects of PUFA on neuronal excitability. We tested the effects of DHA on seizures in vitro, using the zero-Mg^{2+} model (Stafstrom, Wang, and Jensen, unpublished data). Extracellular recordings were made from the CA1 subfield in hippocampal slices of P14 rats. When the external Mg^{2+} was removed from the bathing medium, spontaneous interictal discharges and more prolonged, intense ictal bursts appeared (Fig. 3). Perfusion of DHA (100 μ*M*) eliminated those ictal bursts in a

Table 3
Comparison of Ketogenic and Polyunsaturated Fatty Acid Diets

	Ketogenic diet	*PUFA Diet*[a]
Fat content	Mixture of saturated FA and unsaturated FA of various chain lengths; may be deficient in PUFA	Fish oil (20% menhaden oil diet)
Calories from fat (%)	90	44
Ketosis	Yes	No
Efficacy in epilepsy: humans	Yes	Not studied
Efficacy in animal models: in vitro	Variable	Preliminary study only; Fig. 3
Efficacy in animal models: in vivo	Effective in many, but not all models; study for reviews, see Stafstrom (1999) and Thavendiranathan et al. (2000)	Single preliminary negative (Stafstrom & Sarkisian, 1997)

[a]As used in the study by Stafstrom and Sarkisian, 1997; Zapata-Haynle Corp. (Reedville, VA, USA), RBU-D Product code 548.

reversible manner, suggesting that this PUFA blocked epileptiform events and therefore might be able to exert an antiepileptic action in the brain. These intriguing findings are preliminary and further experiments are ongoing.

In conclusion, PUFAs play multiple important roles in brain development and the regulation of neuronal excitability. Exploration of their possible function in epilepsy treatment is just now beginning. The possibility that a "PUFA diet", analogous to the existing "ketogenic diet" (Table 3), could play a role in the battle against epilepsy is indeed tantalizing.

REFERENCES

Alexander GJ, Kopeloff LM. Induced hypercholesterolemia and decreased susceptibility to seizures in experimental animals. Exp Neurol 1971; 32:134–140.

Al-Mudallal AS, LaManna JC, Lust WD, Harik SI. Diet-induced ketosis does not cause cerebral acidosis. Epilepsia 1996; 37:258–261.

Al-Mudallal AS, Levin BE, Lust WD, Harik SI. Effects of unbalanced diets on cerebral glucose metabolism in the adult rat. Neurology 1995; 45:2261–2265.

Appleton DB, DeVivo DC. An animal model for the ketogenic diet. Epilepsia 1974; 15:211–217.

Ballaban-Gil K, Callahan C, O'Dell C, Pappo M, Moshé S, Shinnar S. Complications of the ketogenic diet. Epilepsia 1998; 39:744–748.

Bazan N, Packard M, Teather L, Allan G. Bioactive lipids in excitatory neurotransmission and neuronal plasticity. Neurochem Int 1996; 2:225–231.

Ben-Ari Y. Limbic seizure and brain damage produced by kainic acid: mechanisms and relevance to human temporal lobe epilepsy. Neuroscience 1985; 14:375–403.

Birkhahn R, McCombs C, Clemens R, Hubbs J. Potential of the monoglyceride and triglyceride of DL-3-hydroxybutyrate for parenteral nutrition: synthesis and preliminary biological testing in the rat. Nutrition 1997; 13:213–219.

Bough K, Yao S, Eagles D. Higher ketogenic ratios confer protection from seizures without neurotoxicity. Epilepsy Res 1999a; 38:15–25.

Bough KJ, Eagles DA. A ketogenic diet increases the resistance to pentylenetetrazole-induced seizures in the rat. Epilepsia 1999; 40:138–143.

Bough KJ, Valiyil R, Han FT, Eagles DA. Seizure resistance is dependent upon age and caloric restriction in rats fed a ketogenic diet. Epilepsy Res 1999b; 35:21–28.

Bourre JM, Bonneil M, Clement M, Dumont O, Durand G, Lafont H, et al. Function of dietary polyunsaturated fatty acids in the nervous system. Prostaglandins Leukotrienes Essential Fatty Acids 1993; 48:5–15.

Bridge EM, Iob LV. The mechanism of the ketogenic diet in epilepsy. Bull Johns Hopkins Hosp 1931; 48:373–389.

Brunengraber H. Potential of ketone body esters for parenteral and oral nutrition. Nutrition 1997; 13:233–235.

Carrie I, Clement M, Javel DD, Frances H, Bourre J. Learning deficits in first generation OF1 mice deficient in (n-3) polyunsaturated fatty acids do not result from visual alteration. Neurosci Lett 1999; 266:69–72.

Cavalheiro EA, Riche DA, Le Gal La Salle GL. Long-term effects of intrahippocampal kainic acid injection is rats: a method for inducing spontaneous recurrent seizures. Electroencephalogr Clin Neurophysiol 1982; 53:581–589.

Couch SC, Schwarzman F, Carroll J, Koenigsberger D, Nordli DR, Deckelbaum RJ, et al. Growth and nutritional outcomes of children treated with the ketogenic diet. J Am Diet Assoc 1999; 99:1573–1575.

Crawford M. The role of essential fatty acids in neural development: implications for perinatal nutrition. Am J Clin Nutr 1993; 57(Suppl):703S–710S.

Cronin J, Dudek FE. Chronic seizures and collateral sprouting of dentate mossy fibers after kainic acid treatment in rats. Brain Res 1988; 474:181–184.

Cullingford TE, Dolphin CT, Bhakoo KK, Peuchen S, Canevari L, Clark JB. Molecular cloning of rat mitochondrial 3-hydroxy-3-methylglutaryl-coA lyase and detection of the corresponding mRNA and of those encoding the remaining enzymes comprising the ketogenic 3-hydroxy-3-methylglutaryl-coA cycle in central nervous system of the suckling rat. Biochem J 1998; 329:373–381.

Dahlquist G, Persson U, Persson B. The activity of D-beta-hydroxybutyrate dehydrogenase in fetal, infant and adult rat brain and the influence of starvation. Biol Neonate 1972; 20:40–50.

Davidian NM, Butler TC, Poole DT. The effect of ketosis induced by medium chain triglycerides on intracellular pH of mouse brain. Epilepsia 1978; 19:369–378.

de Curtis M, Manfridi A, Biella G. Activity-dependent pH shifts and periodic recurrence of spontaneous interictal spikes in a model of focal epileptogenesis. J Neurosci 1998; 18:7543–7551.

Deitmar JW, Rose CR. pH regulation and proton signaling by glial cells. Prog Neurobiol 1996; 48:73–103.

Dekaban AS. Plasma lipids in epileptic children treated with the high fat diet. Arch Neurol 1966; 15:177–184.

DeVivo DC. How to use drugs (steroids) and the ketogenic diet. In: Morselli PL, Pippenger CE, Penry JK, eds. Antiepileptic Drug Therapy in Pediatrics. Raven, New York, 1983, pp. 283–292.

DeVivo DC, Leckie MP, Ferrendelli JS, McDougal DB. Chronic ketosis and cerebral metabolism. Ann Neurol 1978; 3:331–337.

DeVivo DC, Malas KL, Leckie MP. Starvation and seizures. Arch Neurol 1975; 32:755–760.

Dodson WE, Prensky AL, DeVivo DC, Goldring S, Dodge PR. Management of seizure disorders: Selected aspects, Part II. J Pediatr 1976; 89:695–703.

Donevan S, White H, Anderson G, Rho J. Dibenzylamine, a contaminant in L-(+)-beta-hydroxybutyrate, exhibits anticonvulsant actions via blockade of N-methyl-D-aspartate receptors. Epilepsia 1999; 40(Suppl. 7):160.

Dudek FE, Spitz M. Hypothetical mechanisms for the cellular and neurophysiologic basis of secondary epileptogenesis: Proposed role of synaptic reorganization. J Clin Neurophysiol 1997; 14:90–101.

Enslen M, Milon H, Malnoe A. Effect of low intake of n-3 fatty acids during development on brain phospholipid fatty acid composition and exploratory behavior in rats. Lipids 1991; 26:203–208.

Erecinska M, Nelson D, Daikhin Y, Yudkoff M. Regulation of GABA level in rat synaptosomes: fluxes through enzymes of the GABA shunt and effects of glutamate, calcium, and ketone bodies. J Neurochem 1996; 67:2325–2334.

Fay T. The therapeutic effect of dehydration on epileptic patients. Arch Neurol and Psychiatry 1930; 23:920–945.

Freeman JM, Vining, E.P.G., Pillas DJ, Pryzik PL, Casey JC, Kelly MT. The efficacy of the ketogenic diet— 1998: a prospective evaluation of intervention in 150 children. Pediatrics 1998a; 102:1358–1363.

Freeman JM, Kelly MT, Freeman JB. The Epilepsy Diet Treatment. Demos, New York, 1994.

Freeman JM, Vining EPG, Pyzik PL, Gilbert DL. Beta-hydroxybutyrate levels in blood correlate with seizure control in children on the ketogenic diet. Epilepsia 1998b; 39(Suppl. 6):167.

Freeman JM, Vining EPG. Seizures decrease rapidly after fasting: preliminary studies of the ketogenic diet. Arch Pediatr Adolescent Med 1999; 153:946–949.

Ge S, Niesen CE. Beta-hydroxybutyrate potentiates GABA-A mediated inhibitory postsynaptic potentials in immature hippocampal CA1 neurons. Epilepsia 1998; 39 (Suppl 6):135.

Geyelin H. Fasting as a method for treating epilepsy. Med Rec 1921; 99:1037–1039.

Gibson R. Long-chain polyunsaturated fatty acids and infant development. Lancet 1999; 354:1919–1920.

Hamano H, Nabekura J, Nishikawa M, Ogawa T. Docosahexanoic acid reduces GABA response in substantia nigra neuron of rat. J Neurophysiol 1996; 75:1264–1270.

Hawkins RA, Williamson DH, Krebs HA. Ketone-body utilization by adult and suckling rat brain in vivo. Biochem J 1971; 122:13–18.

Helmholz HF, Keith HM. Eight years' experience with the ketogenic diet in the treatment of epilepsy. J Am Med Assoc 1930; 95:707–709.

Hendley CD, Davenport HW, Toman JEP. Effect of acid–base changes on experimental convulsive seizures. Am J Physiol 1948; 153:580–585.

Herzberg GZ, Fivush BA, Kinsman SL, Gearhart JP. Urolithiasis associated with the ketogenic diet. J Pediatr 1990; 117:742–745.

Hori A, Tandon P, Holmes GL, Stafstrom CE. Ketogenic diet: effects on expression of kindled seizures and behavior in adult rats. Epilepsia 1997; 38:750–758.

Horimoto N, Nabekura J, Ogawa T. Arachadonic acid activation of potassium channels in rat visual cortex neurons. Neuroscience 1997; 77:661–671.

Huttenlocher PR. Ketonemia and seizures: metabolic and anticonvulsant effects of two ketogenic diets in childhood epilepsy. Pediatr Res 1976; 10:536–540.

Huttenlocher PR, Wilbourn AJ, Signore JM. Medium-chain triglycerides as a therapy for intractable childhood epilepsy. Neurology 1971; 21:1097–1103.

Hwang D, Rhee SH. Receptor-mediated signaling pathways: potential targets of modulation by dietary fatty acids. Am J Clin Nutr 1999; 70:545–556.

Innis SM. Essential fatty acids in growth and development. Prog Lipid Res 1991; 30:39–103.

Janigro D. Blood-brain barrier, ion homeostasis and epilepsy: possible implications towards the understanding of ketogenic diet mechanisms. Epilepsy Res 1999; 37:223–232.

Jumpsen J, Lien EL, Goh YK, Clandinin MT. Small changes of dietary (n-6) and (n-3)/fatty acid content ratio alter phosphatidylethanolamine and phosphatidylcholine fatty acid composition during development of neuronal and glial cell in rats. J Nutr 1997; 127:724–731.

Kang JX, Xiao Y-F, Leaf A. Free, long-chain, polyunsaturated fatty acids reduce membrane electrical excitability in neonatal rat cardiac myocytes. Proc Natl Acad Sci USA 1995; 92:3997–4001.

Keith HM. Factors influencing experimentally produced convulsions. Arch Neurol and Psychiatry 1933; 29:148–154.

Keros S, McBain CJ. Arachadonic acid inhibits transient potassium currents and broadens action potentials during electrographic seizures in hippocampal pyramidal and inhibitory interneurons. J Neurosci 1997; 17:3476–3487.

Kinsman SL, Vining EPG, Quaskey SA, Mellits D, Freeman JM. Efficacy of the ketogenic diet for intractable seizure disorders: review of 58 cases. Epilepsia 1992; 33(6):1132–1136.

Kraus H, Schlenker S, Schwedesky D. Developmental changes of cerebral ketone body utilization in human infants. Hoppe-Seyler's Zeitschr Physiol Chem 1974; 355:164–170.

Kurlak LO, Stephenson TJ. Plausible explanations for effects of long chain polyunsaturated fatty acids (LCPUFA) on neonates. Arch Dis Child Fetal Neonatal Ed 1999; 80:F148–F154.

Leaf A, Kang JX, Xiao Y-F, Billman GE, Voskuyl RA. The antiarrhythmic and anticonvulsant effects of dietary n-3 fatty acids. J Membr Biol 1999; 172:1–11.

Leaf A, Kang JX, Xiao Y-F, Billman GE, Voskuyl RA. Functional and electrophysiologic effects of polyunsaturated fatty acids on excitable tissues: heart and brain. Prostaglandins Leukotrienes Essential Fatty Acids 1999; 60:307–312.

Lucas A, Stafford M, Morley R, Abbott R, Stephenson T, MacFadyen U, et al. Efficacy and safety of long-chain polyunsaturated fatty acid supplementation of infant-formula milk: a randomised trial. Lancet 1999; 354:1948–1954.

Lustig S, Niesen CE. Beta-hydroxybutyrate suppresses pentylenetetrazol (PTZ)-induced seizures in young adult rats. Epilepsia 1998; 39 (Suppl 6):36.

Mathern G, Bertram E, Babb T, Pretorius J, Kuhlman P, Spradlin S, et al. In contrast to kindled seizures, the frequency of spontaneous epilepsy in the limbic status model correlates with greater aberrant fascia dentata excitatory and inhibitory axon sprouting, and increased staining for N-methyl-D aspartate, AMPA and GABA-A receptors. Neuroscience 1997; 77:1003–1019.

Mathern GW, Cifuentes F, Leite JP, Pretorius JK, Babb TL. Hippocampal EEG excitability and chronic spontaneous seizures are associated with aberrant synaptic reorganization in the rat intrahippocampal kainate model. Electroencephalogr Clin Neurophysiol 1993; 87:326–339.

McGahon B, Martin D, Horrobin D, Lynch M. Age-related changes in synaptic function: analysis of the effect of dietary supplementation with ω-3 fatty acids. Neuroscience 1999; 94:305–314.

McGarry JD, Foster DW. Regulation of hepatic fatty acid oxidation and ketone body production. Ann Rev Biochem 1980; 49:395–420.

McQuarrie I, Keith HM. Experimental study of the acid–base equilibrium in children with idiopathic epilepsy. Am J Dis Child 1929; 37:261–277.

Meves H. Modulation of ion channels by arachidonic acid. Prog Neurobiol 1994; 43:175–186.

Millichap JG, Jones JD, Rudis BP. Mechanism of anticonvulsant action of the ketogenic diet. Am J Dis Child 1964; 107:593–604.

Moore TJ, Lione AP, Sugden MC, Regen DM. Hydroxybutyrate transport in rat brain: developmental and dietary modulations. Am J Physiol 1976; 230:619–630.

Muller-Schwarze AB, Tandon P, Liu Z, Yang Y, Holmes GL, Stafstrom CE. Ketogenic diet reduces spontaneous seizures and mossy fiber sprouting in the kainic acid model. NeuroReport 1999; 10:1517–1522.

Nehlig A. Age-dependent pathways of brain energy metabolism: the suckling rat, a natural model of the ketogenic diet. Epilepsy Res 1999; 37:211–221.

Neuringer M, Anderson GJ, Connor WE. The essentiality of n-3 fatty acids for the development and function of the retina and brain. Ann Rev Nutr 1988; 8:517–541.

Neuringer M, Reisbick S, Janowsky J. The role of n-3 fatty acids in visual and cognitive development: current evidence and methods of assessment. J Pediatr 1994; 125:S39–S47.

Niesen CE, Ge S. The effect of ketone bodies, beta-hydroxybutyrate and acetoacetate on acute seizure activity in hippocampal CA1 neurons. Epilepsia 1998; 39 (Suppl. 6):35.

Nishikawa M, Kimura S, Akaike N. Facilitatory effect of docosahexaenoic acid on N-methyl-D-aspartate response in pyramidal neurons of rat cerebral cortex. J Physiol (Lond) 1994; 475:83–93.

Nordli DR, DeVivo DC. The ketogenic diet revisited: back to the future. Epilepsia 1997; 38:743–749.

Ordway RW, Singer JJ, Walsh JVJ. Direct regulation of ion channels by fatty acids. Trends Neurosci 1991; 14:96–100.

Otani K, Yamatodani A, Wada H, Mimaki T, Yabuuchi H. Effect of ketogenic diet on convulsive threshold and brain monoamine levels in young mice. No To Hattatsu 1984; 16:196–204.

Owen OE, Morgan AP, Kemp HG, Sullivan JM, Herrera MGGF, Cahill J. Brain metabolism during fasting. J Clin Invest 1967; 46:1589–1595.

Pan JW, Bebin EM, Chu WJ, Hetherington HP. Ketosis and epilepsy: ^{31}P spectroscopic imaging at 4.1T. Epilepsia 1999; 40:703–707.

Pellerin L, Pellegri G, Martin J-L, Magistretti PJ. Expression of monocarboxylate transporter mRNAs in mouse brain: support for a distinct role of lactate as an energy substrate for the neonatal vs. adult brain. Proc Natl Acad Sci USA 1998; 95:3990–3995.

Poling JS, Vicini S, Rogawski MA, Salem N. Docosahexaenoic acid block of neuronal voltage-gated K$^+$ channels: subunit selective antagonism by zinc. Neuropharmacology 1996; 35:969–982.

Raiten DJ, Talbot JM, Waters JH. Assessment of nutrient requirements for infant formulas. Am J Clin Nutr 1998; 115:2089–2110.

Rho JM, Kim DW, Robbins CA, Anderson GD, Schwartzkroin PA. Age-dependent differences in flurothyl seizure sensitivity in mice treated with a ketogenic diet. Epilepsy Res 1999; 37:233–240.

Rho JM, Shin D, Robbins CA, Baldwin R, Wasterlain CG, Schwartzkroin PA, et al. GABA levels are increased in the hippocampus of mature mice fed a ketogenic diet. Epilepsia 1999; 40 (Suppl 7):161.

Sankar R, Sotero de Menezes M. Metabolic and endocrine aspects of the ketogenic diet. Epilepsy Res 1999; 37:191–201.

Schuchmann S, Buchheim K, Meierkord H, Heinemann U. A relative energy failure is associated with low-Mg^{++} but not with 4-aminopyridine induced seizure-like events in entorhinal cortex. J Neurophysiol 1999; 81:399–403.

Schwartz RM, Boyes S, Aynsley-Green A. Metabolic effects of three ketogenic diets in the treatment of intractable epilepsy. Dev Med Child Neurol 1989; 31:152–160.

Schwartzkroin PA. Mechanisms underlying the anti-epileptic efficacy of the ketogenic diet. Epilepsy Res 1999; 37:171–180.

Siesjo BK, Ingvar M, Folbergrova J, Chapman AG. Local cerebral circulation and metabolism in bicuculline-induced status epilepticus: relevance for development of cell change. Adv Neurol 1983; 34:217–230.

Sirven J, Whedon B, Caplan D, Liporace J, Glosser D, O'Dwyer J, et al. The ketogenic diet for intractable epilepsy in adults: preliminary results. Epilepsia 1999; 40:1721–1726.

Sokoloff L. Metabolism of ketone bodies by the brain. Ann Rev Med 1973; 24:271–279.

Stafstrom CE, Sarkisian MR. A diet enriched in polyunsaturated fatty acids does not protect against flurothyl seizures in the immature brain. Epilepsia 1997; 38(Suppl. 8):34.

Stafstrom CE, Spencer S. The ketogenic diet: a therapy in search of an explanation. Neurology 2000; 54:282–283.

Stafstrom CE, Wang C, Jensen FE. Effects of beta-hydroxybutyrate on epileptiform activity in immature rat hippocampus in vitro. Epilepsia 1999a; 40(Suppl. 7):81.

Stafstrom CE, Wang C, Jensen FE. Electrophysiological observations in hippocampal slices from rats treated with the ketogenic diet. Dev Neurosci 1999b; 21:393–399.

Stafstrom CE. Animal models of the ketogenic diet: what have we learned, what can we learn? Epilepsy Res 1999; 37:241–259.

Stafstrom CE, Thompson JL, Holmes GL. Kainic acid seizures in the developing brain: status epilepticus and spontaneous recurrent seizures. Dev Brain Res 1992; 65:227–236.

Swink TD, Vining EPG, Freeman JM. The ketogenic diet: 1997. Adv Pediatr 1997; 44:297–329.

Tauck DL, Nadler JV. Evidence of functional mossy fiber sprouting in hippocampal formation of kainic acid-treated rats. J Neurosci 1985; 5:1016–1022.

Thavendiranathan P, Mendonca A, Dell C, Likhodii SS, Musa K, Iracleous C, et al. The MCT ketogenic diet: effects on animal seizure models. Exp Neurol 2000; 161:696–703.

Thio LL, Wong M, Yamada KA. Ketone bodies do not directly alter excitatory or inhibitory hippocampal transmission. Neurology 2000; 54:325–331.

Uauy R, Peirano P, Hoffman D, Mena P, Birch D, Birch E. Role of essential fatty acids in the function of the developing nervous system. Lipids 1996; 31(Suppl):S167–S176.

Uhlemann ER, Neims AH. Anticonvulsant properties of the ketogenic diet in mice. J Pharmacol Exp Ther 1972; 180:231–238.

Valencia P, Carver J, Wyble L, Benford V, Gilbert-Barness E, Weiner D, et al. The fatty acid composition of maternal diet affects the response to excitotoxic neural injury in neonatal rat pups. Brain Res Bull 1998; 45:637–640.

Velíšek L, Velíšková J, Moshé SL. Site-specific effects of local pH changes in the substantia nigra pars reticulata on flurothyl-induced seizures. Brain Res 1998; 782:310–313.

Vining EPG. Clinical efficacy of the ketogenic diet. Epilepsy Res 1999; 37:181–190.

Voskuyl RA, Vreugdenhil M, Kang JX, Leaf A. Anticonvulsant effect of polyunsaturated fatty acids in rats, using the cortical stimulation model. Eur J Pharmacol 1998; 341:145–152.

Vreugdenhil M, Bruehl C, Voskuyl RA, Kang JX, Leaf A, Wadman WJ. Polyunsaturated fatty acids modulate sodium and calcium currents in CA1 neurons. Proc Natl Acad Sci USA 1996; 93:12,559–12,563.

Walker M, Sander J. The impact of new antiepileptic drugs on the prognosis of epilepsy: seizure freedom should be the ultimate goal. Neurology 1996; 46:912–914.

Wheless JW. The ketogenic diet: fa(c)t or fiction. J Child Neurol 1995; 10(6):419–423.

Wilder RM. The effects of ketonemia on the course of epilepsy. Mayo Clinic Proc 1921; 2:307–308.

Withrow CD. The ketogenic diet: mechanisms of anticonvulsant action. In: Glaser GH, Penry JK, Woodbury WM, eds. Antiepileptic Drugs: Mechanisms of Action. Raven, New York, 1980, pp. 635–642.

Woods BT, Chiu T-M. Fatty acid elevation and brain seizure activity. [Letter]. Trends Neurosci 1991; 14:405.

Xiao Y-F, Li X. Polyunsaturated fatty acids modify mouse hippocampal neuronal excitability during excitotoxic or convulsant stimulation. Brain Res 1999; 846:112–121.

Xiong Z-Q, Saggau P, Stringer J. Activity-dependent intracellular acidification correlates with the duration of seizure activity. J Neurosci 2000; 20:1290–1296.

Yehuda S, Carasso RL, Mostofsky DI. Essential fatty acid preparation (SR-3) raises the seizure threshold in rats. Eur J Pharmacol 1994; 254:193–198.

17 Fatty Acid Ethyl Esters
Toxic Nonoxidative Metabolites of Ethanol

Zbigniew M. Szczepiorkowski and Michael Laposata

1. INTRODUCTION

Despite the fact that alcoholism is a major cause of disease in society, little is known about the mechanism by which ethanol abuse induces organ damage. One increasingly compelling hypothesis is that fatty acid ethyl esters (FAEEs), esterified products of fatty acids and ethanol, are at least partly responsible for the observed pattern of organ damage in alcoholics. As shown in Fig. 1, ethyl alcohol can be metabolized by oxidative and nonoxidative pathways. In the oxidative pathway, ethanol can be converted to acetaldehyde through the action of alcohol dehydrogenase, the microsomal oxidizing system, or catalase. Acetaldehyde is then subsequently metabolized to acetate by aldehyde dehydrogenase, either in mitochondria or in the cytoplasm. In one of the nonoxidative pathways of ethanol metabolism, ethanol substitutes for the headgroup of a phospholipid to form phosphatidylethanol. This process occurs through the action of phospholipase D on phosphatidylcholine in the presence of ethanol. The other nonoxidative pathway that leads to the synthesis of FAEEs is the focus of this chapter. This is the enzyme-mediated esterification of fatty acid or fatty acyl-CoA and ethanol.

Through the 1960s and 1970s, naturally occurring long-chain fatty acid ethyl esters were detected in a number of organisms, including insects, fungi, and mammals (Calam, 1971; Laseter, 1971; Skorepa, 1968). In the early 1980s, interest in FAEE metabolism and their pathologic effects increased significantly and information began to emerge on FAEE synthesis and the role of FAEEs in mediating ethanol-induced organ damage. The last 5 yr have witnessed the emergence of in vitro and in vivo evidence that FAEEs contribute to ethanol-induced organ damage, with a variety of different mechanisms proposed for mediation of this toxic effect. Multiple enzymatic activities associated with FAEE formation have now been described. Independent of their role in mediating cell injury, it has very recently been shown that FAEEs are useful short-term and long-term serum markers of ethanol intake, given their appearance in the blood rapidly after ethanol ingestion and their presence when ethanol is no longer detectable.

From: *Fatty Acids: Physiological and Behavioral Functions*
Edited by: D. Mostofsky, S. Yehuda, and N. Salem Jr. © Humana Press Inc., Totowa, NJ

Oxidative   Ethanol   **Non-oxidative**

Microsomal Ethanol Oxidizing System
Alcohol Dehydrogenase
Catalase

FAEE Synthase I, II, III — FA
Ethanol O-Acyltransferase — FA CoA
Carboxylesterase — FA / TG-FA
Lipoprotein Lipase — TG-FA
Pancreatic Triglyceride Lipase — TG-FA
Cholesterol Esterase — FA

Acetaldehyde

Aldehyde Dehydrogenase

Acetate

Fatty Acid Ethyl Ester

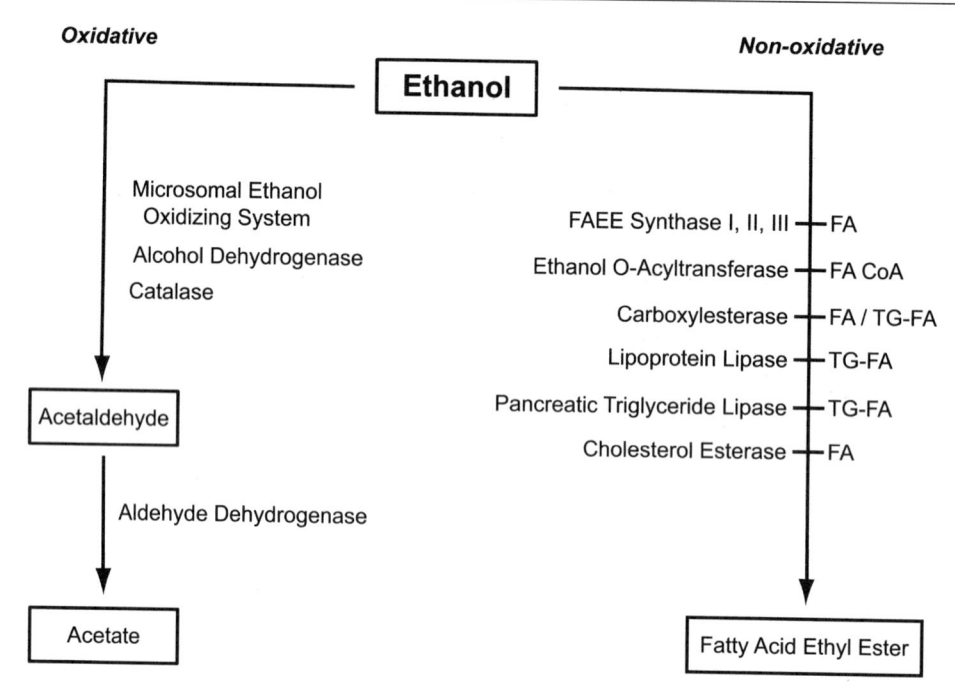

Fig. 1. Oxidative and non-oxidative pathways of ethanol metabolism.

2. THE PRESENCE OF FATTY ACID ETHYL ESTERS IN VITRO AND IN VIVO

The first description of in vitro enzymatic synthesis of fatty acid ethyl esters came from Margolis in 1962 (Margolis, 1962). In experiments with adipose tissue microsomes, he showed that glycerol, ethanol, and several other alcohols form esters with ^{14}C-palmitate when the alcohol is added in millimolar concentration. The esterification of these alcohols required the presence of CoA and ATP or of ^{14}C-palmityl-CoA. Intrigued by this observation, Goodman and Deykin injected rats with ^{14}C-radiolabeled ethanol and tentatively identified long-chain fatty acid ethyl esters in total-body lipid extract (Goodman, 1963). In 1965, Newsome and Rattray reported that "porcine pancreatin" contained a source of enzyme activity to esterify ethanol with fatty acids (Newsome, 1965). Patton and McCarthy in 1966 observed that goat milk converted alcohol and fatty acids into ethyl esters in vitro (Patton, 1966). A prominent spot was observed on the thin-layer chromatography autoradiogram between the solvent front and the position of triglycerides, which was subsequently identified as ethyl palmitate. When alcohol was given to lactating goats, low levels of ethyl palmitate, ethyl oleate, ethyl stearate, and ethyl linoleate were detected in the milk by gas chromatography within 3 h of ethanol ingestion. A 1966 report by Newsome and Rattray documented that normal and postheparin plasma were able to esterify endogenous free fatty acid with ethanol to form ethyl esters (Newsome, 1966). The maximal esterification of oleic acid with ethanol in their experiments occurred at pH 6.0. However, significant activity existed at physiological pH, indicating that the plasma was capable of esterifying ethanol and fatty acids in concentrations that are found in vivo. In 1973, Grigor and Bell reported that ethyl esters of fatty acids are produced when rat liver microsomes are incubated in the presence of ethanol, radioactive fatty acids, CoA, ATP, and Mg^{2+} (Grigor, 1973). They used gas–liquid chromatography to identify the product. They reported that other short-chain alcohols such as methanol, propanol, butanol, pentanol, and hexanol were also esterified. The K_m for the alcohols was found to decrease with increasing chain length from methanol to hexanol. A report in 1976 by Johnson et al. showed that ethyl esters can be formed artifactually during the preparation of methyl esters in vitro if ethanol is not removed from the chloroform used in lipid extraction (Johnson, 1976). They noted, however, that the ethyl esters could be resolved from the corresponding methyl esters by gas–liquid chromatography. This observation was in agreement with earlier reports that indicated that ethanol used in lipid extraction systems can participate in artifactual FAEE synthesis (Gordon, 1970). In 1978, Polokoff and Bell reported in studies with an assay system employing radiolabeled palmitoyl-CoA and ethanol (dispersed in diacylglycerol) that radiolabeled ethyl palmitate was generated by a rat microsomal enzyme (Polokoff, 1976). In 1981, Lange et al. observed the incorporation of ^{14}C-ethanol into a neutral lipid fraction of lipid extracts from isolated and perfused rabbit hearts and from rabbit heart homogenates (Lange, 1981). In this 1981 report, thin-layer chromatography experiments initially suggested that the ^{14}C-ethanol was incorporated into the triacylglycerol fraction. However, it was subsequently determined that it was a metabolic product of ethanol, fatty acid ethyl ester, migrating with triacylglycerol in the neutral lipid fraction in the thin-layer chromatography solvent system. A modification of the solvent system allowed for the subsequent isolation of the FAEE.

3. THE TOXIC EFFECTS OF FATTY ACID ETHYL ESTERS

In 1986, a hypothetical connection was established between FAEE and ethanol abuse. Using tissues obtained postmortem from humans acutely intoxicated at the time of death, the organs commonly damaged by ethanol abuse—pancreas, liver, heart, and brain—were found to have high levels of enzyme activity for the synthesis of FAEEs and the highest concentrations of FAEEs among many different organs and tissues tested (Laposata, 1986). Adipose tissue was also found to have an accumulation of FAEEs and measurable levels of FAEE synthetic activity. The organs not typically damaged by ethanol abuse showed little or no FAEEs and correspondingly little ethyl ester synthase activity. FAEEs were implicated as mediators of ethanol toxicity because FAEEs and the enzyme(s) responsible for their synthesis were distributed primarily in organs damaged by ethanol abuse. In this study, it was also found that in chronic alcoholics, FAEEs were present only in adipose tissue. As an explanation, it was suggested that the FAEEs synthesized in adipose are trapped within a pool of triacylglycerol or that FAEEs synthesized in other organs are transported to adipose for storage. The autopsy study provided the first compelling association between organ damage and the generation of FAEEs. However, there was no causal effect demonstrated between FAEEs and cytotoxicity in this report.

Even before the 1986 autopsy study, the possibility that FAEEs mediate ethanol-induced organ damage had been raised. Using FAEEs solubilized in emulsions and isolated heart muscle mitochondria from rabbits, a report in 1983 indicated that FAEEs inhibited mitochondrial function by uncoupling oxidative phosphorylation (Lange, 1983). It was speculated that FAEEs are hydrolyzed in cells in the mitochondrial membrane, resulting in a high concentration of free fatty acids, which impair mitochondrial function.

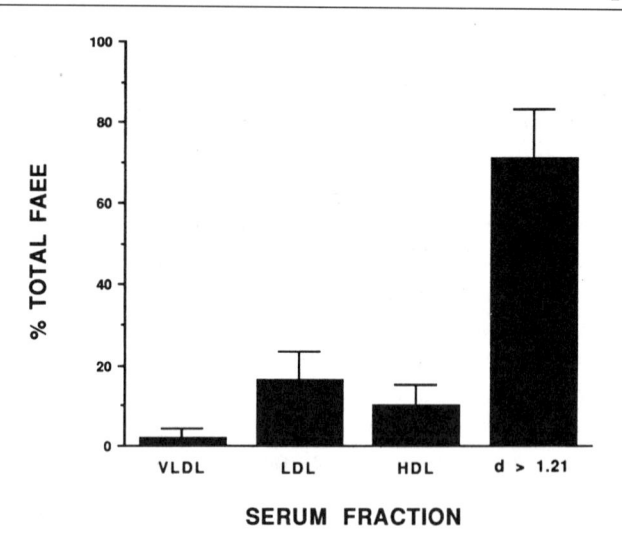

Fig. 2. Percent of total FAEEs in fractions obtained by density gradient ultracentrifugation. Pooled sera from 15 intoxicated emergency-room patients were subjected to density gradient ultracentrifugation and serum fractions were isolated. Lipids were extracted and FAEEs were isolated by thin-layer chromatography and quantitated by gas chromatography. (From Doyle 1994 with permission.)

In 1988, Hungund et al. found ethyl esters, predominantly ethyl oleate and ethyl linoleate, in the livers of mice treated with ethanol by inhalation (Hungund, 1988). They were also able to detect ethyl esters in the brains of these animals, further suggesting a role for FAEEs in impairment of central nervous system (CNS) function. When the FAEEs reached steady-state levels in neural tissues after 3–4 d of alcohol treatment, the authors found that the FAEEs were incorporated into synaptosomal plasma membranes. In vitro studies revealed that the ethyl esters disordered the membrane bilayer, as indicated by a reduction in fluorescence anisotropy. In 1993, Haber et al. demonstrated that rat pancreatic lysosomes incubated for 20 min with ethyl oleate become unstable and leak their enzymes into the surrounding medium (Haber, 1993). The hypothesis was raised that increased pancreatic lysosomal fragility mediated by FAEEs is associated with ethanol abuse. A 1994 study by Ponnappa and colleagues involving the toxic effects of FAEEs in the pancreas showed that isolated rat pancreatic acinar cells incubated with ethanol generated FAEEs endogenously and concomitantly experienced a 20–30% decrease in protein synthesis (Ponnappa, 1994). In support of this finding, Schmidt et al. recently demonstrated that FAEEs, as well as fatty acid methyl esters (FAMEs), are transcriptional regulators of nuclear hormone receptors of the peroxisome proliferator-activated receptor (PPAR) family (Schmidt, 1996). Free fatty acids have been shown to act as transcriptional regulators in the PPAR family. This raises the possibility that hydrolysis of FAEEs in cells produces free fatty acids that regulate transcription following FAEE degradation.

Studies from our research group by Szczepiorkowski et al. provided the first demonstration that FAEEs in a physiologic particle, low-density lipoprotein, exert toxic effects on an intact cell (Szczepiorkowski, 1995). We had previously observed in human subjects that FAEEs are present within low-density lipoproteins (LDL) following ethanol ingestion, as shown in Fig. 2 (Doyle, 1994). Therefore, in the toxicity studies, to deliver

FAEEs to intact cells, we synthesized radiolabeled FAEEs and incorporated them into human LDL particles, which bind to LDL receptors (Bird, 1995). In these studies, ethyl palmitate and ethyl oleate were incorporated into LDL, yielding molar ratios of FAEE to LDL particle of 1614 ± 187 and 3154 ± 303 ($n = 6$, mean ± S.E.), respectively. LDL reconstituted with FAEEs (rLDL) was not oxidatively modified, and native LDL markedly decreased the uptake of ethyl oleate in rLDL by HepG2 cells. Cultured HepG2 cells were incubated with LDL containing a FAEE (either ethyl oleate or ethyl arachidonate), and cell proliferation was measured by [methyl-^{3}H]thymidine incorporation, and protein synthesis was determined using L-[^{35}S] methionine. Incubation of cells with 600 µM ethyl oleate or 800 µM ethyl arachidonate, which permitted us to achieve intracellular FAEE concentrations found in vivo in the 1986 autopsy study, decreased [methyl-^{3}H] thymidine incorporation into HepG2 cells by 30% and 35%, respectively. LDL reconstituted with ethyl oleate (400 µM) decreased protein synthesis in intact HepG2 cells by 40%. Electron microscopy revealed significant changes in cell morphology, with accumulation of intracellular lipids, a distortion of the intracellular organelles and a distortion of the nuclear membrane. FAEEs delivered in reconstituted LDL were rapidly hydrolyzed, and the fatty acids reesterified into phospholipids, triacylglycerols, and cholesteryl esters, with preference for triacylglycerols. These findings provided evidence that FAEEs are toxic for intact human hepatoblastoma cells and that they or their metabolites may be a causative agent in ethanol-induced liver damage.

A report by Gubitosi-Klug and Gross examined the effect of FAEEs on a human brain potassium channel in SF9 cells expressing the recombinant channel (Gubitsi-Klug, 1996). They found that physiologically relevant FAEE concentrations accelerated the kinetics of activation of the channel and raised the possibility that this in vitro observation may be one of the pathologic consequences of ethanol abuse in the central nervous system.

The in vitro toxicity studies still left unanswered the question of whether FAEEs are toxic in vivo. This led us to perform experiments to assess the toxicity of FAEEs in reconstituted LDL in vivo in rats (Werner, 1997). In these studies, rats received FAEEs in reconstituted LDL at FAEE concentrations that are physiologically attainable (10–30 µM) after ethanol ingestion. The FAEEs were delivered as a bolus and then by continuous infusion for 1 h through a cannula placed in the carotid artery and advanced through the aorta to the superior mesenteric artery. This placement was made to maximize the likelihood for observing FAEE-induced cytotoxicity in the pancreas. The rats were sacrificed 3–24 h after infusion of FAEEs and biochemical makers of organ damage were measured. Histology was performed for multiple organs, and the wet/dry ratio of the pancreas was determined to assess pancreatic edema. Control rats received LDL, which was reconstituted with cholesterol ester. The wet/dry ratio for the rats receiving the LDL containing FAEEs was significantly higher than that of the rats receiving LDL reconstituted with cholesterol ester. In addition, by 3 h, the level of the trypsinogen-activating peptide, a biochemical marker for pancreatic cell damage, was threefold higher in the animals receiving the LDL reconstituted with FAEEs relative to controls.

Bora et al. also performed an in vivo experiment in which FAEEs were injected directly into the left ventricle of rats undergoing thoracotomy and significant myocardial cell damage was observed, whereas control animals not receiving FAEEs did not suffer similar damage (Bora, 1996).

Taken together, many in vitro and in vivo studies now show that FAEEs have significant capacity to induce cell injury.

4. ENZYMES ASSOCIATED WITH FATTY ACID ETHYL ESTER SYNTHESIS

In 1984, an FAEE synthase enzyme was purified to homogeneity from rabbit myocardium (Mogelson, 1984). At an early step in the purification, there were two peaks of enzyme activity separable by DEAE–cellulose chromatography. The larger peak of enzyme activity (the second one) was purified more than 5000-fold, with an overall yield of 40%. Sodium dodecyl sulfate–polyacrylamide gel electrophoresis (SDS-PAGE) revealed a single polypeptide with a molecular weight of 26,000, and gel permeation chromatography under nondenaturing conditions indicated a molecular weight of 50,000 for the active enzyme. This result suggested that the enzyme is a soluble dimeric enzyme composed of two identical or nearly identical subunits.

In 1987, a study involving FAEE synthesis was performed using a postmortem human brain (Laposata, 1987). FAEE synthase activity was found in 10 different anatomic locations in the human brain, with gray-matter sites containing approximately twice the activity of white-matter sites. Two forms of the synthase present in cytosol or loosely bound to membrane fractions, in both gray-matter and white-matter homogenates, were isolated by ion-exchange chromatography. Gray and white matter were found to have different proportions of the two forms of the synthase. In this report, it was also shown that higher blood ethanol concentrations are associated with higher FAEE concentrations in the cerebral cortex, suggesting that FAEEs are generated in the human brain during acute intoxication.

In 1989, using human myocardium rather than rabbit myocardium as starting material, Lange's group found a third peak of FAEE synthase activity. This peak eluted before the minor fraction off the DEAE column in the FAEE synthase purification procedure (Bora, 1989a). To clarify the nomenclature of the FAEE synthases at that time, Lange named the FAEE synthase peaks in order of their elution from DEAE–cellulose at pH 8.2 as synthase I, II, and III (Bora, 1989b). Synthase I was a dimer of identical subunits of 26,000 Da each. The enzyme had crossreactivity with synthase III by Western blot. Synthase II was a monomer that dimerized easily, but dimerization was found not to be required for catalytic activity. The molecular weight of the enzyme was 65,000–67,000, and it bore a relationship to cholesterol esterase. Synthase III, like synthase I, was a dimer of identical subunits of 26,000 Da each. Synthase III was reported by Lange to have glutathione *S*-transferase activity (Lange, 1991; Board, 1993), but this conclusion was subsequently seriously challenged (Board, 1993; Sharma, 1991), as described on pg 297.

A study published in 1990 investigated the synthesis and degradation of FAEE by rat hepatoma cells exposed to ethanol in tissue culture (Laposata, 1990). FAEE synthesis by cultured hepatoma cells was found to be linearly associated with the concentration of ethanol added to the culture medium. In addition, the enzymes responsible for the synthesis of ethyl esters were shown to be primarily membrane bound and concentrated in the microsomal fraction of the hepatocytes, rather than in the cytoplasm. It was also observed in this study that FAEEs can be hydrolyzed to free fatty acids and ethanol by membrane-bound enzymes in the microsomal and mitochondrial–lysosomal fractions.

A 1996 report by Treolar et al. provided additional information on the cytosolic and microsomal forms of FAEE synthase (Treloar, 1996). The cytosolic FAEE synthase activity, which uses ethanol and free fatty acid as substrates for FAEE synthesis, was designated as FAEE synthase, and the microsomal FAEE synthase activity, which uses ethanol and fatty acyl-CoA as substrates, was referred to as acyl-CoA : ethanol acyltransferase (AEAT). The data indicated that the activities per gram of liver of these two FAEE synthetic enzymes are comparable.

Further characterization of the role of acyl-CoA : ethanol O-acyltransferase has been performed by Diczfalusy et al. (Diczfalusy, 1999). In her study, she compared the activity of AEAT and FAEE synthase in isolated rat liver microsomes. The baseline activities were determined to be similar. However, the addition of bis-(4-nitrophenyl) phosphate, a serine esterase inhibitor, increased AEAT activity sixfold with complete inhibition of FAEE synthase activity. A cysteine-reacting compound, p-hydroxylmercuribenzoic acid, stimulated AEAT activity fourfold with no appreciable effect on FAEE synthase activity. This result could be explained by significant enzymatic hydrolysis of the substrate (acyl-CoA) and the product (FAEE) by serine esterases, which would decrease the observed formation of FAEE by AEAT. The authors concluded that under optimal conditions, without competing hydrolysis, the capacity of AEAT to synthesize FAEEs is significantly higher than FAEE synthase. This observation and the fact that the ratio of acyl-CoA to nonesterified fatty acids may be high under normal conditions may indeed make AEAT the most important enzyme in fatty acid ethyl ester formation.

In 1991, Lange's group reported the cDNA cloning and sequencing of human myocardial FAEE synthase III (Bora, 1991). They claimed that the sequence of the isolated FAEE synthase III cDNA was highly homologous to the sequence of glutathione S-transferase pi-1 cDNA from multidrug-resistant Michigan Cancer Foundation (MCF) cells (Bora, 1989a). The conclusion that FAEE synthase III is the same enzyme molecule as the one that has glutathione S-transferase activity has been challenged in a number of reports. In 1991, Sharma et al. purified glutathione S-transferase isoenzymes from human pancreas and evaluated the FAEE synthase activity of these enzyme preparations (Sharma, 1991). The results indicated that human glutathione S-transferase isoenzymes belonging to alpha, mu, and pi classes do not express FAEE synthase activity, which led Sharma to conclude that FAEE synthase and glutathione S-transferase are distinct proteins. Subsequently in 1993, Board et al. (Board, 1993) also addressed the question of FAEE synthase III identity with glutathione S-transferase. They constructed a clone that encoded the FAEE synthase III protein described by Lange's group. They concluded that their results provided no evidence to support a relationship between FAEE synthase III and glutathione S-transferase and suggested that the cDNA produced by Lange's group in 1991 may have resulted from a cloning artifact. Using their purified FAEE synthase III, Lange's group performed site-directed mutagenesis to alter two histidine residues in the molecule (Bora, 1992). Inhibition of both FAEE synthase and glutathione S-transferase activity with this site-directed mutagenesis would support their conclusion that one enzyme protein has both activities. However, the mutagenesis resulted in inhibition of only the glutathione S-transferase activity and not the FAEE synthase activity. In 1992, Lange's group reported the expression of FAEE synthase III in cultured neuronal cells (Isenberg, 1992). The presence of FAEE synthase in neuronal cells was consistent with earlier work showing that FAEE synthase activity is present in the human brain (Laposata, 1987). In two reports by Carlson's group, FAEE formation was demonstrated in in vivo and in vitro preparations of rat and rabbit lung (Manautou, 1992; Manautou, 1991), as these organs had not been evaluated in the 1986 autopsy study by Lange's group.

Several enzymes known to catalyze other reactions have also been shown to catalyze FAEE synthesis. These include carboxylesterase from adipose tissue (Tsujita, 1992) and pancreas (Tsujita, 1994b), lipoprotein lipase (Chang, 1997; Tsujita, 1994a), pancreatic triglyceride lipase (Riley, 1990), and possibly cholesterol esterase (Lange, 1982).

In 1982, Lange demonstrated that FAEEs could be generated by the action of cholesterol esterase on the substrates free fatty acids and ethanol (Lange, 1982). Using purified cholesterol esterase and purified radiolabeled substrates, cholesterol esterase catalyzed the formation of FAEEs from ethanol and free fatty acids. This report showed that cholesterol esterase could catalyze the synthesis of FAEEs, but there was no evidence to suggest that this is the mechanism whereby FAEEs are synthesized in vivo.

In studies on the FAEE synthase enzyme, it was demonstrated by Tsujita and Okuda that carboxylesterase is capable of catalyzing the FAEE synthesis reaction (Tsujita, 1992; Tsujita, 1994b). They demonstrated that purified carboxylesterase was able to promote the synthesis of FAEEs in a number of organs. In support of the findings of Tsujita et al., who showed that carboxylesterase can promote the synthesis of FAEE, we reported that purified FAEE synthase also has cocaethylene (an esterification product of ethanol and cocaine) synthetic capacity (Heith, 1995). Cocaethylene synthase, like FAEE synthase, has been shown to have carboxylesterase activity (Dean, 1991), linking FAEE synthesis again to carboxylesterase.

In 1994, it was shown that carboxylester lipase obtained from pig pancreas is associated with FAEE synthase activity (Tsujita, 1994b). It was proposed that the lipase attacks triacylglycerol, forming an acyl-enzyme intermediate, and that during the deacylation process, ethanol binds to fatty acid as an acceptor. It was proposed that if ethanol is present during triacylglycerol degradation in the intestinal lumen, carboxylester lipase contributes to nonoxidative ethanol metabolism.

Tsujita and Okuda demonstrated in vitro that lipoprotein lipase is capable of catalyzing FAEE synthesis (Tsujita, 1994a). Chang and Borensztajn supported this observation by demonstrating that FAEE synthesis in an isolated rat heart perfused with chylomicrons and ethanol is mediated by lipoprotein lipase (Chang, 1997). The basic characteristics of enzymes involved in FAEE synthesis is presented in Table 1.

We have reported that there exists an apparent fatty acid specificity for palmitate and oleate for the FAEEs present in the blood following ethanol intake and that FAEE synthesis increases with higher extracellular concentrations of their corresponding fatty acids for ethyl oleate, ethyl linoleate, and ethyl arachidonate—but not ethyl palmitate (Dan, 1997). Using HepG2 cells incubated with ethanol, we have also quantitated the actual mass of FAEE synthesized and demonstrated that the fatty acid used for FAEE synthesis is derived from a designated intracellular pool of fatty acids (Dan, 1998).

Because cytosolic enzymes such as aspartate aminotransferase, lipase, and amylase appear in the blood after liver or pancreatic damage, we hypothesized that FAEE synthase, which is both cystolic and membrane bound, is released into the blood of patients with liver or pancreatic disease. We demonstrated that patients with liver or pancreatic disease release FAEE synthase into their plasma in amounts proportional to the level of asparate aminotransferase, amylase, or lipase (Aleryani, 1996). The presence of FAEE synthase in plasma may permit nonoxidative ethanol metabolism in the circulation in such individuals, although this remains to be conclusively demonstrated.

In another study, we showed that white blood cells (WBCs) are capable of synthesizing FAEE (Gorski, 1996). This finding is significant because it demonstrates that a blood sample rather than a biopsy can be used to quantitate FAEE synthase activity in populations. We determined that the lymphocyte–monocyte population of WBCs was particu-

Table 1
Enzymes Associated with Fatty Acid Ethyl Ester Synthesis

Name	Cellular distribution	Substrates	Ref.
FAEE synthase I	Cytosolic	FFA	Bora, 1989b
FAEE synthase II	Cytosolic	FFA	Bora, 1989b
FAEE synthase III	Cytosolic	FFA	Bora, 1989b; Isenberg, 1992
Carboxylesterase	Microsomal	MG, DG, TG (short-chain FA), methyl-FA; FA	Tsujita, 1992; Tsujita, 1994b; Chang, 1997
Acyl-CoA:ethanol O-acyltransferase (AEAT)	Microsomal	FA-CoA	Diczfalusy, 1999
Lipoprotein lipase	Extracellular	FFA, TG (both require bile salts)	Tsujita, 1994a; Tsujita, 1999
Cholesterol esterase	Cytosolic	FFA	Lange, 1982
Triglyceride lipase	Cytosolic	FFA	Riley, 1990

Abbreviations: FFA: free fatty acid; FA:-CoA-fatty acid coenzyme A; MG: monoglyceride; DG: diglyceride; TG: triglyceride.

larly high in FAEE synthase activity and that the natural-killer cells had the highest activity among the lymphocytes. This result correlated with a report published at the same time showing that delivery of ethanol to rats previously injected with tumor cells permitted metastasis by inhibition of natural-killer cell function (Ben-Eliyahu, 1996). This observation raises the possibility that upon ethanol ingestion, FAEE synthesis in natural-killer cells is a causative factor in the inhibition of immunologic activity and the promotion of tumor metastasis.

We also demonstrated in this study that FAEE synthase activity could be induced nearly twofold in the WBC fraction of humans ingesting 2 oz of scotch whiskey for 6 d (Gorski, 1996). This supports the conclusion that FAEE synthase is regulated to some extent by the presence of ethanol. The enzyme activity returned to baseline levels despite ingestion of 2 oz of scotch whiskey for an additional 3 d. In this report, it was also shown that alcoholic individuals have approximately half the WBC FAEE synthase activity detected in normal controls. The lower enzyme activity observed in the WBCs of alcoholics in a detoxification center may be the result of years of ethanol abuse, or it may be that alcoholics congenitally have low levels of FAEE synthase. If the latter is true, this finding may explain, in part, the genetic predisposition of many alcoholic individuals to ethanol abuse.

In a 1991 report, the concentration of FAEEs and the activity of FAEE synthase were determined in adipose tissues of rats ingesting ethanol for different periods of time (DePergola, 1991). After 10 wk of ethanol exposure, 300 nmol of FAEE per gram of adipose tissue were detected. Importantly, the ethyl esters disappeared completely after 1 wk of abstinence from ethanol. The half-life of the ethyl esters was determined to be somewhat less than 24 h, similar to an earlier report of 16.6 h (Laposata, 1989). In this study, the majority of the FAEEs were found in a membrane preparation of isolated adipose tissue rather than in the cytoplasm. It was also shown that following 10 wk of exposure to ethanol, FAEE synthase activity in adipose tissue doubled. After the ethanol

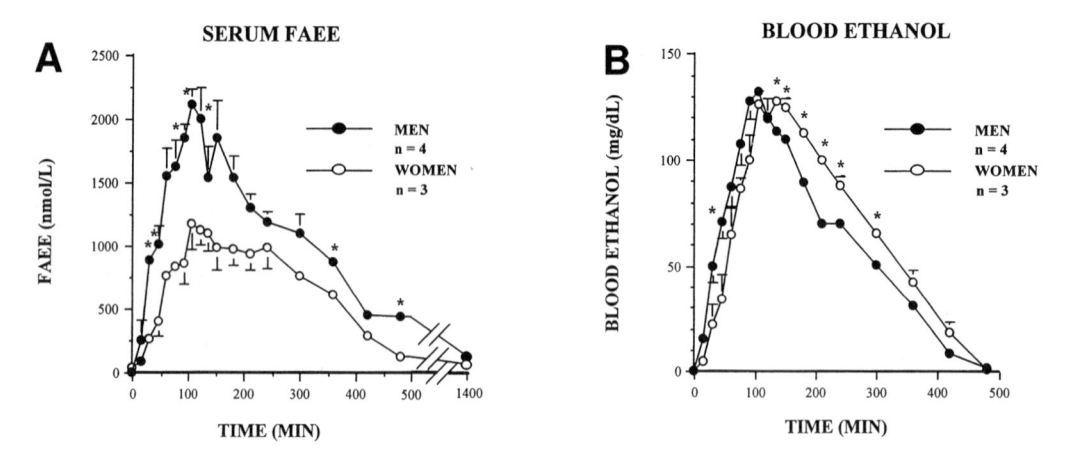

Fig. 3. Time-course for serum FAEEs (**A**) and blood ethanol (**B**) concentrations for men and women consuming alcohol to legal limits of intoxication. The asterisk (*) indicates significant differences ($p \leq 0.05$) between sexes for a given time-point. (From Soderberg 1999 with permission.)

ingestion was stopped, the enzyme activity decreased with a half-life of approximately 1 wk, suggesting that FAEE synthase is regulated to some extent by the presence of ethanol in the exogenous medium.

In a 1992 report by Bearer et al. (Bearer, 1992), it was shown that human and mouse placentas have significant FAEE synthase activity and that they can accumulate FAEEs after maternal ethanol exposure, persisting for up to 7 d in the placentas. These studies raised the possibility that the accumulation of FAEEs in placentas is contributory to the embryopathy of fetal alcohol syndrome.

Hungund and Gokhale reported in 1994 that alcohol administration to pregnant rats leads to significant accumulation of ethyl esters of long-chain fatty acids in both maternal and fetal organs (Hungund, 1994). They also reported that ganglioside GM 1 treatment 1 h and 24 h prior to ethanol exposure on both gestational d 7 and gestational d 14 reduce the accumulation of FAEEs. The mechanism by which GM 1 treatment reduced the accumulation of FAEE was not determined.

Recently, Soderberg et al. have reported on differences in FAEE synthesis between genders. (Fig. 3) (Soderberg, 1999). In this study, four men and three women ingested a weight-adjusted amount of ethanol as a mixture of vodka and fruit juice. All subjects attained a peak blood ethanol concentration of more than 21.7 mmol/L (100 mg/dL). There were no significant differences in the length of time required to achieve the peak ethanol level between genders (Fig. 3B). Interestingly, the peak level of FAEE was twofold higher in men than in women. (Fig. 3A). The observed difference may be attributed to differences in activity of FAEE synthases and different rates of egress of FAEEs into the circulation or activity of enzymes responsible for FAEE degradation. Additional studies with larger groups of subjects are necessary to identify and explain this observation.

5. FATTY ACID ETHYL ESTERS AS FATTY ACID SUPPLEMENTS

There is substantial clinical interest in fatty acid supplements as treatments for a variety of diseases. Oral preparations of FAEEs have been made available for fatty acid supplementation and with the introduction of FAEE capsules for oral intake, there were a number of studies to evaluate ethyl ester absorption from the gastrointestinal (GI) tract. In 1991, Nordoy et al. reported that ethyl esters and triglycerides were equally well absorbed from the GI tract in human subjects (Nordoy, 1991). In 1992, Yamazaki et al. (Yamazaki, 1992) infused emulsions of ethyl eicosapentaenoate (ethyl EPA) into rat veins and demonstrated that the EPA content in the phospholipids of a variety of organs substantially increased. It has also been shown that treatment of normal volunteers with oral n-3 FAEE, with approximately 3 g of ethyl EPA plus ethyl docosahexaenoic acid (ethyl 22:6) per day, results in a marked accumulation of the fatty acids from these ethyl esters in the plasma and in cell lipids within 6 wk (Marangoni, 1993). Harris et al. conducted a randomized, placebo-controlled, double-blind, crossover trial in which 10 mildly hypertriglyceridemic patients were given capsules containing n-3 FAEEs or an olive-oil placebo for two 4-wk treatment periods with a 1-wk washout phase in between (Harris, 1993). They found that the n-3 FAEEs were effective hypotriglyceridemic agents and that they impact lipoprotein metabolism very quickly. However, they also determined that the incorporation of n-3 FAEEs into lipoproteins was associated with an increased susceptibility of the lipoproteins to oxidation. Krokan et al. (Krokan, 1993) reported in a 1993 study that there is enrichment of 20:5n-3 and 22:6n-3 in total plasma lipids and phospholipids in healthy volunteers after oral ingestion of either a concentrated ethyl ester or a natural triglyceride and that the enteral absorption of 20:5 and 22:6 was very similar for FAEEs and triglycerides. Importantly, there were no acute toxic effects associated with oral FAEE intake in any of these studies. There is now evidence of a significant therapeutic effect of n-3 ethyl esters in mice. Paulsen et al. recently demonstrated that the administration of ethyl esters of eicosapentaenoic acid (EPA, 20:5n-3) and docosahexaenoic acid (DHA, 22:6n-3) for 18 wk suppresses the formation and growth of intestinal polyps in a mouse strain genetically predisposed to develop tumors in the gastrointestinal tract (Paulsen, 1997).

We recently tested the hypothesis that orally ingested supplemental FAEEs are rapidly degraded in the GI tract and blood to explain the lack of toxicity (Saghir, 1997). Using rats given FAEEs as an oil directly into the stomach or within LDL particles directly into the circulation, we demonstrated that FAEE hydrolysis in the gastrointestinal tract and blood is rapid and extensive. The fate of the fatty acid from FAEE hydrolysis was highly dependent on the organ or tissue presented with the FAEE.

6. FATTY ACID ETHYL ESTERS AS MARKERS OF ETHANOL INTAKE

Because of the long half-life of FAEEs in adipose tissue, it was suggested that FAEE in adipose tissue could be a laboratory marker for previous alcohol intake, particularly for forensic applications where adipose tissue samples can be readily obtained. In postmortem samples from four chronically intoxicated subjects whose blood ethanol levels were zero, it was demonstrated that prior ethanol ingestion could be established by the presence of FAEEs in the adipose tissue (Laposata, 1989). In this report, a separate series of experiments determined the half-life of FAEEs to be 16.6 h in the adipose tissue of rabbits that received 10% ethanol in their drinking water for 10 mo.

In a 1994 ethanol ingestion study with five volunteers, FAEEs were shown to be bound to lipoproteins and albumin in serum (Fig. 2) (Doyle, 1994). A higher percentage of saturated fatty acids was found in the FAEE pool than in the serum free fatty acid or

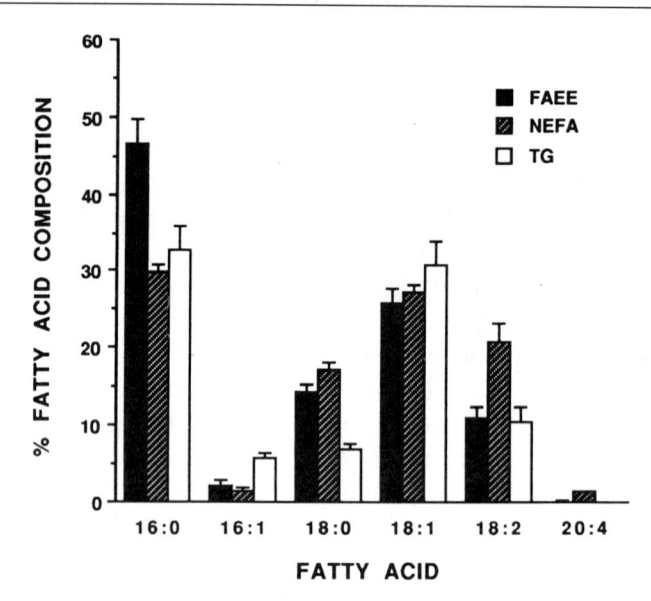

Fig. 4. The fatty acid composition of serum FAEEs and serum triglycerides from hospital emergency-room patients and plasms non-esterified fatty acids (NEFA) of healthy subjects. The values represent the percent of each fatty acid within either FAEE, triglyceride, or NEFA pools. (From Doyle 1994 with permission.)

triacylglycerol pools (Fig. 4). In this study, we also demonstrated that when FAEEs are isolated by density gradient ultracentrifugation from the sera of intoxicated emergency room patients, 68.3% of the FAEEs associate with the $d > 1.21$ fraction, which contains albumin. To assess whether FAEE in serum, like nonesterified fatty acids, are bound to albumin, sera from several emergency-room patients with detectable blood ethanol were pooled, and the serum albumin in the sample was immunoprecipitated. Lipids from the immunoprecipitate were extracted, and FAEEs were found in the precipitates. These data provide evidence to support the conclusion of earlier density gradient studies that a significant fraction of serum FAEEs are bound to albumin.

In a series of nuclear magnetic resonance (NMR)-based studies, we analyzed FAEEs binding to small unilamellar phospholipid vesicles, human LDL, and bovine serum albumin by ^{13}C-NMR spectroscopy using ethyl (1-^{13}C, 99%) oleate (Bird, 1996). The conclusions from this study were that the addition of ethyl oleate to isolated human LDL resulted in ethyl oleate in the core of the lipoprotein, that albumin has a much greater affinity for oleic acid than for ethyl oleate, and that ethyl oleate rapidly transfers between LDL and phospholipid vesicles. In our report by Doyle et al., we found that albumin transports the majority of FAEE in plasma (Doyle, 1994). The accumulated observations suggest that despite its low affinity for FAEEs, albumin transports most of the FAEEs in the blood because it has a much greater plasma concentration than lipoprotein particles. In a separate study, we reported that as the serum FAEE concentration rises, the percentage of FAEEs associated with lipoproteins increases and, correspondingly, the percentage associated with albumin decreases (Bird, 1997). This observation is consistent with a hypothesis that once FAEE binding sites on proteins in the $d > 1.21$ fraction are saturated, additional FAEE molecules are incorporated into lipoproteins.

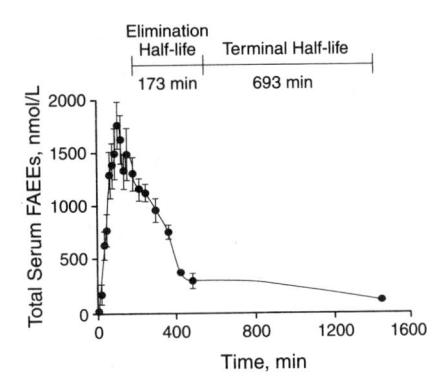

Fig. 5. Composite-time course for total serum FAEE concentration over a 24-h period. Each data point represents the mean FAEE value for subjects 1 through 7 at the indicated time-point; results are shown as ±SEM. Ethanol ingestion occurred during the first 1.5 h of the time-course. (From Doyle 1996, copyrighted 1996, American Medical Association, with permission).

Fatty acid ethyl esters may be clinically important, independent of whether FAEEs induce cytotoxicity, because they can serve as a marker for ethanol intake. We performed a study to determine the clinical utility of FAEE in the blood as a short-term confirmatory marker for ethanol intake and as a long-term marker for ethanol intake after ethanol is no longer detectable (Doyle, 1996). To isolate FAEEs from plasma for quantitation by gas chromatography–mass spectrometry (GC–MS) in these clinical studies, we developed a two-step method using solid-phase extraction with a recovery of 70 ± 3%, using ethyl oleate as a recovery marker (Bernhardt, 1996).

The design for the study was a controlled clinical trial with seven healthy subjects. The subjects ingested a known amount of ethanol at a fixed rate. The concentration of FAEEs in the blood after ethanol intake was determined for up to 24 h. FAEE disappearance from the blood followed a decay curve that initially resembled the decay curve for blood ethanol. However, because of a very slow secondary elimination phase, the FAEEs persisted in the blood for at least 24 h after ethanol intake was completed. Also included in this report was a blinded comparison involving 48 samples that were either positive, negative, or equivocal for blood ethanol. All 20 samples positive for ethanol were positive for ethyl esters; 7 of 7 samples equivocal for ethanol and classified as negative because the amount was too low for accurate quantitation were positive for ethyl esters; and 21 of 21 samples negative for ethanol were negative for ethyl esters. These data allowed us to conclude that FAEEs in the blood can serve as an excellent short-term confirmatory test for ethanol intake as well as a long-term marker of ethanol ingestion (Figs. 5 and 6). A recent study investigated the use of FAEE in meconium of 248 newborn babies to identify ethanol exposed offspring. The presence of ethyl linoleate, as well as total FAEE, was strongly associated with the maternal self-report of ethanol intake during pregnancy (Bearer, 1999; Klein 1999).

There is clinical utility for this FAEE measurement in several situations, as it has been outlined by Laposata in several algorithms for selection of tests to monitor ethanol intake (Laposata, 1997). First, for those involved in vehicular accidents who were drinking ethanol, if the delay until blood collection is too long, the blood ethanol test may be negative because the ethanol usually stays in the blood for about 6 h or less. The FAEE

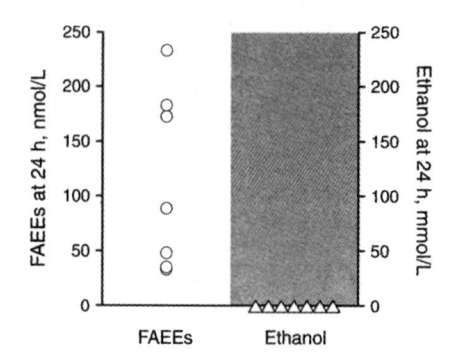

Fig. 6. Serum FAEE and ethanol levels 24 h after the initiation of ethanol ingestion and 22.5 h after cessation. (From Doyle 1996, copyrighted 1996, American Medical Association, with permission.)

test, which stays positive for a full day after ethanol intake, may be able to show that the operator of the vehicle was ingesting ethanol even though the blood ethanol level is zero. Second, because the FAEE test appears to be a more sensitive test for ethanol intake than blood ethanol measurement, the FAEE test may be valuable in confirming within minutes after ethanol intake, that ethanol has been ingested, even if the blood ethanol level is very low.

Third, for forensic cases in which there is a question of poisoning with ethanol, a sample of blood could be analyzed for FAEE quantitation to assess prior ethanol administration. It is not yet known if there are any causes for a false-positive FAEE test or a false-negative FAEE value. It is possible that ingestion of certain medications or foodstuffs will make the FAEE test falsely positive, and it is also possible that interferences exist that block the detection of FAEE. No causes of false-positive or false-negative tests have been yet identified, but there is always a possibility that one will appear as more research is done in the field. In addition, one cannot use the level of blood FAEE 24 h after drinking alcohol to predict the peak blood ethanol concentration much earlier.

8. CONCLUSION

A summary of many of the major observations in the field is shown in Fig. 7 and described in a recent reviews (Laposata, 1998a; Laposata, 1998b). The last several years have seen the rapid growth of our understanding of FAEE synthesis and degradation, FAEE-induced organ damage, and the monitoring of ethanol intake with serum FAEE measurements. The years ahead could bring a mechanistic explanation for FAEE-induced cell injury and a robust clinical assay for detecting ethanol intake.

ACKNOWLEDGMENT

Parts of this chapter are reprinted from *Progress in Lipid Research,* Vol 37, No 5, 1998, pp. 307–316, M. Laposata, Fatty acid ethyl esters: ethanol metabolites which mediate ethanol-induced organ damage and serve as markers of ethanol intake, Copyright (1998), with permission from Elsevier Science.

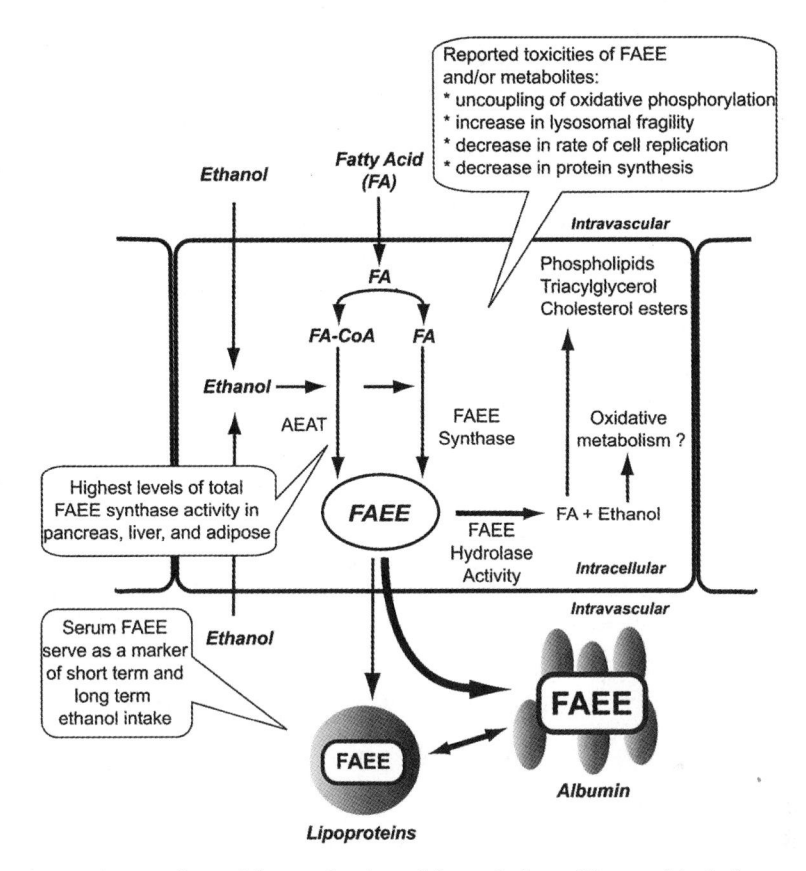

Fig. 7. Current understanding of the synthesis and degradation of fatty acid ethyl esters. FA, fatty acid; FA-CoA, fatty acyl-CoA; FAEE, fatty acid ethyl esters; AEAT, acyl-CoA:ethanol acyltransferase.

REFERENCES

Aleryani S, Kabakibi A, Cluette-Brown JE, Laposata M. Fatty acid ethyl ester synthase, an enzyme responsible for nonoxidative ethanol metabolism, is present in serum following liver and pancreatic injury. Clin Chem 1996; 42:24–27.

Bearer CF, Gould S, Emerson R, Kinnunen P, Cook CS. Fetal alcohol syndrome and fatty acid ethyl esters. Pediatr Res 1992; 31:492–495.

Bearer CF, Lee S, Salvator AE, Minnes S, Swick A, Yamashita T, Singer LT. Ethyl linoleate in meconium: a biomarker for prenatal ethanol exposure. Alcoholism: Clin Exp Res 1999; 23:487–493.

Ben-Eliyahu S, Page GG, Yirmiya R, Taylor AN. Acute alcohol intoxication suppresses natural killer cell activity and promotes tumor metastases. Nature Med 1996; 2:457–460.

Bernhardt TG, Cannistraro PA, Bird DA, Doyle KM, Laposata M. Purification of fatty acid ethyl esters by solid-phase extraction and HPLC. J Chromatogr B 1996; 675:189–196.

Bird DA, Kabakibi A, Laposata M. The distribution of fatty acid ethyl esters among lipoproteins and albumin in human serum. Alcoholism: Clin Exp Res 1997; 21:602–605.

Bird DA, Laposata M, Hamilton JA. Binding of ethyl oleate to low density lipoprotein phospholipid vesicles, and albumin: A ^{13}C NMR study. J Lipid Res 1996; 37:1449–1458.

Bird DA, Szczepiorkowski ZM, Trace VC, Laposata M. Low-density lipoprotein reconstituted with fatty acid ethyl ester as a physiological vehicle for ethyl ester delivery to intact cells. Alcoholism: Clin Exp Res 1995; 19:1265–1270.

Board P, Smith S, Green J, Coggan M, Suzuki T. Evidence against relationship between fatty acid ethyl ester synthase and the pi class glutathione S-transferase in humans. J Biol Chem 1993; 268:15,655–15,658.

Bora PS, Spilburg CA, Lange LG. Identification of a satellite fatty acid ethyl ester synthase from human myocardium as a glutathione S-transferase. J Clin Invest 1989; 84:1942–1946.

Bora PS, Spilburg CA, Lange LG. Metabolism of ethanol and carcinogens by glutathione transferases. Proc Natl Acad Sci USA 1989; 86:4470–4473.

Bora PS, Nalini S, Wu X, Lange LG. Molecular cloning, sequencing, and expression of human myocardial fatty acid ethyl ester synthase-III cDNA. J Biol Chem 1991; 266:16,774–16,777.

Bora PS, Wu X, Lange LG. Site-specific mutagenesis of two histidine residues in fatty acid ethyl ester synthase-III. Biochem Biophys Res Commun 1992; 184:706–711.

Bora PS, Farrar MA, Miller D, Chaitman BR, Guruge BL. Myocardial cell damage by fatty acid ethyl esters. J Cardiovasc Pharmacol 1996; 27:1–6.

Calam DH. Natural occurrence of fatty acid ethyl esters [letter]. Science 1971; 173:78.

Chang W, Waltenbaugh C, Borensztajn J. Fatty acid ethyl ester synthesis by the isolated perfused rat heart. Metabolism 1997; 46:926–929.

Dan L, Cluette-Brown JE, Kabakibi A, Laposata M. Quantitation of the mass of fatty acid ethyl esters synthesized by HepG2 cells incubated with ethanol. Alcoholism: Clin Exp Res 1998; 22:1125–1131.

Dan L, Laposata M. Ethyl palmitate and ethyl oleate are the predominant fatty acid ethyl esters in the blood after ethanol ingestion and their synthesis is differentially influenced by the extracellular concentrations of their corresponding fatty acids. Alcoholism: Clin Exp Res 1997; 21:286–292.

Dean RA, Christian CD, Sample RHB, Bosron WF. Human liver cocaine esterases: ethanol-mediated formation of ethylcocaine. FASEB J 1991; 5:2735–2739.

DePergola G, Kjellstrom C, Holm C, Conradi N, Pettersson P, Bjorntorp P. The metabolism of ethyl esters of fatty acids in adipose tissue of rats chronically exposed to ethanol. Alcoholism: Clin Exp Res 1991; 15:184–189.

Diczfalusy MA, Bjorkhem I, Einarsson C, Alexson SEH. Formation of fatty acid ethyl esters in rat liver microsomes. Evidence for a key role for acyl-CoA : ethanol O-acyltransferase. Eur J Biochem 1999; 259:404–411.

Doyle KM, Bird DA, Al-Salihi S, Hallaq Y, Cluette-Brown JE, Goss KA, et al. Fatty acid ethyl esters are present in human serum after ethanol ingestion. J Lipid Res 1994; 35:428–437.

Doyle KM, Cluette-Brown JE, Dube DM, Bernhardt TG, Morse CR, Laposata M. Fatty acid ethyl esters in the blood as markers for ethanol intake. J Am Med Assoc 1996; 276:1152–1156.

Goodman DS, Deykin D. Fatty acid ethyl ester formation during ethanol metabolism in vivo. Proc Soc Exp Biol Med 1963; 113:65–67.

Gordon SG, Philippon F, Borgen KS, Kern F. Formation of fatty acid ethyl esters during lipid extraction and storage: an important artifact. Biochim Biophys Acta 1970; 218:366–368.

Gorski NP, Nouraldin H, Dube DM, Preffer FI, Dombkowski DM, Villa EM, et al. Reduced fatty acid ethyl ester synthase activity in the white blood cells of alcoholics. Alcoholism, Clin Exp Res 1996; 20:268–274.

Grigor MR, Bell IC Jr. Synthesis of fatty acid esters of short chain alcohols by an acyltransferase in rat liver microsomes. Biochim Biophys Acta 1973; 306:26–30.

Gubitosi-Klug RA, Gross RW. Fatty acid ethyl esters, nonoxidative metabolites of ethanol, accelerate the kinetics of activation of the human brain delayed rectifier K^+ channel, Kv 1.1. J Biol Chem 1996; 271:32,519–32,522.

Haber PS, Wilson JS, Apte MV, Pirola, R.C. Fatty acid ethyl esters increase rat pancreatic lysosomal fragility. J Lab Clin Med 1993; 121:759–764.

Harris WS, Windsor SL, Caspermeyer JJ. Modification of lipid-related atherosclerosis risk factors by ω3 fatty acid ethyl esters in hypertriglyceridemic patients. J Nutr Biochem 1993; 4:706–712.

Heith AM, Morse CR, Tsujita T, Volpacelli SA, Flood JG, Laposata M. Fatty acid ethyl ester synthase catalyzes the esterification of ethanol to cocaine. Biochem Biophys Res Commun 1995; 208:549–554.

Hungund BL, Goldstein DB, Villegas F, Cooper TB. Formation of fatty acid ethyl esters during chronic ethanol treatment in mice. Biochem Pharmacol 1988; 37:3001–3004.

Hungund DL, Gokhale B. Reduction of fatty acid ethyl ester accumulation by ganglioside GM 1 in rat fetus exposed to ethanol. Biochem Pharmacol 1994; 48:2103–2108.

Isenberg KE, Bora PS, Zhou X, Wu X, Moore BW, Lange LG. Nonoxidative ethanol metabolism: expression of fatty acid ethyl ester synthase-III in cultured neural cells. Biochem Biophys Res Commun 1992; 185:938–943.

Johnson AR, Fogerty AC, Hood RL, Kozuharov S, Ford GL. Gas–liquid chromatography of ethyl ester artifacts formed during preparation of fatty acid methyl esters. J Lipid Res 1976; 17:431–432.

Klein J, Karaskov T, Koren G. Fatty acid ethyl esters: a novel biologic marker for heavy in utero ethanol exposure: a case report. Ther Drug Monitor 1999; 21:644–646.

Krokan HE, Bjerve KS, Mork E. The enteral bioavailability of eicosapentaenoic acid and docosahexaenoic acid is as good from ethyl esters as from glyceryl esters in spite of lower hydrolytic rates by pancreatic lipase in vitro. Biochim Biophys Acta 1993; 1168:59–67.

Lange LG, Bergmann SR, Sobel BE. Identification of fatty acid ethyl esters as products of rabbit myocardial ethanol metabolism. J Biol Chem 1981; 256:12,968–12,973.

Lange LG, Sobel BE. Mitochondrial dysfunction induced by fatty acid ethyl esters, myocardial metabolites of ethanol. J Clin Invest 1983; 72:724–731.

Lange LG, Mechanism of fatty acid ethyl ester formation and biological significance. Ann NY Acad Sci 1991; 625:802–806.

Lange LG. Nonoxidative ethanol metabolism: formation of fatty acid ethyl esters by cholesterol esterase. Proc Natl Acad Sci USA 1982; 79:3954–3957.

Laposata EA, Lange LG. Presence of nonoxidative ethanol metabolism in human organs commonly damaged by ethanol abuse. Science 1986; 231:497–499.

Laposata EA, Harrison EH, Hedberg EB. Synthesis and degradation of fatty acid ethyl esters by cultured hepatoma cells exposed to ethanol. J Biol Chem 1990; 265:9688–9693.

Laposata EA, Scherrer DE, Lange LG. Fatty acid ethyl esters in adipose tissue. Arch Pathol Lab Med 1989; 113:762–766.

Laposata EA, Scherrer DE, Mazow C, Lange LG. Metabolism of ethanol by human brain to fatty acid ethyl esters. J Biol Chem 1987; 262:4653–4657.

Laposata M. Fatty acid ethyl esters: short-term and long-term serum markers of ethanol intake. Clin Chem 1997; 43:1527–1534.

Laposata M. Fatty acid ethyl esters: ethanol metabolites which mediate ethanol-induced organ damage and serve as markers of ethanol intake. Prog Lipid Res 1998; 37:307–316.

Laposata M. Fatty acid ethyl esters: nonoxidative metabolites of ethanol. Addict Biol 1998; 3:5–14.

Laseter JL, Weete JD. Fatty acid ethyl esters of Rhizopus arrhizus. Science 1971; 172:864–865.

Manautou JE, Carlson GP. Ethanol-induced fatty acid ethyl ester formation in vivo and in vitro in rat lung. Toxicology 1991; 70:303–312.

Manautou JE, Buss NJ, Carlson GP. Oxidative and non-oxidative metabolism of ethanol by the rabbit lung. Toxicol Lett 1992; 62:93–99.

Marangoni F, Angeli MT, Collis S, Eligini S, Tremoli E, Sirtori CR, et al. Changes of n-3 and n-6 fatty acids in plasma and circulating cells of normal subjects, after prolonged administration of 20:5 (EPA) and 22:6 (DHA) ethyl esters and prolonged washout. Biochim Biophys Acta 1993; 1210:55–62.

Margolis S, Vaughan M. α-Glycerophosphate synthesis and breakdown in homogenates of adipose tissue. J Clin Chem 1962; 237:44–48.

Mogelson S, Lange LG. Nonoxidative ethanol metabolism in rabbit myocardium: purification to homogeneity of fatty acyl ethyl ester synthase. Biochemistry 1984; 23:4075–4081.

Newsome WH, Rattray JBM. Fatty acid ethyl ester formation by plasma in vitro. Can J Biochem 1966; 4:219–227.

Newsome WH, Rattray JBM. The enzymatic esterification of ethanol with fatty acids. Can J Biochem 1965; 43:1223–1233.

Nordoy A, Barstad L, Connor WE, Hatcher L. Absorption of the n-3 eicosapentaenoic and docosahexaenoic acids as ethyl esters and triglycerides by humans. Am J Clin Nutr 1991; 53:1185–1190.

Patton S, McCarthy RD. Conversion of alcohol to ethyl esters of fatty acids by the lactating goat. Nature 1966; 209:616–617.

Paulsen JE, Elvsaas I-KO, Steffensen I-L, Alexander J. A fish oil derived concentrate enriched in eicosapentaenoic and docosahexaenoic acid as ethyl ester suppresses the formation and growth of intestinal polyps in the Min mouse. Carcinogenesis 1997; 18:1905–1910.

Polokoff MA, Bell RM. Limited palmitoyl-CoA penetration into microsomal vesicles as evidenced by a highly latent ethanol acyltransferase activity. J Biol Chem 1976; 253:7173–7178.

Ponnappa BC, Ellingson JS, Hoek JB, Rubin E. Intracellular accumulation of fatty acid ethyl esters inhibits pancreatic protein synthesis. Gastroenterology 1994; 106:316A.

Riley DJS, Kyger EM, Spilburg CA, Lange LG. Pancreatic cholesterol esterases. 2. Purification and characterization of human pancreatic fatty acid ethyl ester synthase. Biochemistry 1990; 29:3848–3852.

Saghir M, Werner J, Laposata M. The rapid and extensive hydrolysis of fatty acid ethyl esters, toxic nonoxidative ethanol metabolites, in the gastrointestinal tract and circulation of rats. Am J Physiol (Gastrointest Liver Physiol 36) 1997; 273:G184–G190.

Schmidt A, Vogel RL, Witherup KM, Rutledge SJ, Pitzenberger SM, Adam M, et al. Identification of fatty acid methyl ester as naturally occurring transcriptional regulators of the members of the peroxisome proliferator-activated receptor family. Lipids 1996; 31:1115–1124.

Sharma R, Gupta S, Singhal SS, Ahmad H, Haque A, Awasthi YC. Independent segregation of glutathione S-transferase and fatty acid ethyl ester synthase from pancreas and other human tissues. Biochem J 1991; 275:507–513.

Skorepa J, Hrabak P, Mares P, Linnarson A. Mass-spectrometric evidence for methyl and ethyl esters of long-chain fatty acids in ox pancreas. Biochem J 1968; 107:318–319.

Soderberg BL, Sicinska ET, Blodget E, Cluette-Brown JE, Suter PM, Schuppisser T, et al. Preanalytical variables affecting the quantitation of fatty acid ethyl esters in plasma and serum samples. Clin Chem 1999; 45:2183–2190.

Szczepiorkowski ZM, Dickersin RG, Laposata M. Fatty acid ethyl esters decrease human hepatoblastoma cell proliferation and protein synthesis. Gastroenterology 1995; 108:515–522.

Treloar T, Madden LJ, Winter JS, Smith JL, de Jersey J. Fatty acid ethyl ester synthesis by human liver microsomes. Biochim Biophys Acta 1996; 1299:160–166.

Tsujita T, Okuda H. Fatty acid ethyl ester synthase in rat adipose issue and its relationship to carboxylesterase. J Biol Chem 1992; 267:23,489–23,494.

Tsujita T, Okuda H. Fatty acid ethyl ester-synthesizing activity of lipoprotein lipase from rat post heparin plasma. J Biol Chem 1994; 269:5884–5889.

Tsujita T, Okuda H. The synthesis of fatty acid ethyl ester by carboxylester lipase. Eur J Biochem 1994; 224:47–62.

Tsujita T, Sumiyoshi M, Okuda H. Fatty acid alcohol ester-synthesizing activity of lipoprotein lipase. J Biochem 1999; 126:1074–1079.

Werner J, Laposata M, Castillo FC, Saghir M, Lozzo RV, Lewandrowski KB, et al. Pancreatic injury induced by fatty acid ethyl ester, a nonoxidative metabolite of alcohol. Gastroenterology 1997; 113:286–294.

Yamazaki K, Hamazaki T. Changes in fatty acid composition in rat blood and organs after infusion of eicosapentaenoic acid ethyl ester. Biochim Biophys Acta 1992; 1128:35–43.

V Psychiatry and Behavior

18 Omega-3 Fatty Acids and Psychiatric Disorders
Current Status of the Field

Joseph R. Hibbeln and Norman Salem Jr.

1. INTRODUCTION

The study of omega-3 essential fatty acids in psychiatric disorders is a field in its infancy. The most compelling aspect of this field is the potential to significantly reduce the burden of major mental illnesses because they are a significant source of disability worldwide. For example, the World Health Organization has identified major depression, which has been strongly linked to low omega-3 status, as the world's single greatest cause of morbidity, defined as years of life lost to a disability (Murray et al., 1996). The magnitude of the effects of low omega-3 status upon populations can be estimated by cross-national comparisons of the dietary intake of seafood and the prevalence rates of the major psychiatric illnesses. It is remarkable that 50- to 60-fold differences in the prevalence rates across countries for three affective disorders, major depression, bipolar depression, and postpartum depression, are each robustly related to the amounts of seafood consumed in each country (Hibbeln 1998, Hibbeln, unpublished data). In contrast, seafood consumption does not appear to predict the prevalence or outcome of schizophrenia consumption, which may indicate a specific relationship to affective and impulsive disorders (Noaghuli, Hibbeln, and Weissman, unpublished data).

Several factors have made this field of study attractive to patients and practicing clinicians, including the nontoxic and nutritionally based approach to treatment and lack of known adverse interactions with other psychotrophic medications. However, the most important factors are that the emerging clinical studies have shown dramatic results and physicians are already using these compounds in their practices, with good results. There is a great diversity in the etiology and prognosis of the various psychiatric disorders, some of which may respond to α-linolenic acid (LNA), eicosapentaenoic acid (EPA), and/or docosahexaenoic acid (DHA), whereas others may not. Finally, the field of omega-3 fatty acids in psychiatric disorders offers a rich diversity of biological mechanisms to explore, which may lead to a better understanding of the pathophysiology of mental illnesses.

Data in this field are only beginning to emerge. Thus, this chapter will review data regarding the relationships of LNA, EPA, and DHA to psychiatric disorders that are available from meeting abstracts, private communications, and unpublished data, in addition to published data, in order to provide as broad a base of information as possible

From: *Fatty Acids: Physiological and Behavioral Functions*
Edited by: D. Mostofsky, S. Yehuda, and N. Salem Jr. © Humana Press Inc., Totowa, NJ

for consideration. Several unresolved fundamental questions will be considered. (1) Has treatment efficacy been demonstrated with multiple independent trials in any specific psychiatric disorder or symptom complex? Depressive symptoms and some clinical aspects of schizophrenia are coming close to meeting those criteria. (2) Is there any specificity of treatment with LNA, EPA, or DHA on specific psychiatric illnesses or symptom complexes? Little is know about which individual fatty acid or combinations of fatty acids are clinically effective, let alone the appropriate doses or treatment courses. (3) What are the possible mechanisms of action of EPA and DHA on nervous system functions related to psychiatric disorders? Are biophysical alterations of synaptic membranes a primary mechanism or should peripheral interactions, for example, immune-neuro-endocrine, mechanisms be considered? (4) What is the impact of a low omega-3 status during early neurological development on the increased risk of later psychiatric disorders? Are treatments effective only in adulthood or are specific stages in the illness in adulthood too late? There is clearly not enough data to appropriately evaluate these questions; however, we do feel that these questions should be posed.

2. MAJOR DEPRESSION IN THE CONTEXT OF EVOLUTION

Fundamental to the understanding of any biological process, including mental illnesses, involves understanding their evolutionary context. There are few readily apparent evolutionary advantages for the development of a high prevalence of major depressive illnesses among *Homo sapiens*. Major depression frequently destroys peoples lives and has a chronic recurring course. Major depression is defined by DSM-IV diagnostic criteria (American Psychiatric Association, 1994) by the loss of the ability to function in family or job life for at least 2 wk, as a a result of disturbances in mood, sleep, concentration, self-esteem, appetite, physical energy, and sexual energy or function. This disease not only directly reduces the likelihood of procreation but also causes significant disruption in social and family interactions. Prior to 1910, the prevalence of major depression may have been nearly 100-fold less than current rates (Klerman et al., 1985). The increased rates of prevalence of affective disorders that have been identified among birth cohorts during the last century in the United States has been well documented (Klerman et al., 1985; Wickramaratne, 1989; Burke et al., 1991; Robins et al., 1994). These increased prevalence rates of depressive disorders during the last century have also been described in several other industrialized nations, including Germany, Taiwan, Canada, and others (Klerman & Weissman, 1989; Cross National Collaborative Group, 1992). The argument that this cohort effect arose from changes in gene frequency over these few decades has been questioned and other potential explanations that have been offered, including nutritional, environmental, and social causes (Bebbington, 1994). Several investigators (Eaton et al., 1998; Broadhurst et al., 1998) have postulated that our genetic patterns evolved in the context of a nutritional milieu that allowed for optimal brain development and psychiatric functioning. Unfortunately, with regard to essential fatty acid intake, the modern diets of postindustrialized societies appear to be discordant with our genetic pattern and may contribute to the increased prevalence rates of major depression (Hibbeln and Salem, 1995). *Homo sapiens* are thought to have evolved consuming diets rich in directly available long-chain omega-3 fatty acids, which may have been permissive for the development of proportionally larger brains (Eaton et al., 1998; Broadhurst et al., 1998; Walter et al., 2000). However, during the last century, diets of industrialized societies have significantly diverged from these traditional diets, reducing the absolute amounts of EPA and DHA intake while increasing the relative intake of omega-6 fatty acids, which compete with omega-3 functions (Eaton et al., 1998). The ratios of omega-6 to omega-3 fatty acids are estimated to be between 0.4 and 2.8 in Paleolithic and evolutionary diets, including models in which either plants and hyper animal butchering were considered as sole sources of food. These models did not include seafood consumption, which would bring the ratios even lower. In contrast, current average ratios are estimated to be approximately 17 in typical Western industrialized societies (Eaton et al., 1998) and we have observed schizophrenics with AA/EPA ratios in red blood cells of more than 70 (unpublished data). These changes in the dietary intake are based not only on decreased seafood consumption but also on the increased production of seed oils (in particular, soybean and corn oils). These seed oils have much higher ratios of linoleic to α-linolenic acids compared to leafy plants, wild game, and seafood (Sinclair et al., 1987; Naughton et al., 1986). Clearly, there have been numerous dramatic changes in human civilization in the last century that may have contributed to the increased prevalence of major depression. However, because changes in the dietary intake of essential fatty acids appear to be able to directly influence central nervous system function, these nutritional factors should be carefully considered (Hibbeln and Salem, 1995). Quantitative assessment of this hypothesis is, unfortunately, difficult because of the uneven quality and paucity of historical dietary data.

3. CROSS-NATIONAL EPIDEMIOLOGY AND MAJOR DEPRESSION

The analyses of cross-national epidemiological data, collected using high-quality modern diagnostic and epidemiological sampling methods, does provide one method of testing the hypothesis that a lower omega-3 fatty acid status is related to higher prevalence rates of affective disorders, psychotic disorders, or aggressive behaviors. Economic data describing seafood consumption has been useful in these cross-national studies. Although economic data on the production and consumption of seafood cannot accurately be used to quantify dietary intake for an individual, these data can be used to describe trends for the populations of entire countries and thus provide a basis for comparing consumption across countries (World Health Organization, 1996). The financial incentive to produce accurate data also adds some confidence to the accuracy of consumption estimates derived from economic data. When compared cross-nationally, greater amounts of seafood consumption were robustly correlated ($r = -0.85, p < 0.0005$) with lower lifetime prevalence rates of major depression (Hibbeln, 1998) (*see* Table 1). The prevalence rates of major depression varied nearly 50-fold across countries as assessed by studies using methods similar to the Epidemiological Catchment Area study, a gold standard of studies in psychiatric epidemiology. The primary data sets of the studies were combined for cross-national analyses and weighted for age, sex, and other demographic differences (Weissman et al., 1996). These data appear to be robust after consideration of potential differences in cultural biases in diagnosis because a structured interview with standardized diagnostic criteria was validated in each country prior to the community sampling. Comparative relationships to prevalence rates of postpartum depression, bipolar affective disorder, homicide mortality, suicide mortality, and schizophrenia will be presented later in this chapter, with the discussion of each of these issues. Although cross-national studies do not provide direct evidence of causal relationships, they do offer a perspective on the potential magnitude of these relationships.

Table 1
Seafood Consumption and Annual Prevalence Rates of Major Depression by Country

Country	Major Depression (annual rate/100 persons)	Seafood consumption (lbs/person/yr)
Japan	0.1	147.7
Taiwan	0.8	81.6
Korea	2.3	105.2
United States	3.0	48.1
Puerto Rico	3.0	67.5
France	4.5	63.9
West Germany	5.0	27.6
Canada	5.2	50.7
New Zealand	5.8	39.0

Note: Seafood consumption is based on disappearance data calculated as catch plus imports minus exports. Prevalence rates of major depression are from the Epidemiological Catchment Area Study, with the exception of Japan, as described in Hibbeln (1998). Results of a simple persons linear regression that includes all countries is ($r = -0.85, p < 0.0005$) and that excludes Japan ($r = -0.74, p < 0.03$). These data do not demonstrate a causal relationship between seafood consumption and lower prevalence rates of major depression.

The results of cross-national analysis comparing prevalence rate of major depression are consistent with results of epidemiological studies conducted within single countries. Tanskanen et al. (in press, 2000) studied a population of 1767 subjects within northern Finland and reported that subjects who consumed fish twice a week or more were at a lower risk of reporting depressive symptoms (odds ratio 0.63) and suicidal thinking (odds ratio 0.57) compared to infrequent fish consumers. Consistent with this report, Silvers and Scott (2000) also found that greater fish consumption predicted improved mental health status by self-report among 4644 subjects in New Zealand. An assessment of essential fatty acid status using direct tissue sampling has also yielded similar results in a community sample of 200 elderly subjects selected from a sample population of 4500, representing 80% of the elderly people in 2 counties in Iowa (Hibbeln et al., unpublished data). Comparisons were made among the 50 most depressed men, the 50 most depressed women, 50 control men, and 50 control women. Depressed women had lower plasma concentrations of DHA, but no other fatty acid, compared to control women. Low plasma concentrations of plasma DHA alone also significantly predicted more severe sleep complaints and reports of anxiety among all groups (Hibbeln et al., unpublished data). DHA was the only plasma fatty acid measure or plasma cholesterol measure that was related to psychiatric symptoms. These cross-national and within-country observational studies are consistent with the hypothesis that low omega-3 status is associated with greater risk of depressive symptoms.

4. MAJOR DEPRESSION AND TISSUE COMPOSITION STUDIES

Since the publication our of initial hypothesis (Hibbeln and Salem, 1995), a series of clinical studies have reported that depressed patients have low tissue concentrations of EPA and/or DHA, and several supplementation trials have reported improvements in depressive or symptoms in the affective spectrum. Although initial reports described elevated plasma concentrations of DHA among subjects with some mixture of depressive symptoms (Ellis et al., 1977; Fehily et al., 1981), unfortunately these studies lacked diagnostic specificity, did not control for alcoholism or smoking, and did not specify the use of psychotrophic medications. Following those initial reports, eight studies have reported that lower concentrations of n-3 fatty acids in plasma or red blood cells (RBCs) predicted depressive symptoms (Adams et al., 1996; Maes et al., 1996; Peet et al., 1998; Edwards et al., 1998a; Edwards et al., 1998b; Peet et al., 1999; Maes et al., 1999; Hibbeln et al., 2000). Adams et al. (1996) were the first to report that lower measures of DHA in the phospholipids of red blood cells ($r = 0.80, p < 0.01$) and a greater aracidonic acid (AA) to EPA ratio ($r = 0.73, p < 0.01$) predicted more severe depressive symptoms. Edwards et al. (1998) carefully controlled common confounding factors that would alter omega-3 status among depressed subjects by controlling for both alcohol consumption and cigarette smoking while also assessing dietary intake. DHA concentrations in RBCs emerged as the most significant predictor of Beck Depression Inventory Scores in a multiple-regression model (β coefficient of -0.92). Because the tissue compositional studies have been observational, it has been difficult to absolutely determine which omega-3 fatty acids, or combination of fatty acids, have the most important mechanistic role. It is important to note that the robust correlational relationships described in these tissue compositional studies are remarkably consistent with the robust cross-national relationships that have been described between lower seafood consumption and higher prevalence rates of major depression ($r = -0.84, p < 0.005$, Hibbeln, 1998). These data do offer strong support that DHA and omega-3 status is compromised among depressed subjects, but these data should not be interpreted as evidence of a metabolic defect in omega-3 essential fatty acid metabolism among depressed subjects. Because most of these studies did not assess dietary intake, the possibility that being depressed has altered dietary preferences and reduced omega-3 intake must also be considered.

5. MAJOR DEPRESSION: TREATMENT

Only one treatment study of omega-3 fatty acids in major depression has been completed and none have been published (Marengell et al., 2000). In contrast to the predictions of the tissue compositional and epidemiological studies, this 6-wk trial of 2 g/d of DHA alone did not document any differences in depressive symptoms among subjects with mild to moderate major depression. This study of medication-free patients with a Hamilton depression rating scale of greater than 17 and no significant comorbid psychiatric diagnoses was well designed and carefully conducted. However, several questions remain before the efficacy of DHA in major depression can be ruled out. First, the trial length may not have been adequate. Second, although the dose of DHA (2 g/d) appeared adequate to change most tissue compositions, this dose may have been excessive if there is a nonlinear or "inverted U"-shaped pattern of response. In other words, response may be seen at low doses, but secondary antagonistic mechanisms may be activated at higher doses. This interpretation is supported by anecdotal reports from some patients who have described better efficacy at doses of 100–300 mg/d (personal observations). Although speculative, it is possible that high doses of dietary DHA may replace EPA or arachidonic acid (AA) from small biologically active phospholipid pools (de la Presa-Owens et al., 1998) and decrease eicosanoid production, an effect not apparent at low-dose regimens. Finally, it is also possible that EPA alone or a combination of EPA plus DHA may be needed for optimal treatment efficacy. More controlled clinical studies will help to resolve these issues. Several studies have described a reduction of depressive symptoms among patients with other primary psychiatric disorders treated with EPA or a combination of EPA plus DHA (Stoll et al., 1999a; Peet et al., 2000). These studies do provide evidence of treatment efficacy for long-chain omega-3 fatty acids in depressive symptoms.

6. BIPOLAR AFFECTIVE DISORDER

One of the most important contributions to the field of omega-3 fatty acids in psychiatric disorders has been the double-blind, placebo-controlled treatment study conducted by Stoll et al. (1999a) among subjects with bipolar affective disorder. Bipolar affective disorder is also commonly known as manic–depressive disorder, which more vividly describes the debilitating clinical course of this illness. Thirty subjects were treated with 14 capsules per day containing either 9.6 g/d of ethyl ester EPA plus DHA or an olive oil placebo. Subjects were studied as outpatients for 4 mo and received the capsules in addition to their regular pharmacological therapies. After 4 mo, there was a significantly reduced relapse to a severe episode of mania or depression in the omega-3-treated group compared to the placebo treated group. Among subjects taking no other medications, four subjects in the EPA plus DHA group remained symptom-free for the length of the study, whereas the four subjects in the placebo group all relapsed. Those results indicate that the EPA plus DHA may function as a sole treatment in some patients with bipolar affective disorder. Significant improvements were seen in the Hamilton Depression Rating scales and the Clinical Global Impressions Scale but not in the Young Mania Depression Scale. These data suggest that depressive symptoms may be more responsive to EPA and DHA than manic symptoms. Further studies in bipolar affective disorder are being conducted by the Stanley Foundation ($n = 240$) through funding by the Center for Complementary and Alternative Medicine at the National Institutes of Health.

The cross-national relationships between lower prevalence rates of bipolar affective disorders and greater seafood consumption (Noaghuli, Hibbeln, and Weissman, unpublished data) are strongly consistent with the clinical intervention trial described by Stoll et al. (1999a) above. Bipolar spectrum disorders have the strongest, most well-defined relationship to seafood consumption in a nonlinear power regression with an apparent threshold of approximately 75 lbs/person/y. Below this level of consumption, the prevalence rates of bipolar disorder rise precipitously from 0.4% in Taiwan (81.6 lbs/person/y) to 6.5% in Germany (27.6 lbs/person/y) a nearly 60-fold difference in prevalence (Noaghuli, Hibbeln, and Weissman, unpublished data). Bipolar affective disorders I and II have similar nonlinear relationships.

7. POSTPARTUM DEPRESSION

Postpartum depression may provide an extremely useful model for testing the hypothesis that a deficiency of omega-3 fatty acids in adulthood, and in particular DHA, increases the predisposition to suffering depressive disorders (Hibbeln and Salem, 1995). Throughout pregnancy, the placenta actively transfers DHA from the mother to the developing fetus (Cambell et al., 1998). Without adequate dietary replenishment, DHA stores in mothers can become depleted (Holman et al., 1991; Al et al., 1995) and may not be replenished for 26 wk (Otto et al., 1997). Given these basic findings, we predicted that the prevalence rates of postpartum depression would be higher in countries with lower rates of seafood consumption. This study evaluated published data on the prevalence of postpartum depression and published data on the DHA content of breast milk (Hibbeln, unpublished). We found that the DHA concentration of mothers' milk predicted prevalence rates of postpartum depression in a simple linear regression model ($r = -0.88$, $p < 0.0001$, $n = 16$), whereas seafood consumption predicted prevalence rates of postpartum depression in a nonlinear logarithmic regression ($r = -0.81$, $p < 0.0001$). These differences comparing seafood consumption data to the breast milk compositional data may be the result of the nonlinear relationship between dietary intake of omega-3 fatty acid and final tissue concentrations (Lands et al., 1992). The observation that the AA composition of mothers' milk did not predict postpartum depression prevalence rates adds confidence to the proposition that DHA and/or other omega-3 fatty acids are the active components. These cross-national relationships remained significant even after the exclusion of all Asian countries and the exclusion of countries where there are extreme differences in the percentage of women of low socioeconomic status or other risk factors that have been identified for postpartum depression (Hibbeln, unpublished). We again note that these cross-national studies are not proof of a causal or protective relationship, but they do provide a strong rationale to conduct interventional trials. To our knowledge, there has been no intervention trial completed using either omega-3 or omega-6 fatty acids to prevent or to treat postpartum depression although one is underway at the University of Arizona (Freeman, personal communication).

8. HOSTILITY AND HOMICIDE MORTALITY

Low omega-3 status has been associated with disorders of impulsivity, including homicide and aggression in cross-national, observational, and interventional studies. Higher homicide mortality rates (World Health Organization, 1995) are correlated to lower rates of seafood consumption (World Health Organization, 1996) across 26 countries in a cross-national analysis ($r = -0.63$, $p < 0.0005$) (Hibbeln, 2001). The diagnosis of death as a result of homicide is subject to fewer cross-national cultural differences than are other behavioral outcomes. These data also are consistent with observational and interventional data for violence and hostility. Virkkunen et al. (1987) reported that impulsive and violent offenders had lower plasma concentrations of DHA and higher concentrations of 22:5n6 than nonimpulsive offenders and healthy controls. Three human, double-blind, placebo-controlled intervention trials have demonstrated the efficacy of omega-3 fatty acids in reducing hostility, an affective state closely related to anger and aggression. One double-blind, placebo-controlled trial was specifically conducted to assess the efficacy of DHA in reducing measures of hostility. Hamazaki et al. (1996) reported that 1.5–1.8 g/d of DHA reduced measures of hostility in a picture frustration test among Japanese students undergoing the stress of university exams, in comparison to a placebo-treated group. Consistent with these reports, Weidner et al. (1992) documented reductions in hostility and depression scores among subjects consuming a high-fish diet. Hamazaki et al. (1999) reported that DHA, administered together with a small amount of EPA, reduced plasma catecholamine levels while subjects were resting, consistent with a hostility-reducing effect. These investigators have failed to reproduce the effects of DHA in reducing hostility under nonstressful conditions (Hamazaki et al., 1998). This failure to replicate may also have been the result of the high baseline intake of seafood among Japanese subjects or to the high variability in the Rosensweig Picture Frustration Test, the dependent measure of hostility. Thienprasert et al. (2000) reported decreases in hostility measures among Thai university employees, but not rural farmers in a 2 mo-long double-blind trial of 1.5 g/d of DHA, compared to a mixture of plant oils. Carefully controlled intervention trials using multiple well-validated dependent measures of hostility, conducted on subjects with a low baseline DHA intake, are needed to evaluate the possible role of EPA and or DHA in reducing hostility or aggression.

9. SUICIDE

Preliminary data are beginning to emerge that suggest a relationship between low-omega-3 status and suicidal behavior, which often has a component of impulsivity. Across 31 countries, greater seafood consumption predicts a lower risk of death as a result of suicide among males in an age-adjusted logarithmic regression ($r = -0.42$, $p < 0.02$) (Hibbeln, unpublished data). This relationship is not as robust as the relationship of seafood consumption to prevalence rates of major depression, postpartum depression, or bipolar affective disorders. However, two epidemiological studies within countries are consistent with these cross-national findings. We examined 1767 subjects in northern Finland and reported that frequent fish consumption (twice per week or more) significantly reduced the risk of reporting depressive symptoms (odds ratio = 0.63, $p < 0.03$) and of reporting suicidal thinking (odds ratio = 0.57, $p < 0.04$) (Tanskanen et al., 2001). In a 17-yr follow-up of 256,118 Japanese subjects (Hirayama, 1990), daily fish eating also had a protective effect in reducing the risk of death from suicide (odds ratio = 0.81 [0.27–0.91], 90% confidence intervals [CIs]) compared to all subjects with a nondaily (less than 7 d/wk) consumption of fish. These observations are also consistent with the assessment of omega-3 status directly among patients. Among suicide attempters, low concentrations of plasma EPA alone were robustly correlated with greater psychopathology on rating scales of impulsivity, guilt, future suicide risk, and most subscales of the Comprehensive Psychopathological Rating Scale (Hibbeln et al., 2000). These correlations were observed only among suicide attempters without a primary diagnosis of depression, which may indicate a selective subgroup response. Death by suicide and suicide attempts may be linked to several different psychiatric disorders, including major depression, personality disorders, schizophrenia, and alcohol and substance abuse, and to a variety of precipitants, including manipulation within personal relationships, psychosis, or traumatic life events. Suicidal behavior often contains elements of impulsivity, depression, anxiety, and hostility; thus, a heterogenous relationship to omega-3 fatty acids may be expected, as suicide or suicide attempts occur among heterogenous groups of subjects.

10. SCHIZOPHRENIA

The use of omega-3 fatty acids also seems promising for some symptoms of schizophrenia (*see* Fenton et al., 2000, for a review). A series of case reports by Rudin (1981; 1982) described significant reductions of psychosis after treatment with flax seed oil. More recent open trials have also indicated possible efficacy (Mellor et al., 1995; Puri and Richardson, 1998; Puri et al. 2000). Three double-blind, placebo-controlled trials (Peet and Mellor, 1998; Peet et al., 2000a; Peet et al., 2000b) have reported clinical improvements among schizophrenics treated with EPA alone, while one study found no clinical improvements (Fenton, et al., unpublished data). Unfortunately, results in the form of full publications are not yet available for any of the double-blind trials. Of these studies, the most striking report is that when treated with 3 g/d of EPA alone, 10 of 30 unmedicated first-episode schizophrenics required no other medications to be free of psychosis (Peet et al., 2000a). In contrast, 29 of 29 patients in the placebo group required medications. In a separate trial of schizophrenics already being treated with medications, Peet and Mellor (1998) reported a 24% improvement in psychosis with EPA, but only a 3% improvement with DHA. Chronic schizophrenics also showed a clear reduction in depression scores in a double-blind, placebo-controlled trial of 2 g/d and 4g/d of EPA (Peet et al., 2000b). However, only the patients with the most severe symptoms showed clinical improvements in psychosis. Fenton, et al. (unpublished data) found no clinical improvements comparing 3 g/d of EPA to 3 g/d of mineral oil among 74 chronic schizophrenics who were ill for an average of 20 yr and already on medications. This study was designed to treat residual symptoms and these patients had low baseline levels of psychosis and depression. The striking differences in treatment response comparing these studies may the result of differences in the severity of symptoms, the length of illness, or the concomitant use of psychotrophic medications. In future studies, subject characterization will continue to be a critical issue in the study of schizophrenics. Some fortunate specific subgroups, such as the subject described by Puri and Richardson (1998) may show dramatic responses, whereas others may show no response. For example, the augmentation studies done among well-controlled chronic schizophrenics have thus far shown no great improvements in residual symptoms or in psychosis, whereas the single study among unmedicated subjects early in their illness showed a dramatic response. Needless to say, more clinical trials are needed.

11. NEGATIVE RESULTS IN CROSS-NATIONAL STUDIES OF SCHIZOPHRENIA

In cross-national analyses, no specific relationship between seafood or fish consumption and schizophrenia prevalence nor outcome has been reported. Among eight countries, Christensen and Christensen (1988), reported that better outcome measures for patients with schizophrenia were correlated with a low percentage of fats from animals ($r = 0.91$–0.95, $p < 0.01$) but not correlated to a higher percentage of dietary fat from vegetable and seafood sources ($r = 0.23$–0.50, $p < 0.10$). These results are also consistent with a recent analysis of seafood consumption and prevalence rates of schizophrenia (Naoghiul, Weissman, and Hibbeln, unpublished) obtained using a gold standard of cross-national psychiatric prevalence data. Across 14 countries, there was no significant relationship between prevalence rates of schizophrenia and seafood consumption utilizing simple and nonlinear models. In summary, these data suggest that there could be a specific relationship of seafood consumption and omega-3 status to prevalence rates of affective and impulsive disorders, but not to schizophrenia.

12. ATTENTION-DEFICIT HYPERACTIVITY DISORDER

Initial reports of deficiencies of essential fatty acids among children with attention-deficit hyperactivity disorder and dyslexia raised the promise of potential nondrug treatments. Convincing treatment data from interventional trials has not yet been reported. (Stevens et al. (1995) found that 53 subjects with attention-deficit hyperactivity disorder had significantly lower concentrations of AA, EPA, and DHA in plasma polar phospholipids when compared to 43 control subjects. (Stordy 1995) has described decreased rod function comparing 10 young dyslexics to 10 controls. In an open trial, supplementation with a fish oil containing 480 mg of DHA/d for 1 mo improved scotopic vision among these dyslexics. Stordy (2000) has also reported that supplementation with a mixture of essential fatty acids improved motor skills in a open trial of 15 children. Two well-controlled, double-blind placebo-controlled trials among children with attention-deficit hyperactivity disorder supplemented with either DHA alone (Voigt, 1998) or a mixture of EPA plus DHA (Burgess, 1998) were presented at an international workshop in 1998. However, to our knowledge, neither positive nor negative results from these studies have appeared in publication. One important confounding factor is that the diagnosis of behavioral disorders among children is often very difficult. For example, many children with bipolar affective disorder may respond to supplementation but may be mistaken for children with attention-deficit hyperactivity disorder. The published data are insufficient to either accept or reject the hypothesis that omega-3 supplementation is useful in treating behavioral disorders among children, such as attention-deficit hyperactivity disorder.

13. DEFICIENCIES IN GESTATION AND PSYCHIATRIC OUTCOMES

An intriguing possibility is that some level of deficiency in omega-3 status either during interuterine, postnatal, or in later development, could contribute to a lifelong risk of suffering psychiatric illnesses through irreversible neurodevelopmental changes (Peet et al., 1999). For example, studies of periods of famine such as the Dutch Hunger War have generated hypotheses that specific psychiatric illness may be related to nutritional deprivation in specific periods of development. Prenatal malnutrition in the first trimester may increase risk for developing schizophrenia (Susser and Lin, 1992) as reflected in increased sulcal sizes (Hulshoff Pol et al., 2000); whereas later gestational famine during the second and third trimester increased the risk for later affective disorders (Brown et al., 2000). One test of the hypothesis that mothers are deficient in essential fatty acids during gestation has been to examine the composition of maternal plasma on the day of birth and the development of psychotic illnesses over the next four decades among the children (Hibbeln et al., 2000b). The total fatty acid content as well as the cholesterol ester 18:3n3 and CE 18:2n6 were significantly higher in the plasma from mother of 27 children who developed psychosis compared to the plasma of the 51 control mothers matched for age, date of birth, and ethnicity. Cholesterol ester DHA and cholesterol ester EPA concentrations were also elevated but were not statistically different. These data are inconsistent with the hypothesis that mothers of children with psychosis are deficient in n-6 or n-3 essential fatty acids.

It is also possible that essential fatty acid deficiencies in the postnatal period may increase the risk of the development of psychiatric or behavioral disorders. Higley et al. (1991) has extensively validated a model for inducing greater risk for lifetime behavioral disturbances and alcohol preference among rhesus monkeys. In this paradigm, infant rhesus monkeys are separated from their mothers at birth, raised in a nursery with a cloth surrogate mother, and fed standard formulas. These infant formulas resembled human infant formulas commercially available in the United States in that they are virtually devoid of DHA and AA (Hibbeln et al., 1999). Several studies have documented that infants fed standard formulas virtually devoid of DHA and AA have suboptimal neurological development (as evidenced by behavior and visual outcome measures), when compared to infants fed formulas supplemented with DHA and AA (Willats et al., 1998;

Birch et al., 2000; Gibson and Makrides, 2000; Carlson and Neuringer, 1999). Thus, we postulated that some portion of the increased risk for aggressive and depressive behaviors among the nursery-raised rhesus monkey described could be the result of the very low levels of dietary AA and DHA. It is known that maternal–infant interactions clearly have profound effects on early development and maturation, so a nutrition deprivation could at best contribute to only a small portion of the behavioral differences observed when mother-raised and nursery-raised infants are compared. In our experiment (Hibbeln et al., 2000c) infant rhesus monkeys were removed from their mothers at birth, and for 6 mo, received one of the two formulas while being raised in a stringently controlled nursery. The DHA/AA group received formulas supplemented with AA (0.8%) and DHA (0.8%) that were similar to the milk of rhesus monkey mothers. The control formulas were similar to commercially produced human infant formulas, as they were virtually devoid of DHA and AA. At 6 mo, all animals were switched to diets rich in omega-3 fatty acids. DHA/AA-fed infants had profoundly improved motor development and visual orientation scores in as little as 7 d (Champoux, et al., 1998). The heart-rate variability remained improved in adolescence at up to 2.3 yr after the dietary intervention had stopped, indicating an enduring developmental effect (Hibbeln et al., 2000c). In cerebrospinal fluid, concentrations of 5-hydroxyindolacetic acid (CSF 5-HIAA, a metabolite of serotonin) were decreased in the DHA/AA group, but only during the 6 mo of formula feeding (Hibbeln et al., 2000c). We cannot determine directly whether the supplementation raised or lowered the brain concentrations of serotonin among these infants. However, the behavioral and physiological improvements noted above were consistent with improved serotinergic function. Long-term follow-up data on behavior differences are still being assessed.

14. NEUROTRANSMITTER ALTERATIONS IN INFANCY

Other investigators have established that neurotransmission in the frontal cortex can also be affected by dietary essential fatty acids during infancy. De la Presa Owens and Innis (1999) fed piglets one of four infant formulas for 18 d. The formulas were either adequate or deficient in 18:2n6 and 18:3n3 or contained supplemental AA (0.2%) and DHA (0.16%). Frontal cortex concentrations of serotonin, tryptophan, dopamine, homovanillic acid, and norepinephrine were nearly doubled in the formulas supplemented with long-chain polyunsaturated fatty acids. It is remarkable that only 18 d of dietary intervention altered concentrations of these fundamental neurotransmitters. Austead et al. (2000) also reported changes in concentrations of both serotonin and CSF 5-HIAA in frontal cortex among piglets given control and DHA/AA-supplemented formulas. They reported that frontal cortex concentrations of serotonin increased while the concentration of the metabolite, CSF 5-HIAA, decreased. One interpretation of these data is that DHA/AA supplementation increases tissue concentrations of serotonin by decreasing degradation of serotonin to its metabolite, CSF 5-HIAA. Several drugs used to treat depression have similar effects of raising tissue concentrations of serotonin and lowering CSF 5-HIAA concentrations. For example, fluoxetine (Prozac® increases the concentration of serotonin in the synapse by inhibiting reuptake into the cell (Paex and Hernandez, 1996). CSF 5-HIAA decreases because the serotonin is not exposed to the degradative enzyme monoamine oxidase. Consistent with this interpretation, DeLion et al. (1997) reported that diets rich in omega-3 fats lead to a decrease in monoamine oxidase activity.

15. SEROTINERGIC METABOLISM AND IMPULSIVITY

Abnormalities in serotinergic neurotransmission caused by low-omega-3 status could potentially increase the risk for impulsive behaviors such as suicide and homicide and may be an important mechanism leading to an increased predisposition toward developing a depressive disorder (Hibbeln et al., 1998a). Abnormalities in serotinergic function are thought to be important in impulsive, suicidal, and depressive behaviors. Most of the commonly used antidepressant medications act to increase serotinergic neurotransmission (Meltzer and Lowey, 1987). One of the best-replicated findings in biological psychiatry is that low concentrations of CSF 5-HIAA are associated with suicide and depression (Roy et al., 1991). Low CSF 5-HIAA concentrations predict impulsive, hostile, and aggressive behaviors (Mann, 1995; Virkkunen et al., 1994; Linnoila et al., 1983) and reflect serotonin turnover in the frontal cortex (Stanley et al., 1985).

16. HUMAN DATA ON OMEGA-3 FATTY ACIDS AND NEUROTRANSMITTER METABOLITES

Correlational data from human studies are consistent with the proposition that omega-3 status is related to cerebrospinal fluid (CSF) neurotransmitter metabolite concentrations. We observed that plasma concentrations of DHA and AA predicted CSF 5-HIAA and CSF homovanillic acid concentrations in 234 subjects (Hibbeln, 1998a; Hibbeln et al., 1998b). In healthy control subjects and late-onset alcoholics, higher concentrations of plasma DHA predicted higher concentrations of CSF 5-HIAA. It is remarkable that this correlational relationship was found between a cerebrospinal fluid measure of a neurotransmitter metabolite and a plasma level of a fatty acid. We have also replicated this finding among 104 adult rhesus monkeys. Higher concentrations of the omega-3 fatty acids DHA and EPA in plasma predicted higher concentrations of CSF 5-HIAA (Hibbeln, et al., unpublished data). Among these animals, higher EPA and DHA plasma concentrations also predicted more functional dominance behaviors. These correlational findings suggest that increasing DHA intake may increase brain serotonin concentrations. Higher brain concentrations of CSF 5-HIAA may reduce depression and aggression. Direct data that demonstrate that increasing EPA or DHA status will change the concentrations among human subjects is sparse. Nizzo et al. (1978) reported that acute intravenous infusions of DHA-containing phospholipids increased CSF 5-HIAA and homovanillic acid concentrations in human subjects over 6 h. Unfortunately, this study was poorly controlled and examined few subjects. Further interventional protocols are needed to determine if there is a causal relationship between increasing peripheral omega-3 status and changes in several neurotransmitter metabolites among humans.

In the animal literature, it is documented that altering the omega-3 fatty acid composition of an the diet can alter neurotransmitter concentrations. It is also important to note that deficiencies in omega-3 fatty acids impact monoaminergic neurotransmission in the frontal cortex, as it is important in regulating impulsive behaviors. Zimmer et al. (2000) documented a significant reduction in the number of dopaminergic synaptic vesicles in the frontal cortex of omega-3-deficient rats. This dietary deficiency in omega-3 fatty acids resulted in a 90% reduction in the quantity of dopamine released after tyramine stimulation (Zimmer, Delion-Vancassel et al., 2000). In comparing diets containing fish oils to diets deficient in omega-3 fats, Chalon et al. (1998) found that dopamine levels were 40% greater in the frontal cortex of rats fed fish oils compared to those fed a control diet. In frontal cortex, there was also a reduction in monoamine oxidase activity and greater binding to dopamine D_2 receptors.

Although animal studies have most strongly documented differences in dopaminergic function, they also indicate that the function of serotinergic neurons may also be altered in adult animals. DeLion et al. (1996) reported that a chronic dietary deficiency in omega-3 fatty acids specifically affected monoaminergic systems in the frontal cortex of rats. They reported that a 40–75% lower level of endogenous dopamine in the frontal cortex occurred in deficient rats according to age. This deficiency also induced a 10% reduction in the density of dopaminergic D_2 receptors and an 18–46% increase in serotonin 5-HT_2 receptor density in the frontal cortex with no change in binding affinity and without variation in serotonin levels. These changes in serotonin 5-HT_2 receptor density were strikingly similar to the abnormalities noted by Stanley and Mann (1983) among victims of suicide; there was a 44% increase in 5-HT_2 receptor density and no change in binding affinity in the frontal cortex. Olsson et al. (1998) reported that a diet low in omega-3 fatty decreased serotonin and 5-HIAA concentrations in several brain regions, including the cortex. In direct biophysical experiments, Heron et al. (1980) found that changing membrane fatty acid composition resulted in markedly altered serotonin receptor binding, an effect they attributed to alterations in the membrane biophysical state. These observations support the assertion that greater omega-3 fatty acid intake may serve to diminish some of the abnormalities of serotinergic neurotransmission, which are associated with impulsive behaviors.

17. OMEGA-3 FATTY ACIDS AND OTHER POTENTIAL MECHANISMS OF ACTION

A discussion of possible mechanisms is difficult and somewhat premature, as the body of clinical data that demonstrates efficacy is still emerging and selective effects of individual fatty acids on specific illness have not been demonstrated. Nonetheless, many of the clinical studies of omega-3 fats in psychiatric disorders have emerged in part from the observation that DHA is selectively concentrated in neuronal membranes (Salem et al., 1986; Salem, 1989). Alterations in the polyunsaturated fatty acid composition of synaptic membranes may exacerbate a predisposition to psychiatric disorders through several mechanisms. (1) Depletion of DHA from synaptic membranes creates biophysical changes in membrane structures that alter the shape and activity of membrane bound proteins, receptors, and enzymes (Mitchell et al., 1992; Huster et al., 1998; Mitchell et al., 1998; Litman et al., 1991; Litman et al., 1996). These biophysical changes may affect neurotransmitter receptor binding, catecholamine metabolism, catabolism and neurotransmitter reuptake because each of these processes are dependent on membrane-bound proteins (Hibbeln and Salem, 1995). (2) Psychiatric disturbances may result from disturbances in signal transduction pathways such as phosphatidylinositol turnover and protein kinase C activity (Stoll et al., 1999b). (3) DHA appears to have an important role in neuronal growth. DHA concentrates in synaptic growth cones during neuronal growth (Martin, 1998), and improves dendritic branching (Wainwright et al., 1998). A dietary deficiency of n-3 fats causes decreased nerve growth factor content in the rat hippocampus (Ikemoto et al., 2000). DHA may be protective for neuronal apoptotic cell death (Kim et al., 2000). Each of these mechanisms has been implicated in the biology of affective disorders (Stoll et al., 1999b). Clearly, many of these putative biochemical mechanisms can be applied to psychiatric disorders other than affective disorders.

18. CONSIDERATIONS OF CENTRAL AND PERIPHERAL MECHANISMS

One paradoxical observation that appears to be emerging from the clinical intervention trials is that EPA appears to be more clinically effective than DHA. In fact, no clinical study has yet demonstrated a clinical effect of DHA (Voigt et al., 1998; Peet et al., 2000; Marengell et al., 2000). Whereas DHA is selectively concentrated in neuronal membranes, EPA is virtually absent from neuronal tissues. Few peripheral mechanisms by which EPA could have central nervous systems effects have been described. For example, it is possible that EPA and/or DHA might increase vascular blood flow through the brain. However, a more plausible and well-substantiated immune-neuro-endocrine mechanism has been proposed by Smith (1991) and examined by Maes et al. (1993). This model has been described extensively in other publications (Smith, 1991; Maes et al 1993; Hibbeln, 1999) and will only be briefly presented here. Hyperactivity of the hypothalamic pituitary axis has repeatedly been described among depressed patients and is thought is be caused by elevated levels of corticotrophic releasing factor (CRF) in the paraventricular nucleus and hypothalamus (Kling et al., 1989). One possible contributing cause of elevations in CRF concentrations is excessive stimulation by interleukin-1β, which is dependent on eicosanoid release (Rivier and Vale 1991). Central interleukin-1β concentrations can be elevated as a result of increased peripheral interleukin-1β release from macrophages (Berkenbosch et al., 1987; Sapolsky et al., 1987) and depressed patients appear to have elevated concentrations of interleukin-1β compared to controls (Maes et al., 1993). EPA and DHA supplementation has been shown to reduce interleukin-1β production by as much as 54% in double-blind clinical trials (Endres et al 1987; Kremer et al., 1990; Meydani et al., 1993). Further testing of this hypothetical mechanism would require careful control of essential fatty acid intake and careful assessment of both immune functions and the hypothalamic pituitary axis.

19. COMMENTS FOR THE DESIGN OF FUTURE STUDIES; TISSUE CONCENTRATION STUDIES

Comparisons of tissue concentrations between controls and patients groups may be useful for exploratory analyses, but the groups should be controlled for confounding influences on fatty acid metabolism and catabolism and care should be take in the interpretation of the results. Comparison studies are of little value without careful diagnostic assessment and patient characterization. Consideration should also be given to the observation that improvement of depressive symptoms among subjects with other diagnoses may confound other outcome measures such as improvements in psychosis or cognition. Typically, comparison studies of the tissue composition of the fatty acids only determine if there is a difference between the groups at single point in time. Thus, it may be difficult to make interpretations concerning differences in basal metabolism or causal relationships. A series of critical questions must be addressed. Does this measure reflect differences in dietary intake? Does the measure reflect, for example, that schizophrenics tend to smoke more and lose polyunsaturates to oxidative stress? Does it reflect that there was greater alcohol consumption among the patient group? Does it reflect the type of neuroleptic medications the patient has been taking? Do these alter lipid metabolism or increase oxidation? Few comparison studies have simultaneously controlled for smoking, alcohol use, and dietary intake, but when controlled, the results were clearly interpretable and had low variability (Peet et al., 1998). Even after known confounding factors have been controlled, it is doubtful that tissue compositional studies can accurately test the hypothesis that the metabolism of essential fatty acids are altered by comparing patient groups to controls. For example, there was a poor correlation between the ratios of precursors to products and the in vivo production of deuterated AA and DHA from deuterated fatty acid precursors in adults (Hibbeln and Salem, unpublished data). A decreased product-to-precursor ratio may not indicate that there is a block in metabolism, it may also indicate that the product is being rapidly degraded. Finally, technical issues should be considered in studies of tissue composition such as the difficult separation and quantification of 24:0 and 24:1 from 22:6n3 and the stability of samples.

20. CONSIDERATIONS FOR FUTURE INTERVENTION TRIALS

The long history of clinical studies in cardiovascular diseases, infant nutrition, and other illnesses have yielded valuable lessons that can help to form the design of future studies of psychiatric disorders. Baseline measurements are essential, as some subjects may enter the study with high tissue concentrations of EPA and DHA and supplementation may produce little additional effect. Biological markers are the most accurate measure of fatty acid status, but a simple dietary questionnaire may be able to screen out subjects with a high omega-3 intake. The baseline diet should be standardized at a low seafood and omega-3 intake. If subjects in the placebo group begin to eat more fish because of their excitement in being part of the study, then treatment effects may be difficult to detect. Consideration should be given to the course of their illness and number of failed treatments. For example, it may be difficult to detect treatment differences among subjects who are treatment resistant and have failed all other therapies. Sample sizes should be calculated differently for prevention and treatment studies. Unfortunately, it is currently unknown what length of treatment will be required to change brain fatty acid composition among psychiatric subjects. It is also unknown what time period is required to observe a clinical effect after a change in tissue composition is achieved. The use of placebos and comparison oils should also be carefully considered. Flavor masking should be considered with the addition of small amounts of fish oils or other flavors to placebo oils. If possible, compounds should be balanced for their degree of unsaturation, delivery form (i.e., triglycerides or ethyl esters or phospholipids), and their caloric content. A positive placebo control may be useful when omega-3 fatty acids are to be compared to other psychotrophic agents that have sexual or other side effects. Study subjects should always be asked if they can identify their treatment group and these data reported. Whenever possible, plasma or red blood cell analyses should be used to confirm treatment and dietary compliance.

21. CONCLUSIONS

As a field, the use of omega-3 fatty acids in psychiatric disorders appears to have significant potential. Cross-national comparisons between seafood consumption and prevalence rates of psychiatric disorders indicate that there is a significant predictive relationship to the 50- to 60-fold range of rates of prevalence of affective disorders. These studies also indicate that there may be a more selective relationship of omega-3 status to affective and impulsive disorders than to schizophrenia. Thus far, a limited number of placebo-controlled intervention trials appears to demonstrate greater efficacy among patients with more severe symptoms (e.g., who are not undergoing current treatment with psychotrophic medications) (Peet et al., 2000). Overall, there appears to be a treatment response to supplements containing at least some EPA, and significant positive treatment results have not yet been described in trials using DHA alone. This clinical observation has resulted in a mechanistic paradox, as EPA is not found in significant concentrations in neuronal tissues, whereas DHA is selectively concentrated in the brain (Salem 1986). Future studies should carefully consider patient diagnosis and stage of illness as well as baseline tissue concentrations and dietary intake of omega-3 fatty acids, both before and during the study. Continued work in this field may yield inexpensive treatments and prevention strategies for these psychiatric illness, which are a significant source of morbidity worldwide.

REFERENCES

Adams PB, Lawson S, Sanigorski A, Sinclair AJ. Arachidonic to eicosapentaenoic acid ratio in blood correlates positively with clinical symptoms of depression. Lipids 1995; 31:S157–S161.

Al MDM, Van Houwelingen AC, Kester ADM, Hasaart THM, De Jong AEP, Hornstra G. Maternal essential fatty acids patterns during normal pregnancy and their relationship to the neonatal essential fatty acid status. Br J Nutr 1995; 75:55–68.

American Psychiatric Association. Diagnostic Criteria from DSM-IV. American Psychiatric Association, Washington, DC, 1994.

Austead N, Innis SM, de la Presa Owens S. Auditory evoked response and brain phospholipids fatty acids and monoamines in rats fed formula with out without arachidonic acid (AA) and/or docosahexaenoic acid (DHA). Brain Uptake and Utilization of Fatty Acids. Conference Organizations: Watkins P, Spector A, Hamilton J, Katz, R. Applications to Peroxisomal Biogenesis Disorders. National Institutes of Health Conference, Bethesda, MD, 2000, p. 3.

Bebbington P, Ramana R. The epidemiology of bipolar affective disorder. Social Pschiatry Psychiatr Epidemiol 1995; 30:279–292.

Berkenbosch F, van Oers J, del Rey A, Tilders F, Besedovsky H. Corticotropin-releasing factor-producing neurons in the rat activated by interleukin-1. Science 1987; 238(4826):524–526.

Birch EE, Garfield S, Hoffman DR, Uauy R, Birch DG. A randomized controlled trial of early dietary supply of long-chain polyunsaturated fatty acids and mental development in term infants. Dev Med Child Neurol 2000; 42(3):174–181.

Broadhurst CL, Cunnane SC, Crawford MA. Rift Valley lake fish and shellfish provided brain-specific nutrition for early *Homo*. Br J Nutr 1998; 79(1):3–21.

Brown AS, van Os J, Driessens C, Hoek HW, Susser ES. Further evidence of relation between prenatal famine and major affective disorder. Am J Psychiatry 2000; 157:190–195.

Burgess JR. Attention deficit hyperactivity disorder, observational and interventional studies. NIH Workshop on Omega-3 Essential Fatty Acids and Psychiatric Disorders, 1998, p. 23.

Burke KC, Burke JD Jr, Regier DA, Rae DS. Age at onset of selected mental disorders in five community populations. Arch Gen Psychiatry 1990; 47(6):511–518.

Cambell FM, Gordon MJ, Dutta-Roy AK. Placental membrane fatty acid-binding protein preferentially binds arachidonic and docosahexaenoic acids. Life Sci 1998; 63(4):235–240.

Carlson SE, Neuringer M. Polyunsaturated fatty acid status and neurodevelopment: a summary and critical analysis of the literature. Lipids 1999; 34:171–178.

Chalon S, Delion-Vancassel S, Belzung C, Guilloteau D, Leguisquet AM, Besnard JC, et al. Dietary fish oil affects monoaminergic neurotransmission and behavior in rats. J Nutr 1998; 128:2512–2519.

Champoux M, Shannon C, Hibbeln JR, Salem N Jr. Essential fatty acid formula supplementation and neuromotor capabilities in nursery reared-reared rhesus monkeys neonates. [abstract]. American Society of Primatologists Annual Meeting, 1998.

Christensen O, Christensen E. Fat consumption and schizophrenia. Acta Psychiatr Scand 1988; 78:587–591.

Cross National Collaborative Group. The changing rate of major depression: crossnational comparisons, JAMA 1992; 268(21):3098–3105.

de la Presa Owens S, Innis SM. Docosahexaenoic and arachidonic acid prevent a decrease in dopaminergic and serotoninergic neurotransmitters in frontal cortex caused by a linoleic and alpha-linolenic acid deficient diet in formula-fed piglets. J Nutr 1999; 129(11):2088–2093.

Delion S, Chalon S, Guilloteau D, Besnard JC, Durand G. Alpha-linolenic acid dietary deficiency alters age-related changes of dopaminergic and serotoninergic neurotransmission in the rat frontal cortex. J Neurochem 1996; 66(4):1582–1591.

Delion S, Chalon S, Guilloteau D, Lejeune B, Besnard JC, Durand G. Age-related changes in phospholipid fatty acid composition and monoaminergic neurotransmission in the hippocampus of rats fed a balanced or an n-3 polyunsaturated fatty acid-deficient diet. J Lipid Res 1997; 38(4):680–689.

Eaton SB, Eaton SB, Sinclair A, Cordain L, Mann NJ. Dietary intake of long-chain polyunsaturated fatty acids during the paleolithic. In: Simopoulos AP, ed. The Return of ω-3 Fatty Acids to the Food Supply I: Land-Based Animal Food Products and Their Health Effects. World Review of Nutrition and Dietetics Vol. 83. Karger, Basal, 1998, pp. 12–23.

Edwards R, Peet M, Shay J, Horrobin D. Omega-3 polyunsaturated fatty acid levels in the diet and in red blood cell membranes of depressed patients. J Affect Disord 1998; 48(2–3):149–55.

Edwards R, Peet M, Shay J, Horrobin D. Depletion of docosahexaenoic acid in red blood cell membranes of depressive patients. Biochem Soc Trans 1998; 26(2):S142.

Ellis FR, Sanders TA. Long chain polyunsaturated fatty acids in endogenous depression. J Neurol Neurosurg Psychiatry 1977; 40(2):168–169.

Endres S, Ghorbani R, Kelley VE, Georgilis K, Lonnemann G, van der Meer JW. The effect of dietary supplementation with n-3 polyunsaturated fatty acids on the synthesis of interleukin-1 and tumor necrosis factor by mononuclear cells. N Eng J Med 1989; 320:265–271.

Fehily AMA, Bowery OAM, Ellis FR, Meade BW, Dickerson JWT. Plasma and erythrocyte membrane long chain fatty acids in endogenous depression. Neurochem Int 1981; 3:37–42.

Fenton WS, Hibbeln J, Knable M. Essential fatty acids, lipid membrane abnormalities, and the diagnosis and treatment of schizophrenia. Biol Psychiatry 2000; 47(1):8–21.

Gibson RA, Makrides M. N-3 Polyunsaturated fatty acid requirements of term infants. Am J Clin Nutr 2000; 71(1):251S–255S.

Hamazaki T, Sawazaki S, Itomura M, Asaoka E, Nagao Y, Nishimura N, et al. The effect of docosahexaenoic acid on aggression in young adults. A placebo-controlled double-blind study. J Clin Invest 1996; 97:1129–1133.

Heron D, Shinitzky M, Hershkowitz M, Samuel D. Lipid fluidity markedly modulates the binding of serotonin to mouse brain membranes. Proc Natl Acad Sci USA 1980; 77(12):7463–7467.

Hibbeln JR, Salem N Jr. Dietary polyunsaturated fatty acids and depression: when cholesterol does not satisfy. Am J Clin Nutr 1995; 62:1–9.

Hibbeln JR. Fish consumption and major depression. Lancet 1998; 351:1213.

Hibbeln JR, Linnoila M, Umhau JC, Rawlings R, George DT, Salem N Jr. Essential fatty acids predict metabolites of serotonin and dopamine in cerebrospinal fluid among healthy control subjects, and early- and late-onset alcoholics. Biol Psychiatry 1998; 44:235–242.

Hibbeln JR, Umhau JC, Salem N Jr, Rawlings R, George DT, Regan P, et al. Essential fatty acids and the neurochemistry of violence II: a replication study of violent and nonviolent subjects. Biol Psychiatry 1998; 44(4):243–249.

Hibbeln JR, Umhau JC, George DT, Shoaf SE, Linnoila M, Salem N Jr. Plasma total cholesterol concentrations do not predict cerebrospinal fluid neurotransmitter metabolites: implications for the biophysical role of highly unsaturated fatty acids. Am J Clin Nutr 2000; 71(1):331S–338S.

Hibbeln JR, DePetrillo P, Higley JD, Schoaf S, Lindell S, Salem N Jr. Improvement in heart rate variability which persist into adolescence using infant formulas containing docosahexaenoic and arachidonic acids. American College of Neuropsychopharmacology Annual Meeting, 1999.

Hibbeln JR, Enstrom G, Majchrzak S, Salem N Jr, Traskman-Benz L. Suicide attempters and PUFAS: lower plasma eicosapentaenoic acid alone predicts greater psychopathology. 4th Congress of the International Society for the Study of Lipids and Fatty Acids, 2000, p. 191.

Hibbeln JR, Buka S, Yolken R, Klebanoff M, Majchrzak S, Salem N Jr. Maternal plasma EFA composition sampled at birth and psychotic illnesses among their children. Brain Reuptake and Utilization of Fatty Acids, Applications to Peroxisomal Disorders, an International Workshop, 2000, p. 11.

Hibbeln JR, DePetrillo P, Higley JD, Schoaf S, Lindell S, Salem N Jr. Neuropsychiatric implications of improvements in heart rate variability among adolescents rhesus monkeys fed formula supplemented with DHA and AA as infants. 4th Congress of the International Society for the Study of Lipids and Fatty Acids, 2000, p. 103.

Hibbeln JR. Homicide mortality rates and seafood consumption: A cross-national analysis. World Rev Nutr Diet 2001; 88:41–46.

Higley JD, Hasert MF, Suomi SJ, Linnoila M. Nonhuman primate model of alcohol abuse: effects of early experience, personality, and stress on alcohol consumption. Proc Natl Acad Sci USA 1991; 88(16):7261–7265.

Hirayama T. Life-Style and Mortality. A Large Census-Based Cohort Study in Japan. Karger, Basal, 1990.

Holman RT, Johnson SB, Ogburn PL. Deficiency of essential fatty acids and membrane fluidity during pregnancy and lactation. Proc Natl Acad Sci USA 1991; 88(11):4835–4839.

Hulshoff Pol HE, Hoek HW, Susser E, Brown AS, Dingemans A, Schnack HG, et al. Prenatal exposure to famine and brain morphology in schizophrenia. Am J Psychiatry 2000; 157(7):1170–1172.

Huster D, Arnold K, Garwrish K. Influence of docosahexaenoic acid and cholesterol on lateral domain organization in phospholipid mixtures. Biochemistry 1998; 37(49):17,299–17,308.

Ikemoto A, Nitta A, Furukawa S, Ohishi M, Nakamura A, Fujii Y, et al. Dietary n-3 deficiency decreases nerve growth factor content in rat hippocampus. Neurosci Lett 2000; 285:99–102.

Kim HY, Akbar M, Lau A, Edsall L. Inhibition of neuronal apoptosis by docosahexaenoic acid (22:6n-3): role of phosphatidylserine in antiapoptotic effect. J Biol Chem 2000; 351(3):709–716.

Kling MA, Perini GI, Demitrack MA, Geracioti TD, Linnoila M, Chrousos GP, Gold PW. Stress-responsive neurohormonal systems and the symptom complex of affective illness. Psychopharmacol Bull 1989; 1989;25(3):312–318.

Klerman RC, Weissman MM. Increasing rates of depression. JAMA 1985; 261:2229–2235.

Kremer JM, Lawrence DA, Jubiz W, DiGiacomo R, Rynes R, Bartholomew LE. Dietary fish oil and olive oil supplementation in patients with rheumatoid arthritis. Clinical and immunologic effects. Arthritis Rheumatology 1990; 33:810–820.

Lands WE, Libelt B, Morris A, Kramer NC, Prewitt TE, Bowen P, et al. Maintenance of lower proportions of (n-6) eicosanoid precursors in phospholipids of human plasma in response to added dietary (n-3) fatty acids. Biochim Biophys Acta 1992; 1180(2):147–62.

Linnoila M, Virkkunen M, Scheinin M, Nuutila A, Rimon R, Goodwin FK. Low cerebrospinal fluid 5-hydroxyindoleacetic acid concentration differentiates impulsive from non-impulsive violent behavior. Life Sci 1983; 33:2609–2614.

Litman BJ, Lewis EN, Levin IW. Packing characteristics of highly unsaturated bilayer lipids: Raman spectroscopic studies of multilamellar phosphatidylcholine dispersions. Biochemistry 1991; 30(2):313–319.

Litman BJ, Mitchell DC. A role for phospholipid polyunsaturation in modulating membrane protein function. Lipids 1996; 31:S193–S197.

Maes M, Smith R, Christophe A, Cosyns P, Desnyder R, Meltzer H. Fatty acid composition in major depression: decreased omega 3 fractions in cholesteryl esters and increased C20: 4 omega 6/C20:5 omega 3 ratio in cholesteryl esters and phospholipids. J Affect Disord 1996; 38(1):35–46.

Maes M, Bosmans E, Meltzer HY, Scharpe S, Suy E. Interleukin-1 beta: a putative mediator of HPA axis hyperactivity in major depression? Am J Psychiatry 1993; 150(8):1189–1193.

Maes M, Christophe A, Delanghe J, Altamura C, Neels H, Meltzer HY. Lowered omega-3 polyunsaturated fatty acids in serum phospholipids and cholesteryl esters of depressed patients. Psychiatry Res 1999; 85(3):275–291.

Mann JJ. Violence and aggression. In: Bloom FE, Kupfer DJ, eds. Psychopharmacology: The Fourth Generation of Progress. Raven, New York, 1995, pp. 1919–1928.

Marengell LB, Zpoyan HA, Cress KK, Benisek D, Arterburn L. A double blind placebo controlled study of docosahexaenoic acid (DHA) in the treatment of depression. 4th Congress of the International Society for the Study of Lipids and Fatty Acids, 2000, p. 105.

Martin RE. Docosahexaenoic acid decreases phospholipase A2 activity in the neurites/nerve growth cones of PC12 cells. J Neurosci Res 1998; 56(6):805–813.

Meltzer HY, Lowey MT. The serotonin hypothesis of depression. In: Meltzer HY, ed. Psychopharmacology: The Third Generation of Progress. Raven, New York, 1987, pp. 233–248.

Meydani SN, Lichtenstein AH, Cornwall S, Meydani M, Goldin BR, Rasmussen H. Immunologic effects of national cholesterol education panel step-2 diets with and without fish-derived N-3 fatty acid enrichment. J Clin Invest 1993; 92:105–113.

Mellor JE, Laugharne JD, Peet M. Schizophrenic symptoms and dietary intake of n-3 fatty acids. Schizophr Res 1995; 18(1):85–86.

Mitchell DC, Straume M, Litman BJ. Role of sn-2-polyunsaturated phospholipids and control of membrane receptor conformational equilibrium: effects of cholesterol and acyl chain unsaturation on the metarhodopsin I-metarhodopsin II equilibrium. Biochemistry 1992; 31:662–670.

Mitchell DC, Litman B. Effect of cholesterol on membrane order and dynamics in highly polyunsaturated phospholipid bilayers. Biophys J 1998; 75:896–908.
Murray CJL and Lopez, AD, eds. Global Burden of Disease: A Comprehensive Assessment of Mortality and Disability from Diseases, Injuries and Risk Factors in 1990 and Projected to 2020. Harvard University Press, Boston, 1996.
Naughton JM, O'Dea K, Sinclair AJ. Animal foods in traditional Australian aboriginal diets: polyunsaturated and low in fat. Lipids 1986; 21(11):684–690.
Nizzo MC, Tegos S, Gallamini A, Toffano G, Polleri A, Massarotti M. Brain cortex phospholipids liposomes effects on CSF HVA, 5-HIAA and on prolactin and somatotropin secretion in man. J Neural Transm 1978; 43:93–102.
Olsson NU, Shoaf S, Salem N Jr. The effect of dietary polyunsaturated fatty acids and alcohol on neurotransmitter levels in rat brain. Nutr Neurosci 1998; 1:133–140.
Otto SJ, Houwelingen AC, Antal M, Manninen A, Godfrey K, Lopez-Jaramillo P, et al. Maternal and neonatal essential fatty acid status in phospholipids: an international comparative study. Eur J Clin Nutr 1997; 51(4):232–242.
Paez X, Hernandez L. Simultaneous brain and blood microdialysis study with a new removable venous probe. Serotonin and 5-hydroxyindolacetic acid changes after D-norfenfluramine or fluoxetine. Life Sci 1996; 58(15):1209–1221.
Peet M, Murphy B, Shay J, Horrobin D. Depletion of omega-3 fatty acid levels in red blood cell membranes of depressive patients. Biol Psychiatry 1998; 43(5):315–319.
Peet M, Mellor J. Double blind placebo controlled trial of n-3 polyunsaturated fatty acids as an adjunct to neuroleptics. 9th Schizophrenia Winter Workshop, Davos Switzerland, 1998.
Peet M, Poole J, Laughaune JDE. Breast feeding neurodevelopment and schizophrenia. In: Peet M, Glen AS, Horrobin D (eds). Phospholipid Spectrum Disorder in Psychiatry. Marius Press, Lancashire, Scotland, 2001.
Peet M, Mellor J, Ramchand CN, Shah S, Vankar GK. Eicosapentaenoic acid (EPA) in the treatment of schizophrenia. 4th Congress of the International Society for the Study of Lipids and Fatty Acids, 2000, p. 108.
Peet M, EPA Treatment Group. Eicosapentaenoic acid (EPA) in the treatment of schizophrenia. Treatment of schizophrenia with ethyl-eicosapentaenoate (EPA): a randomized placebo controlled trial. Brain Reuptake and Utilization of Fatty Acids, Applications to Peroxisomal Disorders, an International Workshop, 2000, p. 11.
Puri BK, Richardson AJ. Sustained remission of positive and negative symptoms of schizophrenia following treatment with eicosapentaenoic acid. Arch Gen Psychiatry 1998; 55(2):188–189.
Puri BK, Richardson AJ, Horrobin DF, Easton T, Saeed N, Oatridge A, et al. Eicosapentaenoic acid treatment in schizophrenia associated with symptom remission, normalisation of blood fatty acids, reduced neuronal membrane phospholipid turnover and structural brain changes. Int J Clin Prac 2000; 54(1):57–63.
Rivier C, Vale W. Stimulatory effect of interleukin-1 on adrenocorticotropin secretion in the rat: is it modulated by prostaglandins? Endocrinology 1991; 129:384–388.
Robins LN, Heltzer JE, Weissman MM, Orvaschel H, Gruenberg E, Burke JD, et al. Lifetime prevalence of specific psychiatric disorders in three sites. Arch Gen Psychiatry 1984; 41:949–958.
Roy A, Virkkunen M, Linnoila M. Serotonin in suicide, violence, and alcoholism. Neuro-Psychopharmacol Biol Psychiatry 1987; 11:173–177.
Rudin DO. The dominant diseases of modernized societies as omega-3 essential fatty acid deficiency syndrome: substrate beriberi. Med Hypotheses 1982; 8(1):17–47.
Rudin DO. The major psychoses and neuroses as omega-3 essential fatty acid deficiency syndrome substrate pellagra. Biol Psychiatry 1981; 16(9):837–850.
Salem N Jr. Omega-3 fatty acids: molecular and biochemical aspects. In: Spiller GA, Scala J, eds. New roles for Selective Nutrients. Alan R. Liss, New York, 1989, pp. 109–228.
Salem N Jr, Kim HY, Yergey JA. Docosahexaenoic acid: membrane function and metabolism. In: Simopoulos A, Kifer RR, Martin R, eds. Health Effects of Polyunsaturated Fatty Acids in Seafoods, Academic, New York, 1986, pp. 263–317.
Sapolsky R, Rivier C, Yamamoto G, Plotsky P, Vale W. Interleukin-1 stimulates the secretion of hypothalamic corticotropin-releasing factor. Science 1987; 238(4826):522–524.
Silvers K, Scott KM. Fish consumption and self-reported physical and mental health status. 4th Congress of the International Society for the Study of Lipids and Fatty Acids, 2000, p. 188.

Sinclair AJ, O'Dea K, Dunstan G, Ireland PD, Niall M. Effects on plasma lipids and fatty acid composition of very low fat diets enriched with fish or kangaroo meat. Lipids 1987; 22(7):523–529.

Smith RS. The macrophage theory of depression. Med Hypotheses 1991; 35(4):298–306.

Stanley M, Mann JJ. Increased serotonin-2 binding sites in frontal cortex of suicide victims. Lancet 1983; 1(8318):214–216.

Stanley M, Traskman-Bendz L, Dorovini-Zis K. Correlations between aminergic metabolites simultaneously obtained from human CSF and brain. Life Sci 1985; 37:1279–1286.

Stevens LJ, Zentall SS, Deck JL, Abate ML, Watkins BA, Lipp SR, et al. Essential fatty acid metabolism in boys with attention-deficit hyperactivity disorder. Am J Clin Nutr 1995; 62(4):761–768.

Stoll AL, Severus WE, Freeman MP, Rueter S, Zboyan HA, Diamond E, et al. Omega-3 fatty acids in bipolar disorder: a preliminary double-blind, placebo-controlled trial. Arch Gen Psychiatry 1999; 56(5):407–412.

Stoll AL, Locke CA, Marangell LB, Severus WE. Omega-3 fatty acids and bipolar disorder: a review. Prostaglandins Leukotreines Essential Fatty Acids 1999; 60(5–6):329–337.

Stordy BJ. Benefit of docosahexaenoic acid supplements to dark adaptation in dyslexics. Lancet 1995; 346(8971):385.

Stordy BJ. Dark adaptation, motor skills, docosahexaenoic acid, and dyslexia. Am J Clin Nutr 2000; 71(1):323S–326S.

Susser, Lin. Schizophrenia after prenatal exposure to the Dutch hunger winter of 1944-1945. Arch Gen Psychiatry 1992; 49:983–988.

Tanskanen A, Hibbeln JR, Hintikka J, Haatainen K, Honkalampi K, Viinamäki H. Fish consumption and depressive symptoms in the population of northern Finland. Psychiatr Services.

Tanskanen A, Hibbeln JR, Hintikka J, Haatainen K, Honkalampi K, Viinamäki H. Fish consumption, depression and suicidality in a general population. Arch Gen Psych 2001: In Press.

Thienprasert A, Hamazaki T, Kheovichai K, Samuhaseneetoo S, Nagasawa T, Watanabe S. The effect of docosahexaenoic acid on aggression/hostility in elderly subjects: a placebo-controlled double blind study. 4th Congress of the International Society for the Study of Lipids and Fatty Acids, 2000, p. 189.

Virkkunen ME, Horroboin DF, Jenkins DK, Manku MS. Plasma phospholipid essential fatty acids and prostaglandins in alcoholic, habitually violent and impulsive offenders. Biol Psychiatry 1987; 22:1087–1096.

Voigt RG, Llorente A, Jensen C, Berretta MC, Boutte C, Heird WC. Preliminary results of a placebo controlled trial in attention deficit hyperactivity disorder. NIH Workshop on Omega-3 Essential Fatty Acids and Psychiatric Disorders, 1998, p. 24.

Wainwright PE, Bulman-Fleming MB, Levesque S, Mutsaers L, McCutcheon D. A saturated-fat diet during development alters dendritic growth in mouse brain. Nutr Neurosci 1998; 1:49–58.

Walter RC, Buffler RT, Bruggemann JH, Guillaume MM, Berhe SM, Negassi B, et al. Early human occupation of the Red Sea coast of Eritrea during the last interglacial. Nature 2000; 405(6782):65–69.

Weidner G, Connor SL, Hollis JF, Connor WE. Improvements in hostility and depression in relation to dietary change and cholesterol lowering. Ann Intern Med 1992; 117:820–823.

Weissman MM, Bland RC, Canino GJ, Faravelli C, Greenwald S, Hwu HG, et al. Cross-national epidemiology of major depression and bipolar disorder. JAMA 1996; 276(4):293–299.

Wickramaratne PJ, Weissman MM, Leaf PJ, Holford TR. Age, period and cohort effects on the risk of major depression: results from five United States communities. J Clin Epidemiol 1989; 42(4):333–343.

Willatts P, Forsyth JS, DiModugno MK, Varma S, Colvin M. Effect of long-chain polyunsaturated fatty acids in infant formula on problem solving at 10 months of age. Lancet 1998; 352(9129):688–691.

World Health Organization. World Health Statistics Annual: Annuaire de Statistiques Sanitaires Mondiales. Switzerland, Geneva, WHO, 1995.

World Health Organization. Fish and Fishery Products. World apparent Consumption Based on Food Balance Sheets (1961–1993). Italy: WHO, Rome, 1996.

Zimmer L, Delpal S, Guilloteau D, Aioun J, Durand G, Chalon S. Chronic n-3 polyunsaturated fatty acid deficiency alters dopamine vesicle density in the rat frontal cortex. Neurosci Lett 2000; 284(1–2):25–28.

Zimmer L, Delion-Vancassel S, Durand G, Guilloteau D, Bodard S, Besnard JC, et al. Modification of dopamine neurotransmission in the nucleus accumbens of rats deficient in n-3 polyunsaturated fatty acids. J Lipid Res 2000; 41(1):32–40.

19 Disorders of Phospholipid Metabolism in Schizophrenia, Affective Disorders, and Neurodegenerative Disorders

David F. Horrobin

1. INTRODUCTION

All neuronal internal and external membranes are composed of phospholipids. Almost all of the brain's proteins are embedded in or attached to phospholipids. There is a tendency to regard these phospholipids as mere structural scaffolding providing support for all the interesting things that happen in the brain, which are dependent on neurotransmitters, proteins, and nucleic acids. However, this is far from the case. Abnormalities in phospholipid metabolism are now known to play central roles in the major psychiatric and neurodegenerative disorders (Peet, Glen, & Horrobin, 1999). Therapeutic approaches based on understanding of phospholipid metabolism are initiating a therapeutic revolution that promises effective treatment with none of the side effects, which, sadly, have come to be accepted as the inevitable norm with drugs for central nervous system (CNS) disorders.

The structural role of phospholipids is far from passive and unimportant. Many proteins are embedded in phospholipid membranes. The quaternary structure of these proteins, and hence their function, is regulated by the phospholipid microenvironment. Tiny changes in that microenvironment, such as the insertion of a single double bond into a fatty acid phospholipid side chain, can double the affinity of a receptor for a ligand (Witt, WesthHansen, Rasmussen, Hastrup, & Nielsen, 1996; Witt & Nielsen, 1994). Many other proteins must be acylated (have a fatty acid side chain attached) before they can function. The side chain is required to anchor the protein in its appropriate placebo on a phospholipid membrane. Without an appropriate side chain and an appropriate phospholipid environment to which that side chain can link, the protein is nonfunctional (Webb, Hermida Matsumoto, & Resh, 2000).

However, the phospholipids are not only structural. Every activation of a receptor, every ion movement, and every stimulus to a cell triggers phospholipid-related signal transduction processes (Horrobin, 1999b; Horrobin, Glen, & Vaddadi, 1994; Horrobin, 1998). These signal transduction mechanisms regulate protein phosphorylation, cyclic nucleotide metabolism, calcium movements, and gene function. They are just as important as and tightly integrated with the protein-based signal transduction mechanisms that receive so much attention.

From: *Fatty Acids: Physiological and Behavioral Functions*
Edited by: D. Mostofsky, S. Yehuda, and N. Salem Jr. © Humana Press Inc., Totowa, NJ

1.1. Phospholipid Structure and Remodeling

All phospholipids have a three-carbon glycerol backbone. To that backbone are attached two fatty acids (acyl groups) at the first and second (so-called Sn1 and Sn2) carbon atoms. To the third carbon atom (Sn3) is attached a phosphorus atom and to the other side of the phosphorus is a water-soluble headgroup, usually consisting of either inositol, choline, serine, or ethanolamine. The phospholipids are thus highly polar with the acyl groups being hydrophobic and the headgroups hydrophilic. This polar nature confers on the phospholipids their property of forming bilayer membranes in aqueous media. The headgroups face out toward the water, and the acyl groups face inward to the core of the membrane.

In standard texts, phospholipid membranes are invariably portrayed as uniform bilayers, with circular blobs for headgroups, and two linear tails for the acyl groups. This consistent line-drawing portrayal of spurious simplicity gives a misleading impression. Phospholipid membranes are as complex as proteins. About 30 different fatty acids may be attached at the Sn1 and Sn2 positions; each combination has somewhat different properties that, when combined with the four different headgroups, confers an extraordinary degree of complexity on brain membranes.

A further degree of complexity is provided by the fact that the fatty acid at the Sn1 position may be attached by either an acyl group or an alkyl group. The alkyl series are known as ether phospholipids and have different properties from the acyl phospholipids. Although some enzymes can act equally well on acyl and on ether phospholipids, most are specific for one or the other. The outlines of acyl phospholipid metabolism, which are given later, can almost all be duplicated by an equivalent set of either phospholipid reactions.

This extraordinary complexity makes phospholipid biochemistry of great subtlety and functional interest. The biochemistry of phospholipids also offers a key site—perhaps *the* key site—in the body where genes and environment interact. This is because the enzymes that define phospholipid metabolism are, of course, genetically determined. However, these enzymes must operate using fatty acids that, to a substantial degree, are provided by the environment. Phospholipids therefore provide extraordinary opportunities for gene–environment interactions. Because most psychiatric and neurological disorders are influenced by more than one gene and also show strong environmental effects, the phospholipids are likely to offer fruitful avenues for exploration of the interaction.

Phospholipids can be synthesized by a number of different routes. However, the primary synthesis route is relatively unimportant because the final structures of brain phospholipids are largely determined not by the structure first synthesized but by the highly active and constant phospholipid remodeling that goes on within the brain. This remodeling plays a central role in several signal transduction cycles, which will be summarized in the following subsections.

1.2. The Phospholipase A_2 Cycle

The phospholipase A_2 (PLA$_2$) series of enzymes remove fatty acids, usually highly unsaturated fatty acids (HUFAs) from the Sn2 position of phospholipids, creating a free HUFA and a lysophospholipid (LyPL), a phospholipid without a fatty acid at the Sn2 position (Fig. 1). There are many PLA$_2$ enzymes, but in the brain the most important ones are the calcium-dependent and calcium-independent cytosolic PLA$_2$s (cPLA$_2$s) (Horrobin

Chapter 19 / Disorders of Phospholipid Metabolism

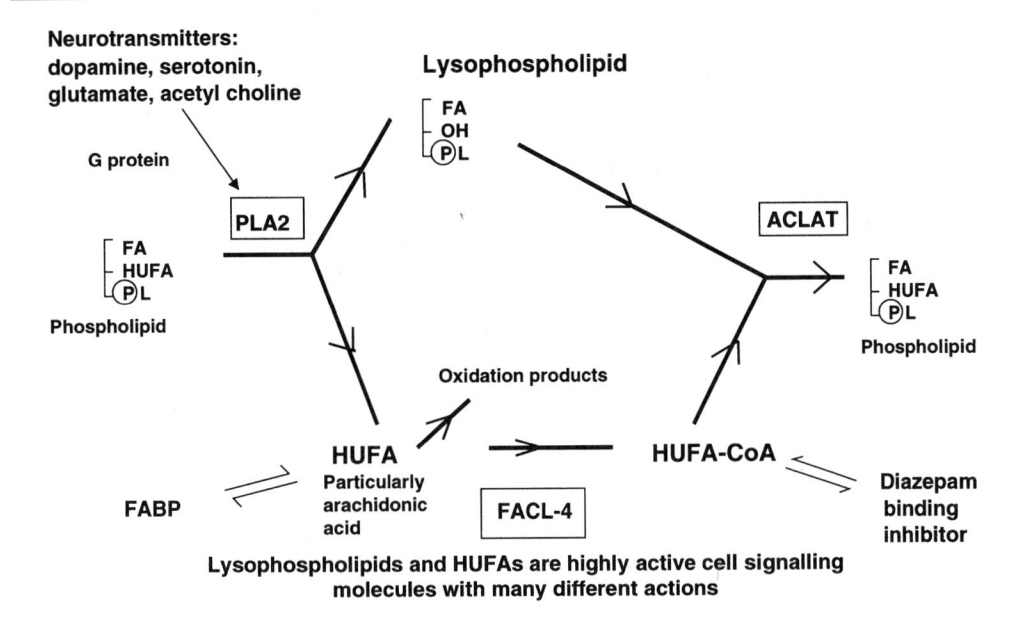

Fig. 1. The phospholipase A$_2$ signal transduction cycle.

& Bennett, 1999b). Most attention has been paid to the calcium-dependent enzyme, which has considerable specificity for arachidonic acid (AA) at the Sn2 position.

Cystostolic PLA$_2$ in neurons is normally inactive. It is regulated by one or more G-proteins, which are functionally lined to many neurotransmitter receptors, including ones for dopamine, serotonin, N-Methyl-D-Aspartate (NMDA), and acetylcholine. Receptor occupation changes the G-protein configuration which activates the PLA$_2$. Both the LyPL and the HUFA that result are highly active cell-signalling molecules that can regulate protein phosphorylation, calcium movements, cyclic nucleotide metabolism, and gene function. If the HUFA is AA, it can be converted to a wide range of eicosanoids, which have other signal transduction functions. It is clearly important that the cycle should be transient and so the activation is stopped, first by conversion of the HUFA to a coenzyme-A derivative by one of a series of enzymes known as fatty acid coenzyme-A ligases (FACL) or acyl-CoA synthetases. Enzymes known as acyl-CoA–lysophospholipid acyl transferases (ACLATs) then close the cycle by linking the HUFA–CoA derivative back to the LyPL, thus generating a stable phospholipid again (Horrobin et al., 1999b).

1.3. The PLC Cycle

There are multiple phospholipase C's (PLCs) that are linked via G-proteins to neurotransmitter receptors. PLCs split the link between the Sn3 carbon and the phosphorus, thus generating two molecules: the headgroup phosphate and diacylglycerol (DAG). PLCs particularly act on inositol phospholipids, generating inositol phosphates and DAG. As in the PLA$_2$ cycle, both of these products are active signal transduction molecules. Again, the cycle must be terminated as shown in Fig. 2, by means of inositol phosphatases and DAG kinases.

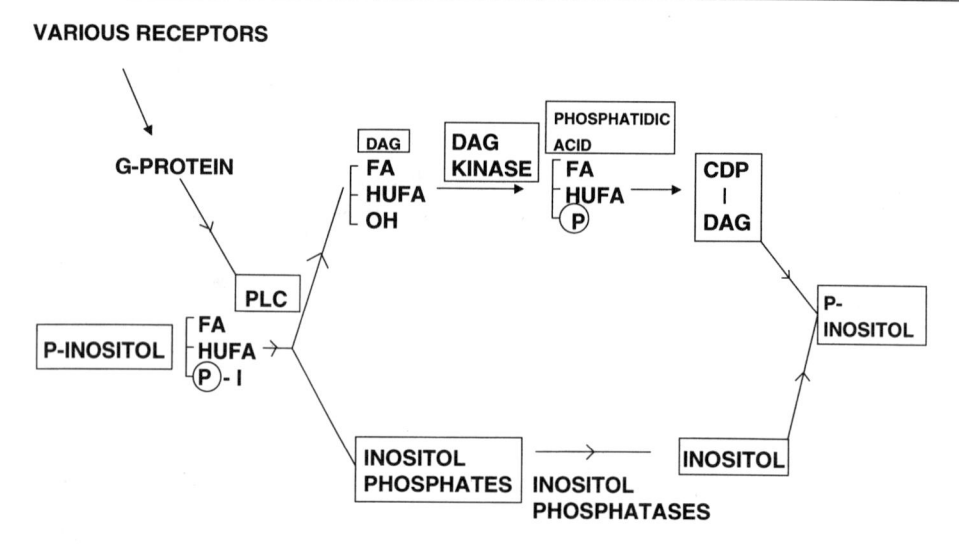

Fig. 2. The phospholipase C signal transduction cycle.

1.4. The PLD Cycle

Phospholipase D (PLD) also is linked to neurotransmitter receptors and, when activated, it splits the phospholipid between the phosphorus and the headgroup generating phosphatidic acid (PA) and the free headgroup. As earlier, the cycle must be terminated as quickly as it is initiated.

1.5. Phospholipase Crosstalk

During neuronal stimulation, the activation of multiple phospholipases seems to be the norm and it can be very difficult to sort out the interactions among them. For example, whereas PA generated by PLD is a potent signal transduction molecule, it can be further metabolized to Lyso-PA, another signal transduction molecule with different effects, by the actions of PLA_2. Lyso-PA can be converted back to PA by lysophosphatidic acid acyl transferase (LPAAT), and PA can be converted to DAG by phosphatidate phosphohydrolase. These are just a few examples of the complex interactions that can occur during neuronal activation.

1.6. Ether Lipids

A substantial proportion of the brain phospholipids are, in fact, ether lipids. They may undergo similar complex interactions, sometimes mediated by the same enzymes that act on standard phospholipids and sometimes mediated by their own specific ether lipid enzymes. As yet, we know relatively little about the specific roles of ether lipids in the brain, but given their abundance, it is not unreasonable to assume that they are substantial.

1.7. Fatty Acid Synthesis

The brain is unusual in being extremely rich in HUFAs at the Sn2 position. These HUFAs come from two classes of fatty acids: the so-called omega-3 and omega-6 essential fatty acids (Fig. 3). These are called essential because at least one of each series, usually linoleic of the omega-6 and α-linolenic of the omega-3 series, must be provided

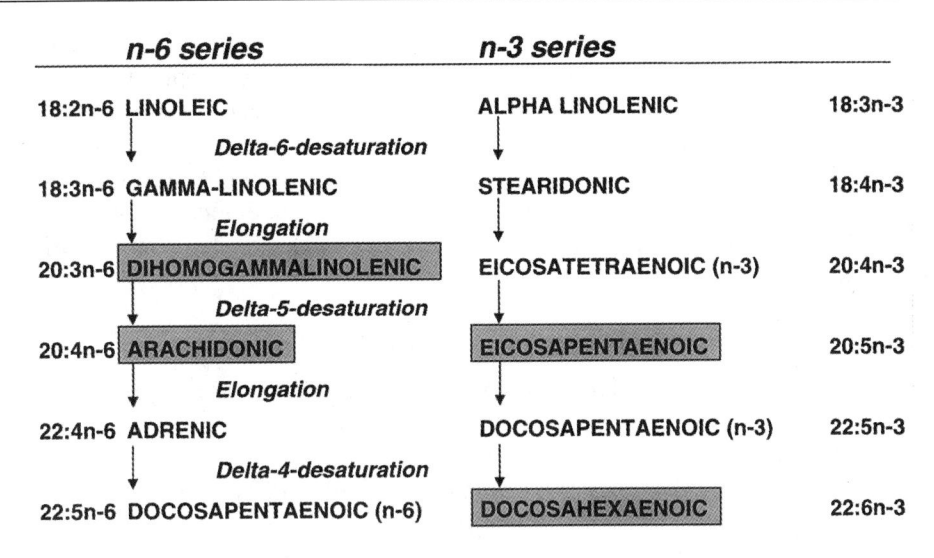

n-6 series		n-3 series	
18:2n-6 LINOLEIC		ALPHA LINOLENIC	18:3n-3
↓ *Delta-6-desaturation*		↓	
18:3n-6 GAMMA-LINOLENIC		STEARIDONIC	18:4n-3
↓ *Elongation*		↓	
20:3n-6 DIHOMOGAMMALINOLENIC		EICOSATETRAENOIC (n-3)	20:4n-3
↓ *Delta-5-desaturation*		↓	
20:4n-6 ARACHIDONIC		EICOSAPENTAENOIC	20:5n-3
↓ *Elongation*		↓	
22:4n-6 ADRENIC		DOCOSAPENTAENOIC (n-3)	22:5n-3
↓ *Delta-4-desaturation*		↓	
22:5n-6 DOCOSAPENTAENOIC (n-6)		DOCOSAHEXAENOIC	22:6n-3

Fig. 3. The metabolism of the omega-6 and omega-3 essential fatty acid.

in the diet. The mammalian body can convert or retroconvert members of each class into other members of the same class, but cannot convert one class to another. Although the brain is rich mainly in AA of the n-6 series and DHA of the n-3 series, most diets only contain limited amounts of these HUFAs. The body must make them from linoleic and α-linolenic acids via reactions that seem inherently slow and that can be impaired by many factors, including diabetes and the stress-related hormones, adrenaline and cortisol.

1.8. Fatty Acid Transport and Binding Proteins

Although fatty acids can relatively readily cross the blood-brain barrier and also cross cell membranes, their transport can be facilitated by fatty-acid-transport proteins (FATPs), and fatty acid binding proteins (FABPs) which bind fatty acids with high affinity and create concentration gradients drawing fatty acids into cells or into particular cellular compartments. There is now substantial evidence that FATPs and FABPs are central to the processes of neural development and synaptic remodeling (Horrobin et al., 1999b).

1.9. Phospholipid and Fatty Acid Metabolism and the Brain

Considering the obvious and central roles that phospholipids and fatty acids play in brain structure, in neuronal modeling and remodeling, and in neurodevelopment and neurodegeneration, it is astonishing how little attention has been paid to them in relation to psychiatric and neurological disorders. This situation is now changing and the research is proving productive both in terms of understanding and therapeutics (Peet, Glen, & Horrobin, 1999; Horrobin, 1999b; Horrobin et al., 1994; Horrobin, 1998; Horrobin et al., 1999b).

2. SCHIZOPHRENIA

2.1. Clinical Observations

Much of the impetus to investigate phospholipid metabolism in psychiatric disorders came from two old clinical observations (Horrobin, 1977; Horrobin, 1979; Horrobin et al.,

1978). The first is that patients with schizophrenia have a reduced risk of developing inflammatory disorders. The second is that some patients with schizophrenia experience a transient reduction in psychosis when they have a fever. Because metabolism of arachidonic acid is involved both in inflammation and in fever, this led me to consider the possibility that arachidonic acid metabolism abnormalities might be involved in schizophrenia. There is now substantial evidence that this broad statement is true and considerable progress has been made in identifying the precise biochemical abnormalities that may be involved.

2.2. Plasma and Red Cell Arachidonic Acid Levels

There is now substantial evidence from several groups that, although plasma levels of AA and other HUFAs are normal or near normal, red cell membrane phospholipid levels are reduced, particularly in patients with the negative syndrome (Peet, Laugharne, Horrobin, & Reynolds, 1994; Yao, van Kammen, & Welker, 1994; Glen et al., 1994). The abnormality may become more evident if red cells are not frozen immediately at very low temperatures, such as $-70°C$. Maintenance for a period at room temperature or for long periods at $-20°C$ rather than $-70°C$ is associated with loss of AA from membrane phospholipids in samples from schizophrenic patients but not from controls. This indicates increased loss of AA from membranes, a process that may continue in stored samples.

2.3. Brain Composition

Brain samples are more difficult to obtain, but two groups have found reduced AA in brain phospholipids in schizophrenic patients (Horrobin, Manku, Hillman, Iain, & Glen, 1991; Yao, Leonard, & Reddy, 2000).

2.4. PLA_2 Bioassays

Gattaz was the first to report increased blood PLA_2 activity by bioassay in schizophrenia (Gattaz, Hubner, & Nevalainen, 1990; Gattaz & Brunner, 1996). Some have failed to replicate this, but it now seems that this can be accounted for by differences in assay technique. Ross has confirmed the Gattaz observations in blood and has also reported that there is increased activity of a calcium-independent PLA_2 in the brain (Ross, Hudson, Erlich, Warsh, & Kish, 1997).

2.5. PLA_2 Immunoassays

Glen and Macdonald have developed an immunoassay for $cPLA_2$ and have identified the protein in red cells (MacDonald et al., 2000). In normal individuals, $cPLA_2$ concentrations in red cells show a normal distribution. In schizophrenic patients, in contrast, the distribution is strongly skewed, with about half the patients showing values above the 95th percentile for normals.

2.6. AA Incorporation Into Membranes

Two groups have studied the incorporation of labeled arachidonic acid into membrane phospholipids, a reaction that depends on the sequential effects of FACL and ACLAT. Excess amounts of LyPL substrate are available in schizophrenic platelets, but in spite of that, the rate of incorporation of AA into phospholipids is impaired (Yao, 1999; Demisch, Heinz, Gerbaldo, & Kirsten, 1992).

2.7. Magnetic Resonance Spectroscopy (MRS)

^{31}P-MRS (magnetic resonance spectroscopy) can be used to probe phospholipid metabolism in the living human brain. The detailed interpretation of the results is open to much controversy, but there is broad agreement that phospholipid metabolism is abnormal in unmedicated schizophrenia patients and that the abnormality is consistent with an increased rate of phospholipid breakdown (Williamson & Drost, 1999; Pettegrew, Keshavan, & Minchew, 1993; Pettegrew et al., 1991; Shioiri, Kato, Inubushi, Murashita, & Takahashi, 1994). There is also evidence from magnetic resonance imaging that, in some patients with schizophrenia, there are expanded lateral ventricles consistent with loss of brain tissue. Because the main bulk of brain tissue consists of phospholipid, the finding is consistent with phospholipid breakdown (Horrobin, 1998).

2.8. Niacin Flushing

Oral or topical administration of niacin generates a flushing response. The flushing depends on the conversion of AA to the potent vasodilator PGD$_2$. In a substantial proportion of schizophrenic patients, the flushing response to both oral and topical niacin is impaired, indicating an abnormality in AA metabolism (Hudson, Lin, Cogan, Cashman, & Warsh, 1997; Ward, Sutherland, Glen, & Glen, 1998; Ward, Glen, Sutherland, & Glen, 1998). The abnormality is greater in unmedicated than in medicated patients and, therefore, is not caused by medication (Peet, Shah, & Ramchand, 1998).

2.9. Overview of the Evidence

The evidence *in toto* is strongly supportive of the idea that there is an abnormality in AA metabolism in schizophrenia, particularly in patients with predominantly negative syndromes. It is unlikely that the observed phenomenon can be explained by a single enzyme deficit. The most likely possibility is that there are two abnormalities, one in a cytosolic PLA$_2$ and one in an enzyme involved in incorporation of AA, most probably one of the FACL group, but also possibly one of the ACLAT group (Horrobin et al., 1999b). The FACL group is likely to be involved because several members of the group are highly expressed in the brain and because deletion of FACL-4 in variant Alport syndrome is associated with severe mental retardation and also with mid-facial hypoplasia, which is also found in schizophrenia (Horrobin et al., 1999b; Piccini et al., 1998). Standard Alport syndrome, which is not associated with mental retardation or deletion of FACL-4, is closely linked to schizophrenia (Shields, Pataki, & DeLisi, 1990).

2.10. Therapeutic Implications

If the biochemical observations are correct, then novel treatments of schizophrenia might be ones that inhibit PLA$_2$ and activate FACL. Eicosapentaenoic acid (EPA) is such a substance and has already generated promising therapeutic results, which are described in Chapter 20 (Peet, 1999).

3. DEPRESSION

3.1. Clinical Observations

Recently, an abundance of clinical observations has accumulated that have shown that depression is associated with a wide range of medical conditions (Horrobin & Bennett, 1999a). These conditions include cardiovascular disorders, stroke, inflammatory disorders, cancer, and osteoporosis. It is remarkable that all of these conditions are also associated with deficits of omega-3 essential fatty acids, suggesting that omega-3 deficits might contribute both to the depression and to the medical disorders (Horrobin et al., 1999a).

Epidemiological observations are consistent with the clinical observations. Both between-country and within-country studies show that there are strong inverse relationships between dietary intake of omega-3 fatty acids and the incidence of both unipolar depression and postpartum depression (Hibbeln, 1999; Hibbeln, 1998; Hibbeln & Salem, 1995). Both sets of observations are thus consistent with the idea that omega-3 deficits are important in depression and in the associated medical conditions.

3.2. Plasma and Red Cell Composition

There are clear abnormalities of fatty acid concentrations in depression that are distinct from those in schizophrenia (Adams, Lawson, Sanigorski, & Sinclair, 1996; Maes et al., 1996; Maes et al., 1999; Peet, Murphy, Shay, & Horrobin, 1998; Edwards, Peet, Shay, & Horrobin, 1998). First, the abnormalities are present in both plasma and in red cells, raising the possibility that the problem may be in fatty acid metabolism in general, rather than membrane phospholipid metabolism in particular. Second, the abnormalities are specifically deficits in the omega-3 fatty acids EPA, DHA, and docosapentaenoic acid (DPA), and particularly in EPA. In contrast to the situation in schizophrenia, AA levels are either normal or elevated.

3.3. Enzyme Abnormalities

These have been much less investigated than is the case with schizophrenia and there are no clear studies that point to a particular enzyme problem. The fatty acid abnormalities suggest that the difficulty may be specifically in the conversion of dietary α-linolenic acid through to EPA, without a corresponding deficit in omega-6 metabolism. There may even be an excess rate of AA formation or an excess rate of incorporation of AA into a signal transduction compartment, which is important in inflammatory and immune responses exaggerated in depression (Horrobin et al., 1999a). Coenzyme-A-independent transacylase (CoAIT) is a possible candidate for such an enzyme (Horrobin et al., 1999a).

3.4. Therapeutics

There is preliminary evidence that omega-3 fatty acids may be beneficial, as discussed in Chapter 20.

4. BIPOLAR DISORDER

There is increasing evidence that abnormalities in phospholipid signal transduction are important in bipolar disorder. To date, most of this evidence has come from investigations of the mechanisms of action of the established drugs for bipolar disorder—lithium and valproate. This evidence suggests that there are abnormalities in the lipid-related protein kinase signal transduction systems and also in the PLA_2 and PLC systems (Manji et al., 1999; Manji & Lenox, 1999; Ikonomov & Manji, 1999; Stoll & Severus, 1996).

4.1. Lithium and PLA_2

Many different actions have been proposed to explain the effects of lithium. Unfortunately, the great majority of these have been demonstrated to occur only at concentrations that are above the normal range.

An exception is the effect of lithium on PLA$_2$-mediated fatty acid release. Over 20 yr ago, lithium at therapeutic concentrations was shown to inhibit AA and dihomogamma-linolenic acid release mediated by vasopressin and prolactin (Manku, Horrobin, Karmazyn, & Cunnane, 1979; Karmazyn, Manku, & Horrobin, 1978). Such hormone-mediated reactions are now known to be mediated by PLA$_2$ (Horrobin, 1999a). More recently, Rapoport and Chang have developed methodology for evaluating PLA$_2$ activity in the brain in vivo (Rapoport, 1999; Robinson et al., 1992). They have shown very specifically that lithium at low therapeutic concentrations can substantially inhibit the activity of cPLA$_2$ in the rat brain. They have also shown that this inhibition is associated with reductions in cPLA$_2$ mRNA and protein levels (Rintala et al., 1999). It seems very likely that this is the primary action of lithium.

Other evidence consistent with this is that in bipolar disorder there is excessive flushing in response to niacin (Hudson et al., 1997; Ward et al., 1998). This is consistent with overactivity of cPLA$_2$ but without any associated abnormality in FACL such as seems to occur in schizophrenia and that reduces the availability of AA for niacin-mediated signal transduction. However, there is no evidence from the enzyme-linked immunosorbent assay (ELISA) assay that, in bipolar disorder, there is any elevation of cPLA$_2$ protein levels in red cells (MacDonald et al., 2000). If cPLA$_2$ is elevated in bipolar disorder, it seems more likely to be related to increased activity of the enzyme, possibly as a result of an activating factor such as one of the S-100 group of protein, rather than to an increased amount of enzyme.

4.2. PLC and Bipolar Disorder

Prior to the recent work by Chang and Rapoport, the most widely accepted hypothesis of the mechanism of action of lithium was that it interfered with the PLC–inositol cycle by blocking inositol phosphatase and thus preventing the completion of the cycle by the regeneration of phosphatidyl–inositol. This effect undoubtedly occurs, but doubt has been cast on its relevance because it only does so at concentrations that are considerably higher than those required to inhibit cPLA$_2$.

4.3. Valproate Actions and Phospholipids

Valproate is a fatty acid that was initially developed as an antiepileptic agent. It seems very likely that it works by interfering with phospholipid-related signal transduction systems, a theme which has been developed by Manji and colleagues (Manji et al., 1999; Manji et al., 1999).

4.4. Therapeutic Implications

Based on the likelihood that bipolar disorder is related to phospholipid signal transduction problems, Severus and Stoll developed the idea that these might be modulated by omega-3 fatty acids (Stoll et al., 1996). They therefore set up a randomized study of high-dose mixed EPA and DHA in partially treatment-resistant patients with rapid-cycling bipolar disorder. The EPA/DHA preparations were found to be highly significant in preventing relapse and in improving depression (Stoll et al., 1999). This initially promising observation is now being followed up in much larger studies.

5. NEURODEGENERATION

Neurodegeneration disorders are among the most important causes of disability. From relatively rare conditions like Huntington's disease and progressive supranuclear palsy, to much commoner ones like Parkinson's disease, Alzheimer's disease, and normal aging, they impose an enormous burden, especially on the aging populations of the developed world.

There is increasing evidence that two common elements are involved in these processes. The first is the deposition of one of a number of abnormal proteins or abnormal protein fragments. The proteins differ, and the most susceptible neurons differ, but, in the end, pathology relates to the accumulation of protein aggregates.

The second common element is the involvement at the end stage of membrane-disrupting proinflammatory processes involving the generation of interleukin-1β by caspase-1 (also known as interleukin-converting enzyme, ICE), the generation of free radicals, and the activation by free radicals and other mechanisms of PLA_2s. These processes seem to be involved in the final common path of many different types of neurodegenerations (Beal, 1992; Alexi et al., 2000; Wellington & Hayden, 2000; Schapira, 1998).

There is exciting new evidence that regulation of phospholipid and fatty acid metabolism may play a crucial role in the regulation of neurodegeneration.

5.1. The Aging Rat Brain

The mechanisms of acting in the rat hippocampus have been investigated by Lynch and colleagues in Dublin (McGahon, Clements, & Lynch, 1997; McGahon, Murray, Horrobin, & Lynch, 1999; McGahon, Martin, Horrobin, & Lynch, 1999a; McGahon, Maguire, Kelly, & Lynch, 1999; McGahon, Martin, Horrobin, & Lynch, 1999b). They have compared 4-mo-old and 22-mo-old rats and have identified a large series of differences. These include loss of long-term potentiation (LTP) and glutamate release, which are believed to be important substration of memory. There is also elevation of caspase-1 activity and of levels of interleukin-1β and lipid peroxidation products (markers of free-radical activity). At the same time, concentrations of the antioxidants, vitamins A and C are reduced, whereas superoxide dismutase (SOD), a marker of pro-oxidant stress, is elevated. Finally, the membrane levels of AA in the hippocampus are substantially reduced, possibly indicating an overactive PLA_2.

It is difficult to know which, if any, of these abnormalities might be primary and causative, which are secondary, or whether they are all secondary to some other problem. However, remarkably, it has now been shown that all of these abnormalities without exception can be corrected by the correction of membrane arachidonic acid levels (McGahon et al., 1997; McGahon et al., 1999; McGahon et al., 1999a; McGahon et al., 1999) (McGahon et al., 1999b). This can be achieved either by the direct administration of AA or by the administration of EPA, which inhibits PLA_2 and therefore indirectly restores brain membrane AA levels to normal.

This is a truly remarkable observation. LTP, glutamate release, caspase-1, interleukin-1β, SOD, vitamin E, vitamin C and MDA levels can all be restored fully to normal by the simple expedient of correcting membrane AA levels. This suggests that the membrane abnormalities may be primary, possibly caused by the brain decline of AA and EPA synthesis that occurs with aging. The observations open up the possibility of a novel approach to neurodegenerative disorders, an approach which has already begun to bear fruit in the management of Huntington's disease.

5.2. Huntington's Disease

Huntington's disease (HD) is a devastating disorder caused by an excess number of CAG repeats in the gene for huntingtin on chromosome 4. The expanded polyglutamine tracts in the protein give rise to abnormal fragments when cleaved by enzymes such as caspase. These fragments progressively accumulate until they reach a critical level, usually between the ages of 30 and 40, when the first clinical signs of neurodegeneration in the basal ganglia and cerebral cortex appear. The clinical syndrome has three main elements: a psychosis, often characterized by outbursts, movement disorders with chorea and severe writhing movements of the face and body, and progressive dementia. Death usually occurs 15–25 yr later. There are no treatments.

There are now several transgenic mouse models of Huntington's disease in which various different fragments of the whole Huntington's gene have been incorporated into the mouse. These models show severe movement disorders and imitate the human disease to varying degrees.

Vaddadi was involved in early attempts to treat schizophrenia by essential fatty acids and decided to try the same approach in patients with HD (Vaddadi, 1999). He obtained some surprisingly positive results in open-label studies and then, with Drago and colleagues, tested this approach in the R6/1 model of HD mice. They were able to prevent completely the movement disorder (Vaddadi et al., 2000).

Puri et al. followed Vaddadi's work with a randomized, double-blind, placebo-controlled study of EPA in patients with end-stage HD (Puri et al., 2000). Seven patients were involved in the study, and over 6 mo, three patients improved and four deteriorated. On breaking the code all three patients who improved were on EPA, whereas all four who deteriorated were on placebo ($p = 0.04$). Moreover, magnetic resonance imaging showed the expected ongoing neurodegeneration with ventricular expansion on placebo, but a partial reversal with reduction of ventricular size on active treatment.

6. CONCLUSIONS

For many years, phospholipid metabolism has been ignored in psychiatric and neurological disorders. This situation is changing rapidly and this approach is poised to become part of the new mainstream in research on brain disorders.

REFERENCES

Adams PB, Lawson S, Sanigorski A, Sinclair AJ. Arachidonic acid to eicosapentaenoic acid ratio in blood correlates positively with clinical symptoms of depression. Lipids 1996; 31:S157–S161.

Alexi T, Borlongan CV, Faull, RLM, Williams CE, Clark RG, Gluckman PD, et al. Neuroprotective strategies for basal ganglia degeneration: Parkinson's and Huntington's diseases. Prog Neurobiol 2000; 60:409–470.

Beal MF. Does impairment of energy metabolism result in excitotoxic death in neurodegenerative illnesses? Ann Neurol 1992; 31:119–130.

Demisch L, Heinz K, Gerbaldo H, Kirsten R. Increased concentrations of phosphatidylinositol (PI) and decreased esterification of arachidonic acid into phospholipids in platelets from patients with schizoaffective disorders or atypic phasic psychoses. Prostaglandins Leukotrienes Essential Fatty Acids 1992; 46:47–52.

Edwards R, Peet M, Shay J, Horrobin D. Omega-3 polyunsaturated fatty acid levels in the diet and in red blood cell membranes of depressed patients. J Affect Disord 1998; 48:149–155.

Gattaz WF, Brunner J. Phospholipase A2 and the hypofrontality hypothesis of schizophrenia. Prostaglandins Leukotrienes and Essential Fatty Acids 1996; 55:109–113.

Gattaz WF, Hubner CVK, Nevalainen TJ. Increased serum phospholipase A2 activity in schizophrenia: a replication study. Biol Psychiatry 1990; 28:495–501.

Glen, A.I.M., Glen EM, Horrobin DF, Vaddadi KS, Spellman M, Morse-Fisher N, et al. A red cell membrane abnormality in a subgroup of schizophrenic patients: evidence for two diseases. Schizophr Res 1994; 12:53–61.

Hibbeln JR. Fish consumption and major depression. Lancet 1998; 351:1213.
Hibbeln JR. Long-chain polyunsaturated fatty acids in depression and related conditions. In: Peet M, Glen I, Horrobin DF, eds. Phospholipid Spectrum Disorder in Psychiatry. Marius, Carnforth, UK, 1999, pp. 195–210.
Hibbeln JR, Salem N. Dietary polyunsaturated fatty acids and depression: when cholesterol does not satisfy. Am J Clin Nutr 1995; 62:1–9.
Horrobin DF. Schizophrenia as a prostaglandin deficiency disease. Lancet 1977; 1:936–937.
Horrobin DF. Schizophrenia: reconciliation of the dopamine, prostaglandin, and opioid concepts and the role of the pineal. Lancet 1979; 1:529–531.
Horrobin DF. The membrane phospholipid hypothesis as a biochemical basis for the neurodevelopmental concept of schizophrenia. Schizophr Res 1998; 30:193–208.
Horrobin DF. Lithium, essential fatty acids (efas) and eicosanoids. In: Birch NJ, ed. Lithium: 50 Years of Psychopharmacology. Weidner, Cheshire, CT, 1999, pp. 154–167.
Horrobin DF. The phospholipid concept of psychiatric disorders and its relationship to the neurodevelopmental concept of schizophrenia. In: Peet M, Glen I, Horrobin DF, eds. Phospholipid Spectrum Disorder in Psychiatry. Marius, Carnforth, UK, 1999, pp. 3–20.
Horrobin DF, Ally AI, Karmali RA, Karmazyn M, Manku MS, Morgan RO. Prostaglandins and schizophrenia: further discussion of the evidence. Psychol Med 1978; 8:43–48.
Horrobin DF, Bennett CN. Depression and bipolar disorder: Relationships to impaired fatty acid and phospholipid metabolism and to diabetes, cardiovascular disease, immunological abnormalities, cancer, aging and osteoporosis: Possible candidate genes. Prostaglandins Leukotrienes Essential Fatty Acids 1999; 60:217–234.
Horrobin DF, Bennett CN. New gene targets related to schizophrenia and other psychiatric disorders: enzymes, binding proteins and transport proteins involved in phospholipid and fatty acid metabolism. Prostaglandins Leukotrienes Essential Fatty Acids 1999; 60:111–167.
Horrobin DF, Glen AI, Vaddadi K. The membrane hypothesis of schizophrenia. Schizophr Res 1994; 13:195–207.
Horrobin DF, Manku MS, Hillman H, Iain A, Glen M. Fatty acid levels in the brains of schizophrenics and normal controls. Biol Psychiatry 1991; 30:795–805.
Hudson CJ, Lin A, Cogan S, Cashman F, Warsh JJ. The niacin challenge test: Clinical manifestation of altered transmembrane signal transduction in schizophrenia? Biol Psychiatry 1997; 41:507–513.
Ikonomov OC, Manji HK. Molecular mechanisms underlying mood stabilization in manic-depressive illness: the phenotype challenge. Am J Psychiatry 1999; 156:1506–1514.
Karmazyn M, Manku MS, Horrobin DF. Changes of vascular reactivity induced by low vasopressin concentrations: interactions with cortisol and lithium and possible involvement of prostaglandins. Endocrinology 1978; 102:1230–1236.
MacDonald DJ, Glen, A.C.A., Boyle RM, Glen, AIM, Ward P, Horrobin DF. An ELISA method for type IV cPLA2 in RBC: a potential marker for schizophrenia. Schizophr Res 2000; 41:259.
Maes M, Christophe A, Delanghe J, Altamura C, Neels H, Meltzer HY. Lowered omega3 polyunsaturated fatty acids in serum phospholipids and cholesteryl esters of depressed patients. Psychiatry Res 1999; 85:275–291.
Maes M, Smith R, Christophe A, Cosyns P, Desnyder R, Meltzer H. Fatty acid composition in major depression: decreased omega 3 fractions in cholesteryl esters and increased C20:4 omega 6/C20: 5 omega 3 ratio in cholesteryl esters and phosopholipids. J Affect Disord 1996; 38:35–46.
Manji HK, Bebchuk JM, Moore GJ, Glitz D, Hasanat KA, Chen G. Modulation of CNS signal transduction pathways and gene expression by mood-stabilizing agents: therapeutic implications. J Clin Psychiatry 1999; 60(Suppl 2):27–39; discussion 40–41 and 113–116.
Manji HK, Lenox RH. Protein kinase C signaling in the brain: molecular transduction of mood stabilization in the treatment of manic-depressive illness. Biol Psychiatry 1999; 46:1328–1351.
Manku MS, Horrobin DF, Karmazyn M, Cunnane SC. Prolactin and zinc effects on rat vascular reactivity: possible relationship to dihomo-gamma-linolenic acid and to prostaglandin synthesis. Endocrinology 1979; 104:774–779.
McGahon B, Clements MP, Lynch MA. The ability of aged rats to sustain long-term potentiation is restored when the age-related decrease in membrane arachidonic acid concentration is reversed. Neuroscience 1997; 81:9–16.
McGahon B, Maguire C, Kelly A, Lynch MA. Activation of p42 mitogen-activated protein kinase by arachidonic acid and trans-1-amino-cyclopentyl-1,3-dicarboxylate impacts on long-term potentiation in the dentate gyms in the rat: analysis of age-related changes. Neuroscience 1999; 90:1167–1176.

McGahon BM, Martin, D.S.D., Horrobin DF, Lynch MA. Age-related changes in LTP and antioxidant defenses are reversed by an alpha-lipoic acid-enriched diet. Neurobiol Aging 1999; 20:655–664.

McGahon BM, Martin DSD, Horrobin DF, Lynch MA. Age-related changes in synaptic function: Analysis of the effect of dietary supplementation with ω-3 fatty acids. Neuroscience 1999; 94:305–314.

McGahon BM, Murray CA, Horrobin DF, Lynch MA. Age-related changes in oxidative mechanisms and LTP are reversed by dietary manipulation. Neurobiol Aging 1999; 20:643–653.

Peet M. New strategies for the treatment of schizophrenia: omega-3 polyunsaturated fatty acids. In Peet M, Glen I, Horrobin DF. (eds.), Phospholipid Spectrum Disorder in Psychiatry. Marius, Carnforth, UK, 1999, pp. 189–192.

Peet M, Glen I, Horrobin DF. Phospholipid Spectrum Disorder in Psychiatry. Marius, Carnforth, UK, 1999.

Peet M, Laugharne JD, Horrobin DF, Reynolds GP. Arachidonic acid: a common link in the biology of schizophrenia? Arch Gen Psychiatry 1994; 51:665–666.

Peet M, Murphy B, Shay J, Horrobin D. Depletion of omega-3 fatty acid levels in red blood cell membranes of depressive patients. Biol Psychiatry 1998; 43:315–319.

Peet M, Shah S, Ramchand CN. Niacin challenge skin flushing in drug-treated, drug-naive schizophrenics and matched controls: A study conducted in India. XXIst CINP Congress, 1998.

Pettegrew JW, Keshavan MS, Minchew NJ. ^{31}P nuclear magnetic resonance spectroscopy: neurodevelopment and schizophrenia. Schizophrenia Bull 1993; 19:35–53.

Pettegrew JW, Keshavan MS, Panchalingam K, Strychor S, Kaplan DB, Tretta MG, Allen M. Alterations in brain high-energy phosphate and membrane phospholipid metabolism in first-episode, drug-naive schizophrenics. Arch Gen Psychiatry 1991; 48:563–568.

Piccini M, Vitelli F, Bruttini M, Pober BR, Jonsson JJ, Villanova M, et al. FACL4, a new gene encoding long-chain acyl-CoA synthetase 4:is deleted in a family with Alport syndrome, elliptocytosis, and mental retardation. Genomics 1998; 47:350–358.

Puri BK, Bydder GM, Counsell SJ, Richardson AJ, Hajnal JV, Appel C, et al. Treatment of Huntington's disease with ethyl-eicosapentaenoic acid: a randomized, placebo-controlled trials. British Association of Psychopharmacology Summer Meeting, Cambridge, UK, 2000.

Rapoport SI. In vivo fatty acid incorporation into brain phospholipids in relation to signal transduction and membrane remodeling. Neurochem Res 1999; 24:1403–1415.

Rintala J, Seemann R, Chandrasekaran K, Rosenberger TA, Chang L, Contreras MA, et al. 85 kDa cytosolic phospholipase A2 is a target for chronic lithium in rat brain. NeuroReport 1999; 10:3887–3890.

Robinson PJ, Noronha J, DeGeorge JJ, Freed LM, Nariai T, Rapoport SI. A quantitative method for measuring regional in vivo fatty-acid incorporation into and turnover within brain phospholipids: review and critical analysis. Brain Res Rev 1992; 17:187–214.

Ross BM, Hudson C, Erlich J, Warsh JJ, Kish SJ. Increased phospholipid breakdown in schizophrenia—evidence for the involvement of a calcium-independent phospholipase A(2). Arch Gen Psychiatry 1997; 54:487–494.

Schapira AH. Mitochondrial dysfunction in neurodegenerative disorders. Biochim Biophys Acta 1998; 1366:225–233.

Shields GW, Pataki C, DeLisi LE. A family with Alport syndrome and psychosis. Schizophr Res 1990; 3:235–239.

Shioiri T, Kato T, Inubushi T, Murashita J, Takahashi S. Correlations of phosphomonoesters measured by phosphorus-31 magnetic resonance spectroscopy in the frontal lobes and negative symptoms in schizophrenia. Psychiatry Res 1994; 55:223–235.

Stoll AL, Severus WE. Mood stabilizers: shared mechanisms of action at postsynaptic signal-transduction and kindling processes. Harvard Rev Psychiatry 1996; 4:77–89.

Stoll AL, Severus WE, Freeman MP, Rueter S, Zboyan HA, Diamond E, et al. Omega 3 fatty acids in bipolar disorder—a preliminary double-blind, placebo-controlled trial. Arch Gen Psychiatry 1999; 56:407–412.

Vaddadi K. Essential fatty acids and movement disorders. In: Peet M, Glen I, Horrobin DF, eds. Phospholipid Spectrum Disorder in Psychiatry. Marius, Carnforth, UK, 1999, pp. 285–296.

Vaddadi KS, Clifford JJ, Waddington JL, Natoli AL, Wong JYF, Drago J. Essential fatty acids prevent the development of motor abnormalities in a transgenic mouse model of Huntington's disease, in press.

Ward PE, Glen E, Sutherland J, Glen A. The response of the niacin skin test in the functional psychoses. XXIst CINP Congress, 1998.

Ward PE, Sutherland J, Glen EMT, Glen AIM. Niacin skin flush in schizophrenia: a preliminary report. Schizophr Res 1998; 29:269–274.

Webb Y, Hermida Matsumoto L, Resh MD. Inhibition of protein palmitoylation, raft localization, and T cell signaling by 2-bromopalmitate and polyunsaturated fatty acids. J Biol Chem 2000; 275:261–270.

Wellington CL, Hayden MR. Caspases and neurodegeneration: on the cutting edge of new therapeutic approaches. Clin Genet 2000; 57:1–10.

Williamson PC, Drost DJ. ^{31}P Magnetic resonance spectroscopy in the assessment of brain phospholipid metabolism in schizophrenia. In: Peet M, Glen I, Horrobin DF, eds. Phospholipid Spectrum Disorder in Psychiatry. Marius, Carnforth, UK, 1999, pp. 45–55.

Witt MR, Nielsen M. Characterisation of the influence of unsaturated free fatty acids on brain GABA/benzodiazepine receptor binding in vitro. J Neurochem 1994; 62:1432–1439.

Witt MR, WesthHansen SE, Rasmussen PB, Hastrup S, Nielsen M. Unsaturated free fatty acids increase benzodiazepine receptor agonist binding depending on the subunit composition of the GABA(A) receptor complex. J Neurochem 1996; 67:2141–2145.

Yao JK. Red blood cell and platelet fatty acid metabolism in schizophrenia. In Peet M, Glen I, Horrobin DF, eds. Phospholipid Spectrum Disorder in Psychiatry. Marius, Carnforth, UK, 1999, pp. 57–71.

Yao JK, Leonard S, Reddy RD. Membrane phospholipid abnormalities in postmortem brains from schizophrenic patients. Schizophr Res 2000; 42:7–17.

Yao JK, van Kammen DP, Welker JA. Red blood cell membrane dynamics in schizophrenia. II. Fatty acid composition. Schizophr Res 1994; 13:217–226.

20 Eicosapentaenoic Acid
A Potential New Treatment for Schizophrenia?

Malcolm Peet and Shaun Ryles

1. INTRODUCTION

Mental illness still carries a sense of stigma, shame, and blame that goes well beyond that found in almost any other type of illness. It is perhaps fair to say that this is no small part because of the difficulty one has with conceptualizing the brain. Unlike the heart or the kidney or the liver, which can be broken down fairly easily into form and function, to most people the brain remains unfathomable. The problem is it does so much. It governs our beliefs, values, emotions, actions, and aspirations. Perhaps unsurprisingly, then, the worst thing that can happen to a person is failure of the brain.

One sometimes wonders if the awesome nature of our perception of the brain and its mysteries works against our attempts at addressing illnesses that affect our emotions and perceptions. Do we get carried away by wanting to map the brain, make it understandable by studying it without reference to the rest of the organism? It seems that we do. There are data that suggest those of us prone to depression are more susceptible to other physical illnesses, including heart disease, hypertension, and diabetes (Peet & Edwards, 1997). There are other data, known for some time, showing that those with schizophrenia seldom suffer from rheumatoid arthritis (Eaton et al 1992). Both sets of data suggest a substantial interplay between brain and body that cannot simply be dismissed as "psychosomatic." Is it fair, therefore, to concentrate our attentions, our finances, our best research brains on receptors in the striatum and limbic system as a means of explaining and solving the mysteries of mental illness? Many researchers throughout the world are now seriously challenging the notion that mental illness is solely a dysfunction of the brain. Is it not true that the intricate system of the brain is inextricably intertwined with those of the whole body? It is this simple question that has lead a number of researchers to begin to investigate more holistic approaches to mental illness. The idea is simple, the science complicated, but the consequences of such ideas being proven may mean that we can begin to position mental illness alongside other, less stigmatized illnesses such as heart disease and diabetes.

One such unifying hypothesis postulates that an abnormality of phospholipid metabolism underlies a spectrum of mental disorders (Peet et al 1999). The study of phospholipids and lipid biochemistry in general is gaining credibility as a legitimate and fruitful

From: *Fatty Acids: Physiological and Behavioral Functions*
Edited by: D. Mostofsky, S. Yehuda, and N. Salem Jr. © Humana Press Inc., Totowa, NJ

area of interest to those studying the etiology of mental illnesses such as depression, bipolar disorder, and schizophrenia (Walker et al., 1999; Freeman, 2000; Maidment, 2000; Fenton et al, 2000). In this area, the most studied of the major mental illnesses is schizophrenia.

2. SCHIZOPHRENIA AND ITS TREATMENT

Schizophrenia is the most devastating of the major mental illnesses. Its prevalence is stable across cultural and national boundaries, affecting between 0.5% and 1.5% of all populations. In the United Kingdom the economic cost of schizophrenia has been put at some £850 million per year, ranking it third behind strokes and learning disabilities as the most costly illness to the health service (Knapp, 1997). In human terms, the damage caused by schizophrenia is incalculable. It is known, for example, that as many as 12% of schizophrenics eventually commit suicide (Brown, 1997).

A cursory review of some of the major symptoms of schizophrenia offers an insight into why it is so devastating. Conceptually, the symptoms can be divided into two broad groups, positive and negative. The positive symptoms, those most readily associated with schizophrenia by the general public, include such things as hallucinations (hearing voices) and delusions. The negative symptoms, not given the attention that perhaps their prevalence deserves, include a loss of emotional responsiveness, social withdrawal, and the neglect of the activities of daily living. In reality, many people suffering from schizophrenia experience a mixture of these symptoms, making their treatment difficult and expensive. People suffering from schizophrenia require a range of interventions, including social outreach and strict medication regimes, that are difficult to administer to a patient group that can be often elusive and suspicious of the world around them.

In terms of treatment, little has changed since the introduction in the 1950s of a range of antipsychotic or neuroleptic medications such as chlorpromazine and haloperidol, which work by blocking dopamine D_2 receptors in the brain. Although these drugs are partially effective, they are far from ideal. Not only do they produce a range of side effects almost as devastating as schizophrenia itself, but they have little impact on negative symptoms and only limited success in treating hallucinations and delusions. It is sobering to reflect on the fact that the long-term outcome of schizophrenia has been no better in the 1990s than it was in the 1940s and 1950s, with only about 40% showing a favorable outcome despite modern treatment (Hegarty et al., 1994). Side effects most readily associated with these medications include drug-induced parkinsonism, comprising muscle stiffness, tremor, shuffling gait and reduced facial expression, and tardive dyskinesia, a potentially irreversible condition comprising involuntary movements affecting the tongue and facial muscles. More recently introduced drugs, such as risperidone and olanzapine, have less of these side effects but are no more efficacious. The only drug to offer greater efficacy is clozapine, and this carries the potentially fatal side effect of agranulocytosis.

In view of these effects, it is then unsurprising to note that as much as 37% of the relapse rate in schizophrenia has been attributed to noncompliance, with the remaining 63% attributed to loss of drug response (Weiden & Olfson, 1995). Despite this relative failure, the current major hypothesis as to the etiology of schizophrenia has been derived from the study of the action of these drugs within the human brain and new treatment approaches have been receptor based.

3. THE ETIOLOGY OF SCHIZOPHRENIA

There are currently two dominant theories as to the etiology of schizophrenia. The most popular postulates abnormal neurotransmitter receptor function within the brain and, to a large extent, concentrates on studying dopamine and serotonin. Although this theory still provides psychiatry with its best tools for dealing with psychosis, namely medications, it has proved sterile ground for developing radical new treatment approaches.

The second most dominant theory is known as the neurodevelopmental hypothesis and concentrates on the way the brain develops during the early months of life. Emphasizing the role of genetics married to environmental conditions, the neurodevelopmental hypothesis is fast gaining credence within the research community. The genetic factor is clear, and the most significant risk factor for individuals developing schizophrenia is having someone within your immediate family who has also been a sufferer. It is also clear that there are deficits in the way those who go on to develop the illness learn language skills and interact as children (Jones et al., 1994). None of this evidence, however, is conclusive and studies have shown that even in identical twins, the concordance rate is only some 40% (Maier & Schwab, 1998) which suggests that substantial environmental factors are operating.

Both hypotheses clearly have something to offer, but present significant challenges when it comes to explaining the way schizophrenia arises and develops. Lately, however, a new hypothesis has begun to be developed that has the potential to bring together both existing theories to create a new and exciting totality.

4. THE PHOSPHOLIPID HYPOTHESIS

The emerging phospholipid hypothesis takes as its basis increasing evidence that phospholipid metabolism is abnormal in schizophrenia. Broadly, it suggests that as a result of this altered metabolism, the cell membranes of schizophrenia sufferers are depleted of certain polyunsaturated fatty acids (PUFAs), including arachidonic acid (AA), docosahexaenoic acid (DHA), and eicosapentaenoic acid (EPA) (Horrobin et al., 1994; Peet et al., 1999). Currently, there are thought to be several possible mechanisms responsible for this. All of them offer an explanation for schizophrenia that brings together the genetic and the environmental, the neurotransmitter, and the lipid all bound up in a package that suggests that diet may have a crucial role to play in ameliorating or exacerbating the symptoms of this much misunderstood illness.

Evidence for the phospholipid hypothesis is arriving from various sources and growing rapidly. Studies at several centers have shown depleted PUFAs in red blood cell (RBC) membrane phospholipids (Peet et al., 1995; Glen et al., 1994; Yao et al., 1994), although not all reports are consistent (Doris, et al., 1998). In drug-treated schizophrenic patients, RBC membranes have been shown to have an apparent bimodal distribution of AA and DHA, where the lower mode is related to negative schizophrenic symptoms (Glen et al., 1994). Reduced PUFA levels have also been reported in cultured skin fibroblasts (Mahadik, et al., 1996). RBC PUFA levels have also been related to tardive dyskinesia, where the depletion of AA and DHA in particular were related to the severity of the abnormal movements (Vaddadi, 1989).

There is, however, a need for direct measurement within the brain. This is notoriously difficult to carry out, but in 1991, Pettegrew et al. used ^{31}P magnetic resonance spectroscopy (MRS) to make a direct examination of brain phospholipid metabolism. Using a group of first-episode drug-free schizophrenic patients, they found significantly raised levels of phosphodiesters (breakdown products of phospholipid) and significantly decreased levels of phosphomonoester (involved in phospholipid synthesis) when compared with controls (Pettegrew et al., 1991). For some, this study, along with others, including those of Stanley et al. (1994, 1995) showing similar results, remains a landmark investigation showing abnormal brain phospholipid metabolism in schizophrenic patients. Most recently, Yao et al. (2000) have shown a significant reduction in PUFAs, particularly AA, within RBC membrane phospholipids in the caudate region in the postmortem brain of a group of schizophrenics compared to well-matched, healthy controls.

A method for indirectly assessing prostaglandin production has come from the application of what has become known as the niacin skin flush (Ward et al., 1998). Previously, it has been shown that normal, healthy controls show a generalized skin flush response when given a high dose of oral niacin. This is caused by a release of prostaglandin D_2 in skin (Morrow et al., 1992). Prostaglandin D_2 is the major cyclo-oxygenase metabolite of AA in skin (Ruzicka & Printz 1982). Any reduction in availability of AA would therefore be expected to affect the subsequent availability of prostaglandin D_2, hence a reduction in skin flush. For many years, it has been observed that people with schizophrenia do not flush in response to oral niacin (Horrobin, 1980). This effect has recently been substantially confirmed (Hudson et al., 1997; Rybakowski & Weterle, 1991; Glen et al., 1996) but confounded by a possible medication effect. Recently, however, studies have been conducted in drug-free schizophrenia sufferers. Using a skin-patch test that utilizes topical aqueous methyl nicotinate developed by the Highland Psychiatric Research Group (Ward et al., 1998), Shah et al. (2000) looked at the flush reaction of a number of medicated and unmedicated schizophrenic patients plus a group of well-matched controls. The results confirm that compared to healthy controls, schizophrenics show a highly significant difference in flush response to topical niacin, with the unmedicated patients showing a similar response to the medicated. This is the most telling evidence accrued thus far that the reduced flush response in schizophrenia sufferers is not the result of any medication effect. It also suggests that metabolic abnormalities in schizophrenia affect the whole body, not just the brain.

Another indirect measure of physiological function relating to PUFA metabolism comes from studies of the electroretinogram (ERG). It is known that adequate n-3 PUFA levels are essential for normal retinal function. Our group has shown that schizophrenic patients show ERG changes similar to those that occur in states of experimental dietary deprivation of n-3 in primates (Warner et al., 1999).

One explanation for this evidence of reduced membrane PUFA may come from studies showing that within the plasma, platelets, and brain of schizophrenic patients, there are elevated levels of phospholipase A_2 (PLA_2) (Gattaz et al., 1990, 1995; Ross et al., 1997, 1999). Existing in several forms, PLA_2 is an enzyme that is involved in the turnover of PUFA in cell membranes. Thus, any overactivity or underactivity of PLA_2 will have a significant effect on phospholipid metabolism. Until recently, studies of PLA_2 activity in schizophrenia have delivered mixed results, suggesting either raised (Gattaz et al., 1987) or normal (Albers et al., 1993) PLA_2 activity. However, Ross et al. (1997) have looked at the possibility that different investigators had studied different subtypes of PLA_2. Subsequently, they have followed up their report showing increased PLA_2 activity in blood, by studying two classes of PLA_2, calcium stimulated and calcium-independent, in the brain of schizophrenics (Ross et al., 1999). These data show a 45% increase in calcium-independent PLA$_2$ in the temporal cortex within the brain of schizophrenics when compared to well-matched controls and a group with bipolar disorder. Furthermore, animal studies by the same group show that this is unlikely to be a medication effect.

As evidence from a host of sources has now firmly posited schizophrenia as being the result of a disease process, the search for genetic markers has grown apace. The evidence associating PLA$_2$ with schizophrenia has indicated that one area to study may be among those genes that regulate PLA$_2$ activity. Indeed, recent findings have demonstrated significant differences in allele distribution between schizophrenic patients and controls on polymorphic sites in the region of the gene for PLA$_2$ (Hudson et al., 1996; Peet et al., 1998; Wei et al., 1998), although there are also negative reports (Strauss et al., 1999; Doris et al., 1998). This raises the possibility that the abnormality of cytosolic PLA$_2$ is genetically determined.

Thus far, we have reviewed evidence demonstrating that a depletion of PUFAs in RBC membranes and an over activity of PLA$_2$ may be responsible for stripping away those PUFA. Other evidence has suggested that sufferers of schizophrenia also show increased levels of oxidative stress. Mahadik et al. (1998) reported elevated plasma lipid peroxides in first-episode nonaffective psychosis. Recently, the Scottish Schizophrenia Research Group (2000) has suggested that such findings could be an artifact caused by smoking, which, in itself, causes oxidative stress. However, work by our group (Zhang, 1999) has demonstrated increased oxidative stress in unmedicated schizophrenic patients who do not smoke. Oxidative stress could be either a cause or a consequence of increased breakdown of membrane PUFA. Yao et al. (1998) showed a reduction in plasma total antioxidant capacity in schizophrenic patients, which was not attributable to antipsychotic drug treatment. Mukerjee et al. (1996) also showed impaired antioxidant defense enzymes at the onset of psychosis. This suggests that there may be a primary abnormality of antioxidant defenses in schizophrenia, possibly of genetic origin (Edgar et al., 2000).

5. DIET AND SCHIZOPHRENIA

Although there have been studies that have shown the importance of dietary essential fatty acids (EFAs) for healthy neurodevelopment (Crawford, 1993), until recently there had been no studies of infant feeding practices in relation to subsequent schizophrenia. It has been shown that the children of women pregnant during the Dutch famine (1944–1945) were significantly more likely to develop schizophrenia in later life (Susser & Lin, 1992). Although these women suffered multiple dietary deficiencies and not just PUFA, it is known that during early development infants require healthy amounts of AA and DHA to maintain normal brain development (Rogers, 1978; Rogan & Gladen 1993; Lucas et al., 1992). What the Dutch data do demonstrate is that dietary impairment *in utero*, presumably by affecting neurodevelopment of the fetus, can have substantial effects in adulthood. Two initial studies of breast feeding in schizophrenia, one by our group and one in Scotland (McCreadie, 1997), produced consistent results. Our study (Peet et al., 1999) used a case-control methodology and the Scottish study compared patients with controls from the general population. Both studies showed that patients who developed schizophrenia were less likely to have been breast-fed as infants. This is consistent with a protective effect of breast-feeding resulting from a better balance of PUFAs, particularly DHA and AA, which were historically lacking in formula feeds. It was suggested that a genetic predisposition to abnormal PUFA metabolism may be modulated by diet in infancy, leading to impaired neurodevelopment and subsequent schizophrenia. However, there are two recent negative studies (Sasaki et al., 2000; Leask et al., 2000) and the issue is unresolved.

In adults with established schizophrenia, two studies suggest that increased consumption of PUFA, and particularly of n-3 PUFAs, is beneficial to the short-term symptoms and longer-term outcome of schizophrenia. It is well recognized that the long-term outcome of schizophrenia is better in undeveloped nations than it is in the West. The reasons for this are unclear, but Christensen and Christensen (1988) made a detailed analysis of World Health Organization data on prevalence and outcome in schizophrenia around the world. They correlated these data with that relating to the national diet and found that a high intake of total fat and saturated fat from land animals and birds and a relatively low intake of PUFAs was significantly associated with an unfavourable outcome of schizophrenia.

Our group conducted a detailed analysis of dietary fatty acid intake in 20 patients hospitalized with chronic schizophrenia (Mellor et al., 1996). All of these patients were receiving optimal dosages of neuroleptic medication but still suffered from schizophrenic symptoms. The normal dietary intake of each patient was assessed over 7 d using weighed intake for set meals and food diaries to record snacks at other times. The dietary data were analyzed with the use of the FOODBASE program (Institute of Brain Chemistry and Human Nutrition, London); this program gives a full fatty acid analysis. Schizophrenic symptomatology was rated using the Positive and Negative Syndrome Scale (PANSS) (Kay et al., 1987) and tardive dyskinesia was rated using the Abnormal Involuntary Movement Scale (AIMS) (Kane et al., 1992). The relationship between dietary fatty acid intake and symptomatology was examined using Pearson's correlation coefficients, followed by multiple regression analysis. The strongest correlation ($r = -0.55, p = 0.01$) was between total n-3 fatty acids in the normal diet and severity of positive symptoms. Thus, the greater the dietary intake of n-3 PUFA, the less severe are the positive schizophrenic symptoms. Within the individual n-3 PUFA, EPA showed a significant negative correlation with both total PANSS score and severity of tardive dyskinsia. Multiple regression analysis confirmed that it was the n-3 fatty acids and particularly EPA that contributed most to the variance in schizophrenic symptoms and tardive dyskinesia. All relationships were negative, so that a greater dietary intake of total n-3 PUFA related to fewer schizophrenic symptoms.

6. TREATMENT OF SCHIZOPHRENIA WITH N-3 PUFA

In the second phase of our study, these same subjects were then supplemented with MaxEPA fish oil (Seven Seas Healthcare Ltd, Hull) daily for 6 wk. Each capsule contained 171 mg of EPA and 114 mg of DHA. The scores on both PANSS and AIMS scales improved significantly (Mellor et al., 1996). Total PANSS score fell from 79 ± 18 to 66 ± 9 ($p < 0.002$). The AIMS score fell from 20 ± 7 to 12 ± 5 ($p > 0.001$). RBC membrane levels were measured by the method of Manku et al. (1983). Laboratory staff were blind to the coding system. RBC membrane levels of EPA as well as other n-3 PUFA further down the metabolic chain and also the total n-3 fatty acid content increased significantly during supplementation. Multiple regression analysis showed that changes in RBC membrane total n-3 PUFA were strongly associated with changes in total PANSS scores.

Previously a small number of studies had attempted to treat schizophrenia by PUFA supplementation. None of them showed the sort of encouraging results that our group achieved. For example, Vaddadi et al. (1986) treated 21 neuroleptic-resistant inpatients with depot neuroleptic medication and dihomo-γ-linolenic acid (DGLA), placebo depot medication and DGLA, or placebo depot medication and placebo DGLA. There were no significant treatment effects on behavior or schizophrenic symptomatology. In a subsequent study, evening primrose oil was given to patients with movement disorders who were predominantly schizophrenic and who had long-term exposure to neuroleptic medication (Vaddadi et al., 1989). The study used a crossover design with 16-wk treatment periods, making the study difficult to interpret because of potential carryover effects. There was a significant treatment effect on schizophrenic symptoms but no clinically important effect on tardive dyskinesia. In another small study, evening primrose oil did not benefit either schizophrenic symptoms or tardive dyskinesia (Wolkin et al., 1986). Supplementation with n-3 in the form of linseed oil (50% α-linolenic acid) was given to five patients in an open study, with some apparent benefit (Rudin, 1981).

Encouraged by our open-pilot study, we embarked upon a double-blind, placebo-controlled trial of n-3 supplementation (Peet et al., 2001). Subjects for the study were schizophrenics diagnosed according to DSM-IV (American Psychiatric Association) criteria who were still significantly symptomatic even with medical treatment that was considered optimal by the responsible psychiatrist. All patients were on stable medication and none suffered from significant physical illness or other psychiatric disorders. Patients were taking a variety of antipsychotic drugs in both oral and depot preparations, and some were also taking anticholinergic medication to treat side effects. Patients continued on their normal medication, which the responsible consultant psychiatrist was asked to keep stable during the 3 mo of treatment. Patients were randomly allocated to treatment with either an EPA-enriched oil (Kirunal, Laxdale Limited, Stirling, UK), a DHA-enriched oil, or a corn oil placebo. Patients were rated on the PANSS scale at the beginning and end of treatment. In addition, blood was taken for the measurement of RBC PUFA levels as described earlier, at the beginning and end of treatment.

Patients were well matched at baseline for age, gender, and severity of illness, with no significant differences among the three treatment groups. Positive symptoms, general psychopathology, and total PANSS score improved most in the EPA-supplemented patients. There was no treatment effect on negative symptoms. The treatment effect for positive symptoms reached statistical significance even with this small group of patients. Using a 25% improvement criterion, EPA was significantly superior to both DHA and placebo ($p < 0.05$).

This was a treatment-resistant group of patients who were substantially symptomatic despite current orthodox clinical treatment. The average 25% improvement in positive schizophrenia symptoms in the EPA group was significant not only statistically but was also clinically meaningful. Furthermore, the EPA treatment was not associated with adverse clinical side effects.

As expected, the highest levels of EPA and of DHA in RBC membranes of 12 patients from each group were found in the appropriate treatment groups. This confirms that the patients as a group were taking their medication and that it was having the predicted effect on membrane PUFA levels. No correlation was found between these changes in membrane PUFA levels and clinical improvement.

The apparent differential response between EPA- and DHA-enriched oils was unexpected and requires explanation. The brain contains only small amounts of EPA relative to the large quantities of DHA (Horrobin et al., 1991). It is therefore uncertain that dietary

EPA would produce any marked or physiologically significant changes in neuronal membrane EPA levels. It is possible that EPA is working through a quite different mechanism than incorporation into membranes. One possibility is through the inhibition of PLA_2. It has been shown in epidermal cells in vitro that PLA_2 is inhibited by EPA but not by DHA (Finnen & Lovell, 1991). However, it is unclear whether this is the same form of PLA_2 that is elevated in schizophrenia. Both EPA and DHA can induce glutathione peroxidase in human endothelial cells. Glutathione peroxidase may then protect cells and tissues from oxidative damage (Crosby et al., 1996). By these two mechanisms, EPA could be preventing both enzymatic membrane breakdown and oxidative damage. This is, however, very speculative.

This initial study has been followed up by a double-blind, placebo-controlled trial of ethyl-EPA at three dosage levels (1, 2, and 4 g) used as an adjunct to existing antipsychotic treatment (Peet et al., unpublished data). Patients were taking either typical antipsychotic drugs such as clopromazine and haloperidol, new atypicals such as olanzapine, or clozapine. The overall analysis so far has suggested that the 2 g dosage of EPA is the most effective. Patients on typical antipsychotic drugs responded poorly to the addition of ethyl-EPA to their treatment, but those patients already taking clozapine showed substantial extra benefit in clinical symptomatology relative to placebo. This is of particular interest because clozapine is currently regarded as the most effective treatment for schizophrenia, and it is used in cases which have failed to respond to other treatments. Clozapine is reserved as a last-resort treatment because of its tendency to produce agranulocytosis, which requires careful blood monitoring of patients taking the drug. Another unwanted effect is elevation of triglyceride levels. In this study, it was found that treatment with 2 or 4 g of ethyl-EPA normalized the elevated triglyceride levels. Also of interest is the effect of the different dosages of ethyl-EPA on RBC membrane fatty acid levels. There was a dose-related increase of membrane EPA levels. The 2-g dose of ethyl-EPA also resulted in elevations of DHA and AA. These changes were particularly marked in the clozapine-treated patients. In contrast, the 4-g dose of ethyl-EPA resulted in a decrease of membrane AA and DHA, presumably because of competition for incorporation into the membrane. Multiple-regression analysis suggested that increased membrane levels of AA were the most important determinant of clinical response.

Further studies have been conducted into the possibility of using EPA as a sole treatment for schizophrenia. Puri et al. (1998) have reported a single case who showed dramatic remission following treatment with EPA. This one case was investigated in great detail. It was found that during EPA treatment, previously low membrane levels of PUFA were normalized (Richardson et al., 2000). Also, progressive enlargement of the cerebral ventricles reversed following EPA treatment (Puri et al., 2000). No other treatment has been shown to have this effect in schizophrenic patients. We have now conducted a double-blind trial of EPA as a sole treatment in schizophrenia. The work was done with collaborators in Baroda, India. The protocol for this study was that subjects were schizophrenic patients who came for treatment at the outpatient clinic but who were not currently on any antipsychotic medication. About a third of the patients had never been previously treated, whereas others had received antipsychotic medication in the past. Thirty patients were treated at random with either EPA or placebo, and the treating clinicians were asked not to prescribe antipsychotic medication unless this was clinically imperative. The main outcome measure was the requirement for antipsychotic drugs. In addition, the PANSS rating scale was carried out at the beginning and end of the study.

When the code was broken after 3 mo of treatment, it was found that all patients in the placebo group required antipsychotic medication by the end of the study, whereas one-third of the patients on EPA completed the 3-mo trial period off antipsychotic medication. Avoiding the use of antipsychotic medication is clinically interesting only if patients did as well as those who received the conventional treatment together with placebo. The PANSS scores demonstrated that there was no significant difference between the patients treated with EPA or placebo at baseline, but by the end of the 3-mo treatment period, the EPA-treated patients had significantly lower PANSS scores than did the placebo group. Thus, the patients taking EPA had a better outcome even though a third of them were being managed without antipsychotic drugs by the end of the trial period (Peet et al., 2000).

Taking the evidence as a whole, it is apparent that the greatest effect of EPA on schizophrenic symptoms is seen in those patients who are either on no antipsychotic treatment or who are taking clozapine. Antipsychotic drugs such as clopromazine are known to have marked effects on some of the enzymes that affect phospholipid metabolism, and high-dose antipsychotic medication can in itself lead to depleted membrane levels of PUFA (Fischer et al., 1992). There is evidence that clozapine does not have this effect (Horrobin, 1999). Thus, the therapeutic effect of EPA may be impaired by concurrent treatment with neuroleptic drugs that damage cell membrane phospholipids.

In summary, existing drug treatments for schizophrenia are of limited efficacy and have substantial side effects. New treatment can arise only on the basis of a new hypothesis. The phospholipid hypothesis of schizophrenia provides the theoretical basis for treatment with PUFA supplementation. Previous studies using n-6 supplementation have had mixed results. We now have evidence from a double-blind, placebo-controlled trial that EPA, but not DHA, is effective in reducing the symptoms of schizophrenia. It is possible that the response to EPA is impaired by concomitant treatment with antipsychotic drugs that damage membrane phospholipids. The best treatment effects of EPA have been seen in patients who are otherwise unmedicated or who are currently taking clozapine. This remains to be explored further.

REFERENCES

Albers M, Meurer H, Marki F, et al. Phospholipase A2 activity in the serum of neuroleptic-naïve psychiatric inpatients. Pharmacopsychiatry 1993; 26:94–98.

Brown S. Excess mortality of schizophrenia. Br J Psychiatry 1997; 171:741–4773.

Crawford MA. The role of essential fatty acids in neural development: implications for perinatal nutrition. Am J Clin Nutr 1993; 57(Suppl):703–710.

Christensen O, Christensen E. Fat consumption and schizophrenia. Acta Psychiatr Scand 1988; 78:587–591.

Crosby AJ, Wahle KWJ, Duthi G.G. (1996). Modulation of glutathione peroxidase activity in human vascular endo-thelial cells by fatty and the cytokine interleukin-1. Biochim Biophys Acta 1303:187–192.

Doris AB, Wahle K, MacDonald A, Morris S, Coffey I, Muir W, et al. Red cell membrane fatty acids cytosolic phospholipase A2 and schizophrenia. Schizophr Res 1998; 31:185–196.

Eaton WW, Haywood C, Ram R. Schizophrenia and rheumatoid arthritis: a review. Schizophr Res 1992; 6:181–192.

Edgar PF, Douglas JE, Cooper GJ, Dean B, Kydd R, Faull RL. Comparative proteome analysis of the hippocampus implicates chromosome 6q in schizophrenia. Mol Psychiatry 2000; 5:85–90.

Fenton WS, Hibbeln J, Knable M. Essential fatty acids, lipid membrane abnormalities, and the diagnosis and treatment of schizophrenia. Biol Psychiatry 2000; 47:8–21.

Fischer S, Kissling W, Kuss HJ. Schizophrenic patients treated with high dose pheniothiazine or thioxanthine become deficient in polyunsaturated fatty acids in their thrombocytes. Biochem Pharmacol 1992; 44:317–323.

Freeman MP. Omega3 fatty acids in psychiatry: a review. Ann Clin Psychiatry 2000; 12:159–165.

Finnen MJ, Lovell CR. Purification and characterisation of phospholipase A2 from human epidermis. Biochem Soc Trans 1991; 19:91S.

Gattaz WF, Kollisch M, Thuren T, et al. Increased phospholipase A2 activity in schizophrenic patients: reduction after neuroleptic therapy. Biol Psychiatry 1987; 22:421–426.

Gattaz WF, Hubner CK, Nevalainen TJ, Thuren T, Kinnunen PKJ. Increased serum phospholipase A2 activity in schizophrenia: a replication study. Biol Psychiatry 1990; 28:495–501.

Gattaz WF, Schmitt A, Maras A. Increased platelet phospholipase A2 activity in schizophrenia. Schizophr Res 1995; 16:1–6.

Glen AIM, Cooper SJ, Rybakowski J, Vadadi K, Brayshaw N, Horrobin DF. Membrane fatty acids, niacin flushing and clinical parameters. Prostaglandins Leucotrienes Essential Fatty Acids 1996; 55:9–15.

Glen AIM, Glen EMT, Horrobin DF, Vaddadi, KS, Spellman M, Morse-Fisher N, et al. A red cell membrane abnormality in a subgroup of schizophrenic patients: evidence for two diseases. Schizophr Res 1994; 12:53–61.

Hegarty JD, Bladessarini RJ, Tohen M, Waternaux C, Oepen G. One hundred years of schizophrenia: a meta-analysis of the outcome literature. Am J Psychiatry 1994; 151:1409–1416.

Horrobin DF, Manku MS, Hillman H, Glen AIM. Fatty acid levels in the brains of schizophrenics and normal controls. Biol Psychiatry 1991; 30:795–805.

Horrobin DF, Glen AIM, Vaddadi K. The membrane hypothesis of schizophrenia. Schizophr Res 1194; 13:195–207.

Horrobin DF. Schizophrenia: a biochemical disorder. Biomedicine 1980; 32:54–55.

Horrobin DF. The effects of antipsychotic drugs on membrane phospholipids: a possible novel mechanism of action of clozapine. In: Peet M, Glen I, Horrobin DF, eds. Phospholipid Spectrum Disorder in Psychiatry. Marius Press, Carnforth, UK 1999, pp. 113–120.

Hudson CJ, Lin A, Cogan S, Cashman F, Warsh JJ. The niacin challenge test: clinical manifestation of altered transmembrane signal transduction in schizophrenia? Biol Psychiatry 1997; 41:507–513.

Hudson CJ, Kennedy JL, Gatoweic A, et al. Genetic variant near cytosolic phospholipase A_2 associated with schizophrenia. Schizophr Res 1996; 21:111–116.

Jones P, Rodgers B, Murray R, Marmot M. Child developmental risk factors for adult schizophrenia in the British 1946 birth cohort. Lancet 1994; 344:1398–1402.

Kane JM, Jeste DV, Barnes TE, Casey DE, Cole JO, et al. Tardive dyskinesia: a task force report of the American Psychiatric Association. American Psychiatric Association, Washington, DC, 1992.

Kay SR, Fiszbein A, Opier LA. The positive and negative syndrome scale (PANSS) for schizophrenia. Schizophr Bull 1987; 13:261–276.

Knapp M. Costs of schizophrenia. Br J Psychiatry 1997; 171:509–517.

Leask SJ, Jones PB, Done DJ, Crow TJ, Richards M. No association between breast-feeding and adult psychosis in two national birth cohorts. Br J Psychiatry 2000; 177:218–221.

Lucas A, Morley R, Cole TJ, Lister G, Leeson-Payne C. Breast milk and subsequent intelligence quotient in children born pre-term. Lancet 1992; 339:261–264.

Mahadik SP, Mukherjee S, Horrobin DF, Jenkins K, Correnti EE, Scheffer RE. Plasma membrane phospholipid fatty acid concentration of cultured skin fibroblasts from schizophrenic patients: comparison with bipolar patients and normal subjects. Psychiatry Res 1996; 63:133–142.

Mahadik SP, Mukherjee S, Scheffer R, Correnti EE, Mahadik JS. Elevated plasma lipid peroxides at the onset of nonaffective psychosis. Biol Psychiatry 1998; 43:674–679

Maidment ID. Are fish oils an effective therapy in schizophrenia: an analysis of the data. Acta Psychiatr Scand 2000; 102:3–11.

Maier W, Schwab S. Molecular genetics of schizophrenia. Curr Opin Psychiatry 1998; 11:19–25.

Manku, MS., Horrobin DF, Huang YS, Morse N. Fatty acids in plasma and red cell membranes. Lipids 1983; 18:906–908.

McCreadie RG. The Nithsdale Schizophrenia Surveys 16. Breastfeeding and schizophrenia: preliminary results and hypotheses. Br J Psychiatry 1997; 170:334–347.

Mellor JE, Laugharne JDE, Peet M. Omega-3 fatty acid supplementation in schizophrenic patients. Hum Psychopharmacol 1996; 11:39–46.

Morrow JD, Oates JA, Roberts LJ. Identification of the skin as a major site of cutaneous vasodilatation after oral niacin. J Invest Dermatol 1992; 98:812–815.

Mukerjee S, Mahadik SP, Scheffer R, Correnti EE, Kelkar H. Impaired antioxidant defence at the onset of psychosis. Schizophr Res 1996; 19:19–26.

Peet M, Edwards RHW. Lipids, depression and physical diseases. Curr Opin Psychiatry 1997; 10:477–480.
Peet M, Laugharne JDE, Rangarajan N, Horrobin DF, Reynolds G. Depleted red cell membrane essential fatty acids in drug treated schizophrenic patients. J Psychiatr Res 1995; 29:227–232.
Peet M, Ramchand CN, Lee KH, Telang SD, Vankar GK, Shah S. Association of the Ban I dimorphic site at the human cytosolic phospholipase A2 gene with schizophrenia. Psychiatr Genet 1998; 8(3):191–192.
Peet M, Brind J, Ramchand CN, Shah S, Vankar GK. Two double-blind placebo-controlled pilot studies of eicosapentaenoic acid in the treatment of schizophrenia. Schizophr Res, in press.
Peet M, Glen I, Horrobin DF, eds. Phospholipid Spectrum Disorder in Psychiatry. Marius, Carnforth, UK, 1999.
Peet M, Poole J, Laugharne JDE. Breastfeeding, neurodevelopment and schizophrenia. In: Peet M, Glen I, Horrobin, DF, eds. Phospholipid Spectrum Disorder in Psychiatry. Marius, Carnforth, UK, 1999, pp. 159–166.
Pettegrew JW, Keshavan MS, Panchalingam K, Strychor S, Kaplan DB, Tretta MG, et al. Alterations in brain high-energy phosphate and membrane phospholipid metabolism in first-episode, drug-naïve schizophrenics. Arch Gen Psychiatry 1991; 48:563–568.
Puri BK, Steiner R, Richardson AJ. Sustained remission of positive and negative symptoms of schizophrenia following treatment with eicosapentaenoic acid. Arch Gen Psychiatry 1998; 55:188–189.
Puri BK, Richardson AJ, Horrobin DF, Easton T, Saeed N, Oatridge A, et al. Eicosapentaenoic acid treatment in schizophrenia associated with symptom remission, normalisation of blood fatty acids, reduced neuronal membrane phospholipid turnover and structural brain changes. Int J Clin Pract 2000; 54:57–63.
Richardson AJ, Easton T, Puri BK. Red cell and plasma fatty acid changes accompanying symptom remission in a patient with schizophrenia treated with eicosapentaenoic acid. Eur Neuropyschopharmacol 2000; 10:189–193.
Rogan WJ, Gladen RC. Breast feeding and cognitive development. Early Hum Dev 1993:31, 181–193.
Rogers B. Feeding in infancy and later ability and attainment: a longitudinal study. Dev Med Child Neurol 1978; 20:421–426.
Ross BM, Hudson C, Erlich J, Warsh JJ, Kish SJ. Increased phospholid breakdown in schizophrenia. Evidence for the involvement of a calcium-independent phospholipase A2. Arch Gen Psychiatry 1997; 54:487–494.
Ross BM, Turenne S, Moszczynska A, et al. Differential alteration of phospholipase A2 activities in brain of patients with schizophrenia. Brain Res 1999; 821:407–13.
Rudin DO. The major psychoses and neuroses as omega-3 essential fatty acid deficiency syndrome: substarte pellagra. Biol Psychiatry 1981; 30:795–805.
Ruzicka T, Printz MP. Arachidonic acid metabolism in guinea pig skin. Biochim Biophys Acta 1982; 711:391–397.
Rybakowski J, Weterle R. Niacin test in schizophrenia and affective illness. Biol Psychiatry 1991; 29:834–836.
Sasaki T, Okazaki Y, Akaho R, Masui K, Harada S, Lee I, et al. Type of feeding during infancy and later development of schizophrenia. Schizophr Res 2000; 42:79–82.
Scottish Schizophrenia Research Group (2000) Smoking habits and plasma lipid peroxidation and vitamin E levels in neuro-treated first-episode schizophrenia patients. Br. J Psychiatry, 176; 290–293.
Shah SH, Vankar GK, Ramchand CN, Peet M. Unmedicated schizophrenic patients have a reduced skin flush in response to topical niacin. Schizophr Res 2000; 43:163–164.
Shepherd M, Lader M, Rodright R. Clinical Psychopharmacology. English Universities, London, 1968, p. 87.
Stanley JA, Williamson PC, Drost DJ, Carr, TJ, et al. Membrane phospholipid metabolism and schizophrenia: an in vivo 31 PMR spectroscopy study. Schizophr Res 1994; 13:209–215.
Stanley JA, Williamson PC, Drost DJ, Carr TJ, Rylett RJ, Malla A, et al. An in vito study of the prefrontal cortex of schizophrenic patients at different stages of illness via phosphorus magnetic resonance spectroscopy. Arch Gen Psychiatry 1995; 52:399–406.
Strauss J, Zhang XR, Barron Y, Ganguli R, Nimgaonkar VL. Lack of association between schizophrenia and pancreatric phospholipise-A2 gene (PLA2G1B) polymorphism. Psychiatr Genet 1999; 9:153–155.
Susser ER, Lin SP. Schizophrenia after prenatal exposure to the Dutch Hunger Winter of 1944–1945. Arch Gen Psychiatry 1992; 49:985–988.
Vaddadi KS, Courtney T, Gilleard CJ, Manku MS, Horrobin DF. A double-blind trial of essential fatty acid supplementation in patients with tardive dyskinesia. Psychiatry Res 1989; 27:313–323.
Vaddadi KS, Gilleard CJ, Mindham RHS, Butler RA. A controlled trial of prostaglandin E1 precursor in chronic neuroleptic-resistant schizophrenic patients. Psychopharmacology 1986; 88:362–367.
Walker NP, Fox HC, Whalley LJ. Lipids and schizophrenia. Br J Psychiatry 1999; 174:101–104.

Ward PE, Sutherland J, Glen EM, Glen AI. Niacin skin flush in schizophrenia: a preliminary report. Schizophr Res 1998; 29:269–274.
Warner R, Laugharne J, Peet M, Brown L, Rogers N. Retinal function as a marker for cell membrane omega-3 fatty acid depletion in schizophrenia: a pilot study. Biol Psychiatry 1999; 45:1138–1142.
Weir J, Lee KH, Hemmings GP. Is the cPLA$_2$ gene associated with schizophrenia? Mol Psychiatry 1998; 3:480–481.
Weiden PJ, Olfson M. Cost of relapse in schizophrenia. Schizophr Bull 1995; 21:419–429.
Wolkin A, Jordan B, Peselow E, Rubenstein M, Petrosen J. Essential fatty acids supplementation in tardive dyskinesia. Am J Psychiatry 1986; 143:912–914.
Yao JK, van Kammen DP, Welker JA. Red blood cell membrane dynamics in schizophrenia II. Fatty acid composition. Schizophr Res 1994; 13:217–226.
Yao JK, Reddy R, McElhinny LG, van Kammen DP. Reduced status of total antioxidant capacity in schizophrenia. Schizophr Res 1998; 32:1–8.
Yao JK, Leonard S, Reddy RD. Membrane phospholipid abnormalities in postmortem brains from schizophrenic patients. Schizophr Res 2000; 42:7–17.
Zhang ZJ. Investigation into membrane lipid peroxidation and antioxidant defence enzymes in schizophrenia. PhD thesis, University of Sheffield, UK, 1999.

21

The Importance of DHA in Optimal Cognitive Function in Rodents

Claus C. Becker and David J. Kyle

1. INTRODUCTION

The mammalian brain is a lipid-rich organ. Approximately 60% of the dry weight of the brain is lipid (Crawford, 1993). Most of the lipid in the brain is present in the form of phospholipids comprising the complex array of neural fibers that make up the central nervous system. A unique aspect of the lipid composition of all mammalian neurological tissues is the extraordinarily high concentration of docosahexaenoic acid (DHA) and arachidonic acid (AA). Indeed, DHA is the most abundant fatty acid building block of brain lipids and represents over 30% of the fatty acids of the phosphatidylinositol (PI), phosphatidylethanolamine (PE), and phosphatidylserine (PS) in the neuron (Salem, et al., 1986). DHA is primarily concentrated in the neuronal endings and synaptosomes and is also associated with the neurite growth cones, where it has been shown to promote neurite outgrowth (Ikemoto, et al., 1997). Its unique conformational characteristics (twenty-two carbons and 6 double bonds; C22:6) allow this fatty acid to have a structural as well as a functional role in biological membranes.

Although the fatty acid composition of the brains of various species of mammals are remarkably similar, some species (e.g., cats) require DHA in their diet, whereas others (e.g., mice) appear to be able to make DHA quite well from dietary precursors (Pawlosky, et al., 1994, Pawlosky, et al., 1996). In this chapter, we will explore the consequences of modulation of brain fatty acid composition on cognitive and behavioral functions in animals. The rodent model provides the opportunity to manipulate brain composition by controlling the diet. Such studies can provide insight into the neurological effects of DHA dietary insufficiency in humans, such as in the case of infants who are fed conventional infant formulas rather than their mother's milk.

2. LIPID COMPOSITION OF THE BRAIN

The mammalian brain to body weight ratio (the encephalization quotient; EQ) is generally larger than that of egg-laying amphibians, reptiles, and fish. Among mammals, however, there is a inverse relationship between adult body weight and brain to body weight ratio (Crawford, 1990). In other words, small mammals such as a mouse have a much larger EQ than larger mammals such as the rhinoceros. It has been proposed that

From: *Fatty Acids: Physiological and Behavioral Functions*
Edited by: D. Mostofsky, S. Yehuda, and N. Salem Jr. © Humana Press Inc., Totowa, NJ

Table 1
Concentration of Selected Fatty Acids from the Brain Cortex,
Olfactory Bulb, and Liver of Long–Evans Rats $(n = 6)^a$

	Brain cortex		Olfactory bulb		Liver	
	Adequate	Deficient	Adequate	Deficient	Adequate	Deficient
16:0	20.6	20.3	24.9	25.8	22.7	23.0
18:0	21.7	21.6	20.1	18.4	17.4	18.4
18:1n-9	16.6	15.4	16.3	16.8	12.1	10.8
20:4n-6	10.7	11.4	9.6	9.1	16.9	18.6
22:4n-6	3.0	3.9	2.6	2.9	0.3	0.6
22:5n-3	0.1	Not detected	0.14	Not detected	0.5	Not detected
22:5n-6	0.6	12.6	1.9	13.0	0.5	5.8
22:6n-3	15.3	2.5	15.7	2.2	7.1	0.5

[a]Significantly different from n-3 adequate group (Student's *t*-test, two-sided, $p < 0.001$).
Source: Hamilton et al., 2000.

as the body masses of the evolving terrestrial mammal species increased, the EQ dropped off precipitously as a result of the unavailability DHA in the diet. These differences are so great that the availability of DHA could be considered as a limiting factor in the evolution of the brain. The notable exceptions to this generalization are marine mammals and man. Marine mammals have an abundant supply of DHA in their diet, and it has been speculated that the movement to a DHA-rich environment (i.e., the land–water interface) approx 200,000 yr ago allowed the rapid development of the large brain of *Homo sapiens* (Broadhurst, et al., 1998).

Most rodents are herbivorous and, therefore, do not have access to preformed DHA in their diet. Consequently, in order to produce a brain with the same DHA content of other mammals, they have developed an efficient system for converting the n-3 precursors of DHA (e.g., linolenic acid; LNA), which are commonly found in fruits, nuts, and other dietary vegetable sources, into DHA (Pawlosky, et al., 1996). In addition, they also have an efficient system for the transfer of preformed DHA from the mother to the growing fetus *in utero* (placental transfer) or postnatally to the infant through their mother's milk. Because there is very little brain growth in the rodent after weaning, the requirement for DHA in the diet of the adult animal is reduced.

The lipid composition of the brain varies slightly depending on the particular section of the brain studied (Table 1). The "white matter" of the brain consists primarily of myelinated neurons and myelin is predominantly sphingolipid (e.g., sphingomyelin) and cholesterol (Fig. 1). The fatty acid composition of myelin is mostly saturated and contains relatively little DHA. The "gray matter" of the brain, on the other hand, has very little myelin and the phosphatides PS, PE, and PI of the active neuronal membranes make up the bulk of the lipid (Salem, et al., 1986). These lipids are very unsaturated and DHA and AA are the predominant building blocks thereof.

The biosynthesis of long-chain polyunsaturated fatty acids (PUFAs), like DHA, involves a succession of elongase and desaturase activities that act in concert on both n-6 and n-3 precursors. Work by Sprecher and colleagues (Voss, et al., 1991) has demonstrated that the final steps in DHA biosynthesis involves an elongation of C22:5 to C24:5, and the action of a delta-6 desaturase to produce C24:6 (Fig. 2). This 24-carbon

Chapter 21 / DHA and Optimal Cognitive Function

Complex Lipids

Triglyceride

Phosphatidylcholine

Ethanolamine   Serine   Inositol

Sphingomyelin

Fatty acids (building blocks)

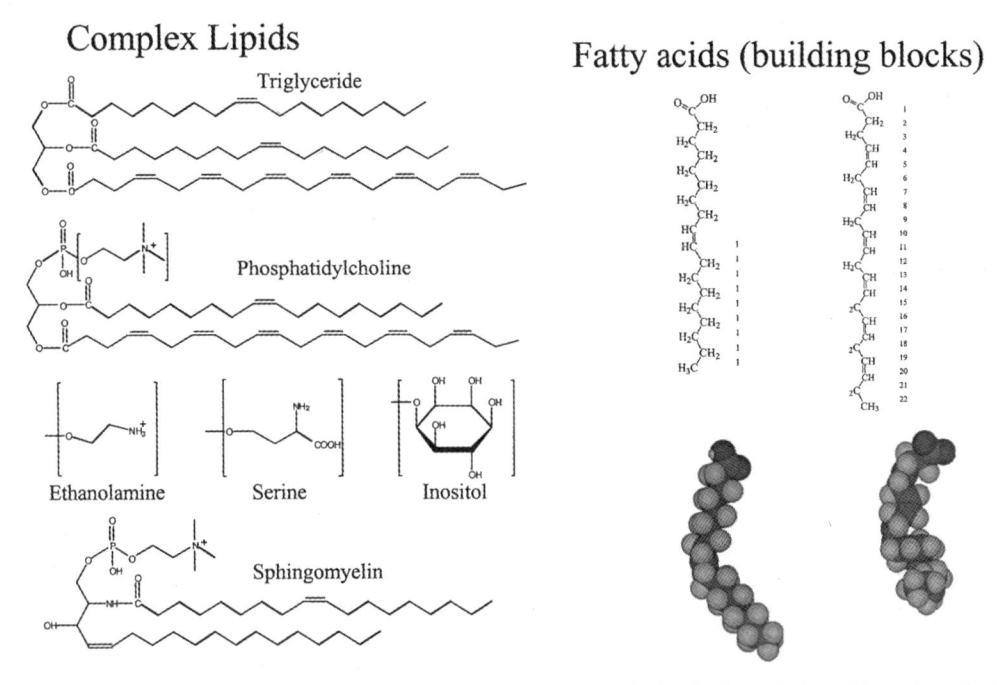

Fig. 1. Complex lipid forms and fatty acids in mammalian cells, including triglycerides, phosphatides with either choline, ethanolamine, serine or inositol, and shphigomyelin. Fatty acid building blocks include oleic acid (C18:1) and DHA (C22:6) in molecular structure form.

Fatty Acid Biosynthesis

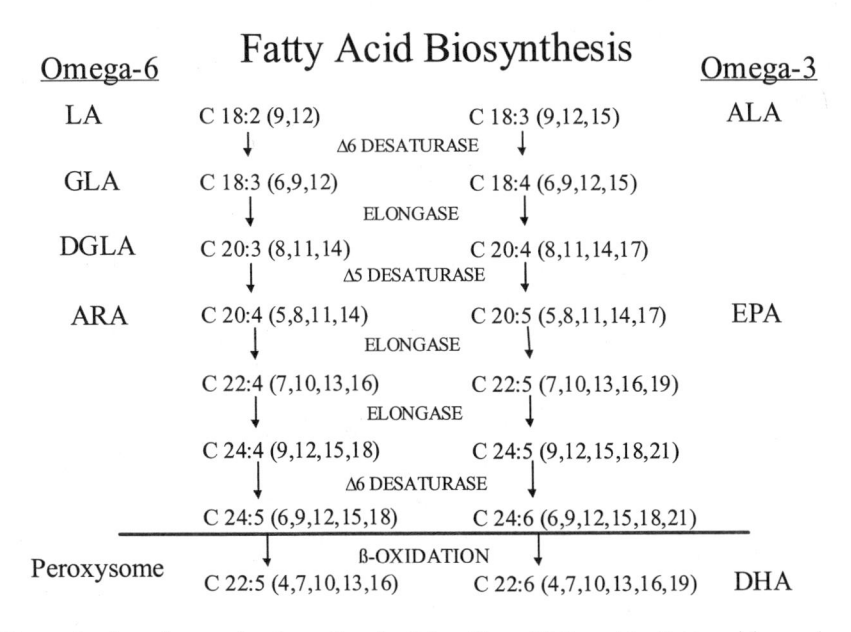

Omega-6			Omega-3
LA	C 18:2 (9,12)	C 18:3 (9,12,15)	ALA
	↓ Δ6 DESATURASE ↓		
GLA	C 18:3 (6,9,12)	C 18:4 (6,9,12,15)	
	↓ ELONGASE ↓		
DGLA	C 20:3 (8,11,14)	C 20:4 (8,11,14,17)	
	↓ Δ5 DESATURASE ↓		
ARA	C 20:4 (5,8,11,14)	C 20:5 (5,8,11,14,17)	EPA
	↓ ELONGASE ↓		
	C 22:4 (7,10,13,16)	C 22:5 (7,10,13,16,19)	
	↓ ELONGASE ↓		
	C 24:4 (9,12,15,18)	C 24:5 (9,12,15,18,21)	
	↓ Δ6 DESATURASE ↓		
	C 24:5 (6,9,12,15,18)	C 24:6 (6,9,12,15,18,21)	
Peroxysome	↓ β-OXIDATION ↓		
	C 22:5 (4,7,10,13,16)	C 22:6 (4,7,10,13,16,19)	DHA

Fig. 2. Biosynthetic pathways for the n-6 and n-3 families of fatty acids. Fatty acids are shown with the number of carbon atoms, followed by the number of double bonds with the positions of the double bonds in parentheses. LA, linoleic acid; GLA, γ-linolenic acid; DGLA, dihomogamma linolenic acid; ARA, arachidonic acid; LNA, linolenic acid; EPA, eicosapentaenoic acid; DHA, docosapentaenoic acid.

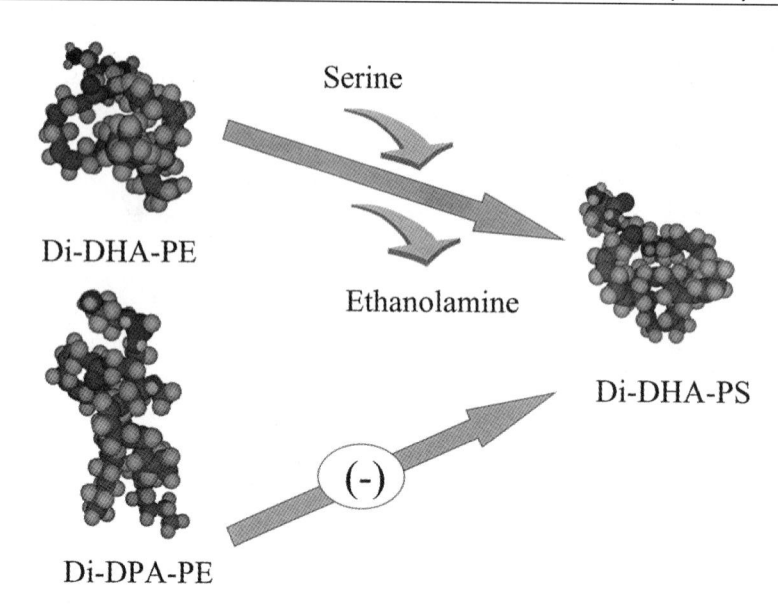

Fig. 3. Molecular models for the conversion of di-DHA phosphatidyl ethanolamine or di-DPA (n-6) phosphatidylethanolamine (PE) into phosphatidylserine (PS) by base exchange catalyzed by PS synthase. Molecular models were computed by energy minimization of the phospholipid structure given the di-DHA PE structure.

intermediate product is then converted to DHA (C22:6) by one step of β-oxidation in the peroxisomes. Isolated neurons in culture cannot undertake this last step, but the glial cells (helper cells) of the brain complete this reaction and provide preformed DHA directly to the neurons (Moore, et al., 1991).

Docosahexaenoic acid is not present as a free acid but is incorporated into a membrane phospholipid such as phosphatidyl serine (PS), typically along with stearic acid (18:0). PS is usually found on the inner leaflet of the plasma membrane and is a key structural and regulatory lipid in the brain. PS is synthesized from phosphatidylethanolamine (PE) or phosphatidylinositol (PI) by an enzyme-catalyzed, base-exchange reaction of serine with either choline or ethanolamine (Hamilton, et al., 2000) (Fig. 3). Under conditions of DHA depletion, there is a substantial reduction of brain PS. This was recently explained by the observation that the base-exchange reaction was favored only when DHA was present as one of the acylmoieties on the substrate PE or PI (Garcia, et al., 1998). In conditions of DHA depletion, where docosapentaenoic acid (n-6 DPA) is found in high concentrations in the PE and PI, the base-exchange reactions are inhibited and PS levels in the brain are reduced. Presumably, this differential activity is due to different conformations of PE when n-6 DPA is present compared to DHA (Fig. 3). It is quite possible, therefore, that the presence of DHA in the membrane phospholipids allows for the correct interconversion of the various phospholipid types in order to maximize the operational efficiency of the neurons.

3. THE EFFECT OF DIET ON BRAIN COMPOSITION

Although isotope-tracer studies have shown that most mammals can synthesize DHA from the essential fatty acid LNA, there is considerable debate over the capacity of

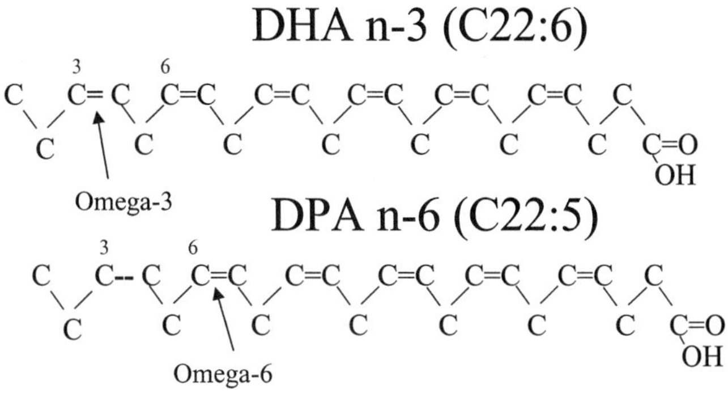

DHA n-3 (C22:6)

$$\text{C–C=C–C=C–C=C–C=C–C=C–C=C–C=C–COOH}$$

Omega-3

DPA n-6 (C22:5)

$$\text{C–C–C–C=C–C=C–C=C–C=C–C=C–C=C–COOH}$$

Omega-6

Fig. 4. Carbon skeleton models of DHA (n-3) and DPA (n-6) noting the position of the methyl terminal double bond at the n-3 and n-6 positions respectively.

various animal species (including man) to make DHA. In humans, the conversion rate of LNA to DHA is estimated to be 2.5–4.0% (Gerster, 1998) and that depends on the diet of the individual. In infants, this can even be as low as 1% conversion (Salem, et al., 1996). This means that most humans living on a Western diet will have insufficient in vivo production of DHA from LNA and will depend on preformed dietary DHA such as is found in some fish, algae, and animal organ meats (or breast milk for infants) to maintain adequate DHA stores. Rodents are more efficient at converting LNA into DHA, and at dietary LNA levels in excess 0.4% of energy the brain DHA does not increase (Bourre, et al., 1989). Cessation of DHA supplementation leads to a gradual decline in tissue and blood concentrations of DHA.

There have been a number of studies clearly demonstrating that the rodent brain fatty acid composition can be significantly altered by dietary modifications. Because the primary n-3 fatty acid in the brain is DHA, virtually all of these studies have involved some sort of n-3 restriction from the diet. The most remarkable change in the brain as a consequence of n-3 restriction is the near total replacement of brain DHA with the equivalent product from the n-6 pathway, namely n-6 DPA (Bourre, et al., 1993, Garcia, et al., 1998). DHA and n-6 DPA vary only by a single double bond at a position 19 carbons from the carbonyl function, or at the n-3 position (Fig. 4). It appears that 22-carbon, highly unsaturated fatty acids are critical elements for brain function, and in the absence of DHA or precursors with which to make DHA, the body substitutes DPA (n-6).

Depletion kinetic studies have shown that although some organs like the liver can become depleted in DHA relatively quickly, DHA is highly conserved in the brain (Greiner, et al., 1999). In rats, for example, it takes about three generations of n-3 deficient diets to almost totally replace the brain DHA with n-6 DPA (Fig. 5). When diets are n-3 restricted, the DHA levels in the brain are the most highly conserved. Since smaller mammals (e.g., mice) have a larger brain to body weight ratio, they generally have more DHA per unit body mass and may deplete more quickly from one generation to the next when n-3 fatty acids are not included in the diet.

Once an animal is depleted in brain DHA, one can look at the rates of repletion of DHA in various organs when provided either DHA *per se* or the parent precursor LNA. Salem and co-workers (personal communication) have demonstrated that the rates of repletion of tissues in mature animals vary with the organ. The liver will replete very quickly (2 wk), whereas the brain will replete more slowly (6–9 wk). When DHA is added back to the diet

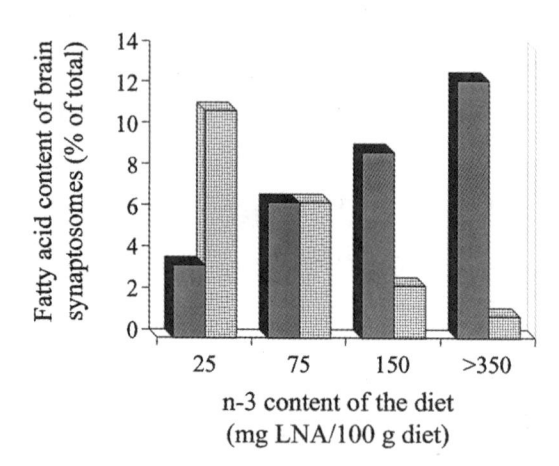

Fig. 5. The effect of n-3 deficiency on the ratio of DHA (solid bars) to n-6 DPA (stippled bars) in the brain synaptosomes of the rat. Different levels of n-3 deficiency were obtained by raising rats for three generations on diets containing various levels of linolenic acid (LNA). (Modified from Bourre et al., 1989.)

of deficient animals, however, the body appears to "recognize" the newly available DHA and use it to quickly replace n-6 DPA in the brain.

Tam and co-workers showed that if n-6 DPA is provided in the diet of rats, it can be incorporated into brain lipids even when three times higher levels of DHA are present (Tam, et al., 2000). The n-6 DPA in the liver increased from a virtually undetectable level before the supplementation to levels of about 7% of the total fatty acid. The n-6 DPA levels in the brain and testes also increased, but by a much smaller amount (i.e., about 15% in both cases). Because it is not known whether the relationship between performance deficit and high brain n-6 DPA (seen in n-3 deficiency) is incidental or causal, it is not known if the observed transfer of dietary n-6 DPA to brain lipids would have a detrimental effect on cognition.

Repletion studies may be undertaken using either preformed DHA or the parent precursor LNA. If LNA is provided, one can get a measure of the animal's ability to convert the LNA to DHA. As indicated previously, different species have different abilities to make this conversion (Salem and Pawlosky, 1994). Most rodents, however, are believed to be similar in this ability compared to other mammals.

Finally, it is important to note that there is another experimental paradigm that results in depletion of brain DHA levels other than n-3 dietary restriction. Chronic consumption of high levels of alcohol results in a specific depletion of DHA from the brain. When cats are given 1.2 g ethanol/kg/d (a value not different than is the case with chronic human alcoholics), their brain DHA levels are reduced by about 20% (Pawlosky and Salem, 1995). About a 30% reduction in brain DHA levels has also been observed in elderly patients with Alzheimer's disease compared to normal geriatric controls (Kyle, et al., 1999, Prasad, et al., 1998). It is interesting to note that human infants fed standard infant formulas without supplemental DHA also have about a 30% lower level of brain DHA at about 4 mo of life compared to infants fed their mother's DHA-containing breast milk (Farquharson, et al., 1992).

In the remaining sections of this chapter, we will discuss the cognitive and behavioral consequences of altering the brain fatty acid composition (i.e., replacing DHA with n-6 DPA) using rodent models.

4. EVALUATION OF COGNITIVE FUNCTION

Domains of behavior may be classified as arousal, cognitive, social, motor, sensory, and motivational. Webster's dictionary defines "cognition" as "the act or process of knowing, including both awareness and judgment." As such, it includes observation of the world and mental treatment of these inputs. Cognition can be subdivided into a number of components such as perception (e.g., odor, visual), action (e.g., exploration), learning (problem solving), and memory (long term and short term or working memory). In humans, some of these components can be further subdivided. For instance, long-term memory can be explicit (conscious storage and retrieval of information) or implicit (information is filed in the absence of conscious effort).

5. TESTS FOR COGNITIVE FUNCTION

5.1. Morris Water Maze

The Morris water maze is useful for the study of learning, adaptation to change, and loss of memory. It consists of a circular tub that can be anywhere from 60 to 240 cm in diameter and that is filled with water that has been rendered opaque by the addition of a white pigment. In the tub is a small platform, which may or may not be submerged. The animal will explore by swimming until it finds the platform where it will rest. Visual cues in the room will help the animal guide itself to the platform when the test is repeated. Collected data may include initial heading, path length, speed of swimming, and time to find platform (also known as escape latency). If the position of the platform is signaled by an attachment, the trial is called "cue learning". If no visual help is provided, it is called "place learning."

5.2. Radial Arm Maze

The radial maze consists of eight dead-end arms arranged like spokes around a central chamber (Fig. 6a). The spatial memory and foraging behavior of the animal can be studied after it has been released in the maze. In one variant, hungry rats are presented with one food pellet at the end of each arm and given a limited time to feed. Parameters typically recorded include number of arms repeatedly visited, or omitted, and time to complete the maze. Rats with impaired short-term memory function will do poorly in this maze.

5.3. T-Maze

In the T-maze the rat must correctly choose between the left or right arm to obtain food pellets (Fig. 6b). Audio, olfactory, or visual cues may guide the choice, and provide a test for efficacy of working memory. The data collected can be time to goal and percent correct choices. One arm may be covered to assess the effect of learning versus the normal tendency of the rodent to seek the darker passage.

5.4. Rotating Rod

In this experimental device, animals are tested for their ability to maintain themselves upright on a rod rotating around a horizontal axis. This is a measure of the agility of the animal and of the desire to escape. By increasing the distance the animal falls, the desire to escape can be reduced.

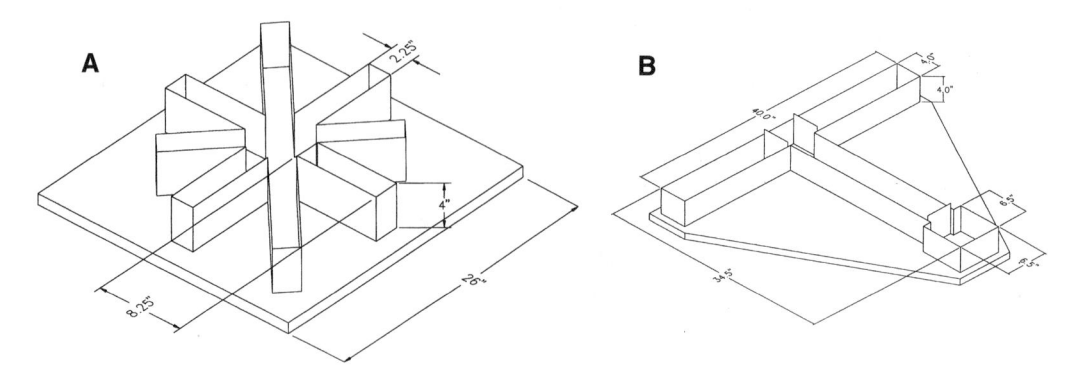

Fig. 6. Radial arm maze (**a**) and T-maze (**b**). Mazes are classic tools for the assessment of behavior, learning and memory. Modern mazes are often automated using photocells and video monitoring. (From: http://www.colinst.com with permission.)

5.5. Elevated Plus-Maze

The plus-maze has four arms attached at right angles to a central platform. High walls enclose each of two arms, and the two other arms have no restrictions. This maze can be used as a test for memory similar to the above-described mazes, or as a test of anxiety because rodents prefer enclosed spaces.

5.6. Skinner Box

The Skinner box consists of a small cage with a lever that the rodent can operate to obtain a reward when certain conditions are met. This reinforces behavior. Where mazes are suitable for the study of discrete behaviors, the Skinner box allows the study of a string of successive behaviors.

5.7. Passive/Active Avoidance or Shuttle Box Test

The avoidance test takes place in a two-compartment cage (shuttle box) with a central connecting opening. An electrical grid on the floor can deliver a mild shock to one of the compartments. A stimulus (sound or light) will announce the start of the experiment. In the "active avoidance" variant, the rat must move from one compartment to the other in order to avoid a shock. In passive avoidance, the shock will be delivered if the rat does not stay in the compartment where it is at the time of the stimulus. Data recorded may be "avoidance," "passage with shock," and "nonpassage." "Avoidance" is when the animal escapes shock by moving to the correct room in time. "Nonpassage" is when an animal remains in the wrong room in spite of the electric stimulus. "Passage with shock" occurs when the animal moves to the right room, but not quickly enough to avoid shock.

5.8. Open Field

The open field is a cage, typically 50 cm × 50 cm, which is sound and light attenuated. Inside the cage, there are no visual cues. An animal put in the cage will be monitored and simultaneous measurement of locomotion and rearing will be conducted. The data collected may include number of head movements, number of rearings against the wall (exploratory behavior), and distance covered (a combination of exploratory behavior and anxiety).

5.9. Escape Tendency

Rodents have a natural aversion to being exposed. By putting a rodent inside an inverted glass beaker with no access to food or drink, they will be motivated to escape. The number of attempts to escape is monitored. The animal will attempt to do so by rearing or scratching the surface of the floor.

5.10. Actimeter

A cage with two infrared photocells centered along one of the short and one of the long sides is used. The rodent is let loose in the cage and the number of times it crosses into a new square is tallied.

5.11. Forced Swimming Test

One of the most fundamental forms of learning is habituation. As rodents become familiar with a stimuli or environment, the behavior elicited will decrease. Habituation can be measured by repeatedly placing the animal in the same environment and recording exploratory behavior and duration of immobility. One example of habituation is the forced swimming test: On two separate occasions, an animal is placed in a Plexiglas cylinder partially filled with water. There is no escape, so the animal is forced to swim. At first, the animal will swim around, but later it will tend to only expend enough effort to keep its head out of the water. During the second trial, duration of immobility is measured. Longer periods of immobility are often interpreted as a measure of learning, but could also be an expression of satisfaction with the current situation, or an absence of anxiety or motivation.

5.12. Electroretinogram

The electroretinogram, or ERG, is the recording of the electric impulses generated by neurons and other cells in the retina and pigment epithelium when this is presented with a short but intense strobe of white light. Because the concentration of DHA in the healthy retina may exceed 50%, it is not surprising that this tissue is affected by n-3 fatty acid depletion. The two major components of an ERG are the fast negative a-wave followed by the fast positive b-wave (Fig. 7). Because the ERG only reveals information about the retina and not about the subsequent processing, it does not tell us anything about the animal's perceived quality of eyesight. The latter quality depends on how well the retina operates.

6. EFFECT OF DIETARY n-3 FATTY ACIDS ON COGNITIVE FUNCTION

Animal models are useful because greater depletion of DHA is possible than in human studies. Typically, a depleted condition is achieved by feeding several generations of rodents a diet essentially free of n-3 fatty acids. A swifter protocol for experimental n-3 deficiency consists of removing the infant rat from the dam, gastrostomizing it, and feeding it synthetic formula via a pump (Ward, et al., 1996). This leads to over 50% reduction in brain DHA in just 8 wk and as much as 90% reduction in the offspring.

As long as sufficient n-6 fatty acids are supplied, n-3 deficiency does not result in conspicuous changes as one might see if the animals were depleted in certain vitamins. As a result, it took 50 yr longer to recognize the essentiality in mammals of n-3 fatty acids than it did for n-6 fatty acids. Because the growth rate, the total amount of lipid in the

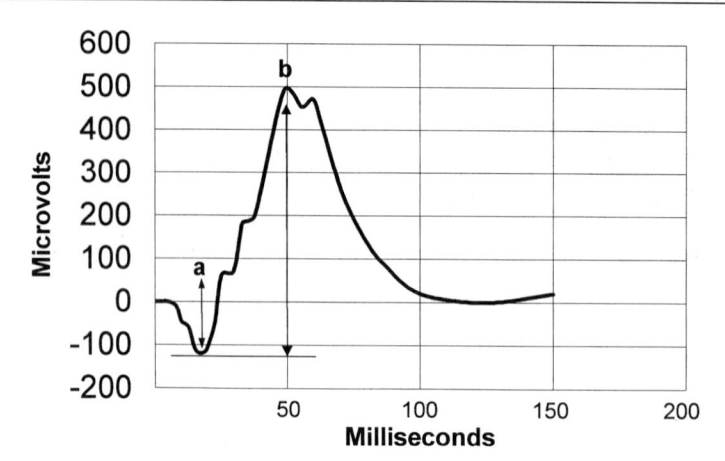

Fig. 7. A typical electroretinogram is recorded by placing an electrode on the cornea and monitoring the succession of electric impulses which follow the brief illumination of the retina with white light.

brain, and most other parameters remain approximately unchanged, these are not confounding factors and the effects of n-3 deficiency are generally assumed to be the result of the change in fatty acid composition of organs such as the brain. Table 2 lists oils commonly used in dietary depletion studies according to whether they are rich or poor sources of n-3 fatty acids.

7. DIET DEPLETION STUDIES

There is a large body of evidence, mainly from mice and rats, suggesting that low consumption of dietary n-3 fatty acids results in impaired cognitive and/or visual performance. Because of the differential abilities of mammals to synthesize DHA from LNA, a second body of data addresses the issue of providing preformed DHA to animals that are not n-3 deficient.

7.1. Dietary n-3 Deficiency in the Rat

7.1.1. VISUAL ACUITY AND n-3 DEFICIENCY

Perhaps the most consistent effect of n-3 deficiency is found on vision. For instance, when Wistar rats that had been fed sunflower-oil-based chow for three generations were compared to a similar soybean-oil-fed group, there were striking and highly significant differences in their electroretinograms (Bourre, et al., 1989). In 4-wk-old rats, the amplitudes of the a- and b-waves in the LNA-rich group were 39% and 80% higher, respectively, compared to the deficient group. Furthermore, for the a- and b-waves to reach the detection limit, the intensity of the light stimulus had to be 10 times higher in the LNA-deficient group than in the sufficient group. This effect diminished with time, but even between adult rats, a statistically significant difference of 20% between the a-wave amplitudes of the two dietary groups remained.

7.1.2. BEHAVIOR AND n-3 DEFICIENCY

In a two-generation study (Lamptey, 1976) rats were fed diets containing 10% soybean oil or 10% safflower oil. The effect on brain lipids was primarily that the first group was

Table 2
Dietary Sources of Dietary Fat Commonly Used in Animal Experiments

Good dietary sources of n-3	Poor dietary sources of n-3
Rapeseed oil	Safflower oil
Soybean oil	Sunflower oil
Perilla oil	Lard or tallow
Linseed oil or flax seed oil	Coconut oil
Fish oil from menhaden, salmon, tuna	Hydrogenated oil
Algal oil (e.g., DHASCO®)	Palm oil
	Peanut oil

found to deposit DHA, whereas the second group deposited n-6 docosapentaenoic (DPA) in the brain of the rat. There were no detectable outward physical effects of the difference in diet, but the group fed the LNA-rich diet had superior learning and motivation in the Y-maze. This may have been the first study to show an effect on behavior of an n-3-deficient diet.

When spontaneously hypertensive rats and Wistar/Kyoto rats were fed perilla or safflower oil, Yamamoto and colleagues (Yamamoto, et al., 1987) found the same differences in brain lipid as Lamptey (Lamptey and Walker, 1976) in the second-generation of n-3 deficient rats. These rats showed a significantly lower probability of returning a correct response when compared to the perilla-fed animals in a Skinner box with light as a stimulus. The lower performance was the result of inferior learning ability and not inactivity on the part of the rat, as deficient rats responded relatively more frequently to the stimuli. These results were later reproduced in the Donryu strain of rat (Yoshida, et al., 1997).

Three generations of Wistar rats were fed diets deficient or sufficient with respect to LNA (Bourre, et al., 1989). In one group, the dietary source of fat was sunflower oil, whereas in the other group, the dietary fat contained the same amount of soybean oil. There was little outward physical evidence of the difference in polyunsaturated fatty acid (PUFA) quality, but cognition was highly perturbed. When 7 male 60-d-old rats from each group were studied in the shuttle box (30 repetitions), the LNA-deficient group had significantly lower avoidance (2.8 vs 7.2) and a significantly higher number of nonpassages (10.5 vs 2.8) compared to the LNA-rich group. At the second session, the difference was still significant but smaller and became insignificant in the following two sessions. This seems to indicate that the main effect on the animals was on the speed of learning, not the capacity for learning.

The above-cited studies provide evidence to support the premise that diets very low in n-3 will have a negative impact on learning. Fewer studies have tested diets that are n-3 adequate against diets that are a rich source of LNA. Yamamoto (Yamamoto, et al., 1988) studied rats fed standard n-3-sufficient laboratory chow against a diet enriched with perilla oil and found that the latter group had a significantly higher rate of correct responses in the brightness discrimination test. Enslen did not find a difference between lab chow and soybean-oil-enriched groups in a two-generation rat study as gauged by the open-field test (Enslen, et al., 1991). Both groups of male Sprague–Dawley rats showed superior outcomes, however, including increased exploratory activity, relative to a group fed safflower oil.

In one of the few short-term studies published, Coscina studied 3 dietary groups, each consisting of 16 young male hooded rats that had previously been fed lab chow (Coscina, et al., 1986). One group was fed 20% soy bean oil, one group was fed 20% lard, and a third group was maintained on lab chow containing 4.5% mixed fat. After as little as 3 wk, testing began in the Morris water maze. The soybean oil group displayed significantly superior learning. After the first 3 trials, this LNA-enriched group had significantly shorter latency for 21 consecutive trials. The apparent learning abilities of the lard and the lab chow group were not significantly different, but in most trials, it took an average rat in these groups at least 30% longer to find the platform than the average rat from the LNA-rich group. Unfortunately, attempts years later to reproduce these results failed (Coscina, 1997). After the first 24 trials, the platform was moved and another 6 trials were conducted. It is interesting that it took the soybean oil group some time to adjust, but from the third trial onward, this group again had the shortest latency.

7.1.3. Effect of Dietary DHA During n-3 Sufficiency on Behavior in the Rat

Few studies in rats have supplied preformed DHA to the diets of the experimental animals. In one such study (Wainwright, et al., 1999), four groups of male Long–Evans rats were weaned and raised on a diet containing 10% LA and 1% LNA with or without 2.5% of DHA and/or AA (in a 2x2 factorial scheme). After 6–9 wk, these rats were tested in the Morris water maze, but there were no significant differences between their performances, nor were they different from a control group suckled by dams fed standard lab chow or a group fed 13.5% lard.

This is in contrast with results reported in Wistar rats that were fed a n-3-sufficient diet with varying levels of DHA for two generations (Yonekubo, et al., 1993). A control diet containing 2% LNA (n-6/n-3 ratio of 10.4) was compared to a diet with sardine oil added to give 7.0% long-chain n-3 fatty acids (n-6/n-3 ratio of 1.8). The behavior of 6-wk-old rats was tested in the Morris water maze. After 8 trials with 20 rats in each group, the mean time-to-escape for the control group was 26% longer than for the DHA-enriched group. The time to find the platform fell by approximately 40% in both groups, but the difference remained statistically highly significant. When the platform was moved, the DHA-fed group initially took longer to find the platform, as they were searching significantly longer in the area where the platform used to be. However, the superior learning ability of the DHA group was applied to the new situation, and after the third trial, the escape latency of this group was again significantly shorter than that of the control. It is interesting that there was a demonstrable advantage of adding long-chain n-3 to a diet already considered adequate with respect to n-3. This reflects the way terms like "adequate" and "sufficient" are often defined based on physical measures and not behavior.

7.2. Dietary n-3 Deficiency in the Mouse

7.2.1. Mouse Visual Acuity and n-3 Deficiency

As seen in rats, there is no evidence of a strong correlation between visual acuity and learning in the mouse. The second generation of mice (F1) fed a diet poor in LNA was compared to a group fed laboratory chow. In adult mice, there was a significant difference in the retinal concentration of DHA, but the difference in b-wave electroretinograms ceased to be significant at the seventh week (Carrie, et al., 1999). The ability to learn was examined using the passive-avoidance test and the LNA-deficient group continued to significantly underperform the chow group, even as adults. Normalization of visual performance was not accompanied by an improvement in cognitive performance.

7.2.2. MOUSE BEHAVIOR AND n-3 DEFICIENCY

The data on behavior for the mouse is less consistent than the data on visual acuity, even if most studies point to beneficial effects of a diet enriched in DHA and its precursor LNA. Wainright (Wainright, et al., 1994) fed mice one of three diets through pregnancy, lactation, and weaning. They were the same basic diet but with (1) saturated fat, (2) adequate LA but deficient in LNA, and (3) sufficient LNA and LA (n-6/n-3 ratio of 3.7). Six weeks after weaning, two females from each litter were tested in the Morris water maze and the open field. Initially, mice fed the saturated diet were a little slower in the water maze, but that effect was transient. No difference was found in the open field or during reverse learning in the water maze.

In contrast, Bourre's group (Frances, et al., 1996a) compared OF1 mice fed a peanut-oil-based diet containing trace amounts of LNA with a group receiving 200 mg of LNA/ 100 g of feed from rape seed oil and found significant performance differences. Both groups had the same total fat and LA content in the diet and had been reared by dams receiving the same diet as their pups. There were 10 female mice in each group. The LNA-deficient group improved its performance over time, but in the Morris water maze (place-test version), it took them significantly longer to find the platform after 16 practice trials. The difference was not significant in the cued version of the test. When mice raised by the same scheme were tested (Frances, et al., 1996b) it was found that the n-3-sufficient mice but not n-3-deficient mice showed reduced exploration in a photocell Actimeter. In the forced swimming test, the n-3-sufficient mice were significantly less active, and the same was found ($p < 0.061$, one-sided test) for the number of escape attempts from a confined space as well as for activity in an open field (pooled data for 2 d). The authors concluded that n-3-sufficient mice showed greater habituation. There were 14 mice per group.

Frances et al. (Frances, et al., 1996b) used the same protocol to rear LNA-deficient and LNA-sufficient pups. Animals on the n-3-deficient diet showed significantly slower learning. Mice raised on a sufficient diet were quicker to escape the Rotarod. Paw coordination, muscular function, defensive behavior, and anxiety was not different between the groups.

Suzuki et al. (Suzuki et al., 1998) found that mice fed sardine oil for 12 mo were better at finding their way though a maze. They negotiated it more quickly and strayed down wrong paths less frequently when compared to a n-3-deficient group

7.2.3. EFFECT OF DIETARY DHA DURING n-3 SUFFICIENCY ON BEHAVIOR IN THE MOUSE

Recently, Carrie and colleagues (Carrie, et al., 2000) studied the effect of addition of sardine oil (a source of long-chain n-3 fatty acids) to a diet containing adequate n-3 and n-6 fatty acids. The control was a diet with equivalent amounts of palm oil. Both diets contained 14% fat of which 31% and 2.0%, respectively, were n-3. The sardine oil diet contained 0.94 g DHA and 1.6 g of EPA per 100 g of feed. Groups of 12 female OF1 mice from each dietary treatment were tested in the open-field, Morris water maze, and avoidance tests as young adults, mature, and old. To avoid memory effects, mice were not subjected to the same test twice. The effect of the n-3 diet on young mice was a significant increase in exploration and locomotor activity. In older mice, exploration, as measured in the open-field test was the same in both groups. However, the locomotor activity of old mice was significantly lower in the fish-oil group. Perhaps this is an effect of superior memory in this group. If the hidden platform is removed from the Morris maze, the time that a mouse spends searching in that quadrant is a measure of memory. By this criterion mature mice from the fish-oil group were significantly better than the control group at remembering the location of the hidden platform. In old mice, the fish-oil group spent 45% more time in the training quadrant than the palm-oil group, but this did not reach significance at the 5% level. In active avoidance, the fish-oil-fed mice were significantly different from the control on the first day of training. Young fish-oil-fed mice had higher avoidance, whereas in the two other ages, mice from the palm-oil group were better at escaping before the electric chock. This could be a consequence of the increased locomotor activity. Mice that move about less may take longer to realize that the room of the cage they are in makes a difference. Also, n-3 fatty acids may make the mouse less averse to the electric stimulus. Yehuda and Carasso (Yehuda and Carasso, 1993) reported that rats injected with 25 mg/kg/d of LA and LNA in a ratio of 4 : 1 had a latency to lick the paw, after being placed on a 58°C hotplate, which was twice as long as a saline-injected control.

8. REPLETION STUDIES

Repletion studies allow the assessment of whether the nutritional effects of n-3 fatty acids are solely a function of the concentration of the fatty acids in organs, or if a period during which n-3 intake is low can have a lasting impact irrespective of later nutritional status. We have seen earlier that n-3 deprivation may lead to cognitive impairment relative to n-3-sufficient animals, but what happens if the n-3 deprivation is ended?

Dunkin–Hartley guinea pigs were born to a sow that had been fed standard chow (Weisinger, et al., 1999). Following weaning, they were fed a safflower-oil-based diet deficient in LNA and a canola-oil-based diet with adequate LNA. The first-generation deficient animals showed a reduction in retinal phospholipid DHA levels from 22.1% to 9.3% relative to the LNA-sufficient group at wk 16. The reciprocal increase in n-6 DPA was accompanied by electroretinographic deficiencies such as significant reductions in a- and b-wave amplitudes (35% and 60%, respectively, at wk 16) and a significantly reduced sensitivity to light stimuli (42% at 16 wk). Some animals were switched from the safflower-oil diet to the canola-oil diet at 6 wk for 5 and 11 wk, or at 11 wk for 5 wk. The effect of the n-3 repletion was to reduce the functional deficit of the n-3-deficient group. After 10 wk of repletion, the differences in function between the canola-oil group and the repleted group were no longer significant, even if the DHA content of retinal phospholipids was still significantly lower in the latter group. Animals fed canola oil for 5 wk were more sensitive to light stimulus if the repletion started early. The study did not go on for long enough to resolve whether this was because animals that had been on the n-3-deficient diet longer had smaller DHA stores or whether repletion must happen before a certain critical period in development. This is a particularly important result, as guinea pigs and humans have more extensive postnatal neurological development than rats and mice, and as such the guinea piglet is more analogous to the condition of the human infant.

A recent study in rats found that behavior too can be normalized by repleting a n-3-deficient animal (Moriguchi, 2000). For three generations, Long–Evans hooded rats were fed a diet that was either adequate (n-6/n-3 ratio of 4.1 with 1.3% preformed DHA and 2.6% LNA) or deficient (n-6/n-3 ratio of 378). The third generation of male deficient rats were switched to the adequate diet either at birth (wk 0), weaning (wk 3), or adulthood (wk 7) and their performance in the Morris water maze was studied at wk 9 and 13. As expected, the deficient group had significantly reduced brain DHA and greater difficulty in learning the task and took longer and swam farther to find the platform. Furthermore, rats that were tested during repletion performed significantly better the longer they had been consuming the n-3-adequate diet. The total escape latency was directly and statistically significantly correlated to the concentration of DHA ($r^2 = 0.88$) and negatively correlated with the n-6 DPA ($r^2 = -0.95$). This means that almost all the variance in the time it took to find the platform can be predicted just from the concentration of n-6 DPA or DHA in the brain. The higher the level of n-6 DPA in the brain, the worse was the performance deficit. It also illustrates how closely n-6 DPA and DHA are regulated. After 9 wk of feeding an n-3-adequate diet, the DHA concentration in the brain was no longer different from that of rats that had always been fed an adequate diet.

9. BIOCHEMICAL BACKGROUND FOR COGNITION

It is clear from the above studies that diet can affect both the brain fatty acid composition as well as the visual and cognitive performance in rodents. One is now left to speculate about the causal relationship between brain fatty acid composition and visual and cognitive performances. Is it plausible that a seemingly minor change primarily in two fatty acids (i.e., the substitution of n-6 DPA for the n-3 DHA) could result in such a significant effect?

Structural changes in the brain itself do appear to be correlated with the diet and performance. Yoshida and co-workers found that poor learning in rats as a consequence of n-3-deficient diets was associated with low average densities of hippocampal synaptic vesicles (Yoshida, et al., 1997). Salem has recently reported that n-3-deficient diets that result in poorer performance scores on various tests was also accompanied by lower densities of neuronal cell bodies in both the frontal cortex and the hippocampus (personal communication). Heinz and co-workers have identified a DHA-specific binding protein that is expressed during the migration of the neurons from the ventricles to the cortical plate during development in the rat brain (Xu, et al., 1996). The inactivation of this latter DHA-binding protein prevented the normal migration of cells and disrupted the development of the brain cortex. Finally, the recent work of Salem's group (Hamilton, et al., 2000) has demonstrated that the synthesis of phosphatidyl serine is dramatically (sixfold) reduced when the membranes are enriched with the n-6 DPA instead of DHA. PS is not only an important structural lipid in the brain, but it is also specifically involved in the function of certain protein kinases (e.g., Raf-1 kinase, which is important in growth factor signaling) and the control of apoptosis.

Fatty acid compositional changes in the retina of the rodent eye can also impact the development of vision, which, in turn, will affect behavior of the animal. Recent biochemical and biophysical studies have demonstrated how the response of rhodopsin to different membrane lipids may significantly affect vision. DHA is primarily associated with membrane proteins of the 7-TM motif (i.e., seven transmembrane spanning regions). Such 7-TM proteins are also associated with a G-protein-coupled event (activation of a phosphodiesterase which cleaves GTP to GDP). Rhodopsin is a typical 7-TM G-protein-coupled "receptor" and recent work by Litman and colleagues have shown that if the rhodopsin is buried in a membrane rich in n-6 DPA instead of DHA, the activity of the phosphodiesterase is reduced by about sixfold (Nui, et al., 2000). Using very sensitive analytical tools and nuclear magnetic resonance measurements, Gawrisch and colleagues have demonstrated that the internal membrane order parameter is unfavorable to rhodopsin if the membrane contains n-6 DPA instead of n-3 DHA (Gawrisch, et al., 2000). These types of detailed analyses are now helping to explain how a simple change in the structure of a key fatty acid in the brain can result in dramatic changes in neurological functions from visual perception to cognitive abilities.

The critical importance of these findings are amplified by data from nonrodent studies and the resulting understanding of our own neurological development. Within the brain, there are many other G-protein-coupled receptors such as the neurotransmitter receptors with similar membrane spanning regions. This may be why the DHA concentration is very high in the synaptic junctions of the neurons. Innis has recently reported that when infant pigs are fed formula diets deficient in DHA and ARA, there is a significant alteration in the levels of various neurotransmitters in the brain (de la Presa Owens and Innis, 1999). Hibbeln has demonstrated a correlation between DHA status in certain individuals and serotonin levels in the cerebral spinal fluid (Hibbeln, et al., 1998). Hibbeln went on to validate this correlation with observational data indicating that human populations consuming the highest quantities of DHA in their diet also had the lowest incidence of depression (Hibbeln, 1998). Thus, there is a plausible link among the diet, the fatty acid composition of the brain, and the concentrations of certain neurotransmitters that can affect behavior.

10. CONCLUSION

Rodents need DHA to meet their full developmental potential. If this fatty acid or its precursor LNA is absent from the diet, delayed or reduced cognitive development, visual impairment, and abnormal electroretinograms are often observed. In some cases, these differences disappear with time. It cannot be concluded from this, however, that there are no long-term functional effects of hypodocosahexaenemia. The behavioral tests used are generally not specific enough to permit this. For instance, a maze is a test not just of the memory of an animal but also its motivation, eyesight, and physical condition.

Controlled dietary depletion studies in humans are difficult at best and have serious ethical issues in most cases. Therefore, we must rely on animal data, particularly the rodent data, to extrapolate to the human condition. Given the above behavioral effects of dietary fatty acids and the associated biochemical plausibility establishing a causal relationship between DHA in the diet and optimal brain function, we need to seriously consider the effects of dietary DHA insufficiency in man. In no place is this as critical as in the feeding of infants. From an evolutionary point of view, the best source of nutrition for an infant of any species is its mother's milk. In humans, mother's milk is a rich source of DHA as well as AA that is needed for the rapid brain growth that takes place in the first 2 yr of postnatal life. Most artificial infant formula mixtures provided for human infants today do not contain DHA or AA. Consistent with all the animal and biochemical data, the long-term neurological outcomes of infants fed these formulas are significantly poorer than the outcomes for breast-fed babies (Anderson, et al., 1999). The animal data provided in this chapter gives us strong arguments to continue to encourage mother's to breast feed their infants as long as possible to attain the maximum genetic potential of the next generation.

ACKNOWLEDGMENTS

We are grateful to Joi Rivas for her assistance in the preparation of this document and to Columbus Instruments International, Columbus, OH, for supplying the illustrations of the behavior instruments.

REFERENCES

Anderson JW, Johnstone BM, Remley DT. Breast-feeding and cognitive development: a meta-analysis [see comments]. Am J Clin Nutr 1999; 70(4):525–535.

Bourre JM, Dumont O, Pascal G, Durand G. Dietary alpha-linolenic acid at 1.3 g/kg maintains maximal docosahexaenoic acid concentration in brain, heart and liver of adult rats. J Nutr 1993; 123(7):1313–1319.

Bourre JM, Francois M, Youyou A, Dumont O, Piciotti M, Pascal G, et al. The effects of dietary alpha-linolenic acid on the composition of nerve membranes, enzymatic activity, amplitude of electrophysiological parameters, resistance to poisons and performance of learning tasks in rats. J Nutr 1989; 119(12):1880–1892.

Broadhurst CL, Cunnane SC, Crawford MA. Rift Valley lake fish and shellfish provided brain-specific nutrition for early Homo [see comments]. Br J Nutr 1998; 79(1):3–21.

Carrie I, Clement M, De Javel D, Frances H, Bourre JM. Learning deficits in first generation OF1 mice deficient in (n-3) polyunsaturated fatty acids do not result from visual alteration. Neurosci Lett 1999; 266(1):69–72.

Carrie I, Guesnet P, Bourre JM, Frances H. Diets containing long-chain n-3 polyunsaturated fatty acids affect behaviour differently during development than ageing in mice [see comments]. Br J Nutr 2000; 83(4):439–447.

Coscina DV. Polyunsaturated fats and learning, old data, new questions. In: Yshuda S, Mostofsky M, eds. Handbook of Essential Fatty Acid biology, Humana, Totowa, NJ, 1997, pp. 215–244.

Coscina DV, Yehuda S, Dixon LM, Kish SJ, Leprohon-Greenwood CE. Learning is improved by a soybean oil diet in rats. Life Sci 1986; 38(19):1789–1794.

Crawford MA. The early development and evolution of the human brain. Upsala J Med Sci 1990; 48(Suppl):43–78.

Crawford MA. The role of essential fatty acids in neural development: implications for perinatal nutrition. Am J Clin Nutr 1993; 57(5 Suppl):703S–709S.

de la Presa Owens S, Innis SM. Docosahexaenoic and arachidonic acid prevent a decrease in dopaminergic and serotoninergic neurotransmitters in frontal cortex caused by a linoleic and alpha-linolenic acid deficient diet in formula-fed piglets. J Nutr 1999; 129(11):2088–2093.

Enslen M, Milon H, Malnoe A. Effect of low intake of n-3 fatty acids during development on brain phospholipid fatty acid composition and exploratory behavior in rats. Lipids 1991; 26(3):203–208.

Farquharson J, Cockburn F, Patrick WA, Jamieson EC, Logan RW. Infant cerebral cortex phospholipid fatty-acid composition and diet [see comments]. Lancet 1992; 340(8823):810–813.

Frances H, Coudereau JP, Sandouk P, Clement M, Monier C, Bourre JM. Influence of a dietary alpha-linolenic acid deficiency on learning in the Morris water maze and on the effects of morphine. Eur J Pharmacol 1996; 298(3):217–225.

Frances H, Monier C, Clement M, Lecorsier A, Debray M, Bourre JM. Effect of dietary alpha-linolenic acid deficiency on habituation. Life Sci 1996; 58(21):1805–1816.

Garcia MC, Ward G, Ma YC, Salem N Jr, Kim HY. Effect of docosahexaenoic acid on the synthesis of phosphatidylserine in rat brain in microsomes and C6 glioma cells. J Neurochem 1998; 70(1):24–30.

Gawrisch K, Holte LL, Koenig BW, Polozov IV, Safley AMK, Teague WE. NMR Investigations of docosahexaenoic acid structure and flexibility. PUFA in Maternal and Child Health Conference, 2000.

Gerster H. Can adults adequately convert alpha-linolenic acid (18:3n-3) to eicosapentaenoic acid (20:5n-3) and docosahexaenoic acid (22:6n-3)? Int J Vitam Nutr Res 1998; 68(3):159–173.

Greiner RS, Moriguchi T, Hutton A, Slotnick, BM, Salem N Jr. Rats with low levels of brain docosahexaenoic acid show impaired performance in olfactory-based and spatial learning tasks. Lipids 1999; 34(Suppl):S239–S243.

Hamilton L, Greiner R, Salem N Jr, Kim HY. n-3 Fatty acid deficiency decreases phosphatidylserine accumulation selectively in neuronal tissues [In Process Citation]. Lipids 2000; 35(8):863–869.

Hibbeln JR. Fish consumption and major depression [letter] [see comments]. Lancet 1998; 351(9110):1213.

Hibbeln JR, Linnoila M, Umhau JC, Rawlings R, George DT, Salem N Jr. Essential fatty acids predict metabolites of serotonin and dopamine in cerebrospinal fluid among healthy control subjects, and early- and late-onset alcoholics. Biol Psychiatry 1998; 44(4):235–242.

Ikemoto A, Kobayashi T, Watanabe S, Okuyama H. Membrane fatty acid modifications of PC12 cells by arachidonate or docosahexaenoate affect neurite outgrowth but not norepinephrine release. Neurochem Res 1997; 22(6):671–678.

Kyle DJ, Schaefer E, Patton G, Beiser A. Low serum docosahexaenoic acid is a significant risk factor for Alzheimer's dementia. Lipids 1999; 34:S245.

Lamptey MS, Walker BL. A possible essential role for dietary linolenic acid in the development of the young rat. J Nutr 1976; 106(1):86–93.

Moore SA, Yoder E, Murphy S, Dutton GR, Spector AA. Astrocytes, not neurons, produce docosahexaenoic acid (22:6 omega-3) and arachidonic acid (20:4 omega-6). J Neurochem 1991; 56(2):518–524.

Moriguchi T, Loewke J, Salem N Jr. Spatial task performance depends upon the level of the brain docosahexaenoic acid. PUFA in Maternal and Child Health Conference, 2000.

Nui S, Mitchell DC, Litman BJ, Effect of DHA and cholesterol on receptor–G protein coupling. PUFA in Maternal and Child Health Conference, 2000.

Pawlosky R, Barnes A, Salem N Jr. Essential fatty acid metabolism in the feline: relationship between liver and brain production of long-chain polyunsaturated fatty acids. J Lipid Res 1994; 35(11):2032–2040.

Pawlosky RJ, Salem N Jr. Ethanol exposure causes a decrease in docosahexaenoic acid and an increase in docosapentaenoic acid in feline brains and retinas. Am J Clin Nutr 1995; 61(6):1284–1289.

Pawlosky RJ, Ward G, Salem N Jr. Essential fatty acid uptake and metabolism in the developing rodent brain. Lipids 1996; 31(7):S103–S107.

Prasad MR, Lovell MA, Yatin M, Dhillon H, Markesbery WR. Regional membrane phospholipid alterations in Alzheimer's disease. Neurochem Res 1998; 23(1):81–88.

Salem N Jr, Pawlosky RJ. Arachidonate and docosahexaenoate biosynthesis in various species and compartments in vivo. World Rev Nutr Dietics 1994; 75:114–119.

Salem N Jr, Wegher B, Mena P, Uauy R. Arachidonic and docosahexaenoic acids are biosynthesized from their 18-carbon precursors in human infants. Proc Natl Acad Sci USA 1996; 93(1):49–54.

Salem NJ, Kim H-Y, Yergey JA. Docosahexaenoic acid: membrane function and metabolism. In Health Effects of Polyunsaturated Fatty Acids in Seafoods. Simopoulos AP, Kiter RR, Martin RE, eds. Academic, New york, 1986, pp. 263–317.

Suzuki H, Park SJ, Tamura M, Ando S. Effect of the long-term feeding of dietary lipids on the learning ability, fatty acid composition of brain stem phospholipids and synaptic membrane fluidity in adult mice: a comparison of sardine oil diet with palm oil diet. Mech Ageing Dev 1998; 101(1–2):119–128.

Tam PS, Umeda-Sawada R, Yaguchi T, Akimoto K, Kiso Y, Igarashi O. The metabolism and distribution of docosapentaenoic acid (n-6) in rats and rat hepatocytes. Lipids 2000; 35(1):71–75.

Voss A, Reinhart M, Sankarappa S, Sprecher H. The metabolism of 7,10,13,16,19-docosapentaenoic acid to 4,7,10,13,16,19-docosahexaenoic acid in rat liver is independent of a 4-desaturase. J Biol Chem 1991; 266(30):19,995–20,000.

Wainwright PE, Huang YS, Bulman-Fleming B, Levesque S, McCutcheon D. The effects of dietary fatty acid composition combined with environmental enrichment on brain and behavior in mice. Behav Brain Res 1994; 60(2):125–136.

Wainwright PE, Xing HC, Ward GR, Huang YS, Bobik E, et al. Water maze performance is unaffected in artificially reared rats fed diets supplemented with arachidonic acid and docosahexaenoic acid. J Nutr 1999; 129(5):1079–1089.

Ward G, Woods J, Reyzer M, Salem N Jr. Artificial rearing of infant rats on milk formula deficient in n-3 essential fatty acids: a rapid method for the production of experimental n-3 deficiency. Lipids 1996; 31(1):71–77.

Weisinger HS, Vingrys AJ, Bui BV, Sinclair AJ. Effects of dietary n-3 fatty acid deficiency and repletion in the guinea pig retina. Invest Ophthalmol Vis Sci 1999; 40(2):327–338.

Xu LZ, Sanchez R, Sali A, Heintz N. Ligand specificity of brain lipid-binding protein. J Biol Chem 1996; 271(40):24,711–24,719.

Yamamoto N, Hashimoto A, Takemoto Y, Okuyama H, Nomura M, Kitajima R, et al. Effect of the dietary alpha-linolenate/linoleate balance on lipid compositions and learning ability of rats. II. Discrimination process, extinction process, and glycolipid compositions. J Lipid Res 1988; 29(8):1013–1021.

Yamamoto N, Saitoh M, Moriuchi A, Nomura M, Okuyama H. Effect of dietary alpha-linolenate/linoleate balance on brain lipid compositions and learning ability of rats. J Lipid Res 1987; 28(2):144–151.

Yehuda S, Carasso RL. Modulation of learning, pain thresholds, and thermoregulation in the rat by preparations of free purified alpha-linolenic and linoleic acids: determination of the optimal omega 3-to-omega 6 ratio. Proc Natl Acad Sci USA 1993; 90(21):10,345–10,349.

Yonekubo A, Honda S, Okano M, Takahashi K, Yamamoto Y. Effects of dietary fish oil during the fetal and postnatal periods on the learning ability of postnatal rats. Biosci Biotechnol Biochem 1993; 58(5):799–801.

Yoshida S, Yasuda A, Kawazato H, Sakai K, Shimada T, Takeshita M, et al. Synaptic vesicle ultrastructural changes in the rat hippocampus induced by a combination of alpha-linolenate deficiency and a learning task. J Neurochem 1997; 68(3):1261–1268.

22 The Role of Omega-3 Polyunsaturated Fatty Acids in Body Fluid and Energy Homeostasis

Richard S. Weisinger, James A. Armitage, Peta Burns, Andrew J. Sinclair, Algis J. Vingrys, and Harrison S. Weisinger

1. INTRODUCTION

The control of body fluid, sodium, and energy homeostasis is fundamental to mammalian life. The mechanisms involved are complex and rely heavily on receptor-mediated events. Given the ubiquity of n-3 fatty acids in mammalian cell membranes and the reliance of receptor protein activity on membrane lipid composition, we felt it timely to review the mechanisms responsible for body fluid, sodium, and energy homeostasis and to examine evidence suggesting modulation of these processes by n-3 polyunsaturated fatty acids (n-3 PUFAs).

This chapter will describe the role of the n-3 PUFAs in modulating some of the mechanisms that control body fluid, sodium, and energy homeostasis. The chapter is divided into three major sections. The first section is dedicated to the dietary intake and metabolism of the n-3 PUFAs, as well as the interrelated n-6 PUFAs. The second section describes the major mechanisms responsible for the maintenance of body fluid and sodium homeostasis. The third section describes the major mechanisms responsible for the maintenance of body energy homeostasis. The mechanisms involved in these homeostatic functions are complex and interactive; thus, they will first be introduced in terms of normal function. Then, evidence regarding the role of n-3 PUFAs in these homeostatic mechanisms will be described. Interestingly, although it is well known that n-3 PUFAs can have a profound influence on receptor-mediated events and that the control mechanisms involved in the homeostatic processes under consideration utilize such processes, there has been surprisingly little investigation into a possible role for the n-3 PUFAs in the maintenance of body fluid, sodium, or energy homeostasis.

2. SUPPLY AND METABOLISM OF POLYUNSATURATED FATTY ACIDS

Mammals do not produce $\Delta 12$ and $\Delta 15$ desaturase enzymes and therefore cannot produce either linoleic acid (LA) or α-linolenic acid (LNA), precursors of the n-6 and n-3 families, respectively, *de novo*. LNA is metabolized, in a series of desaturation and elongation reactions, to produce long-chain n-3 PUFA, the most prevalent of which are eicosapentaenoic acid (EPA) and docosahexaenoic acid (DHA). This metabolism occurs primarily in the liver and cerebral microvasculature of the blood-brain barrier, but also by astrocytes and the cerebral endothelium (Moore, Yoder, & Spector, 1990). EPA is further metabolized by the cyclo-oxygenase and lipoxygenase pathways to produce the eicosanoids (prostaglandins, thromboxanes, and leukotrienes), which are potent, short-acting, local hormones or second messengers. The eicosanoids are derived from EPA as well as from arachidonic acid (20:4n-6). These fatty acids are liberated from the cell membrane by the phospholipase A_2. Phospholipase A_2 has equal affinity for both EPA and arachidonic acid, therefore, the proportion of n-3 : n-6 hydrolyzed by phospholipase A_2 is determinant on the tissue n-3 : n-6 profile (Calder, 1996). DHA is primarily sequestered into the phospholipid membranes of cells within the brain and central nervous system. The predominant PUFA in the brain is DHA, which comprises up to 30% of the phospholipid fatty acids in the gray matter (Sinclair, 1975). DHA is a determinant in the regulation of synaptic transmission, ultrastructural changes in neural synapses and changes in dopamine and N-methyl D-Aspartate (NMDA) receptor function (for reviews, *see* Fenton, 2000; Kurlack & Stephenson, 1999). Furthermore, n-3 PUFA deficiency early in life may result in irreversible damage to biochemical processes in the central nervous system (Uauy, Mena, & Rojas, 2000).

The metabolism of n-3 and n-6 PUFAs is interlinked, as they compete for enzymes and metabolic substrates at all levels. Therefore, relative as well as absolute dietary intake is relevant in the determination of tissue n-3 and n-6 fatty acid levels. The Western diet typically contains high levels of n-6 fatty acids, as these are components of most animal and vegetable fats. Dietary sources of n-3 PUFA are varied. The most plentiful sources are fish, shellfish, and marine products, which contain large amounts of EPA and DHA. Certain plant oils, such as rapeseed (canola), soybean, and perilla contain large amounts of LNA (Crawford & Sinclair, 1972; Sinclair, 1975). Although beef and lamb do contain n-3 PUFAs, both the absolute content and the n-3 : n-6 ratio of PUFAs within these meats is low.

Burr and Burr (1929) highlighted the importance of certain fats in the human diet and demonstrated that dietary deprivation of the essential fatty acids was deleterious to health. They concluded that the n-6 PUFAs were most important to function in mammals because their omission resulted in overt systemic dysfunction. Conversely, it has been shown that deprivation of n-3 PUFAs results in subtle dysfunction across a range of behavioral and physiological modalities.

The n-3 PUFAs are important components of cell membranes throughout the body, as they are incorporated into the phospholipids that form cell membranes. Each phospholipid molecule is comprised of a headgroup to which fatty acid esters are bound. There are two binding positions, termed *sn*-1 and *sn*-2. The acyl chains that bind to these headgroups interact with other chains in neighboring phospholipid molecules within the bilayer, and the level of chain interaction determines the biophysical properties of the membrane. The *sn*-1 position is usually occupied by a saturated fatty acid; the *sn*-2 position is usually occupied by a relatively unsaturated fatty acid chain.

Lipid bilayers comprised mainly of saturated fat are relatively stable due to low acyl chain interactions. These stable membranes have higher viscosity and produce rigid cell membranes. An example of cells with this type of membrane is myelinated nerve fibres.

Conversely, membranes composed of phospholipids with highly unsaturated acyl chains are less rigid, because of interactions between such bulky acyl chains. This state of membrane disorder produces a physical fluidity and it is proposed that such membranes offer minimal hindrance to the function of proteins (e.g., receptors or channels) that lie within the membrane. Metabolically active membranes, such as those found in neuron cell bodies, are comprised of phospholipids containing unsaturated *sn*-2 chains. Work by Brown (1994), Dratz (Dratz & Holte, 1993), and Litman (Littman & Mitchell, 1996) support this notion, as recombinant membranes that comprise higher levels of unsaturated PUFA, particularly the long-chain n-3, DHA, mimic, more closely, biological activity when measured in vitro.

Western diets typically contain low levels of n-3 PUFAs (Simopoulos, Leaf, & Salem, 2000). Furthermore, relative to the diet of early man, today's diet is not only higher in saturated fats but has an altered ratio of the two major groups of PUFAs. Modern-day humans are, most probably, descendants of coastal-dwelling hunter–gatherers, with a primary diet of fish, shellfish, and plant matter. This omnivorous, and opportunistic, diet was varied and most probably low in saturated fat. The ratio of n-6 to n-3 fatty acids during the early part of human evolution was close to 1, whereas, today, the ratio is 10:1 or more (Eaton, Eaton, Sinclair, Cordain, & Mann, 1998; Simopoulos, 1998). A recent study in Australia (Sinclair, unpublished) indicates that DHA intake is 0.1% of fat intake, or about 100 mg/d. The recommended intake is 0.5–1% (Simopoulos et al., 2000). Furthermore, recent data from our group (Sinclair, unpublished) has shown that reduced DHA levels in neural tissue occur in animals that are maintained on high saturated fat diets similar to that of a Western diet.

3. OMEGA-3 PUFA IN THE BRAIN AND CENTRAL NERVOUS SYSTEM

A significant proportion of neural development occurs postnatally; hence, early nutrition is vital to support optimal neural development (Connor, Neuringer, & Reisbick, 1992). The large demand for DHA during pregnancy and early development is usually met via the placenta and then by maternal milk (Neuringer, Connor, S, Barstad, & Luck, 1986; van Houwelingen et al., 1996). Maternal dietary LA and LNA serve as precursors for n-3 and n-6 long-chain PUFAs by the maternal liver. The ability to obtain sufficient amounts of DHA to ensure normal growth and development, however, may be compromised by the decreased ability to transform precursor into DHA during pregnancy and lactation (van Houwelingen et al., 1996). There is a progressive enrichment of DHA within circulating lipids of the fetus during the third trimester, a time when fetal demands for vascular and neural growth are greatest (Clandinin et al., 1980; van Houwelingen et al., 1996). DHA is concentrated in synapses, and most synaptogenesis occurs during perinatal life.

4. BODY FLUID AND SODIUM HOMEOSTASIS

The regulation of body fluid and sodium homeostasis requires central nervous system mechanisms to coordinate information regarding body fluid status with the appropriate physiological and behavioral responses. Mechanisms have evolved to ensure that the physiological and behavioral responses to a body deficit are appropriate. For example, during periods of water deprivation, renal mechanisms minimize loss of body water. Secretion of the pituitary hormone, arginine vasopressin (AVP), is increased and acts on the kidney to ensure that urinary excretion of water is maximally decreased. In addition, the animal will seek out and ingest water when encountered. Analogously, with sodium deficiency, as would occur with diarrhea or profuse sweating, renal mechanisms minimize loss of body sodium. Secretion of the adrenal hormone, aldosterone, is increased and acts on the kidney to ensure that urinary excretion of sodium is maximally decreased. Furthermore, the animal will seek out and ingest sodium-containing substances or fluids when encountered. Behaviorally, thirst and sodium appetite motivate the seeking out and ingestion of water and sodium, respectively.

Fluid in the body is distributed into two fluid compartments, the intracellular fluid (i.e., the fluid within the cells) and the extracellular fluid (i.e., the fluid that is contained in the circulatory system [intravascular fluid or plasma] and that which surrounds the cells [interstitial fluid]). The sensation of thirst, and in the main, the secretion of AVP, is stimulated by depletion of the intracellular and/or the extracellular fluid compartments (Andersson, Leksell, & Rundgren, 1982; Fitzsimons, 1972; Fitzsimons, 1998). The ingestion of water that occurs with cellular dehydration, as would occur with the ingestion of salty foods, is mediated by both cerebral sodium sensors and osmoreceptors (Andersson et al., 1982; Blair-West et al., 1994; Fitzsimons, 1972; McKinley, Denton, Leksell, Tarjan, & Weisinger, 1980; McKinley, Denton, & Weisinger, 1978; Osborne, Denton, & Weisinger, 1987; Park, Denton, McKinley, Pennington, & Weisinger, 1989; Weisinger, Denton, McKinley, Muller, & Tarjan, 1985a; Weisinger, Denton, McKinley, & Nelson, 1985b; Weisinger, Denton, McKinley, Simpson, & Tarjan, 1986). Thirst caused by extracellular dehydration, as would occur with hemorrhage, is mediated by stretch or pressure receptors located in the circulatory system and mechanisms responsive to the increased levels of angiotensin II (ANG II) in the circulation caused by the depletion of the extracellular fluid compartment (Andersson et al., 1982; Fitzsimons, 1972; Fitzsimons, 1998). ANG II formed in the brain, acting as a neurotransmitter, may also be involved. The stimulation of water intake following dehydration or water deprivation is the result of the depletion of both the intracellular and the extracellular fluid compartments. Water intake following water deprivation is greatly attenuated by preventing elevation of cerebral sodium concentration (Osborne et al., 1987) or by blocking brain mechanisms responsive to ANG II (Blair-West, Carey, Denton, Weisinger, & Shade, 1998; Blair-West, Denton, McKinley, & Weisinger, 1988; Blair-West, Denton, McKinley, & Weisinger, 1997; Weisinger, Blair-West, Denton, & Tarjan, 1997). In general, increased levels of ANG II, formed in the circulatory system or in the brain, as well as increased cerebral sodium concentration/osmolality in the brain contribute to the expression of thirst and to the secretion of AVP.

Depletion of body sodium results in a reduction in plasma volume and thereby stimulates production of ANG II. Subsequently, increased blood levels of ANG II stimulate the secretion of aldosterone. The stimulation of sodium intake that occurs in sodium-deplete animals is thought to be mediated, in the main, by central mechanisms responsive to the high blood levels of ANG II (Tarjan, Ferraro, May, & Weisinger, 1993; Thunhorst & Fitts, 1994; Weisinger et al., 1996; Weisinger et al., 1990a; Weisinger, Denton, Di Nicolantonio, & McKinley, 1988). ANG II formed in the brain (Blair-West et al., 1998; Blair-West et al., 1997; Sakai, Chow, & Epstein, 1990; Weisinger et al., 1997; Weiss, Moe, & Epstein, 1986), possibly in synergy with the elevated blood levels of aldosterone (Epstein, 1992; Sakai, Nicolaidis, & Epstein, 1986) could be involved. In addition, sodium intake is stimulated by a decrease in cerebral sodium concentration acting on sodium sensors involved in sodium appetite (Blair-West et al., 1987; Weisinger, Considine, Denton, & McKinley, 1979; Weisinger et al., 1985a; Weisinger et al., 1986). A decrease in cerebral sodium concentration may also contribute to the increase in aldosterone secretion (Coghlan et al., 1980).

Angiotensin II is an 8 amino acid peptide that does not readily cross the blood-brain barrier (Epstein, 1981; Reid, 1984). Thus, circulating ANG II works by influencing brain structures lacking a blood-brain barrier (e.g., the circumventricular organs such as the subfornical organ [SFO] and the organum vasculosum of the lamina terminalis [OVLT]). ANG II generated within the brain works by influencing brain structures with (such as the median preoptic nucleus, MnPO) or without a blood-brain barrier (McKinley, Badoer, Vivas, & Oldfield, 1995). The activities of ANG II are mediated primarily by ANG II, type 1 (AT1) receptors (Beresford & Fitzsimons, 1992; Blair-West et al., 1997; McKinley et al., 1996). These receptors utilize guanine nucleotide binding protein (G-protein) coupled signal transduction mechanisms. The AT1 receptors are coupled to stimulation of inositol phospholipid hydrolysis and mobilization of Ca channels (Lenkei, Palkovits, Corvol, & Llorens-Cortes, 1998).

Angiotensin II has been identified in a number of different brain structures, including the hypothalamus (e.g., paraventricular nucleus [PVN] and supraoptic nucleus [SON]), the amygdala and bed nucleus of the stria terminalis (BNST), and the brain areas located on the midline of the front wall of the third brain ventricle [i.e., the SFO, MnPO, and OVLT (McKinley et al., 1990; Oldfield, Ganten, & Mckinley, 1989)]. ANG II may act as a neurotransmitter or neuromodulator in these brain areas (Ferguson & Washburn, 1998; Li & Ferguson, 1993). AT1 receptors have also been identified in these above-mentioned brain structures (McKinley et al., 1996; Song, Allen, Paxinos, & Mendelsohn, 1992).

Evidence consistent with an involvement in thirst and sodium appetite of the brain areas noted above has been reported. For example, water intake (Lind & Johnson, 1983; McKinley et al., 1984; Vivas & Chiaraviglio, 1992) as well as AVP secretion (McKinley et al., 1984) caused by water deprivation is reduced or eliminated in animals with lesion of a brain area that includes the MnPO and the OVLT. Lesion of either the SFO (Thunhorst, Ehrlich, & Simpson, 1990; Weisinger et al., 1990b) or the OVLT (Chiaraviglio & Perez Guaita, 1984; Fitts, Tjepkes, & Bright, 1990) reduces or eliminates sodium-depletion-induced sodium appetite. The amygdala and bed nucleus of the stria terminalis appear to have a role in sodium appetite caused by aldosterone (mineralocorticoid) or by sodium depletion (Reilly, Maki, Nardozzi, & Schulkin, 1994; Schulkin, Marini, & Epstein, 1989; Zardetto-Smith, Beltz, & Johnson, 1994). See Fig. 1.

Also, a number of other neuropeptides have also been shown to interfere with the expression of thirst subsequent to water deprivation and/or sodium appetite subsequent to sodium depletion. In some instances, ANG II, presumably in the neural system subserving thirst, stimulates the release of a neuropeptide that inhibits the expression of sodium appetite. ANG II stimulates the release of oxytocin (Lang et al., 1981) and oxytocin appears to be one of the main inhibitory factors that limit the expression of sodium appetite (Stricker & Verbalis, 1987; Verbalis, Blackburn, Hoffman, & Stricker, 1995). Oxytocin is a 9-amino acid peptide synthesized in PVN or SON neurons that project either to the neurohypophysis or to sites within the central nervous system. Oxytocin receptors are part of the 7 transmembrane, G-protein-coupled receptor family (Barberis, Mouillac, & Durroux, 1998) and are found in the BNST, nucleus of the solitary

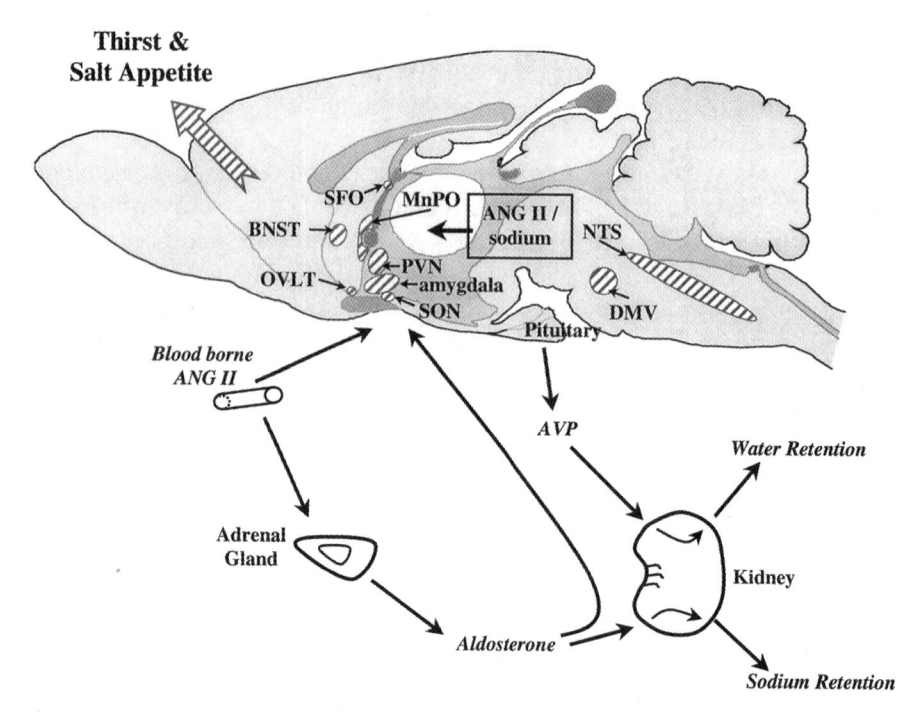

Fig. 1. Model showing mechanisms controlling water and sodium homeostasis. Major brain nuclei and peripheral pathways are shown. See text for abbreviations.

tract (NTS), and dorsal motor nucleus of the vagus (DMV). Another peptide, somatostatin, may have actions similar to oxytocin. Central administration of somatostatin decreases sodium intake caused by sodium depletion but does not alter water intake caused by water deprivation (Weisinger, Blair-West, Denton, & Tarjan, 1991). Somatostatin, a 14- or 28-amino acid peptide, was originally isolated from the ovine hypothalamus (Brazeau et al., 1973). Somatostatinergic neurons (Crowley & Terry, 1980; Krisch & Leonhardt, 1980) and somatostatin receptors (Leroux & Pelletier, 1984; Patel, Baquiran, Srikant, & Posner, 1986) have been identified in the circumventricular organs as well as other areas implicated in sodium appetite and thirst. Somatostatin receptors are members of the 7 transmembrane, G-protein-coupled receptors (Raulf, Perez, Hoyer, & Bruns, 1994).

Another peptide, atrial natriuretic peptide, does not appear to be as specific as oxytocin and somatostatin. Central administration of atrial natriuretic peptide has been shown to attenuate both the increase in water intake caused by water deprivation and the increase in sodium intake caused by sodium depletion (Nakamura, Katsuura, Nakao, & Imura, 1985; Tarjan, Denton, & Weisinger, 1988; Weisinger, Blair-West, Denton, & Tarjan, 1992). Atrial natriuretic peptide, a 28-amino acid peptide, and its receptors are distributed widely throughout the brain including many of the areas thought to be involved in body fluid homeostasis (e.g., many of the brain areas that contain ANG II receptors such as the SFO, MnPO, OVLT, and PVN) (Mendelsohn, Allen, Chai, Sexton, & Figdor, 1987; Quirion et al., 1984).

Overall, the evidence points to the conclusion that neuropeptides, both excitatory (ANG II) and inhibitory (oxytocin, atrial natriuretic peptide, and somatostatin), have a

Table 1
Influence of n-3 PUFA Deficiency on Intake of Water and 0.5 *M* NaCl Under Basal (No Treatment) and Stimulated (Water Deprivation and Sodium Depletion, Respectively) in Rats

	n-3 Deficient diet for perinatal period only	n-3 Deficient diet for entire life	n-3 Sufficient diet for perinatal period only	n-3 Sufficient diet for entire life
Water intake				
Basal	Normal	Normal	Normal	Normal
Stimulated	↓	↓	Normal	Normal
Sodium intake				
Basal	Normal	Normal	Normal	Normal
Stimulated	↑	↑	Normal	Normal
DHA	Normal	↓↓	↓	Normal

role in the maintenance of body fluid homeostasis. Interestingly, the receptors of these neuropeptides are members of the 7 transmembrane G-protein-coupled receptor family.

4.1. Influence of Omega-3 PUFAs on Body Fluid and Sodium Homeostasis

The above-described homeostatic functions are, in large part, dependent on proper neural functioning and the n-3 PUFAs are crucial to neural function. Receptor-driven processes are also crucial to these homeostatic functions. n-3 PUFAs can influence mechanisms known to be involved in body fluid homeostasis. Vaskonen (1996) showed that supplementation of a high-sodium (6%) diet with n-3 PUFA-rich fish oil increased the ability of stroke-prone–spontaneously hypertensive rats to excrete sodium and prevented the increase in blood pressure and water intake normally caused by the ingestion of 6% NaCl. Fish oil has also been shown to attenuate the increase in blood pressure caused by subcutaneous infusion of ANG II in normotensive rats (Hui, St.-Louis, & Falardeau, 1989). In normotensive humans (Kenny et al., 1992), ingestion of n-3 PUFAs (including 1200 mg/d DHA and EPA) for 7 days decreased plasma triglycerides and attenuated the increase in vascular resistance caused by ANG II. The n-3 PUFAs can influence mechanisms responsive to changes in sodium or osmolarity and ANG II, mechanisms involved in the regulation of body fluid and sodium regulation.

In preliminary experiments, we evaluated the effect of maintaining rats during their early development (gestation through to 2 mo of age = perinatal period) on an n-3-PUFA-deficient diet (safflower-oil diet) or an n-3 PUFA-replete or control diet (canola-oil diet, containing 0.8 g LNA per 100 g of diet) on subsequent ingestive behavior. From 2 mo of age, the rats continued to receive the same diet or were switched to the reverse diet for 4 wk. Diet did not appear to affect plasma sodium, potassium, osmolality, or hematocrit. However, it was observed that animals raised on n-3-PUFA-deficient diets underdrank water subsequent to water deprivation and overdrank sodium subsequent to sodium depletion. See Table 1.

These results are consistent with an aberration in central osmoreceptor or sodium sensor or ANG II-sensitive mechanisms that influence body fluid and sodium homeostasis. The results of this study suggest a potential requirement for n-3 PUFAs, early in life, for normal development and neural function. Evidence that early dietary n-3 PUFA deprivation can interfere with body fluid homeostasis was also observed in nonhuman primates. In contrast to our observations, however, Reisbick and co-workers demonstrated that n-3-PUFA-deficient animals have increased intake of both water and NaCl solutions under basal conditions (Reisbick, Neuringer, Connor, & Barstad, 1992; Reisbick, Neuringer, Connor, & Iliff-Sizemore, 1991; Reisbick, Neuringer, Hasnain, & Connor, 1990). Clearly, further studies are required to determine the role of n-3 PUFAs in body fluid and sodium homeostasis.

5. BODY ENERGY HOMEOSTASIS

Many factors are involved in the control of food intake. Some of the most important factors controlling the amount of food that we eat include environmental factors such as food availability, the characteristics of the food itself (e.g., smell, taste, our eating habits, learned preferences and aversions) as well as other psychological and social factors, including our lifestyle. Although these psychosocial factors are extremely important to the food intake patterns of humans, this section will concentrate on the physiological factors, primarily the role of fat in food intake.

The regulation of body energy homeostasis, the balance between energy input and expenditure, is crucial to an animal's growth and survival. Complex central nervous system mechanisms have evolved to ensure that the animal's needs and its behavioral and physiological responses are coordinated. A starving animal, motivated by hunger, seeks out and ingests food when encountered. Simultaneously, mechanisms act to conserve body energy when possible (e.g., body temperature and activity may be decreased). In the long term, the interplay between the various environmental and physiological factors controlling energy input and expenditure determines the individual's body weight.

The importance of the hypothalamic brain area to the control of food intake and body weight has been known for at least 50–60 yr (Anand & Brobeck, 1951; Hetherington & Ranson, 1940). Initially, evidence demonstrated that the lateral hypothalamus (LH) and the ventromedial nucleus of the hypothalamus (VMN) were crucial. Lesion of the LH caused animals to lose body weight. Animals with a lesion of the LH decrease their food intake and increase their body temperature. Experiments suggested that the lesion of the LH reduced the animal's body weight "set point"; that is, when the animal's body weight was decreased by food deprivation prior to lesion of the LH, the normally observed undereating did not occur after the lesion, but, rather, the new low body weight was maintained. In contrast, lesion of the VMN caused animals to increase food intake and gain body weight. When an animal's body weight was increased prior to lesion of the VMN, the normally observed overeating did not occur after lesion, but, rather, the new elevated body weight was maintained. These results suggest that the "set point" is determined, at least in part, by hypothalamic mechanisms (Friedman & Stricker, 1976; Hoebel & Teitelbaum, 1966; Keesey & Hirvonen, 1997; Keesey, Mitchel, & Kemnitz, 1979; Keesey & Powley, 1975; Mrosovsky & Powley, 1977). Clearly, however, the "set point" can be influenced by various environmental and physiological factors, as evidenced by obesity. Interestingly, even "obese" humans appear to have a body weight set point (e.g., following a weight-loss regimen, most obese patients regain all of their lost weight within 9 yr) (Johnson & Drenick, 1977).

Evidence consistent with the idea of a body-weight "set point" is observed in animals with normal body weight as well. For example, like others (Keesey & Hirvonen, 1997),

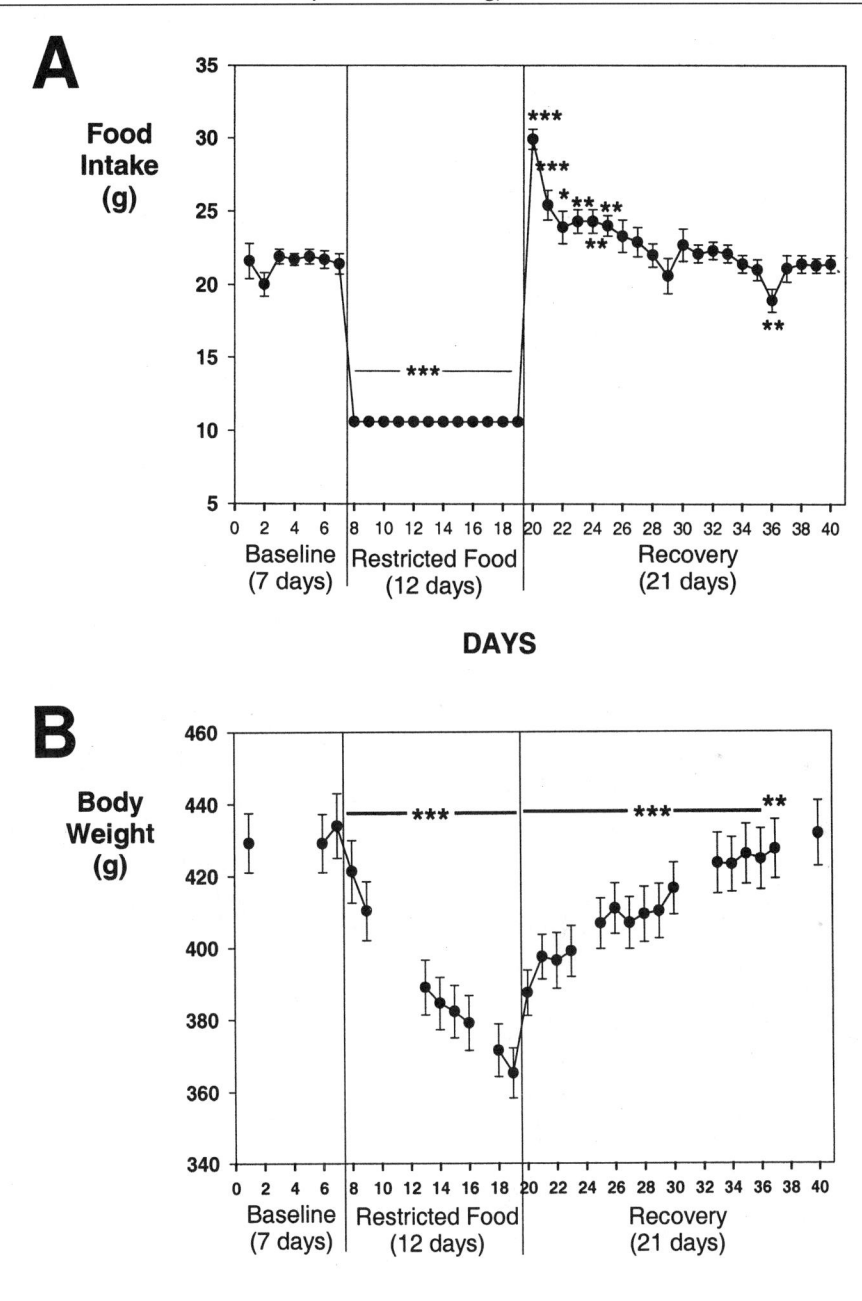

Fig. 2. Effect of food restriction on subsequent food intake (**A**) and body weight (**B**) in rats. Following a 7-d baseline period, male Sprague–Dawley rats ($n=7$) were restricted to 50% of their daily food intake for 12 d (to achieve a 15% decrease in body weight). Unlimited food was then made available and food intake and body weight was monitored for a further 21 d. $* = p < 0.05$, $** = p < 0.01$, $*** = p < 0.001$ versus baseline.

we have observed that when animals are given free access to food after a period of food restriction, the loss in body weight is restored. Interestingly, overeating may only be evident during the first week following the return of free food access. *See* Fig. 2.

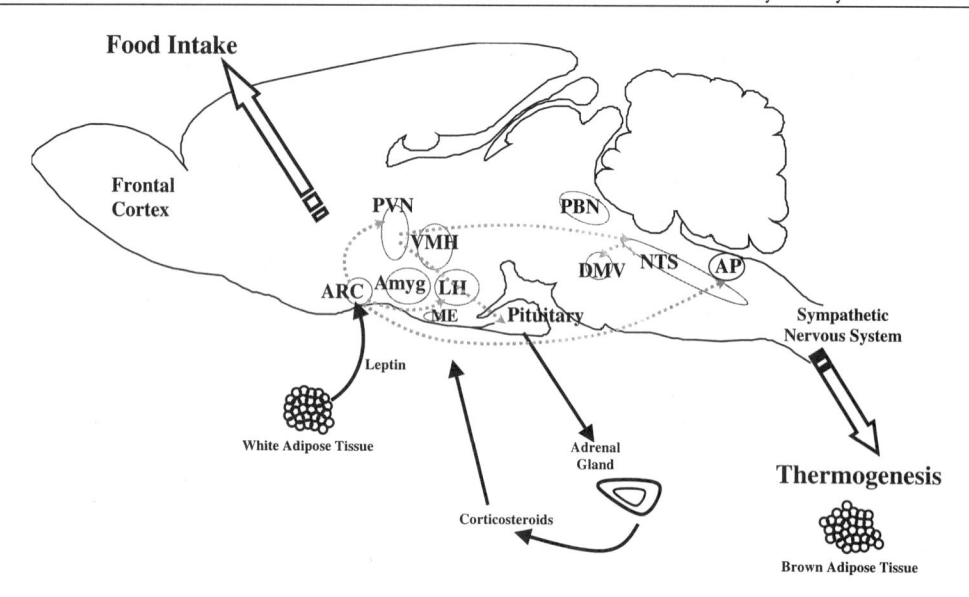

Fig. 3. Model showing mechanisms controlling body energy homeostasis. Major brain nuclei and central and peripheral pathways are shown. See text for abbreviations.

Today, other brain areas are known to be important to the control of food intake and body weight. Lesion of the arcuate nucleus (ARC) or PVN result in obesity (Bray, 1993; Dallman et al., 1993; Holzwarth-McBride, Hurst, & Knigge, 1976; Kirchgessner & Sclafani, 1988; Rothwell, 1990). One of the important factors contributing to obesity appears to be an increase in the activity of the hypothalamic–pituitary–adrenal axis. Adrenalectomy inhibits or prevents development of obesity subsequent to lesion of the ARC or PVN (Dallman et al., 1993). Other brain areas, such as the amygdala, area postrema (AP), NTS, parabrachial nucleus (PBN), DMV and frontal cortex are also involved (Bray & York, 1998; Dallman et al., 1993; Holzwarth-McBride et al., 1976; King et al., 1999; Kirchgessner & Sclafani, 1988; Ritter & Hutton, 1995; Woods, Seeley, Porte, & Schwartz, 1998). Disruption of the neural pathways between PVN, NTS, and DMV simulate aspects of the obesity caused by lesion of the VMN (Kirchgessner & Sclafani, 1988). Thus, these results are consistent with neural networks rather than neural centers controlling food intake and body weight. Of course, although many key brain areas have been identified, there is much to be elucidated in terms of determining the relevant pathways. *See* Fig. 3.

Glucose and fat are the two major sources of energy for the body. Early theories proposed that food intake was controlled by central mechanisms sensing changes in the level or utilization of these fuels. Mayer (1952; 1955) suggested that food intake was controlled by blood glucose levels or by levels of glucose utilization. Glucostats in the hypothalamus were thought to sense the changes in glucose levels. Some evidence suggests that a small decrease in plasma glucose precedes hunger and food intake in humans (Campfield & Smith, 1986).

Another theory, the "lipostatic" theory proposed by Kennedy (1950), suggested that food intake varied so as to maintain body fat stores (i.e., the "set point"). Changes in body fat stores, reflected in signals dependent on the size of those stores (e.g., blood levels of fatty acids), controlled food intake. Indeed, it has been shown that animals with lesion of the LH and VMN have pronounced disturbances in fat metabolism (Friedman & Stricker, 1976). Animals with a VMN lesion show increased fat storage such that circulating levels of the metabolic products of fat metabolism are decreased. It has been proposed (Friedman & Stricker, 1976) that it is this decreased availability of the metabolic products of fat metabolism that stimulate food intake.

Presently, in line with the idea that fat metabolism is crucial to body energy homeostasis, evidence suggests that peptides produced in the body and correlated with body fat mass are essential to the long-term control of food intake and body weight. Leptin is one such peptide. Leptin is the protein product of the *ob* gene. It is primarily produced in white adipose tissue, and the plasma level of leptin accurately reflects total-body adiposity (Friedman, 1997; Porte et al., 1998). It enters the brain from the circulation by a saturable transport mechanism (Friedman, 1997; Seeley & Schwartz, 1999; Woods et al., 1998). Leptin expression is stimulated by cortisol, insulin, and fasting and is attenuated by long-chain fatty acids (Houseknecht, Baile, Matteri, & Spurlock, 1998; Trayhurn, Hoggard, Mercer, & Rayner, 1999). It has been shown that central administration of leptin decreases body weight, primarily by causing the loss of fat (Chen et al., 1996; Halaas et al., 1997). Interestingly, although decreased food intake normally causes a reduction in energy utilization, this does not occur during administration of leptin. Indeed, body weight decreases during administration of leptin, even when food intake has returned to normal (Halaas et al., 1997).

The decrease in food intake caused by leptin is mediated by its actions on central peptidergic systems involved in food intake. For example, leptin stimulates peptidergic systems that inhibit food intake. Leptin stimulates corticotropin releasing factor (CRF) release in the PVN. CRF and urocortin, a newly discovered member of the CRF family, decrease food intake and increase energy expenditure (Richard, 1993; Rothwell, 1990; Weisinger et al., 2000). The decreased food intake caused by CRF/urocortin is mediated by CRF-R2 receptors (Smagin, Howell, Ryan, De Souza, & Harris, 1998; Spina et al., 1996), found in the VMN, amygdala, and PVN (Baram, Chalmers, Chen, Koutsoukos, & De Souza, 1997; Eghbal-Ahmadi, Hatalski, Avishai-Eliner, & Baram, 1997; Gray & Bingaman, 1996). It has been shown that the influence of leptin on food intake is blocked by a CRF receptor antagonist (Gardner, Rothwell, & Luheshi, 1998). CRF is thought to decrease food intake by its influence on the release of another peptide that inhibits food intake (i.e., oxytocin) (Olson, Drutarosky, Stricker, & Verbalis, 1991), or by its ability to decrease gastric emptying (Tache, Maeda-Hagiwara, & Turkelson, 1987). The increase in energy expenditure caused by CRF/urocortin has been attributed to their activation of the sympathetic nervous system (Arase, York, Shimizu, Shargill, & Bray, 1988; Rothwell, 1990; Smagin et al., 1998; Spina et al., 1996). The influence of leptin and CRF on food intake and body weight are clearly very similar and it seems likely that the actions of leptin are, at least in part, mediated via a system involving CRF and its receptors.

In addition, leptin acts to inhibit the activity of peptidergic systems that stimulate food intake. Leptin receptors have been found on many of the peptidergic neurons thought to increase food intake and body weight. Such neurons are found throughout the brain and include the melanin-concentrating hormone (MCH)-containing neurons of the LH, the neuropeptide Y (NPY)-containing neurons of the ARC and VMN, and the galanin-containing neurons of the PVN (Hakansson, de Lecea, Sutcliffe, Yanagisawa, & Meister, 1999; Meister, 2000). Much seems to be known about how these various peptidergic neurons influence food intake. MCH is synthesized in the LH and MCH-containing neurons project to various brain areas involved in energy balance. Stimulation of MCH (Chambers et al., 1999) pathways causes increased food intake. NPY is found in neurons in the ARC that project to the PVN, LH, and NTS. Stimulation of NPY pathways is followed by increased food intake, lipogenesis, and decreased sympathetic nervous system activity and energy expenditure in brown fat (Billington & Levine, 1992; Grundemar & Hakanson, 1994; Levine & Billington, 1997; Richard, 1995; White, 1993). Stimulation of food intake by NPY is thought to be mediated by neurons expressing MCH, via an ARC–LH pathway (Broberger, De Lecea, Sutcliffe, & Hokfelt, 1998). NPY neurons appear to be more responsive to changes in carbohydrate metabolism than to fat metabolism (Leibowitz, 1995). Galanin-expressing neurons and neurons with galanin receptors are found in many brain areas involved in food intake such as LH, PVN, and amygdala (Merchenthaler, Lopez, & Negro-Vilar, 1993). Activation of galaninergic neurons that project from the anterior PVN to the median eminence (ME) causes increased intake and accumulation of fat (Leibowitz, 1995; Leibowitz & Alexander, 1998). Interestingly, not only does leptin prevent food intake caused by NPY, MCH, and galanin, but also leptin appears to block the formation of these peptides (i.e., leptin decreases hypothalamic NPY, MCH, and galanin gene expression) (Sahu, 1998). The decrease in MCH caused by leptin is, at least in part, mediated by α-melanocyte stimulating hormone (αMSH). Leptin increases expression of αMSH, a peptide that acts on neurons with melanocortin-4 (MC-4) receptors and stimulation of these receptors decreases MCH (Broberger, 1999; Hanada et al., 2000). αMSH may also be involved in the decrease in food intake caused by CRF (Rothwell, 1990).

Leptin also influences body weight by interacting with peptides that control short-term food intake (i.e., peptides that affect meal size and/or meal frequency). Secretion of the gut-peptide cholecystokinin (CCK) produces short-term satiety. However, although CCK can decrease meal size, meal frequency is increased such that body weight is not altered (West, Fey, & Woods, 1984). One of the mechanisms by which CCK decreases food intake is by decreasing gastric emptying (Moran & McHugh, 1982). Interestingly, it has been shown (Matson, Reid, Cannon, & Ritter, 2000) that when leptin (given centrally) and CCK (given peripherally) are given simultaneously, body-weight loss is greater than that observed when leptin is given alone. The synergy between leptin and CCK cannot be explained by decreased food intake alone. Presumably, increased energy expenditure is involved. It should be noted that most of the peptides that influence food intake also influence, inversely, the activity of the sympathetic nervous system controlling thermogenesis in the brown adipose tissue. For example, NPY increases food intake and decreases activity of the sympathetic nervous system, whereas CRF decreases food intake and increases activity of the sympathetic nervous system (Bray, 1993).

5.1. Influence of n-3 PUFAs on Acquisition and Distribution of Body Fat, Food Intake and Body Weight

Addition of long-chain n-3 PUFAs to the diet significantly reduces body adiposity and increases lean body mass. For example, in a long-term (6 months) study, 3- to 4-mo-old rats were maintained on semisynthetic diets in which the fat content of the diet comprised 45% of the energy. Relative to the rats maintained on a diet containing lard (high in saturated fats) or corn oil, rats maintained on a diet containing n-3-PUFA-rich fish oil had less adipose tissue in epididymal, inguinal, mesenteric, and retroperitoneal stores (Hill et al., 1993). Body weights were not different between groups. Parrish (1991) maintained male Wistar rats on a lard or MaxEPA (high in long-chain n-3 PUFA) diet in which the fat content of the diet comprised 40% of the energy. By 5 wk, fat in epididymal and perirenal tissues, as well as plasma triglycerides, were lower in the MaxEPA group. Given that nearly 100% of the cell volume of adipose tissue is triglycerides, factors that influence their synthesis and breakdown are crucial. It has been suggested that n-3 PUFAs restricts adipose tissue growth by limiting the amount of triglyceride in each cell (Parrish et al., 1991). Altered expression of genes encoding lipogenic (fatty acid synthase and lipoprotein lipase), lipolytic (hormone sensitive lipase), and glyceroneogenic enzymes because of ingestion of n-3 PUFAs appear to be important in reducing body fat stores.

Several studies have suggested that the reduction in body fat mass is not universal and that some adipose tissues are more susceptible than others. For example, Raclot (1997) offered 50-d-old male Wistar rats diets containing n-3-PUFA-rich fish oil, purified DHA, purified EPA, a mixture of DHA and EPA, or a diet containing lard and olive oil (control). These semisynthetic diets were formulated so that the caloric content was matched and that 40% of caloric content was derived from fat. Diets containing n-3 fatty acids produced a site-specific antiadipogenic effect. Decreased retroperitoneal fat was noted, but subcutaneous adipose tissue remained constant. Belzung (1993) obtained a similar result. In this study, the addition of n-3 PUFA to a diet high in saturated fat limited hypertrophy of retroperitoneal and epididymal adipose tissues, but had no effect on subcutaneous or mesenteric fat stores. Further work will be needed to establish the factors responsible for reduction of fat in the specific adipose tissues.

Several studies have reported that diets containing n-3 PUFAs can decrease body weight or body-weight gain. McGahon (1999) reported that after 22 mo on a long-chain n-3 PUFA (DHA) or short-chain n-6 diet (corn oil), male Wistar rats maintained on the DHA diet were significantly lighter. Long-chain n-3 PUFA diets were also observed to decrease body weight (but not food intake) of diabetic (streptozotocin-treated) rats. Such diets increased responsiveness of muscle glucose transport to insulin (Sohal, Baracos, & Clandinin, 1992). Cunnane (1986) compared a diet high in n-6 PUFA (evening primrose oil = 18:2n-6 and 18:3n-6) with a diet high in n-3 PUFA (cod-liver oil, EPA, and DHA). The diets were matched for caloric content whereby fat provided 20% of energy. In neither lean (*ln/ln*) nor obese (*ob/ob*) mice, were differences in food intake observed. However, although not very large, the body weight of obese mice on the n-3 PUFA diet was less than those obese mice on the n-6 PUFA diet. *See* Fig. 4. This finding was attributed to alterations in prostaglandin synthesis produced by the high n-3 PUFA intake. In a study in which C57 mice received 58% of calories as either saturated fat or long-chain n-3 PUFA (Wang, Storlien, & Huang, 1999), animals maintained on the high-saturated-fat diet gained the most body weight and accumulated the most body fat. In animals fed the high-saturated-fat diet, neural activity, as measured by an increase in fos immunoreactivity, was observed to be increased in the LH ("hunger" area) and decreased in the VMN ("satiety" area). Conversely, the activity in the VMN but not the LH was increased in the fish-oil-fed group. These results are consistent with the proposition that relative to diets with long-chain n-3 PUFAs, diets high in saturated fats cause obesity not only because they are less satiating but also because they act to increase activity in brain areas associated with increasing food intake and body weight.

In summary, dietary fat determines plasma and cellular lipid composition and, thus, effects cellular function and metabolism. This, in turn, determines the animal's food intake and energy expenditure and, thus, determines body weight and body composition.

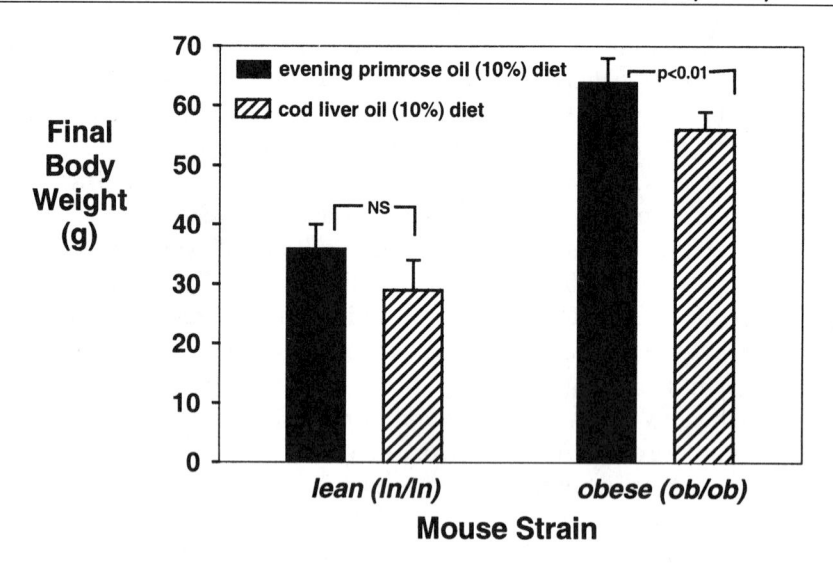

Fig. 4. Effect of fat content of diet on body weight of lean and obese mice. Lean (ln/ln) and obese *(ob/ob)* mice were fed a n-6-PUFA-rich (evening primrose oil, closed bars) or n-3 PUFA-rich (cod-liver oil, hatched bars)-based diet for 16 wk and the effect on final body weight determined. (Adapted from Cunnane et al., 1986)

However, not all studies have shown that the addition of n-3 PUFAs (long or short chain) to the diet of normal rats or mice has an influence on food/energy intake or body weight (e.g., Awad, Bernardis, & Fink, 1990; Belzung et al., 1993; Chicco, D'Alessandro, Karabatas, Gutman, & Lombardo, 1996; Cunnane et al., 1986; Hill et al., 1993; Parrish et al., 1991; Raclot et al., 1997; Russell, Amy, & Dolphin, 1991; Rustan, Hustvedt, & Drevon, 1993; Sohal et al., 1992). Clearly, further work needs to be done to establish the influence on body weight of the addition of long chain n-3 PUFA to the diet. In particular, physiologically relevant amounts (e.g., 0.5–1.0% of fat intake) of the n-3 PUFAs have to be examined.

5.2. Influence of n-3 PUFAs on Fat Metabolism: Oxidation and Thermogenesis

Jones and Schoeller (1988) showed that when added to a saturated-fat diet, n-3 PUFA increased basal metabolic rate and total energy expenditure. Within minutes of ingestion, n-3 PUFAs upregulate genes involved in lipid oxidation and downregulate genes involved in lipogenesis. Hepatic oxidation of fatty acids increases within 3 d when the diet contains 12–15% fish oil, but does not increase for several weeks when the diet contains n-6 PUFAs. Evidence suggests that unlike intake of a high-fat diet, intake of a diet rich in n-3 PUFAs may promote fat utilization rather than storage (Price, Nelson, & Clarke, 2000).

Thermogenesis is one of the major components of energy expenditure and therefore a major aspect of body-weight control. In addition to their ability to decrease lipogenesis, n-3 PUFAs also decrease fat deposition by inducing genes involved in thermogenesis, thereby increasing total-body heat production. Thermogenesis in brown adipose tissue is mainly regulated by the sympathetic nervous system. Increased thermogenesis can be the result of increased responsiveness to noradrenaline or increased release of noradrenaline. Thus, n-3 PUFAs could influence the effectiveness of the sympathetic nervous system because it increases the sensitivity of brown adipose tissue to noradrenaline (Ohno, Ohinata, Ogawa, & Kuroshima, 1996).

5.3. High-Fat Diets and Obesity: Possible Influence of n-3 PUFAs

Obesity is one of the major health risks for a number of diseases, particularly heart disease and diabetes. It is well known that ingestion of a diet high in saturated fats is one of the major causes of obesity. There are two explanations for this observation. First, diets high in saturated fats do not seem to be as satiating as either high-carbohydrate or high-protein diets (Doucet et al., 1998), even when the high-fat diet is less palatable (Warwick, 1996). Second, whereas increased intake of either carbohydrate or protein causes a concomitant increase in energy expenditure (e.g., nonshivering thermogenesis), increased intake of saturated fat does not cause a similar increase in energy expenditure. Individuals maintained for 1 wk on a high-carbohydrate or high-protein diet have, whereas individuals maintained on a diet high in saturated fat did not have, increased body temperature and ingested fat was mostly sequestered to adipose tissue; that is, increased intake of saturated fat does not promote the use of fat as a metabolic fuel (Schutz, Flatt, & Jequier, 1989; Thomas et al., 1992). Thus, diets high in saturated fats are more obesity producing than diets high in carbohydrate or protein because of their lesser ability to satiate appetite and to increase fat oxidation. With the ingestion of more energy than expended, storage of fat is increased and obesity results (Tremblay, Plourde, Despres, & Bouchard, 1989).

The peptide hormone insulin is produced in the pancreas and secreted in proportion to the degree of adiposity. Similar to leptin, insulin levels are correlated with amount of abdominal fat (Porte et al., 1998; Woods et al., 1996; Woods, Figlewicz Lattemann, Schwartz, & Porte, 1990; Woods et al., 1998). It is transported into the brain where it acts to decrease food intake and body weight (Schwartz, Figlewicz, Baskin, Woods, & Porte, 1992; Woods et al., 1996). High insulin resistance is a characteristic of obesity, hypertension, and non-insulin-dependent diabetes mellitus. There is an inverse relationship between insulin action and triglyceride content. With the ingestion of fat, insulin secretion is increased. Insulin stimulates fatty acid synthase, an enzyme that catalyzes all reactions involved in lipogenesis, and thereby results in the accumulation of triglycerides (Sul, Latasa, Moon, & Kim, 2000). Monounsaturated fatty acids (such as oleate) and saturated fatty acids (such as palmitate), fatty acids with little or no influence on fatty acid synthase, are incorporated into triglycerides in adipose tissue (Parrish, Pathy, & Angel, 1990; Parrish et al., 1991; Su & Jones, 1993) and thereby increase insulin resistance. In contrast, n-3 PUFAs markedly inhibit fatty acid synthase and are, therefore, preferentially incorporated into phospholipids for cell membrane remodeling; that is, n-3 PUFAs decrease triglyceride production, resulting in low levels of tissue and plasma triglycerides (*see* Fig. 5), and thereby decrease insulin resistance (Awad et al., 1990; Chicco et al., 1996; Hill et al., 1993; Parrish et al., 1991; Russell et al., 1991; Rustan et al., 1993; Sohal et al., 1992). Clearly, ingestion of diets high in n-3 PUFAs should be beneficial for treatment of obesity and diabetes. For maximal benefit, ingestion of diets with appropriate amounts of n-3 PUFAs should begin at an early age or possibly prolonged ingestion may be required.

5.4. Cancer Cachexia: Influence of n-3 PUFAs

In patients with cancer, weight loss indicates a poor prognosis and a shorter survival time. Cancer cachexia involves a massive loss of body weight, with extensive breakdown of both body fat and skeletal muscle, often, but not always, accompanied by anorexia (DeWys, 1985). Metabolic studies have shown that increased free fatty acid mobilization

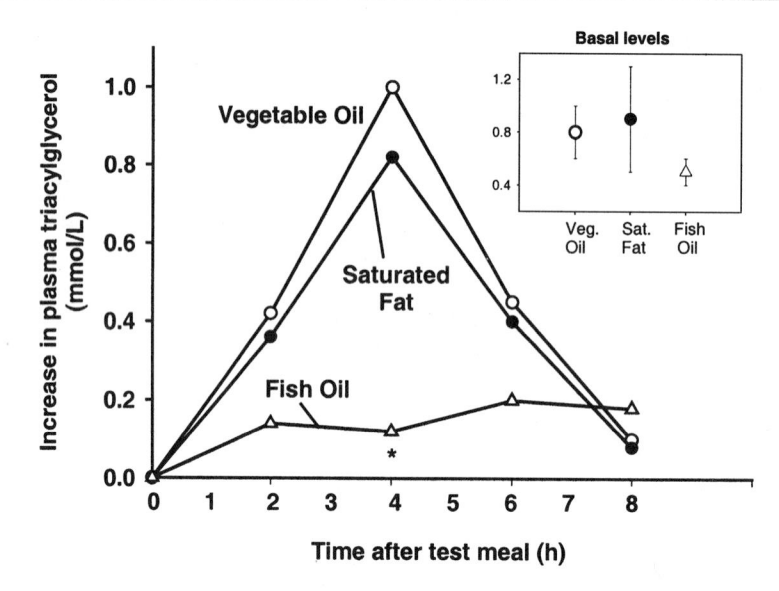

Fig. 5. The increase in plasma triacylglycerol concentrations after ingestion of a 50-g test meal. Individuals were maintained on saturated fat, vegetable-oil- or fish-oil-based diets. A test meal of saturated fat was given to the individuals maintained on the saturated-fat-based diet (●), a vegetable oil test meal was given to individuals maintained on the vegetable-oil diet (○), and a salmon-oil test meal was given to individuals maintained on the salmon-oil diet (△). Mean basal triacylglycerol concentrations are also shown. (Adapted from Harris, Connor, Alam, & Illingworth, 1988)

occurs prior to weight loss in cancer patients (Costa, Bewley, Aragon, & Siebold, 1981) and weight loss is not reversed by parenteral nutrition; thus, the weight loss associated with cancer cachexia is different from simple starvation (Brennan, 1977).

Although not fully understood, it appears that the most likely model for the development of cancer cachexia is based on increased cytokine production. Interleukin-1β (IL-1β), an endogenous pyrogen, is a cytokine released by activated macrophages and monocytes that mediates local and systemic responses to inflammation. Peripheral or central administration of IL-1β produces decreased food and water intake and increased body temperature. Such a decrease in food intake caused IL-1β can be attenuated by an MC-4 receptor antagonist (Huang, Hruby, & Tatro, 1999) or by depletion of neuronal histamine (Sakata, Kang, Kurokawa, & Yoshimatsu, 1995). Activation of the MC-4 receptor (Broberger, 1999; Hanada et al., 2000), as noted earlier, and the release of histamine (Morimoto, Yamamoto, & Yamatodani, 2000) contribute to the decrease in food intake caused by leptin.

In a mouse model of cancer (MAC16 mice), cachexia was reversed by a semisynthetic diet in which a high proportion of the calories were derived from n-3-PUFA-rich fish oil [19% EPA, 13% DHA, (Tisdale & Dhesi, 1990)], suggesting that EPA and DHA interfered with catabolic activity. Fish-oil supplements also elevated the normally reduced blood glucose levels observed in MAC16 mice, suggesting a decrease in glucose utilization by the tumor. Free fatty acids were unaltered. An n-6-PUFA-rich diet had no effect on weight loss or tumor growth and none of the diets influenced food intake. The weight gain by fish-oil-treated mice was shown to be the result of a reduction in loss of fat and muscle, not to increased water retention (Tisdale & Dhesi, 1990).

In a subsequent study, Tisdale and Beck (1991) investigated the inhibition of lipid mobilization by PUFAs in MAC16 mice. Ingestion (by gavage) of EPA, but not of DHA or LNA, prevented weight loss and minimized tumor growth. None of the n-6 PUFAs (e.g., LA, arachidonic acid) were effective in inhibiting the mobilization of lipid. EPA was also effective in inhibiting the lipolytic effect of salbutamol (β-adrenergic agonist) or adrenocorticotrophin. In another mouse model of cancer (C57 mice, Lewis lung carcinoma), treatment with either DHA or EPA prevented the decrease in body weight (Ohira et al., 1996), but did not alter food or water intake or tumor size.

Given that weight loss in cancer patients occurs prior to decrease in food intake, the problem appears to be, at least initially, the result of disruption of the sympathetic nervous system and the regulation of energy expenditure. It is possible that some byproduct of the cancer causes damage to the LH. A cachectic individual with multiple sclerosis was shown to have damage to the LH (Kamalian, Keesey, & ZuRhein, 1975). Another possibility is that the factors that mediate the cachectic process stimulate inhibitory factors such as leptin or inhibit stimulatory factors such as NPY (Inui, 1999). Furthermore, it has been shown that patients with surgically inoperable pancreatic adenocarcinoma normally lose body weight, but this is reversed by eating n-3-PUFA-rich fish oil. This reversal is the result of both an increase in appetite and a decrease in energy expenditure (Barber, Ross, Voss, Tisdale, & Fearon, 1999). Weight loss is a potent stimulus to food intake; therefore, the failure of patients with cancer to increase their intake of food and not learn to ingest diets containing large amounts of n-3-PUFA-rich fish oil, which appears to be effective in reducing cancer cachexia, seems to require further investigation.

6. CONCLUSION

The evidence presented in this chapter shows that n-3 PUFAs can influence body fluid and energy homeostasis. In regard to thirst and sodium appetite, there is some evidence in rats that early n-3 PUFA deprivation causes some aberrant responses in later life. If applicable to the human situation, the results suggest that the ingestion of n-3 PUFAs should begin early, probably with the pregnant mother, but certainly with the infant. It is conceivable that ingestion of n-3 PUFAs early in life may prevent many of the fluid disorders that are known to occur in the elderly (McAloon Dyke et al., 1997; Stachenfeld, DiPietro, Nadel, & Mack, 1997). However, although suggestive, much work is still needed to confirm this recommendation. In regard to the intake of food and regulation of body weight, the possibility that a critical period exists is minimal. Clearly, n-3 PUFAs can influence the accumulation and utilization of fat, both crucial to the regulation of body weight. Ingestion of n-3 PUFAs have been shown to reduce the size of adipose tissue, to increase thermogenesis, and, in some instances, to reduce body weight. In this regard, ingestion of n-3 PUFAs as a replacement for saturated fat or monounsaturated fatty acids might be important in overweight individuals. Ingestion of the n-3 PUFAs could facilitate the loss of body fat via its effect on increasing fat metabolism and thermogenesis. n-3 PUFAs could also improve glucose utilization by its effects on reducing cellular triglycerides and thus increasing insulin sensitivity. Interestingly, in animal models of cancer cachexia and in patients with cancer cachexia, n-3 PUFAs can act to increase body weight. The mechanisms involved seem to involve interference with the catabolic influence of toxins produced by the tumor and are clearly different from those mechanisms by which the n-3 PUFAs act to reduce body fat in individuals without cancer. Thus, the ingestion of n-3 PUFAs can influence fat mass and body weight via several different mechanisms.

Although our current understanding of the role of n-3 PUFAs in fluid and food intake and body-weight maintenance requires further work, enough evidence from behavioral experiments exists to suggest that these metabolites might be important in a multitude of cellular functions. Certainly, n-3 PUFAs are crucial for normal neural function and it is possible that deprivation of such metabolites might alter cell membrane properties in crucial areas of the brain that monitor appetite. Perhaps coincidentally, there seems to be a commonality in the neural functions altered by n-3 PUFA deprivation. More often than not, affected processes are mediated by receptors belonging to the 7-transmembrane domain, G-protein-coupled receptor family. G-Protein-coupled receptors are essential in the integration of feeding and drinking control mechanisms (Plata-Salaman, Wilson, & Ffrench-Mullen, 1998). For example, in regard to feeding, NPY receptors (Blomqvist & Herzog, 1997; Du et al., 1997), galanin receptors (Kask, Berthold, & Bartfai, 1997) and MC-4 receptors (Huszar et al., 1997) belong to G-protein-coupled receptor superfamily. In regard to drinking, ANG II receptors (Lenkei et al., 1998), OT receptors (Barberis et al., 1998) and somatostatin receptors (Raulf et al., 1994) belong to G-protein-coupled receptor superfamily. The function of this receptor superfamily may be more susceptible to alteration by decreased n-3 PUFA because of their morphology. These large and complex membrane-bound proteins alter conformation after ligand binding, and it is possible that a decrease in membrane fluidity would hamper conformational change such that the active conformation is no longer energetically favorable, following binding of the ligand.

Over the past 50 yr, there has been an increase in diabetes, heart disease and cancer, thought to be caused, in part, by changes in environmental factors. The nutrients we eat may be the most influential environmental stimuli, and fat is a strong determinant of cell differentiation, growth, and metabolism. Although the effects of n-3 PUFAs on several behavioral indices have been widely reported, there is still much work to be done if we are to understand the mechanism by which this family of fatty acids exerts its many actions.

ACKNOWLEDGMENTS

We would like to thank Brett Purcell for his expert technical assistance. This work was supported by a block grant to the Howard Florey Institute of Experimental Physiology and Medicine from the National Health and Medical Research Council of Austrlalia (Grant No. 98 3001) and grants from the Robert J. Kleberg, Jr. and Helen C. Kleberg Foundation, the G. Harold and Leila Y. Mathers Charitable Foundation, the Rebecca L. Cooper Medical Research Foundation, the Windermere Foundation, and the Philip Bushell Foundation.

REFERENCES

Anand BK, Brobeck JR. Hypothalamic control of food intake in rats and cats. Yale J Biol Med 1951; 24:123–140.

Andersson B, Leksell LG, Rundgren M. Regulation of water intake. Ann Rev Nutr 1982; 2:73–89.

Arase K, York DA, Shimizu H, Shargill N, Bray GA. Effects of corticotropin-releasing factor on food intake and brown adipose tissue thermogenesis in rats. Am J Physiol 1988; 255:E255–E259.

Awad AB, Bernardis LL, Fink CS. Failure to demonstrate an effect of dietary fatty acid composition on body weight, body composition and parameters of lipid metabolism in mature rats. J Nutr 1990; 120:1277–1282.

Baram TZ, Chalmers DT, Chen C, Koutsoukos Y, De Souza EB. The CRF1 receptor mediates the excitatory actions of corticotropin releasing factor (CRF) in the developing rat brain: in vivo evidence using a novel, selective, non-peptide CRF receptor antagonist. Brain Res 1997; 770:89–95.

Barber MD, Ross JA, Voss AC, Tisdale MJ, Fearon KC. The effect of an oral nutritional supplement enriched with fish oil on weight-loss in patients with pancreatic cancer. Br J Cancer 1999; 81:80–86.

Barberis C, Mouillac B, Durroux T. Structural bases of vasopressin/oxytocin receptor function. J Endocrinol 1998; 156:223–229.

Belzung F, Raclot T, Groscolas R. Fish oil n-3 fatty acids selectively limit the hypertrophy of abdominal fat depots in growing rats fed high-fat diets. Am J Physiol 1993; 264:R1111–R1118.

Beresford MJ, Fitzsimons JT. Intracerebroventricular angiotensin II-induced thirst and sodium appetite in rat are blocked by the AT1 receptor antagonist, Losartan (DuP 753), but not by the AT2 antagonist, CGP 42112B. Exp Physiol 1992; 77:761–764.

Billington CJ, Levine AS. Hypothalamic neuropeptide Y regulation of feeding and energy metabolism. Curr Opin Neurobiol 1992; 2:847–851.

Blair-West JR, Burns P, Denton DA, Ferraro T, McBurnie MI, Tarjan E, Weisinger RS. Thirst induced by increasing brain sodium concentration is mediated by brain angiotensin. Brain Res 1994; 637:335–338.

Blair-West JR, Carey KD, Denton DA, Weisinger RS, Shade RE. Evidence that brain angiotensin II is involved in both thirst and sodium appetite in baboons. Am J Physiol 1998; 275:R1639–R1646.

Blair-West JR, Denton DA, Gellatly DR, McKinley MJ, Nelson JF, Weisinger RS. Changes in sodium appetite in cattle induced by changes in CSF sodium concentration and osmolality. Physiol Behav 1987; 39:465–469.

Blair-West JR, Denton DA, McKinley MJ, Weisinger RS. Angiotensin-related sodium appetite and thirst in cattle. Am J Physiol 1988; 255:R205–R211.

Blair-West JR, Denton DA, McKinley MJ, Weisinger RS. Central infusion of the AT1 receptor antagonist losartan inhibits thirst but not sodium appetite in cattle. Am J Physiol 1997; 272:R1940–R1945.

Blomqvist AG, Herzog H. Y-receptor subtypes—how many more? Trends Neurosci 1997; 20:294–298.

Bray GA. The nutrient balance hypothesis: peptides, sympathetic activity, and food intake. Ann NY Acad Sci 1993; 676:223–241.

Bray GA, York DA. The MONA LISA hypothesis in the time of leptin. Recent Prog Horm Res 1998; 53:95–117.

Brazeau P, Vale W, Burgus R, Ling N, Butcher M, Rivier J, Guillemin R. Hypothalamic polypeptide that inhibits the secretion of immunoreactive pituitary growth hormone. Science 1973; 179:77–79.

Brennan MF. Uncomplicated starvation versus cancer cachexia. Cancer Res 1977; 37:2359–2364.

Broberger C. Hypothalamic cocaine- and amphetamine-regulated transcript (CART) neurons: histochemical relationship to thyrotropin-releasing hormone, melanin-concentrating hormone, orexin/hypocretin and neuropeptide Y. Brain Res 1999; 848:101–113.

Broberger C, De Lecea L, Sutcliffe JG, Hokfelt T. Hypocretin/orexin- and melanin-concentrating hormone-expressing cells form distinct populations in the rodent lateral hypothalamus: relationship to the neuropeptide Y and agouti gene-related protein systems. J Compar Neurol 1998; 402:460–474.

Brown MF. Modulation of rhodopsin function by properties of the membrane bilayer. Chem Phys Lipids 1994; 73:159–180.

Burr G, Burr M. A new deficiency disease produced by the rigid exclusion of fat from the diet. J Biol Chem 1929; 82:345–367.

Calder PC. Immunomodulatory and anti-inflammatory effects of n-3 polyunsaturated fatty acids. Proc Nutr Soc 1996; 55:737–774.

Campfield LA, Smith FJ. Functional coupling between transient declines in blood glucose and feeding behavior: temporal relationships. Brain Res Bull 1986; 17:427–433.

Chambers J, Ames RS, Bergsma D, Muir A, Fitzgerald LR, Hervieu G, et al. Melanin-concentrating hormone is the cognate ligand for the orphan G-protein-coupled receptor SLC-1. Nature 1999; 400:261–265.

Chen G, Koyama K, Yuan X, Lee Y, Zhou YT, O'Doherty R, et al. Disappearance of body fat in normal rats induced by adenovirus-mediated leptin gene therapy. Proc Natl Acad Sci USA 1996; 93:14,795–14,799.

Chiaraviglio E, Perez Guaita MF. Anterior third ventricle (A3V) lesions and homeostasis regulation. J Physiol 1984; 79, 446–52.

Chicco A, D'Alessandro ME, Karabatas L, Gutman R, Lombardo YB. Effect of moderate levels of dietary fish oil on insulin secretion and sensitivity, and pancreas insulin content in normal rats. Ann Nutr Metab 1996; 40:61–70.

Clandinin MT, Chappell JE, Leong S, Heim T, Swyer PR, Chance GW. Intrauterine fatty acid accretion rates in human brain: implications for fatty acid requirements. Early Hum Dev 1980; 4:121–129.

Coghlan JP, Blair-West JR, Butkus A, Denton DA, Hardy JJ, Leksell LG, et al. Factor regulating aldosterone secretion. In: Cumming I, Funder J, Mendelsohn F, eds. Endocrinology. Australian Academy of Science, Lanberra, Australia, 1980, pp. 385–388.

Connor WE, Neuringer M, Reisbick S. Essential fatty acids: the importance of n-3 fatty acids in the retina and brain. Nutr Rev 1992; 50:21–29.

Costa G, Bewley P, Aragon M, Siebold J. Anorexia and weight loss in cancer patients. Cancer Treat Res 1981; 65:3–7.

Crawford MA, Sinclair AJ. Nutritional influences in the evolution of the mammalian brain. In: CIBA Foundation Symposium on Lipids, Malnutrition and the Developing Brain. Associated Scientific Publishers, Amsterdam, 1972, pp. 267–287.

Crowley WR, Terry LC. Biochemical mapping of somatostatinergic systems in rat brain: effects of periventricular hypothalamic and medial basal amygdaloid lesions on somatostatin-like immunoreactivity in discrete brain nuclei. Brain Res 1980; 200:283–291.

Cunnane SC, McAdoo KR, Horrobin DF. n-3 Essential fatty acids decrease weight gain in genetically obese mice. British J Nutr 1986; 56:87–95.

Dallman MF, Strack AM, Akana SF, Bradbury MJ, Hanson ES, Scribner KA, et al. Feast and famine: critical role of glucocorticoids with insulin in daily energy flow, Frontiers in Neuroendocrinology. Raven, New York, 1993, Vol. 14, pp. 303–347.

DeWys W. Management of cancer cachexia. Semin Oncol 1985; 12:452–460.

Doucet E, Almeras N, White MD, Despres JP, Bouchard C, Tremblay A. Dietary fat composition and human adiposity [see comments]. Eur J Clin Nutr 1998; 52:2–6.

Dratz EE, Holte LL. The molecular spring model for the function of docosahexaenoic acid (22:6n-3) in biological membranes. In: Sinclair AJ, Gibson RA, eds. The Third International Congress on Essential Fatty Acids and Eicosanoids. AOCS, Champaign, IL, 1993, pp. 122–127.

Du P, Salon JA, Tamm JA, Hou C, Cui W, Walker MW, et al. Modeling the G-protein-coupled neuropeptide Y Y1 receptor agonist and antagonist binding sites. Protein Eng 1997; 10:109–117.

Eaton SB, Eaton SB 3rd, Sinclair AJ, Cordain L, Mann NJ. Dietary intake of long-chain polyunsaturated fatty acids during the paleolithic. World Rev Nutr Diet 1998; 83:12–23.

Eghbal-Ahmadi M, Hatalski CG, Avishai-Eliner S, Baram TZ. Corticotropin releasing factor receptor type II (CRF2) messenger ribonucleic acid levels in the hypothalamic ventromedial nucleus of the infant rat are reduced by maternal deprivation [published erratum appears in Endocrinology]. Endocrinology 1997; 138:5048–5051; erratum: 1998; 139(i):136.

Epstein AN. Angiotensin-induced thirst and sodium appetite. In: Martin JB, Reichlin S, Bick KL, eds. Neurosecretion and Brain Peptides. Raven, New York, 1981, pp. 373–387.

Epstein AN. Control of salt intake by steroids and cerebral peptides. Pharmacol Res 1992; 25:113–124.

Fenton WS, Hibbeln, JR, Knable M. Essential fatty acids, lipid membrane abnormalities, and the diagnosis and treatment of schizophrenia. Biol Psychiatry 2000; 47:8–21.

Ferguson AV, Washburn DL. Angiotensin II: a peptidergic neurotransmitter in central autonomic pathways. Prog Neurobiol 1998; 54:169–192.

Fitts DA, Tjepkes DS, Bright RO. Salt appetite and lesions of the ventral part of the ventral median preoptic nucleus. Behav Neurosci 1990; 104:818–827.

Fitzsimons JT. Thirst. Physiol Rev 1972; 52:468–561.

Fitzsimons JT. Angiotensin, thirst, and sodium appetite. Physiol Rev 1998; 78:583–686.

Friedman JM. Leptin, leptin receptors and the control of body weight. Eur J Med Res 1997; 2:7–13.

Friedman MI, Stricker EM. The physiological psychology of hunger: a physiological perspective. Psychol Rev 1976; 83:409–431.

Gardner JD, Rothwell NJ, Luheshi GN. Leptin affects food intake via CRF-receptor-mediated pathways. Nature Neurosci 1998; 1:103.

Gray TS, Bingaman EW. The amygdala: corticotropin-releasing factor, steroids, and stress. Crit Rev Neurobiol 1996; 10:155–168.

Grundemar L, Hakanson R. Neuropeptide Y effector systems: perspectives for drug development. Trends Pharmacol Sci 1994; 15:153–159.

Hakansson M, de Lecea L, Sutcliffe JG, Yanagisawa M, Meister B. Leptin receptor- and STAT3-immunoreactivities in hypocretin/orexin neurones of the lateral hypothalamus. J Neuroendocrinol 1999; 11:653–663.

Halaas JL, Boozer C, Blair-West J, Fidahusein N, Denton DA, Friedman JM. Physiological response to long-term peripheral and central leptin infusion in lean and obese mice. Proc Natl Acad Sci USA 1997; 94:8878–8883.

Hanada R, Nakazato M, Matsukura S, Murakami N, Yoshimatsu H, Sakata T. Differential regulation of melanin-concentrating hormone and orexin genes in the agouti-related protein/melanocortin-4 receptor system. Biochem Biophys Res Commun 2000; 268:88–91.

Harris WS, Connor WE, Alam N, Illingworth DR. Reduction of postprandial triglyceridemia in humans by dietary n-3 fatty acids. J Lipid Res 1988; 29:1451–1460.
Hetherington AW, Ranson SW. Hypothalamic lesions and adiposity in the rat. Anat Rec 1940; 78:149–172.
Hill JO, Peters JC, Lin D, Yakubu F, Greene H, Swift L. Lipid accumulation and body fat distribution is influenced by type of dietary fat fed to rats. Int J Obesity Related Metab Disord 1993; 17:223–236.
Hoebel BG, Teitelbaum P. Weight regulation in normal and hypothalamic hyperphagic rats. Journal of Comparative Physiology. A. Sensory Neural Behav Physiol 1966; 61:189–193.
Holzwarth-McBride MA, Hurst EM, Knigge KM. Monosodium glutamate induced lesions of the arcuate nucleus. I. Endocrine deficiency and ultrastructure of the median eminence. Anat Rec 1976; 186:185–205.
Houseknecht KL, Baile CA, Matteri RL, Spurlock ME. The biology of leptin: a review. J Anim Sci 1998; 76:1405–1420.
Huang QH, Hruby VJ, Tatro JB. Role of central melanocortins in endotoxin-induced anorexia. Am J Physiol 1999; 276:R864–R871.
Hui R, St.-Louis J, Falardeau P. Antihypertensive properties of linoleic acid and fish oil omega-3 fatty acids independent of the prostaglandin system. Am J Hypertens 1989; 2:610–617.
Huszar D, Lynch CA, Fairchild-Huntress V, Dunmore JH, Fang Q, Berkemeier LR, et al. Targeted disruption of the melanocortin-4 receptor results in obesity in mice. Cell 1997; 88:131–141.
Inui A. Cancer anorexia–cachexia syndrome: are neuropeptides the key? Cancer Res 1999; 59:4493–4501.
Johnson D, Drenick EJ. Therapeutic fasting in morbid obesity. Arch Int Med 1977; 137:1381–1382.
Jones PJ, Schoeller DA. Polyunsaturated:saturated ratio of diet fat influences energy substrate utilization in the human. Metabolism 1988; 37:145–151.
Kamalian N, Keesey RE, ZuRhein GM. Lateral hypothalamic demyelination and cachexia in a case of "malignant" multiple sclerosis. Neurology 1975; 25:25–30.
Kask K, Berthold M, Bartfai T. Galanin receptors: involvement in feeding, pain, depression and Alzheimer's disease. Life Sci 1997; 60:1523–1533.
Keesey RE, Hirvonen MD. Body weight set-points: determination and adjustment. J Nutr 1997; 127:1875S–1883S.
Keesey RE, Mitchel JS, Kemnitz JW. Body weight and body composition of male rats following hypothalamic lesions. Am J Physiol 1979; 237:R68–R73.
Keesey RE, Powley TL. Hypothalamic regulation of body weight. Am Sci 1975; 63:558–565.
Kennedy GC. The hypothalamic control of food intake in rats. Proc R Soc London 1950; 137:535–549.
Kenny D, Warltier DC, Pleuss JA, Hoffmann RG, Goodfriend TL, Egan BM. Effect of omega-3 fatty acids on the vascular response to angiotensin in normotensive men. Am J Cardiol 1992; 70:1347–1352.
King BM, Rollins BL, Stines SG, Cassis SA, McGuire HB, Lagarde ML. Sex differences in body weight gains following amygdaloid lesions in rats. Am J Physiol 1999; 277:R975–R980.
Kirchgessner AL, Sclafani A. PVN–hindbrain pathway involved in the hypothalamic hyperphagia–obesity syndrome. Physiol Behav 1988; 42:517–528.
Krisch B, Leonhardt H. Luliberin and somatostatin fiber-terminals in the subfornical organ of the rat. Cell Tissue Res 1980; 210:33–45.
Kurlack L, Stephenson T. Plausible explanations for effects of long chain polyunsaturated fatty acids on neonates. Arch Dis Child 1999; 80:148–154.
Lang RE, Rascher W, Heil J, Unger T, Wiedemann G, Ganten D. Angiotensin stimulates oxytocin release. Life Sci 1981; 29:1425–1428.
Leibowitz SF. Brain peptides and obesity: pharmacologic treatment. Obesity Res 1995; 3(Suppl 4):573S–589S.
Leibowitz SF, Alexander JT. Hypothalamic serotonin in control of eating behavior, meal size, and body weight. Biol Psychiatry 1998; 44:851–64.
Lenkei Z, Palkovits M, Corvol P, Llorens-Cortes C. Distribution of angiotensin type-1 receptor messenger RNA expression in the adult rat brain. Neuroscience 1998; 82:827–841.
Leroux P, Pelletier G. Radioautographic localization of somatostatin-14 and somatostatin-28 binding sites in the rat brain. Peptides 1984; 5:503–506.
Levine AS, Billington CJ. Why do we eat? A neural systems approach. Ann Rev Nutr 1997; 17:597–619.
Li Z, Ferguson AV. Subfornical organ efferents to paraventricular nucleus utilize angiotensin as a neurotransmitter. Am J Physiol 1993; 265:R302–R309.
Lind RW, Johnson AK. A further characterization of the effects of AV3V lesions on ingestive behavior. Am J Physiol 1983; 245:R83–R90.
Littman BJ, Mitchell DC. A role for phospholipid polyunsaturation in modulating membrane protein function. Lipids 1996; 31:S-193–S-197.

Matson CA, Reid DF, Cannon TA, Ritter RC. Cholecystokinin and leptin act synergistically to reduce body weight. Am J Physiol 2000; 278:R882–R890.

Mayer J. The glucostatic mechanisms of regulation of food intake. Bull N Eng Med Center 1952; 14:43–49.

Mayer J. Regulation of energy intake and the body weight: the glucostatic theory and the lipostatic hypothesis. Ann NY Acad Sci 1955; 63:15–42.

McAloon Dyke M, Davis KM, Clark BA, Fish LC, Elahi D, Minaker KL. Effects of hypertonicity on water intake in the elderly: an age-related failure. Geriatr Nephrol Urol 1997; 7:11–16.

McGahon BM, Martin DS, Horrobin DF, Lynch MA. Age-related changes in synaptic function: analysis of the effect of dietary supplementation with omega-3 fatty acids. Neuroscience 1999; 94:305–314.

McKinley MJ, Badoer E, Vivas L, Oldfield BJ. Comparison of c-fos expression in the lamina terminalis of conscious rats after intravenous or intracerebroventricular angiotensin. Brain Res Bull 1995; 37:131–137.

McKinley MJ, Congiu M, Denton DA, Park RG, Penschow J, Simpson JB, et al. The anterior wall of the third cerebral ventricle and homeostatic responses to dehydration. J Physiol 1984; 79:421–427.

McKinley MJ, Denton DA, Leksell L, Tarjan E, Weisinger RS. Evidence for cerebral sodium sensors involved in water drinking in sheep. Physiol Behav 1980; 25:501–504.

McKinley MJ, Denton DA, Weisinger RS. Sensors for antidiuresis and thirst—osmoreceptors or CSF sodium detectors? Brain Res 1978; 141:89–103.

McKinley MJ, McAllen RM, Mendelsohn FA. O., Allen AM, Chai SY, Oldfield BJ. Circumventricular organs: neuroendocrine interfaces between the brain and the hemal milieu, Frontiers in Neuorendocrinology. Raven, New York, 1990, Vol. 11, pp. 91–127.

McKinley MJ, McAllen RM, Pennington GL, Smardencas A, Weisinger RS, Oldfield BJ. Physiological actions of angiotensin II mediated by AT1 and AT2 receptors in the brain. Clin Exp Pharmacol Physiol 1996; 3(Suppl):S99–S104.

Meister B. Control of food intake via leptin receptors in the hypothalamus. Vitam Horm 2000; 59:265–304.

Mendelsohn FA, Allen AM, Chai SY, Sexton PM, Figdor R. Overlapping distributions of receptors for atrial natriuretic peptide and angiotensin II visualized by in vitro autoradiography: morphological basis of physiological antagonism. Can J Physiol Pharmacol 1987; 65:1517–1521.

Merchenthaler I, Lopez FJ, Negro-Vilar A. Anatomy and physiology of central galanin-containing pathways. Prog Neurobiol 1993; 40:711–769.

Moore SA, Yoder E, Spector AA. Role of the blood-brain barrier in the formation of long-chain omega-3 and omega-6 fatty acids from essential fatty acid precursors. J Neurochem 1990; 55:391–402.

Moran TH, McHugh PR. Cholecystokinin suppresses food intake by inhibiting gastric emptying. Am J Physiol 1982; 242:R491–R497.

Morimoto T, Yamamoto Y, Yamatodani A. Leptin facilitates histamine release from the hypothalamus in rats. Brain Res 2000; 868:367–369.

Mrosovsky N, Powley TL. Set points for body weight and fat. Behav Biol 1977; 20:205–223.

Nakamura M, Katsuura G, Nakao K, Imura H. Antidipsogenic action of alpha-human atrial natriuretic polypeptide administered intracerebroventricularly in rats. Neurosci Lett 1985; 58:1–6.

Neuringer M, Connor WE, Linn DS, Barstad L, Luck S. Biochemical and functional effects of prenatal and postnatal omega-3 fatty acid deficiency on retina and brain in monkeys. Proc Natl Acad Sci USA 1986; 83:4021–4025.

Ohira T, Nishio K, Ohe Y, Arioka H, Nishio M, Funayama Y, et al. Improvement by eicosanoids in cancer cachexia induced by LLC-IL6 transplantation. J Cancer Res Clin Oncol 1996; 122:711–715.

Ohno T, Ohinata H, Ogawa K, Kuroshima A. Fatty acid profiles of phospholipids in brown adipose tissue from rats during cold acclimation and repetitive intermittent immobilization: with special reference to docosahexaenoic acid. Jpn J Physiol 1996; 46:265–270.

Oldfield BJ, Ganten D, Mckinley MJ. An ultrastructural analysis of the distribution of angiotensin II in the rat brain. J Neuroendocrinol 1989; 1:121–128.

Olson BR, Drutarosky MD, Stricker EM, Verbalis JG. Brain oxytocin receptor antagonism blunts the effects of anorexigenic treatments in rats: evidence for central oxytocin inhibition of food intake. Endocrinology 1991; 129:785–791.

Osborne PG, Denton DA, Weisinger RS. Inhibition of dehydration induced drinking in rats by reduction of CSF Na concentration. Brain Res 1987; 412:36–42.

Park R, Denton DA, McKinley MJ, Pennington G, Weisinger RS. Intracerebroventricular saccharide infusions inhibit thirst induced by systemic hypertonicity. Brain Res 1989; 493:123–128.

Parrish CC, Pathy DA, Angel A. Dietary fish oils limit adipose tissue hypertrophy in rats. Metabolism 1990; 39:217–219.

Parrish CC, Pathy DA, Parkes JG, Angel A. Dietary fish oils modify adipocyte structure and function. J Cell Physiol 1991; 148:493–502.

Patel YC, Baquiran G, Srikant CB, Posner BI. Quantitative in vivo autoradiographic localization of [125I-Tyr11]somatostatin-14-and [Leu8,D-Trp22–125I-Tyr25]somatostatin-28-binding sites in rat brain. Endocrinology 1986; 119:2262–2269.

Plata-Salaman CR, Wilson CD, Ffrench-Mullen JM. In vivo IL-1beta-induced modulation of G-protein alphaO subunit subclass in the hypothalamic ventromedial nucleus: implications to IL-1beta-associated anorexia. Mol Brain Res 1998; 58:188–194.

Porte D Jr, Seeley RJ, Woods SC, Baskin DG, Figlewicz DP, Schwartz MW. Obesity, diabetes and the central nervous system [see comments]. Diabetologia 1998; 41:863–881.

Price PT, Nelson CM, Clarke SD. Omega-3 polyunsaturated fatty acid regulation of gene expression. Curr Opin Lipidol 2000; 11:3–7.

Quirion R, Dalpe M, De Lean A, Gutkowska J, Cantin M, Genest J. Atrial natriuretic factor (ANF) binding sites in brain and related structures. Peptides 1984; 5:1167–1172.

Raclot T, Groscolas R, Langin D, Ferre P. Site-specific regulation of gene expression by n-3 polyunsaturated fatty acids in rat white adipose tissues. J Lipid Res 1997; 38:1963–1972.

Raulf F, Perez J, Hoyer D, Bruns C. Differential expression of five somatostatin receptor subtypes, SSTR1-5, in the CNS and peripheral tissue. Digestion 1994; 55:46–53.

Reid IA. Actions of angiotensin II on the brain: mechanisms and physiologic role. Am J Physiol 1984; 246:F533–F543.

Reilly JJ, Maki R, Nardozzi J, Schulkin J. The effects of lesions of the bed nucleus of the stria terminalis on sodium appetite. Acta Neurobiol Exp 1994; 54:253–257.

Reisbick S, Neuringer M, Connor WE, Barstad L. Postnatal deficiency of omega-3 fatty acids in monkeys: fluid intake and urine concentration. Physiol Behav 1992; 51:473–479.

Reisbick S, Neuringer M, Connor WE, Iliff-Sizemore S. Increased intake of water and NaCl solutions in omega-3 fatty acid deficient monkeys. Physiol Behav 1991; 49:1139–1146.

Reisbick S, Neuringer M, Hasnain R, Connor WE. Polydipsia in rhesus monkeys deficient in omega-3 fatty acids. Physiol Behav 1990; 47:315–323.

Richard D. Involvement of corticotropin-releasing factor in the control of food intake and energy expenditure. Ann NY Acad Sci 1993; 697:155–172.

Richard D. Exercise and the neurobiological control of food intake and energy expenditure. Int J Obesity Related Metab Disord 1995; 19(Suppl 4):S73–S79.

Ritter S, Hutton B. Mercaptoacetate-induced feeding is impaired by central nucleus of the amygdala lesions. Physiol Behav 1995; 58:1215–1220.

Rothwell NJ. Central effects of CRF on metabolism and energy balance. Neurosci Biobehav Rev 1990; 14:263–271.

Russell JC, Amy RM, Dolphin PJ. Effect of dietary n-3 fatty acids on atherosclerosis prone JCR:LA-corpulent rats. Experim Mol Pathol 1991; 55:285–293.

Rustan AC, Hustvedt BE, Drevon CA. Dietary supplementation of very long-chain n-3 fatty acids decreases whole body lipid utilization in the rat. J Lipid Res 1993; 34:1299–1309.

Sahu A. Leptin decreases food intake induced by melanin-concentrating hormone (MCH), galanin (GAL) and neuropeptide Y (NPY) in the rat. Endocrinology 1998; 139:4739–4742.

Sakai RR, Chow SY, Epstein AN. Peripheral angiotensin II is not the cause of sodium appetite in the rat. Appetite 1990; 15:161–170.

Sakai RR, Nicolaidis S, Epstein AN. Salt appetite is suppressed by interference with angiotensin II and aldosterone. Am J Physiol 1986; 251:R762–R768.

Sakata T, Kang M, Kurokawa M, Yoshimatsu H. Hypothalamic neuronal histamine modulates adaptive behavior and thermogenesis in response to endogenous pyrogen. Obesity Res 1995; 3(Suppl 5):707S–712S.

Schulkin J, Marini J, Epstein AN. A role for the medial region of the amygdala in mineralocorticoid-induced salt hunger. Behav Neurosci 1989; 103:179–185.

Schutz Y, Flatt JP, Jequier E. Failure of dietary fat intake to promote fat oxidation: a factor favoring the development of obesity [see comments]. Am J Clin Nutr 1989; 50:307–314.

Schwartz MW, Figlewicz DP, Baskin DG, Woods SC, Porte D Jr. Insulin in the brain: a hormonal regulator of energy balance. Endocr Rev 1992; 13:387–414.

Seeley RJ, Schwartz MW. Neuroendocrine regulation of food intake. Acta Paediatr Scand 1999; 88(Suppl):58–61.

Simopoulos AP. Overview of evolutionary aspects of w3 fatty acids in the diet. World Rev Nur Diet 1998; 83:1–11.

Simopoulos AP, Leaf A, Salem N Jr. Workshop statement on the essentiality of and recommended dietary intakes for omega-6 and omega-3 fatty acids. Prostaglandins Leuktorienes Essential Fatty Acids 2000; 63:119–121.

Sinclair AJ. Long chain polyunsaturated fatty acids in the mammalian brain. Proc Nutr Soc 1975; 34:287–291.

Smagin GN, Howell LA, Ryan DH, De Souza EB, Harris RB. The role of CRF2 receptors in corticotropin-releasing factor- and urocortin-induced anorexia. NeuroReport 1998; 9:1601–1606.

Sohal PS, Baracos VE, Clandinin MT. Dietary omega 3 fatty acid alters prostaglandin synthesis, glucose transport and protein turnover in skeletal muscle of healthy and diabetic rats. Biochem J 1992; 286:405–411.

Song K, Allen AM, Paxinos G, Mendelsohn FA. Mapping of angiotensin II receptor subtype heterogeneity in rat brain. J Compar Neurol 1992; 316:467–484.

Spina M, Merlo-Pich E, Chan RK, Basso AM, Rivier J, Vale W, et al. Appetite-suppressing effects of urocortin, a CRF-related neuropeptide. Science 1996; 273:1561–1564.

Stachenfeld NS, DiPietro L, Nadel ER, Mack GW. Mechanism of attenuated thirst in aging: role of central volume receptors. Am J Physiol 1997; 272:R148–R157.

Stricker EM, Verbalis JG. Central inhibitory control of sodium appetite in rats: correlation with pituitary oxytocin secretion. Behav Neurosci 1987; 101:560–567.

Su W, Jones PJ. Dietary fatty acid composition influences energy accretion in rats. J Nutr 1993; 123:2109–2114.

Sul HS, Latasa MJ, Moon Y, Kim KH. Regulation of the fatty acid synthase promoter by insulin. J Nutr 2000; 130:315S–320S.

Tache Y, Maeda-Hagiwara M, Turkelson CM. Central nervous system action of corticotropin-releasing factor to inhibit gastric emptying in rats. Am J Physiol 1987; 253:G241–G245.

Tarjan E, Denton DA, Weisinger RS. Atrial natriuretic peptide inhibits water and sodium intake in rabbits. Regul Petpides 1988; 23:63–75.

Tarjan E, Ferraro T, May C, Weisinger RS. Converting enzyme inhibition in rabbits: effects on sodium and water intake/excretion and blood pressure. Physiol Behav 1993; 53:291–299.

Thomas CD, Peters JC, Reed GW, Abumrad NN, Sun M, Hill JO. Nutrient balance and energy expenditure during ad libitum feeding of high-fat and high-carbohydrate diets in humans. Am J Clin Nutr 1992; 55:934–942.

Thunhorst RL, Ehrlich KJ, Simpson JB. Subfornical organ participates in salt appetite. Behav Neurosci 1990; 104:637–642.

Thunhorst RL, Fitts DA. Peripheral angiotensin causes salt appetite in rats. Am J Physiol 1994; 267:R171–R177.

Tisdale MJ, Beck SA. Inhibition of tumour-induced lipolysis in vitro and cachexia and tumour growth in vivo by eicosapentaenoic acid. Biochem Pharmacol 1991; 41:103–107.

Tisdale MJ, Dhesi JK. Inhibition of weight loss by omega-3 fatty acids in an experimental cachexia model. Cancer Res 1990; 50:5022–5026.

Trayhurn P, Hoggard N, Mercer JG, Rayner DV. Leptin: fundamental aspects. Int J Obesity Related Metab Disord 1999; 23(Suppl 1):22–28.

Tremblay A, Plourde G, Despres JP, Bouchard C. Impact of dietary fat content and fat oxidation on energy intake in humans. Am J Clin Nutr 1989; 49:799–805.

Uauy R, Mena P, Rojas C. Essential fatty acids in early life: structural and functional role. Proc Nutr Soc 2000; 59:3–15.

van Houwelingen AC, Foreman-van Drongelen MM, Nicolini U, Nicolaides KH, Al MD, Kester AD, et al. Essential fatty acid status of fetal plasma phospholipids: similar to postnatal values obtained at comparable gestational ages. Early Hum Dev 1996; 46:141–152.

Vaskonen T, Laakso J, Mervaala E, Sievi E, Karppanen H. Interrelationships between salt and fish oil in stroke-prone spontaneously hypertensive rat. Blood Pressure 1996; 5:178–189.

Verbalis JG, Blackburn RE, Hoffman GE, Stricker EM. Establishing behavioral and physiological functions of central oxytocin: insights from studies of oxytocin and ingestive behaviors. Adv Exp Med Biol 1995; 395:209–225.

Vivas L, Chiaraviglio E. The effects of reversible lidocaine-induced lesion of the tissue surrounding the anterior ventral wall of the third ventricle on drinking in rats. Behav Neural Biol 1992; 57:124–130.

Wang H, Storlien LH, Huang XF. Influence of dietary fats on c-Fos-like immunoreactivity in mouse hypothalamus. Brain Res 1999; 843:184–192.

Warwick ZS. Probing the causes of high-fat diet hyperphagia: a mechanistic and behavioral dissection. Neurosci Biobehav Rev 1996; 20:155–161.

Weisinger RS, Blair-West JR, Burns P, Denton DA, McKinley MJ, Purcell B, et al. The inhibitory effect of hormones associated with stress on Na appetite of sheep. Proc Natl Acad Sci USA 2000; 97:2922–2927.

Weisinger RS, Blair-West JR, Burns P, Denton DA, McKinley MJ, Tarjan E. The role of angiotensin II in ingestive behaviour: a brief review of angiotensin II, thirst and Na appetite. Reg Peptides 1996; 66:73–81.

Weisinger RS, Blair-West JR, Denton DA, McBurnie M, Ong F, Tarjan E, et al. Effect of angiotensin-converting enzyme inhibitor on salt appetite and thirst of BALB/c mice. Am J Physiol 1990; 259:R736–R740.

Weisinger RS, Blair-West JR, Denton DA, Tarjan E. Central administration of somatostatin suppresses the stimulated sodium intake of sheep. Brain Res 1991; 543:213–218.

Weisinger RS, Blair-West JR, Denton DA, Tarjan E. Central administration of atrial natriuretic peptide suppresses sodium and water intake of sheep. Brain Res 1992; 579:113–118.

Weisinger RS, Blair-West JR, Denton DA, Tarjan E. Role of brain angiotensin II in thirst and sodium appetite of sheep. Am J Physiol 1997; 273:R187–R196.

Weisinger RS, Considine P, Denton DA, McKinley MJ. Rapid effect of change in cerebrospinal fluid sodium concentration on salt appetite. Nature 1979; 280:490–491.

Weisinger RS, Denton DA, Di Nicolantonio R, Hards DK, McKinley MJ, Oldfield B, et al. Subfornical organ lesion decreases sodium appetite in the sodium-depleted rat. Brain Res 1990; 526:23–30.

Weisinger RS, Denton DA, Di Nicolantonio R, McKinley MJ. The effect of captopril or enalaprilic acid on the Na appetite of Na-deplete rats. Clin Exp Pharmacol Physiol 1988; 15:55–65.

Weisinger RS, Denton DA, McKinley MJ, Muller AF, Tarjan E. Cerebrospinal fluid sodium concentration and salt appetite. Brain Res 1985; 326:95–105.

Weisinger RS, Denton DA, McKinley MJ, Nelson JF. Dehydration-induced sodium appetite in rats. Physiol Behav 1985; 34:45–50.

Weisinger RS, Denton DA, McKinley MJ, Simpson JB, Tarjan E. Cerebral Na sensors and Na appetite of sheep. In: deCaro G, Epstein AN, Massi M, eds. The Physiology of Thirst and Sodium Appetite. Life Sciences Series Vol. 105, Plenum, New York, 1986, pp. 485–490.

Weiss ML, Moe KE, Epstein AN. Interference with central actions of angiotensin II suppresses sodium appetite. Am J Physiol 1986; 250:R250–R259.

West DB, Fey D, Woods SC. Cholecystokinin persistently suppresses meal size but not food intake in free-feeding rats. Am J Physiol 1984; 246:R776–R787.

White JD. Neuropeptide Y: a central regulator of energy homeostasis. Reg Peptides 1993; 49:93–107.

Woods SC, Chavez M, Park CR, Riedy C, Kaiyala K, Richardson RD, et al. The evaluation of insulin as a metabolic signal influencing behavior via the brain. Neurosci Biobehav Rev 1996; 20:139–144.

Woods SC, Figlewicz Lattemann DP, Schwartz MW, Porte D Jr. A re-assessment of the regulation of adiposity and appetite by the brain insulin system. Int J Obesity 1990; 14:69–73; discussion pp. 74–76.

Woods SC, Seeley RJ, Porte D Jr, Schwartz MW. Signals that regulate food intake and energy homeostasis. Science 1998; 280:1378–1383.

Zardetto-Smith AM, Beltz TG, Johnson AK. Role of the central nucleus of the amygdala and bed nucleus of the stria terminalis in experimentally-induced salt appetite. Brain Res 1994; 645:123–134.

23 PUFA: Mediators for the Nervous, Endocrine, and Immune Systems

Shlomo Yehuda, Sharon Rabinovitz, and David I. Mostofsky

1. INTRODUCTION

Three major systems in the body mediate "information" traffic, namely the nervous system, the endocrine system, and the immune system. Chemical or electrical signals to or from other cells are generated and received, signals are modified, and tissues, organs, and muscles are activated as a result of their activities. Conventional wisdom of the past considered each system as independent from the others and able to interact only with component members of its own domain. More recently, this notion has been challenged, and studies have shown that each system is able to interact with external signals and stimuli and, moreover, that each system could interact with the other two systems. Studies of individual differences in autonomic reactivity not only may help identify a disposition to long-term health changes but may better clarify the convergence of the cardiovascular, neuroendocrine, and psychoneuroimmunological systems (Kiecolt-Glaser et al., 1992). Representative reviews of the bidirectionalities within these systems have been provided elsewhere (Dunn, 1989;1995; Song & Leonard, 2000). The biochemical modulation and mediation of the various activities of these systems and, by extension, the behavioral consequences have received less attention than deserved among behavioral neuroscientists. In this chapter, we will attempt to focus our attention on the role of the lipids as they impact the mutifactorial world of the body economy. In addition, although the bidirectionality that exists globally between behavior and physiology is largely beyond the scope of this chapter, the considerations of stress effects on the dynamics of PUFA regulation and associated behaviors are particularly relevant.

One group of molecules in particular, the polyunsaturated fatty acids (PUFAs), seems to play an important role in each of these three systems. Previous studies by us and others have shown that PUFA has a profound effect in all three systems, such as altering neuronal membrane fluidity (Yehuda, et al., 1993) and the production and function of neurotransmitters (Yehuda, et al.,1997). PUFA also was shown to modify the production and activity of releasing factors and certain hormones and the activity of certain cytokines (Yehuda, et al., 2000).

From: *Fatty Acids: Physiological and Behavioral Functions*
Edited by: D. Mostofsky, S. Yehuda, and N. Salem Jr. © Humana Press Inc., Totowa, NJ

The central hypothesis of this chapter is that PUFA occupies a major role in being able to modify the activity of each system and the interactions among them. Psychological functions such as cognition, learning, mood, stress, pain, and brain-related states or clinical neurological disorders can be best understood by appreciating the unifying elements of these systems. In this chapter, we will first review some basic aspects of the chemistry and structure of PUFA, followed by a review of some fundamental aspects of the immune system, and conclude with selected examples involving pain and stress. A glossary of common terms encountered in the PUFA literature is provided as an appendix to this volume.

2. ESSENTIAL FATTY ACIDS

Few topics in nutrition cause so much controversy and concern and are as misunderstood as fat. The dominant message from the medical profession has been to dramatically reduce the amount of fat we eat, in the interest of combating risks associated with cardiovascular disorders, diabetes, and other chronic disorders. At the same time, deficiencies in fat intake are equally likely to contribute to health hazards, including increased risk of infection, dysregulation of chronobiological activity, and impaired cognitive and sensory functions (especially in infants). A consensus has emerged from recent research that suggests that it is not so much the amount of fat we eat as the balance of the different types of fats. The type of dietary fat affects the behavior of each cell, that is, how well it can perform its vital functions and the ability to resist disease.

Two polyunsaturated fatty acids (PUFAs), linoleic and α-linolenic acid. are necessary for good health. Referred to as "essential fatty acids" (EFAs) because they cannot be manufactured by the body, but depend on being provided by nutritional intake, EFAs have beneficial effects when available in moderation. Excesses of the otherwise beneficial fatty acids may, however, exert harmful effects, with high intakes of saturated and hydrogenated fats being linked to an increase in a number of health risks, including degenerative diseases, cardiovascular disease, cancer, and diabetes.

2.1. PUFA: Omega-6 and Omega-3 Fatty Acids (n-6 and n-3 FA)

Linoleic acid is a member of the family of omega-6 (n-6) fatty acids and α-linolenic acid is an omega-3 (n-3) fatty acid. These terms refer to characteristics in the chemical structure of the fatty acids. Other omega-6 fatty acids can be manufactured in the body using linoleic acid as a starting point. These include γ-linoleic acid (GLA), dihomo-γ-linoleic acid (DHGLA), and arachidonic acid (AA). Similarly, other omega-3 fatty acids manufactured in the body using α-linolenic acid as a starting point include eicosapentaenoic acid (EPA) and docosahexaenoic acid (DHA).

Essential fatty acids are involved in energy production, the transfer of oxygen from the air to the bloodstream, and the manufacture of hemoglobin. They are also involved in growth, cell division, and nerve function. Essential fatty acids are found in high concentrations in the brain and are essential for normal nerve impulse transmission and brain function (Salvati et al., 1993).

Among the significant components of cell membranes are the phospholipids, which contain fatty acids. The type of fatty acids in the diet determine the type of fatty acids that are available to the composition of cell membranes. A phospholipid made from a *saturated* fat has a different structure and is less fluid than one that incorporates an *essential fatty acid.* In addition, linoleic and α-linolenic acids *per se* have an effect on the neuronal membrane fluidity index. They are able to decrease the cholesterol level in the membrane, which would otherwise decrease membrane fluidity, which, in turn, would make it difficult for the cell to carry out its normal functions, and increase the cell's susceptibility to injury and death. These consequences for cell function are not restricted to absolute levels of FAs; rather, the *relative* amounts of omega-3 fatty acids and omega-6 fatty acids in cell membranes have also been shown to affect cellular function.

At least five categories of PUFA effects on brain functions have been noted and discussed elsewhere (Yehuda, et al., 1997), namely (1) modifications of membrane fluidity, (2) modifications of the activity of membrane-bound enzymes, (3) modifications of the number and affinity of receptors, (4) modifications of the function of ion channels, and (5) modifications of the production and activity of neurotransmitters.

Symptoms of essential fatty acid deficiency may include fatigue, skin problems, immune weakness, gastrointestinal disorders, heart and circulatory problems, growth retardation, and sterility (Belzung et al., 1998). In addition to these symptom conditions, a lack of dietary essential fatty acids has been implicated in the development or aggravation of breast cancer, prostate cancer, rheumatoid arthritis, asthma, preeclampsia, depression, schizophrenia, and ADHD (attention-deficit/hyperactivity disorder). The list is neither exhaustive nor conclusive.

2.2. PUFA and the Immune System

Repeated demonstrations that PUFA can modify the production and activity of various components of the immune system have left unexplained the mode of action by which it exerts its effects. Several mechanisms had been proposed, including the following: *membrane fluidity*—changes that might effect the capability of cytokines to bind to their respective receptors on the cell membrane; *lipid peroxidation*—decrease in free-radical-induced tissue damage; *prostaglandin production*—an indirect mechanism whereby prostaglandins that are derivatives of PUFA modify cytokine activity; *regulation of gene expression*—PUFA influences on the signal transduction pathways and on mRNA activity. The role of PUFA in immune function is complicated by the fact that n-3 and n-6 have differential effects on various immune components. A recent review (Zimmer et al., 2000) indicated that n-3 fatty acids induce a decrease in lymphocyte proliferation in humans and rats, a decrease in interleukin-1 (IL-1) production and a decrease in IL-2 production in both humans and animals. In addition n-3 FA decreases tumor necroses factor-α (TNF-α) production in humans, but increases it in mice macrophages, and also decreases natural-killer (NK) cell activity (McGahon et al., 1998). On the other hand, n-6 increases the production of IL-2 in mice and decreases production of TNF-α production and NK cell activity. Still other studies have shown that linoleic acid (n-6) decreases the activity of IL-2 (Moussa, et al., 2000), and increases IL-1 production and tissue response to cytokines, whereas n-3 FA generally decreases IL-1 production and activity (Grimble and Tappia, 2000, Rizzo et al., 1996). Despite some disagreements among studies, it seems that n-3 FA (α-linolenic acid, DHA, EPA) decreases the production and activity of the proinflammatory cytokines (IL-1, IL-6, TNF-α) (Blok, et al.,1997; Chavali, et al., 1998, Hughes and Finder, 1997, Yano, et al., 2000) and that n-6 has the opposite effect (Caughey, et al., 1996; James, et al., 2000; Grimble and Tappia, 1998). The ability of n-3 PUFA to reduce proinflammatory cytokines and prostaglandin (Chavali and Forse, 1999) led to the proposal for the use of fish oil to relieve pain. Fish oil (rich in n-3 PUFA) has been shown to decrease IL-6, IL-10, IL-12, TNF, and prostaglandin E_2 (PGE_2) (Chavali and Forse, 1999).

Increasingly, the effects of PUFA are being examined not only with respect to their absolute level in diet, supplementation, or serum and tissue levels, but also with respect to the proportional relationship to other FAs. The level of anti-inflammatory IL-2 production is increased following treatment by a mixture of n-3 to n-6 FA in a ratio of 1 : 3 (Wu, et al., 1996), whereas it induces an increase in n-3 in the tissue (James, et al., 1999). These observations will be discussed in greater detail later in this chapter.

2.3. Stress

In psychology and biology, the term "stress" is applied to describe a strain or interference that disturbs the functioning of an organism. Organisms respond to physical and psychological stress with a combination of behavioral and physiological defenses. If the stress is too powerful or the defenses inadequate, a psychosomatic or comparable dysfunction may result in both animals and humans.

Outside the laboratory, stress is accepted as an unavoidable effect of living and is an especially complex phenomenon in the modern technological society. There is little doubt that an individual's success or failure in controlling potentially stressful situations can have a profound effect on the ability to function. The ability to "cope" with stress has figured prominently in anxiety and psychosomatic research. Stress has also figured prominently in the world of Health Psychology or Behavioral Medicine. Reports of a statistical link between coronary heart disease and individuals with a particular personality profile that is characterized by a behavioral pattern that manifests a lifestyle of impatience, a sense of time urgency, hard-driving competitiveness, and a preoccupation with vocational and related deadlines ("Type A Personality") has been replicated numerous times. Correlations with other profiles have suggested potential links to cancer, diabetes, and other chronic medical conditions. Although different types of stress can be identified, this chapter will refer only to psychological stress.

3. EFA AND STRESS

As early as 1964, Back and Bogdnoff (1964) reported elevations of free fatty acids and cholesterol among stressed peoples. Rosenman (1997) summarized many years of research on the increased level of cholesterol among Type A behavior subjects. Subsequent studies confirmed the correlation between stressful situations and an increased level of cholesterol and free fatty acids (Arbogast, et al. 1994; Brennan, et al., 1996; Clark, et al., 1998; Morrow, et al., 1994). On the other hand, stressed medical students were found to exhibit lower levels of linoleic and arachidonic acids (n-6), with no change in n-3 fatty acids (Onho, et al., 1996; Williams, et al., 1992). Stress was shown to be able to modify several key steps in fatty acid and lipid metabolism (Matsmoto, et al., 1999; Milles, et al., 1994). More specifically, during stress, the cardiac uptake of free FA was reduced (Bagger, 1997). Administration of DHA (an n-3 derivative) improved cardiac response to stress (Rouseau, et al., 1998), decreased the level of aggression (Hamzaki, et al., 1996; Sawazaki, 1996), decreased stress responses (Hamazaki, et al., 1999; Sawazaki, 1999; Singer, et al., 1991), and decreased the level of prostaglandin E_2 (Deutch, 1995; Rossetti, et al., 1997). Not surprising, therefore, was the finding that dietary intake of soybean oil and fish oil has stress-reduction properties (Furukawa, 1999; Ulak, et al., 1995). It should be useful to speculate on the biological bases of these correlations.

3.1. Stress and the Endocrine System

One function of the central nervous system is to evaluate and identify situations that can qualify as stress situations. The major components in the brain that are involved are the cortex, the limbic system, and the hypothalamus. The hypothalamus is the brain structure that bridges the nervous and the endocrine systems. In stress situations ("fight or flight"), the body mobilizes all of its energy to deal with the stress. The sympathetic nervous system, which is part of the autonomic nervous system, coordinates the effort to cope with the stress. The signal to activate the sympathetic nervous system is provided by the hormone epinephrine that is released from the medulla of the adrenal gland. A simplified description of the chain of events can be represented as follows: Beginning with encountering a stress situation, hypothalamic production of a cortico-releasing factor (CRF) signals the pituitary to release the hormone ACTH (adreno-corticotrophic hormone), which, in turn, acts on the adrenal cortex to release *cortisol*, which releases the hormone *epinephrine*. It is epinephrine, the "stress hormone," that activates the sympathetic nervous system. The release of CRF that initiates the cycle is controlled by the neurotransmitter dopamine. The reciprocal relationships between various brain neurotransmitters and releasing factors have been summarized elsewhere (Yehuda, et al., 1998; Song & Leonard, 2000).

3.2. Examples of Fatty Acid–Peptide Interaction

Several factors influence the release of CRF, including IL-2 and naloxon release (Jorgensen and Suny, 1994, Raber, et al., 1995, Ye, et al., 1994). Stress may increase the level of CRF directly or indirectly by action of the lutenizing hormone releasing hormone (LHRH) (Karanth, et al., 1995, Panajotova et al., 1997), while β-endorphin and α-MSH inhibit CRF. Release of CRF induces a higher level of prostaglandin E_2 (Faletti, et al., 1995). Polyunsaturated fatty acids improve the binding of β-endorphin to its receptor (Hargreaves and Clandinin, 1990; Murphy et al., 1987; Schwyzer, 1988). Similar findings were found for enkephalin binding (Myers, et al., 1995). Delta sleep-inducing peptide (DSIP) inhibits lipid peroxidation (Catalan, et al., 1986; Shmalwko, 1988), and most probably induces sleep through phospholipids (Yukhananov et al., 1991) and by modification of ionic channel activity (Sazontova, et al., 1996).

3.3. Prostaglandins

Essential fatty acids are a special class of unsaturated fatty acids, which also act as precursors of yet other FA. Most of the prostaglandins are derivatives of arachidonic acid and all of them have a high physiological, hormonelike, activity level. They are involved in numerous brain functions, such as regional blood flow and permeability of various biological membranes. It has been suggested that prostaglandins are also involved in the functional level of the activity of cAMP (Joo, 1993). The behavioral and physiological effects of an n-3/n-6 compound (in a ratio of 1 : 4, which we have designated as SR-3 in our formulations) correlate with changes in the fatty acid profile and with changes in the cholesterol level. It may well be that SR-3 has an effect on the prostaglandin system as well and that this change mediates the behavioral and biochemical changes that we observed in the rat. There is evidence that prostaglandin D_2 has a profound effect on sleep (Fadda, et al., 1995, 1996; Gelfand, et al., 1995; Ongini, 1993). Prostaglandins enhance CRF activity (Behan, et al., 1966; Lacroix, et al., 1996; Thompson, et al., 1991; Wanabe, et al.,

1994), and CRF induces release of prostaglandins (Petraglia, et al., 1995). Prostaglandins enhance TRH release and stimulate the dopaminergic and noradrenergic receptors activity (Murphy, et al., 1987; Yamasugi and Hame, 1993), whereas β-endorphin inhibits prostaglandin synthesis (Gelfand, et al., 1995).

3.4. Cholesterol and Fatty Acids

Cholesterol is involved in many functions of the membrane. It is well established that cholesterol decreases the membrane fluidity index, with consequences on the activity of ion channels, and receptor functions and is involved in dopamine release. Moreover, cholesterol is a key molecule in the end product of the CRF–ACTH axis. Steroids are derivatives of cholesterol and it is, therefore, interesting that various fatty acids have differential effects on cholesterol metabolism.

Huang et al. (1993) cite many studies that demonstrate that n-6 fatty acid are able to reduce the level of cholesterol in the blood serum. The data of Horrocks and Harder (1993) showed that n-6 fatty acids and n-3 fatty acids differ in their mode of action in cholesterol reduction, such that n-6 fatty acids redistribute cholesterol and n-3 fatty acid actually reduces the level of cholesterol. Davis (1992) demonstrated that n-3 essential fatty acids are more effective in reducing cholesterol levels in macrophages than n-6 essential fatty acids, most probably by the differential effect on the enzyme acyl-CoA (cholesterol acyltransferase). However, Horrocks and Harder (1983) indicated that cholesterol-esterifying enzymes that incorporate free fatty acids into cholesterol esters without the participation of CoA are also present in the rat brain.

The mechanism by which n-3 fatty acids are able to reduce the cholesterol level is still unclear, although some hypotheses have been advanced. For example, Bourre (1991) claimed that α-linolenic acid controls the composition of nerve membranes. Salem (1995) proposed that docosahexaenoic acid [DHA or (22:6n-3)] controls the composition and functions of the neuronal membrane. We recently reviewed a number of studies that provided support for a preferred n-3/n-6 ratio of 1 : 4 (Yehuda, et al., 1993). It is possible that such a ratio optimizes uptake into the brain and its eventual incorporation into the neuronal membranes.

3.5. Specific Fatty Acid and the Ratio Among Various Fatty Acids

Various fatty acids have a different role in the nervous system and in the body. It has been suggested that the nervous system has an absolute molecular configuration requirement for proper functioning (Salem, 1995). Studies in our laboratory offer confirmation for this suggestion and provide an added qualifying requirement, namely the need for a proper ratio between the essential fatty acids. We experimentally tested our hypothesis that the ratio of n-3 and n-6 may be a key factor in modulating behavioral and neuropharmacological effects of polyunsaturated fatty acid, and we attempted to identify the optimal ratio (Yehuda, et al., 1993). To avoid changes in the composition of fatty acids in commercially available oils and in order to exclude the confounding effects of other fatty acids or lipids, we used highly purified α-linolenic and linoleic acids. We tested a wide range of ratios of α-linolenic/linoleic acid including 1 : 3, 1 : 3.5, 1 : 4, 1 : 4.5, 1 : 5, 1 : 5.5, 1 : 6 (vol/vol). We found that a mixture of α- linolenic and linoleic acids with a ratio of 1:4 (which we referred to as SR-3) was the most effective in improving learning performance (as assessed by the Morris water maze), elevating pain threshold, improving sleep, and improving thermoregulation (Yehuda, et al., 1986, 1997, 1998). The mixture with this ratio was also able to alleviate learning deficits induced by the neurotoxins AF64A and 5,7-dihydroxytryptamine (Yehuda, et al., 1995), and provided protection from seizures induced by PTZ (Yehuda, et al., 1994). In addition, a preliminary study showed that SR-3 administration exerts beneficial effects in rats given diluted doses of the experimental allergic encephalomyelitis (EAE) toxin. The EAE rats showed learning and motor deficits as well as major changes in the fatty acids profile and the cholesterol level in frontal cortex synaptosomes. The SR-3 treatment was able to rehabilitate the changes induced by EAE to a significant degree, yet not completely to the normal level of the controls (unpublished observation).

The importance of the differentiation among the various types of fatty acids may be appreciated by noting the effects of immunological factors on peptides: namely n-3 fatty acids suppress the synthesis of interleukin-1 and interleukin-6, and enhance the synthesis of interleukin 2, whereas n-6 fatty acids have the opposite effect. It should be recalled that interleukin-1 and interleukin-6 promote CRF release via arachidonic acid (Cambromero, et al., 1997; Canonico, et al., 1996; Lyso, et al., 1992; Rivier, 1995). However, CRF inhibits the stimulating effect of interleukin-1 on prostaglandin synthesis (Fleisher-Berkovich and Danon, 1995; Oka, et al., 1993). Interleukin-2 is also capable of releasing CRF, but to a lesser degree (Karanth, et al., 1995; Kerttula, et al., 1994).

The study of the fatty acid–peptide interaction is not sufficiently developed to permit a definitive differentiation among the effects of various fatty acids. However, it appears that n-3 and n-6 fatty acids are the most active fatty acids in fatty acid–peptide interactions.

3.6. Peptide Interaction with P450, Prostaglandins, Cholesterol, and Fatty Acids

In light of the above-reviewed studies, it seems that the various fatty acids and lipids play a major role in the syntheses, release, and function of several peptides, mainly the peptides that are concerned with releasing factors. For example, prostaglandins are derivatives of fatty acids, but their metabolism is modulated by P450, a collective name for a group of enzymes that convert cholesterol to steroids. In addition to the effect on steroids, P450 is involved in the metabolism of neurotransmitters (including dopamine), CRF, and ACTH. These enzymes are crucial for the production of stress-related molecules (Makita et al., 1996). The interaction between prostaglandins and ACTH is not yet clear. Cholesterol is a major molecule in the membrane, and an elevated level of cholesterol results in a decrease in the membrane fluidity and in a disturbance of the membrane function. In addition, steroids are derivatives of cholesterol. Many studies have demonstrated the role of specific fatty acids (mainly essential fatty acids of n-3 and n-6 groups) on all aspects of synthesis, release, and receptor functions of dopamine, CRF, ACTH, and steroids.

3.7. EFA Ratio and Stress

Our own study examined the effects of a mixture of fatty acids on cortisol and cholesterol levels in laboratory stress situations. (Yehuda, al el, 2000). A compound of free nonesterified unsaturated fatty acids α-linolenic and linolenic acids in a ratio 1 : 4 was administered for 3 wk prior to injection of cortisone (10 mg/kg) or prior to immersion of rats in a 10°C saline bath. The results confirmed the expected elevation of cortisol and cholesterol level in stress, but, more importantly, the treatment prevented elevation of blood levels of cortisol and cholesterol and protected against deficits in the Morris water maze learning that usually accompany such stressful conditions. Differences from controls on all behavioral and biochemical measures were statistically significant ($p < 0.05$). It is proposed that induction of intense stress and its associated increase in cortisol, cholesterol, and other corticosteroids may damage hippocampal structures and may help account for cognitive decline witnessed in Alzheimer's disease and other age-related phenomena. The modulation of these consequences by the fatty acid mixture may provide an alternative strategy for the study of stress markers and for the development of other intervention options in humans.

3.8. The Immune System and Stress

The immune system is the collection of organs, tissues, and cells responsible for the organism to resist attack by antigens or invasive foreign bodies, particularly microbes. In light of the research of the past few decades, the classic definition and conceptualization of the immune system has changed. The immune system, once considered a "closed system" (i.e., it reacts only to internal body events), is now recognized to be "open" and subject to activation by the nervous system. There are established strong relationships between the components of the immune system and behavior with reciprocal influences on each other. The recognition of these interactions gave rise to the label "psychoneuroimmunology" (PNI), a term probably first introduced by (Solomon, 1989). The popularity of PNI is easily attributable to the classic studies and writings of Ader and his associates (Ader et al., 1987) which have provided an impressive body of evidence that the nervous system is capable of modulating the immune response. Molecular research from many disciplines have confirmed that receptors for neuromodulators and neurohormones may be found on human T-lymphocytes and that activation of these receptors can be stimulatory or inhibitory, depending on the neuroactive substance. The immune system may be able to communicate with the nervous system using neuromodulators and neurohormones secreted by lymphocytes.

3.9. Psychological Stress and the Immune System

Early studies in the field examined the effects of stress on general components of the immune system, such as T-cells and natural-killer (NK) cells. More recent studies chose to measure another component of the immune system—the cytokines, a system that is composed of molecules called interleukin. They are a group of naturally occurring proteins that are important in the activation of lymphocytes of the immune system. They were discovered in the 1970s, and several known types of interleukin (IL) are recognized as crucial constituents of the body's immune system. The most studied interleukins are interleukin-1 (IL-1) and interleukin-2 (IL-2). IL-1 is considered a proinflammatory and a pyrogenic agent, whereas IL-2 is considered an anti-inflammatory agent.

Recent studies showed that stress is able to cause an increase in the level of norepinephrine, CRF, ACTH, cortisol (the "stress" hormones), IL-1, IL-2, and natural killer cells, and an increase in the level of PGE_2 (Chavali, et al., 1998; Van Doornrn, et al., 1998). On the other hand, IL-2 induces an increase in CRF, ACTH, and cortisol (Denicoff, et al., 1989; Hanisch, et al; 1994, Karanath, et al., 1995). Stress and IL-2 induces an increase of β-endorphin level (the natural painkiller opiate peptide).

The EFA has a differential effect on the production and the activity of ILs. For example, a diet based on n-3 PUFA abolishes the anorexia response to IL-1 (Endres, 1997, Calder, 1997, DePablo et al., 2000). DHA (n-3) administration causes a decrease in IL-1, IL-2, and TNF-α (Mohan and Das, 1997). The complex relationships between the endocrine and the immune system mediating stress is diagrammed in Fig. 1.

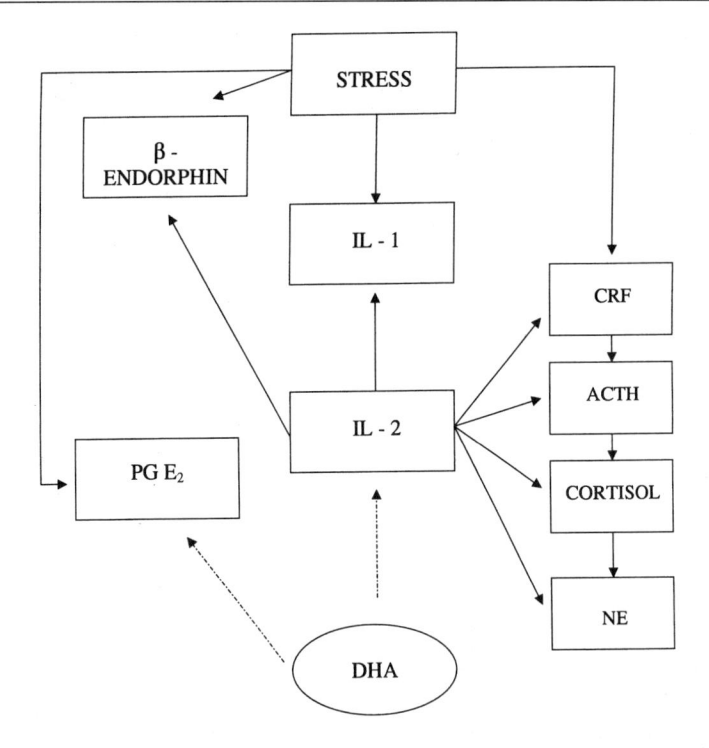

Fig. 1. Possible effects of EFAs in the stress situation on endocrine, immunological, and opiate molecules. Solid line indicates a stimulatory effect; broken line indicates an inhibitory effect.

Other relationships between fatty acids and cytokines and stress have been found. Recently, Zimmer et al. (2000) found that a chronic n-3-PUFA-deficient diet induced a marked decrease in dopamine vesicle density in the rat frontal cortex. This is an interesting finding because dopamine modifies the cell-mediated immune-response and mediates stress responses in mice (Inglis and, Konstandi, et al., 2000; Moghaddam, 1999; Nizoguchi, et al., 2000; Wu, et al., 1999).

More direct connections have been demonstrated by Meas (et at, 2000). Previous studies showed that psychological stress induces the production of proinflammatory cytokines, such as interferon (IFN), TNF-α, and IL-6 (Feng et al., 1990, Hankenson et al., 2000). Responses to stress are mediated by blood serum fatty acids. Subjects with a lower serum n-3 PUFA level (or with a higher n-6/n-3 ratio) experienced a stronger stress response and their IFN and TNF-α levels were higher, compared to subjects with higher serum n-3 levels or with lower n-6/n-3 ratios. These findings suggest that an imbalance in n-6/n-3 PUFA predisposes subjects to overreact in situations that pose psychological stress (Maes, 1997).

The cytokines IL-1 and IL-6 inhibit fatty acids oxidation; IL-1 is somnogenic and pyrogenic (Broughton and Morgan, 1994); IL-2 acts as a stimulant; and IL-6 is pyrogenic but not somnogenic (Opp et al. 1989). PUFAs that belong to the n-3 group strongly influence the production and activity of cytokines in that they inhibit the production and activity of IL-1 and IL-6 and stimulate the production and activity of IL-2 (De-Caterina et al. 1994). Diets containing vegetable oil (mainly n-6 FAs) have significantly different effects on the cytokine system than diets with fish oil that contain mainly n-3 FAs (Alexander, 1995).

4. PAIN AND EFA

"Pain" can be described as a complex experience consisting of a physiological response to a noxious stimulus followed by an affective (emotional) response to that event. Pain, in the context of a homeostatic mechanism, is a warning signal that helps to protect an organism by influencing it to withdraw from harmful stimuli; it is primarily associatez with injury, or the threat of injury, to the body tissue.

Pain is not an emotion, but painful sensations that can undoubtedly elicit stress. Like emotion, pain usually energizes the organism into action. Just as fear prepares for fight or flight, pain signals to do something in order to break contact with the potentially damaging agent and to begin restoring the injured tissue. On the one hand, many pain situations are associated with inflammation. The first event after injury to a tissue that occurs is a narrowing of the nearby arterioles (i.e., the tiny blood vessels that carry blood from the larger arteries to the network of microscopic vessels called capillaries). This constriction soon wears off, however, and is replaced by dilation, or widening, of the arterioles, and the resulting increase in blood flow flushes the entire network of capillaries with blood. White blood cells (leukocytes) then leave their normal position in the center of the arterial bloodstream and adhere to the walls of the blood vessels. Somehow, these leukocytes then penetrate the walls of the vessels, pass through to the exterior, and come to rest in the injured tissues near the vessel walls. The measurement and dynamics of pain, however, is not reducible to the immune system alone, which, although responsible for the inflammatory process, is complicated by added variables, such as stress.

The study of the relationships between pain situation and fatty acids is not well developed. However, several studies indicate a strong relationship between fatty acids and pain. Menstrual pain in Danish women correlated with low n-3 PUFA intake (Deutch, 1995). More than 50 scientific papers recommended various n-3 fatty acids to treat rheumatoid arthritis (RA) (e.g., Fagarasen, et al., 1991; Nordstrom, et al., 1995; Rothman, et al., 1995; Volker, et al. 1996). Recently, it has been reported that oleamide (a new potent fatty acid molecule) resulted in an elevation of the pain threshold (Tannenbaum et al., 1997). The clinical and biological implications of such relationships is worth examining a bit closer.

Fatty acid involvement with pain has been reported often, with DHA and other n-3 EFA having been proposed for various pain syndromes, notably for RA. It must be recalled that prostaglandins play a major role in pain, with PGE_2 able to increase reported pain (Eguchi, et al., 1999; Minami, et al., 1996, 1997, 1999; Sanyal, et al., 1990), whereas PGD_2 decreases pain (Eguchi, et al., 1999; Hosoi, et al., 1999). Similarly, β-endorphin (the brain painkiller) induces a decrease in the level and function of PGE_2, (Hawranko and Smith, 1999), and PGE_2 upregulates β-endorphin activity.

The immune system plays a major role in pain. Under pain situations, IL-1 induces a decrease in the level of CRF, cortisol, and β-endorphin (Bessler, et al., 1996; Parsadaniatz, et al., 1997). CRF alone releases β-endorphin. IL-1 alone has the similar effect—releasing β-endorphin. n-3 Fatty acids protected the rats from opiate side effects (Frances, et al., 1996, Hargreaves et al., 1990, Jones et al., 1999). IL-1, on the other hand, increases the level of PGE_2 (Hori, et al., 1998) to increase pain sensation.

A simplified diagram of the complex relationships between the molecules is given in Fig. 2.

Pain increases the levels of CRF, IL-1, IL-6, TNF-α, and PGE_2 (Cabot, et al., 1997; Hori, et al., 1998; Shafer, et al., 1996). TNF increases pain (Ignatouwski, et al., 1999;

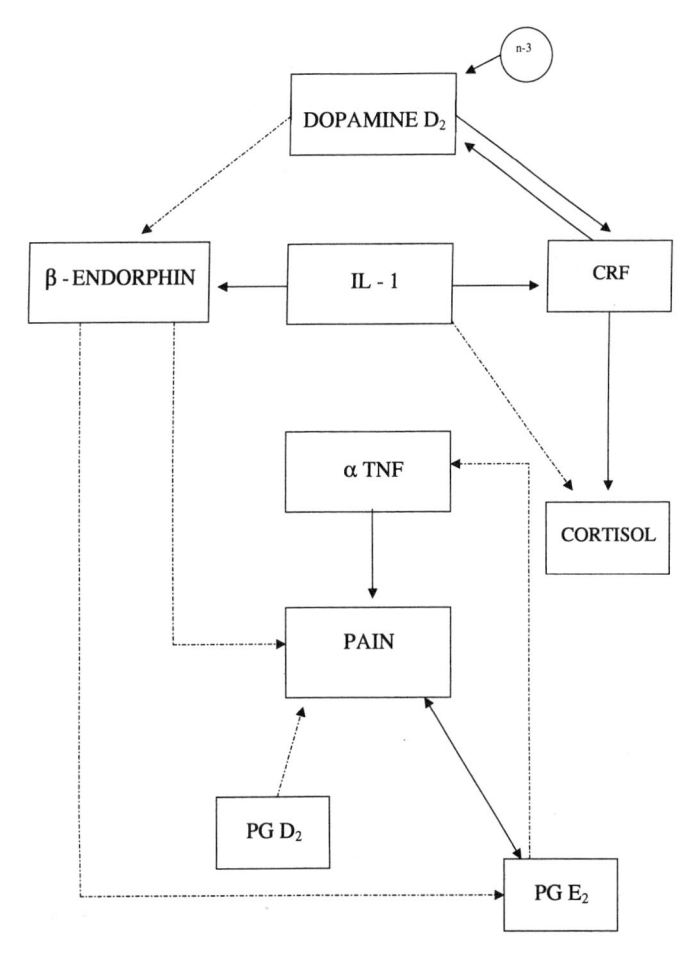

Fig. 2. Possible effects of EFAs in the pain situation on endocrine, immunological, and opiate molecules. Solid line indicates a stimulatory effect; broken line indicates an inhibitory effect.

Junger and Sorkin, 2000; Wagner and Myers, 1996). PGE$_2$ increases the production of TNF (Hori, et al., et al., 1998). The keen interest in TNF-α is explained by the fact that this molecule is a key factor in RA and other immunological disorders. TNF, a naturally occurring protein that is produced in the human body by the phagocyte cells known as macrophages, has also been found to play a much broader and more positive role in regulating inflammatory and immune responses throughout the body. Recent studies showed that n-3 and n-6 fatty acids had a differential effect on TNF-α (Chavali, et al., 1998; Raina, et al., 1995; Venkatraman and Chu, 1998).

4.1. n-3 and n-6 FAs and Pain Threshold and Pain Tolerance

Animal studies showed that the FA mixture has an analgesic effect on rats tested in hot plate and tail flick (Yehuda. 1993). Our unpublished results showed that a mixture of n-3/n-6 fatty acids is able to reduce the amount of nonsteroidal anit-inflammatory drug medication in the RA group. An unpublished study with human subjects showed that the FA mixture is able to increase the pain threshold of various pains (e.g., ischemic, thermal, etc.) but not the pain tolerance of all types of pains. These results may indicate that the "analgesic" effects of the FA mixture are mediated via the opiate system.

5. STRESS-INDUCED ANALGESIA

Stress-induced analgesia (SIA) is the phenomenon that certain stressors can induce changes in the pain threshold following the stress situation. The biological basis of SIA is not clear. Some studies showed that SIA is mediated by the opiate system (Altier and Stewart, 1996; Starec, et al., 1006; Vaccarino, et al., 1999; Yamada, et al., 1995), by β-endorphin (Hawanko, et al., 1994; Herz, 1995, Nakagawassi et al., 1999), substance P (Altier and Stewart, 1998), corticosterone (Filaretou, et al., 1996), or by CRF and cytokines (Laiviere and Melzack, 2000). Other studies found that the effect of SIA gradually decreases over aging (Laiviere and Melzack, 2000), implicating the hippocampus in SIA. Fish oil protects from SIA (Sutton, et al., 1994), and stressors that induced SIA also induced an increase in cholesterol level(Bernnan, et al., 1992). It is of interest to note that PTSD (posttraumatic stress disorder)-induced analgesia was shown to be mediated by the opiate system, but not by cortisol (Baker, et al., 1997, Pitman, et al., 1990). It has also been found that the level of cholesterol that is elevated in PTSD Vietnam veterans (Kagan, 1990) reflects the long term effects of the stress situation.

6. EPILOGUE

An understanding of the pervasive effects of PUFA on the complex interactions among the nervous, endocrine, and immune systems is not yet complete. However, it is already possible to speculate on the directions that much of the future research and clinical developments will pursue in these areas. Not least among such concerns are the needs to clarify the role of the individual fatty acids, to properly explain the profile of the fatty acids in both the brain and serum, and, in particular, to delineate the importance and functional consequences of the n-6/n-3 ratio. Perhaps most important is to recognize the *mediating* function that PUFAs provide and, thereby, to advance basic research in both behavior and the life sciences, and to develop clinical applications for the treatment and prevention of a wide range of chronic disorders.

REFERENCES

Ader R, Cohen N, Felten DL. Brain, behavior and immunity. Brain, Behav Immun 1987; 1:1–6.
Alexander JW. Specific nutrients and the immune response. Nutrition 1995; 11:229–232.
Altier N, Stewart J. Opiate receptors in the ventral tegmental are contribute to stress-induced analgesia in formalin test for tonic pain. Brain Res 1996; 718:203–206.
Altier N, Stewart J. Dopamine receptor antagonists in the nucleus accumbens attenuate analgesia induced by ventral tegmental area substance P or morphine and by nucleus accumbens amphetamine. J Pharmacol Exp Ther 1998; 285:208–215.
Arbogast BW, Neumann, J.K, Arbogast LY, Leeper SC, et al. Transient loss of serum protective activity following short-term stress: A possible biochemical link between stress and atherosclerosis. J Psychosom Res 1994; 38:871–884.
Back KW, Bogdanoff MD. Plasma lipid response to leadership, conformity and deviation. In: Leiderman PH, Shapiro D, eds. Psychological Approach to Aocial Behavior. Tavistock, London, 1964.
Bagger JP, Botker HE, Thomassen A, Nielsen TT. Effects of ranolizine on ischemic threshold, coronary sinus blood flow, and myocardial metabolism in coronary artery disease. Cardiovasc Drugs Ther 1997; 11:479–484.

Baker DG, West SA, Orth DN, Hill KK, Nicholson WE, Ekhator NN, Bruce AB, Wortman MD, Keck PE, Geracioti TD. Cerebrospinal fluid and plasma beta-endorphin in combat veterans with post-traumatic stress disorder. Pschoneuroendocrinology, 22:517–529.

Behan DP, Grigoriadis DE, Lovenberg T, Chalmers D, Heinrichs S, Liaw C, et al. Neurobiology of corticotropin releasing factor (CRF) receptors and CRF-binding protein: implications for the treatment of CNS disorders. Mol Psychiatry 1996; 1:265–277.

Belzung C, Leguisquet AM, Barreau S, Doelion-Vacassal S, Chalon S. Alpha-linoleic acid deficiency modifies distractibility but not anxiety and locomotion in rats during aging. J Nutr 1998; 128:1537–1542.

Bernnan FX, Job RS, Watkins LR, Maier SF. Total plasma cholesterol levels of rats are increased following only three sessions of tailshock. Life Sci 1992; 50:945–950.

Bessler H, Sztein MB, Serrate SA. Beta-endorphin modulation of IL-1-induced IL-2 production. Immunopharmacology 1990; 19:5–14.

Blok WL, Deslypere JP, Demacker PN, van der an Jongekrijg J, Hectors MP, van der Meer JW, et al. Pro- and anti-inflammatory cytokines in healthy volunteers fed various doses of fish oil for 1 year. Eur J Clin Invest 1997; 27:1003–1008.

Bourre JM, Bonnet M, Halbone G, Lafont H, Pascal G, Durand G. Essentiality of w3 fatty acids for brain structure and function. World Rev Nutr 1991; 66:103–117.

Brennan FX, Cobb CL, Silbert LE, Watkins LR, Maier SF. Peripheral beta-adrenoreceptors and stress-induced hypercholesterolemia in rats. Physiol Behav 1996; 60:1307–1310.

Broughton KS, Morgan LJ. Frequency of (n-3) polyunsaturated fatty acid consumption induces alterations in tissue lipid composition and eicosanoid synthesis in CD-1 mice. J Nutr 1994; 124:1104–1111.

Cabot PJ, Carter L, Gaiddon C, Zhang Q, Schefer M, Loeffler JP, et al. Immune cell-derived beta-endorphin: Production, release, and control of inflammatory pain in rats. J Clin Invest 1997; 100:142–148.

Calder PC. N-3 polyunsaturated fatty acids and immune cell function. Adv Enzyme Regul 1997; 37:197–237.

Cambromero JC, Rivas FJ, Borrell J, Guaza C. Role of arachidonic acid metabolism on corticotropin-releasing factor (CRF)-release induced by interleukin-1 from superfused rat hypothalami. J Neuroimmunol 1992; 39:57–66.

Canonico PL, Speciale C, Sortino MA, Cronin MJ, MacLeo RM, Scapagnini U. Growth hormone releasing factor (GRF) increases free arachidonate levels in the pituitary; a role for lipoxygenase products. Life Sci 1986; 38:267–272.

Catalan RE, Martinez AM, Aragones MD, Robles A, Miguel BG. Substance P induces alterations on cerebral lipids involved in membrane fluidity. Biochem Biophys Res Commun 1987; 144:232–237.

Caughey GE, Mantzioris E, Gibson RA, Cleland LG, James MJ. The effect on human tumor necrosis factor alpha and interleukin-1 beta production of diets enriched in n-3 fatty acids from vegetable oil or fish oil. Am J Clin Nutr 1996; 63:116–122.

Chavali SR, Zhong WW, Forse RA. Dietary alpha-linolenic acid increases TNF-alpha, and decreases IL-6, IL-10 in response to LPS: effects of sesamin on the delta-5 desaturation of omega-6 and omega-3 fatty acids in mice. Prostaglandins Leukotrienes Essential Fatty Acids 1998; 58:185–191.

Chavali SR, Forse RA. Decreased production of interleukin-6 and prostaglandins E2 associated with inhibition of delta-5 desaturation of omega-6 fatty acids in mice fed safflower oil diets supplemented with sesamol. Prostaglandins Leukotrienes Essential Fatty Acids 1999; 61:347–352.

Clark VR, Moore CL, Adams JH. Cholesterol concentration and cardiovascular reactivity to stress in African American college volunteers. J Behav Med 1998; 21:505–515.

Davis P.J. n-3 and n-6 polyunsaturated fatty acids have different effects on acyl-CoA:cholesterol acyltransfered in fatty macrophages. Biochem Cell Biol 1992; 70:1313–1318.

De-Caterina R, Cybulsky MI, Clinton SK. Gimbrone MA, Libby P. The omega-3 fatty acid decosahexaenoate reduces cytokine-induced expression of proatherogenic and proinflammatory proteins im human endothial cell. Arteriosclerosis Thromb Vasc Biol 1994; 14(11):1829–1836.

De Pablo MA, De Cienfuegos GA. Modulatory effects of dietary lipids on immune system functions. Immunol Cell Biol 2000; 78:31–39.

Denicoff KD, Durkin TM, Lozte MT, Quinlan PE, Davis CL, Listwak SJ, et al. The neuroendocrine effects of interleukin-2 treatment. J Clin Endocrinol Metab 1989; 69(2):402–410.

Deutch B. Menstrual pain in Danish women correlated with low n-3 polyunsaturated fatty acid intake. Eur J Clin Nutr 1995; 49:508–516.

Dunn AJ. Psychoneuroimmunology for the psychoneuroendocrinologist. A review of animal studies of nervous system-immune system interactions. Psychoneuroendocrinology 1989; 14:251–274.

Dunn AJ. Interactions between the nervous system and the immune system: implications for psychopharmacology. In: Bloom FE, Kupfer DJ, eds. Psychopharmacology: The Fourth Generation of Progress. Raven, New York, 1995, pp. 719–731.

Eguchi N, Minami T, Shirafuji N, Kanaoka Y, Tanaka T, Nagata A, et al. Lack of tactile pain (allodynia) in lipocalin-type prostaglandin D synthase-deficient mice. Proc Natl Acad Sci USA 1999; 96:726–730.

Endres, S. n-3 Polyunsaturated fatty acids and human cytokines synthesis. Lipids 1997; 31:5239–5242.

Fadda P, Martellotta MC, Gessa GE, Fratta W. Dopamine and apodes interactions in sleep deprivation. Prog Neuropsychopharmacol Biol Psychiatry 1993; 17:269–278.

Fadda P, Martellotta MC, De Montis MG, Gessa GE, Fratta W. Dopamine D1 and opioid receptor binding changes in the limbic system of sleep deprived rats. Neurochem Int 1992; 20:153S–156S.

Fagarasan MO, Arora PK, Axelrod J. Interleukin-1 potentiation of beta-endorphin secretion and the dynamics of interleukin-1 internalization in pituitary cells. Prog Neuropsychopharmacol Biol Psychiatry 1991; 15:551–560.

Faletti A, Vagina JM, Gimeno MA. Beta-endorphin inhibits prostaglandin synthesis in rat ovaries and blocks induced ovulation. Prostaglandins 1995; 49:93–103.

Feng C, Keisler DH, Fritsche KL. Dietary omega-3 polyunsaturated fatty acids reduce IFN-gamma receptor expression in mice. J Interferon Cytokine Res 1999; 19:41–48.

Filaretov AA, Bogdanov AI, Yarushkina NL. Stress-induced analgesia. The role of hormones produced by the hypophyseal-endocortical system. Neurosci Behav Physiol 1996; 26:527–578.

Fleisher-Berkovich S, Danon A. Effect of corticotropin-releasing factor on prostaglandin synthesis in endothelial cells and fibroblasts. Endocrinology 1995; 136:4068–4072.

Frances H, Coudereau JP, Sandouk P, Clement M, Monier C, Bourre JM. Influence of dietary alpha-linolenic acid deficiency on learning in the Morris water maze and on the effects of morphine. Eur J Pharmacol 1996; 18:217–225.

Furukawa K, Tashiro T, Yamamori H, Takagi K, Morishima Y, Sugiura T, et al. Effects of soybean oil emulsion and eicosapentaenoic acid on stress response function after a severely stressful operation. Ann Surg 1999; 229:255–261.

Gelfand RA, Wepsic HT, Parker LN, Jadus MR. Prostaglandin E2 induces up-regulation of murine macrophage beta-endorphin receptors. Immunol Lett 1995; 45:143–148.

Grimble RF. Dietary lipids and the inflammatory response. Proc Nutr Soc 1998; 57:535–542.

Hamazaki T, Sawazaki S, Nagasawa T, Nagao Y, Kanagawa Y, Yazawa K. Administration of docosahexaenoic acid influence behavior and plasma catecholamine levels at times of psychological stress. Lipids 1999; 34:S33–S37.

Hamzaki T, Sawazaki S, Itomura M, Asaoka E, Nagao Y, Nishimura N, et al. The effect of docosahexaenoic acid on aggression in young adults: A placebo-controlled double-blind study. J Clin Invest 1996; 97(4):1129–1133.

Hanisch UK, Rowe W, Sharma S, Meaney MJ, Quirion R. Hypothalamic-pituitary-adrenal activity during chronic central administration of interleukin-2. Endocrinology 1994; 135(6):2465–2472.

Hankenson KD, Watkins BA, Schoenlein IA, Allen KG, Turek JJ. Omega-3 fatty acids enhance ligament fibroblast collagen formation in association with changes in interleukin-6 production. Proc Soc Exp Biol Med 2000; 223:88–95.

Hargreaves KM, Flores CM, Dionne RA, Mueller GP. The role of pituitary beta-endorphin in mediating corticotropin-releasing factor-induced antinociception. Am J Physiol 1990; 258(2):E235–E242.

Hargreaves K, Clandinin MT. Dietary lipids in relation to postnatal development of the brain. Upsala J Med Sci 1990; 48s:79–95.

Hartmann G, Endres S. n-3 Polyunsaturated fatty acids and human cytokine synthesis. In: Yehuda S, Mostofsky DI, eds. Handbook of Essential Fatty Acids Biology, Humana, Totowa, NJ, 1997, pp. 103–113.

Hawranko AA, Smith DJ. Stress reduces morphine's antinociceptive potency: dependence upon spinal cholecystokinin processes. Brain Res 1999; 824:251–257.

Hawranko AA, Monro PL, Smith DJ. Repetitive exposure to the hot-plate test produces stress induced analgesia and alters beta-endorphine neuronal transmission within the periaqueductal gray of the rat. Brain Res 1994; 667:283–286.

Herz A. Role of immune processes in peripheral opiate analgesia. Adv Exp Med Biol 1995; 373:193–199.

Hori T, Oka T, Hosoi M, Aou S. Pain modulatory actions of cytokines and prostaglandin E_2 in the brain. Ann NY Acad Sci 1998; 840:269–281.

Horrocks LA, Harder HW. Fatty acids and cholesterol. In: Lajtha A, ed. Handbook of Neurochemistry. Plenum, New York, 1983, pp. 1–16.

Hosoi M, Oka T, Abe M, Hori T, Yamamoto H, Mine K, et al. Prostaglandins E(2) has antinociceptive effect through EP (1) receptor in the ventromedial hypothalamus in rats. Pain 1999; 83:221–227.

Huang YS, Koba K, Horrobin DF, Sugano, M. Interrelationship between dietary protein, cholesterol and n-6 polyunsaturated fatty acid metabolism. Prog Lipid Res 1993; 32:123–137.

Hughes DA, Pinder AC. N-3 Polyunsaturated fatty acids modulate the expression of functionally associated molecules on human monocytes and inhibit antigen-presentation in vitro. Clin Exp Immunol 1997; 110:516–523.

Ignatowski TA, Covey WC, Knight PR, Severin CM, Nickola TJ, Spengler RN. Brain-derived TNF-alpha mediates neuropathic pain. Brain Res 1999; 841:70–77.

Inglis FM, Moghaddam B. Dopaminergic innervation of the amygdala is highly responsive to stress. J Neurochem 1999; 72:1088–1094.

James MJ, Gibson RA, Cleland LG. Dietary polyunsaturated fatty acids and inflammatory mediator production. Am J Clin Nutr 2000; 71:343s–348s.

Jones DR, Pettitt TR, Sanjuan MA, Mrida I, Wakelam MJ. Interleukin-2 causes an increase in saturated/monounsaturated phosphatidic acid derived from 1,2-diacylglycerol and 1-o-alkyl-2-acyglycerol. J Biol Chem 1999; 274:16846–16852.

Joo F. Brain microvascular cyclic nucleotides and protein phosphorylation. In: Pardridge WM, ed. Blood Brain Barrier: Cellular and Molecular Biology. Raven, New York, 1993, pp. 267–287.

Jorgensen C, Sany J. Modulation of the immune response by the neuro-endocrine axis in rheumatoid arthritis. Clin Exp Rheumatol 1994; 12:435–551.

Junger H, Sorkin LS. Nociceptive and inflammatory effects of subcutaneous. Pain 2000; 85:145–151.

Kagan BL, Lesskin G, Hans B, Willkins J, Foy D. Elevated lipid levels in Vietnam veterans with chronic posttramutic stress disorder. Biol Psychiatry 1990; 45:374–377.

Karanath S, Lyson K, Aguila MC, McCann SM. Effects of lutenizing-hormone-releasing hormone, alpha-melanocyte-stimulating hormone, naloxon, dexamethasone and indomethacin on interleukin-2-induced corticotropin-releasing factor release. Neuroimmunomodulation 1995; 2(3):166–173.

Kerttula T, Kaukinen S, Riutta A, Seppala F, Mucha I, Vapaatalo H, Alanko J. Effects of noradrenaline and dopamine infusions on arachidonic acid metabolism in man. Thromb Res 1995; 80:169–178.

Kiecolt-Glaser JK, Cacioppo JT, Malarkey WB, Glaser R. Acute psychological stressors and short term immune changes: What, why, for whom and to what extent? Psychosom Med 1992; 54:680–685.

Konstandi M, Johnson E, Lang MA, Malamas M, Marselos M. Norarenaline, dopamine, serotonin: different effects of psychological stress on brain biogenic amines in mice and rats. Pharmacol Res 2000; 41:341–346.

Kremer JM. N-3 fatty acid supplements in rheumatoid arthritis. Am J Clin Nutr 2000; 71:349s–351s.

Lacroix S, Rivest S. Role of cyclo-oxygenase pathways in the stimulatory influence of immune challenge on the transcription of a specific CRF receptor subtype in the rat brain. J Chem Neuroanat 1996; 10:53–71.

Laiviere WR, Melzack R. The role of corticotropin-releasing factor in pain and analgesia. Pain 2000; 84:1–12.

Lyson K, McCann S. Involvement of arachidonic acid cascade in interleukin-6-stimulated corticotropin-releasing factor release in vitro. Neuroendocrinology 1992; 50:708–709.

Maes M. The immunoregulatory effects of antidepressants. Hum Psychophar Clin Exp 2001; 16(1):95–103.

Maes M, Christophe A, Eugene, Bosmas E, Lin A, Neels H. In humans, serum polyunsaturated fatty acid levels predict the response of proinflammatory cytokines to psychological stress. Biol Psychiatry 2000; 47(10):910–920.

Makita K, Falck JR, Capdevile JH. Cytochrome P450: the arachidonic acid cascade, and hypertension: new vistas for an old enzyme system. FASEB J 1996; 10:1456–1463.

Matsmoto K, Yobimoto K, Huong NT, Abdel-Fattah M, Van Hien T, Watanabe H. Psychological stress-induced enhancement of brain lipid peroxidation via nitric oxide systems and its modulation by anxiolytic and anxiogenic drugs in mice. Brain Res 1999; 839:74–84.

McGahon B, Murray CA, Clements MP, Lynch MA. Analysis of the effect of membrane arachidonic acid concentration on modulation of glutamate release by interleukin-1: an age-related study. Exp Gerontol 1998; 33:343–354.

Meas et al., 2000.

Milles DE, Huang YS, Narce M, Poisson JP. Psychosocial stress catecholamines, and essential fatty acid metabolism in rats. Proc Soc Exp Biol Med 1994; 205(1):56–61.

Minami T, Okuda-Ashitaka E, Hori Y, Sakuma S, Sugimoto T, Sakimura K, et al. Involvement of primary afferent C-fibres in touch-evoked pain (allodynia) induced by prostaglandin E2. Eur J Neurosci 1999; 11:1849–1856.

Minami T, Okuda-Ashitaka E, Nishizawa M, Mori H, Ito S. Inhibition of nociceptin-induced allodynia in conscious mice by prostaglandin D2. Br J Pharmacol 1997; 122. 605–610.

Minami T, Okuda-Ashitaka E, Mori H, Ito S, Hayaishi O. Prostaglandin D2 inhibits prostaglandin E2-induced allodynia in conscious mice. J Pharmacol Exp Ther 1996; 278:1146–1152.

Mizoguchi K, Yuzurihara M, Ishige A, Sasaki H, Chubi DH, Tabira T. Chronic stress induces impairment of spatial working memory because of prefrontal dopaminergic dysfunction. J Neurosci 2000; 20:1568–1574.

Mohan IK, Das UN. Oxidant stress, anti-oxidants and essential fatty acids in systemic lupus erythematosus. Prostaglandins Leukotrienes Essential Fatty Acids 1997; 56:193–198.

Morrow LE, McClellan JL, Klir JJ, Kluger MJ. The CNS site of glucocorticoid negative feedback during LPS and psychological stress induced fevers. Am J Psychol 1996; 271:732–737.

Moussa M, Le Boucher J, Garcia J, Tkaczuk J, Ragab J, Dutot G. et al. In vivo effects of olive oil-based lipid emulsion on lymphocyte activation in rats. Clin Nutr 2000; 19:49–54.

Murphy MG, Moak CM, Rao BG. Effects of membrane polyunsaturated fatty acids on opiate peptide inhibition of basal and prostaglandin E1-stimulated cyclic AMP formation in intact NIE-115 neurometabolism cells. Biochem Pharmacol 1987; 36:4079–4084.

Myers M, Freire E. Calorimetric and fluorescence characterization of interactions between enkephalins and liposomal and synaptic plasma membranes containing gangliosides. Biochemistry 1995; 24:4076–4082.

Nakagawassi O, Tadano T, Tan No K, Niijima F, Endo Y, Kisara K. Change in beta-endorphin and stress-induced analgesia in mice after exposure to forced waking stress. Methods Findings Exp Clin Pharmacol 1999; 21:471–476.

Nordstrom DC, Honkanen VE, Nasu Y, Antila E, Frimen C, Konttinen YT. Alpha-linoleic acid in the treatment of rheumatoid arthritis. A double-blind, placebo-controlled and randomized study: flaxseed vs. safflower seed. Rheumatol Int 1995; 14:231–234.

Ohno T, Ohinata H, Ogawa K, Kuroshima A. Fatty acid profiles of phospholipids in brown adipose tissue from rats during cold acclimation and repetitive intermittent immobilization: with special reference to docosahexaenoic acid. Jpn J Physiol 1996; 46:265–270.

Oka T, Aou S, Hori T. Intracerebroventriculaer injection of interleukin-1 beta induces hyperalgesia in rats. Brain Res 1993; 624:61–68.

Ongini E, Bonizzoni E, Ferri N, Milani S, Trampus M. Differential effects of dopamine D-1 and D-2 receptor antagonist antipsychotics on sleep–wake patterns in the rat. J Pharmacol Exp Ther 1993; 266:726–731.

Opp M, Obal, J.R, Cady AB, Johannsen L, Krueger JM. Interleukin-6 is pyrogenic but not somnogenic. Physiol Behav 1989; 45 1069–1072.

Panajotova V. The effect of dopaminergic agents on cell-mediated immune response in mice. Physiol Res 1997; 46:113–118.

Parsadaniantz SM, Batsch E, Gegout-Pottie P, Terlain B, Gillet P, Netter P, et al. Effects of continuous infusion of interleukin-1 beta on corticotropin-releasing hormone (CRH) receptors, proopiomelanocortin gene expression and secretion of corticotropin, beta-endorphin and corticosterone. Neuroendocrinology 1997; 65:53–63.

Peterson LD, Jeffery NM, Thies F, Sanderson P, Newsholme EA, Calder PC. Eicosapentaenoic and docosahexaenoic acids alter rat spleen leukocyte fatty acid composition and prostaglandins E_2 production but have different effects on lymphocyte functions and cell-mediated immunity. Lipids 1998; 33:171–180.

Peterson LD, Thies F, Calder PC. Dose-dependent effects of dietary gamma-linolenic acid on rat spleen lymphocyte functions. Prostaglandins Leukotriens Essential Fatty Acids 1999; 61:19–24.

Petraglia F, Benedetto C, Florio P, D'Ambrogio G, Genazzani AD, Marozio L, et al. Effect of corticotropin-releasing factor-binding protein on prostaglandin release from cultured maternal decide and on contractile activity of human myometrium in vitro. J Clin Endocrinol Metab 1995; 80:3073–3076.

Pitman RK, van der Kolk BA, Orr SP, Greenberg MS. Naloxon reversible analgesia: response to combat related stimuli in posttraumatic stress order. A pilot study. Arch Gen Psychiatry 1990; 47:541–545.

Raber J, Koob GF, Bloom FE. Interleukin-2 induces corticotropin-releasing factor (CRF) release from the amygdala and involves a nitric oxide-mediated signaling; comparison with the hypothalamic response. J Pharmacol Exp Ther 1995; 272:815–824.

Raina N, Matsui J, Cunnane SC, Jeejeebhoy KN. Effect of tumor necrosis factor-alpha on triglyceride and phospholipid content and composition of liver and carcass in rats. Lipids 1995; 30:713–718.

Rivier C. Influence of immune signals on the hypothalamic–pituitary axis of the rodent. Front Neuroendocrinol 1995; 16,151–182.

Rizzo MT, Carlo-Stella C. Arachidonic acid mediates interleukin-1 and tumor necrosis factor-alpha induced activation of the c-jun amino terminal kinases in stroma cells. Blood 1996; 10:3792–3800.

Rosenman RH. Do environmental effects on human emotions cause cardiovascular disorders? Acta Physiol Scand 1997; 640(Suppl):133–136.

Rossetti RG, Seiler CM, DeLuca P, Laposata M, Zurier RB. Oral administration of unsaturated fatty acids: Effects on human peripheral blood lymphocyte proliferation. Leukotrienes Biol 1997; 62(4):438–443.

Rothman D, Allen H, Herzog L, Pilapil A, Seiler CM, Zurier RB. Effects of unsaturated fatty acids on interleukin-1 beta production by human monocytes. Cytokine 1997; 9:1008–1012.

Rothman D, DeLuca P, Zurier RB. Botanical lipids: effects on inflammation, immune responses, and rheumatoid arthritis. Semin Arthritis Rheum 1995; 25:87–96.

Rouseau D, Moreau D, Raederstorff D, Sergiel, J.P, Muggli R, Grynberg A. Is a dietary n-3 fatty acid supplement able to influence the cardiac effect of the psychological stress? Mol Cell Biochem 1998; 178:353–366.

Salvati S, Cambegei LY, Sorcinni M, Olivieri A, DiBase A. Effects of propylthiouracial induced hypothyroidism on membranes of adult rat brain. Lipids 1993; 28:1075–1078.

Salem N Jr. The nervous system has an absolute molecular species requirement for proper function. Mol Memb Biol 1995; 12:131–134.

Sanyal AK, Srivastava DN, Bhattachayarya SK. The antinociceptive effect of intracerebroventricularly administered prostaglandin E-sub-1 in the rat. Psychopharmacology 1990; 60:159–163.

Sawazaki S, Hamazaki T, Yazawa K, Kobayashi M. The effect of docosahexaenoic acid on plasma catecholamine concentrations and glucose tolerance during long-lasting psychological stress: a double blind placebo-controlled study. J Nutr Neurosci Vitaminol 1999; 45:655–665.

Sazontova TG, Golantsova NE, Kolmykova SN, Arkhangel'Skaia MI, Koshatskaia IL, Arkhipenko. Effect of delta sleep-inducing peptide on the status of the calcium transport system of the sarcoplasmic reticular and activity of antioxidant defense enzymes in the myocardium. Bull Exp Biol Med 1996; 121:248–251.

Schafer M, Mousa SA, Zhang Q, Carter L, Stein C. Expression of corticotropin-releasing factor in inflamed tissue is required for intrinsic peripheral opiod analgesia. Proc Natl Acad Sci USA 1996; 93:6096–6100.

Schwyzer R. Estimated membrane structure and receptor subtype selection of an opioid alkaloid-peptide hybrid. Intl J Peptides Protein Res 1988; 32,476–483.

Shmalwko IP, Mikhaleva II. Antimetastatic effect of the delta-sleep peptide during stress in mice with Lewis lung carcinoma. Exp Oncol 1988; 10:57–60.

Shmalwko IP, Mikhaleva II. Antimetastatic effect of the delta-sleep peptide during stress in mice with Lewis' lung carcinoma. Exp Oncol 1988; 11:37–61.

Singer P, Richterheinrich E. Stress and fatty liver—possible indications for dietary long-chain n-3 fatty acids. Med Hypotheses 1991; 36:90–94.

Solomon GF. Psychoneuroimmunology and human immunodeficiency virus infection. Psychiatr Med 1989; 7(2):47–57.

Song C, Leonard BE. Fundamentals of Psychoneuroimmunology. Wiley, New York, 2000.

Starec M, Rosina J, Malek J, Krsjak M. Influence of dynorphin A (1-13) and dynorphin A (1-10) amide on stress-induced analgesia. Physiol Res 1996; 45:433–438.

Sutton LC, Fleshner M, Mazzeo R, Maiser SF, Watkins LR. A permissive role of corticosterone in an opioid form of stress-induced analgesia—blockade of opiate analgesia is not due to stress-induced hormone-release. Brain Res 1994; 663:19–29.

Tannenbaum BM, Brindley DN, Tannenbaum GS, Dallman MF, McArthur MD, Meaney MJ. High fat feeding alters both basal ans stress-induced hypothalamic–pituitary–adrenal activity in the rat. Am J Physiol 1997; 273:6 Pt 1 E1 168–177.

Thompson ABR, Keelan M, Clandinin MT. Feeding rats a diet enriched with saturated fatty acids presents the inhibitory effects of acute and chromic ethanol exposure on the in vitro uptake of hexoses and lipids. Biochem Biophys Acta 1991; 1084:122–128.

Ulak G, Cinek R, Sermet A, Guzel C, Ulak M, Denli O. Protective effect of fish oil against stress induced gastric injury in rats. Arzneimittelforschung 1995; 45:1174–1175 (in German).

Vaccarino AL, Olson GA, Olson RD, Kastin AJ. Endogenous opiates: 1998. Peptides 1999; 20:1527–1574.

Van Doornen LJ, Snieder H, Boomsma DI. Serum lipids and cardiovascular reactivity to stress. Biol Psychol 1998; 47:279–297.

Venkatraman JT, Chu WC. Effects of dietary omega-3 and omega-6 lipids and vitamin E on serum cytokines, lipid mediators and anti-DNA antibodies in a mouse model for rheumatoid arthritis. J Am Coll Nutr 1999; 18:602–613.

Volker D, Garg M. Dietary n-3 fatty acid supplementation in rheumatoid arthritis—mechanisms, clinical outcomes, controversies, and future directions. J Clin Biochem Nutr 1996; 20:83–97.

Wagner R, Myers RR. Endoneural injection of TNF-alpha produces neuropathic pain behaviors. NeuroReport 1996; 25:2897–2901.

Watanabe T, Clarck WG, Ceriani G, Lipton JM. Elevation of plasma ACTH concentration in rabbits made febrile by systemic injection of bacterial antitoxin. Brain Res 1994; 652:201–206.

Williams LY, Kiecolt Glaser JK, Horrocks LA, Hillhouse JT, Glaser R. Quantitative association between altered plasma esterfied omega-6 fatty-acid proportions and psychological stress. Prostaglandins Leukotrienes Essential Fatty Acids 1992; 47:165–170.

Wu D, Meydani SN, Meydani M, Hayek MG, Huth P, Nicolosi RJ. Immunologic effects of marine- and plant-derived n-3 polyunsaturated fatty acids in nonhuman primates. Am J Clin Nutr 1996; 63:273–280.

Wu YL, Yoshida M, Emoto H, Tanaka M. Psychological stress selectively increases extracellular dopamine in the "shell", but not in the "core" of the rat nucleus accumbens: a novel dual-needle probe simultaneous microdialysis study. Neurosci Lett 1999; 275:69–72.

Yamada K, Nabeshima T. Stress-induced behavioral responses and multiple opiod systems in the brain. Behav Brain Res 1995; 67:133–145.

Yamaguchi K, Hama H. Evaluation for roles of brain prostaglandins in the catecholamine-induced vasopressin secretion in conscious rats. Brain Res 1993; 607:149–153.

Yano M, Kishida E, Iwasaki M, Shosuke K, Masuzawa Y. Docosahexaenoic acid and vitamin E can reduce human monocytic U937 cell apoptosis induced by tumor necrosis factor. J Nutr 2000; 130:1095–1101.

Ye H, Wolf RA, Kurz T, Corr PB. Phosphatidic acid increases in response to noradrenaline and endothelin-1 in adult rabbit ventricular myocytes. Cardiovasc Res 1994; 28:1828–1834.

Yehuda S, Carraso RL. Modulation of learning pain thresholds, and thermoregulation in the rat by preparations of free-purified α-linolenic and linoleic acids: determination of the optimal w3-to-w6 ratio. Proc Natl Acad Sci 1993; 90:10,345–10,349.

Yehuda S, Carasso RL, Mostofsky DI. Essential fatty acid preparation (SR)-3 raises the seizure threshold in rats. Eur J Pharmacol 1994; 254:193–198.

Yehuda S, Carasso RL, Mostofsky DI. Essential fatty acid preparation (SR-3) rehabilitates learning deficits induced by AF64A and 5,7-DHT. NeuroReport 1995; 6:511–515.

Yehuda S, Leprohon-Greenwood CE, Dixon LM, Coscina DV. Effects of dietary fat on pain threshold, thermoregulation and motor activity in rats. Pharmacol Biochem Behav 1986; 24:1775–1777.

Yehuda S, Rabinovitz S, Carasso RL, Mostofsky DI. Fatty acids and brain peptides. Peptides 1998; 19:407–419.

Yehuda S, Rabinovitz S, Carasso RL, Mostofsky DI. Mixture of essential fatty acids rehabilitates stress effects on learning, and cortisol and cholesterol level. Intl J Neurosci 2000; 101:73–87.

Yehuda S, Rabinovitz S, Mostofsky DI. Effects of essential fatty acid preparation (SR-3) on brain lipids, biochemistry and behavioral and cognitive functions. In: Yehuda S, Mostofsky DI, eds. Handbook of Essential Fatty Acid Biology, Biochemistry Physiology, and Behavioral Neurobiology. Humana, Totowa, NJ, 1997, pp. 427–452.

Yehuda S, Rabinovitz S, Mostofsky DI. Essential fatty acids and sleep: mini review and hypothesis. Med Hypothesis 1998; 50:139–145.

Young VM, Toborek M, Yang F, McClain CJ, Hennig B. Effect of linoleic acid on endothelial cell inflammatory mediators. Metabolism 1998; 47:566–572.

Yukhananov R, Rebrov I, Tennila T, Maisky A. Effects of ethanol and DSIP on the 36Cl-flux in synaptosomal vesicles. Mol Chem Neuropathol 1991; 15:235–248.

Zimmer L, Delpal S, Guiloteau D, Aioun J, Durand G, Chalon S. Chronic n-3 polyunsaturated fatty acid deficiency alters dopamine vesicle density in the rat frontal cortex. Neurosci Lett 2000; 284:25–28.

GLOSSARY*

Alpha-linolenic acid (LNA or 18:3n-3): An 18-carbon, three double-bond fatty acid in the omega-3 family. This polyunsaturated fatty acid is produced in the chloroplast of terrestrial and marine plants. In animals, alpha-linolenic acid is metabolized into the longer-chain omega-3 fatty acids, eicosapentaenoic acid, and docosahexaenoic acid.

Arachidonic acid (AA or 20:4n-6): A 20-carbon, four double-bond fatty acid of the omega-6 family. This long-chain polyunsaturate is produced in animals through the metabolism of linoleic acid. It is considered an essential fatty acid, as it can support the physiological functions associated with EFAs when substituted for linoleic acid.

Chain elongation: A process that occurs during fatty acid metabolism which yields the addition of two carbon atoms to the chain on the carboxyl end of the molecule.

Desaturation: A process that occurs during fatty acid metabolism which involves removing hydrogen atoms from the carbon chain and the subsequent formation of a double bond.

Dihomo-gamma linolenic acid (DGLA or 20:3n-6): A 22-carbon, six double-bond fatty acid of the omega-3 family produced in animals via metabolism of alpha-linolenic acid. DHA is a major component of neural membranes and may be an essential fatty acid during infancy when the newborn's capacity to convert LNA appears limited. DHA is found in fish and fish oils.

Eicosanoids: Hormone-like compounds, including prostaglandins, thromboxanes, lipoxins, and leukotrienes, which are involved in many important biological processes in the human body (i.e., central nervous function, regulation of blood pressure, regulation of other hormones, inflammatory reactions, and immune response). They are produced mainly through the metabolism of arachidonic acid and eicosapentaenoic acid, although other fatty acids may also be similarly metabolized.

Eicosapentaenoic acid (EPA or 20:5n-3): A 20-carbon, five double-bond fatty acid of the omega-3 family produced in animals through the metabolism of alphalinolenic acid. One of the principal fat components of fish.

*With permission from Roche Vitamins, Inc. and Norman Salem, Jr., Introduction to polyunsaturated fatty acids. Backgrounder 1999; 3(1):7.

From: *Fatty Acids: Physiological and Behavioral Functions*
Edited by: D. Mostofsky, S. Yehuda, and N. Salem Jr. © Humana Press Inc., Totowa, NJ

Essential fatty acid (EFA): A fatty acid which can prevent a constellation of pathophysiological effects associated with fat deficiency, including impaired growth, dermatitis, transdermal water loss, impaired reproductive function, and susceptibility to bacterial infections.

Fat: A nutrient that is a major source of energy and plays a key role in the absorption of the fat-soluble vitamins A, D, E, and K. Fat components are the primary building blocks of cell membranes and are precursors to hormone-like compounds involved in a number of important physiological processes. Fats also provide foods with flavor and texture.

Fatty acid: A basic component in the chemical structure of fat composed of a hydrocarbon chain and an acid group. Fatty acids can be either saturated, monounsaturated, or polyunsaturated, depending on the number of double bonds. Commonly occurring mammalian fatty acids have from 12 to 24 carbons and from 0 to 6 double bonds.

Gamma-linolenic acid (GLA or 18:3n-6): An 18-carbon, three double-bond fatty acid of the omega-6 family produced in animals via the metabolism of linoleic acid.

Linoleic acid (LA or 18:2n-6): An 18-carbon, two double-bond fatty acid. It is the most predominant PUFA in the Western diet. It is found in mayonnaise, salad dressings, and in the seeds and oils of most plants, with the exception of coconut, cocoa, and palm. Linoleic acid is metabolized into longer-chain fatty acids, such as arachidonic acid and gamma-linolenic acid, in animals through a process of chain elongation and desaturations.

Monounsaturated fatty acids (MUFA): A fatty acid molecule that contains only one double bond along the carbon chain. Oleic acid (18:1n-9) is the most common monounsaturated fatty acid. All plants and animals, including humans, can synthesize oleic acid.

Omega-3 fatty acid: A polyunsaturated fatty acid whose first double bond is located three carbons from the methyl end of the fatty acid molecule. Alpha-linolenic acid is the precursor or "parent" compound from which other longer-chain omega-3 PUFA are synthesized, including EPA and DHA. Sources of alpha-linolenic acid include green leafy vegetables, linseed and rapeseed oils, as well as phytoplankton, algae, and fish. Scientists may also refer to this family of fatty acids as n-3 PUFA.

Polyunsaturated fatty acids (PUFA): A fatty acid molecule that contains two or more double bonds along the carbon chain. Two classes of PUFA exist: omega-6 and omega-3. Humans, like all mammals, are unable to totally synthesize omega-3 or omega-6 PUFA and must obtain them from dietary sources.

Saturated fatty acids: Fatty acids whose carbon chain contains as many hydrogen atoms as is chemically possible and thus contains no double bonds. The most common saturated fatty acids include myristic acid (14:0), palmitic acid (16:0), and stearic acid (18:0).

Unsaturated fatty acids: Fatty acids whose carbon atoms do not contain the maximum potential number of hydrogen atoms, resulting in the formation of at least one carbon–carbon double bond.

Abbreviations:

AA	Arachidonic acid
DHA	Docosahexaenoic acid
DGLA	Dihomo-gamma linolenic acid
EFA	Essential fatty acid
EPA	Eicosapentaenoic acid
GLA	Gamma-linolenic acid
LA	Linoleic acid
LNA	Alpha-linolenic acid
MUFA	Monounsaturated fatty acid
n-3 PURA	Omega-3 polyunsaturated fatty acid
n-6 PUFA	Omega-6 polyunsaturated fatty acid
PUFA	Polyunsaturated fatty acid

Principal Polyunsaturated Fatty Acids

Trivial name	Shorthand designation	Abbreviation
Omega-6 PUFA		
Linoleic acid	18:2n6	LA
Gamma-linolenic acid	18:3n6	GLA
Dihomo-gamme linolenic acid	20:3n6	DGLA
Arachidonic acid	20:4n6	AA
Docosatetraenoic acid	22:4n6	DTA
Docosapentaenoic acid	22:5n6	DPA (n-6)
Omega-3 PUFA		
Alpha-linolenic acid	18:3n3	LNA
Eicosapentaenoic acid	20:5n3	EPA
Docosapentaenoic acid	22:5n3	DPA (n-3)
Docosahexaenoic acid	22:6n3	DHA

Index

A

Acetylcholine, docosahexaenoic acid supplementation effects, 228
N-Acetylcysteine (NAC), diabetic neuropathy management, 251
ACTH, *see* Adrenocorticotropic hormone
ADHD, *see* Attention-deficit hyperactivity disorder
Adrenocorticotropic hormone (ACTH), lipid interactions, 409
Aggression, omega-3 fatty acid intake effects, 317, 318
Aging, *see* Immunosenescence
Alzheimer disease, autoradiography of cholinergic signaling, 136
γ-Aminobutyric acid (GABA),
　docosahexaenoic acid effects, 75, 273
　ketogenic diet effects, 282, 283
Angiotensin II,
　brain distribution, 381
　function, 380, 381
　omega-3 fatty acid synergism, 383
　receptors, 381, 394
　regulation of secretion, 380
Arachidonic acid,
　brain development,
　　accretion, 161, 162
　　dietary studies of neuronal cell composition and function, 164–166
　　metabolism, 163, 164
　　synthesis, 162–164
　　transport, 163
　　visual system development, 162
　breast milk composition, 99
　diabetic neuropathy levels, 242, 243
　essentiality, 177
　function, 99
　gene expression effects, 13–16
　lithium effects on turnover, 132
　recommended dietary intakes,
　　adults, 20, 21
　　infants, 20, 22
　　pregnancy and lactation, 20, 22
　synthesis, 99, 100, 177, 193, 219
Arachidonic acid, schizophrenia dysfunction,
　brain composition, 336
　evidence for metabolism dysfunction, 335–337
　membrane incorporation, 336
　niacin flushing response, 337
　phospholipase A_2 assays, 336
　plasma and red cell levels, 336
　therapeutic implications, 337
Arginine vasopressin (AVP),
　body fluid homeostasis, 379, 380
　regulation of secretion, 380
Atopic dermatitis, essential fatty acid balance, 230
Attention-deficit hyperactivity disorder (ADHD),
　essential fatty acid balance, 230
　omega-3 fatty acid intake effects, 319, 320
AVP, *see* Arginine vasopressin

B

Basal metabolic rate, omega-3 fatty acid effects, 390
Bipolar disorder,
　lithium effects on phospholipase A_2 function, 338, 339
　omega-3 fatty acid intake effects, 316
　phospholipase C function, 339
　therapeutic implications of omega-3 fatty acid effects, 339
Body energy homeostasis, *see* Energy balance
Brain development,
　essential fatty acids, *see* Arachidonic acid; Brain docosahexaenoic acid
　functional connections, 160, 161
　gliogenesis, 160
　lipid composition, 161, 178, 357, 358, 360
　neurogenesis, 159, 160
　prospects for study, 170, 171
　size versus weight, 159

stages, 159
Brain docosahexaenoic acid,
   animal models for synthesis studies, 101, 102, 119, 120
   development role,
      accretion, 161, 162
      dietary studies of neuronal cell composition and function, 164–166
      metabolism, 163, 164
      synthesis, 162–164
      transport, 163
      visual system development, 162
   dietary preformed docosahexaenoic acid and α-linolenic acid studies,
      compositional studies, 102–104
      efficiency of synthesis, 361
      tracer studies,
         accretion in baboons, 120
         materno-fetal conversion in primates, 106–108
         neonatal conversion in primates, 108, 109
         neonatal conversion in rodents, 104, 105
         perinatal conversion in primates, 105, 106
         rat diet studies, 361, 362
   essentiality in diet, 334, 335
   transport and binding proteins, 335
   turnover, 190
Brain phospholipid metabolism,
   dietary omega-3 fatty acid effects on monkey phospholipid species,
      deficiency effects, 183
      diet compositions, 183
      fish oil effects, 183, 184
      turnover of phospholipids, 186
   ether lipids, 334
   lipid composition of brain, 161, 178, 357, 358, 360
   α-linolenic acid deficiency effects, 227, 228
   mental illness theories, 345, 346
   pathological conditions, 125
   pathways, overview, 126–128
   phosphorous-31 magnetic resonance spectroscopy in schizophrenia, 337, 347, 348
   remodeling, 332, 335
   signal transduction cycles,
      crosstalk, 334
      phospholipase A_2, 332, 333
      phospholipase C, 333
      phospholipase D, 334
   structural overview, 182, 183, 331, 332
   tracer studies,
      autoradiography of phospholipase A_2 response,
         cholinergic signaling in Alzheimer disease model, 136
         cholinergic stimulation with arecoline, 133, 135, 136
      compartments in model, 128, 129
      dilution factor, 129
      dopaminergic signaling in Parkinson disease model, 137, 138
      lithium effects on arachidonic acid turnover, 132
      neocortex remodeling with nucleus basalis lesion, 137
      positron emission tomography of fatty acid incorporation in human brain, 138, 139
      regional incorporation coefficient determination, 126, 129
      rodent model, 125, 126
      specificity of phospholipid labeling, 130
      turnover and half-life determination, 126, 130, 132, 139
      unlabeled fatty acid incorporation rate, 129, 130

C

Calcium ATPase, polyunsaturated phospholipid bilayer effects on function,
   curvature strain, 33
   reconstituted protein system, 32
   thickness of bilayer, 32, 33
Calcium channel,
   anticonvulsant drug effects, 70
   polyunsaturated fatty acid effects, 70, 74, 273, 274
Calcium flux, docosahexaenoic acid effects on leukocyte signaling, 55
Cancer cachexia,
   cytokine roles, 392
   free fatty acid mobilization, 391, 392
   omega-3 fatty acid effects, 392, 393
Carbamazepine, polyunsaturated fatty acid synergism, 76
Carbon recycling, *see* α-Linolenic acid
CD28, loss in aging, 81

Index

Cereal grains,
    dietary evolution, 5
    fatty acid composition, 6
    risks in diet, 5, 6
Cheese, fatty acid composition, 8, 11
Cholesterol,
    depression correlation with serum levels, 24
    omega-3 fatty acid effects on levels, 408
    polyunsaturated phospholipid bilayer interactions, 30, 31, 37, 47, 48
Cognition, docosahexaenoic acid role in rodents,
    biochemical background for cognition, 371, 372
    brightness–discrimination learning behavior studies of long-term α-linolenic acid deficiency,
        biochemical studies, 227–229
        blindness to dim light control, 226, 227
        diet compositions, 221
        docosahexaenoic acid supplementation effects, 224
        negative response effects, 221–224, 231
        reversibility and omega-6 fatty acid effects, 224, 225
        rod outer segment turnover, 225, 226
    lipid composition of brain,
        diet effects, 360–362
        overview, 357, 358, 360
    memory types, 363
    omega-3 fatty acid depletion studies,
        diet characteristics, 365, 366
        mouse,
            behavior, 366–369
            docosahexaenoic acid-specific effects, 369, 370
            visual acuity, 368, 369
        rat,
            behavior, 366–368
            docosahexaenoic acid-specific effects, 368
            visual acuity, 366
        repletion studies, 370, 371
    tests,
        actimeter, 365
        electroretinogram, 365
        elevated plus-maze, 364
        escape tendency, 365
        forced swimming test, 365
        Morris water maze, 363
        open field, 364
        radial arm maze, 363
        rotating rod, 363
        shuttle box test, 364
        Skinner box, 364
        T-maze, 363
Corticotropin-releasing factor (CRF),
    leptin effects, 387, 388
    lipid interactions, 409
    omega-3 fatty acid intake effects, 324
    regulation of release, 409
    stress response, 407
CRF, *see* Corticotropin-releasing factor
Cytokines, *see also* specific cytokines,
    pain role, 412, 413
    polyunsaturated fatty acid effects on levels, 405, 406
    stress response, 410, 411

D

Depression, *see also* Bipolar disorder,
    cholesterol, correlation with serum levels, 24
    comorbid conditions, 337, 338
    essential fatty acid deficiency effects, 229, 230
    lipid enzyme abnormalities, 338
    omega-3 fatty acid intake effects,
        bipolar affective disorder, 316
        major depression,
            cross-national epidemiology and seafood intake, 313, 314
            dietary evolution effects, 312–313
            omega-3 fatty acid supplementation studies, 315, 316
            tissue composition of fatty acids, 314, 315
        postpartum depression, 316, 317
    plasma and red cell fatty acid composition, 338
DHA, *see* Docosahexaenoic acid
Diabetes, omega-3 fatty acid effects, 391, 394
Diabetic neuropathy,
    biosynthesis,
        eicosanoids, 241, 242
        fatty acids, 241, 252
    clinical presentation, 239
    fatty acids,
        metabolism in nerves, 241–243
        supplementation effects,
            fish oil studies, 246
            linoleic acid, 244

γ-linolenic acid, 244, 245, 252
  rationale, 243, 244
nitric oxide formation, 247, 252
onset, 239
oxidative stress,
  antioxidant defenses, 250
  cultured neuron studies, 248
  lipid peroxidation, 250, 251
  reactive carbonyls, 248, 249
  reactive oxygen species sources, 247, 248
pathogenesis, 239–241, 251
polyol pathway activity and essential fatty acids, 246, 247
treatment,
  N-acetylcysteine, 251
  γ-linolenic acid synergism, 252
  pyridoxamine, 251, 252
Diet, see Hunter–gatherer diet; Western diet
Docosahexaenoic acid (DHA),
  breast milk composition, 99
  cardiomyocyte supplementation effects, 32
  cognitive function, see Cognition, docosahexaenoic acid role in rodents
  deficiency, see Omega-3 fatty acid deficiency
  essentiality, 177
  gene expression effects, 13–16
  leukocyte modulation, see Leukocyte, docosahexaenoic acid modulation
  membrane composition effects, see Membrane lipid bilayers
  metabolism in blood versus brain, 188, 189
  peroxisome biogenesis disorder therapy, see Peroxisome biogenesis disorder
  recommended dietary intakes,
    adults, 20, 21
    infants, 20, 22
    pregnancy and lactation, 20, 22
  serotonin correlation, 24, 37, 321, 322
  sodium channel effects, see Sodium channel, voltage-regulated
  synaptosome enrichment, 115
  synthesis, 99, 100
    animal species ability, 357
    brain, see Brain docosahexaenoic acid
    central nervous system, 117
    development, 118, 119, 154, 155
    dietary factors, 116, 117
    genetic factors, 116, 117
    liver, 116, 117, 194
  mathematical modeling, 121
  precursor, see α-Linolenic acid
  retina, see Retinal docosahexaenoic acid
  tissue distribution, 42
  Western diet intake, 9
Dopamine, omega-3 fatty acid effects,
  deficiency, 228, 229
  intake, 322, 323

E

Egg, fatty acid composition dependence on chicken feed, 8, 10
Eicosapentaenoic acid (EPA),
  deficiency, see Omega-3 fatty acid deficiency
  gene expression effects, 13–16
  metabolism in blood versus brain, 188, 189
  psychiatric disorder supplementation efficacy, 324, 326
  recommended dietary intakes,
    adults, 20, 21
    infants, 20, 22
    pregnancy and lactation, 20, 22
  schizophrenia therapy, see Schizophrenia
  sodium channel effects, see Sodium channel, voltage-regulated
  synergistic effects with antiepileptic drugs, 75, 76
  Western diet intake, 9
Electroretinogram (ERG),
  component overview, 199–201, 365
  docosahexaenoic acid deficiency,
    amplitudes, 207, 208, 212, 225
    early receptor potential effects, 201, 202
    fast PIII, 202, 203, 205, 206
  photoreceptor response modeling, 203–206
  retinal adaptation and morphology, 199
  signal origins, 199
  twin-flash response and inactivation studies, 206
Energy balance,
  appetite factors, 384
  arcuate nucleus regulation, 386
  food restriction effect on subsequent intake, 385
  glucose reguation, 386
  hypothalamic control, 384
  leptin function, 387, 388
  lipostatic theory, 386, 387

Index

omega-3 fatty acid effects,
    adiposity reduction, 388, 389, 393
    basal metabolic rate, 390
    body weight reduction, 389
    total energy expenditure, 390
EPA, *see* Eicosapentaenoic acid
Epilepsy,
    anticonvulsant criteria, 285
    antiepileptic actions of polyunsaturated fatty acids,
        cell culture studies, 285
        evidence, 63, 71, 72, 274
        hippocampal slice studies, 285, 286
        ketogenic diet comparison, 286
        mechanisms, 73–75, 284
        rationale for therapy, 274, 285
        synergistic effects with antiepileptic drugs, 75, 76
    classification, 63, 64
    epileptiform discharges, 64
    ion channels, *see* Calcium channel; Sodium channel, voltage-regulated
    ketogenic diet control, *see* Ketogenic diet
    prevalence, 63
ERG, *see* Electroretinogram
Ethanol,
    fatty acid metabolites, *see* Fatty acid ethyl esters
    metabolic pathways, 291, 292
Exercise,
    aging effects on capacity, 84, 85, 90, 91
    immunosenescence effects,
        cytokine production, 91, 92
        leukocytosis induction, 85, 86, 91
        natural killer cell activity, 85, 86, 92
        oxidative stress and antioxidants, 86, 87
        proinflammatory effects, 91
        stress response, 86

F

FAEEs, *see* Fatty acid ethyl esters
Fatty acid ethyl esters (FAEEs),
    alcohol specificity for formation, 293
    biosynthesis,
        acyl-CoA:ethanol *O*-acyltransferase, 297
        carboxylester lipase, 298
        cholesterol esterase, 298
        lipoprotein lipase, 298
        overview, 291–293
        placenta, 300
        sex differences, 300

        sites in brain, 296
        synthases,
            cytosolic synthase, 296, 297
            ethanol induction, 299, 300
            microsomal synthase, 296, 297
            purification, 296
            sequencing, 297
        white blood cells, 298, 299
    ethanol intake marker,
        advantages, 303, 304
        gas chromatography–mass spectrometry study, 303
        half-life, 301
        protein binding, 301, 302
    supplement utilization, 300, 301
    toxicity,
        alcoholic autopsy studies, 293
        cell studies, 294, 295
        central nervous system, 294, 295
        membrane disruption, 294
        mitochondria, 293
        pancreas, 294
        rat studies, 295

G, H

GABA, *see* γ-Aminobutyric acid
GLA, *see* γ-Linolenic acid
HD, *see* Huntington's disease
Hostility, omega-3 fatty acid intake effects, 317, 318
Hunter–gatherer diet,
    characteristics, 4
    lifestyle features, 4
    omega-6:omega-3 fatty acid balance, 9, 11, 12
Huntington's disease (HD),
    animal models, 341
    clinical presentation, 341
    eicosapentaenoic acid therapy, 341
    polyglutamine repeats, 340, 341

I

IL-1b, *see* Interleukin-1b
IL-2, *see* Interleukin-2
IL-6, *see* Interleukin-6
Immune function,
    aging, *see* Immunosenescence
    overview of polyunsaturated fatty acid effects, 405, 406
    psychoneuroimmunology, 410
    stress response, 410, 411

Immunosenescence,
    aging, immunologic theory, 80
    apoptosis of lymphocytes, 82, 83
    cell-mediated immunity changes, 80, 81
    consequences, 79
    exercise effects,
        cytokine production, 91, 92
        leukocytosis induction, 85, 86, 91
        natural killer cell activity, 85, 86, 92
        oxidative stress and antioxidants, 86, 87
        proinflammatory effects, 91
        stress response, 86
    interleukin-2 production, 81, 82
    neuroendocrine system changes, 82
    nutritional intervention,
        omega-3 fatty acids, 88–90, 92
        omega-6 fatty acids, 90, 92
        undernutrition in the elderly, 87, 88
        vitamin E, 88, 89
        zinc, 88
    signal transduction abnormalities, 83
    telomerase shortening, 83
    thymus,
        atrophy reversibility, 79, 84
        changes, 83, 84
Infantile Refsum disease, see Peroxisome biogenesis disorder
Interleukin-1b (IL-1b),
    cancer cachexia role, 392
    neurodegeneration role, 340
    omega-3 fatty acid intake effects, 324
    stress response, 411
Interleukin-2 (IL-2),
    aging effects on production, 81, 82
    receptor, docosahexaenoic acid effects on leukocyte signaling, 54
    stress response, 411
Interleukin-6 (IL-6), stress response, 411

K

KD, see Ketogenic diet
Ketogenic diet (KD),
    animal studies, 276, 280
    anticonvulsant mechanisms,
        acid–base changes, 280, 281
        cerebral energy metabolism, 282
        chronic effects, 284
        criteria, 277–279
        ketone levels and seizure control, 280
        ketosis role, 278
        latency of onset, 278, 279
        lipid effects, 281
        neuronal excitability effects, 282, 283
        overview, 279
        phases of control, 280
        study design, 277, 278
    biochemical aspects of seizure control, 277
    calorie restriction, 275
    children, 275, 276
    composition, 274, 275
    historical perspective of use for epilepsy, 274
    risks, 275
    seizure control efficacy, 275, 276

L

LA, see Linoleic acid
Leptin, function, 387, 388
Lettuce, fatty acid composition, 8
Leukocyte, docosahexaenoic acid modulation,
    adhesion proteins,
        membrane composition effects, 51, 52
        types, 51
    calcium flux, 55
    CD4 expression, 52
    CD8 effects, 52
    cell types, 50, 51
    interleukin-2 receptor signaling, 54
    major histocompatibility complex proteins,
        classes, 52
        MHC I expression, 53
        MHC II expression, 53, 54
        structures, 52, 53
    protein kinase C stimulation, 54, 55
Linoleic acid (LA),
    deficiency effects, 206, 207, 219, 220
    diabetic neuropathy, supplementation effects, 244
    essentiality, 177, 193, 377, 404
    gene expression effects, 13–16
    heart attack risks with intake, 230
    recommended dietary intakes,
        adults, 20, 21
        infants, 20, 22
        pregnancy and lactation, 20, 22
α-Linolenic acid (LNA),
    bioavailability, 146
    brightness–discrimination learning behavior studies of long-term deficiency,
        biochemical studies, 227–229
        blindness to dim light control, 226, 227

Index

diet compositions, 221
docosahexaenoic acid supplementation effects, 224
negative response effects, 221–224, 231
reversibility and omega-6 fatty acid effects, 224, 225
rod outer segment turnover, 225, 226
carbon recycling studies,
    fetal primate IRMS studies, 152
    prospects for study, 152, 153
    suckling rat,
        GC-IRMS, 151, 152
        nuclear magnetic resonance, 151
        tracer studies, 150, 151
dietary sources, 193
docosahexaenoic acid synthesis, *see* Brain docosahexaenoic acid; Retinal docosahexaenoic acid
essentiality, 177, 193, 377, 404
gene expression effects, 13–16
human behavior effects, 229, 230
metabolic fates, 109–111, 121, 219, 378
β-oxidation,
    carbon recycling role, 146
    docosahexaenoic acid synthesis impact, 153, 154
    energy deficit studies, 149, 150
    prospects for study, 152, 153
    tracer analysis, 147
    whole-body balance analysis, 148
recommended dietary intakes,
    adults, 20, 21
    infants, 20, 22
    pregnancy and lactation, 20, 22
synthesis, 145
tissue distribution, 109
g-Linolenic acid (GLA), diabetic neuropathy supplementation effects, 244, 245, 252
Lithium, effects on arachidonic acid turnover, 132, 338, 339
LNA, *see* α-Linolenic acid
Long-term potentiation (LTP), loss in aging, 340
LTP, *see* Long-term potentiation

M

Major histocompatibility complex (MHC) proteins,
    classes, 52
    docosahexaenoic acid effects,
        MHC I expression, 53
        MHC II expression, 53, 54
    structures, 52, 53
Membrane lipid bilayers,
    classification, 195, 378, 379
    docosahexaenoic acid,
        benefits in disease, 41, 42
        cholesterol interactions, 30, 31, 37, 47, 48
        conformation in membranes, 43
        effects on membranes,
            curvature stress, 46, 47
            fluidity, 43, 44
            fusion, 46
            hydration, 44, 45
            membrane order, 44, 115
            packing, 45, 46
            permeability, 45
            phase transitions, 46
        function, 23, 210, 211
        leukocyte modulation, *see* Leukocyte, docosahexaenoic acid modulation
        lipid microdomain induction,
            detergent-resistant membranes, 48
            model membranes, 49
            trans-membrane domains, 49
        tissue distribution, 42
        α-tocopherol interactions, 48
    fatty acid ethyl ester disruption, 294
    fluidity, factors affecting, 196, 210, 211, 378, 379, 405
    membrane protein function effects of composition, *see also* specific proteins,
        acyl chain packing, 34–36
        curvature strain, 33, 34, 50
        reconstituted protein systems, 32, 50
        thickness of bilayer, 32, 33
    phospholipid composition, 195
    polyunsaturated phospholipid bilayer features,
        acyl chain packing, 26–30, 196
        cholesterol interactions, 30, 31
        curvature strain, 26
        elastic area compressibility modulus, 25
        model systems, 24, 25
        phase transition temperature, 25
        thickness, 26, 210
        water permeability, 26

Memory, *see* Cognition, docosahexaenoic acid role in rodents
MHC proteins, *see* Major histocompatibility complex proteins
Mustard, fatty acid composition, 8
Mutation rate, human DNA, 3

NAC, *see* N-Acetylcysteine
Natural killer cell,
    exercise effects on activity, 85, 86, 92
    omega-3 fatty acid effects, 405
Neonatal adrenoleukodystrophy, *see* Peroxisome biogenesis disorder
Nerve growth factor (NGF), omega-3 fatty acid deficiency effects, 229
Neuropeptide Y (NPY), food intake pathways, 387, 388
NF-κB, *see* Nuclear factor-κB
NGF, *see* Nerve growth factor
Nitric oxide (NO), diabetic neuropathy pathogenesis, 247, 252
NMR, *see* Nuclear magnetic resonance
NO, *see* Nitric oxide
NPY, *see* Neuropeptide Y
Nuclear factor-κB (NF-κB), antioxidant effects on activity, 87
Nuclear magnetic resonance (NMR), α-linolenic acid carbon recycling studies, 151

O

Obesity,
    fat intake role, 391
    insulin resistance, 391
    omega-3 fatty acid effects, 388, 389, 391
Omega-3 fatty acid deficiency,
    brain phospholipid effects, 183, 184, 186
    cognition studies,
        diet characteristics, 365, 366
        mouse,
            behavior, 366–369
            docosahexaenoic acid-specific effects, 369, 370
            visual acuity, 368, 369
        rat,
            behavior, 366–368
            docosahexaenoic acid-specific effects, 368
            visual acuity, 366
        repletion studies, 370, 371
    psychiatric illness prevalence studies,
        attention-deficit hyperactivity disorder, 319, 320
        bipolar affective disorder, 316
        epidemiologic methodology, 311
        hostility and aggression, 317, 318
        intervention trial design, 325
        major depression,
            cross-national epidemiology and seafood intake, 313, 314
            dietary evolution effects, 312–313
            omega-3 fatty acid supplementation studies, 315, 316
            tissue composition of fatty acids, 314, 315
        mechanisms of action, 323, 324
        neurotransmitter alterations, 321–323
        post-gestation psychiatric outcomes, 320, 321
        postpartum depression, 316, 317
        schizophrenia, 318, 319
        suicide, 318
        tissue concentration studies, 324, 325
    rhesus monkey studies of reversibility,
        behavioral changes, 189
        cerebral cortex lipid response, 180–182, 187
        diet compositions, 178, 179
        erythrocyte lipid response, 179, 180
        fatty acid turnover, 182, 187
        implications for humans, 189, 190
        plasma lipid response, 179
    rodent brain recovery studies, 187, 188
    testes fatty acid composition effects, 185, 186
    vision effects, 23, 24, 36, 99, 116, 165, 178, 188
    water and sodium balance effects, 383, 384
Omega-6:omega-3 fatty acid balance,
    agribusiness effects on foods, 8–10, 378
    cognition effects, 408, 409
    dietary evolution, 9, 11, 12
    gene expression effects, 13–16
    geographic differences in diets, 9, 10, 13
    health consequences, 11–13
    ideal balance, 16
    immunologic mediation, 409
    stress supplement formulation, 409, 410

P

Pain,
    affective response, 412
    fatty acid involvement, 412–414
    stress-induced analgesis, 414

threshold ad fatty acid supplement studies, 413, 414
Parkinson disease, autoradiography of dopaminergic signaling, 137, 138
PBD, *see* Peroxisome biogenesis disorder
Peroxisome biogenesis disorder (PBD),
animal models, 262
clinical manifestations, 258, 259
docosahexaenoic acid replacement therapy,
animal studies, 267
nonrandomized trials,
biochemical effects, 263
clinical effects, 263–265
diet and administration, 263
magnetic resonance imaging of brain, 265, 266
randomized trial, 266, 267, 268
rationale, 257, 258, 262, 267, 268
recommendations, 268, 269
Zellweger syndrome deficiency features, 262
gene mutations, 259–261, 269
genotype–phenotype correlations, 251
peroxisomal reactions, 257
PEX5 knockout mouse studies, 267
single-enzyme defects in differential diagnosis, 261
Peroxisome proliferator-activated receptors (PPARs), fatty acid ligands, 245
PET, *see* Positron emission tomography
PGE1, *see* Prostaglandin E1
Phospholipase A2 (PLA2),
aging brain activity, 340
autoradiography of response,
cholinergic signaling in Alzheimer disease model, 136
cholinergic stimulation with arecoline, 133, 135, 136
lithium effects on arachidonic acid turnover, 132, 338, 339
schizophrenia abnormalities, 336, 348, 349
signal transduction cycle, 332, 333
Phospholipase C (PLC), signal transduction cycles, 333
Phospholipase D (PLD), signal transduction cycles, 334
Phospholipid metabolism, *see* Brain phospholipid metabolism
Pineal gland, aging changes, 82
PKC, *see* Protein kinase C

PLA2, *see* Phospholipase A2
PLC, *see* Phospholipase C
PLD, *see* Phospholipase D
Positive health, Hippocratic concept, 3
Positron emission tomography (PET), fatty acid incorporation in human brain, 138, 139
PPARs, *see* Peroxisome proliferator-activated receptors
Pregnancy,
docosahexaenoic acid demands, 379
fatty acid intake requirements, 20, 22
omega-3 fatty acid intake effects,
post-gestation psychiatric outcomes, 320, 321, 349
postpartum depression, 316, 317
Prostaglandin E1 (PGE1), γ-linolenic acid induction of synthesis, 245
Protein kinase C (PKC),
aging changes in function, 83
docosahexaenoic acid effects on leukocyte signaling, 54, 55
polyunsaturated phospholipid bilayer effects on function,
curvature strain, 33, 50
reconstituted protein system, 32
Purslane, fatty acid composition, 8
Pyridoxamine, diabetic neuropathy management, 251, 252

R

Retina, *see also* Electroretinogram,
cell types, 196
inactivation of phototransduction, 198, 199, 206
rod outer segment phototransduction, 197, 198
Retinal docosahexaenoic acid, *see also* Rhodopsin,
animal models for synthesis studies, 101, 102
composition, 99
deficiency effects,
early receptor potential, 201, 202
electroretinogram amplitudes, 207, 208, 212, 225
fast PIII, 202, 203, 205, 206
repletion following depletion, 209
retinal fatty acid profile, 207
vision, 23, 24, 36, 99, 116, 165, 178
developmental role, 100

dietary preformed docosahexaenoic acid and a-linolenic acid studies,
compositional studies, 102–104
tracer studies,
materno-fetal conversion in primates, 106–108
neonatal conversion in primates, 108, 109
neonatal conversion in rodents, 104, 105
perinatal conversion in primates, 105, 106
diet,
breast-feeding impact on vision, 168
modulation of fatty acid composition and function, 167, 168
rhodopsin content modulation, 168
very-long-chain fatty acid modulation, 168–170
infant formula composition effects, 100, 101
a-linolenic acid, brightness–discrimination learning behavior studies of long-term deficiency,
biochemical studies, 227–229
blindness to diim light control, 226, 227
diet compositions, 221
docosahexaenoic acid supplementation effects, 224
negative response effects, 221–224, 231
reversibility and omega-6 fatty acid effects, 224, 225
rod outer segment turnover, 225, 226
maternal diet manipulation studies, 208, 209
membrane effects,
expansion, 210
fluidity, 210, 211
photoactivity role, 210
polyunsaturated fatty acid profile, 194, 195
recycling, 120, 121
retinal pigment epithelium phagosome effects, 211, 212, 225, 226
supply and conservation in development, 166
Rhizomelic chondrodysplasia punctata, *see* Peroxisome biogenesis disorder
Rhodopsin,
dietary fat effects on retinal content, 169
docosahexaenoic acid modulation, 209, 210
light response, 196–198
polyunsaturated phospholipid bilayer effects on function,
activation, 116
acyl chain packing, 34–36
curvature strain, 33, 34
reconstituted protein system, 32

S

Salmon, fatty acid composition in wild vs cultured fish, 8, 9
Schizophrenia,
arachidonic acid,
brain composition, 336
evidence for metabolism dysfunction, 335–337
membrane incorporation, 336
niacin flushing response, 337, 348
phospholipase A_2 assays and activity, 336, 348, 349
plasma and red cell levels, 336
therapeutic implications, 337
eicosapentaenoic acid therapy,
ethyl ester study, 352, 353
fish oil double-blind studies, 351, 352
membrane fatty acid levels, 351
resistance, 351
symptom response, 350–353
essential fatty acid deficiency effects, 229, 349
etiology theories,
neurodevelopmental hypothesis, 347
neurotransmitter theory, 347
phospholipid hypothesis, 347–349
medications and compliance, 346
omega-3 fatty acid intake effects, 318, 319, 349, 350
phosphorous-31 magnetic resonance spectroscopy of phospholipid metabolism, 337, 347, 348
prevalence, 346
symptoms, 346
Serotonin, omega-3 fatty acid intake effects, 24, 37, 321–323
SIA, *see* Stress-induced analgesia
Sodium channel, voltage-regulated,
action potential generation, 64, 65
blocking mechanisms,
anesthetics and anticonvulsants, 66, 67
polyunsaturated fatty acids, 67, 74
opening and closing, 65, 66
phosphorylation, 66

polyunsaturated fatty acid effects in hippocampal CA1 neurons,
concentration–effect relationship, 70
model sysem, 67, 68
sodium current activation, 70
voltage dependence of sodium current inactivation, 68, 70
voltage clamping, 66
Sodium homeostasis,
atrial natriuretic peptide role, 382
omega-3 fatty acid deficiency effects, 383, 384
overview, 380–382
oxytocin role, 381, 382
receptors, 380, 381
somatostatin role, 382
Spinach, fatty acid composition, 8
Stress, see also Corticotropin-releasing factor,
definition, 406
endocrine response, 407
fatty acid supplementation, 409, 410
fatty acid–peptide interactions, 407
immune response, 410, 411
lipid profile effects, 406
prostaglandin response, 407–409
Stress-induced analgesia (SIA),
features, 414
fish oil prevention, 414
Suicide, omega-3 fatty acid intake effects, 318

T

Thirst, regulation, 380, 381
TNF-α, see Tumor necrosis factor-α
α-Tocopherol,
docosahexaenoic acid interactions, 48
immunosenescence intervention, 88, 89
Total energy expenditure, omega-3 fatty acid effects, 390
Tumor necrosis factor-α (TNF-α), pain role, 412, 413

V

Valproic acid,
phospholipid interactions, 339
polyunsaturated fatty acid synergism, 76
Vegetable oil,
fatty acid composition, 6, 7
hydrogenation, 7
Vitamin E, see α-Tocopherol
Voltage-regulated sodium channel, see Sodium channel, voltage-regulated

W–Z

Western diet,
characteristics, 4, 6
disease association, 230, 231
lifestyle features, 4
major depression, evolutionary effects, 312, 313
omega-6:omega-3 fatty acid balance, 9, 12, 378, 379
Zellweger syndrome, see Peroxisome biogenesis disorder

About the Editors

David I. Mostofsky is Professor of Experimental Psychology at Boston University, and holds associate appointments at several major medical institutions in the Boston area. A Charter Fellow of the Academy for Behavioral Medicine Research, he has been active in Behavioral Medicine and Psychopharmacology for many years, and has published numerous articles and books on a variety of topics in this area, especially with regard to epilepsy, pain, essential fatty acids, and psychoneuroimmunology. He has been the recipient of many awards including, a Marshal Fund Award, Einstein Award, NIH/Fogarty Fellowship, and a Fulbright Fellowship.

Shlomo Yehuda has a PhD in Psychology and Brain Science from M.I.T. He is a Professor in the Department of Psychology and is the Director of the Psychopharmacology Laboratory at Bar Ilan University. He currently holds the Ginsburg Chair for Alzheimer research and is active at the Farber Center. He was a research Associate in Professor Wurtman's Lab at M.I.T., a visiting Professor at Professor Kastin's Peptide Laboratory (Tulane Medical School, New Orleans, LA), and Rosenstadt Professor at Toronto Medical School–Nutrition Department. He has published over 150 scientific papers and 4 books in the following fields: Brain Biochemistry, Effects of Nutrients on Brain and Behavior (mainly brain iron and essential fatty acids), and Aging of the Brain and Animal Models of Neurological Disorders.

Norman Salem Jr. is Chief of the Laboratory of Membrane Biochemistry & Biophysics within the National Institutes on Alcohol Abuse and Alcoholism at the National Institutes of Health. He is also a Research Professor at the F. Edward Hebert School of Medicine, Uniformed Services University of the Health Sciences and Adjunct Professor at the Georgetown University Medical School. He has worked on studies of DHA composition, metabolism and biological function for more than 25 years, with an emphasis on the nervous system. His current research is focused upon the role of DHA in neural development and the means by which DHA is obtained by the brain.

His professional affiliations include the American Society for Biochemistry & Molecular Biology, the American Society for Neurochemistry, the International Society for the Study of Fatty Acids and Lipids, the American Oil Chemists Society and the Research Society on Alcoholism. He is an author of about 150 scientific papers and book chapters. Dr. Salem serves as an Associate Editor of Lipids and is on the editorial board of *Nutritional Neuroscience*.

About the Series Editor

Dr. Bendich is Associate Director, New Product Research at GlaxoSmithKline, where she is responsible for leading research initiatives for products that include TUMS, Os-Cal and Geritol, as well as other nutrient-containing dietary supplements marketed in and outside the US. Prior to joining SmithKline Beecham, Dr. Bendich was Assistant Director of Clinical Nutrition at Roche where she participated in many of the groundbreaking clinical studies involving antioxidants and folic acid.

Dr. Bendich is an internationally recognized authority on antioxidants, nutrition and immunity and pregnancy outcomes, vitamin safety and the cost-effectiveness of vitamin/mineral supplementation. Author of over 100 scientific papers, editor of seven books, including *Preventive Nutrition: The Comprehensive Guide For Health Professionals* and Series Editor of *Nutrition and Health* for Humana Press, Dr. Bendich also serves as Associate Editor for *Nutrition: The International Journal* and is a member of the Board of Directors of the American College of Nutrition. Dr. Bendich was recently a recipient of the Burroughs Wellcome Visiting Professorship in Basic Medical Sciences, 2000–2001, and holds academic appointments as Adjunct Professor at Columbia University, UMDNJ, and Rutgers University.